ESTATES GAZETTE LAW REPORTS

Estates Gazette Law Reports

1994

Volume 1

Edited by
Barry Denyer-Green LLM PhD FRICS
of the Middle Temple, barrister

Audrey Boyle BA
Publisher's editor

Estates Gazette

Routledge
Taylor & Francis Group
LONDON AND NEW YORK

EG Books is an imprint of Routledge
2 Park Square, Milton Park, Abingdon, Oxon OX14 4RN
711 Third Avenue, New York, NY 10017, USA
Routledge is an imprint of the Taylor & Francis Group, an informa business

ISBN for complete set of 2 volumes: 0 7282 0211 5
ISBN for this volume 0 7282 0212 3
ISSN 0951-9289

For information on all EG Books publications
visit our website www.routledge.com

CONTENTS

SUPPLEMENT

Cases not reported in "Estates Gazette"

TABLE OF CASES

Titles of cases shown in bold type
Names of parties reversed shown in ordinary type

INDEX OF SUBJECT-MATTER

LEASEHOLD ENFRANCHISEMENT — See **Landlord and tenant** (Leasehold enfranchisement)

MILK QUOTA

Breach of Milk Marketing Board scheme

Power of disciplinary committee

Alleged breach of scheme relating to "withholder" monthly returns — Whether disciplinary committee of Milk Marketing Board has power to impose penalties where breach also involves ingredients of a crime — Held, committee has power to hear allegations and impose fines and penalties where breach of scheme and criminal offence overlap, but crime has additional ingredients.

***R* v *Milk Marketing Board (CA)*, 7**

Negligent advice

Claim

Plaintiff purchasing farm not used for dairy purposes in April 1984 — Adverse decision of Dairy Produce Quota Tribunal on claims for wholesale and exceptional hardship quota — Whether tribunal erred in law — Whether defendant failed to advise judicial review of tribunal's decision — Held, tribunal erred in applying regulations and failing to give reasons — Defendant in breach of duty care.

***Hood* v *National Farmers' Union (CA)*, 1**

MORTGAGE

Mortgage loan report — See **Negligence** (Mortgage loan report)

NEGLIGENCE — See also **Milk quota**

Duty of care and liability

Estate agents

Negligent misstatement of area of gardens to house — Whether duty of care owed to purchaser — Held, duty of care owed because ingredients of proximity, foreseeability of loss and just and reasonableness present.

***McCullagh* v *Lane Fox & Partners Ltd (QB)*, 48**

Valuers — See also **Negligence** (Mortgage loan report)

Turnover valuation of restaurant premises for mortgage purposes — Use of figures for turnover provided by borrowers — Whether valuer negligent in use of chosen figures and variables — Held, no negligence found.

***Craneheath Securities Ltd* v *York Montague Ltd (Ch)*, 159**

Damages: measure and quantification

Estate agents

Negligent misstatement of area of gardens to house — Valuation — Whether plaintiff paid more than market value — Held, on the evidence plaintiff did not pay more than market value.

***McCullagh* v *Lane Fox & Partners Ltd (QB)*, 48**

Mortgage loan report

Damages

Valuer failed to observe defect in concrete purlin in Bison system terrace house otherwise observable in roof space — Plaintiffs lost sale, suffered job dismissal, temporary housing accommodation before obtaining council tenancy — Substantial arrears owing to building society and loss of equity — Held, damages awarded included worry, distress and physical inconvenience.

***Ezekiel* v *McDade (Off Ref)*, 255**

Duty of care

Mortgage valuation of country house with prospects of planning permission for conversion — Whether valuation report for mortgage purposes may include element of hope value for planning permission — Held, where mortgage valuation is to open market value, hope value may be included — Valuer entitled to follow RICS guidance notes on valuation.

***Allied Trust Bank Ltd* v *Edward Symmons & Partners (QB)*, 165**

Whether valuation of £155,000 for mortgage purposes in 1988 overvaluation — Premises sold in 1991 for £75,000 — Whether defendant valuers negligent — Held, valuation at top end of valuation bracket and not negligent.

***Axa Equity & Law Home Loans Ltd* v *Goldsack & Freeman (QB)*, 175**

Valuers — See **Negligence** (Duty of care and liability — Valuers)

NOTICES — See **Agricultural holdings** (Notice to quit)

OPTION TO PURCHASE — See **Vendor and purchaser** (Option)

PETROL FILLING STATION — See **Landlord and tenant** (Rent review clause — Construction: general)

PRACTICE AND PROCEDURE

Address for service — See **Landlord and tenant** (Forfeiture)

REPAIRS — See **Landlord and tenant** (Repairs)

RENT ACTS — See **Landlord and tenant** (Rent Acts)

RENT REVIEW — See **Landlord and tenant** (Rent review clause)

RESTRICTIVE COVENANT

Construction

Use of property for schizophrenics

Property subject to covenant against trade or pursuit which would cause nuisance, annoyance or danger to adjoining properties — Whether use by charity for schizophrenics in quiescence stage a breach — Held, no breach of covenant — Covenant did not control identity or personal characteristics of occupiers — No danger to neighbours.

***National Schizophrenia Fellowship* v *Ribble Estates SA (Ch)*, 181**

RICS GUIDANCE NOTES — See **Negligence** (Mortgage loan report — Duty of care)

"RIGHT TO BUY" — See **Housing** (Secure tenancy)

RIGHT TO LIGHT — See **Easement** (Right to light)

SCHOOL SITES

Reverter

Claim by grantor's successor in title

School site, which had been conveyed to trustees under School Sites Act 1841, sold after closure of school — Whether grantor's successor entitled to proceeds of sale under 1841 Act — Held, site did not revert to adjoining land by proceeds held in trust for grantor's successor.

***Marchant* v *Onslow (Ch)*, 187**

SECURE TENANCY — See **Housing** (Secure tenancy)

SITES OF SPECIAL SCIENTIFIC INTEREST

Management agreement

Assessment of compensation

Award of arbitrator — Correct test for assessing compensation under Circular 4/83 — Whether net profits forgone difference between profits which would have been earned and maximum income farm will produce subject to restrictions — Held, proper basis for loss is same as assessment of damages at common law.

***Thomas* v *Countryside Council for Wales (QB)*, 17**

SURETY

Covenant — See **Landlord and tenant** (Assignment — Surety covenant)

TELECOMMUNICATION CODE — See **Compensation for telecommunication cables**

TOWN AND COUNTRY PLANNING

Advertisements

Criminal offence

Whether landowner charged with offence of display of advertisement guilty where advertisement removed immediately on having knowledge of the same — Whether defence in s224(5) of Town and Country Planning Act 1990 that advertisements displayed without owners' "knowledge or consent" has conjunctive or disjunctive meaning — Held, "knowledge or consent" must be construed disjunctively — Defence available if owner has knowledge but has not given consent.

***Wycombe District Council* v *Michael Shanly Group Ltd (QB)*, 200**

Listed building

Compensation

Listed building consent refused for proposed works involving demolition of chimney breasts and creation of additional office space — Whether works of demolition or alteration for purposes of s27 of Planning (Listed Buildings and Conservation Areas) Act 1990 — Held, whether works demolition or alteration question of fact — Relevant how planning authority treated application — On evidence works consisted of alteration — Potential for compensation.

***Shimizu (UK) Ltd* v *Westminster City Council (LT)*, 214**

Planning permission

Compensation for refusal

Claim for £363,000 with agreed cost of works of £92,000 and existing value of £645,000 on valuation date of June 29 1988 — Authority supporting compensation liability of £91,998 — Decision, compensation of £263,000 — Use of building society indices unreliable — Floor area analysis not practical approach to valuation of houses.

***Toye* v *Kensington and Chelsea Royal London Borough Council (LT)*, 204**

Conditions

Proposed condition related to land not owned by applicant — Whether planning permission can be refused on ground that proposed condition has no reasonable prospect of fulfilment — Held, mere fact that desirable condition appears to have no reasonable prospects of fulfilment does not mean that planning permission must necessarily be refused.

***British Railways Board* v *Secretary of State for the Environment (HL)*, 197**

Sites of special scientific interest — See **Sites of special scientific interest** (Management agreement)

VALUATION

Sale transfer — See **Vendor and purchaser** (Sale transfer — Hope value)

VENDOR AND PURCHASER

Option

Validity

Option to purchase leased subjects in contract for lease at current open market value included words referring to mutually agreed price — Whether option was agreement to agree and unenforceable — Held, option clause must be read as a whole and parties intention clear — Option enforceable with price to be determined objectively by court in absence of agreement.

***Scottish Wholefoods Collective Warehouse Ltd* v *Raye Investments Ltd (Ct Sess)*, 245**

Sale transfer

Hope value

Transfer of land provided for additional consideration upon grant of planning permission — Whether valuation of land prior to grant of planning

AGRICULTURE

Court of Appeal

April 21 1993

(Before Sir Stephen BROWN, President, Lord Justice EVANS and Lord Justice PETER GIBSON)

HOOD AND ANOTHER v NATIONAL FARMERS' UNION

Estates Gazette January 8 1994

[1994] 01 EG 109

Negligence — Milk quota — Adverse decision of Milk Quota Tribunal — Whether tribunal erred in finding against plaintiffs — Whether defendant negligent in failing to advise remedy

On September 29 1983 the plaintiffs, who had previously run a diary herd of 100 cows on a 112-acre farm in Wiltshire, purchased Greys Green Farm, Oxfordshire, a 420-acre farm that had not been used for diary purposes for 10 years. Between November 1983 and March 1984 the plaintiffs constructed a milking parlour, silage plant and water supply at Greys Green Farm. Although some stock were moved to the farm in December 1983, the milking herd of 102 cows and followers were moved on April 4 1984 and the sale of the Wiltshire farm was completed on May 1 1984. The Diary Produce Quotas Regulations 1984 (SI 1984 No 1047) gave effect to the EEC Council Regulations 856/84 and 804/84, which imposed levies on milk production above prescribed quotas as from April 1 1984. The plaintiffs received a provisional assessment of primary milk quota calculated on the basis of milk production on the Wiltshire farm, but none for Greys Green Farm. The Diary Produce Quota Tribunal dismissed the reference to them of the plaintiffs' claim for a development claim and an exceptional hardship claim for quota. The plaintiffs obtained judgment on liability against the defendant in a claim for damages for negligence by the defendant in failing to advise them that they had the right to issue judicial review proceedings to challenge the decision of the tribunal. On appeal the defendant contended that the judge misconstrued the 1984 regulations and regulation 857/84; made findings contrary to the weight of the evidence; and misinterpreted his role in determining whether the plaintiffs had suffered the loss of any chance to pursue a successful judicial review.

Held: **The appeal was dismissed. On the judge's findings of fact the defendant was in breach of its duty of care in failing to advise about the possibility of judicial review and the urgency of such application. The tribunal erred in not giving reasons for its conclusions and for applying a cut-off date of March 31 1985 for an exceptional hardship claim. The plaintiffs were producers to whom article 4.1(c) of EEC Regulation 857/84 applied and were not otherwise ineligible under those regulations to receive exceptional hardship quota. The tribunal erred in law in finding that para 17(3)(*a*) and (*b*) of Schedule 2 to the 1984 regulations were not satisfied in relation to the plaintiffs' exceptional hardship claim; there was no cut-off date of March 31 1985 to exceptional hardship claims. Had an application for judicial review been made the Divisional Court would have exercised its discretion to remit the case to the tribunal. The judge was wrong to have suggested a 60% chance of establishing a claim to an exceptional hardship quota, this being a matter for the quantification of damages.**

The following cases are referred to in this report.

Poyser and Mills' Arbitration, In re [1964] 2 QB 467; [1963] 2 WLR 1309; [1963] 1 All ER 612; *sub nom Poyser* v *Mills* [1963] EGD 421; (1963) 185 EG 609

R v *Dairy Produce Quota Tribunal for England and Wales, ex parte Caswell* [1989] 2 EGLR 1; [1989] 26 EG 130, QB

R v *Dairy Produce Quota Tribunal for England and Wales, ex parte Caswell* [1989] 1 WLR 1089; [1989] 3 All ER 205, CA

Stubbs v *Hunt & Wrigley* [1992] 1 EGLR 17; [1992] 20 EG 107

This was an appeal from the judgment of Judge Bates QC (sitting as a judge of the High Court), who on December 20 1991 gave judgment on liability in a claim for damages by the plaintiffs, Alexander Robert Hood and Ann Marie Hood, against the defendant, the National Farmers Union, for negligence: see [1992] 1 EGLR 175; [1992] 25 EG 135.

Stuart Isaacs QC and Richard McManus (instructed by Lloyd Cooper) appeared for the appellant; Alan Moses QC and Richard Gordon (instructed by Dawson & Co) represented the respondents.

Giving the judgment of the court, **Peter Gibson LJ** said: This is an appeal by the defendant, the National Farmers' Union ("the NFU"), from the order of Judge Bates QC, sitting as a judge of the High Court in the Queen's Bench Division, who in an action brought by the plaintiffs, Mr and Mrs Hood, held on the trial of a preliminary issue on the question of liability only that the NFU was liable to them in negligence. Mr and Mrs Hood are dairy farmers and sought the advice of the NFU in 1985 after they had been disappointed with the decision of the Dairy Produce Quota Tribunal ("the tribunal") rejecting their exceptional hardship claim to an additional milk quota. The learned judge held that the NFU was negligent in respect of the advice which it had given them and that thereby Mr and Mrs Hood had lost a 60% chance of establishing a claim to that quota.

Mr and Mrs Hood have at all material times carried on their dairy farming business as partners. Throughout the 1970s and until November 1983 they owned a farm of 112 acres, Archers Farm, Wiltshire, with about 100 dairy cows. The milk they produced was sold to the Milk Marketing Board ("the MMB"). Early in 1983 Mr Hood came into a substantial inheritance and he and Mrs Hood looked for a larger farm on which to continue their business. In June 1983 they contracted to purchase, and on September 29 1983 they purchased, Greys Green Farm, near Henley, Oxfordshire. This was a farm of some 420 acres for which they paid £640,500 exclusive of the farmhouse. It had not been used for dairy farming for some 10 years, but between November 1983 and March 1984 they spent £49,621.25 in constructing a milk parlour, silage plant and water supply and a further £12,800 in sowing over 100 acres of grass leys and inseminating cows to produce 40 heifers. All with a view of expanding their herd; initially, they planned to have 200 milking cows, but with longer-term plans to double that number. In November 1983 they contracted to sell Archers Farm, but continued to keep their herd there until the building work permitted the moving of the herd. On

A December 27 1983, eight dry dairy cows and 20 calves were moved to Greys Green Farm and the milking herd of 102 cows and 37 followers were moved on April 4 1984. The sale of Archers Farm was completed on May 1 1984. In the middle of these major changes to their dairy farming business, Mr and Mrs Hood, like all the other dairy farmers in the European Community, suddenly found themselves at the beginning of April 1984 subject to drastic new rules intended to reduce milk production.

On March 31 1984 the council of the European Communities adopted Council Regulation (EEC) No 856/84 ("Regulation 856/84"). This recited that the market in milk products was suffering from
B structural surpluses as a result of an imbalance between supply and demand and that although a uniform coresponsibly levy had been applied, quantities of milk delivered were increasing at such a rate that disposal of surpluses was imposing financial burdens and market difficulties, which were jeopardising the very future of the Common Agricultural Policy. Accordingly, the regulation, which came into force the next day, introduced an a additional levy for an initial period of five years on quantities of milk delivered by producers to purchasers in excess of what were called reference quantities, but we would describe as quotas. A new article 5(c) was inserted in Council Regulation (EEC) No 804/68 by article 1 of Regulation 856/84:

C 1. During five consecutive periods of 12 months beginning on 1 April 1984, an additional levy payable by producers or purchasers of cows' milk shall be introduced. The object of the said levy shall be to curb the increase in milk production while at the same time permitting the structural developments and adjustments required, having regard to the diversity of the situations among individual Member States, regions and collection areas in the Community. However, the first period shall start on 2 April 1984.

It was provided that the legislation should be implemented in accordance with one of two formulas, of which Formula B is the formula relevant to this case:

D — A levy shall be payable by every purchaser of milk or other milk products on the quantities of milk or milk equivalent which have been delivered to him by a producer and which, during the 12 months concerned, exceed a reference quantity to be determined.

— The purchaser able to the levy shall pass on the burden in the price paid to those producers who have increased their deliveries, in proportion to their contribution to the purchaser's reference quantity being exceeded

The sum of the reference quantities was not to exceed what was called a guaranteed total quantity equal to the sum of quantities of milk being delivered to undertakings treating or processing milk or other milk products in each member state during 1981 plus 1% and the
E guaranteed total quantity for the UK was specified as in 15,338,000 tonnes, which was increased to 15,698,000 tonnes for the period April 2 1984 to March 31 1985. Also on April 31 1984 the Council of the European Community adopted a further Regulation (EEC) No 857/84 ("Regulation 857/84") specifying rules for the application of the additional levy. This regulation recited that the member states should be enabled to adapt the reference quantities to take into account the special situations of certain producers and to establish for this purpose, as necessary, a reserve within the guaranteed total quantity and that the scheme must "as a matter of overwhelming public interest" enter into force on April 2 1984. By article 1 the levy was fixed at no less than
F 100% of the target price for milk where Formula B was applied. By article 2 the reference quantity of a purchaser of milk delivered to him by a producer where Formula B applied could if the member state chose (and as the UK in fact chose) be equal to the quantity of milk or milk equivalent purchased by the purchaser during the 1983 calendar year weighted by a specified percentage. Article 3 provided that for the determination of the reference quantities certain special situations should be taken into account. Producers who adopted milk production development plans (in accordance with a particular directive (72/159/EEC)) lodged before March 1 1984 might obtain special reference quantities. The final paragraph of article 3.1 also provided:

Investments carried out without a development plan can also be taken into account if the Member State has sufficient information. G

Article 4, so far is relevant, provided:

1. In order to complete the restructuring of milk production at national or regional level or at the level of the collecting areas, the Member States may, in connection with the application of formulas A and B:

. . .

(b) grant an additional reference quantity to producers realising a milk production development plan approved after the entry into force of this Regulation under Directive 72/159/EEC, on condition that this plan meets the criteria referred to in Article 1(2) of Regulation (EEC) No 1946/81;

(c) grant producers undertaking farming as their main occupation an additional reference quantity, whether their herd fulfils the conditions set out in paragraph (b) or not. H

But by article 5 additional reference quantities could be granted only within the guaranteed total quantity and were to be drawn from a reserve to be constituted by the member state within the guaranteed total quantity. By article 12(c) "producer" was defined as:

a natural or legal person or group of natural or legal persons farming a holding located within the geographical territory of the Community:

— selling milk or other milk products directly to the consumer, and/or

— supplying the purchaser

By article 12(d) "holding" was defined as: J

all the production units operated by the producer and located within the geographical territory of the Community.

Both regulations were expressed to be directly applicable in all member states. These regulations can be seen to have laid down the basic framework for the new milk quota and additional levy system, imposing on each member state a maximum amount of milk that could be produced annually by producers without incurring a penal levy, but, subject to that maximum, leaving the member state some freedom in the detailed applications of that system.

To implement these regulations within the UK the Dairy Produce Quotas Regulations 1984 (SI 1984 No 1047), ("the 1984 regulations") K were made on June 23 1984 and came into operation the next day. Under these regulations three types of quotas might be available to a producer: (1) primary wholesale quota (typically, equal to the milk production in 1983 from the producer's holding less 1%); (2) secondary wholesale quota (so as far as relevant, a quota based on a producer's expected amount of milk production consequent on the producer's reliance on an investment of a capital nature made or contracted to be made before March 1 1984 and essential to the increase of milk intended to be delivered by him from his holding); and (3) exceptional hardship quota. This quota could be allocated in L satisfaction of "exceptional hardship claims". Such a claim was defined by para 17(3) of Schedule 2 to the 1984 regulations as follows:

(3) An exceptional hardship claim shall be a claim by a person to whom Article 4(i)(*c*) of Council Regulation 857/84 (which deals with producers undertaking farming as their main occupation) applies that —

(*a*) before 2nd April 1984 he has entered into, or become obliged to enter into, a transaction or made an arrangement —

. . . or

(ii) the reasonably expected outcome of which is a level of wholesale delivery of dairy produce in respect of which, or a substantial part of which, wholesale quota is not otherwise capable under these regulations of being allocated to him M

(*b*) at the time of entering into or being obliged to enter into that transaction or making that arrangement, he intended to go into or remain in business as a producer,

(*c*) he has not received, and will not receive or become entitled to receive, as a result of that transaction or argument, benefit reasonably commensurate with the want of wholesale quota to which his exceptional hardship claim relates,

(*d*) as a result of the matters specified in the preceding paragraphs of this subparagraph, he has suffered or will suffer exceptional hardship in comparison with producers in general, and (*e*) it is fair and reasonable that he should be allocated wholesale quota as a result of his exceptional hardship claim.

When Mr and Mrs Hood heard of the introduction of the new scheme of milk quotas, they found themselves in very great uncertainty, committed as they were to moving their dairy farming from the smaller Archers Farm to the much larger Greys Green Farm in the purchase of which and on the development of which for increased milk production they had incurred substantial expenditure. In April 1 1984 Mr and Mrs Hood received a provisional assessment of primary wholesale quota of 439,222 litres calculated on the basis of their 1983 milk production from 90 cows at Archers Farm with an average yield of 5,352 litres per cow. Had the new purchasers of that farm been producers within the meaning of Regulation 857/84 and had they claimed in time, that quota would have belonged to them (see article 7 of Regulation 857/84 and article 5.1 of Commission Regulation (EEC) No 1371/84 Regulation 1371/84), but because those purchasers were not producers and made no claim in time, Mr and Mrs Hood by an agreement with the MMB pursuant to regulation 8 of the 1984 regulations retained that quota in connection with Greys Green Farm. In August 1984 they made a claim for secondary wholesale quota relying on the final paragraph of article 3.1 of Regulation 857/84 and on their purchase and development of Greys Green Farm to which they had committed themselves before the quota scheme came into force. In the language of the 1984 regulations, this was a special case development claim. They claimed a quota of 750,000 litres based on their expected milk sales to the MMB between April 2 1984 and March 31 1985 from a herd which they expected to be 150 cows at March 31 1984. They also made an application for exceptional hardship quota.

A local panel of the Dairy Produce Quota Tribunal on October 12 1984 assessed their secondary wholesale quota at 220,164 litres. The exceptional hardship claim was passed to the tribunal itself. Unfortunately for Mr and Mrs Hood they decided to appeal to the tribunal against the local panel's assessment of secondary wholesale quota. They put in detailed written submissions alleging that they would suffer exceptional hardship if they were not allocated quotas of 750,000 to 1,000,000 litres for a herd of 150 to 200 dairy cows. They appeared in person before the tribunal at a brief oral hearing of both their appeal from the local panel and of their application for exceptional hardship quota on December 12 1984. They were asked about the number of milking cows which they then expected they would have on March 31 1985 and told the tribunal they would have 115 cows. On January 13 1985 the tribunal by a written decision reduced the secondary wholesale quota assessment to 85,633 litres, being the increase in the number of cows from the 1983 figure of 90 to the March 31 1985 figure of 115, *viz* 25, multiplied by the 1983 average yield of 5,352 litres per cow. That increase, the tribunal held, was justified by the investment made by Mr and Mrs Hood. By a further written decision on February 18 1985 the tribunal rejected the exceptional hardship claim. It referred to their primary wholesale quota and the reduced secondary wholesale quota, which it said was in respect of their herd of 115 cows as forecast for March 31 1985. It referred to their claim for an additional hardship quota and continued:

3. The tribunal considers that the evidence does not establish that [Mr and Mrs Hood] are suffering, or will suffer, exceptional hardship in comparison with milk producers in general as a result of the imposition of milk quotas. On the basis of their 1983 average yield they have sufficient quota for some 98 dairy cows.

4. The tribunal hold that, on the evidence before it, none of the requirements of a claim for additional quota as set out in provisions (*a*) to (*e*) of subparagraph 17(3) of the second schedule to the [1984 regulations] is satisfied in this case.

Mr and Mrs Hood were unhappy with the way the oral hearing before the tribunal had been conducted and even before the tribunal's decisions were received they wrote to the Country Landowners' Association, giving details of their claims and of the hearing before the tribunal and seeking advice on how to challenge a decision of the tribunal.

On January 17 1985, they were advised by the association that the only possible way of challenging the tribunal's decision would be to apply for an order from the High Court on the ground that the decision of the tribunal was not within its statutory powers or that the requirements of the regulations had not been complied with. But they were warned that the fact was that many dairy farmers were angry and frustrated about not being able to get the quotas that they expected to get, but that it did not follow that the tribunal had failed to act within its statutory powers or that the requirement of the regulations had not been complied with.

Discouraged by that response but still dissatisfied, Mr and Mrs Hood then turned to the NFU, which they had recently joined. They provided their local NFU group secretary, Mr John Davis, with a copy of their letter to the Country Landowners' Association. He referred the matter to the NFU's county secretary, Mr William Goldsworthy, who in turn took advice from a barrister in the NFU's headquarters, Miss Diane Smith. On February 6 Mr Goldsworthy wrote to Mr Davis, repeating Miss Smith's advice, save in one respect, and Mr Davis on February 12 1985, sent to Mr and Mrs Hood Mr Goldsworthy's letter.

Miss Smith had referred in her advice to an earlier advisory memorandum of which she was the co-author and which had been sent to county secretaries. In that memorandum the possibility of judicial review had been discussed and the comment was made: "time is very much of the essence in judicial review cases". Mr Goldsworthy in his letter raised the possibility of judicial review, but made no mention of any time constraints. He mentioned that the NFU was taking one or two cases through the courts to try and produce some kind of clarification. Mr Goldsworthy was encouraging to the extent that he said that certain arguments could be put forward by Mr and Mrs Hood. Mr Goldsworthy, on February 22 1985, asked Miss Smith to re-examine the papers "in connection with the procedure to a judiciary [*sic*] review", which, he told her, Mr and Mrs Hood had decided that they wanted. He said: "any help and guidance which we can give them in that regard would be appreciated". She replied on March 1 1985 that their case was not suitable for judicial review and her advice was passed to Mr and Mrs Hood on March 12 1985, as was a minute of Mr Goldsworthy in which he again referred to the several cases the NFU was pursuing. But on the evidence it cannot be doubted that Mr and Mrs Hood would have commenced judicial review proceedings promptly, had they been told of the time-limits.

The main contact between the NFU and Mr and Mrs Hood was through Mr Davis. As the learned judge recorded:

Mr Davis said that the impression he was left with from reading the opinions from the legal division and from his discussions with Mr Goldsworthy was that there was no immediate action which [Mr and Mrs Hood] could take to obtain redress but that the NFU was taking a number of cases through the courts which might set a precedent to enable [Mr and Mrs Hood] to make a successful claim at some time in the future. This, he said, would have been the message which he would have conveyed to the Hoods when he spoke to them, which he did at regular intervals.

And again:

Right up until the time when Mr Hood consulted his own solicitors in June 1989, Mr Davis said that he continued to be under the impression that there were still cases going through the courts which could set a precedent for [Mr and Mrs Hood] to follow. That is a message he said he would have continued to convey to Mr Hood and, on their evidence, that was the impression and the message which [Mr and Mrs Hood] received.

No mention was ever made to Mr and Mrs Hood of the requirement (in RSC, Ord 53, r4) that an application for leave to apply for judicial review should be made promptly and in any event within three months from the decision sought to be impugned.

In June 1989 Mr Hood saw an article in a farming magazine about the Court of Appeal decision in *R* v *Dairy Produce Quota Tribunal for England and Wales, ex parte Caswell* [1989] 1 WLR 1089 and because certain facts in that case bore a similarity to the present one, he instructed Mr Albert Caswell's solicitors. On July 14 1989 Mr and Mrs Hood applied for leave to apply for judicial review out-of-time, but that application was refused on paper by Rose J and a renewed application for leave was refused at an interpartes hearing by Pill J who held that there was no good reason for extending time.

On July 31 1990, this action was commenced. On April 18 1991 Master Hodgson, by consent, ordered that the question of liability be tried as a preliminary issue before the question of *quantum.* At the trial of the preliminary issue the learned judge held that Mr and Mrs Hood relied on the NFU for advice, but that the NFU failed to advise them that to challenge the tribunal's decision on exceptional hardship quotas they had to proceed by way of judicial review within three months of the date of the decision. He found that the NFU was thereby in breach of its duty to exercise reasonable care and skill in the advice which it gave and was liable in contract and in tort.

He further held that Mr and Mrs Hood were entitled to make an exceptional hardship claim because they were producers for the purposes of article 4.1(c) of Regulation 857/84 and he rejected the argument of the NFU that none of the requirements of para 17(3)(*b*), (*c*), (*d*) and (*e*) was satisfied. He found that the tribunal erred in law in three respects: (1) in failing to give reasons, (2) in looking only to the period ending March 31 1985, and (3) in not giving Mr and Mrs Hood a fair hearing.

He concluded:

> Looking at all the circumstances and putting myself in the position of the divisional court, I have to ask whether the tribunal might not on a remitter have reached a different conclusion properly directing themselves as to the law. The answer is, that I think it would.
>
> Further, I consider that, on the remitter, there is notwithstanding the complexity of the regulation, a substantial chance (say, 60%) of establishing a claim to exceptional hardship quota.

The NFU appeals on the ground essentially that the learned judge:

(1) misconstrued the 1984 regulations,

(2) misconstrued Regulation 857/84,

(3) in some instances (and in particular in relation to the finding of negligence and in assessing at 60% the chance of success) made findings which were contrary to the weight of the evidence, and

(4) misappreciated his role in determining whether Mr and Mrs Hood had suffered the loss of any chance to pursue a successful judicial review, in particular in assuming without explanation that the court on an application for judicial review would have exercised its discretion to intervene and to remit the matter to the tribunal.

Mr and Mrs Hood, by their respondents' notice, sought to support the learned judge's conclusion on additional grounds, but it has not proved necessary to go into those further matters.

We have had the benefit of admirable argument from Mr Stuart Isaacs QC, for the NFU, and Mr Alan Moses QC, for Mr and Mrs Hood, and we are much indebted to them for guiding us so skilfully and economically through the considerable complexities of the community and domestic regulations. We confess that we are left with a strong feeling of sympathy both with dairy farmers like the Hoods, understandably concerned as they were when the new quota and levy system, to which they had been subjected without warning, came into force, but left without authoritative guidance, and also with the NFU, inundated as it was with requests from farmers for assistance and advice at a time when so much was uncertain.

Numerous issues emerged in the course of argument, but it is convenient at this stage to identify those which may call for decision for the resolution of this appeal. The first issue is whether the NFU was negligent in not advising Mr and Mrs Hood of the time-limits for judicial review. If it was negligent, that conclusion would not avail the Hoods unless they establish that:(a) on a timeous application by them for judicial review the divisional court would have acceded to that application and remitted the exceptional hardship claim for a rehearing by the tribunal; and (b) there is more than a negligible chance that the tribunal would have awarded exceptional hardship quotas to them.

Mr Moses submitted that the application for judicial review would have succeeded because of the number of errors of law by the tribunal including: (i) a failure to give reasons; (ii) the adoption of a cut-off date of March 31 1985 for an exceptional hardship award; and (iii) the holding that none of the requirements of subparas (*a*) to (*e*) of para 17(3) of Schedule 2 to the 1984 regulations are satisfied.

Mr Isaacs challenged the factual basis of (i) and (iii), and submitted that Mr and Mrs Hood were not eligible for exceptional hardship quota, first because they were not "producers" to whom article 4.1(c) of Regulation 857/84 applied for the purposes of para 17(3) and, second because they had no entitlement to primary wholesale quota which was, he said, a prerequisite to the grant of exceptional hardship quota. The first submission and, unless the Hoods had a primary wholesale quota entitlement, the second, raise questions of interpretation of Regulation 857/84 and it is common ground that unless this court is of the view that the interpretation is clear and free from doubt, the possibility of the necessity of a reference to the European Court of Justice under article 177 of the Treaty of Rome arises.

Mr Isaacs further submitted that the tribunal was right to conclude that none of subparas (*a*) to (*e*) of para 17(3) was satisfied. He also submitted that it was a separate question, not addressed by the learned judge, whether the divisional court would have exercised its discretion to order a remittal.

Finally, both sides presented submissions on whether the learned judge was right to make an assessment in percentage terms of the chance which the Hoods claim to have lost in consequence of the NFU's negligence.

We propose to deal with the issues in the following order:

(1) Was the NFU negligent?

(2) Did the tribunal —

(a) fail to give reasons, and/or

(b) apply a cut-off date of March 31 1985 and so err in law?

(3) Is it clear and free from doubt that the Hoods were or were not eligible under Regulation 857/84 and, if not, should this court refer the question to the European Court of Justice?

(4) Did subparas (*a*) and (*b*) of para 17(3) (which largely turn on a question of law) apply to the circumstances of Mr and Mrs Hood?

(5) Did Mr and Mrs Hood have no chance or a chance which is more than negligible of establishing before the tribunal at a rehearing that subparas (*c*), (*d*) and (*e*) of para 17(3), (which depend largely on value judgment by the tribunal) apply to their circumstances?

(6) Would the divisional court have remitted the case to the tribunal for a rehearing?

(7) Was the learned judge correct in assessing that Mr and Mrs Hood lost a 60% chance of establishing a claim to exceptional hardship quota?

1 *Negligence*

It was common ground between the parties that: (1) the NFU held itself out as competent to advise Mr and Mrs Hood as to the consequences and legal effect of the tribunal's decision and as to whether it could and should be challenged and knew that they relied on such advice; (2) the NFU owed them a duty to so advise them with reasonable care and skill; but (3) the NFU was not acting as their solicitors. The crucial question is as to the scope of the duty assumed by the NFU: did it extend to advising Mr and Mrs Hood on the procedure for alleging the decision of the tribunal on the exceptional hardship claim? Mr Isaacs submits that it did not. He contends that it is Miss Smith's perception of what she was asked to advise (*viz* on whether secondary wholesale quota was available in respect of development not completed by March 31 1985) and of the likely outcome of proceedings for judicial review (*viz* that no more quota would have been allocated) which is relevant and that unless it can be established that that perception on the material out before her was negligent, it is immaterial that her advice about time being of the essence in judicial review proceedings was not communicated to the Hoods. He further relied on what he described as Mr Goldsworthy's honest impression that Mr Hood knew that applications for judicial

A review had to be made within three months and Mr Goldsworthy's honest belief that the Hoods had solicitors acting for them.

We cannot accept these submissions. The perception of Miss Smith, whom no one has criticised, and the impression and belief of Mr Goldsworthy seem to me to be nothing to the point, and in any event the learned judge did not accept the evidence of Mr Goldsworthy on his impression of Mr Hood's knowledge of judicial review. The question is whether the NFU was negligent. Mr Hood, who was accepted by the learned judge as a witness of truth, referred in his witness statement to telling Mr Davis of his dissatisfaction with the tribunal's decision on the exceptional hardship claim and said:

B I told Mr Davis that we needed the expertise of the NFU to guide and advise us on how to challenge the tribunal's decision.

Mr Davis in his witness statement (and it must be remembered that he was the NFU official who had direct contact with the Hoods) said that it was clear to him that so far as their milk quota problem was concerned they were expecting the NFU to advise them on their rights and remedies.

We have already referred to Mr Goldsworthy's minute of February 22 1985 to Miss Smith in which he reveals his awareness of what and guidance Mr and Mrs Hood wanted. And yet the NFU, while accepting the role of adviser to give that help and guidance, failed to C advise them of the time-limits for judicial review. Indeed, the NFU by raising false hopes that there were cases which it would be pursuing through the courts and which might assist the Hoods implied that they need not get on with judicial review proceedings. In our judgment, the learned judge's findings of fact, to which we have already referred, leading to his conclusion that the NFU in breach of its duty failed to exercise reasonable care and skill in the advice which it gave, are unassailable, as is that conclusion.

2 *Error of law by the tribunal*

It is common ground that there was a duty on the tribunal to give D reasons (see section 12(1) of Tribunals and Inquiries Act 1971) as Megaw J said *In Re Poyser and Mills' Arbitration* [1964] 2 QB 467 at p478:

Parliament provided that reasons shall be given, and in my view that must be read as meaning that proper, adequate reasons must be given. The reasons that are set out must be reasons which will not only be intelligible, but which deal with the substantial points that have been raised.

Mr and Mrs Hood had set out in a detailed document supplied to the tribunal why in their submission each of the conditions for exceptional hardship quota was satisfied. Mr Isaacs submitted that E adequate reasons were given for the rejection of their claim because in each of paras 3 and 4 of the tribunal's decision, which we have already cited, it is made clear that the basis of the decision is the evidence before the tribunal. For our part we think that Mr Moses was entirely correct in saying that the tribunal stated its conclusions but, save possibly for the second sentence in para 3, no reasons for those conclusions. We do not doubt that the tribunal was under heavy pressure of work at the time, but we regret to have to say that it erred in failing to give an adequate explanation of its decision and to deal with the substantial points raised.

It is also common ground that, as Popplewell J held in *R* v *Dairy* F *Produce Quota Tribunal for England & Wales, ex parte Caswell* November 23 1988*, on the true construction of the 1984 regulations is for the purposes of an exceptional hardship claim there was no applicable cut-off date of March 31 1985. Mr Isaacs disputes that the tribunal did apply such a cut-off date in the present case, and it is correct that the only express reference to that date is in relation to the factual recitation of the forecast size of the herd which explained the secondary wholesale quota assessment. But there is nothing to explain why the planned increase in the herd after March 31 1985 had no effect on the outcome of the application and it is clear from the *Caswell* case, decided as it was by the tribunal only five days before the decision on the exceptional hardship claim in the present case, that G the tribunal had taken the view that it was appropriate to apply a cut-off date of March 31 1985 for such a claim. It is beyond belief that the tribunal would apply a different interpretation of the 1984 regulations so shortly after it had decided the *Caswell* case. We are prepared to infer that the tribunal erred in law on this point also.

Eligibility under Regulation 857/84

To be eligible for exceptional hardship quota, para 17(3) requires that the claimant must be a person to whom article 4.1(c) of Regulation 857/84 applies. Mr Isaacs submitted, and Mr Moses was content to accept, that that requirement needed to be satisfied at April H 2 1984, when the milk quota and additional levy scheme came into force.

Mr Moses contended that Mr and Mrs Hood were "producers undertaking farming as their main occupation" and that by reason of the definitions of "producer" and "holding on" in article 12 they were such producers as being persons farming a production unit, *viz* Archers Farm, operated by them on that day and located within the community. That is the literal construction which found favour with the learned judge.

Mr Isaacs' contention was that to be a person to whom article 4.1(c) applies, the producer must be farming the production unit in J respect of which the additional reference quota is claimed. He submitted that this is the construction of article 4.1(c) which should be adopted in accordance with the purposive approach applicable to the interpretation of community legislation. The purpose of the scheme was, he said, to curb the increase in milk production and he argued that the literal construction ran counter to that purpose as it would have the effect of increasing the acreage devoted to dairy production and would therefore encourage rather than discourage milk production, both the vendor and the purchaser of a holding being able to base a claim for quota on the same holding. He referred us to a number of decisions of the European Court to illustrate the adoption of K that purposive approach in construing community provisions restrictively, but we were shown no authority that is directly in point on the provisions we are asked to construe.

Both Mr Moses and Mr Isaacs contended that his construction was so clearly correct that if a decision on this question of community law is necessary to enable the court to give judgment, it is nevertheless not necessary to refer the question to the European Court of Justice. We have considered anxiously whether the point raised is so clear and free from doubt that this court itself can properly decide it without a reference, but we have reached the firm conclusion that the literal construction is manifestly correct.

L The construction advanced by Mr Isaacs seems to me to involve implying a qualification on the words used in article 4.1(c) when read with article 12, for which we can see no justification. The literal construction does not lead to an increase in milk production. Those who are eligible under article 4.1(c) for additional reference quantities, if a member state chooses to grant them, can only receive quotas out of the reserve referred to in article 5 and that will not raise the guaranteed total quantity of the member state. The stated purpose of article 4 was to complete the restricting of milk production and that article recognised that there were exceptional situations when it might be appropriate to allow a member state to grant additional M quotas.

In our view, it would be anomalous and grossly unfair if dairy farmers like the Hoods, who had produced milk for years (including the year the milk production in which was taken into account for the purpose of the guaranteed total quantity), who were producing milk on April 2 1984 and who, well before the scheme came into force, had incurred significant expenditure with a view to continuing and increasing milk production, should be wholly ineligible for the additional quotas contemplated. The question we are concerned with at this stage is eligibility, not with the allocation of quotas. The need for specific rules to govern transfers of holdings and reallocations was recognised and met: see articles 7 and 8 of Regulation 857/84 and

*Editor's note: Reported at [1989] 2 EGLR 1.

article 5 of Regulation 137 1/84. Under these (and the 1984 regulations) there is no possibility of two quotas being granted in respect of the same holding (in the sense that the quota for the same holding could be enjoyed at one and the same time by a transferor and a transferee). But we see no difficulty in eligibility being founded for both a transferor and a transferee in respect of the same land, as that is what justice may demand in an unusual set of circumstances such as that which obtains here. In our judgment therefore, the literal construction is plainly correct and we can so decide without requesting a ruling of the European Court of Justice.

Is it necessary as a matter of construction of Regulation 857/84, for a producer to be entitled to primary wholesale quota if he is to be eligible to receive an "additional reference quantity"? The argument turns on the meaning of "additional". Mr Isaacs, relying on the decision of Phillips J in *Stubbs* v *National Farmers Union* (June 11 1991)* to the effect that the producer must be so entitled, answers that question in the affirmative. Mr Moses answers it in the negative, contending that *additional* means no more than "another". We see considerable force in this contention of Mr Moses, but it is unnecessary to decide this question of construction, as in any event the Hoods were entitled to, albeit initially defeasibly, and did receive primary wholesale quota.

In our judgment, therefore, it is clear that Mr and Mrs Hood were producers to whom article 4.1(c) applied and were not otherwise ineligible under Regulation 857/84 to receive exceptional hardship quota.

4 *Para 17(3), (a) and (b)*

Mr and Mrs Hood claim that before April 2 1984 they entered into a transaction or made an arrangement, *viz* the purchase and development of Greys Green Farm, the reasonably expected outcome of which was a level of wholesale delivery of dairy produce, *viz* between 750,000 and 1m litres from 150 to 200 cows, in respect of a substantial part of which, *viz* that quantity less the primary and secondary wholesale quota as totalling 524,865 litres, wholesale quota was not otherwise capable upper the 1984 regulations of being allocated to them.

Mr Isaacs submitted that as the transaction or arrangement is the same transaction or arrangement relied on by the Hoods for their secondary wholesale quota claim in respect of which they had received quota, quota was capable of being allocated in respect of the entire expected level of wholesale delivery, and so the claim for exceptional hardship quota should have been rejected *in limine*. But that, in our view, deprives the words *a substantial part of which* of any meaning. Subpara (*a*) must contemplate that there being be a situation in which quota is allocated or is capable of being allocated in respect of milk production consequent on a development plan and yet the same development plan can found a claim to exceptional hardship quota in respect of further amounts of milk production.

On Mr Moses' construction of the regulations, it is explicable on the basis that there is a cut-off date of March 31 1985 for secondary wholesale milk quota whereas there is no such cut-off date for the exceptional hardship quota. For this he relies on what was said by Popplewell J in the *Caswell* case to that effect.

To understand this point, we must refer to the relevant parts of the 1984 regulations. Secondary wholesale quota can be allocated pursuant to a special case claim made by an applicant in accordance with para 7 of Schedule 2 to the 1984 regulations. By para 8(1) each special case claim is to be examined by the minister who, shall accept or reject it in accordance with the other provisions of para 8. By subpara (2) he is to accept it only if, *inter alia*, it is a development claim and he is satisfied that the applicant may be entitled to secondary wholesale quota. A development claim is defined in regulation 2(1) of the 1984 regulations as meaning a special case claim based on article 3.1 of Regulation 857/84 submitted by a producer in respect of investments of a capital nature on his holding calculated to achieve an average monthly quantity of milk production on his holding greater by a specified certificates adjourn than the average of his milk production taken from each month of the period with reference to which his primary wholesale quota was estimated. By Schedule 2, para 8(4):

A development claim shall be accepted by the Minister only if it contained the following particulars from the applicant —

(*a*) the subparagraph of Article 3(1) of . . . Regulation 857/84 on which he relies;

(*b*) evidence that his reliance is based on an investment of a capital nature (which was made or in respect of which an obligation was entered into before March 1st 1984) essential to the increase of dairy produce intended by him to be delivered by wholesale delivery from his holding;

(*c*) the total quantity of dairy produce which he expects to sell by direct sale or deliver by wholesale delivery from his holding in the first quota year; and

(*d*) his secondary wholesale quota claim, being a specified quantity of dairy produce equal to that percentage of the quantity of dairy produce claimed by him as his development quantity in relation to his holding which he does not claim as secondary direct sales quota.

By regulation 2 (1)

"development quantity" means the quantity of milk calculated, by reference to the increase from a producer's annualised quantity (that is to say his average monthly quantity of milk production, in any base period applicable to him, multiplied by 12) to the quantity of milk reasonably expected to be produced on his holding in the first quota year, by taking, in respect of any expected increased quantity

— certain specified amounts, special case claims which are not rejected have to be reserved for further examination and by Schedule 2, para 9(3):

The estimate of a secondary wholesale quota of each applicant shall be —

(*a*) such quantity of dairy produce (if any) as is considered by the further examination body to be justified by his special case claim, or

(*b*) his secondary wholesale quota claim —

Whichever is the less.

The claimant can cause his claim to be determined by the tribunal and by para 10(1) the tribunal shall determine the secondary wholesale quota in accordance with provisions similar to those in para 9(3).

It will be noted that by para 8(4)(*c*) and (through the references to "development quantity"), (*d*) there are two references to the first quota year and it was after alluding to the definition of development quantity that Popplewell J commented in respect of the development claim: "it is obviously geared to the year ending 31 March 1985". With that he contrasted the provisions relating to exceptional hardship quota.

Mr Isaacs, in his reply, pointed out that Popplewell J was wrong to suggest that there was such a cut-off date for a development claim. He said that it did not follow that because particulars relating to the first quota year had to be supplied to the minister, attention had to be confined to that period and he submitted that there was nothing in para 9(3) or 10(1) to limit the further examination body or the tribunal to consider the period up to March 31 1985 only and instead they were required to consider that what was justified by the development plan.

We cannot accept Mr Isaac's submissions. It would be remarkable if the particulars to be supplied to the minister did not contain all the information needed to determine the secondary wholesale quota and there is nothing in para 8(4) to suggest that the applicant should supply information relating to what was expected to happen after the end of the first quota year. Equally for the purpose of paras 9(3) and 10(1) it would be surprising if what could be justified by the special case claim was on a different basis from that on which the secondary wholesale quota claim had been made. In our judgment, therefore, a cut-off date of March 31 1985 applies to secondary wholesale quota. We are the happier to have reached this conclusion for two reasons. First, it appears to have been the practice of the tribunal not to look beyond March 31 1985 and the practice will have been confirmed by the *Caswell* case. It is very likely therefore, that numerous determinations will have been made on that basis and we would be reluctant at this late stage in the history of secondary wholesale quota to declare the

*Editor's note: *sub nom Stubbs* v *Hunt & Wrigley* [1992] 1 EGLR 17.

A law to be other than that which it has previously been understood to be. Second, it is singularly unattractive for the NFU now to advance the argument propounded on its behalf, the implication of it being that the tribunal was wrong in its approach to the assessment of secondary wholesale quota for the Hoods and yet the NFU failed to advise the Hoods to seek a judicial review of that decision by the tribunal.

We therefore conclude that subpara (*a*) of para 17(3) applies to the circumstances of Mr and Mrs Hood, and, as we understand him, Mr Isaacs accepts that if the requirements of subpara (*a*) are satisfied, so on the facts are the requirements of subpara (*b*). It follows that the
B tribunal erred in law in its holding to the contrary.

5 *Para 17(3), (c), (d) and (e)*

These subparagraphs impose conditions requiring the the tribunal to make a value judgment as to whether the Hoods became entitled to receive benefit "reasonably commensurate" with the want of wholesale quota to which their exceptional hardship claim related, whether they suffered "exceptional hardship in comparison with producers in general" and whether it is "fair and reasonable" that they should be allocated wholesale quota as a result of that claim. At this stage the court is only considering the preliminary issue of liability and it is sufficient to say that we are satisfied on all the evidence put
C before the court that Mr and Mrs Hood had more than a negligible chance that they would have established at a rehearing those subparagraphs applied to their circumstances.

In view of the tribunal's errors: (i) in its construction of subparas (*a*) and (*b*) and (ii) in its application of a cut-off date of March 31 1985 to exceptional hardship claims, its consideration of subparas (*c*), (*d*) and (*e*) would appear to have been vitiated, and certainly the way the tribunal expressed itself in the decision is insufficient to establish the contrary.

6 *Remittal*

In the light of the conclusions that the tribunal erred in law in
D failing to give reasons, in applying a cut-off date of March 31 1985 to the exceptional hardship claim and in its construction and application of para 17, there being be no question but that the Divisional Court, had an application for judicial review been made in time, would have exercised its discretion to remit the case to the tribunal for a rehearing.

7 *Assessment of the lost chance*

Both sides made submissions on the meaning and effect of the learned judge's holding that there was "a substantial chance (say 60%) of establishing a claim to exceptional hardship quota". With all respect
E to the learned judge, he was ill-advised to suggest any figure, that being a matter for the later stage of the action when damages will be quantified. Both Mr Isaacs and Mr Moses helpfully agreed that if the court concluded that the Hoods had lost a more than negligible chance of establishing a claim to exceptional hardship quota, it should so declare without quantifying the chance. Indeed, for the avoidance of doubt we would wish to make clear that in assessing the lost chance as more than negligible we do not thereby intend to trench in any way on the freedom of the trial judge to quantify the damages as he sees fit in the light of the evidence put before him.

We dismiss this appeal.

F *Appeal dismissed.*

Court of Appeal G

November 17 1993

(Before Lord Justice BUTLER-SLOSS, Lord Justice STUART-SMITH and Sir Tasker WATKINS)

R v MILK MARKETING BOARD

Estates Gazette March 12 1994

[1994] 10 EG 106

Milk quotas — Alleged breach of scheme relating to "withholder" monthly returns — Whether disciplinary committee of Milk Marketing Board has power to impose penalties where breach also involves ingredients of a crime H

On June 6 1990 the disciplinary committee of the Milk Marketing Board decided that the applicant registered milk producer should pay penalties of £3,200 under para 77(1)(c)(i) of the Milk Marketing Scheme and repay losses to the board of £18,264 under section 10 of the Agricultural Marketing Act 1958. The committee had found proved 16 charges arising out of the monthly returns which the applicant, as a "withholder", was required to make to the board for the period January 1988 to March 1989. The judge in the court below J granted an application by way of judicial review of that decision and quashed it on the ground that its decision involved making findings which also constituted criminal offences, which could be tried only by the criminal courts. The board appealed.

Held: **The appeal was allowed. The committee has power to hear allegations of breaches of the board's scheme where, as here, (1) the allegations amount to a criminal offence, which includes all the ingredients of the breach of the scheme, plus extra crime requirements; and (2) if the crime and the breach of the scheme share some ingredients, but each has separate ingredients. The board had to prove that the applicant was under an obligation to** K **make a return and that in doing so he knowingly made a false statement as to the quantity of milk produced by him; dishonesty is not an ingredient of this alleged breach of the scheme although it is under section 2(1)(*c*) of the Theft Act 1978.**

No cases are referred to in this report.

This was an appeal by the appellant, Milk Marketing Board, from a decision of Hutchison J, who on October 16 1992 granted the applicant/respondent, Stephen Brook, judicial review of a decision of the Milk Marketing Board by its disciplinary committee to impose L fines and penalties on the applicant.

Stephen Coward QC and Simeon Maskrey (instructed by Wedlake Bell) appeared for the appellant; Cherie Booth (instructed by Beviss & Beckingsale, of Chard) represented the respondent.

Giving judgment, SIR TASKER WATKINS said: This is an appeal by the Milk Marketing Board from the decision of Hutchison J on October 16 1992 granting an application for judicial review by the applicant, Stephen Brook, to remove into the High Court and quash a decision of the disciplinary committee of the board on June 6 1990 that the applicant should pay penalties of £3,200 under para 77(1)(c)(i) of the Milk Marketing Scheme (as amended) and repay losses to the M board of £18,264.20 under section 10 of the Agricultural Marketing Act 1958, "the 1958 Act".

A number of grounds were relied upon by the applicant and before the learned judge. He declined to give the applicant relief on the grounds of natural justice and Wednesbury unreasonableness, but he accepted the submissions of Miss Cherie Booth, on behalf of the applicant, that the disciplinary committee had no jurisdiction to entertain the complaints preferred by the board against him. Accordingly, he quashed the findings of the disciplinary committee and the penalties which it had imposed upon the applicant.

The board appeals against those decisions on the bases that:

A (1) The judge was wrong in law in holding that para 77(1)(d) of the scheme and/or section 9(2) of the 1958 Act prevents the disciplinary committee from entertaining any charge brought under para 77(1)(c)(i) of the scheme where, even though the charge under the paragraph does not embrace the elements of some other statutory offence, the real nature of what is alleged against a milk producer constitutes some offence under a statutory provision, namely section 2(1)*(c)* of the Theft Act 1978.

(2) The judge was wrong in any event in finding that what was alleged against the applicant was a dishonest falsification of his records in order to avoid payment of contributions due to the board B because dishonesty and/or any such intention were not elements of the offence under para 77(1)(c)(i) of the scheme and were not so alleged against the applicant.

(3) That the judge ought to have found that the disciplinary committee had the power to find contraventions under para 77(1)(c)(i) of the scheme (and to impose fines and penalties accordingly), because on its terms the contravention under the paragraph does not constitute an offence under any other statute and, in particular, does not constitute an offence under section 2(1)(*c*) of the 1978 Act, dishonesty and/or intention to obtain any exemption and/or abatement of liability to make a payment and/or the obtaining of any such exemption and/or C abatement forming no part of the contravention under the paragraph, but being essential conditions of liability under the statutory offence.

The applicant was charged, *inter alia*, as follows:

> That you knowingly made a false statement in reply to the Board's demand under paragraph 75 of the Scheme for a return of the quantity of milk of your own production which (i) you sold by retail or semi/retail, and (ii) you sold under your Producer Processor Agreement in the month of June 1988.

There were 37 such charges levelled against the applicant arising out of the monthly returns of his milk production that he was required to make to the board during the period January 1998 to March 1989. The disciplinary committee found the case proved in respect of the D first 16 charges and dismissed the balance of them. It fined the applicant £200 upon each of the 16 charges found proved and ordered him to repay losses in the sum already mentioned under section 10 of the 1958 Act which provides:

> Where a contravention of a provision of a Scheme (being a contravention for which the disciplinary committee have the duty to impose, and the board the duty to recover, a penalty) causes loss to the board, the disciplinary committee may, if the written statement of the charged served on the producer in accordance with the provisions of subsection (4) of the last foregoing section embodies an estimate by the board of the amount of the loss, together with a statement of the grounds on which that estimate is based, require by their E decision that the producer shall pay to the board, in addition to any penalty imposed, such sum, to be stated in the decision, as the committee may think fit, not exceeding the sum which appears to the committee justly to represent the amount of the loss;
>
> provided that the sum so stated in the decision shall not exceed the amount in the said estimate.

Prior to the proceedings before the disciplinary committee, the applicant had been served with an estimate of the board's loss, which was calculated by assuming that all the milk which the applicant has sold to the roundsmen, from whom the board had obtained statements, was milk of his own production which had not been declared on his F monthly return.

In his usual clear judgment, the judge, as he said, set the scene in a brief summary which I gratefully adopt.

Under the scheme all milk producers in England and Wales must be registered with the board and are, in general, obliged to supply all their milk to the board, unless it agrees otherwise. The board sells the milk thus supplied at differing prices, depending on the use to which it is to be put. Milk destined for drinking commands the highest price, known as the board selling price. Milk intended for other uses is sold at lower prices.

The board pools the proceeds of sales and, after deducting expenses and any moneys required for reserves, distributes the proceeds to producers, who therefore receive what is known as the G average wholesale producers price ("AWPP"). However, the board permits certain producers to sell their milk otherwise than to the board, such persons are commonly referred to as "withholders", and fall into two categories — producer retailers and producer processors. Obviously such persons receive prices in excess of the AWPP. A mechanism, therefore, exists for correcting the unfairness that would otherwise result, by requiring withholders to pay to the board sums calculated on the difference between the receipts on their direct sales and what they would have received at the AWPP rate had they not been withholders.

It is of the essence of this equalisation scheme that withholders H should keep accurate records of their direct sales and make accurate returns to the board. A failure to do so may enable the withholder to secure an unfair advantage over the producers who sell direct to the board.

The board's case against the applicant was that he had not made proper returns of his sales, thereby securing an advantage by depriving the board of sums to which it was entitled.

Para 71(1) of the scheme, as material, states:

> (c) If any registered producer
>
> (i) fails without reasonable excuse to comply with any demand made by the Board under paragraph 75, or knowingly makes any false statement in J reply thereto . . . then the Disciplinary Committee shall impose upon him and the Board shall recover from him such monetary penalty (not exceeding £200) as that Committee think just; . . .
>
> (d) no such penalty shall be imposed in respect of such a contravention as aforesaid which constitutes an offences under any Act other than the Act of 1958.

The provision in (d) the judge compared with section 9(2) of the 1958 Act, which states:

> . . . every Scheme shall require the disciplinary committee to impose on, and the board to recover from, any registered producer who contravenes any provision of the Scheme . . . such monetary penalties as may be specified by K the scheme, so however that no such penalty shall be imposed in respect of a contravention of the scheme which constitutes an offence under this or any other Act.

Section 2(1)(*c*) of the 1978 Act states:

> . . . where a person by any deception
>
> (*c*) dishonestly obtains any exemption from or abatement of liability to make a payment . . . he shall be guilty of an offence.

The judge, in addressing the submissions to him made by Miss Booth, which are succinctly set out in her skeleton argument and were forcibly also submitted by her to us, said that she had, in his view L rightly, concentrated on the provisions of section 9(2), rather than those of para 77(1)(d).

He summarised her submissions in this way.

The board alleged against the applicant that he had deliberately and dishonestly failed to make full and accurate returns of his sales, with the object and effect of depriving the board of the payments to which it was entitled and retaining those moneys for himself. On the facts of this case, the board's allegations plainly imported dishonesty on the part of the applicant. Accordingly, the charges in respect of which the disciplinary committee found the applicant guilty fell squarely within section 2(1)(*c*) of the 1978 Act. M

Miss Booth also drew attention to the way in which the case was presented on behalf of the board to the disciplinary committee, which involved the assertion that the applicant had over a period of years deliberately made false statements as to his milk production in order to avoid the levy. The case was never put on the basis of incompetence or inadvertence, but that the applicant had been caught in deliberately and dishonestly evading his liability to pay what was due.

The object of section 9(2) of the 1958 Act is to ensure that criminal conduct is dealt with by the criminal courts where all the safeguards of criminal procedure and jury trial are available, and that the prohibition that it contains recognises that the disciplinary committee is not an appropriate body to deal with criminal offences. It is an informal domestic body and Parliament clearly envisaged that it should have power to deal only with minor infringements of the Act, but that in cases where serious offences were involved an accused producer should be entitled to the full protection of the law.

Miss Booth further submitted that the substance of the charge before the board should be looked at rather than its form, and that if at any time during the proceedings it becomes apparent that a criminal offence is comprised in the allegation made, then the disciplinary committee must decline jurisdiction and abort the proceedings.

Finally, Miss Booth argued that the prohibition against the imposition of a penalty contained in section 9(2) means, in effect, that in cases where the contravention alleged constitutes an offence under another statutory provision, the committee does not have jurisdiction to try the charge at all.

The judge was persuaded that Miss Booth was correct in submitting that the prohibition in section 9(2) and in subpara 71(1)(d) prevents the disciplinary committee from entertaining charges where, even though the formulation of the charge does not embrace the elements of some other statutory offence, the real nature of what is alleged is ("constitutes") some offence under a statutory provision. That, he found, was the position in this case. So he held that the applicant had established the jurisdiction point.

Earlier in his judgment, the judge posed the question which he regarded as the crucial question in the following way:

> However, when the prohibition in section 9(2) has to be considered in relation to other Acts the question, does the contravention of the Scheme with which the producer is charged constitute an offence under that Act? — presents real difficulties. Does "constitutes" mean that one must have regard merely to the way in which the Board framed the charge to those elements which by their formulation the board accept the obligation to prove; or is it necessary to consider the substance of the charge, posing the question whether in that particular case proof of the elements of the charge which the board accept the obligation to discharge in fact involves proof of a further element or further elements which constitutes some statutory offence?

I agree with him that the matter of construction of subpara 77(1)(d) and section 9(2) of the 1958 Act is crucial to the question as to whether the board had jurisdiction to hear the charges against the applicant. True it is that the judge analysed the impact of the provisions of section 6 of the 1958 Act upon the overall issue of jurisdiction, but for reasons which will appear I think that he was misguided into doing so.

Admirably presented though they were, in my judgment, the judge was in error in accepting the arguments of Miss Booth and in the ultimate conclusion he reached. I have been entirely persuaded of this by the submissions of Mr Stephen Coward QC, on behalf of the board, which clearly emerged from his skeleton argument which is accompanied by a what he calls a "pictorial analysis" which lends further clarity, if that be necessary, to his illumination of the error into which the judge fell.

Mr Coward's examination of any conceivable impact which the provisions of section 6 of the 1958 Act can have upon the matter of jurisdiction results, in my judgment, inevitably with the conclusion that it does not bear upon the essential matter of construction, save to say that the 1958 Act, in section 6 and elsewhere, creates criminal offences, each one of whom obviously has to be strictly proved according to the ingredients of it, otherwise the board will have failed to establish any charge of that kind brought against a milk producer. Section 9(2) and para 71(1)(d) have to be looked at independently of all that and strictly construed.

Mr Coward submitted, and I accept, that if a matter before the disciplinary committee potentially overlaps with a criminal liability, but does not in fact or of necessity require the proof of each and every ingredient of the criminal offence, it does not constitute an offence under "any other Act". The word "constitutes", he said, in para 77(1)(d) means that a tribunal is debarred from imposing a penalty on a registered producer if, but only if, the board, in order to succeed, must establish each and every ingredient of the criminal offence.

In the "pictorial analysis", aided by very useful illustrations, the following propositions, all of which I accept, appear. First, it is stated that if the ingredients of the crime and the breach, that is to say the breach of Act or scheme are identical, there is no power to hear the charge before the disciplinary committee. Second, that if the breach includes all the ingredients of the crime, plus extra breach ingredients, there is no power to hear. Third, if the crime includes all the ingredients of the breach, plus extra crime requirements, there is power to hear. And, fourth, if the crime and the breach share some ingredients, but each has distinct ingredients, there is power to hear.

Having regard to the charge which the applicant faced, and of which in 16 instances the board convicted him, it cannot possibly be said that the first two propositions applied. It can rightly be said, in my judgment, that propositions three or four, or both, did apply. The essential requisite as to construction was to examine what it was the board had to prove in the charge it made. It had to prove that the applicant was under an obligation to make a return to the board that in doing so he knowingly made a false statement as to the quantity of milk produced by him and which he had sold by retail, or semi-retail, and under his producer processor agreement. Dishonesty is not an ingredient of this alleged breach of the scheme. It is, of course, under section 2(1)(*c*) of the 1978 Act. Nothing which was said on behalf of the board to the disciplinary committee with reference to the applicant's conduct can serve, in my view, to alter in any way what it is Parliament has clearly laid down as having to be proved as a breach of the scheme. All that was necessary for the board to do was to establish the ingredients of the charge arising out of the legislation, no more and no less.

The judge observed that it was a remarkable fact that, although a provision such as one finds in section 9(2) of the 1958 Act has existed for some 60 years, so far as the researches of counsel have been able to discover, there is no decision which bears upon the point at issue in this case. I feel bound to say that I do not find that at all surprising.

For these reasons, I would allow the appeal.

Stuart-Smith and **Butler-Sloss LJJ** agreed and did not add anything.

Appeal allowed; leave to appeal to House of Lords refused.

Court of Appeal

December 16 1993

(Before Lord Justice NEILL, Lord Justice BELDAM and Lord Justice HENRY)

HANNAFORD v SMALLACOMBE

Estates Gazette April 16 1994

[1994] 15 EG 155

Agricultural holdings — Arbitration — Whether arbitrator appointed when he received form of appointment or when he notified parties — Whether tenant complied with a notice to pay rent by a cheque received timeously but dishonoured on first presentation after time-limit and later honoured on second presentation but before notice to quit served

The respondent tenant held an agricultural tenancy from the appellant landlord. The tenant failed to pay the gale of rent due on March 25 1990 and on March 29 1990 the landlord's agent served on the tenant a notice to pay having regard to Case D of Schedule 3 to the Agricultural Holdings Act 1986; the last date for receipt of payment under the notice was May 29 1990. On May 22 or 23 the landlord's agent received a cheque from the tenant for the rent; it was presented for payment on May 25 but returned marked "refer to drawer, please re-present" on June 1 1990. The cheque was subsequently re-presented and honoured on June 5 1990. On June 8 1990 the tenant was served with a notice to quit relying on Case D. The tenant sought arbitration of the notice to quit and the parties agreed upon an arbitrator who, by letter dated August 31, accepted the appointment subject to the completion of a form of appointment signed by both parties. A form was executed by both parties by September 12 and received by the arbitrator on September 14, although he did not acknowledge it until September 18 when he reminded them that, in accordance with para 7 of Schedule 11 to the 1986 Act, their respective statements of case were to be submitted to him within 35 days of his appointment. The landlord's case was submitted by October 10, but the tenant's case had not been submitted by October 19. On that date the arbitrator informed the tenant's solicitors that, because no case had been received within the time-limit, the tenant would be precluded from putting forward any evidence at the arbitration. The tenant purported to submit a case contending that the 35-day period ran from September 18, alternatively that the arbitrator had a discretion to extend the time-limit. The arbitrator was asked to state a case on a number of questions to the county court directed to the two principal issues: (1) whether the arbitrator was appointed when he received the executed form of his appointment, or when he informed the parties that he had been appointed; and (2) whether the notice to pay rent had been complied with. The landlord appealed from the decision of the judge, who had decided the principal issues in favour of the tenant.

***Held*: The appeal was allowed.**

1. Having regard to the advance notice the parties had of the arbitrator's consent to act upon receipt of a form of appointment (a condition subsequent), the receipt of that form by the arbitrator on September 14 completed the tripartite agreement: the arbitrator was appointed and the tenant did not have to be informed of that. The tenant's statement of case was therefore not delivered within the statutory time-limit of 35 days from the appointment of the arbitrator. The arbitrator did not have power to extend the statutory time-limit.

2. Upon the dishonouring of the cheque on first presentation, the tenant failed to comply with the notice to pay rent and the landlord became entitled to serve a notice to quit relying on Case D in Schedule 3 to the 1986 Act. The second presentation of the cheque before service of the notice to quit did not waive the right to serve such a notice: *Stoneman* v *Brown* could not be distinguished.

The following cases are referred to in this report.

Beevers v *Mason* (1978) 37 P&CR 452; 248 EG 781, CA
Cohen v *Hale* (1878) 3 QBD 371
Hadley (Felix) & Co v *Hadley* [1898] 2 Ch 680
Luttenberger v *North Thoresby Farms Ltd* [1993] 1 EGLR 3; [1993] 17 EG 102
Nicks & Son Ltd v *Taylor's Bakery Ltd* [1962] NZLR 286
Romer & Haslam, Re [1893] 2 QB 286
Stoneman v *Brown* [1973] 1 WLR 459; [1973] 2 All ER 225, CA
Tradax Export SA v *Volkswagenwerk AG* [1970] 1 QB 537; [1970] 2 WLR 339
University College, Oxford (Master & Fellows) v *Durdy* [1982] Ch 413; [1982] 3 WLR 94; [1982] 1 All ER 1108, CA

This was an appeal against the decision of Judge Darwall-Smith, who in Exeter County Court made determinations on a number of questions in a case stated from the arbitrator, Mr John Neason FRICS FAAV, in an arbitration on a notice to quit served by the appellant, Mrs Betsy Pym Smallacombe, upon the respondent, Vivian Reginald Hannaford.

Barry Denyer-Green (instructed by Horwood & James, of Aylesbury) appeared for the appellant; Dirik Jackson (instructed by Bond Pearce & Co, of Plymouth) represented the respondent.

Giving the first judgment at the invitation of Neill LJ, HENRY LJ said: Mr Vivian Reginald Hannaford was, at the material time, the tenant of an agricultural holding known as Lower West Kimber, Northlew, Devon. Mrs Betsy Pym Smallacombe was his landlord. The rent was £1,700 pa, payable in arrears by half-yearly instalments of £850 on September 29 and March 25 in each year of the tenancy.

Section 26(1) of the Agricultural Holdings Act 1986 places restrictions on the ability of a landlord of an agricultural holding to terminate the tenancy thereof. Generally speaking, the tenant has the right to serve a counternotice in response to any notice to quit, and if he does so the notice to quit is not to have effect until the Agricultural Land Tribunal consents to its operation (which it may do only in certain limited circumstances).

However, section 26(2) of the Act makes an exception in certain cases which are specified in Part I of Schedule 3. One of these cases is Case D, which applies where:

> At the date of the giving of the notice to quit the tenant had failed to comply with a notice in writing served on him by the landlord, being . . .
>
> (a) a notice requiring him within two months from the service of the notice to pay any rent in respect of the agricultural holding to which the notice to quit relates.

Here the tenant did not pay his half year's rent on March 25 1990, the date when it was due. The following day the landlord issued the statutory notice requiring the tenant to pay the rent due within two months from the date of service of the notice on him. That notice was served on March 29 1990. Therefore, the last date for receipt of payment was May 29.

What in fact happened was this. On May 21 the tenant sent the landlord's agent a cheque for £850 dated May 18. That was received on behalf of the landlord on May 22 or 23. It was paid into the landlord's agent's client account at National Westminster Bank in Tavistock on May 25. That bank, by notice dated June 1 1990, notified the landlord's agents that the cheque had been returned marked "refer to drawer, please re-present". That cheque was subsequently re-presented by the bank and honoured on June 5 1990.

However, the landlord's agents issued a notice to quit dated June 8 1990 relying on Schedule 3, Part I, Case D of the Agricultural Holdings Act 1986, that at the date of the giving of the notice the tenant had failed to comply with the notice dated March 26 requiring payment of the outstanding rent within two months of service thereof.

Where the tenant seeks to challenge the effect of such a notice he must proceed by way of arbitration under the provisions of Agricultural Holdings (Arbitration on Notices Order) 1987. On June 20 1990 the tenant gave the landlord notice that he wished to contest "all matters arising out of the reasons stated in the notice to quit" and that he required such questions to be determined by arbitration. The parties agreed that Mr John Neason [FRICS FAAV], the senior partner of Colvilles, chartered surveyors of Exeter, be appointed arbitrator, and he was so appointed. The provisions of para 7 of Schedule 11 to the 1986 Act require the parties to the arbitration to submit their respective statement of cases to him within 35 days "from the appointment of the arbitrator".

Here the parties prepared a document agreeing to the appointment of Mr Neason as arbitrator, the execution of that document was completed on September 12 and it was posted to the arbitrator the next day. He received it on September 14, but did not acknowledge receipt of it until his letter dated September 18, sent to both parties. In this letter he said:

> I would remind you that your Statements of Case should arrive with me not later than 35 days from the date of appointment and this is a mandatory requirement.

Thirty-five days from September 14 would expire at midnight on October 19, but if his appointment was not completed until his acknowledgment of it, the time would not expire until midnight on October 23.

The landlord submitted her case in good time, it being dated October 10. The tenant had not submitted his by October 19, and on that date the arbitrator wrote to the tenant's solicitors, Bond Pearce & Co, stating:

Messrs Bond Pearce had not submitted a statement of case and they should note that they will therefore be precluded from putting forward any evidence at the hearing.

On receipt of this letter, the tenant's solicitors denied that they were out of time, and enclosed a case stated dated October 22 1990.

There were then two separate categories of issue before the arbitrator. The first related to the statutory rules, and the consequences of breach of them.

Question 1: had the tenant submitted his statement of case in time?

Question 2: if he had not, had the arbitrator power to extend that time?

Second, and quite separately, there were the various issues arising under the arbitration proper, relating to the question whether the delivery of a cheque for rent was payment on the date of delivery if that cheque were dishonoured on the first presentation, but subsequently met before the notice to quit was served.

In these circumstances, the arbitrator sensibly decided to state a case on agreed facts for the opinion of the court. Such a case was prepared with the assistance of the parties and is dated May 13 1992. It came for hearing before Judge Darwall-Smith in Exeter County Court, judgment was given on January 26 1993. His findings on the 10 questions posed are now the subject of appeal to us.

Question 1 reads, "Was the tenant's Statement of Case delivered within the statutory period allowed by para 7 of Schedule 11 to the 1986 Act?"

As will already be clear, the answer to this question depends on whether the date of appointment is the date when the arbitrator received the document appointing him (September 14), or the date when he notified the parties that he had received it and was acting on that appointment (his letter of September 18).

The facts as disclosed by the documents are these. Bond Pearce & Co, solicitors, were acting for the tenant and Kivells, estate agents, were acting for the landlord. By letter of August 16 Bond Pearce put three possible names forward and by letter of August 17 Kivells agreed to the appointment of Mr Neason. Bond Pearce then telephoned Mr Neason and he confirmed that he would accept the appointment as arbitrator. They notified Kivells of this. Mr Neason then wrote (August 31) a letter stating:

> I confirm that I would be willing to act as arbitrator in this case, subject to receiving a properly completed form signed by both parties within the appropriate time-limit.
>
> My charges will be based upon a rate of £65 per hour exclusive of VAT, travelling costs and disbursements, and I reserve the right to appoint a legal assessor if this is necessary having seen the respective parties' Statement of Case. . . .
>
> Upon receipt of a properly completed form of appointment, I will write to the parties further specifying arrangements for the conduct of the case.

Bond Pearce then prepared a draft form of appointment which they sent to Kivells. Kivells duly signed that document and the executed copy, dated September 12, was sent to Mr Neason by Bond Pearce in a letter dated September 13. The document reads:

> We Kivells . . . agents for . . . the landlord . . . and Bond Pearce . . . agents for . . . the tenant . . . appoint J S Neason . . . as sole arbitrator under the above Acts . . . to settle and determine all claims questions and differences arising between the landlord and the tenant . . . concerning the matters specified in the schedule below.
>
> We further agree that no objection shall be raised by the issue of the arbitrator of one award determining all matters so referred to him which shall be final and binding on the landlord and the tenant and the persons claiming under them respectively in accordance with the above provisions.
>
> We undertake and agree to sign any further submissions appointments or other documents and to any other acts or things which may be required by the arbitrator in his opinion to empower him legally and effectively to hear settle and determine the above matters or any of them.

The arbitrator received this document on September 14 and on September 18 wrote to the parties in the following form:

> I acknowledge with thanks receipt of your form of appointment, dated 12th day of September 1990. I would remind you that your Statements of Case should arrive with me not later than 35 days from the date of appointment and that this is a mandatory requirement.

He then went on to deal with arrangements for the arbitration. Those then were the facts.

Both the learned judge and this court were referred to the authorities. Certain general rules were laid down in the *La Loma (Tradax Export SA* v *Volkswagenwerk AG* [1970] 1 QB 537). That was dealing with a very different factual situation, namely the appointment by each side of their own arbitrator. Closer to this case is the case of *University College, Oxford (Master & Fellows)* v *Durdy* [1982] Ch 413. That concerned an arbitration under this legislation. But there the parties could not agree on the arbitrator. Where this occurs the parties must now apply to the president of the RICS to make an appointment. Then it was to the minister. He must then seek the consent of his proposed appointee to act. The question then arose as to whether that appointment was complete with the minister's execution of the document of appointment, or his notification of the appointment to the parties. This court there held that the former was the correct construction, but sought to temper the potential injustice of the time running against a party who knew nothing of the event which had started time running by adopting

> a benevolent construction in accordance with natural justice . . . by reading into the paragraph the requirement that a party must have notice before time [for the delivery of his statement of case] begins to run against him

— see Eveleigh LJ at p418.

But by a subsequent amendment to the Schedule, resort to such a construction was prevented, the new para 31 of the Schedule making it clear that in cases of third-party appointments time for the purposes of paras 7 and 14 shall run from the time of execution of the instrument of appointment. Any "benevolence" shown was by the extension of the time from 28 days to 35 days.

Both the judge and the editor of *Muir Watt on Agricultural Holdings* distinguish that case from the appointment of an arbitrator by agreement. The judge quoted a passage from p278 of the 13th ed of *Muir Watt*:

> An agreement that an arbitrator should be appointed in the future, however, is not the same thing as the actual appointment of an arbitrator, a point which may be important where there are time limits in question . . . the appointment of an arbitrator by agreement, as distinct from an appointment by the President, requires co-operation and communication between the parties. It clearly involves agreement between the parties as to the arbitrator, the arbitrator's consent to act and the parties' knowledge that he has consented. It would seem, therefore, that the date of appointment must be the date when the following requirements have been fulfilled: (a) the agreed appointment has been made in writing; (b) the proposed arbitrator has been notified of the appointment and has consented to act (he might have signified consent before the appointment); and (c) the arbitrator's acceptance has been notified to the parties (unless they had advance notice of acceptance before the formal appointment). It is desirable that the arbitrator should accept his appointment in writing, although this is not prescribed by the Act, and the appointment will be effective if he does in fact proceed under it without the formality of writing.

In my judgment, that passage rightly identifies the necessity for the parties' knowledge that the arbitrator has consented to act, *unless they had advance notice of his acceptance before the formal appointment*. This was the question that had to be asked in the light of the letter of August 31 in which he indicated his acceptance of the post "subject to receiving a properly completed form signed by both parties within the appropriate time limit".

The learned judge found that the arbitrator's appointment was complete only when the parties received his letter of September 18 acknowledging receipt of the form of appointment. He found this for the following reasons:

(i) the arbitrator was not bound to act until he had received and approved the written form of appointment; *and*

(ii) the tenant was informed of that fact.

In my judgment, the judge was right on the first proposition in (i): undoubtedly the arbitrator was not bound until he received the notice

A of appointment — that was the condition subsequent on which he had insisted in his letter of August 31. But what he then received was in fact a properly completed appointment form. I doubt whether his formal approval of it was necessary.

But the learned judge was, in my opinion, wrong on (ii): the parties had notice from the arbitrator's letter of August 31 at the time at which he would consider himself bound. They had bound themselves by signing the notice of appointment and sending it to him — his receipt of that document completed the tripartite agreement.

It is true that that conclusion produced the untidy result that the parties to the arbitration will (unless the document is delivered by fax)
B be unaware of the precise time when their 35 days begin to run against them. However, they could avoid this untidiness by making the appointment complete only on the arbitrator's faxed acceptance. This they did not do.

Accordingly, in my judgment, the tenant's statement of case was not delivered within the statutory period and question 1 should therefore have been answered "no".

Question 2 relates to whether the arbitrator has power to extend that 35-day time-limit. The general rule is that a statutory time-limit cannot be extended unless either the statute expressly so provides or unless the requirement can be considered to be merely directory and not mandatory: even though the word "shall" is used. In *Durdy*'s case
C (*supra*) Griffiths LJ (at p923G) held (albeit *obiter*) that the time-limits were fixed. That has been the view of both leading text books, and has been the conventional view taken in the county court over the history of this Act and its predecessor, a view allowed to stand through various amendments of the Act and rules. And, in my judgment, the wording of the rule makes that long-accepted conclusion clear beyond doubt; under that rule the parties *shall* deliver their statement of case within the statutory period and no amendment of it *shall* be allowed after the expiry of that period *except* with the consent of the arbitrator. This makes it abundantly clear that the arbitrator cannot consent to an extension of the original time. The learned judge answered this
D question "no", the tenant reserving their submissions for this court. The judge was clearly right in his answer.

The second major legal question relates to the sequence of events whereby the cheque for the rent was presented within the two-month period, dishonoured after expiry of that period, and then was re-presented and honoured on the second presentation before service of the notice to quit relied on.

The starting point is the payment of the rent by cheque. The law is set out in a decision of this court dealing (coincidentally) with the tenancy of an agricultural holding — *Beevers* v *Mason* (1978) 37 P&CR 452*, where Shaw LJ said [at p458]:

E

> On general principles, the landlord should have the rent in cash in his hand by the due date. This requirement may, however, be waived by express arrangement, or by necessary implication where the facts are sufficiently strong to establish that the landlord has shown that he is content to accept payment by cheque posted by the date of payment.

In this case it is conceded that the landlord was content to accept payment by cheque.

The authorities make it clear that payment by cheque is conditional payment. Thus, Lord Esher MR in *Re Romer & Haslam* [1893] 2 QB 286 said at p296:

F

> It is perfectly well-known law, which is acted upon in every form of mercantile business, that the giving of a negotiable security by a debtor to his creditor operates as a conditional payment only, and not as a satisfaction of the debt, unless the parties agree so to treat it. Such a conditional payment is liable to be defeated on non-payment of the negotiable instrument at maturity, . . .

Where such a conditional payment is made, the creditors' remedies are suspended pending the resolution of that condition — see *Cohen* v *Hale* (1878) 3 QBD 371 *per* Cockburn CJ at p373:

> It is very true that a man who takes a cheque may be estopped from proceeding to enforce payment of the debt until presentment of the cheque, and if the cheque is ultimately paid the debt is extinguished. All that happens in the meantime is that the right of action is suspended. But when the cheque G
> is presented and dishonoured, the debt, the remedy for which was suspended until presentment of the cheque, may be treated as a debt subsisting all along, just as if the cheque had not been given.

*Editor's note: Also reported at (1978) 248 EG 781.

And, where the cheque is honoured, the date of payment of that cheque relates back to the time of the giving of the cheque: *Felix Hadley & Co* v *Hadley* [1898] 2 Ch 680 at p682 *per* Byrne J:

> In this case I think what took place amounted to a conditional payment of the debt; the condition being that the cheque or bill should be duly met or honoured at the proper date. If that be the true view, then I think the position is exactly as if an agreement had been expressly made that the bill or cheque
> should operate as payment unless defeated by dishonour or by not being met; H
> and I think that that agreement is implied from giving and taking the cheques and bills in question . . . Of course, had these bills and cheques not been duly met, then the debts would have revived, and would have been debts due to the vendor; but, inasmuch as they have been met, I consider that, the condition having been observed, there was a payment as from the time of giving the bills and cheques.

Applying those authorities to the facts of this case, it is clear that, had the tenant's cheque been cleared on its first presentment, then the date of payment would have related back to the posting of the cheque (see *Beevers* v *Mason, supra*), and so would have been made within
the two-month period. J
And that same principle was perhaps more controversially applied where payment of the cheque was not stopped, but was delayed for a short time at the request of the drawer (see *Nicks & Son Ltd* v *Taylor's Bakery Ltd* [1962] NZLR 286, but note at p288):

> . . . that there was ever only the one lodgment and presentation of the cheque which was never "stopped" by being marked to that effect or returned to the payee so as to require re-presentation.

The situation where a rent cheque was posted within the two-month period but required to be re-presented after expiry of the period was dealt with at first instance in *Luttenberger* v *North
Thoresby Farms Ltd* [1993] 1 EGLR 3 . There, Ferris J identified the K
ratio of *Beevers* v *Mason* (*supra* at p460):

> When the cheque was put into the post, then, subject only to its being honoured, the rent was paid.

There the rent cheque had not been signed in the way required by the mandate. It needed another signature. That signature was added after the period had expired, and after the collecting bank had been told that the paying bank would not honour it until that signature had been added, and indeed after the collecting bank had asked for its return. In fact, the extra signature was added, and the cheque therefore
honoured, but both occurred after the two-month period had expired. L
Ferris J held this to be too late, and explained and distinguished *Beevers* v *Mason* as follows [at p9A]:

> It seems to me that in saying "subject only to [the cheque] being honoured" the Court of Appeal in *Beevers* v *Mason* must have been envisaging that the cheque would be honoured without further intervention on the part of the tenant. If, for example, a tenant sends by post an unsigned cheque which the landlord presents for payment, not noticing the absence of a signature, it seems to me that that does not constitute any kind of payment on any date. If, when asked to pay the cheque, the tenant's bank notices the absence of a signature and, instead of returning the cheque, contacts its customer and gets him to sign the cheque, which it then pays, the resulting transfer of money is
> undoubtedly a payment, but in my view it is a payment made only when the M
> tenant signs the cheque or when the cheque is then paid.

He then found that the transfer of the money when the cheque was honoured was not as a result of the mandate constituted by the original cheque, but resulted from the new mandate given by the second signature after the period had expired. That case was rightly decided in my view.

In the case before us, at the moment when the cheque was returned marked "refer to drawer, please re-present", the position is quite clear. There had been no payment within the statutory period, and a notice to quit served then by the landlord would be effective. And so the judge

A correctly found. But then the learned judge continued:

> However, on discovering that it was not honoured, instead of serving a notice to quit, the landlord re-presented the cheque which was then honoured. He did that before he served a notice to quit. In so doing, he is still accepting the cheque as conditional payment, and once honoured, it becomes actual payment *ab initio*, ie the date of payment reverts back to the time he received the cheque. The time he received the cheque was May 23, and therefore within the two-month limit set by the notice to pay rent, dating from March 29. It seems to me that if the landlord had re-presented the cheque after he had served his notice to quit, he would be entitled to say that the tenant had failed to pay within the two months; therefore the notice to quit was valid and the subsequent re-presentation of the cheque, which was honoured, was simply B payment he was entitled to for rent owing to him. By serving the notice to quit, he has demonstrated that he has not accepted the cheque as payment under the notice to pay rent, which if dishonoured, he is entitled to do. However, in re-presenting the cheque before serving the notice to quit, he is accepting the cheque as conditional payment in satisfaction of the notice to pay rent.

There the learned judge is saying that all depended on the order of events: if the notice to quit was served before the cheque was re-presented, the notice to quit was valid, but if the notice to quit was served after then, he was accepting payment as if the cheque had been honoured on its original presentation.

C It is necessary to go back to the statute. Under section 26(1) the general rule is that a notice to quit an agricultural holding shall not have effect unless the tribunal consent. But that rule does not apply in any of the cases set out in Part 1 of Schedule 3. This was Case D.

That case deals with the situation where:

> At the date of the giving of the notice to quit the tenant *had* failed to comply with . . . a notice requiring him within two months from the service of the notice to pay any rent [due]. . .

That was this case: the cheque had not been honoured on its original presentation and so there had been a breach. That was not in D contention. What was in contention was what happened when the rent was accepted before service of the notice to quit. This court considered that point in *Stoneman* v *Brown* [1973] 1 WLR 459. There Lord Denning MR said [at p462]:

> In an attempt to avoid the notice to quit, the tenant claimed that he had paid the rent on September 1 when he posted his letter with the cheque in it. He said that the landlord had authorised him to make payment by cheque by post. So on this occasion he said that the rent was paid on September 1 1970 and this payment was before notice to quit was served; because the notice to quit did not get to the tenant until September 2. I will assume that the tenant paid the rent on September 1 when he sent the cheque. Nevertheless I am afraid E that was too late to save him. The two months had already expired on August 29 1970. Once the tenant allowed the two months to expire without payment, the payment thereafter could not save him. Even if he paid the rent after the two months, but before the notice to quit was given, he would still have to F quit. The words of the statute are plain.

Then, by way of further explanation, he said:

> If the tenant does not pay his rent within the two months required by the notice, then the tenant has no further right to be there; the landlord can serve his notice to quit; and even though the tenant should pay the rent before the landlord actually serves his notice to quit, still it does him no good; it does not get him out of his troubles, because he was already in breach, and the landlord's right to serve a notice to quit has accrued. If the landlord serves a notice to quit, the tenancy is ended. It does not survive even if the tenant pays the rent for a subsequent period unless the proper inference is that there was an implied agreement for a new tenancy. G

Here, on the agreed facts and on the material before the arbitrator, there was no basis for inferring an implied agreement for a new tenancy.

Nor, on that material, was there any legal basis for finding that the landlord had, by the mere acceptance of the rent due more than two months earlier, waived his entitlement to claim possession by his notice to quit. It is trite law that forfeiture is not waived by the acceptance of rent accrued due before that forfeiture. Therefore, there are no grounds for distinguishing this case from *Stoneman* v *Brown* and the fact that the rent was paid before the notice to quit was served does not assist the court. Against that background I turn to consider H the specific questions asked.

Question 4: "Is evidence by the tenant of discussions and dealings between him and his bank relevant and admissible on the issue as to whether or not the tenant failed to comply with the notice to pay rent?"

My answer to this question would be "no". As a matter of principle, the debtor must seek out his creditor and pay cash. Here the landlord had softened that general rule by permitting payment of the rent by cheque sent in the post. That was conditional payment, conditional upon the cheque being cleared in the ordinary course of banking. The risk of the bank wrongly refusing to honour that cheque is borne by the debtor and not by the creditor in such circumstances. J

Question 5: "Is delivery of a cheque for rent by a tenant to a landlord payment on the date of delivery if the cheque is afterwards presented for payment and dishonoured on the first presentation but met on the subsequent presentation?"

The answer to this question is "no". When the cheque has been dishonoured, the condition on which it was accepted has not been met. No subsequent presentation and honouring of the cheque could relate back to the date of its original delivery unless the landlord was waiving his right which had accrued on the first dishonour. Such a finding is not possible on the agreed facts alone. And it follows from what I have said that questions 8 and 10 (which the learned judge had K answered in the negative) should both be answered "yes".

Beldam and **Neill LJJ** agreed and did not add anything.

ARBITRATION

Chancery Division

March 26 1993

(Before Judge RICH QC, sitting as a judge of the High Court)

AMEGO LITHO LTD v SCANWAY LTD

Estates Gazette January 15 1994

[1994] 02 EG 110

Arbitration — Rent review — Determination of rent above figure spoken to by the landlord's surveyor — Whether arbitrator had made error of law in handling comparables — Whether misconduct in disregarding evidence

Following a reference to determine the rent at review of Chart House, 16 Chart Street, London N1, on July 15 1992 the arbitrator made his award in the sum of £32,700 pa based upon a rental value in relation to the net internal floor area of £12.07. The tenant applied for leave to appeal under section 1 of the Arbitration Act 1979 and to set aside the award under section 23 of the Arbitration Act 1950 on the ground that the conclusion at which the arbitrator arrived is contradicted by the only true and reasonable conclusion that can be drawn from the findings of fact that the arbitrator made. The arbitrator stated that he relied on a comparable in respect of which the landlord's surveyor had concluded that the rental value was £10.40. Further, the arbitrator stated that he was disrgarding certain improvements he had been advised by counsel to disregard, but that such disregard did not affect the rental value he had already determined.

Held: **1. The application for leave to appeal was dismissed. There was no error of law on the part of the arbitrator within the principles of *Edwards* v *Bairstow* [1956] AC 14 even if he fell into error in the handling of the comparables.**

2. However, the award was set aside and the arbitrator removed. The arbitrator had misconducted the proceedings in rejecting a piece of evidence put forward by the tenant notwithstanding that the evidence could have been objected to by the landlord, and was not, as inadmissible evidence.

The following cases are referred to in this report.

Edwards v *Bairstow* [1956] AC 14; [1955] 3 WLR 410; [1955] 3 All ER 48, HL

Ipswich Borough Council v *Fisons plc* [1990] Ch 709; [1990] 2 WLR 108; [1990] 1 All ER 730; [1990] 1 EGLR 17; [1990] 04 EG 127, CA

Land Securities plc v *Westminster City Council* [1993] 1 WLR 286; [1993] 4 All ER 124

Trayfoot v *Lock* [1957] 1 WLR 351; [1957] 1 All ER 423, CA

This was an application for leave to appeal under section 1 of the Arbitration Act 1979 and to set aside an arbitrator's award under section 23 of the Arbitration Act 1950 made by the plaintiff, Amego Litho Ltd, as tenant under a lease held from the defendant landlord, Scanway Ltd.

Thomas Jefferies (instructed by Burstows, of Crawley) appeared for the plaintiff; Kirk Reynolds QC (instructed by Max Bitel Greene) represented the defendant.

Giving judgment, **JUDGE RICH QC** said: This is an application on behalf of a tenant for leave to appeal under the Arbitration Act 1979 and an application under section 23 of the Arbitration Act 1950 to set aside the award of an arbitrator made on July 15 1992. The award was made by a surveyor, Mr Alistair Voaden [FRICS], who was appointed as arbitrator to determine the rent under a rent review clause in a lease of premises at Chart House, 16 Chart Street, London N1. He gave the reasons for his award in a supplementary document dated July 30 1992. His award was in the sum of £32,700 a year as the review rent from September 29 1990. The supplementary document, in which he set out his reasons, explained that this was based, as to a significant element in that valuation, on a rental value assessed by him as £12.07 per sq ft for the agreed net internal floor area of the premises.

The application for leave to appeal has, of course, to be founded on an allegation of an error of law, and the nature of the error of law which is put forward by Mr Jefferies, on behalf of the applicant tenant, is that the conclusion that the arbitrator arrived at is contradicted by the only true and reasonable conclusion that can be drawn from the findings of fact that the arbitrator made, and that therefore the determination involved an error of law such as was identified by the House of Lords in the case of *Edwards* v *Bairstow* [1956] AC 14.

I assume, for the purposes of the consideration of this submission, that there is indeed such a category of error of law, relevant to the granting of leave to appeal and to the successful appealing against an award under the Arbitration Act of 1979, notwithstanding the provisions of section 1 of the 1979 Act which were, as I accept, clearly directed to restricting the previous jurisdiction of the High Court to set aside or remit awards on the ground of error of fact on the face of the award. I make that point specifically because I have been referred to some observations of Steyn LJ in a case which is available only in an transcript dated November 5 1992, the name of the case being *Geogas SA* v *Trammo Gas*, where Steyn LJ set out some highly persuasive reasons why the courts should not interfere in matters of fact presented, however attractively (and nothing can be more attractive than the presentation of Mr Jefferies) as matters of law.

I approach the question under the 1979 Act, therefore, with some caution. But, nevertheless, if I were persuaded that there was indeed an obvious error here in the sense identified in the case of *Edwards* v *Bairstow,* I would certainly grant leave. I put the test in terms of obvious error because this particular case of a rent review appears to me to be a one-off case, notwithstanding the distinctions which were made by the Court of Appeal in the case of *Ipswich Borough Council* v *Fisons plc* [1990] 1 EGLR 17, where it was pointed out that a determination of law which would be determinative of the position on subsequent reviews is more than a one-off case.

This is, in my judgment, a one-off case because the nature of the error of law arises from the particular handling of the evidence of a particular arbitrator and is therefore not determinative upon the construction of the rent review clause on subsequent reviews. The fact that it is determinative of the rent for a period which will, at least as the minimum rent, extend beyond the immediate review does not make it any the less a one-off determination for the residue of the lease.

I do not think that there is an obvious error of law within the *Edwards* v *Bairstow* principle, although I do think that there is an extreme probability that there has been an error of fact by the arbitrator in the handling of the evidence in this case.

I say that for this reason: the arbitrator indicated in his reasons for his decision that he had had the greatest regard to the open market letting six months prior to the review date of some premises on the third floor of 2-4 Vestry Street, London N1. The analysis of that transaction by the landlord's surveyor derived a rental value of those premises of £10.40 per sq ft and it is not immediately obvious how, without error of fact, the arbitrator could have placed greatest reliance upon that transaction and concluded that the appropriate rental value of the subject premises was £12.07.

The next comparable transaction to which the arbitrator purported to have regard was a rent review of premises at New Roman House, 10 East Road, which on analysis by the landlord's surveyor showed a rental value of those premises of £11.98. This might more nearly have supported the valuation at which the arbitrator arrived of the subject premises, because the landlord's surveyor was able to say on the basis of those two comparable properties that his, the landlord's, surveyor's valuation of the subject premises was £13 per sq ft.

That being the position, I strongly suspect that the arbitrator has, as I indicated, fallen into some error in the handling of the comparable transactions on which he placed reliance, because he placed greatest reliance, on the less highly priced of the two, but I cannot go so far as to say, as is effectively necessary, in order to bring the case within the principles of *Edwards* v *Bairstow*, that no reasonable valuer could, on the evidence, have arrived at a conclusion which was, after all, 93p per sq ft less than that to which the landlord's surveyor had spoken without his reasonableness being impugned.

Mr Jefferies urges, however, that a discount of that sort from the figure put forward by the landlord's surveyor is unsupportable, and therefore an error amounting to an error of law must have crept in because the landlord's surveyor had not in his computation disregarded certain improvements which, as a matter of law, the arbitrator was advised by counsel he had to disregard and which the arbitrator purported to disregard. The arbitrator set out the list of relevant improvements that he was proposing thus to disregard in a letter which he addressed to the parties dated June 30 1992. It referred back to an earlier letter of his of May 26 in which he had indicated, before advice by counsel, an intention not to disregard any improvements undertaken by the tenant. But, having had the advice to disregard certain of them, the arbitrator wrote that he would:

> advise that counsel's opinion and the consequences in "regards" and "disregards" referred to above do not affect the rental value referred to in [his] letters of 26th May, nor [his] previous award on the head lease review.

It is difficult to understand how the arbitrator was rightly applying the principles of law, to which he paid at least lip service, in that letter saying that he proposed to disregard the value of the improvements set out in that letter, and yet did not think it appropriate in any way to adjust a valuation prepared on a different basis which did not disregard such improvements.

Nevertheless, I cannot go so far as to say either that he has in those letters indicated any error of law, nor can I say that it is obvious that he could not have reached a conclusion, that disregarding certain improvements, or the effect of certain improvements upon rental value in fact had no effect upon the rental value at which he had arrived without doing so, because it is perfectly possible logically, although the arbitrator has not spelled this out, for a tenant to make improvements which none the less do not affect the rental value. I am prepared moreover to accept that that is at least more probable, with regard to improvements whose nature is tenants' fixtures, than in respect of other improvements, and I note that the improvements which the arbitrator was advised to disregard were improvements whose nature was tenants' fixtures.

I have therefore come to the conclusion that I am not entitled and therefore do not grant leave to appeal against this award under the Arbitration Act 1979. But my examination of the grounds do leave me with the unhappy feeling that an injustice has in fact been introduced into the determination of the rent of this case. I mean that, by reason of error, a conclusion has been arrived at which is almost certainly unfair to the tenant.

That is not, of itself, a ground upon which the court can interfere. Indeed, on the contrary, the nature of arbitration is that the parties submit to the determination of the arbitrator, certainly so far as matters of fact are concerned, be it right or wrong. They sacrifice perhaps the perfection, as we might hope, that arises out of the court system subject to the safeguard of a fuller system of appeal, for the finality which arises in the submission to arbitration by an agreed arbitrator.

That finality, however, is subject to the control of the courts under the provisions of the Arbitration Act 1950, which include the provision of section 23 of that Act by which the court is given power to set aside the award where an arbitrator has misconducted the proceedings. I am satisfied that in this case, in a technical sense, the arbitrator has indeed misconducted the proceedings, albeit in a manner which I suspect did not, and indeed I would be prepared to find "could not", on the approach that he adopted have had any effect upon his actual determination.

The reasons which he gave for his decision recited the way in which he had dealt with comparable properties. I have already indicated that he has said that he had the greatest regard to the transaction of 2-4 Vestry Street. He went on to say that he had also had regard to the transaction at New Roman House. His third sentence under the head of "Valuation" observed:

> The parties did not provide other open market letting or rent review comparables.

It was from this that he derived his conclusion that the agreed net internal floor area is 2,648 sq ft, to which he said he had attributed a rental value of £12.07 — that is to say, of course, per square foot.

The assertion that the parties did not provide other open market letting or rent review comparables is true only if a somewhat specialised meaning is given to rent review comparables because there had been submitted, other than the two to which the arbitrator had referred, evidence in regard to a rent review determination in fact undertaken by Mr Voaden himself in respect of premises at 53-55 East Road, where he had determined a rental value which was significantly out of line with the rental value which he ultimately attributed to the subject premises. I confess that I am not quite clear at which figure it was analysed, but I think it is a figure of £6.50 per sq ft in respect of one floor and of £7.39 per sq ft in respect of another floor of those premises. No doubt adjustments would have been appropriate to be made if that comparable had been treated as being appropriate to be relied upon, but there is no sign of how it was considered or how it was relied upon.

It may be that, had the arbitrator considered that evidence, he would have come to the conclusion that it was of little weight as it represented only a record of his previous opinion on some other premises and was not a market transaction. Indeed, it is in this sense that it might be said, although I suspect that it was only my suggestion that led to this way of interpreting the phrase, that it was not strictly a rent review comparable at all. But I do not think that that is the sense in which the arbitrator was using the language at this stage in his account of his reasons. No doubt because he did indeed place weight only on the two premises or two transactions to which he made reference, he was at that stage overlooking that there was indeed relied upon by the tenants' surveyor another comparable transaction, which the tenants' surveyor was at least at that time entitled to have considered because it had been introduced to the proceedings and had not been challenged or objected to as being inadmissible.

There is now authority that the mere opinion of the arbitrator is to be treated only as an expert's opinion and is not to be treated therefore as evidence of a market rent and therefore would be inadmissible, as

A held by Hoffmann J in the case of *Land Securities plc* v *Westminster City Council* [1993] 1 WLR 286.

Nevertheless, as I have indicated, it appears to me that there is a technical misconduct in the rejection in effect of a piece of evidence which had been admitted and not objected to when put forward on behalf of the tenant. That is, as it appears to me, to be a technical misconduct such as was identified by the Court of Appeal in the case of *Trayfoot* v *Lock* [1957] 1 All ER 423, where a document was looked at by the arbitrator but was then excluded from consideration so that, in effect, he rejected it and excluded material evidence. In this case also it appears to me that, in his reasoning, the arbitrator has
B shown that he has, probably inadvertently, rejected evidence which was technically material evidence for the purposes of this arbitration.

Having so found it would not seem to me that this would be a sensible reason for setting aside the award if the effect of setting it aside was no more than to require the arbitrator to look at a piece of evidence, whose value he would be well able to appraise himself, and which he apparently from his determination has not valued highly and to which, on remission of the matter to the arbitrator, the landlord might now pertinently and properly object to being considered by reason of the decision to which I have just made reference. Therefore,
C if the sole effect of my setting the arbitration aside was merely to enable the present arbitrator to reconsider that particular piece of evidence, I would not exercise such discretion as I think I have under section 23 to set the award aside.

I do, however, think that it is fair and proper in the circumstances of this case to take advantage — and I use those words advisedly as I am conscious that that is what I am doing — of the technical misconduct of the arbitrator to allow what I hope may be nearer to fairness to be achieved between the parties by a renewed arbitration.

In those circumstances I would not only set aside this award, but
D would remove the arbitrator but only upon these terms, which were submitted to by Mr Jefferies on behalf of the applicant tenant, that the costs thrown away in the arbitration so far should be treated as costs of any renewed arbitration so that they may be in the discretion of the newly appointed arbitrator, who may then have the opportunity to ensure that the opportunity which this decision gives to the tenant to reopen proceedings, which appear to me have gone wrong, although through no fault at all of the landlord, should not impose upon the landlord any unfair burden in costs. I will make orders accordingly, if they can be agreed by counsel. I will hear argument as to whether I should put the tenant upon any terms in respect of the existing
E awarded rent.

The order to remove the arbitrator is conditional upon there being paid to the landlord within 28 days of today, or such longer period as may be agreed between the parties, or determined on application by the court, the difference between the rent which has been paid since the review date and rent at the rate of £24,000 a year and upon rent at the rate of £24,000 a year being paid on the usual quarter days to the landlord on account of liability for rent to be awarded under the renewed arbitration. Such rent be accepted by the landlord subject to the landlord's undertaking to the court that in the event of an award for
F rent making a less sum than has been paid due from the tenant to the landlord the surplus will forthwith become repayable to the tenant.

I make no provision for payment of interest upon these sums which are paid on account of the rent being advanced because in the calculation of the total sums will be due from the tenant to the landlord it will be possible for the landlord to bring into calculation interest that is due under the lease.

Application for leave dismissed; application to set aside the award allowed.

Queen's Bench Division G

July 23 1993

(Before Mr Justice ROUGIER)

THOMAS AND ANOTHER v COUNTRYSIDE COUNCIL FOR WALES

Estates Gazette January 29 1994

[1994] 04 EG 138

Site of special scientific interest — Management agreement restrictions — Correct test for compensation for loss — Arbitration — Appeal — Whether arbitrator applied correct test — Whether he gave reasons H

The appellants farm 371 acres on the Gower Peninsula and have grazing rights on a further 117 acres of cliff land. In 1986-87 the appellants had a sheep flock of 1,100 ewes and were planning to expand to 2,000 ewes having built a covered shelter at a cost of £29,000. The cliff land was designated a site of special scientific interest and on October 23 1986 the respondent authority notified the appellants under section 28(1) of the Wildlife and Countryside Act
1981 that sheep farming must be drastically reduced. Being J dissatisfied with the terms of compensation offered by the respondents for net profits foregone and other losses for an interim management agreement to control sheep farming on the cliff land, the appellants referred the issue to arbitration. In assessing the net profits foregone under the guidelines in Circular 4/83, the arbitrator first estimated the profit which the appellants would have earned had they been allowed to continue as before and then assessed what was the maximum income the farm could produce subject to the restrictions. The appellants appealed under section 1 of the Arbitration Act 1979 contending that the arbitrator had applied the wrong formula under
para 16 of the guidelines, wrongly rejected various heads of claim and K had failed to give reasons.

Held: **The appeal was allowed and the award remitted to the arbitrator. The correct test under para 16 to the guidelines is that the question of loss is to be looked at as a matter of causation in accordance with the principles of law applicable to damages for tortious acts or breaches of contract. Accordingly, the loss is the pecuniary loss which directly and naturally flows from the restriction. Although the appellants owe no duty to mitigate, they cannot expect the respondents to pay for losses which the respondents might have avoided had the appellants taken
reasonable steps to avoid such losses. The matter is one of the L causation — that is to say, the assessment of the losses which have naturally flowed as a result of the restrictions. In relation to the claim for capital losses the arbitrator was erroneous in dismissing such claims in so far as he had applied the wrong test in assessing loss of profits. The arbitrator, in rejecting the claim for loss of capital value of the holding, failed to appreciate that the appellants were claiming for residual capital loss being diminution of the capital value of the land only after the management agreement had run its course. The arbitrator had failed to give reasons for his award.**

M

The following cases are referred to in this report.

British Westinghouse Electric & Manufacturing Co Ltd v *Underground Electric Railways Co of London Ltd* [1912] AC 673; (1912) 81 LJKB 1132; 107 LT 325

Lloyds & Scottish Finance Ltd v *Modern Cars & Caravans (Kingston) Ltd* [1966] 1 QB 764; [1964] 3 WLR 859; [1964] 2 All ER 732

Sotiros Shipping Inc v *Sameiet Solholt ("The Solholt")* [1981] 2 Lloyd's Rep 574; [1983] 1 Lloyd's Rep 605, CA

Universal Petroleum Co v *Handels- und Transportgesellschaft GmbH* [1987] 1 WLR 1178; [1987] 2 All ER 737, CA

This was an appeal under section 1 of the Arbitration Act 1979 by the appellants, Mr and Mrs Thomas, against an award made by an arbitrator on the appellants' claim for compensation against the respondents, Countryside Council for Wales, under the Wildlife and Countryside Act 1981.

Colin Brodie QC and Keith Lindblom (instructed by Lawrence Graham, for John Owen & Co of Llandeilo) appeared for the appellants; Keith Bush (instructed by the Treasury Solicitor) represented the respondents.

Giving judgment, **ROUGIER J** said: This appeal is brought by Mr and Mrs Thomas with leave of the court against an arbitration award made pursuant to section 50(3) of the Wildlife and Countryside Act 1981. Principally the appeal concerns the question of the proper measure of compensation to be paid to a landowner whose use of land is affected by powers exercised by what was originally the Nature Conservancy Council and which has now been subdivided into various regional councils, the respondents, as their name suggests, being that responsible for the principality.

Mr and Mrs Thomas are farmers, whose main enterprise is sheep farming. They are experienced and have a good reputation — a factor remarked upon by the arbitrator in making his award. Their farm is called Paviland Manor Farm, comprising 371 inland acres, and they have grazing rights over a further 117 acres of cliff land all at Rhossili, which is at the south-west corner of the Gower Peninsula. Whether one calls the adjoining sea the Bristol Channel or the Atlantic matters little; the wind and the rain sweep in from the west and shelter is at a premium. This fact is of importance since strangely enough it was the cliff land which provided shelter for the Thomases' flock of sheep and thereby had particular value in the management of their sheep-farming activity. At the relevant time, which was in 1986-87, Mr and Mrs Thomas had expanded their flock to some 1,100 ewes and were planning to increase the numbers to something like 2,000. To cater for the increase they had recently built a bothy on the inland part of the farm at a cost of some £29,000.

The respondents may broadly be described as the authority responsible for conservation and the protection of wildlife in the area. Their powers are derived from the Wildlife and Countryside Act 1981, a measure passed with the objective, among others, of preventing undue exploitation and consequent damage to the flora and fauna of this country. In order to be able to carry out this objective, the relevant authority is given powers under section 28 to declare certain areas to be those of special scientific interest and, by appropriate notification, to ban or limit any operations which appear to them to be likely to damage the countryside unacceptably. The cliff land at Rhossili had been declared an area of special scientific interest and it comes as something of a surprise to learn that the activity which incurred the displeasure of the respondents was that of pasturing sheep. But so it was; on October 23 1986 they gave formal notification to the Thomases pursuant to section 28(1) of the Act that this time-honoured form of husbandry had to be drastically limited. Owing to the crucial importance of the cliff land, clearly compliance with the notification would entail drastic alteration to the Thomases' sheep farming and their plans for the future.

It is necessary briefly to describe the various steps which led to the arbitration, all of which are provided for by the statute. First, any owner or occupier of land who has been served with notification is prevented from carrying out any of the operations specified for four months, and then only if he has served a counternotice that he proposes to carry out the operations proscribed: see section 28(5). The method whereby the council can enforce any ban or limitation is described as a management agreement and the Act provides for the parties to enter into such an agreement by negotiation. Before a long-term management agreement can be finalised, the parties will enter into short-term management agreements, and if no long-term agreement is negotiated the council or other relevant authority next makes a formal offer to enter into a management agreement by virtue of section 50(1). This offer has to specify the terms including the terms as to payment which the council are prepared to make in order to compensate the owner or occupier for the restrictions being imposed on his land. By subsection (2) the payments are stated to be determined in accordance with guidelines issued by the ministry. These guidelines are described as Circular 4/83. If the owner/occupier is dissatisfied with the terms of payment offered, he can require the matter to be referred to arbitration by virtue of subsection (3), and that is what happened in this case.

During the course of the unsuccessful negotiations the respondents commissioned a report from a Mr Hollington, which they disclosed to the Thomases. This report is in the nature of a feasibility study and addresses the problem of how the land might be farmed subject to the restrictions. In the relevant part of the report, it will be seen that Mr Hollington considered two possibilities; the first to reduce the existing flock to some 640 ewes and convert a part of the land to arable. The second was to reduce the flock to a mere 220 ewes with a consequently greater conversion to arable. In terms of cost, Mr Hollington gave it as his opinion that the first alternative would limit the loss to be sustained by approximately £2,000 more than would the second alternative. It is necessarily implicit in his report that he considered both options to be viable.

The relevant terms of the guidelines are as follows: first, by para 13 it is provided that:

For the purposes of the long term agreement, owners and owner-occupiers may choose *either* lump sum payment *or* annual payments . . . only annual payments are available to *tenants*.

Para 14:

Payment, at the commencement of the agreement of a single *lump sum* for the management agreement over a 20-year period . . . *The amount should be equal to the difference between the restricted and unrestricted value of the owner or owner-occupier's interest* calculated having regard to the rules for assessment in respect of the compulsory acquisition of an interest in land as set out in Section 5 of the Land Compensation Act 1961.

Para 16, headed "Annual Payments (owners or owner-occupiers)" provides:

Payment of *annual sums* for an agreement over a 20-year period, . . . The *payments should reflect net profits foregone because of the agreements.*

By para 18 it is stated that individual assessment will be appropriate in most cases to calculate the sums payable.

Later in the document under a heading entitled "Other Factors determining Payment", para 29 provides:

Fees — on completion of a management agreement the offeror should pay the reasonable costs of the offeree incurred in retaining professional advisors to assist him in connection with the agreement.

Para 30 — "Other Costs":

A management agreement may also provide for payment:
(a) For expenditure reasonably incurred within the previous 12 months of the date of notification which has been rendered abortive or in undertaking work rendered abortive by the agreement (subject to minimising loss — see paragraph 11 above);
(b) For any loss or damage directly attributable to the agreement insofar as . . . no relevant payment is available under other provisions of these guidelines.

Under the heading "Minimising Costs", para 11 states:

From the date of initial notification of his proposed operation, an owner or occupier should not take any action which may increase the sum eventually payable to him under these guidelines eg entering into any contract or commitment with third parties relating to the proposed work.

This paragraph earns few prizes for drafting.

The appeal of Mr and Mrs Thomas is brought pursuant to section 1 of the Arbitration Act 1979 and is advanced on two basic grounds:

1. That in determining the measure of net profits foregone because of the agreements within the terms of para 16 of the guidelines, the arbitrator, as a matter of law, applied the wrong

A formula when he calculated the loss by reference to the difference between what the appellants could have earned had they continued with the proposed expansion of their sheep farming activity compared with what they could earn by switching to a farming policy which would maximise profits, without taking into account any considerations other than those which were purely financial. It is contended that this is an incorrect interpretation of the combined effect of para 11 and para 16 of the guidelines.

2. That the arbitrator erred in rejecting various heads of claim for lost capital expenditure on the basis that, the appellants having opted for annual payments, they were not entitled to payments under paras
B 29 and 30 of the guidelines.

Contained in the appeal is also an application pursuant to subsection (5) of section 1 of the Arbitration Act 1979 that the matter be remitted to the arbitrator for him to make further primary findings of fact which are necessary and to give reasons which both parties had asked him to do. To apply to remit in this fashion at the same time as appealing the findings seems to me to be putting the cart, if not in front of, at least beside the horse, but the appellants were given leave, so I am told, under both subsections, and Mr Keith Bush for the respondents, after some initial hesitation, does not seek to argue that I should not deal with both aspects at the same time.

C I now turn to consider the relevant terms of the arbitrator's award. At the outset it should be mentioned that, although this is an award made in relation to a proposed long-term agreement, it is referable only to the three years from September 1 1987 to August 1990. As a further background fact it should be recorded that Mr and Mrs Thomas, faced with the restrictions which were being imposed upon their use of the cliff land, made a considered judgment that any substantial sheep-farming operation would not be viable, and therefore altered their farming policy radically and in essence adopted the second possibility which had been described by Mr Hollington in his report. Nobody has suggested that Mr and Mrs Thomas took their
D decision other than in good faith and backed by their own not inconsiderable experience of farming in the area. More important, the respondents accepted that this decision was not to be regarded as an action which might "increase the sum eventually payable to them" within the terms of para 11. The parties appeared to have agreed that the wording of para 11 was intended solely to deal with actions taken in bad faith with the deliberate intention of increasing the eventual compensation. Whether that is the correct interpretation must abide some later dispute. Fortunately, for present purposes, it is not necessary for me to construe the proper meaning of para 11. I content myself with the prophecy that it is likely to give trouble in the future.

E The formula adopted by the arbitrator for assessing the net profits foregone can be deduced from the following passages in the award:

First, para 12, where he states:

> In reaching my decision I first considered whether the farming policy adopted by the claimants was the optimum in the circumstances and secondly what might the claimants have achieved from the sheep unit had the restrictions not been imposed.

Later, in para 15, we find the following:

> It must be assumed that any farmer managing his holding properly will be looking to a farming policy that will maximise profits. The farmer can, and is
> F free to, farm the holding as he wishes provided he can afford the losses that may result but the respondents are not obliged to work out the profits foregone on other than their view properly established as to the optimum farming system resulting from the imposition of the restrictions. The accounts for the years 1989 and 1990 show very large losses which I consider show that the policy adopted was not the optimum.

And finally in the second of the paras numbered 17:

> The claimants are free to adopt any farming policy they wish to choose but the respondents are required only to consider under the heading "Profit Foregone" the effect on those profits arising from the restrictions imposed . . . I consider that a viable, although reduced, flock could have been maintained using the inland areas after the imposition of the restrictions.

From these passages it is clear that in order to assess the net profits G foregone, the arbitrator first estimated the profit which the Thomases would have earned had they been allowed to continue as before and as the second half of the equation he assessed what was the maximum income the farm could produce, the sole criterion being that of finance; no other considerations entered the picture.

In my judgment, that approach is wrong in law. It would have appealed to Mr Gradgrind, but it does not appeal to me. There are several reasons which can be advanced for this, but the following will suffice:

1. Para 16 makes it clear that the question of loss is to be looked at as a matter of causation; this is in accordance with the general H principles of law applicable to damages for tortious acts or breaches of contract and the equation adopted by the arbitrator goes far beyond any principle of damages applicable to such cases.

2. Farming activities can vary in profitability from year to year; one of the principal variables being the level of subsidies on particular activities. Bearing in mind that this was a calculation designed to fix the level of annual payments, it might well be that one particular strategy would be the commercial optimum one year, but not the next.

3. In determining the proper meaning to be ascribed to the words of para 16 I consider it permissible to look at the entire compensation structure provided for. That which is found in the short-term J agreement is: "to pay the owner for any reasonable financial loss arising as a consequence of this agreement"; it would, in my judgment, be strange indeed if compensation under the long-term agreement was computed on a formula which was likely to be more disadvantageous to the owner/occupier.

4. Para 18 of the guidelines makes it clear that individual assessment will be appropriate in most cases. This seems to me to be incompatible with any proposition that the only yardstick is purely commercial. Even the respondents, albeit in an alternative proposition, were prepared to concede that they must take their farmer as they found him with his preferences and perhaps in some cases his K lack of competence. A basis of computation of loss which involves the notional transmutation of perhaps a somewhat inefficient farmer to one of maximum competence would obviously work great injustice. I must not lose sight of the fact that this is a situation where a statutory body is given arbitrary power to interfere with the manner in which the ordinary citizen is going about a perfectly lawful occupation on his own land. Such interference, while doubtless well-intentioned, is always susceptible to somewhat overzealous application, even, at times, to abuse.

Take the case of a farmer who farms sheep and nothing else because that is what he wishes to do. Restrictions are placed upon him which will entail a very considerable diminution of the number of L sheep he is able to farm; bowing to the inevitable he takes advice and is told that rather than continue with a limited herd he will make more money if he converts his entire farm to turnips and mangoldwurzels. Is he not to be allowed to say: "I don't like turnips and mangoldwurzels; I don't like their smell, I don't like the mud that they produce, and I fail to see why this intrusive body, drest in a little brief authority, should come and radically alter the amenities and tenor of the life that I have led for many years." Are not such arguments to be heard? If the answer is "No" then the world will have grown grey indeed. For these reasons the matter must be remitted to the arbitrator for him to apply the correct test. M

This naturally leads to the question: What is the correct test?

The appellants' first contention was that any obligation on the part of an owner/occupier was limited to and exhaustively defined by the terms of para 11 of the guidelines. Freed from the necessity of determining just what para 11 meant, Mr Colin Brodie QC boldly submitted that his clients were entitled to be compensated in full, whatever altered farming policy they chose to adopt, provided they did not fall foul of the restriction imposed by para 11. As an alternative, he submitted that the whole question was to be equated *pari passu* with the standard law governing damages for breach of contract or for tort, and mitigation.

A I have to reject his first submission; I think that the wording of the guidelines, although far short of ideal, was intended to equate the manner of assessing compensation with the normal law applicable to cases of contractual or tortious liability. To that extent, para 11 is in fact surplusage. A further consideration that has influenced me is that, whereas the parties during the course of the arbitration adopted an interpretation of para 11 which was extremely sensible, I am far from persuaded that in the face of the actual wording it was correct. It seems to me that on a somewhat less sympathetic construction of the words any course of action by a farmer which subsequently turned out to be uneconomic might well fall foul of the paragraph, in which case B the farmer would be worse off than he would be under the ordinary law.

Under the normal law of contract and tort the fundamental basis for the measure of damages is compensation for pecuniary loss which directly and naturally flows from the breach. The authority, so far as breach of contract is concerned, is to be found in *British Westinghouse Electric & Manufacturing Co Ltd* v *Underground Electric Railways Co of London Ltd* [1912] AC 673 at p689. There is, however, a qualification that a plaintiff suing for breach of contract or for that matter for tort cannot call upon a defendant to pay the full direct consequences unless he himself has acted reasonably to mitigate the C loss. It is sometimes loosely described as a plaintiff's duty to mitigate. In *MacGregor on Damages* 15th ed, para 275, it is stated that:

> the first and most important rule is that the plaintiff must take all reasonable steps to mitigate the loss to him consequent upon the defendant's wrong and cannot recover damages for any such loss which he could thus have avoided but has failed, through unreasonable action or inaction, to avoid. Put shortly, the plaintiff cannot recover for avoidable loss.

With respect to the learned author, the word "must", implying some sort of duty, may not be strictly accurate. If he wishes to claim the full measure of his loss, a plaintiff must act reasonably, but as was recently pointed out in the case of *The Solholt* [1983] 1 Lloyd's Rep D 605*, a plaintiff is under no duty to mitigate his loss. He is completely free to act as he judges to be in his best interests. The significance of his failure to act in a reasonable manner is merely that he cannot then call upon the defendant to pay for losses which he might have avoided had he taken reasonable steps to do so. For the purposes of the present remission to the arbitrator I would stress that fundamentally the matter is one of causation — that is to say the assessment of the loss which has naturally flowed as a result of the restrictions.

The questions to be asked are:

1. What could the appellants have achieved? — this is one which the arbitrator has already determined.

E 2. What have the appellants achieved?

3. In between those two poles, has any act or omission of the appellants been unreasonable and of such a nature as to leave one to say that the difference between 1 and 2 can no longer be said to be caused by the restrictions? Put another way, was the appellants' decision to adopt the farming system which they did a reasonable one? That question is not to be answered solely in terms of the commercial optimum. Obviously profitability is a factor, and an important one, but in an occupation such as farming any test of reasonableness should take some account of other circumstances including individual personal factors of amenity, even of aesthetic preference. The question F of what is reasonable is entirely one of fact for the arbitrator. However, in view of the contents of the report by Mr Hollington, some assistance as a matter of law may be derived from the concluding remarks of the judgment in *Lloyds & Scottish Finance* v *Modern Cars & Caravans (Kingston)* [1966] 1 QB 764 at p782 where, without stating a definite proposition of law, the court expressed grave doubt whether, if a plaintiff acted upon the suggestion of the defendant, he could ever be said to have acted unreasonably. Whether or not the two possibilities described by Mr Hollington amounted to a suggestion made by the respondents will again be a matter of fact for the arbitrator.

*Editor's note: *Sotiros Shipping Inc* v *Sameiet Solholt* (*"The Solholt"*).

I next turn to deal with the appellants' second ground of appeal G concerning the capital costs disallowed by the arbitrator. These are to be found, together with such reasons as the arbitrator gave, at para 18 of the award.

Herein the appellants' primary submission was that the award demonstrated on its face that, in disallowing the various items listed under para 18, the arbitrator had done so on the basis that lump sums could not be claimed if the claimants had exercised the option open to them to select annual payments rather than a lump sum payment.

Had this been the basis of the arbitrator's reasoning, I should have had no hesitation in holding that he was in error; paras 29 and 30 of the guidelines seem to me to be a clear indication that, over and above the annual payments, or for that matter the limited basis of the lump H sum recoverable under para 14, certain further items of capital expenditure are recoverable. But I think that Mr Bush is correct when he submits that para 18 of the award, taken as a whole, does not indicate that the arbitrator was at fault in that regard; it was only item (c), described as "Loss in Capital value of the Holding", which he disallowed on that basis. It is nevertheless necessary to consider the various items listed under para 18 of the award to see if, as the plaintiffs contend, there are errors of law on their face. I will deal with them in turn, leaving item (c) until last.

Subpara (a) — this is in a separate category.

Subpara (b) — the capital costs resulting from the need to purchase J machinery and improvements have been disallowed on the basis that these costs would not have arisen if the appellants had retained sheep on the inland areas, that is to say had adopted the first of the two possibilities mentioned by Mr Hollington in his report. Since I have held that the arbitrator applied the wrong test in deciding that the appellants could not claim any losses over and above those which would have been sustained had they adopted that suggestion, it necessarily follows that his approach to this item was erroneous also. If, on the remission, the arbitrator comes to the conclusion that the appellants acted reasonably, it will be necessary for him to examine this item of claim afresh. K Further, contrary to what is set out in the last sentence under subpara (b), it is not necessary to incorporate any award under this head in the annual calculation of profit foregone since, as I have already tried to indicate, paras 29 and 30 of the guidelines allow for capital costs of this nature as separate items.

Subpara (d) — here again the arbitrator seems to give two reasons; the first, namely that reflected by the words "this building would be essential if a flock of sheep was retained", reflects the same error of approach as that described under subpara (b). The arbitrator continues, however, by saying: "even without the flock the building is of such a construction that it adds value to the holding as it could be adopted for other uses of value to the farm". L Again, the wrong test has been applied; the arbitrator should ask himself whether the appellants would be unreasonable if they did not use the building for other uses of value; this distinction, however, may well turn out to be more academic than real.

Subpara (e) — the forced sheep sale. The arbitrator states:

> in a letter dated 22nd July 1987 from Mr DJ Harry Thomas to Mr Thompson, land agent to the NCC, it is made quite clear that the advice of their auctioneer was that because of the number of ewes involved a sale on the farm was likely to produce better prices.

The arbitrator continues that he considers that the advice was M understandable because of the quality of the stock and the reputation of the appellants. But that again is a wholly incorrect reason for disallowing the claim. This matter too must be considered in the light of whether the appellants acted reasonably in changing to the farming policy which they did. It is implicit in the arbitrator's remarks under this subparagraph that he considers that the appellants acted reasonably in selling on their farm rather than at auction, so that if, on remission, he comes to the conclusion that the appellants did act reasonably in their selection of an alternative farming strategy, this too is a claim which must be entertained.

Subpara (f) — this has not been pursued in this appeal.

A Subpara (g) — here the arbitrator does not deal with this on the basis that it is a claim which does not fall within his remit. In view of the terms of para 29 of the guidelines, this again must be classed as an error.

Subpara (h) — this has not been contended for.

I return therefore to subpara (c) which, in the way in which it has been dealt with, gives rise to some difficulty. It is necessary to set out the arbitrator's findings verbatim, as follows:

> Loss in capital value of the holding — no evidence to substantiate the valuations put forward on behalf of the claimants was put forward. The valuation was therefore not tested before me and in any case the guidelines are B quite explicit about the claim to be either profit foregone or lump sum payment.

Two criticisms are made of this, both, as it seems to me valid. To start with, I am told, and there is no dispute about this, that in fact evidence was put before the arbitrator in the form of a report from a Mr John Jeremy [FRICS], whom the appellants did not call. Granted that the respondents objected to the report being put in evidence, nevertheless the arbitrator allowed it. A reading of it reveals that the arbitrator appears to have been under a misapprehension. What was being claimed was not the loss in capital value of the land by reason of C the restrictions from the very moment the restrictions were imposed — that indeed would be attempting to claim both annual payments and lump sum payments, and would be impermissible. On the contrary, what was being claimed was the diminution of the capital value of the land after the management agreement had run its course. For the sake of convenience this has been referred to as residual capital loss. The report of Mr Jeremy places this at £55,000, which was the figure claimed in the appellants' original notice of claim.

This gives rise to the following problem: as expressed, there is no discernable error of law or for that matter of approach in the way in which the arbitrator has dealt with the loss in capital value under D para 18 subpara (c). Am I entitled to go behind the curtain as it were in order to satisfy myself that the arbitrator's finding was based on a misapprehension as to the nature of the claim being made? Mr Bush has drawn my attention to the judgment of the Court of Appeal in *Universal Petroleum Co* v *Handels- und Transportgesellschaft GmbH* [1987] 2 All ER 737. This report contains a definitive statement of the limitations of the right to challenge an arbitrator's award imposed by section 1 of the 1979 Arbitration Act. In particular, the court stressed that primary findings in an award could not be challenged unless there was misconduct by the arbitrator, or a lack of jurisdiction or he had failed to accede to the request of the parties on a specific matter. It was stressed that the arbitrator's primary findings of fact were immune E from review apart from those circumstances and in particular that evidence could not be introduced in order to challenge those findings. Any error open to challenge had to be visible in the award itself rather than arise from the arbitration. Having considered the entirety of the judgment, however, I have reached the conclusion that the case is not strictly in point where the present issue is concerned. The Court of Appeal were dealing with the strict limitations on the power to challenge findings of primary fact by an arbitrator. Subpara (c) of para 18 of the award makes no primary findings of fact other than the incorrect one that no evidence was put forward in support of the claim. For my part, I do not consider that looking at the way in which the F original claim was framed comes under the heading of taking account of matters which are extrinsic to the award. The claim is, in essence, the progenitor of the award itself.

In my judgment, it surely must be permissible to consider precisely what was being claimed in order to go one further and consider whether the arbitrator in fact dealt with that particular claim. To put it another way, though one cannot query whether the arbitrator got his facts right, it must surely be open to a party to contend that he failed to perform his task. In this instance, although the claim was rather baldly stated in its pleaded form as a claim for the capital value of the land, nevertheless the sum of £55,000 was specifically mentioned and reference was made to Mr Jeremy's report. This, I consider, was sufficient to bring that report within the ambit of the claim, and once G that position is arrived at, it is quite clear that the appellants were claiming only for the residual capital loss. This is a matter which the arbitrator failed to appreciate and accordingly this again must be submitted to him for reconsideration.

Finally, I come to the third issue raised in the appeal, namely the contention that the matter should be remitted to the arbitrator for him to make further findings of fact and to give his reasons for arriving at the award which he did. I can deal with this fairly shortly. It should be said at once that, since the arbitrator was specifically asked to state his reasons, sadly, it is apparent that he has not done so. Apart from the bare statement that he has considered this, that or the other factor, he H has merely given figures for the three years in question. It has been quite impossible for either party to determine just how the arbitrator has arrived at those three figures. While accepting Mr Bush's submission that an arbitrator cannot be expected to make every single minute finding of fact involved in a fairly complex award, yet where the whole purpose of the arbitration is to calculate and arrive at a specific sum or sums of money payable by way of compensation, then the steps and calculations by which those sums are arrived at, in my judgment, must amount either to primary findings of fact or at least to reasons. In the present award virtually no reasons have been given; Mr Bush himself, when asked point blank how one could determine J the method whereby the arbitrator had arrived at the three figures referred to, was unable to supply an answer, nor could anyone else.

But that is by no means conclusive. Section 1(5) of the Arbitration Act 1979 has also drastically limited the circumstances in which an award can be remitted for further findings. The subsection states, in its relevant terms:

> If an award is made and, on an application made by any of the parties to the reference it appears to the High Court that the award does not or does not sufficiently set out the reasons for the award, the court may order the arbitrator or umpire concerned to state the reasons for his award in sufficient detail to enable the court, should an appeal be brought under this section, to consider any question of law arising out of the award. K

The foregoing words envisage a two-stage process; first, the court has to determine whether the reasons given for the award are sufficient. That I have already done and held them insufficient. But, second, the court cannot order the matter to be remitted for further findings or reasons unless satisfied that these would enable it to determine whether an error of law had been committed. Despite Mr Brodie's submissions on the point, I remain quite unconvinced that even if the arbitrator gave the precise reasons for the award at which he arrived, which principally would be a matter of mathematics, any question of law would be likely to arise. L

For these reasons I do not think that the application under subsection (5) can succeed. But, for the reasons which I have endeavoured to express, I take the view that this appeal must succeed, and the matter must be remitted to the arbitrator for his further consideration after applying what I have stated to be the correct test when interpreting the meaning of the words "profits foregone because of the restrictions" in para 16 of the guidelines.

Appeal allowed and award remitted.

M

Chancery Division

July 12 1993

(Before Mr Justice EVANS-LOMBE)

SECRETARY OF STATE FOR THE ENVIRONMENT v REED INTERNATIONAL PLC

Estates Gazette February 12 1994

[1994] 06 EG 137

Arbitration — Applications for leave to appeal and to remit — Rent review — Whether arbitrator had evidence to award uplift for overbid of actual tenant of adjoining building — Whether arbitrator failed to consider parties' submissions in awarding uplift

The applicant Secretary of State was the subtenant under two subleases dated July 28 1988 held from the respondent sublessor in respect of two adjoining premises consisting of contiguous multi-storey blocks, which are occupied for the purposes of the Department of the Environment. In accordance with the provisions for rent review under both subleases, an arbitrator was appointed to determine the rent becoming due on May 7 1991. In making his award the arbitrator stated his opinion that, in assessing the rent under each sublease, and faced with the reality of a large organisation occupying the other half of the building and appearing to be a potential active bidder, the hypothetical tenant would have taken all steps to secure a tenancy: this hypothesis envisaged the possibility of two bidders for each unit and therefore the hypothetical tenant would have made a small overbid of £1 per sq ft. The applicant sought leave to appeal under section 1(3) of the Arbitration Act 1979, on the grounds that there was no evidence to support the overbid, and to remit the award under section 22 of the Arbitration Act 1950, on the same grounds and further that the route by which the arbitrator applied the uplift in rent was not one advanced by either party in their respective surveyors' written representations.

***Held*: Both applications were dismissed. (1) The application for leave to appeal under section 1(3) of the 1979 Act was dismissed because the basis of the application, namely that there was no evidence to support the arbitrator's findings on the uplift, is insufficient; there was sufficient evidence before the arbitrator, namely that the department occupied the other property in each case, the proximity of each property and the ease of communication between each. (2) The award should not be remitted because there was evidence before the arbitrator. Further, in deciding to giving his reasons for uplifting the rent, the arbitrator was exercising pure valuation experience on his part.**

The following cases are referred to in this report.

Baleares, The [1993] 1 Lloyd's Rep 215

King v *Thomas McKenna Ltd* [1991] 2 QB 480; [1991] 2 WLR 1234; [1991] 1 All ER 653, CA

This was the hearing of an application by the Secretary of State for the Environment under section 1(3) of the Arbitration Act 1979 for leave to appeal and under section 22 of the Arbitration Act 1950 to remit an award of an arbitrator determining rents payable under subleases held by the applicant from the respondent, Reed International plc.

Paul Morgan QC (instructed by the Treasury Solicitor) appeared for the applicant; David Neuberger QC (instructed by D J Freeman) represented the respondent.

Giving judgment, **EVANS-LOMBE J** said: I have before me two applications. The first is for leave to appeal under section 1(3) of the Arbitration Act 1979, leave being required solely for questions of law as are set out in the summons arising from the arbitrator in this case applying an uplift to his award under two rent review clauses of which he had been appointed by the parties as arbitrator. The second application, after concessions, is to remit under section 22 of the Arbitration Act 1950 the arbitrator's award so that the arbitrator can review that award as to the questions of uplift only.

The situation on the ground is that the applicant, the Secretary of State for the Environment (to whom I will refer as "the department"), is the subtenant under two subleases from the respondent, Reed International plc (to whom I will refer as "Reed"), in respect of two adjoining premises, 83-85 Piccadilly (to which I will refer as "no 83") and 29 Bolton Street (to which I will refer as "no 29"). These premises are contiguous multi-storey blocks and have been occupied by the department for the purpose of that department's administration. In the award the arbitrator said this with relation to the premises:

> The valuations relate to two contiguous but hypothetically separate units within a single structure each of about 42,000 sq ft with some services shared, but in accordance with the lease provisions, separate rentals need to be assessed for each demise.
>
> In reality the entire property is occupied by a Government department and there are no physical signs in the building to indicate the separate hereditaments.

The underlease of no 83 is dated June 28 1988 and, under it, rent is payable for the initial period of the lease down to May 7 1991 of £997,450. The relevant rent review clause is clause 5. That clause provides for a first review date in respect of the rent at May 7 1991.

The provision for arbitration is contained in subclause 3 of that clause, and provides as follows:

> Unless within a period of three months after the relevant review date the parties have agreed in writing upon the amount of the current market rental value then the same shall be determined by an independent chartered surveyor (who shall act as an arbitrator pursuant to the Arbitration Acts 1950 to 1979 or any statutory modification or re-enactment thereof for the time being in force) such surveyor to be agreed upon in writing by the landlord and the tenant or in the event of failure so to agree for a further period of one month to be nominated at the instance of either party by the President for the time being of the Royal Institution of Chartered Surveyors and the decision of such surveyor shall be binding on the landlord and the tenant and the fee payable in respect of any such decision made by such surveyor shall be borne and paid by the parties in such shares and in such manner as he shall determine.

The current market rental value spoken of in that subclause is defined in subclause 1(ii) in the following terms:

> "The current market rental value" at the review dates means the yearly rent obtainable without taking any fine premium at which the demised premises might at the relevant review date be reasonably expected to be let in the open market for a term equivalent to the then unexpired term hereby created with vacant possession under a lease on the same terms and conditions in all other respects as in this present lease (but upon the assumption that the tenant has complied with all its covenants and agreements herein contained) but disregarding —
>
> (a) any goodwill that shall have attached to the demised premises by the carrying on thereat of any business by the tenant and any lawful subtenants of the tenant;
>
> (b) any effect on the rent of the demised premises of any improvements thereto that shall have been carried out by the tenant or any lawful subtenant during the term hereby created other than pursuant to an obligation to the landlord (not being an obligation to comply with statutes or other legislation or the directions of the public);
>
> (c) any effect on rent of the fact that the tenant and the tenant's predecessors in title have been in occupation of the demised premises.

The sublease in respect of no 29 was also dated June 28 and provided for a rent of £742,200 for the first period of the tenancy, also to May 7 1991. The review provisions contained in that sublease were to all intents and purposes precisely the same as those in respect of the sublease of no 83. The parties were not able to agree on a new rent becoming due on May 7 1991. Accordingly, Mr Samuel Levy [FRICS], of Jones Lang Wootton, was appointed an arbitrator. The arbitration was to be conducted in writing. On June 30 a Mr Spencer King [BSc FRICS], of Debenham Tewson & Chinnocks, for Reed, made submissions to the arbitrator. At para 9.12 he refers to:

A (c) any effect on rent that the tenant has been in occupation of *the demised premises* . . .

And the submission is:

9.13 It is accepted therefore that the tenant's occupation of the actual premises is to be ignored. There is no requirement however that the occupation of the tenant in the *adjoining property* is to be ignored. In respect of either review, therefore, the tenant of the adjoining building will be assumed to be in the market with a requirement for accommodation suited by the subject property demised under either lease, which it is assumed will be vacant.

9.14 Clearly, the tenant of the adjoining building may be expected to bid
B very competitively for the subject premises. This is particularly so having regard to the ease of intercommunication, as demonstrated by the existing arrangements, and the ability to secure the arrangements for the provision of services and recovery of costs between the two buildings.

9.15 It is my experience that a bid from the adjoining owner may not be precluded by the assumption of a reasonable rent. This situation seems analogous to the bid for vacant space within a multi-occupied building from a tenant already within the complex.

9.16 I attach at Appendix I an extract from the reasoned award by Mr Graham Plumbe in the Millbank arbitration in which both you and I were involved. He specifically rejected the proposition that such bid represented that of a special purchaser and as such should be disregarded in the assessment
C of a reasonable rent.

9.17 The extent of the overbid is difficult to quantify, although I believe this is a factor which should be taken into account in view of the inter-relationship between these two buildings and their current occupation.

At para 12.2.6, Mr King says:

Having calculated the rental values of each building, I have made a nominal increase of 5%. This is to reflect the bid that could be anticipated from the tenant of the adjoining building who would form part of the market and was known to have a space requirement at the relevant time.

He then went on to apply an uplift to his assessment of the rent,
D which amounted to some £82,850 in respect of no 83 and £87,700 in respect of no 29.

The counter-submission from the department was made by a Mr Adrian Smith [ARICS]. What Mr Smith says starts at para 9.13 and is:

I can make no sense of the suggestion that the tenant of the adjoining building will be assumed to be in the market with a requirement for accommodation suited by the subject property demised under either lease. All we are required to assume is that there will be a hypothetical willing lessee for the vacant premises. This is not the actual occupier as Mr King seems to suggest.

The approach to the possibility of a bid being made by the occupier of the premises adjoining the subject premises is dealt with by the decision in the
E Chancery Division of *First Leisure Trading Ltd* v *Dorita Properties Ltd* [1991]* and I attach a copy of the case . . . The conclusion from this case is that the arbitrator is not required to assume either that the adjoining occupier is or is not a hypothetical tenant. The matter must rest on the facts.

Mr King has offered no evidence to support his contention in this paragraph.

There is nothing unusual about the physical layout of each building when considered separately and there is, accordingly, no reason why the tenant of the adjoining building (whoever he might be) would put in a higher bid to secure the subject premises; nor for that matter is there any reason to suppose that the tenant of the adjoining building would be the only potential bidder for the subject premises, thus enabling him to put in a *lower* bid.
F The same applies in relation to the question of service charge recovery. The provisions are workable . . .

He deals with that, and then continuing at para 9.15:

I attach at Appendix 2 an extract from the Reasoned Award of Mr Stanton [in another case].

In this case a similar "special purchaser" bid argument arose out of the Secretary of State for the Environment's position as tenant in the building under a number of leases. There, as here, the issue was not the question of whether there was such an argument admissible in law but whether in fact "precise evidence was introduced to indicate that the Secretary of State for the Environment would have required these floors so badly as to consider making an overbid". G

Mr King had not introduced any evidence to this effect.

Attached to Mr Smith's submission was a letter. This is a letter addressed to Mr Smith from a Mr Glenn, deputy head of Division of Property Holdings, the Government agency, and materially it says:

You ask me if there was a demand at the rent review dates from any arm of Government for approximately 40,000 sq ft of office space.

There was not. However, in the latter part of June, Savills were instructed to identify market options, to be considered with properties available on the Government Estate for meeting a potential requirement of some 35,000 to
50,000 sq ft. The search covered an area bounded by Baker Street to the H
West, Oxford Street/Holborn to the South, City Road to the East and King's Cross/Euston to the North

— And I notice in parenthesis that that does not include the area of either of these buildings —

Further, at that time, it was known that as a result of Government policy in relocating staff from London some 2m square feet of space in the Government Estate was likely to be surplus to requirements and available within 4-5 years of the review date. This had imported a strong policy presumption that new demands would, wherever possible, be met from the existing Estate. The Government has few holdings in Mayfair/St James's, because the location J
tends not to suit Department's operational requirements, and rents are high. Indeed the occupying Government Department in the subject buildings is itself actively planning to leave Mayfair.

In the circumstances there would have been no possibility of a commitment to take a building the size of either no 83 Piccadilly or no 29 Bolton Street in Mayfair or St James's on the date in question.

I do, however, note in passing that there appears to have been no attempt by the department to dispose of either of the subleases of no 83 or no 29 although the arbitrator makes no reference to this fact in his award.

There was a counter-submission on behalf of Reed by Mr King K
also dated February 17 1992 where he says at para 4.1.2:

The potential occupier of 83 Piccadilly would not be concerned by the implications on security if the escape rights from no 29 Bolton Street are held by an associated organisation. It is therefore worthwhile repeating, when considering the rent review on 83 Piccadilly, the Secretary of State for the Environment is assumed to be in 29 Bolton Street with a requirement suited by the subject premises, from which he is assumed to have vacated. The hypothetical tenant would therefore have to ignore any problems as a consequence of the adjoining building if he is to compete on equal terms with the adjoining occupier.

Subsequent to those submissions, there was correspondence L
between the parties and the arbitrator, in the course of which an opinion of counsel, which had been prepared for the department, was submitted to the arbitrator with the consent of Reed. That matter is dealt with in the award of the arbitrator, which followed on June 16 1992. What the arbitrator said was this, at para 3, under the heading:

Proximity of Adjoining Property

Mr King argues that in each case the actual tenant of the adjoining unit should be considered as a "special bidder" who, requiring additional offices, would make a higher bid, and he adopts an uplift of 5%, which I find unsupported in his evidence. Mr Smith, for his part, produces evidence that the actual lessee of both units had no requirement for additional offices at the M
date of the review.

In the circumstances of these two units held on separate leases but within the reality of a single structure, I find that this particular matter has caused concern to me as I have been uncertain whether, in law, I am entitled to assume that both units were vacant and hypothetically available for letting at the review date or, when considering the rent for just the one property, that the other was let and therefore the occupiers (whoever they shall be) could be considered as a potential bidder for the subject unit.

I now have the benefit of seeing a copy of a counsel's opinion sent to me by Mr Smith but with Mr King's approval which illustrates the legal position and I understand that the two surveyors are at idem that I should not assume that both units were vacant at the review date. This point is no longer valid and has been disregarded.

*Editor's note: Reported at [1991] 1 EGLR 133.

A Moving on to the other quasi legal point as to whether the hypothetical lessee of one unit can be regarded as a special bidder for the other (and vice versa for the second unit of space), I have considered the additional comments put to me by both surveyors.

Although I have received no evidence to show that the actual tenant of each unit would have been prepared to overbid the market rental for the other section of the property, I am of the opinion that faced with the reality of the substantial organisation occupying the "other half" of the building and appearing to be a potential active bidder, the hypothetically willing tenant would have taken all steps to secure a tenancy.

This hypothesis envisages the possibility of two bidders for each unit — the hypothetical tenant and the actual tenant of the other unit — and in these
B circumstances I believe that the hypothetical tenant would have made a small overbid which I measure in the terms of £1 per sq ft for prime office space in each unit.

And he adds in respect of no 83, therefore, a figure of some £39,014, and in respect of no 29 a figure of some £41,345 to the rent that he otherwise would arrive at as being the ordinary market rent.

I will deal with the application for leave first. Section 1 of the Arbitration Act 1979 provides materially as follows:

> (2) Subject to subsection (3) below, an appeal shall lie to the High Court on any question of law arising out of an award made on an arbitration agreement; and on the determination of such an appeal the High Court may by
> C order —
> (*a*) confirm, vary or set aside the award; or
> (*b*) remit the award to the reconsideration of the arbitrator or umpire together with the court's opinion on the question of law which was the subject of the appeal;
> and where the award is remitted under paragraph (*b*) above the arbitrator or umpire shall, unless the order otherwise directs, make his award within three months after the date of the order.
> (3) An appeal under this section may be brought by any of the parties to the reference —
> (*a*) with the consent of all the other parties to the reference; or
> (*b*) subject to section 3 below, with the leave of the court.

D The provisions of section 3 are immaterial so far as the issues before me are concerned.

> (4) The High Court shall not grant leave under subsection (3)(*b*) above unless it considers that, having regard to all the circumstances, the determination of the question of law concerned could substantially affect the rights of one or more of the parties to the arbitration agreement; and the court may make any leave which it gives conditional upon the applicant complying with such conditions as it considers appropriate.

It was submitted for the department, first, that there was no evidence to support the conclusion that an allowance by way of rental
E uplift was proper because the department can be assumed to be prepared to pay the market rent so that an overbid was necessary for any outsider to obtain possession. Second, it was submitted that, there being no evidence to support the arbitrator's conclusion, there was raised a point of law within the provisions of section 1(2). Third, that the arbitrator's award granting the uplift was a matter which substantially affected the interests of the department as a party to the arbitration within the meaning of section 1(4).

I will deal with the third submission first by accepting it. It seems to me that, notwithstanding that the uplift represented a small percentage of the total rent awarded, it was none the less a material
F matter representing rather more than £80,000 pa in respect of both sets of premises.

I now turn to deal with the second submission. In the course of argument I was referred to the decision of the Court of Appeal in *The Baleares* [1993] 1 Lloyd's Rep 215. I do not need to refer to the facts of this case because, in the passage from the judgment of Steyn LJ, which appears at p231, he deals generally with the question of what constitutes a point of law for the purposes of section 1(3), and I start reading from his judgment at p231 in the right-hand column, the last paragraph on that page, where he says:

> The search for a plausible question of law proved elusive. Mr Goldsmith suggested at one stage that the appropriate question of law might be that there was no evidence to support the finding of fact that there was a loss. Mr G
> Goldsmith displayed little enthusiasm for this submission but he never formally abandoned it. And Mr Rokison QC found it necessary to make a contrary submission in reply. In view of the economical way in which Counsel dealt with the point I will be equally brief. One difficulty in Mr Goldsmith's way was that the arbitrators had never been asked to state in their award the evidence on the relevant finding of loss: see *Mondial Trading Co GmbH* v *Duffus Zuckerhandelsgesellschaft* [1980] 2 Lloyd's Rep 376, at p379, col 1. But it seems to me that the difficulty is more fundamental. After a careful analysis of the history and purpose of the Arbitration Act 1979 *Mustill & Boyd* concluded in the first edition of their book, at p541, that this rule should no longer be recognised. In *Universal Petroleum Co Ltd* v *Handels & Transport GmbH* [1987] 1 Lloyd's Rep 517 Lord Justice Kerr tentatively commented that it may be going too far to say that the jurisdiction no longer exists. He H
> thought that the way to deal with the matter is for the Commercial Judge to refuse leave to appeal on such a ground. In the second edition of their book, at p596, *Mustill & Boyd* maintain their view. I respectfully agree with their view. The concept of a jurisdiction which ought never to be exercised is not attractive. Moreover, the problem cannot be solved by the Commercial Judges stifling such applications at the leave to appeal stage. More often than not the Commercial Judge's order giving leave to appeal will not exclude the possibility of an argument that there is no evidence to support a particular finding of fact. The order will be in general terms. If leave is given on one discrete point of law it will then be possible also to argue that there is no evidence to support a finding of fact thereby compelling the court to review the relevance and weight of the evidence before the arbitrators. The power to J
> review a finding of fact of a tribunal on the ground that there is no evidence to support it, and that there is therefore an error of law, is a useful one in certain areas of the law, notably in the administrative law field. But in the limitation appellate jurisdiction of the Court under s1 of the Arbitration Act 1979 this concept has no useful role to play. It is inconsistent with the filtering system for the granting of leave to appeal which was created by the Arbitration Act 1979. In my judgment it has not survived the changes introduced by the reforming measure of 1979.

In his judgment in this case Neill LJ said at p227:

> In conclusion I should make a brief reference to the question discussed by Lord Justice Steyn: Is it still permissible to review as an error of law a finding of fact by arbitrators which is challenged on the ground that there was no K
> evidence to support it. Lord Justice Steyn has concluded in his judgment that such an "error of law" has not survived the coming into force of the Arbitration Act 1979. For my part I do not feel able without further argument to reach a final conclusion on this point, but I am impressed by the argument that it is inconsistent with the thrust and purpose of the 1979 Act to allow findings of fact to be reviewed by means of any examination which involves reference being made to the evidence which was before the arbitrators, where the object of the examination is to establish that there was no or no sufficient evidence to support the finding.

Ralph Gibson LJ agreed with the judgment of Neill LJ.

In my judgment, I should be guided by what Steyn LJ has said and L
I find that the basis of the application for leave, namely on the submission that there was no evidence to support the arbitrator's findings on uplift, is insufficient for this court to grant leave. If I am wrong in that, it seems to me that there was in fact evidence upon which the arbitrator could have based his conclusion. There are really three aspects to that evidence. The fact that the department was in occupation of adjacent premises which was admitted, the nature of those premises or the two sets of premises and their proximity to each other and the apparent ability, according to the arbitrator, to communicate easily between the two and the fact that the department was looking for premises elsewhere in London, if not in this area. So M
it seems to me that, even if my reliance on what must be treated as *obiter dicta* by Steyn LJ is wrong, there was evidence before the arbitrator upon which he could have based his findings.

So I turn to the application to remit under section 22. It was submitted for the department again that there was no evidence to support the conclusion as to uplift. In the light of what I have already said when I dealt with the application for leave to appeal, in my judgment this cannot be a matter which could support an application to remit.

It was further submitted that neither party in their written submissions, from which I have quoted extensively, suggested that

A route to allowing an uplift which was ultimately selected by the arbitrator. It was submitted that not only did he find his own route to that uplift but he did not warn them that he was proposing to do so, and that he should have done so as to enable the parties and, in particular, the department to make submissions that that approach was wrong.

I turn to the notes in the White Book vol 2, p1772 under para 5793, and the latter part of that paragraph where reference is made to the decision in *King* v *Thomas McKenna Ltd* [1991] 1 All ER 653:

> The jurisdiction of the court under s22 is unlimited and extends beyond the traditional categories to any case where, notwithstanding the arbitrator has
> B acted with complete propriety, due to mishap or misunderstanding some aspect of the dispute the subject of the reference has not been considered or adjudicated upon fully or in the manner the parties are entitled to accept, and where it would be inequitable to allow any award to take effect without some further consideration by the arbitrator.

This part of these two applications is the only part which has caused me any very great difficulty, but I have finally concluded that I should not remit this award. It seems to me that what the arbitrator was here doing was the exercise of pure valuation experience on his part. The point with which he was dealing was a pure valuation point, and what the arbitrator was doing was really partially accepting the
C submission of Reed that there should be an uplift, but not accepting it in full. It also seems to me that to remit the award for reconsideration at this point to this arbitrator would be highly unlikely to serve any useful purpose. The arbitrator has come to this conclusion. It is hard to see how the parties can, on the basis of the current evidence, submit other than that his approach was technically wrong, which is likely to be a submission that the arbitrator will reject. I also bear in mind the submissions made to me relating to the modern trend in the legislation and the purpose behind that legislation dealing with arbitration procedure. The modern trend is, where possible, to require parties who have elected to accept arbitration to treat any award made by an
D arbitrator as final — the trend towards finality, as it has been described to me. It seems to me in these circumstances that, where I am in doubt as to whether one of the parties can justly complain that he has not had an opportunity to make submissions with relation to a particular and not very important point in the arbitration, I should resolve that doubt by leaving the arbitrator's award to stand. Accordingly, I propose to take that course. It follows, therefore, that each of these applications fails.

E

Court of Appeal

November 24 1993

(Before Lord Justice DILLON, Lord Justice MANN and Lord Justice HIRST)

COHEN AND OTHERS v BARAM

Estates Gazette April 2 1994

[1994] 13 EG 111

F

Arbitration — Whether successful party under arbitration award entitled to defend claim by arbitrators for fees and expenses on ground that arbitrators' award of costs wrong in law — Whether arbitrators have a cause of action for recovery of fees

In accordance with Jewish law the Beth Din of the Federation of Synagogues, acting as arbitrators under the Arbitration Acts 1950 and 1979, made an award in favour of the appellant in respect of a claim by A and L. Following the usual practice, the Beth Din made no award as to the parties' respective legal costs, but in respect of their own fees and expenses made an unreasoned award that the appellant should pay half of these. In appealing against judgment against him under Ord 14 of the RSC in a claim by the trustees of the Beth Din for its fees, the G appellant contended that: (1) the trustees were the wrong claimants; (2) there had not been a proper judicial exercise of the arbitrators' discretion, because the award of costs did not follow the event and no reasons for the award had been given; (3) the appellant's counterclaim had not been taken into account.

Held: **The appeal was dismissed. The arbitrators were employees of the federation and therefore the federation's trustees were the proper plaintiffs. The claim to the arbitrators' fees and expenses lay under an implied contract that parties to an arbitration will pay such costs. The appellant's remedy in respect of the alleged failure by the arbitrators properly to exercise their H judicial discretion on costs should have been raised by appeal under section 1 of the Arbitration Act 1979, or for remission under section 22 of the Arbitration Act 1950 to seek reasons. The time-limits and procedure were prescribed by Ord 73 of the RSC. He could not challenge the award by way of a defence to claim against him on behalf of the arbitrators. The counterclaim was implausible.**

The following cases are referred to in this report.

Antaios Compania Naviera SA v *Salen Rederierna AB* [1985] AC 191; [1984] 3 WLR 592; [1984] 3 All ER 229; [1984] 2 Lloyd's Rep 235, HL J
Blexen Ltd v *Trentham (G Percy) Ltd* [1990] 2 EGLR 9; [1990] 42 EG 133
Crampton & Holt v *Ridley & Co* (1887) 20 QBD 48
Foster v *Great Western Railway Co* (1882) 8 QBD 515
King v *Thomas McKenna Ltd* [1991] 2 QB 480; [1991] 2 WLR 1234; [1991] 1 All ER 653, CA
Pioneer Shipping v *BTP Tioxide Ltd ("The Nema")* [1982] AC 724; [1981] 3 WLR 292; [1981] 2 All ER 1030; [1981] 2 Lloyd's Rep 239, HL
Smeaton Hanscomb & Co Ltd v *Sassoon I Setty Son & Co (No 1)* [1953] 1 WLR 1468; [1953] 2 All ER 1471
Wandsworth London Borough Council v *Winder* [1985] AC 461; [1984] 3 WLR 1254; [1984] 3 All ER 976; 83 LGR 143, HL

This was an appeal by David Baram from an order of Mr Graham K Hamilton QC (sitting as a deputy judge of the Queen's Bench Division), who had dismissed with costs the appellant's appeal from an order of Master Prebble refusing leave to defend in an action brought by the respondents, Arnold Cohen and four others.

Jack Denbin (instructed by Davison Stone & Lewis, of Sheffield) appeared for the appellant; Stephen Shaw (instructed by Franks Charlesly & Co) represented the respondents.

Giving the first judgment at the invitation of Dillon LJ, **HIRST LJ** said: This is an appeal from the order of Mr Graham Hamilton QC, sitting as a deputy judge of the Queen's Bench Division, made on L March 1 1993, dismissing with costs the appeal of the defendant, Mr David Baram, from an order of Master Prebble dated January 26 1993, by which the learned master refused to give the defendant leave to defend in this action, brought by the plaintiffs, Mr Arnold Cohen and four others, who are the trustees of the Federation of Synagogues.

The appeal centres entirely on an order for costs made by the Beth Din of Federation of Synagogues, London, in an arbitration between the two claimants (Mr Landau and Mr Adler) and the present appellant, who was the respondent in the arbitration.

The factual background and history of the course of proceedings is as follows. The respondents in this appeal are, as I have already M mentioned, the trustees of the Federation of Synagogues, which is one of the two main orthodox synagogual institutions in this country. The institution has its own religious court, called a "Beth Din", to which parties who so desire can refer disputes. The court sits as a court of arbitration.

In February 1991, the appellant submitted to the jurisdiction of the Beth Din by a written agreement for submission, dated February 21 1991, in respect of a dispute involving himself, in this case as respondent, and Mr Landau and Mr Adler as claimants.

The submission was in the following terms:

> In the matter of the Arbitration Acts 1950 and 1979 and in the matter of an

Arbitration between Mr Landau and Mr Adler (claimants) and Mr Baram (respondent).

Whereas a dispute or difference has arisen and still exists between the claimants and the respondents, and the parties have failed to come to terms;

And whereas it is the desire of the parties to refer such dispute or difference by way of Din Torah to the arbitration and final decision of the Beth Din of the Federation of Synagogues, London;

Now therefore the parties agree as follows:—

1. The parties hereby agree to refer to the arbitration and final decision of the Beth Din of the Federation of Synagogues, London, all disputes or difference between them, and all claims which either party alleges that he has against the other party, for determination by way of Din Torah according to the rules of procedure customarily employed in arbitrations before the Beth Din, and according to principles of Halachah and/or general principles of equity customarily employed in arbitrations before the Beth Din.

2. The parties hereby agree each on their part to accept and perform the award of the said Beth Din touching all disputes, differences and claims between the parties, which award shall be final and binding . . .", — the critical words — ". . . and to pay such costs as the Beth Din may determine within the period specified in the award.

3. The parties hereby agree that should either party, after the preliminary hearing has been heard *inter partes*, subsequently fail without good cause to attend any subsequent hearing, the Beth Din may proceed to determine the matter *ex parte*.

Then the document is signed by the two claimants and the present appellant. It is dated February 21 1991.

It should be noted that it is common ground between the parties that so far as the parties' own legal costs are concerned in a Beth Din arbitration it is an established principle of Jewish law that each should bear their own costs. It is therefore also common ground that the reference which I highlighted, that is the agreement to pay such costs as the Beth Din may determine, refers to the arbitrator's own costs and/or fees and not to the parties' legal costs.

There is a parallel submission to arbitration, dated the same date, of another dispute, with a reference in identical terms, although the roles of claimant and respondent are reversed. That second dispute (in what I will call the second arbitration) has not yet been adjudicated upon.

The Beth Din sat on four occasions, with three arbitrators, during the course of 1991, and published its award on the merits in March 1992 in favour of the present appellant, the respondent in the arbitration. In addition, the Beth Din ordered that the costs, that is their costs, should be divided equally between the two sides, with the present appellant being ordered to pay the sum of £2,000 by way of costs within 30 days and Messrs Landau and Adler being ordered to pay the same amount within the same period. No reason was given by the arbitrators for the order which they made in that form, nor were the costs particularised, nor were they broken down between fees and expenses.

The award was sent to the appellant under cover of a letter dated March 12 1992. However, despite subsequent demands for payment, he declined to pay. In consequence this action, seeking recovery of the £2,000 costs awarded against him, has been brought.

The plaintiffs issued Ord 14 proceedings on August 17 1992, which resulted in a summary judgment in the plaintiffs' favour by the master, to which I have already referred, which was upheld by the learned judge.

The appellant raised with the master and the judge, and repeated in this court, a number of grounds which he contends entitle him to unconditional leave to defend, all of which were rejected by the learned judge.

They fall conveniently under three headings, which I shall outline before considering them separately.

The first heading is "respondents' status". The appellant challenged the eligibility of the present respondents, as trustees of the federation, to bring these proceedings, contending that the arbitrators themselves were the proper plaintiffs. There was also a question as to the *locus standi* of the respondents to sue as plaintiffs for recovery of their fees and expenses. Both points are disputed by the respondents.

The second heading is "proprietary of the award of costs and Ord 73". The appellant contends that the award was flawed on a number of separate grounds. First, he contends that, as a matter of law, the arbitrators' decision on costs was not a judicial exercise of their discretion, because they failed to apply the general rule that costs follow the event *a fortiori* since they failed to give any reasons for their decision to depart from that general rule.

Second, he contends that the arbitrators' award of costs, and the procedure adopted in relation to it, conflicted in a number of respects with Jewish law and the principles of Halachah. I do not need to detail the various contentions under this head.

The respondents contend that it is not open to the appellant to raise these matters by way of defence in the action on the ground that, by virtue of Ord 73, r2 of the Rules of the Supreme Court, the only available challenge on these grounds is via an application for leave to appeal under section 1(2) of the Arbitration Act 1979, or by an application for an order for remission under section 22 of the Arbitration Act 1950 or section 1(5) of the Arbitration Act 1979 for reconsideration and/or the furnishing of reasons, and that in default of such application the arbitration award is final, valid and unassailable.

The appellant contends that he cannot be prevented from raising these matters by way of defence, irrespective of the requirements of Ord 73, and that, in any event, he is precluded on religious grounds from making any application of this type to the court.

This is the major issue in the case and is one of significant importance.

The third heading is "counterclaim". The appellant contends that he has a valid counterclaim which on ordinary principles should be treated as a set-off and should entitle him to unconditional leave to defend. This is rebutted by the respondents.

I shall deal with these three issues under the same headings in the same order.

Respondents' status

I am completely satisfied that the present respondents, as trustees of the federation, are the correct plaintiffs. The arbitration took place under their aegis and the expenses which were incurred were incurred by them. The arbitrators themselves are employees of the federation and would not in any event, in my judgment, have been appropriate plaintiffs. The submission as to the *locus standi* of the respondents to sue as plaintiffs for their fees and/or expenses was not eventually pursued by the appellant before us, rightly in my judgment, since it is necessary, in order to give business efficacy to the arrangements between the arbitrators and the respective parties, to imply a contract that the parties respectively will honour the arbitrators' award of costs in consideration of the arbitrators on their part proceeding with the arbitration: *Mustill and Boyd, Commercial Arbitration* 2nd ed, p233. *Crampton & Holt* v *Ridley & Co* (1887) 20 QBD 48 at p54 *per* AL Smith J.

On this contract they are plainly entitled to sue, though of course normally the problem does not arise because it is customary for arbitrators to exercise a lien on the award pending payment of their costs: *Mustill and Boyd*, p235.

Propriety of the award of costs and Ord 73

It is convenient first to deal with the appellant's reliance on his own religious objections to litigating in a secular court. While, of course, fully respecting his religious scruples they are, in my judgment, entirely a matter for his own conscience and cannot be relied upon to exonerate him from compliance with the general law. Ord 73 and the Arbitration Acts thus apply to this case completely unimpaired.

Ord 73, r2(1) and (2) provide as follows:

(1) Every application to the Court —

(*a*) to remit an award under section 22 of the Arbitration Act 1950, or

(*b*) to remove an arbitrator or umpire under section 23(1) of that Act, or

(*c*) to set aside an award under section 23(2) thereof, or

. . .

(*e*) to determine under section 2(1) of the Arbitration Act 1979, any question of law arising in the course of a reference, must be made by originating motion to a single judge in court.

(2) Any appeal to the High Court under section 1(2) of the Arbitration Act

A 1979 shall be made by originating motion to a single judge in court.

The verbs "must" and "shall", in those two subrules respectively, should be particularly noted.

Ord 73, r5 (1) and (2) lay down time-limits for such summonses or motions.

Section 1(1) of the Arbitration Act 1979 provides, so far as relevant, that the previous procedure for the statement of a special case by arbitrators to the High Court shall cease to have effect.

Subsection (2) stipulates as follows:

> Subject to subsection (3) below, an appeal shall lie to the High Court on any
> B question of law arising out of an award made on an arbitration agreement; and on the determination of such an appeal the High Court may by order —
> (*a*) confirm, vary or set aside the award; or
> (*b*) remit the award to the reconsideration of the arbitrator or umpire together with the court's opinion on the question of law which was the subject of the appeal;
> and where the award is remitted under paragraph (*b*) above the arbitrator or umpire shall, unless the order otherwise directs, make his award within three months after the date of the order.
> (3) An appeal under this section may be brought by any of the parties to the reference —
> (*a*) with the consent of all the other parties to the reference;
> or
> C (*b*) . . . with the leave of the court.
> (4) The High Court shall not grant leave under subsection (3)(*b*) above unless it considers that, having regard to all the circumstances, the determination of the question of law concerned could substantially affect the rights of one or more of the parties to the arbitration agreement; and the court may make any leave which it gives conditional upon the applicant complying with such conditions as it considers appropriate.
> (5) . . . if an award is made and, on an application made by any of the parties to the reference, —
> (*a*) with the consent of all the other parties to the reference,
> or
> (*b*) . . . with the leave of the court,
> D it appears to the High Court that the award does not or does not sufficiently set out the reasons for the award, the court may order the arbitrator or umpire concerned to state the reasons for his award in sufficient detail to enable the court, should an appeal be brought under this section, to consider any question of law arising out of the award.

As is very well known, the scope for leave to appeal against an arbitration award under section 1 of the 1979 Act has been very narrowly circumscribed by the decisions of the House of Lords in *The Nema* [1982] AC 724 and *The Antaios* [1985] AC 191.

Section 22 of the Arbitration Act 1950, which has not been repealed, also gives power to the court to remit an award.

E The effect of section 1 of the 1979 Act on appeal against an order for costs in an arbitration have been considered by the Court of Appeal in two recent cases, upon which Mr Shaw strongly relies. In *Blexen Ltd* v *G Percy Trentham Ltd* [1990] 42 EG 133*, the Court of Appeal, consisting of Lloyd and McCowan LJJ and Sir John Magaw, were concerned with an appeal concerning an arbitration award of costs.

At p134 Lloyd LJ stated:

> Mr Thornton argues that the wrongful exercise of the arbitrator's discretion as to costs does not give rise to any question of law. If it does, it is not a question of law arising out of an award on an arbitration agreement within the meaning of section 1(2) of the 1979 Act and therefore section 1(1) does not apply.
> F I find that argument difficult to follow. The arbitrator has explained his reasons for making his order as to costs in great detail in his final award. The award was made pursuant to an arbitration agreement. If the arbitrator had erred in his approach the error must be one of fact or law. If it was an error of fact his finding of fact is conclusive. If it was an error of law then it would be open to the claimants to apply for leave to appeal on that question of law. I can see no difference between an error of law in a reasoned award as to costs and any other error of law.
> Then it was said that even if there was an error of law on the face of the award there is a parallel jurisdiction to set aside or remit for misconduct which survives the 1979 Act. The claimants can therefore choose their remedy. I do not agree. It is of course perfectly true that Parliament did not in 1979 repeal section 23 of the 1950 Act. The power to set aside or remit for misconduct G survives. But the question is whether that power exists when the only misconduct alleged is an error of fact or law in a reasoned award. I am quite clear that it does not. To hold otherwise would defeat one of the main purposes of the 1979 Act. Whatever may have been the position as to costs in the old days the only course open today, where an arbitrator states his reasons, is to challenge those reasons by seeking leave to appeal on a question of law under section 1(3) of the 1979 Act. This the claimants never did.
>
> I say nothing about what would be the appropriate procedure where an arbitrator states no reasons for his award as to costs.

The doubt, expressed in the last sentence I have just quoted, was laid to rest in the subsequent case of *King* v *Thomas McKenna Ltd* H [1991] 2 QB 480, where the Court of Appeal, consisting of the Master of the Rolls (Lord Donaldson of Lymington) Ralph Gibson and Nicholls LJJ, considered another appeal concerned with an arbitrator's award of costs, this time in a case where no reasons had been given. As a result of the Master of the Rolls' judgment, with which Ralph Gibson and Nicholls LJJ agreed, the court clearly held that the same principle as had been laid down in the earlier case also applied where the arbitrators had failed, as in the present case, to give any reasons (at pp494 to 495).

Mr Jack Denbin submits that these cases, and indeed the whole scheme of the Arbitration Act 1979, are not applicable, because he seeks to raise the question as to the propriety of the award of costs by J way of a defence to the action brought by the present respondents as plaintiffs.

In support of this submission he cited the case of *Wandsworth London Borough Council* v *Winder* [1985] AC 461, where the House of Lords held that Ord 53 of the Rules of the Supreme Court, dealing of course with judicial review, did not shut out a private citizen's right to challenge a decision of a local authority, under their powers laid down by the Housing Act, when that private citizen was defending an action by the local authority for arrears of rent.

In my judgment, the very important principle laid down in the *Wandsworth* case has no bearing whatsoever on the present case. K

Section 1 of the Arbitration Act 1979, as interpreted by the Court of Appeal in the *Blexen* and *King* cases, lays down mandatorily that any challenge to an award of costs by arbitrators on grounds of misconduct, or on the grounds of misapplication by the arbitrators of some general legal principle or of Jewish law, must be brought by way of an application for leave to appeal under the 1979 Act. Ord 73, r2(2) prescribes that such application must be made by originating motion to a single judge in court and Ord 73, r5(2) lays down the time-limits.

No such application having been made in the present case, the arbitrators' award stands as a final, valid and unassailable award and the appellant is, in my judgment, precluded from raising these issues L by way of defence in these proceedings or by any other route save that prescribed by the 1979 Act and Ord 73.

As an alternative he could, of course, have sought a remission for reconsideration, or for the giving of reasons under section 22 of the 1950 Act, or perhaps preferably now, under section 1(5) of the 1979 Act, on the ground that the award of costs was devoid of reasons and *prima facie* contrary to the general law that costs should follow the event in arbitrations as in court proceedings: *Smeaton Hanscomb & Co Ltd* v *Sassoon I Setty Son & Co (No 1)* [1953] 1 WLR 1468. This he never did; nor, most regrettably, did he even trouble to apply to the arbitrators direct for an explanation of their reasons. Had he done so I M think it more than likely that he would have received an explanation which would have averted this litigation.

It follows from the above that, in my judgment, Mr Stephen Shaw's submissions, based on Ord 73 and the Arbitration Act 1979, are soundly based and the appellant is debarred from raising the issues concerning the propriety of the costs award by way of defence in this action.

It follows that the respondents, as plaintiffs, were entitled to summary judgment under Ord 14, subject only to the counterclaim.

Counterclaim

This is dated September 23 1992. It claims that the respondents

*Editor's note: Also reported at [1990] 2 EGLR 9.

A have, in breach of the second arbitration agreement, declined to proceed to determine the reference. The counterclaim is as follows:

> Accordingly in the premises and by virtue of the foregoing the plaintiffs are in breach of the arbitration agreement and the defendant has suffered vexation, distress, loss, damage and special damage.

Under "particulars" it says:

> The Particulars will be served separately in the form of a schedule.

No particulars have been forthcoming for this somewhat ambitious claim under those various headings, though more than a year has B elapsed since the counterclaim was served.

It is well settled, and Mr Denbin was perfectly entitled to submit as a matter of general law, that where a defendant in Ord 14 proceedings has a *bona fide* and plausible counterclaim, the court will generally, in the exercise of its discretion, treat it as a set-off and grant unconditional leave to defend, or at least order a stay of execution until the counterclaim is determined.

However, in my judgment, this counterclaim is extremely implausible. When asked when and how the arbitrators declined to proceed with the second arbitration, in other words, when asked to identify his cause of action on the counterclaim, Mr Denbin relied C solely on two letters dated August 15 1991 and October 15 1992, in each of which the arbitrators stated expressly, and in terms, that they were ready to proceed with the second arbitration when requested for a hearing date, thus taking a posture the very opposite of that which is suggested against them, namely that they declined to proceed.

Mr Denbin then sought to rely on a letter dated June 11 1993 where the arbitrators indicated that they were not willing to go ahead with the second arbitration until the present dispute about costs in the first arbitration was resolved.

It might be said, in passing, that it seems somewhat unjustifiable to criticise the arbitrators for declining to embark on the second arbitration and incur further expenses while the costs of the first D arbitration were still in the melting pot.

Be that as it may, this letter, in any event, cannot possibly be relied upon to found a cause of action on this counterclaim, since it is dated a long time afterwards.

Quite apart from this consideration, as Dillon LJ pointed out in argument, the proper course for a party in the position of this appellant, if the arbitrators did indeed decline to proceed, is to apply to the court either for an order that the arbitration should proceed forthwith, or for an order that fresh arbitrators should be appointed.

By neglecting to pursue this course, the appellant has failed to E mitigate the damage which he alleges, not very convincingly, he has suffered as a result of this delay. Such mitigating steps taken punctually would have eliminated, or at least reduced to an insignificant amount, any such damage.

In consequence, in the exercise of my discretion, I would not be prepared to treat this counterclaim as a set-off against the respondents' established claim though, of course, the appellant is free, if he sees fit to do so, to pursue the counterclaim as a separate action.

For all these reasons I would dismiss this appeal.

I should add that there was an application to adduce fresh evidence before us of correspondence subsequent to the learned judges' hearing. F One of the letters was the one of June 11 1993, to which I have referred. That application was allowed.

Mann LJ agreed and did not add anything.

Agreeing, **Dillon LJ** said: I add a few comments only on three minor aspects.

The first is that we admit the new evidence that the appellant sought to have admitted on the appeal because, and only because, one of the letters referred to in that was the letter which he claimed showed that the trustees had refused improperly to proceed with the second arbitration. It is therefore necessary, as Hirst LJ has explained, to consider the submissions founded on the letter.

The second point is as to the point of the *locus standi* of the G trustees to sue for arbitrators' costs which the Beth Din ordered Mr Baram to pay. It would be more accurate to say that the point in its broadest form was initially taken by the court, rather than by Mr Denbin. Mr Denbin asserted that the trustees had no *locus standi*, and was not greatly concerned to consider whether anyone else had a *locus standi*. Indeed, on some of his submissions he was concerned to set up that there was a contract and that the Beth Din had been in breach of it. It had occurred to the members of the court that there might be a more formidable difficulty in the way of the trustees, in that the normal procedure is for an arbitrator to rely on not releasing his award until his fees have been paid, and there was no clear authority that an arbitrator had a right to sue in implied contract for payment of the H arbitrators' fees.

To my mind the reasoning of AL Smith J, in *Crampton & Holt* v *Ridley & Co* (1887) 20 QBD 48, on this point is convincing. It is the right approach that the trustees can sue on an implied contract for the £2,000.

The final point is this. It was claimed in the notice of appeal that the award against Mr Baram of £2,000 arbitrators' costs was made by the Beth Din in excess of jurisdiction.

In his skeleton argument Mr Denbin relied, for that proposition, on a decision in *Foster* v *Great Western Railway Co* (1882) 8 QBD 515. He said that it was beyond the jurisdiction of the Beth Din to award J costs against the successful party in the arbitration. But *Foster's* case is trenchantly criticised by Devlin J in *Smeaton Hanscomb & Co Ltd* v *Sassoon I Setty Son & Co*, to which my lord has already referred.

In my judgment, it is inapplicable under the new regime established under the 1979 Arbitration Act, even if it was still valid before that.

I would, as indicated, dismiss this appeal.

Hirst LJ added: I fully agree with Dillon LJ's judgment and, particularly, with his important reference to the old case in the last section of his judgment. K

Appeal dismissed with costs. Leave to appeal refused.

Chancery Division

December 8 1993

(Before Mr Justice WARNER)

GLADESMORE INVESTMENTS LTD v CARADON HEATING LTD L

Estates Gazette April 16 1994
[1994] 15 EG 159

Arbitration — Leave to appeal arbitrator's award — Whether greater weight in holding parties to award where arbitrator relied on opinion of counsel at parties' request

By a lease dated August 26 1987 the respondent tenant holds a term of 20 years from August 14 1987 of a warehouse and office premises from the applicant landlord; neither party being the original M parties to the lease. The building was constructed shortly before the grant of the lease and by agreement the original tenant improved the building at its expense. In determining the rent at review the arbitrator incorporated the legal opinion of counsel in his award to the effect that in correspondence prior to the grant of the lease the original parties agreed, *inter alia*, that the effect of certain items of expenditure by the original tenant was to be disregarded at the first review and the effect of other items was to be disregarded for the purposes of all rent reviews, including the first. That effect of the correspondence was not reflected in the formal agreement for the lease. The landlord applied under section 1(3)(*b*) of the Arbitration Act 1979 for leave to appeal

A the arbitrator's award on the grounds that: (1) the correspondence did not result in an agreement; and (2) if there was an agreement it was personal to the original parties and did not bind the landlord.

***Held*: The application was refused. Where parties have agreed that questions of law should be referred to counsel by the arbitrator, this weighs in favour of the view that, having chosen to have their dispute referred in this way as part of the arbitration, the parties should be held to that unless there are strong reasons for not doing so. In order to obtain leave the landlord has to show a *prima facie* case that the arbitrator was wrong near the top of the scale; a moderately strong case will not do. The opinion of**
B counsel advising the arbitrator that there was a concluded agreement in the correspondence was a tenable view. The parties anticipated that their agreement would last for the complete term of the lease and must have intended it to bind successors.

The following case is referred to in this report.

Ipswich Borough Council v *Fisons plc* [1990] Ch 709; [1990] 2 WLR 108; [1990] 1 All ER 730; [1990] 1 EGLR 17; [1990] 04 EG 127, CA

This was an application under section 1(3)(*b*) of the Arbitration Act 1979 by the applicant landlord, Gladesmore Investments Ltd, to appeal an arbitrator's award made in a rent review; the respondent
C tenant was Caradon Heating Ltd.

James Thom (instructed by Wayne Leighton Berger) appeared for the applicant; Nicholas Dowding (instructed by McKenna & Co) represented the respondent.

Giving judgment, **Warner J** said: This is an application under section 1(3)(*b*) of the Arbitration Act 1979 for leave to appeal against an arbitration award on questions of law. The arbitrator was a Fellow of the Royal Institution of Chartered Surveyors and the arbitration was under a rent review clause in a lease. The applicant, the landlord, is Gladesmore Investments Ltd, and the respondent, the tenant, is
D Caradon Heating Ltd.

The lease was dated August 26 1987 and was made between R E Hatfield (Contractors) Ltd (as landlord) and Stelrad Group Ltd (as tenant). I shall refer to them as "the original parties". The applicant is an assignee, and not the first assignee, of the freehold and the respondent is an assignee of the lease. The lease was for a term of 20 years from August 14 1987. The demised premises were a mixed warehouse and office building called Accord House in Hull. The rent under the lease was reviewable at the end of the first five years of the term, at the end of the first 10 years and at the end of the first 15 years. There was a break clause in the lease entitling the tenant to break at the expiration of the first 10 years.

E The building in question was being built immediately before the grant of the lease and it was agreed between the original parties that certain works to improve it relative to what was provided in the specification should be carried out at the cost of Stelrad Group Ltd. The issues in the arbitration related largely to the treatment of the effect of those works on the rent review.

The arbitration proceeded on written representations, including not only written representations by the parties' surveyors but also legal submissions by Mr Nicholas Dowding, for the tenant, and Mr James Thom, for the landlord. The arbitrator was authorised by the parties to take, and did take, the opinion of counsel, Mr John Male, on a number
F of points of law that were in issue. The arbitrator completely accepted Mr Male's opinion and in effect incorporated that opinion in his award. I think it is common ground that the effect of that was that Mr Male's reasons became the arbitrator's reasons so far as legal questions were concerned.

The works to which I referred were particularised in a letter dated February 27 1987 from Stelrad Group Ltd to its own solicitors, Payne & Payne, as being works which "should be excluded at the rent review stage". They were listed as 17 items, the total cost of which was estimated at £128,650.

The tenant's contention before the arbitrator was that, in correspondence between the original parties' solicitors which took place between February 1987 and August 1987, an agreement was G
reached that the effect of items 1, 2, 3, 4 and 14 in that list was to be disregarded for the purposes of the first rent review and the effect of the other items was to be disregarded for the purposes of all rent reviews, including the first. That contention, in accordance with the advice of Mr Male, was accepted by the arbitrator.

The landlord seeks leave to appeal against the arbitrator's decision on two grounds. The first is that the correspondence on which the tenant relies did not on its true construction result in any agreement and the second is that, if that correspondence did result in an agreement, it was an agreement that was personal to the original parties and not binding on the landlord, which is a successor in title to H
the original landlord and was not a party to the correspondence.

I am told that there was a formal agreement for a lease, which was made in February 1987 and which incorporated the form of lease that was eventually executed on August 26 1987. The effect of the correspondence was not reflected in the terms of any formal agreement or in the terms of the lease as executed.

I am also told that before the arbitrator the tenant argued that the effect of the works should be disregarded because of the express disregard provisions in the lease. But that argument was rejected by Mr Male and by the arbitrator and there is no cross-appeal on that point. J

There is, of course, a certain amount of authority on the question as to how the court should exercise its discretion to grant leave to appeal under section 1(3)(*b*) — authority which culminates in the decision of the Court of Appeal in *Ipswich Borough Council* v *Fisons plc* [1990] Ch 709*. I was referred by both counsel to the judgment of Lord Donaldson MR in that case. The principles on which the court should act in the light of those authorities and, in particular, of that judgment are common ground. There is a presumption in favour of the finality of arbitral awards. That is because the parties themselves have chosen arbitration rather than resorting to the courts as the means of resolving differences between them. That presumption can never be rebutted by K
showing merely that there is a doubt as to whether the arbitrator was right on a point of law. In what is called a "one-off case", the judge hearing the application for leave to appeal must be satisfied before he grants that leave that it is obvious that the arbitrator was wrong on a point of law. In a case that is not one-off — that is to say where, for instance, the case gives rise to a question of construction of a common form clause or where it gives rise to a general question of law or, in the case of an arbitration under a rent review clause, where the decision will affect subsequent rent reviews between the same parties — leave should be granted only if the judge is satisfied that there is a strong *prima facie* case that the arbitrator was wrong. Mr Thom agreed L
to a suggestion, that I made to him, that that meant that the court must be satisfied that the arbitrator was more probably wrong than right.

How strong a *prima facie* case has to be shown will depend upon all the circumstances. One such circumstance is the nature of the dispute and the nature of the expertise of the arbitrator. At p724 in the *Ipswich* case Lord Donaldson said:

> So, how strong is strong? No meter can be applied or indeed devised. It is a matter of relative values. If the chosen arbitrator is a lawyer and the problem is purely one of construction, the parties must be assumed to have had good reason for relying upon his expertise and the presumption in favour of finality or, to put it the other way round, the strength needed to rebut it will be greater. M
> So too if the dispute really centres upon an issue calling for non-legal expertise, albeit with some underlying question of law, and the chosen arbitrator has that expertise. But if the chosen arbitrator is not a lawyer and the whole dispute centres upon a difficult question of law, less strength may be required.

Here, the chosen arbitrator was not a lawyer, but the questions at issue between the parties are questions of law. On the other hand, as I have said, the parties agreed that those questions of law should be referred to a lawyer — Mr Male — and that, as Mr Dowding suggested, weighs in favour of the view that, having chosen to have

*Editor's note: Also reported at [1990] 1 EGLR 17.

their dispute referred to a lawyer as part of the arbitration, the parties should be held to that unless there are strong reasons for doing otherwise.

Mr Dowding submitted that the present case was to some extent a hybrid. So far as the items 1 to 4 and 14 in the list of February 27 1987 are concerned, it was a one-off case because they were to be disregarded only on the first rent review. So far as the other items in that list are concerned, it was not a one-off case. It is difficult on the wording of the award to assess the relative importance of items 1 to 4 and 14 and the remaining items. Both counsel attempted such an assessment, but I found those attempts somewhat inconclusive. One also has to bear in mind the break clause which means that there might never be another rent review.

Taking it all in all, it seems to me that this is a case where, in order to obtain leave to appeal, the landlord has to show a *prima facie* case that the arbitrator was wrong near the top end of the scale; by which I mean that a moderately strong case will not do.

I must now go through the correspondence on which the tenant relies and which the landlord says does not evince an agreement. It starts, so far as material, with a letter of March 3 1987 from Payne & Payne to the original landlord's solicitors, Gosschalks Weldon Chambers Thomas, whom I shall call for short Gosschalks Weldon. That letter enclosed the letter of February 27 1987 to which I have already referred and I understood it to be common ground that it embodied an offer that all the items in the list in the letter of February 27 should be disregarded at any rent review.

The letter of March 3 was followed by a letter of March 26 from Payne & Payne to Gosschalks Weldon, which was merely a chaser. The reply to that from Gosschalks Weldon to Payne & Payne was dated March 30 and stated that they had referred the matter to their client's agents, who in turn had written to the agents acting for the proposed freeholders, and they said:

> As soon as we receive the appropriate confirmation we will write to you.

That is a reference to the fact that at the time, as was known to all the parties, R E Hatfield (Contractors) Ltd were going to sell the freehold of the property to a company called Kennel Ride (Properties) Ltd as soon as the lease was entered into.

There were then further letters from Payne & Payne to Gosschalks Weldon, of May 29 and July 7, asking for a reply on the matter. On July 7 Gosschalks Weldon wrote to Payne & Payne saying that they were still waiting for their client's future landlord's solicitors to confirm the matters referred to in their letter of March 30. Again, on July 15 Gosschalks Weldon wrote to Payne & Payne saying that they were waiting for written confirmation from the solicitors acting for Kennel Ride (Properties) Ltd.

Then came an important letter from Gosschalks Weldon to Payne & Payne in which they said:

> We have now received confirmation from Kennel Ride (Properties) Ltd's solicitors that the matters set out in your letter of the 27th February 1987 are in order, in that items 1, 2, 3, 4 and 14 are to be excluded only from the first rent review. The further numbered items in the letter are to be treated as standard tenants' improvements.

It is common ground that that letter is to be regarded as a counter-offer to the offer contained in the letter from Payne & Payne of March 3.

On August 20 1987 there was a bizarre letter from Payne & Payne to Gosschalks Weldon, which said:

> We would refer to your letter of 23 July in connection with the above. The only item with which our client and its surveyor disagrees is number 15, in that it is their opinion that this should be excluded from the first rent review. Our clients have improved the quality of the works and as such feel entitled on the first review to exclude the effect from the rent. We should be obliged if you would take your client's further instructions thereon. Subject to the above, all outstanding items seem to be agreed and we expect to be in receipt of the rent tomorrow to enable us to effect completion.

That was followed by a further letter from Payne & Payne on August 24 1987 to Gosschalks Weldon, saying:

> With reference to our client's letter dated 27 February detailing items 1 to 17 inclusive, we confirm the agreement that these shall be excluded from all rent reviews with the exception of items 1 to 4 inclusive, and 14, which are to be excluded only from the first rent review.

The last letter in the correspondence was a letter dated August 25 1987 from Gosschalks Weldon to Payne & Payne saying:

> Thank you for your letter of the 24th of August, 1987. The position is clearly set out in our letter to you of the 23rd of July, 1987.

On that correspondence, Mr Thom takes two points. One, which he referred to as "the counter-offer point", was that the letter of August 20 was itself a counter-offer which destroyed the offer in Gosschalks Weldon's letter of July 23, so that there was no offer in existence available for acceptance by Payne & Payne's letter of August 24. Mr Thorn's second point, which he called "the formulation point", was that the parties were never *ad idem* as to whether the items other than items 1 to 4 and 14 were, in the words of Gosschalks Weldon's letter of July 23, to be "treated as standard tenant's improvements" or were, in the words of Payne & Payne's letter of August 24, to be "excluded from all rent reviews".

I deal first with the formulation point. Mr Thom submits that it is not clear what the phrase "to be treated as standard tenants' improvements" means. It is not clear, he says, that they mean the same as "to be disregarded on all rent reviews", which was what Mr Male held that they meant and was the result, of course, that was contended for on behalf of the tenant.

Mr Thom points out that in the letter of July 23 the words used by Gosschalks Weldon are to the effect that the matters set out in the letter of February 27 are "in order" — that may be taken to mean are agreed — "in that items 1, 2, 3, 4 and 14 are to be excluded only from the first rent review", which, of course, was not what was proposed in the letter of February 27; and then in a new sentence, quite separately, they say:

> The further numbered items in the letter are to be treated as standard tenants' improvements

instead of saying that those further numbered items were, as envisaged in the letter of February 27, to be excluded from all rent reviews, or disregarded in all rent reviews. Nor, when Gosschalks Weldon wrote on August 25 in reply to Payne & Payne's letter of August 24, did they simply say:

> The position as set out in your letter of the 24th of August is agreed

— they went back to their letter of July 23.

Mr Male advised that the effect of the correspondence was, as I have said, in accordance with the tenant's contentions. He said that the effective reply to the letter of March 3 was that dated July 23 and he said:

> In the language of rent review, I read this letter as saying that the specified items were to be disregarded upon the first rent review and that all the other items were to be disregarded upon all rent reviews. By the expression "treated as standard tenants' improvements", I take this to mean works falling within the standard disregard of tenants' improvements. Thus, in terms of offer and acceptance, the letter dated the 23rd of July 1987 was, in my opinion, a counter-offer to that contained in the letter dated the 3rd of March 1986. The reply to the counter-offer was on the 20th August 1987 when Payne & Payne indicated that the only item on which there was disagreement was item 15. However, this disagreement was not pressed by Payne & Payne because on the 24th August 1987 that firm wrote in relation to items 1 to 17 in the letter dated the 27th of February, 1987:
>
> "We confirm that these shall be excluded from all rent reviews with the exception of items 1 to 4 inclusive, and 14, which are to be excluded only from the first rent review."
>
> In my opinion, this letter corresponded with the letter dated 23rd of July 1987 in that both sides were saying that items 1 to 4 inclusive and 14 were to be treated as falling within the disregard of tenant's improvements for the first rent review and all the other items to be so treated for the first and all subsequent rent reviews. The parties were therefore *ad idem*. There was, in my

A opinion, an offer in the letter dated the 23rd of July 1987 and an acceptance in the letter dated the 24th of August 1987.

It seems to me that that is an eminently tenable view. I listened to the arguments of Mr Thom and of Mr Dowding on the point and in the result it seemed to me that they were fairly evenly balanced. As Mr Dowding pointed out, one has to see what a reasonable recipient of the letter of July 23 in the circumstances would have taken it to mean, rather than make a close semantic analysis of the correspondence. It is perfectly true that, as both counsel said, the letter of August 20 is bizarre. But the obscurity of it does not, in my judgment, tip the scale in favour of the view that there is a strong *prima facie* case that Mr
B Male's opinion was wrong. It seems clear to me, as I think it must have been to Mr Male, that the parties thought that they had reached agreement because the lease was signed on August 26.

Mr Thom introduced a refinement to this argument which was to the effect that, where parties differ in their formulation of a particular term in a contract, there may be no effective agreement between them even though on their true construction the two formulations mean the same thing. That is an argument which, of course, would be very strong in a case where, for instance, solicitors were negotiating on the basis of a draft and they had not reached agreement on the precise wording to be inserted in the draft. But I do not think that the
C argument has the same force in a case like this of a contract by correspondence, where the use of different phrases to mean the same thing (if they do) is not necessarily indicative that the parties are not *ad idem*.

Mr Thorn's counter-offer point was a novel one in the sense that it is common ground that no reliance was placed on it in the arguments put forward to the arbitrator and to Mr Male. I thought that, on this point also, the arguments of counsel were fairly evenly balanced. It depends upon whether the letter of August 20 can be regarded as impliedly rejecting the offer contained in the letter of July 23. That, in the end, must be a question of construction. Moreover, as Mr Dowding
D pointed out, although on an appeal the landlord would not be precluded from relying on this argument by reason only that it had not relied on it during the arbitration, it does not sit well with the presumption of finality in the case of arbitral awards to allow an appeal to be brought on the basis of a new point. That, I think, supports the view that leave should not be given to appeal on the counter-offer point.

The second main ground on which the landlord seeks leave to appeal is that the agreement was personal to the parties to the correspondence and not binding on the applicant, who was not a party to that correspondence. On that issue both counsel relied before me on
E the arguments that they had put forward in writing to the arbitrator and, indeed, they read those arguments to me. Mr Male in substance F
accepted Mr Dowding's argument and I am bound to say that it seems to me he was right to do so, the main points being that, with a term as long as 20 years, the parties must have contemplated that the reversion might change hands during the term and, indeed, the correspondence itself shows that there was to be an assignment of it as soon as the lease was entered into, an assignment to Kennel Ride (Properties) Ltd.

Mr Thom in his written submissions suggested that in some way, which I found very difficult to follow, Kennel Ride was in a different position from subsequent assignees of the reversion. Before me, Mr Thom went so far as to suggest that R E Hatfield (Contractors) Ltd were acting as Kennel Ride's agents. I certainly do not find that from G
the correspondence. Mr Thom also suggested that the purpose of the agreement was to remove the unfairness of the tenant paying the landlord to make improvements and then paying the landlord again for them in rental terms. Although the effect on the tenant is the same, the degree of unfairness in paying an enhanced rent to a new landlord is markedly less, he said. Mr Male said of that that it seemed to him that the answer to the point was that if the agreement was intended to prevent a perceived unfairness to the tenant in paying the landlord more than once for improvements, then that unfairness to the tenant could not change when the landlord changed. Mr Thom also argued that it could be a complicated matter to value disregards and if there H
was to be a sale on by Kennel Ride, the purchaser, if aware that he would face a disregard, would have to discount his bid to reflect not only the reduction in rent but also the increased likelihood of disputes on rent review with attendant delay and expense. If a purchaser would be bound, the value of the reversion would be diminished, he said. To that Mr Male answered that the lease already contained a disregard of improvements, the effect of which a purchaser would have to consider in formulating his bid. I cannot see any possibility of saying that there is a strong *prima facie* case that Mr Male was wrong on this issue.

There was in the notice of motion reference to another ground of appeal which was that, even if the agreement was intended to bind J
successors to the original landlord, the present landlord had no notice of it. But Mr Thom told me, quite rightly, that he did not pursue that ground because on reflection he did not feel that there had been an error of law by the arbitrator about it plain enough to warrant leave to appeal.

Having reached those conclusions, I need not express any view on the question whether the effect of the arbitrator's or Mr Male's alleged errors was substantial. Certain calculations were put forward by Mr Thom to show that it was and, had I thought that there was a strong *prima facie* case, I would have said that the effect of the errors was substantial enough. However, for the reasons I have attempted to K
formulate, I think that this application should be dismissed.

COMPULSORY PURCHASE AND COMPENSATION

A
Court of Session

First Division

October 15 1993

(Before Lord President HOPE, Lord MAYFIELD and Lord McCLUSKEY)

EMSLIE & SIMPSON LTD v ABERDEEN CITY DISTRICT COUNCIL

B Estates Gazette May 7 1994

[1994] 18 EG 136

Compulsory acquisition — Whether long lease containing landlord's option to break for redevelopment which was implemented by notice before general vesting declaration took effect compensatable as a lease with less than a year to run — Whether loss of profits in year before acquisition caused by blight compensatable

The appellant claimants trade in shop premises under the name of Target Discount in terms of a lease from the heritable proprietors, W
C Ltd; the period of the lease was from May 28 1975 to May 28 2000. Clause 21 of the lease permitted the landlord to terminate it after May 28 1980 on giving 12 months' notice in the event of the landlord intending to demolish, develop, reconstruct, modernise or refurbish. Following a successful application to quash the decision of the Secretary of State to confirm a compulsory purchase order made in 1981, W Ltd then entered into a "linking agreement" by which it undertook to make its lands, including the premises, available to the council. The compulsory purchase order was then confirmed and on March 17 1985 the council served on the claimants and W Ltd notice of making a general vesting declaration which vested the properties in
D the council on April 15 1987. On March 26 1987 W Ltd served a notice to quit on the claimants under clause 21 requiring them to vacate by March 28 1988. The claimants in fact vacated before April 15 1987. The claimants appealed from the Lands Tribunal for Scotland, which had held that: (1) the notice to quit was valid and their tenancy was a long tenancy about to expire to which para 8 of Schedule 24 to the Town and Country (Scotland) Act 1972 applied by virtue of para 38 by reason of the existence of the landlord's option to break in clause 21 whether or not that option was exercised; and (2) that the claimants were not entitled to loss of profits suffered by Target Discount in the year prior to the vesting date. The claimants submitted that clause 21 was incapable of having any effect because of the
E linking agreement as W Ltd were not able to carry on the development and that their tenancy was a long tenancy which vested in the council by para 7 of Schedule 24 and was appropriately compensatable.

Held: **(1) The option under clause 21 was still available at the date of the general vesting and declaration because W Ltd was still the landlord at that date and the effect of serving the notice to quit was to alter the lease from a long tenancy to a long tenancy about to expire compensatable as a lease with less than a year to run under section 114 of the Lands Clauses (Consolidation) Act 1854.**

(2) *Per Lord President Hope and Lord Mayfield*: A causative link between the taking of land from the claimants and their trading losses in the year before acquisition was not established and accordingly loss of profits attributable to the indiscriminate effects of blight was not recoverable. F

Lord McCluskey dissenting: **The tribunal found as a fact that the loss was caused by the overall blighting effect of the scheme of acquisition. The fact that the claimants would have had no title to claim had their land not been taken and that others suffered loss through blight who had no claim to compensation because no land was taken from them was unsound reasoning. The real cause of the loss was the scheme involving the compulsory purchase order** G **and the threat of compulsory acquisition of the affected land. There was a direct link between the loss and the dispossession of the claimants. The test for compensation was one of equity.**

The following cases are referred to in this report.

Aberdeen City District Council v *Sim* 1983 SLT 250; (1982) 264 EG 621
Evans v *Glasgow District Council* 1978 SLT (Lands Tr) 5
Harvey v *Crawley Development Corporation* [1957] 1 QB 485; [1957] 2 WLR 332; [1957] 1 All ER 504; (1957) 55 LGR 104; 8 P&CR 141, CA
Lanarkshire and Dunbartonshire Railway Co v *Mair* (1895) 22 R 912; 21 R 1018; 31 SC LR 239; 2 SLT 258
Prasad v *Wolverhampton Borough Council* [1983] Ch 333; [1983] 2 WLR 946; [1983] 2 All ER 140; (1983) 82 LGR 265; 47 P&CR 252; [1983] EGD 627; 265 EG 1073, CA H
Venables v *Department of Agriculture for Scotland* 1932 SC 573
Wordie Property Co Ltd v *Secretary of State for Scotland* 1984 SLT 345

This was an appeal by Emslie & Simpson Ltd from the decision of the Lands Tribunal for Scotland, which had awarded the appellants £75,400.86 in a claim for compensation arising out of the compulsory acquisition of a lease by the acquiring authority, Aberdeen City District Council.

Roy Martin QC and Ian Miller (instructed by Ledingham Chalmers, of Aberdeen) appeared for the claimants; Brian Gill QC and J J Gordon Reid QC (instructed by Bennett & Robertson, of Edinburgh) represented the acquiring authority.

Giving his opinion, **LORD PRESIDENT HOPE** said: We have before us in this case an appeal by the claimants, Emslie & Simpson Ltd, against a decision of the Lands Tribunal for Scotland dated August 12 1992 in a reference for the determination of disputed compensation by which the tribunal determined the amount of the compensation to be paid in respect of the compulsory purchase of shop premises at 18 and 20 Loch Street, Aberdeen. There is also an appeal by the acquiring authority, Aberdeen City District Council, against a decision by the K tribunal dated November 4 1992 finding the acquiring authority liable to the claimants in their expenses of the reference. The claimants' appeal is under section 13(1) and (6) of the Tribunals and Inquiries Act 1971, while the appeal by the acquiring authority is under section 11 of the Tribunals and Inquiries Act 1992, a consolidating enactment which came into force on October 1 1992. It was agreed in the course of the hearing that it would be more convenient for the appeal in regard to expenses to be dealt with after a decision has been taken on the claimants' appeal and it is with the claimants' appeal only that this opinion is concerned.

The tribunal determined the compensation to be paid by the acquiring authority in the sum of £75,400.86. This sum was awarded to the claimants, who had been in occupation as tenants of the shop premises, under section 114 of the Lands Clauses Consolidation (Scotland) Act 1845. That section provides that a tenant who has no greater interest in the lands than as a tenant for a year or from year to year is entitled to compensation for the value of his unexpired term, for any allowance which ought to be made to him by an incoming tenant and for any loss or injury which he may sustain. A number of items of claim were agreed, and the tribunal made no award in respect of other items which were disputed by the acquiring authority. Among the disputed items for which an award was made were a claim for extinguishment of the claimants' business and a claim for expenses incurred from the date of vesting. A sum was also included in the award for the value of the leasehold interest. The only items with which we are now concerned are those which are dependent on the tribunal's decision that, following the service of a notice to quit by their landlords, the claimants' lease had less than a year to run at the date of vesting, and a claim for loss of profits in the year prior to vesting as compensation for disturbance which the tribunal disallowed because they held that it had not been shown that this was a consequence of the taking of the claimants' interest in the land.

General background

The claimants had traded in the premises at 18 and 20 Loch Street under the name of Target Discount in terms of a lease from the heritable proprietors, Wordie Property Co Ltd, dated February 23 and March 9 1979. The date of entry was May 28 1975 and the period of the lease was from that date until May 28 2000. The subjects were situated in a part of the central area of Aberdeen which had been under consideration for redevelopment for many years. From the mid-1960s Wordie had held discussions with the planning authority about proposals for redevelopment. This was on the basis that either they would develop their own property by themselves or there would be redevelopment of their land together with other neighbouring land in co-operation with the local authority. In 1979 the Secretary of State for Scotland approved the definition of a comprehensive development area, which included within area B of that area the land on which the premises at 18 and 20 Loch Street were situated. The local authority, which by this stage was Aberdeen City District Council, then entered into a formal agreement with a property developer, Bredero Vast Goed NV, for the development of area B. They then proceeded to make compulsory purchase orders for that area and the other part of the comprehensive development area known as area A and Bredero made an application for outline planning permission to develop area B. Wordie, who at this stage wished to proceed independently with their own development, objected to the compulsory purchase order and to Bredero's application for planning permission relating to area B. They made a planning application in outline for the development by themselves of that part of area B which was substantially in their ownership.

These events took place against the background of a clause in the lease between Wordie and the claimants, which was clause 21, in these terms:

> The Landlords shall be entitled to terminate this Lease, and resume the premises from the Tenants, with effect from or after the Twenty eighth day of May Nineteen hundred and eighty, provided they give to the Tenants written notice to that effect at least twelve months prior to the date of intended termination, in the event of the Landlords wishing to carry out (i) demolition and development or (ii) a re-construction, modernisation or refurbishment, of the premises, or of, *inter alia*, the premises, in the case of (ii) of a kind which would necessitate the Tenants completely vacating the premises. In the event of the Landlords terminating this Lease as aforesaid, the Tenants, provided the development, re-construction *et cetera* is completed before the Twenty eighth day of May Two thousand, and the Landlords so intimate to the Tenants, shall have an exclusive option (to remain open for exercise by the Tenants for not more than three months from the date of said intimation) of obtaining from the Landlords a Lease, to expire at the Twenty eighth day of May Two thousand, of accommodation comparable to the subjects of this Lease and forming part of the property presently owned by the Landlords at Schoolhill, Harriet Street, George Street and Loch Street, Aberdeen, on full repairing and insuring terms and on similar lines to the terms of this Lease but at a then current fair market rent fixed, *mutatis mutandis*, in accordance with Clause Twentieth hereof with rent reviews every fifth year. In the event of the Landlords terminating this Lease as aforesaid no compensation shall be payable to the Tenants or any other parties who may have rights of occupation of the premises or any part thereof.

During the 1960s Wordie had gradually acquired all the properties in a block of approximately 2 acres formed by Schoolhill, Harriet Street, Loch Street and George Street, Aberdeen, with a view to redeveloping them. It was thought likely that the redevelopment would take place in the 1980s and the leases of all the properties in the block contained a provision in the same terms as clause 21 in the claimants' lease, all the leases being terminable on or after the same date in 1980.

In 1981, after a public inquiry, the Secretary of State confirmed the compulsory purchase order, refused Wordie's planning application and approved the application for planning permission by Bredero. Wordie then made an application to the Court of Session under section 233 of the Town and Country Planning (Scotland) Act 1972 and para 15 of Schedule 1 to the Acquisition of Land (Authorisation Procedure)(Scotland) Act 1947 to quash these decisions on the ground that they were *ultra vires*. Their application was successful and the decisions were all quashed: see *Wordie Property Co Ltd* v *Secretary of State for Scotland* 1984 SLT 345. Wordie then entered into an arrangement with the council and Bredero by which Wordie undertook to make their land, excluding the properties on Schoolhill, available to the council for the redevelopment. They were to be entitled to a share of the ground rent payable by Bredero, which was to be determined by a formula which related to rents obtained for the leases of the developed property. They were also to retain a heritable interest consisting of a percentage of the whole property as redeveloped. This arrangement was formalised in an agreement, known as the linking agreement, between the three parties dated May 1 1984 and January 8 and August 7 1985.

Following the making of this agreement Wordie withdrew their application for planning permission together with their objections to the council's application for confirmation of the compulsory purchase order and to Bredero's application for outline planning permission. The public inquiry was then reopened and in 1985 the Secretary of State granted an amended application for planning permission to Bredero and confirmed the compulsory purchase order for the area proposed for the redevelopment. This compulsory purchase order covered all the subjects in the block owned by Wordie, including the premises at 18 and 20 Loch Street. The design of the development was such, however, that the properties belonging to Wordie fronting on to Schoolhill did not need to be acquired. These properties were later refurbished by Wordie themselves in terms of the linking agreement.

On March 15 1985 Bredero served notice on the council and on Wordie in terms of clause 4 of the linking agreement listing all of Wordie's properties which were to be acquired by the council for the redevelopment. By letter dated March 17 1985 the council, as the acquiring authority, served on Wordie and on the claimants notice of the making of a general vesting declaration within the meaning of section 278 of, and Schedule 24 to, the Town and Country Planning (Scotland) Act 1972 in respect of all the properties situated within area B, including the premises at 18 and 20 Loch Street. The effect of the general vesting declaration, in terms of para 7 of Schedule 24 to the 1972 Act, was to vest the properties in the council on April 15 1987, on which date the council were to be entitled to enter upon and take possession of the land. In terms of para 8 of Schedule 24, however, it was necessary, if the claimants' tenancy was a long tenancy about to expire as defined by para 38 of the Schedule, for the council to serve a notice of intended entry on the claimants if they wished to take possession of the premises leased to them on April 15 1987. But the claimants vacated the premises without waiting for such a notice to be

A served on them and the council duly entered into possession of the premises on that date.

In the meantime Wordie served on the claimants a notice to quit dated March 26 1987. The notice was in these terms:

> We hereby give you notice that we require to resume possession of the above premises leased by you because it is wished to carry out demolition and development works which will affect the premises. You are therefore to remove from the premises at 28th March 1988, on which date your lease is terminated.

In a letter to the claimants on the same date which accompanied the
B notice Wordie stated:

> You will have received from the City cf Aberdeen District Council Notice of the making of a General Vesting Declaration affecting the lease to subjects and requiring vacant possession thereof on 15th April 1987. Notwithstanding the removal date stated in this company's enclosed notice this does not prejudice the Council's requirement for vacant possession on 15th April 1987. Our notice may however affect any statutory rights you may have to compensation from the Council from having to relinquish possession on 15th April 1987, rather than the removal date stated in the enclosed notice.

It appears from the terms of that letter that Wordie did not expect that their notice to quit would be enforced, because the council by
C reason of the general vesting declaration were in a position to recover possession of the subjects much earlier by serving on the claimants a notice of intended entry. But the effect of their serving the notice to quit was to entitle them to obtain the benefit of a provision in clause 6(c) of the linking agreement which was in these terms:

> In the event that the Council is required to pay compensation in respect of any interest in the Wordie Subjects other than that of Wordie, Wordie shall pay to the relief of the Council and of Bredero such compensation together with any interest thereon; PROVIDED ALWAYS THAT (One) where in any notice served by Bredero on the Council and on Wordie pursuant to the provisions of Clause Fourth hereof, the date therein specified is earlier than thirteen months after
> D the service of said notice, and where Wordie shall within fourteen days after its receipt of said notice, give to the owner of any such interest the requisite notice to terminate that interest at the earlier lawful date following said date the foregoing obligation upon Wordie to pay to the relief of the Council and Bredero compensation (together with any interest thereon) in respect of that interest shall not apply . . .

The precondition in respect of the date of the notice served by Bredero was satisfied for the purposes of this clause, and Wordie were obliged to serve notice to quit on the claimants if they were to avoid liability to relieve the council and Bredero of any compensation which those parties might have to pay to the claimants in respect of the
E compulsory acquisition. Any liability on Wordie themselves to pay compensation to the claimants was also avoided by giving such notice to quit, provided that the notice could be regarded as having been validly given and the lease could be regarded as having been duly terminated in terms of clause 21 of the lease.

Notice to quit

It was submitted to the tribunal on the claimants' behalf that the notice to quit was invalid and of no effect, on the ground that the lease could be terminated under clause 21 only in the event of Wordie themselves wishing to carry out work of the kind specified in that clause. In the events which had happened that could not have been
F their wish, as the council had already served a general vesting declaration on them and they would cease to have any interest in the premises when that declaration took effect. It was said that the real purpose of the notice was to limit the claimants' right to compensation and to relieve Wordie of any liability for compensation to be paid by the acquiring authority and not to bring the lease to an end. In the opinion of the tribunal, however, the notice to quit was a valid notice given under clause 21. In their view it could reasonably be said that it was given "in the event of the landlords wishing to carry out" demolition and development in terms of that clause, even although this was now to be done not by Wordie alone, but by them in partnership with the council and with Bredero. They considered that the giving of the notice to quit, after the making of the general G
vesting declaration but before the vesting date, had the effect of converting what would otherwise have been a long tenancy on that date into a tenancy for less than a year. The result was that, subject to the giving of a notice of intended entry in terms of para 8 of Schedule 4 to the 1972 Act, the claimants' occupation after that date would continue to be as tenants by virtue of the tenancy. Accordingly, the claimants' only claim for compensation for the loss of their interest in the land under section 114 of the Lands Clauses Consolidation (Scotland) Act 1845 was on the basis that the lease had less than one year to run.

The claimants contend in their grounds of appeal that, in holding H
that the notice to quit was a valid notice, the tribunal erred in law. They also contend, in a separate ground of appeal, that the tribunal erred in law in holding that the provisions of para 8 of Schedule 24 to the 1972 Act applied to their tenancy. This ground of appeal is directed to a passage in the tribunal's opinion in which they state that, if at the date of the general vesting declaration the claimants' tenancy was not in fact a short tenancy or a long tenancy which was about to expire, it was nevertheless a tenancy to which para 8 applied because of the provisions of para 38 of the same Schedule in which definitions of the expressions "short tenancy" and "long tenancy about to expire" are set out. Plainly this was not a short tenancy as defined in that J
paragraph, because it was not a tenancy for a year or from year to year or any lesser interest. But para 38(2) provides that, in determining for the purposes of that paragraph whether a tenancy granted for an interest greater than a short tenancy is, at the date of a general vesting declaration, a long tenancy which is about to expire it shall be assumed that the landlord will exercise any option to terminate the tenancy then or thereafter available to him. In the tribunal's view Wordie still had a valid option to terminate the tenancy under clause 21, so it was to be assumed that that option would be exercised, whether or not that was in fact the case.

Counsel recognised that the points taken in these two grounds of K
appeal must stand or fall together. This was because the claimants' argument is that the notice to quit was invalid because the option given to Wordie by clause 21 of the lease was no longer available to them to exercise. As they were unable to show by that date that they themselves wished to carry out demolition and development of the premises, the conditions necessary to a valid exercise of the option could no longer be satisfied. So the notice to quit which was in fact served was invalid and in any event the option to which para 38(2) applied was not available. The practical effect of these two points is the same also. Unless the tenants' tenancy was a long tenancy about to expire within the meaning of para 38, their interest in the lease would L
require to be treated as having vested in the council at the end of the period specified in the general vesting declaration in terms of para 7. This would have the appropriate consequences as regards the assessment of the compensation to be paid to them for the value of the leasehold interest. The critical question, therefore, is whether this was a long tenancy which was about to expire. The effect of the definition in para 38 is that it is to be assumed that Wordie would exercise the option under clause 21 if it were still available to them, even if that option was not in fact exercised.

The claimants' argument was that by entering into the linking agreement Wordie put it beyond themselves to take advantage of M
clause 21. They submitted that by the date of the general vesting declaration clause 21 was incapable of having any effect as a means of terminating the lease, and that it had to be assumed for the purposes of para 38(2) of Schedule 24 that the option to terminate in terms of that clause was no longer available. The purpose of clause 21, it was said, was to regulate the private rights of the parties to the lease. It had no place where the development was to be taken in the exercise of compulsory powers. It was essential to a valid exercise of the option that the landlords wished to carry out the work described in that clause. The wish which the clause described had to satisfy two requirements. It had to be a wish to carry out demolition and development of the premises let to the tenants or, *inter alia*, those

premises, this being a development over which the landlords themselves would retain control. And it had to be a wish to terminate the lease in order to undertake that development. But by clause 4 of the linking agreement it was agreed that the premises together with other premises were to be developed by the use of the statutory procedure of compulsory acquisition. This meant that the contractual method of terminating the lease was no longer to be used. Under reference to other clauses in the linking agreement it was submitted that the works which that agreement contemplated involved the relinquishing by Wordie of control over the form of the development. This was to be a development of an entirely different kind from that contemplated by clause 21 of the lease. The actings of the parties were, it was submitted, consistent with this approach. The letter of March 26 1987 which accompanied the notice to quit seemed to accept that that notice was without prejudice to the council's right to obtain vacant possession on the date specified in the general vesting declaration. That would be so only if the notice to quit was invalid, because its effect if valid was that para 8 and not para 7 of Schedule 24 to the 1972 Act was to apply. Mr Martin accepted that if the notice was a valid notice it would still have been open to the council to serve a notice of intended entry in terms of para *(a)* of the Schedule if they wished to obtain entry before the date as from which the lease was to terminate. But no such notice was served and the claimants left the premises without waiting for this to be done. All of this was consistent with this having been a long tenancy, the claimants' interest in which was extinguished on April 15 1987. The use of the words "it is wished" in the notice to quit was also significant because there was here a recognition by Wordie that they could not state that they themselves wished to carry out development.

In my opinion, no significance can be attached to the subsequent actings of the parties or to the words used in the notice to quit. The question whether the option under clause 21 was still available to be exercised does not depend on the view taken by the parties of its effect, nor does it depend on the way in which Wordie expressed themselves when they elected to exercise it. It appears from the letter of March 26 1987 that there may have been some misunderstanding, at least on Wordie's part, about the effect of the notice to quit, if it was valid, in regard to the council's right to obtain vacant possession of the subjects on April 15 1987. It can also be said that Wordie's use of the words "it is wished" in regard to the carrying out of the development rather than "we wish" implies a recognition on their part that, as others were now involved in the project, those words were more appropriate. But these points relate only to the way in which the option was exercised, not to question what is central to this part of the case, which is whether the option was still capable of being validly exercised.

On this point I consider that the Lands Tribunal were entitled, on the facts set out in their opinion, to hold that the option under clause 21 was still available. Wordie were still the landlords of the premises at the date of the general vesting declaration and it was still open to them as landlords to entertain a wish in regard to development of the premises. It is clear from the tribunal's findings that for a long time they had had it in mind that there would be a redevelopment of this area. This was why clause 21 had been inserted in this lease, as it had been in leases relating to the other properties in the block. The original plan had been for Wordie to carry out their own development but events had overtaken them. The purpose of entering into the linking agreement was to resolve their long-running dispute with the council and with Bredero. Its effect was that all three parties were to combine together to effect the redevelopment. Mr Martin accepted that the tribunal had correctly understood the effect of the linking agreement when they stated that when Wordie eventually agreed to support the scheme it was on the basis that they would themselves have a share in the development, and that their involvement included the retention by them of a heritable interest in the land subject to development and a right to a share in returns from it when it was carried out.

As Mr Brian Gill QC for the council put it, Wordie had in effect become codevelopers of the subjects with the council and with Bredero, with a significant control over the development and a significant share in the title to the land and in the receipts. In my opinion, it was still open to Wordie to say, as partners in the co-operative venture, that they wished to demolish and develop the claimants' premises together with other premises. The manner of achieving that development had no doubt altered from that which was envisaged when the lease was entered into. But the essential purpose of the wish to recover possession of the subjects was the same.

As for the argument that, since the premises were to be the subject of a compulsory acquisition in terms of the linking agreement, the service of a notice to quit was no longer a necessary part of the project of the development, the answer lies in the fact that there was nevertheless a substantial purpose to be served by giving a notice to quit at this stage. One of the purposes of clause 21 had all along been, as the last sentence of it makes clear, to relieve Wordie of the obligation to pay compensation to the tenants for terminating the lease in terms of that clause. The effect of serving a notice to quit prior to April 15 1987 was to alter the lease from a long tenancy, the tenants' interest in which would be extinguished on the vesting date, to a long tenancy about to expire. The landlords' interest in the lease, subject to the notice to quit, would then transmit to the council, with a consequent reduction in the council's liability to the claimants by way of compensation for the termination of their interest. The council would have the option of terminating the tenancy by serving a notice of intended entry under clause 8 of Schedule 24 or of enforcing the notice to quit and recovering possession of the premises in due course. In my opinion, therefore, there were sound practical reasons for wishing to serve a notice to quit at this stage. It cannot in any event be said that the option to serve such a notice in terms of clause 21 was no longer available, because a notice in terms of that clause was still capable of affecting the rights of the parties to the lease.

For these reasons I consider that the Lands Tribunal reached a sound decision on this point, and that they were right to award compensation to the claimants under section 114 of the 1845 Act on the basis that the lease had less than one year to run at the date of vesting.

Loss of profits for disturbance

The claimants submitted, as a claim for disturbance, a claim for loss of profits sustained by them under the trading name "Target Discount" in the year prior to the vesting date. They averred that from, at latest 1982, the area which included their premises began to be blighted by public knowledge of the development proposals and that customers were discouraged by the closure or relocation of other traders, the commencement of works on area A and the diversion of bus routes. But the tribunal rejected the claim on the ground that that loss, which they accepted was caused by the blight on trading in the surrounding area owing to the scheme, could not be said to have been a consequence of the acquisition of the claimants' interest in land or of their dispossession from it. In their opinion the compensation for disturbance to which the claimants were entitled did not include compensation for business losses which had not been shown to have been a consequence of the taking of their interest in the land.

The claimants contend in their grounds of appeal that the tribunal erred in law in holding that they were not entitled to their loss of profits for compensation for disturbance prior to the vesting date.

There are, I think, inherent in the tribunal's decision on this point two findings, one of which is a finding in fact and the other a finding in law. So far as the facts were concerned, the tribunal were satisfied that the claimants had suffered a loss of profit during the year prior to the vesting date which was due to the overall blighting effect of the scheme. This finding was in accordance with the claimants' averments on this point, and I think that it is significant that it was not suggested in the averments, nor has it been said to have been demonstrated by the evidence, that the claimants' loss of profits was caused by their having to leave the premises. The cause of this loss was not said to be the taking of the claimants' premises in the exercise of compulsory powers but was a blight on trading generally which affected the whole area, owing to the scheme. Among those affected by it were Wordie's

A tenants in Schoolhill, whose interests in land were not acquired although they were within the designated area of the compulsory purchase order. Also affected by it were occupiers of retail premises on the other side of Loch Street, which were outside the area of the compulsory purchase order and were not, therefore, under threat of being compulsorily acquired. The tribunal's findings on these matters are not in dispute.

The question of law which arises on these facts is whether a loss of this character can be recognised as part of a claim for disturbance. The tribunal held, after a review of the authorities, that a claim for compensation for a loss occasioned by disturbance may be valid, B although the loss was incurred prior to dispossession, provided it was caused by the dispossession.

There was no dispute in the argument before us as to whether a loss incurred prior to dispossession may be recovered in a claim for disturbance. The point at issue is whether the loss must be shown to have been caused by the dispossession, that it is to say by the taking of the premises from the claimant in the exercise of compulsory powers, or whether it is sufficient for the loss to be recoverable that it was caused by the overall effects of the scheme of acquisition.

The precise limits of a claim for disturbance have never been defined in the statutes. Section 12 of the Land Compensation C (Scotland) Act 1963, which replaced section 2 of the Acquisition of Land (Assessment of Compensation) Act 1919, set out the rules for the assessment of compensation in respect of a compulsory acquisition. Rule (2) provides for the assessment of the value of the land, and rule (6) states that the provision of rule (2) shall not affect the assessment of compensation for disturbance or any other matter not directly based on the value of land. The foundation for an award of compensation of disturbance lies in the earlier statutes. Section 114 of the Lands Clauses Consolidation (Scotland) Act 1845, in terms of which the tribunal made their award of compensation in the present case, states simply that the tenant "shall be entitled to compensation D . . . for any loss or injury which he may sustain". That section should be read together with section 17 of the same Act, which provides that the promoters of the undertaking must state in the notice to treat that they are willing to treat for the purpose of the lands "and as to the compensation to be made to all parties for the damage that may have been sustained by them by reason of the execution of the works". A reference to compensation for any damage sustained by reason of the execution of the works appears also in section 19 of that Act.

In *Venables* v *Department of Agriculture for Scotland* 1932 SC 573 it was held that rules (2) and (6) of the 1919 Act did not have the effect of displacing in any way the principle recognised by the 1845 Act that the person dispossessed should get compensation for all his E loss in addition to the value of the lands to be taken. Lord Justice Clerk Alness said at p581:

> The sound principle would seem to be that the person dispossessed shall get compensation for all loss occasioned to him by reason of his dispossession. The Act of 1845 recognises that; the text-books recognise it; judicial authority recognises it; and the Act of 1919 continues to the evicted owner all claims formerly open to him, including that claim.

In *Aberdeen City District Council* v *Sim* 1983 SLT 250*, at pp252-3 emphasis was placed on the generality with which this proposition was expressed, and it was noted that the decision in F *Venables* had been cited with approval and followed in England in *Harvey* v *Crawley Development Corporation* [1957] 1 QB 485. The issue in *Sim* was whether expenditure prior to the notice to treat was recoverable. The claim was for reimbursement of solicitors' fees which the claimants had incurred in the purchase of another house in the belief that their house was to be compulsorily acquired in the near future. The argument to the contrary was based on the use of the word "consequence" by Romer LJ in *Harvey* at p494, where he said:

> It seems to me that the authorities to which our attention was drawn do establish that any loss sustained by a dispossessed owner (at all events one who occupies his house) which flows from a compulsory acquisition may G properly be regarded as the subject of compensation for disturbance, provided, first, that it is not too remote and, secondly, that it is the natural and reasonable consequence of the dispossession of the owner.

*Editor's note: Also reported at (1982) 264 EG 621.

Denning LJ, as he then was, used a similar expression in the same case at p492 where he said that the law on the principle of compensation was settled both in England and in Scotland:

> Those cases show that, in addition to the actual market value of the house, when sold with vacant possession, Mrs Harvey is entitled to "compensation for disturbance," which is specifically preserved by section 2, rule (6), of the Act of 1919, and includes all damage directly consequent on the taking of the house under statutory powers. H

In *Sim* it was held that the word "consequent" was to be understood in the causal, not the temporal, sense, and that decision has now been followed in England in *Prasad* v *Wolverhampton Borough Council* [1983] Ch 333*. Mr Roy Martin QC, for the claimants, accepted, however, that the principles as developed by these decisions did not take the matter as far as he needed to go in order to show that the loss of profits which the claimants had sustained in this case was recoverable.

The claimants' submission was that their loss was recoverable in terms of section 17 of the 1845 Act, as damage sustained by them "by reason of the execution of the works". It was accepted that a causative J link had to be established between the loss and the actions of the promoters of the scheme. It was submitted, however, that the word "dispossession" which appears in the authorities ought not to be construed too narrowly. There was a direct and natural relationship between the dispossession of the owner or occupier of the lands and the works for the purpose of which the land was being compulsorily acquired. It was the carrying out of the works which gave rise to the right to compensation, and it was appropriate to take into account the effect of the scheme for the carrying out of these works. Reference was made to *Evans* v *Glasgow District Council* 1978 SLT (Lands Tr) 5 at p12 in which, in a claim for a disturbance payment under section K 35(1)(*b*) of the Land Compensation (Scotland) Act 1973, the tribunal left open the question whether anterior loss of profits consequential upon pending displacement could be claimed for loss sustained by the claimant "consequent upon his having to quit the land". In principle, it was said, there was no reason why a loss incurred before acquisition resulting from action taken by the claimant as a result of the threat of acquisition, as in *Sim*, should be recoverable while other losses resulting directly from that threat, such as from the blight to which the loss was attributable in this case, should be held not to be recoverable.

In my opinion, however, the claimants' loss cannot be brought within the principles which have been established by the authorities. I L think that it is significant that the distinction which the claimants seek to draw, between the expression "by reason of the execution of the works" in sections 17 and 19 of the 1845 Act and the word "dispossession" in the authorities, has never been seen as a point of importance in any previous discussion of the subject. The law as developed in the authorities, against the background of the provisions of the 1845 Act, is that compensation is given for loss caused by the taking of the lands. In *Lanarkshire and Dunbartonshire Railway Co* v *Mair* (1895) 22 R 912 at p919, Lord Kinnear put the matter in this way:

> M . . . It is a well-settled rule in the construction of the Lands Clauses Act that when lands have been taken in the exercise of powers of compulsory purchase, the owner or occupier, as the case may be, is entitled not only to the market value of his interest but to full compensation for all the loss which he may sustain by being deprived of his land.

The phrase used by Denning LJ in *Harvey* at p492 is "all damage directly consequent on the taking of the house under statutory powers". This interpretion of the concept of dispossession has received statutory expression in section 35(1)(*b*) of the 1973 Act, which refers to the loss which the claimant will sustain by reason of

*Editor's note: Also reported at (1983) 265 EG 1073.

A the disturbance to his trade or business "consequent upon his having to quit the land". It is dispossession caused by the taking of the lands which gives rise to the right to compensation, not the threat of dispossession or the effects of publication of plans for the execution of the works.

Where, as in *Sim* and in *Prasad*, loss incurred under the threat of dispossession has been held to be recoverable, this is because the dispossession has followed and the loss has been shown to have been caused by the dispossession. The loss for which compensation was awarded in these cases was for expense which might have been incurred after the dispossession but which the claimants had chosen to B incur beforehand. In *Sim* the court said at p253 that it would be inequitable if a claim which satisfies the tests described by Romer LJ in *Harvey* at p494, that it is not too remote and is the natural and reasonable consequence of the dispossession of the owner, should be denied the right to compensation because the expenditure was incurred prior to the service of the notice to treat, when similar expenditure incurred subsequent to the service of the notice would be admitted. In *Prasad* Stephenson LJ recognised at pp350-351 that, if the person threatened with inevitable dispossession acts reasonably in removing to other accommodation before he is given notice to treat or before his land is actually acquired by compulsory purchase, he can properly be C said to have been displaced in consequence of the acquisition and the loss which he will sustain by reason of his trade or business will be loss consequent upon his having to quit the land.

I can find nothing in these authorities to support the proposition that a loss which cannot be shown to have been directly caused by the taking of the lands is recoverable. As Mr Gill pointed out, the categories of what may be recovered under a claim for disturbance cannot be treated as a closed list. It may be that under certain circumstances a loss of business profits may be regarded as a valid head of disturbance. This may be because the loss is due to steps taken by the claimant to anticipate the effects of the dispossession, such as closing the business for a while in order to make arrangements for its D relocation. An appropriate test is whether the action or the expense incurred to anticipate the effects of dispossession could reasonably have been done or incurred by the claimant after receipt of the notice to treat. But loss which is held to be due to the effects of blight on trading generally cannot be said to have been caused by the dispossession of the claimant from the land. The two events are quite separate. Blight on trading is the result of the way the scheme is perceived by the public, and it is indiscriminate in its effects, as the tribunal found in the present case. It affects all traders in the area irrespective of whether their land is to be taken from them under the scheme. Dispossession, on the other hand, is the result of action taken E directly against the claimant by the promoter, and it is particular to the person whose interest is acquired from him by the taking of the land. A causative link between the taking of the land from the claimants and their trading losses has not been established in this case, and I think that the tribunal were bound in these circumstances to refuse this part of the claim.

Conclusion

For these reasons I consider that the tribunal were right to determine the compensation payable in this case on the basis that the lease had less than one year to run and that the loss of profits in the F business of Target Discount was not allowable prior to the date of vesting. I would uphold the determination and refuse the appeal.

In agreeing, **Lord Mayfield** said: Your lordship in the chair has set out the findings of the tribunal, and the grounds of appeal and has summarised the contentions presented by both parties. I agree with your lordship's reasoning and conclusion in relation to the first two grounds of appeal, namely that the tribunal were right to award compensation to the claimants under section 114 of the 1845 Act on the basis that the lease had less than one year to run at the date of vesting.

In relation to the third ground of appeal I agree with your lordship's conclusion that the claimants' loss cannot be brought within the principles set out in the various authorities which have been G referred to in this case. There is nothing in the authorities to support the contention that a loss which cannot be shown to have been directly caused by the taking of the lands is recoverable. In this case the tribunal held that the loss was due to the effects of blight on trading generally. The blight on trading was indiscriminate in its effects. As your lordship has pointed out, it affected all traders in the area irrespective of whether their land was to be taken from them under the scheme. I consider that the tribunal were entitled to conclude that the necessary causative link between the taking of the land from the claimants and the trading losses had not been established. In the circumstances I agree with your lordship in the chair that the tribunal had not erred in law in holding that the claimants were not entitled to H loss of profits (sustained by them under the trading name "Target Discount") for compensation for disturbance prior to the vesting date.

In the circumstances I agree with your lordship in the chair that the appeal be refused.

In agreeing on the first issue, but dissenting on the second, **Lord McCluskey** said: As your lordship in the chair has made clear there are two distinct issues raised in this appeal. On the first, that raised by the first and second grounds of appeal, I agree entirely with your lordship's opinion and there is nothing that I would wish to add. On the third issue, I have the misfortune to differ from your lordship. I J shall therefore attempt to explain my reasons for reaching a different conclusion on this issue.

In advancing their third ground of appeal, what the appellants claim is compensation for loss caused to their business in the final trading year before the vesting date given in the general vesting declaration. This loss was caused, they maintain, by the blighting effect of the impending acquisition of the premises from which their business was conducted. The tribunal have found in fact that the loss was "caused . . . by the overall blighting effect of a scheme of acquisition". This finding is not challenged; and it was taken to mean that the whole scheme of acquisition by compulsory purchase of all K the designated premises in area B (including the premises owned by Wordie and tenanted by the appellants) had a blighting effect which was not confined to the appellants' business and other businesses also compulsorily acquired, but which extended to other businesses carried on both in premises within area B which were not compulsorily acquired and in premises located "on the other side of Loch Street" outwith but on the fringe of area B; they, of course, were not compulsorily acquired either. The appellants' loss, which is conveniently referred to as a "loss of profits", was measured by reference to a diminution of the expected profits for the period in question, and this diminution resulted directly from, that is to say was L caused by, the blighting effect of the whole general scheme upon the whole area within and adjacent to area B.

Your lordship in the chair has summarised the arguments presented by both sides and I need not attempt to make my own summary. But it is clear that counsel for both parties adopted the same general approach to the question and founded upon the same reported authorities. In particular, they were agreed that a tenant who lost his business premises through compulsory acquisition could properly claim for loss caused by the "dispossession", provided the loss was not too remote and was the reasonable and natural consequence of the dispossession. They were further agreed that such loss was M recoverable even if sustained or incurred before the date of the actual dispossession. Even a loss of profits, incurred during a period before actual dispossession, might be recoverable, although counsel were not agreed as to the kind of circumstances that would have to obtain before such an item of loss might be recovered.

In essence, counsel for the respondents submitted that, for any such claim to be allowed, the claim would have to be one which arose in a way similar to the way in which the claim arose in *Aberdeen City District Council* v *Sim*: there, it was submitted, the loss could be seen as an inevitable consequence of the dispossession, for the claimant had simply taken steps in anticipation of the dispossession so that a loss

A which was properly to be regarded as an inevitable consequence of the dispossession had simply been incurred, or realised, earlier rather than later, before dispossession instead of after it. By way of illustration in relation to a claimable loss of profits, counsel offered the example of a businessman who, prior to actual dispossession, deliberately spent time and effort searching elsewhere for alternative premises in which to relocate his business, with the result that he could not devote that time and effort to running the existing business in the premises from which he was going to have to move and had thereby suffered an immediate diminution of trading with consequent loss of profits. Such a loss, albeit a loss of profits, could properly be regarded as an
B inevitable loss which the claimant had simply anticipated. Counsel for the respondents also submitted that in no case had compensation for loss of profits, caused prior to dispossession by the overall blighting effect of a scheme involving compulsory purchase, been allowed; indeed, no example was cited of such an item being claimed.

The Lord Justice Clerk in *Venables* v *Department of Agriculture for Scotland* and the court in *Aberdeen City District Council* v *Sim* treated the claim for disturbance as one of equity: in *Aberdeen City District Council* v *Sim* that was stated to be "the proper test". It is the test which I seek to apply; the equitable principle is that if a public authority uses compulsory powers to dispossess a person of his
C interest in land and causes him to suffer loss thereby the loss should not be allowed to fall upon the person dispossessed; he should be compensated for that loss by the acquiring authority. In so formulating the principle I recognise that it must be applied in such a way as to guard against spurious claims — *cf* Lord Hunter in *Venables*. But the principle itself is not to be restricted just because it has to be cautiously applied. And clearly the causal link must be proved. That leads me to consider the matter of cause and effect.

In my opinion, it is essential in each case to discover as a matter of fact what it is that has caused the loss giving rise to the claim. In this case the tribunal have identified the cause as the overall blighting
D effect of the scheme. But they, and the respondents support their position, distinguish between the overall blighting effect of the scheme, on the one hand, and the dispossession, on the other. Counsel for the appellants, however, says that this is to take too narrow a view. The claim is one of equity; the disturbance is that occasioned or sustained by reason of the execution of "the works" — *cf* Lands Clauses Consolidation (Scotland) Act 1845, section 17. In the light of the authorities, it is submitted, that means occasioned by the carrying forward of the scheme to completion by dispossessing owners and tenants; but the appellants do not dispute that it must be shown that the loss was caused by the dispossession.

E It is clear that the question of causation is one for the tribunal. Causation raises questions of fact rather than questions of law. But, on the other hand, where the tribunal of fact, as in the present case, have made their approach to the determination of the causation issue absolutely clear and precise it is open to this court, as a matter of law, to consider whether the tribunal have misdirected themselves in the way they have approached the question of causation or have drawn an unwarrantable conclusion as to cause and effect from the primary facts and circumstances which they have found established. It thus becomes necessary to examine closely what it is that the tribunal have said on this issue. I consider that the pertinent reasoning is to be found in the
F passage reading:

> That loss, which it is claimed was caused by the blight on trading in the surrounding area to the scheme, cannot be said to have been a consequence of the acquisition of Emslie and Simpson's interest in land or of their concurrent dispossession. Losses in the year to 31 March 1987, or in any previous trading year in which there may have been a blight on trade, were not caused by the taking of the land. The same losses would have incurred if the land had not been taken. Before the vesting date Emslie and Simpson were in the same position as Wordie's tenants in Schoolhill, whose interests in land were not acquired although they were within the designated area of the compulsory purchase order, and whose trade would also be affected by blight caused by the scheme, as would the trade of the occupiers of retail premises on the other side of Loch Street.

The first strand of this reasoning is that the losses in the year to G
March 31 1987 "were not caused by the taking of the land. The same losses would have occurred if the land had not been taken". I do not regard this as a persuasive and valid reason.

In *Aberdeen City District Council* v *Sim* the claimants incurred their expenditure — and therefore suffered their loss — in 1973; it was then that they bought the alternative house. The resolution to make a compulsory purchase of their original house was not passed until 1976 and it was a year later before the compulsory purchase order was confirmed. The general vesting declaration and the related notice to treat did not follow until 1978. So in *Sim's* case also it can properly be said that the same losses would have occurred even if the H
house had not been taken; there was in 1973 no commitment or obligation to take the house and no power to acquire it compulsorily. That did not prevent the court in *Sim* from concluding that the loss was properly to be regarded as one occasioned by reason of the Sims' dispossession. The true cause operated to produce, in 1973, a loss. That cause could not change its character because of something happening or not happening years later; nor did any later event create a new cause which replaced the cause which operated in 1973. It is obvious, of course, that had the local authority in that case eventually decided not to proceed by way of compulsory purchase of the original house the Sims could not have recovered their alternative expenditure J
as part of a disturbance claim. But in that case, as in the present, there was dispossession eventually under a compulsory purchase order and the procedures consequent on it; so they were entitled to recover. Their 1973 loss was occasioned by and consequent upon their subsequent dispossession. *Prasad* v *Wolverhampton Borough Council* [1983] Ch 333 is also a case in which the expenses constituting the loss were incurred before the date of the notice to treat. Indeed, as all the modern cases make clear, the requirement that the loss be consequent upon or occasioned by reason of the dispossession does not bar a claim in respect of loss incurred before the dispossession. Causation is a matter of fact. If it be proved that the dispossession K
process caused the loss, then the actual dispossession under compulsory purchase powers merely gives the person who has suffered the loss a title to claim under section 114 of the 1945 Act. In my opinion, it is clear that the fact that the claimants would have had no title to claim for this item if the land had not been taken is not a fact which bears upon the causation issue. The tribunal clearly thought otherwise. In this respect, therefore, I am of opinion that the reasoning of the tribunal is unsound.

The second strand of the reasoning is that others were also affected by blight, namely other tenants in Schoolhill, within area B, whose interests in land were not compulsorily acquired and also traders on L
the other side of Loch Street, who were outside the scheme altogether. Thus, what the tribunal is saying is that these people suffered loss without ever being dispossessed and therefore their loss could not have been occasioned by reason of their dispossession; as their loss had the same cause, namely overall blight, it followed that Emslie & Simpson's loss was not caused by dispossession either. In my view, this is very similar reasoning to that underlying the first strand in the tribunal's argument. It does not look directly at the chain of causation. It does not discover the true cause of the loss. It is saying, in effect, the same losses would have occurred if the land had not been taken, because the land on the other side of Loch Street was not taken yet M
those in business there suffered exactly the same kind of loss. The first point of difference is again that the appellants' interest in land *was* taken and the result is that the appellants here had a title to claim under section 114 which the others whose land was not acquired did not have. But the crucial point is that this passage in the reasoning also begs the question as to the cause of the loss sustained by each of the three groups, those within the area whose land was acquired, those within the area whose land was not acquired, and those outwith the area. In my view, the cause of the loss sustained by those on the other side of Loch Street was not, of course, their actual physical dispossession; they were not dispossessed. The cause of their loss was

the whole process of dispossessing of those, including the appellants, who *were* dispossessed by the scheme.

Let me illustrate that by looking again at the example which Mr Brian Gill QC, for the council, offered of the businessman who, faced with a real, impending threat of dispossession under a compulsory purchase order, takes time off from running his business to try to find an alternative location and thereby suffers an immediate loss of profits for which, as Mr Gill rightly concedes, he can claim after he is actually dispossessed. Let us suppose that he is a doctor in busy private practice. Let us suppose further that next door, in premises which are *not* to be compulsorily acquired, a pharmacist carries on a business which is heavily dependent upon the trade from the doctor's surgery. The pharmacist realises that, when the doctor goes, his dispensing business will go and his whole business will suffer seriously; so he too shuts up shop temporarily and follows the doctor around looking for suitable premises, preferably next door to the doctor's new premises. He too will suffer loss of profits for the period while he neglects his business. What is the cause of his loss? The answer is that it is effectively the *same* cause that operated in relation to the doctor, which *ex concessu* is the dispossession of the doctor under the compulsory purchase order. No doubt the pharmacist's own voluntary action in deciding to take steps to counter the apprehended effects of the blight is also a causal factor (as with the doctor who has also chosen voluntarily to anticipate his problems); but the real cause of his action and of his loss is the blight occasioned by the dispossession resulting from the compulsory purchase. The pharmacist, of course, has no claim for disturbance; he has no title to claim under section 114. But the fact that he has no claim for disturbance does not mean that the cause of his loss is not the same cause which resulted in the doctor's loss, being a loss for which the doctor can claim. The effective, underlying and prime cause of both losses is the same. If the same thinking is applied in relation to the persons affected by blight in and around area B the immediate cause of their losses is, as the tribunal correctly divined, the same, namely the blight. But the mere fact that they have no claim, not having had their land taken, does not mean that the present appellants, whose land was taken, have no claim.

What, in my opinion, the tribunal have failed to do in this paragraph is to follow through the search for the full and real cause of the diminution of profits which the present appellants have suffered. The blight is the immediate cause but it is also the result, the effect of another cause. The tribunal have not sought to discover and identify that cause. But the tribunal make it abundantly plain that the paragraph quoted contains the whole of their reasoning, because that passage is followed by the words:

> For the foregoing reasons the Tribunal is satisfied that the compensation for disturbance to which Emslie and Simpson are entitled does not include compensation for business losses occurred before the vesting date which have not been shown to have been a consequence of the taking of their interest in the land.

The phrase in the last 19 words is plainly not part of their reasoning; it is merely a restatement of their conclusion as expressed in the second sentence of the paragraph which I have quoted in full.

In these circumstances, considering as I do that the whole reasoning of the tribunal on causation is flawed, I consider that the matter of causation needs to be looked at again. I do not see the need to obtain any further facts because it appears to me that all the necessary facts have been found by the tribunal to enable us to ascertain the real cause. The immediate cause, as the tribunal has found, is the overall blighting effect of the scheme. But what was the real cause of the overall blighting effect of the scheme? The answer is, in my view, the bringing forward of the scheme including a compulsory purchase order for area B. That compulsory purchase order clearly involved the eventual physical dispossession of the present appellants; that was an integral part of the scheme. In my view, it is plain that the cause of the loss in respect of which the appellants now claim under this head of claim was the threat, ultimately realised, of compulsory acquisition of premises in area B under the scheme. In the light of that conclusion in fact, which I think cannot be challenged, the only question which remains is whether the loss sustained by the appellants, whose interest in land was compulsorily acquired, was a loss occasioned "by reason of" their dispossession. I consider that the claim must be a proper claim if the scheme, of which an integral part is their dispossession, causes the blight, which in turn causes the loss, provided that the threatened dispossession actually occurs. It may be thought that to look at the scheme broadly rather than to look narrowly at the actual dispossession is to go beyond what the decided cases permit. In my opinion, that is not so. What is striking about the case of *Aberdeen City District Council* v *Sim*, in line with the other cases which remove the temporal or sequential element from causation, is that an event happening in 1978 is treated as the event which occasions a different event, which precedes it by five years. In other words, once there is a threat of dispossession followed some time later by dispossession the threat can operate as the cause of the loss which is incurred as a direct consequence of the threat. All that is then needed to give the claimant a title to sue is that the dispossession should actually take place. It was, in my view, the natural and reasonable consequence of the dispossession of the owners of their interests in various lands in area B which caused the blight, which, in turn, caused the loss. In the light of the authorities, especially *Aberdeen City District Council* v *Sim* which is binding on the court, the notions of dispossession, or the taking of the land, or "the works" have been expanded to include those factors integral with the process of dispossession or taking even though the process precedes the actual dispossessing in point of time. I consider that here there is a direct link between the loss and the dispossession and that the loss sustained is in no sense too remote. It therefore appears to me that in accordance with the principles applied in *Aberdeen City District Council* v *Sim* and *Prasad* v *Wolverhampton Borough Council* the appellants are entitled to succeed in respect of this item of claim.

In my opinion, furthermore, it is important to recognise that the test of the claim for disturbance caused by dispossession, as that notion has been developed, is the test of equity. If a person in fact suffers loss because his property is designated in a compulsory purchase order and the loss takes the form of loss of profits through the blighting of the business between the date of the first appearance of the order and the actual physical dispossession of the claimant then who is to bear that loss? The respondents' argument in the present case was that there could be no compensation for that loss. In other words, the person who suffers the loss which is directly caused by the compulsory purchase has no remedy against the compulsory purchaser; so he, the "victim" of the scheme, must bear the loss himself. That appears to me to be a total departure from the test of equity. It does not appear to me that any of the cases to which we were referred requires such a result. In argument, Mr Gill pointed to the fact that others than the present appellants were dispossessed; the scheme was one which related to a substantial area, not just to this particular property. I accept that that is so, but, in my opinion, it makes no difference to the reasoning or the result. If there is only one property affected by a proposed scheme and the result of the scheme is to blight the business carried on in the property then, for the reasons I have considered, the owner of the property and of the business would be able to say that the cause of his loss was the scheme including his actual dispossession. Why should the result be different if there are two, or 20, discrete sets of premises with different businesses carried on in them, and all, or most of them, are affected by blight directly attributable to the threat of impending compulsory purchase? There can be no safety in numbers for an acquiring authority. There can be no equity in saying that as between the promoters of a scheme of compulsory purchase and those who are dispossessed by it it is the dispossessed who must suffer the loss caused by the promotion of the scheme.

Your lordship in the chair points out that the distinction which the claimants sought to make between the expression "by reason of the execution of the works" in sections 17 and 19 of the 1845 Act and the word "dispossession" in the authorities has never been seen as a point of importance in any previous (judicial) discussion of the subject. I agree that that is so; but looking at the facts and the issue in the cases to which we were referred I do not think that the point ever arose for determination. The observations of Stephenson LJ on pp350-1 in *Prasad* are clearly *obiter* and appear to have derived from a concession made by the *amicus curias* in relation to a point that the court there did not have to decide. No doubt the authorities do not, in the light of their own facts, expressly govern the point at issue here; but it appears to me that the principles they enunciate and apply do.

In the whole circumstances I consider that the appellants are entitled to succeed on this branch of their claim.

Appeal dismissed.

Court of Appeal

February 23 1994

(Before Sir Thomas BINGHAM MR, Lord Justice STEYN and Lord Justice ROSE)

BLUE CIRCLE INDUSTRIES PLC v WEST MIDLANDS COUNTY COUNCIL

Estates Gazette May 21 1994
[1994] 20 EG 149

Lands tribunal — Case stated — Whether notes of evidence of member should be produced to parties and court

In December 1988 the Lands Tribunal determined that the respondent county council should pay the appellants just over £1m compensation for the compulsory acquisition in 1981 of 18 ha of disused sand and gravel quarry land. The award was remitted by the Court of Appeal in 1991 to the tribunal for redetermination of the figure to be deducted from the overall sum as the annual operating costs identified in the valuations. The original decision had been that of two members, one of whom had retired; the surviving member consulted the retired member and determined that the same deduction as before should be made. The appellants appealed that second decision on the grounds that the tribunal wrongly and in breach of the rules of natural justice did not give the parties an opportunity of being heard and that the redetermination was wrong in law. The respondents agreed that there had been a breach of procedure. In the present application, the respondents sought production by the tribunal of the members' notes of evidence to seek to resist the appellants' attack upon the decision. In January 1994 the registrar of the tribunal informed the parties of the president's views that: both members were retired; neither had personally retained any notes of evidence; any notes in the tribunal's files were personal to the members and could not be released save by order of the court.

Held: **The application was allowed. It would not be appropriate and standard practice for notes of evidence to be annexed to the case stated. In the circumstances of the present case, where there had been long delays and where both parties were hoping to avoid a further remission, it was proper to order that the members' notes now in the possession of the tribunal are made available to the parties and to the court in advance of the hearing. In future it would be unnecessary for a full court to consider such applications. There was some doubt whether it was proper to regard the members' notes as the personal property of the member in question.**

No cases were referred to in this case.

This was an interlocutory application by the respondents, West Midlands County Council, for an order directing the Lands Tribunal to produce notes of evidence in an appeal by case stated of the appellants, Blue Circles Industries plc, against a decision of the Lands Tribunal awarding compensation to the appellant.

Robin Purchas QC and Andrew Tait (instructed by the solicitor to Birmingham City Council) appeared for West Midlands County Council; Jeffrey Burke QC (instructed by Pickering Kenyon) represented Blue Circle Industries plc.

Giving judgment, **SIR THOMAS BINGHAM MR** said: This is an application made by the West Midlands County Council, otherwise apparently known as Birmingham City Council, and it arises out of a compulsory purchase of land which the council made in 1981. The land comprised 18 ha of a disused sand and gravel quarry. The former owners of the land were Blue Circle Industries plc. There was a difference between the council and the former owners as to the compensation payable on the compulsory acquisition and that was a difference which in due course came before the tribunal. The tribunal consisted of two members, who heard evidence and argument on this question on a series of dates beginning on February 15 1988 and continuing intermittently until June.

In December 1988 the tribunal gave a decision in which they decided that a sum of a little over £1m was payable by way of compensation. The claimants challenged that assessment and it immediately became clear that there were certain errors in the figure at which the tribunal had arrived, some of which the tribunal were able to correct under the slip rule, but others of which they felt unable to correct.

In September 1989 a case stated was settled for the claimants' appeal on a point of law under section 3(4) of the Lands Tribunal Act 1949. That came before the Court of Appeal in July 1991. The Court of Appeal's decision then was that the matter should be remitted to the tribunal for redetermination (on the evidence which had been given before the tribunal) of the figure to be deducted from the overall sum of compensation as the annual operating cost attributable to this operation. The tribunal had originally fixed this figure at £225,000, but there were grounds for criticising the way in which that determination had been reached and accordingly that question was remitted.

The tribunal, which had originally consisted of two members, had altered in composition by this stage in that one of the two members who had previously dealt with this matter had retired. The surviving member, however, addressed himself to the question which had been remitted, discussed it with the original other member and issued a decision on May 30 1992 in which the same annual operating figure of £225,000 was given, and details were given of how that figure had been reached. Unhappily that decision also has in its turn been the subject of challenge by the claimants on the ground that it was flawed in law. So we now have a second appeal to this court by way of case stated.

The first basis upon which the appeal is put is that there was a breach of natural justice in the procedure by which the tribunal reached its second decision. Put very shortly, the complaint is that representations were made to the tribunal that the claimants wished to be heard with reference to the calculation of the annual operating figure and indications were given by the tribunal that a hearing would be arranged. That did not, however, take place and the decision was issued without any opportunity for the claimants or the council to make representations to the tribunal. It appears to be common ground between the parties that there was a breach of the procedure which should have been followed and that on that basis the claimants have good ground for challenging this second decision of the tribunal. Although, therefore, the appeal to this court has not yet been heard, there appears to be no issue between the parties, but that the claimants are entitled to succeed on that ground.

There is, however, a second ground of challenge and it is that which gives rise to this application. The claimants contend that the figure of £225,000, which appeared in the first decision and reappeared in the second decision with reasons given as to how it had been reached, is open to attack on grounds of law. The legal ground upon which it is sought to base that challenge has not been at all fully investigated on the hearing of this application. For my part, I am very far from satisfied from the very brief, and necessarily brief, introduction which Mr Jeffrey Burke QC has given to this ground of appeal that it is in truth a ground of law at all. Therefore, for my part, I would wish to leave open any decision as to whether this second ground is properly open to the claimants as a ground of law under the Act.

There is, however, a dispute as to the basis upon which that figure was calculated. The council say that the evidence to which the tribunal refer in their second decision is accurate and correct and they seek to resist any attack upon it. In support of that ground they further seek to obtain possession and sight of the notes of the evidence taken by the tribunal, since there was evidently no transcript made of the evidence as it was given. The claimants for their part do not resist that application, but neither do they support it. When the approach was originally made to the tribunal for the supply of the notes made by the members, a letter was written on behalf of the tribunal, over the signature of the registrar, dated January 5 1994. In the letter it was said that the notes taken by the sitting members at the hearing were their personal property although it made clear that certain of those notes were in fact retained in the tribunal's file. The penultimate paragraph of the letter reads as follows:

> The President takes the view that these steps

— that is the obtaining of the notes and the approach to the members of the tribunal —

> would not be justified on the basis of a unilateral request from one party. It is of course his wish to be co-operative in placing this matter back before the Court of Appeal expeditiously and economically, but for the reasons indicated he is not prepared to comply with your request unless and until *either* the appellants concur *or* the Court of Appeal so directs.

A copy of that letter was sent to the claimants' solicitors and they did concur in the request which the council made for the release of the members' notes in the possession of the tribunal. Accordingly, the council wrote again on January 12 indicating that concurrence had been received and repeating the request for the notes. The tribunal answered on January 14 1994 and reported that steps were now being taken to ascertain from the required members whether they had retained their notes and whether they were prepared to make them available. The letter continued:

> I am to make it plain that the tribunal has not undertaken to produce any notes, and would not be prepared to do so save by order of the Court of Appeal. As was pointed out earlier, any such notes are the personal property of the member concerned, and it is not the normal practice for the Court of Appeal to consider any material upon a case stated apart from the written decision of the tribunal. The letter of 5 January 1994 was written upon the President's directions, and as there stated, was limited to requesting information from the members concerned if the parties wished this to be done.

A further letter followed on January 17 1994 in which it was reported:

> . . . the President has asked me to let you know that he has now been able to contact the members previously concerned in this case, both of whom, are now in retirement. Neither Mr Mallett nor Mr Rees has retained any notes in his personal possession. As to such notes as may have been deposited in the tribunal's files, I am to repeat that they remain the personal property of the member in question and were made only for his own use. In line with previous invariable practice, both Members decline to make their notes available to the parties, unless required to do so by order of the Court of Appeal.

I have to say that, for my part, I entertain some doubt whether it is proper to regard the members' notes as the personal property of the member in question. It would seem to me that they fall very much within the same category as the notes made by a county court judge in the days when hearings in the county court were not transcribed, and it was standard practice for those notes to be made available, either in a photostat copy of the original or in typed form. However, I am quite sure that it would not be the appropriate and standard practice for notes of evidence to be annexed to a case stated. Indeed, in cases stated from magistrates' courts, for example, that practice has always been frowned upon. I would not, therefore, wish to countenance the suggestion that as a regular course notes should be made available. Moreover, for reasons that I have already indicated, I have real doubt whether the notes in this particular case would bear on any issue which it is proper for the court to entertain. This is, however, a matter which has dragged on for an appallingly lengthy period. Both parties, I think, are anxious, if it can be achieved, to ensure that the forthcoming hearing before the Court of Appeal is the last hearing in this matter and that there is not a remission and the need for a further hearing by the tribunal. Accordingly it seems to me proper to order that the members' notes now in the possession of the tribunal are made available both to the parties and to the court in advance of that hearing, while reserving to the court the decision as to whether it regards it as appropriate to look at the notes or pay attention to them.

The application in this case has been brought before the full court since it seems to be a novel question and one that called for some consideration, but in any future case, should, improbably, any similar application be made, it would seem to me quite unnecessary for the full court to have to consider the matter in advance of an appeal. The standard rule should be that notes of evidence made by members of the tribunal should not be required for purposes of this court's ruling on an appeal by way of case stated and if, in an exceptional case, it is thought necessary for the notes to be made available, then a direction by the Registrar of Civil Appeals should be made to that effect. I wish to emphasise that in the ordinary way the notes will not be produced.

Steyn and **Rose LJJ** agreed and did not add anything.

Application allowed. Costs in the appeal.

For a further case on this subject see p 229

ESTATE AGENTS

A
Court of Appeal

November 8 1993

(Before Lord Justice BUTLER-SLOSS, Lord Justice STUART-SMITH and Lord Justice KENNEDY)

LETGAIN LTD v SUPER CEMENT LTD

Estates Gazette February 19 1994

[1994] 07 EG 192

B *Estate agents* — ***Whether estate agent negligently failed to display "for sale" boards and failed to advertise or market the property properly***

In September 1988 the plaintiff estate agent agreed to find a purchaser for a site owned by the defendants for a commission of 1.25% and in a letter set out how the site would be marketed, including a board, mailing other agents and advertising. At a meeting between the parties and because the defendants were involved in litigation with a past director of the company, and wished to keep confidential the sale of the site, the plaintiffs alleged that they were instructed not to put up a board or to advertise. Particulars were
C circulated to other agents and eventually an offer was received from D Ltd, who purchased the site for £300,000 and shortly afterwards entered into an option to resell at £550,000, which was exercised in June 1990. In the court below, Tudor Evans J allowed the plaintiffs' claim for unpaid commission in the sum of £4,312.50 and dismissed the defendants' counterclaim for £250,000 for breach of contract for negligence in the marketing of the site and for advising the defendants to accept an offer below market value. The defendants appealed.

Held: **The appeal was dismissed. The trial judge was entitled, on the evidence before him, to prefer the account of the date and terms of the plaintiffs' instructions put forward by its negotiator
D on its behalf. The plaintiffs had received specific instructions not to use a board. They had discharged their contractual duty to market the property properly given their instructions about limited advertising. Having regard to other valuations of the site at the time and later, the plaintiffs were not in breach of any duty in leaving the matter of the acceptance of the offer to the defendants to make up their mind one way or the other.**

The following cases are referred to in this report.

Hontestroom, The v The Sagaporack [1927] AC 37

Powell v Streatham Manor Nursing Home [1935] AC 243

E

This was an appeal by the defendants, Super Cement Ltd, from the decision of Tudor Evans J, who on July 31 1992 gave judgment to the plaintiffs, Letgain Ltd, on their claim for unpaid commission and dismissed the defendants' counterclaim for damages for breach of contract.

Augustus Ullstein QC (instructed by Crellins) appeared for the appellants; Simon Brown (instructed by Kennedys) represented the respondents.

Giving the first judgment at the invitation of Butler-Sloss LJ, STUART-SMITH LJ said: This is an appeal from a judgment of Tudor F
Evans J given on July 31 1992 by which he gave judgment for the plaintiffs in the sum of £4,312.50 on the claim, being the value of their charges for services rendered to the defendants in connection with the sale at a price of £300,000 of premises at Spelthorne Lane, Ashford, Middlesex, that being a freehold property which was then owned by the defendant company. He dismissed the defendants' defence and counterclaim, the defence being that the value of the plaintiffs' services were nil with a counterclaim in the sum of about £250,000 for breach of contract. The purchasers of the property from the defendant company was a firm called Doyle Brothers (Bedfont) Ltd, a company which was controlled by a Mr Payne. G

The property in question was somewhat odd-shaped. The judge described it as "funnel-shaped with a relatively narrow entrance where the factory was placed and to the rear of which there was some unused land". It was about half an acre in extent. At the material time the defendants were manufacturing and sold epoxy-resin-based flooring material from the premises. A contract was made in 1988 between the plaintiffs and the defendants by which it was agreed that the plaintiffs would provide their services in exchange for 1.25% commission on the sale price. That, as the judge pointed out, was how the amount of the claim was calculated.

The defendants denied that they were in breach of any of the H
express or implied terms of the contract and, as a result, the plaintiffs thereby lost the chance of selling the property at a higher sum, that being the £250,000 difference which was claimed. That came about in this way, that Mr Payne subsequently sold the property for the price of £550,000 to a company called Ideal Vehicle Rentals.

It was alleged by the defendants in their defence that the plaintiffs were in breach of the contract between themselves and the plaintiffs in a number of respects: first, that they failed to cause a "for sale" board to be erected at the premises; second, that they failed to advertise the property properly; third, that they failed to circularise all London and local agents; fourth, that they failed properly to market the property to J
potential owner-occupiers; fifth, that they failed to attempt to market the property to persons and organisations desirous of purchasing the property; and, finally, that they advised the defendants to accept an offer which was below the true market value. It was the plaintiffs' case that they were expressly instructed not to put up a board and not to advertise the property. They denied that they were in breach of the contract in other respects and the judge found in their favour.

The introduction between the parties came about in this way. In June and August 1988 the plaintiffs wrote two letters to the defendants inquiring whether they were interested in selling the property, because at that stage they had a client who might be interested. That came to K
nothing but Mr Goddard, the managing director of the defendant company, subsequently telephoned the plaintiffs and expressed an interest in instructing the plaintiff company to market the property for them. A meeting was arranged which took place on September 21 between Mr Goddard, on behalf of the defendants, and Mr John Kingston [MA(Cantab) ARICS], a director of the plaintiffs. At that meeting Mr Goddard told Mr Kingston that the company might want to sell either the property or the business. One of the reasons was that there was litigation between Mr Little, a former managing director of the company, and the company for breach of fiduciary duty on the

ground that he had set up in business in competition with the defendants and was soliciting their customers. Mr Goddard on that occasion told Mr Kingston that the discussion was highly confidential and that they were involved in that litigation. As a result of that meeting, it was left that Mr Kingston would write, setting out his opinion as to how the property could be marketed. Mr Goddard told Mr Kingston that there was to be a board meeting of the defendant company on October 25 and that he would thereafter contact him.

Mr Kingston then caused inquiries to be made at the planning office. The planning officer was against residential development and reasonably positive that a B1 use could be developed. B1 use comprises office development, hi-tech development and light industrial use. On September 26 Mr Kingston wrote a letter to Mr Goddard. The letter indicated these points and that Mr Kingston thought that a price of £750,000 could be obtained on a development basis. He said that his optimism was based on a buoyant market and the fact that considerable sums were being offered for commercial developments. He suggested that the planners would not be too strongly opposed to the offer of small two-storey business units for owner-occupiers at the site. He said that vacant possession would have to be given within a reasonable time, and he thought that that would fit in with the litigation which he understood was expected to be resolved within a few weeks. He then set out in some detail how it was proposed that the plaintiffs would market the property, and I read from the letter:

> (a) *Board*
> I feel it would be beneficial to place a board on the premises (perhaps 5'x 4') so that the property could be easily identifiable to interested parties.
> (b) *Mailing*
> I would also offer our in-house service, whereby we would write to all investors/developers who we believe would be interested in such a proposition. This method of marketing is found to be particularly successful and economical.
> (c) *Preparation of Particulars*
> Details of the site would be prepared on our usual A4 details paper and sent to all the London and local agents for their perusal.
> (d) *Advertising*
> It is also recommended that a limited amount of advertising be undertaken and I would advise that perhaps a half page (black and white) advertisement be placed in the *Estates Gazette*, which is a much read document within the property world.

Then he said that they would expect to dispose of the site within a three-month period. In summarising, he put the position as follows:

> . . . it is my considered opinion that to ensure the highest possible value for your property, there is only one suitable course of action. I believe that the site should be marketed at the earliest opportunity as a redevelopment situation.
>
> No doubt there will be aspects of this report which you will wish to take up with me and discuss at greater length and I would be delighted to hear from you at your convenience.

Mr Goddard telephoned Mr Kingston after the board meeting on October 25. According to Mr Kingston, Mr Goddard told him that the company wanted to sell the property and asked him to attend a meeting between himself and Mr Broomfield, who was the company chairman, to discuss the sale. He was not instructed in that telephone conversation to go ahead and market it on the basis of the letter of September 26, and he was not given authority to act at that time. According to Mr Goddard, however, on receipt of the letter they had the board meeting. He telephoned Mr Kingston on October 26 and instructed him to sell the property. He said it was arranged for Mr Kingston to come to a meeting on November 1 to measure up the premises.

A meeting took place on November 1 attended by Mr Goddard, Mr Broomfield and Mr Kingston. As the judge described it, there was a head-on conflict between Mr Kingston and Mr Goddard as to what took place at that meeting. Mr Kingston's evidence was that there was discussion about the proposals contained in the letter of September 26, and that he said the proposals with regard to the use of the sale board were not acceptable. Both the defendants' representatives were concerned that the other person involved in the litigation would become aware of the sale and there was a discussion about the use of the sale board, but eventually he was instructed not to use it. As to the advertising in the *Estates Gazette*, again it was Mr Kingston's evidence that that was not acceptable because Mr Goddard and Mr Broomfield felt that national advertising would jeopardise their desire to keep the matter secret and, accordingly, Mr Kingston was instructed not to advertise at all. Again there is a conflict of evidence about that. Mr Goddard said that Mr Kingston was instructed to place an advertisement in the *Estates Gazette*, but not to undertake any other advertisement in relation to the property. Various other matters were discussed. The closing date for the offers was to be December 9, but that was subsequently changed to December 16.

On the next day, November 2, Mr Kingston went to see the planning officer. Mr Kingston said he was told that he could not proceed with residential development and that he could not proceed with office development, but that, although the planners were pessimistic, they were not closing the door to B1(b) and B1(c) development — that is the other type of development to which I have referred. Mr Kingston said that shortly after that meeting he telephoned Mr Goddard, told him what had happened, said that redevelopment would not be easy, but it was agreed between them that they would go ahead and attempt to market on the basis agreed at the meeting, ie that it was suitable for redevelopment.

Thereafter particulars of the property were prepared on November 8. They consisted of five pages, but the last three pages were plans and directions, which were not sent out unless requested. The particulars were described as "excellent redevelopment potential", and the potential purchasers were invited to consult the attitude of Mr Sterney, the planning officer, with regard to potential development. There was no price indicated in the particulars. According to Mr Kingston, the particulars were sent out, first, to developers who were kept on an in-house list by the plaintiffs (there were about 300 developers on that list); second, to central London estate agents, using an estate agency service which, according to Mr Kingston, mailed about 400 to 500 agents practising in central London, and also to 65 to 70 local agents, using the *Estates Gazette* agents' directory, which contains a list of agents which is updated every month. Mr Kingston said that the particulars were sent only to the commercial agents in locations which he thought were suitable.

According to Mr Goddard, he said that about a week after November 1 he had taken up the fact that there was no sale board up on the premises, but that Mr Kingston had told him that, on reflection, he thought it best not to have a board. The judge did not accept that evidence.

By the closing date of the offers, December 16, there had in fact been no offers at all, and on December 19 Mr Kingston wrote to Mr Goddard in these terms:

> Unfortunately, I have to advise you that we did not receive *one* offer in relation to the above property. Clearly this is most disturbing and, in part, may be explained by the closeness to Christmas.
>
> I have subsequently spoken to the planners again, however, and feel that it is something more fundamental. When I met Spelthorne Council, they indicated that they would be somewhat reluctant to see any form of redevelopment close to the residential area. Many of the interested parties have contacted me advising that the planners have totally "stone walled" any redevelopment plans put to them. I strongly believe that this is the main reason for such a response.
>
> With regard to where to go from here, I believe you have two alternatives:
>
> 1. Re-market the site in the New Year and within this programme, include wholesale advertising showing that it could be of interest to *both* developers and owner/occupiers. You still have sufficient time for this plan to take place satisfactorily.
>
> 2. Consider your idea of simply selling your business, as discussed in our early meetings.

Shortly after receiving that letter Mr Kingston received a letter from Mr Jeff Tapping [ARICS], who was a partner in a local firm of estate agents [Bryant & Partners], making an offer for the premises of

A £265,000 on behalf of Mr Payne. Mr Kingston immediately wrote to Mr Goddard informing him of the offer. He said in the letter:

> The reduced level obviously illustrates that it is being bought by an owner/occupier but I am not sure that it should be dismissed, as the developers may find the planning difficulties insurmountable.
>
> Perhaps you might call me to discuss the situation further, but I strongly suggest that a letter be sent back to this interested party indicating that we would like to interview them. It may be possible to then "wind" up the offer level from them slightly.

A meeting was fixed to take place on January 4 1989 and it was arranged that on that day Mr Kingston would first meet Mr Goddard B and Mr Broomfield to discuss the offer, and then they would see Mr Tapping and Mr Payne. According to Mr Kingston, at any rate in his evidence in chief, there was a fourth person present at the preliminary meeting. He said he was told by Mr Goddard that this was a local agent who had been asked to express his views. The agent said that he envisaged a value in excess of £300,000 for an occupier. Mr Kingston said that he thought there would be a joint marketing between himself and the other agent. He also thought that Mr Goddard told him that the other agent was from a firm called Shepherd Dear & Co. Mr William Shepherd [FRICS], of Shepherd Dear & Co, had in fact advised Mr Goddard about the premises. He had written in 1987 and shortly C before this meeting. It was put to Mr Kingston in cross-examination that there was no other person present, but when Mr Goddard gave evidence he said that his nephew had been present, a man called Mr Denham, and that he had remembered that shortly before the trial.

The offer of £265,000 made by Mr Tapping was discussed. Mr Kingston said that he felt that an offer of £350,000 would be attractive and that the present offer of £265,000 did not reflect the true value. On the other hand, the defendants could by this time wait until June 1990 because vacant possession was not to be given to a purchaser before that date. On the other hand, the existence of the litigation hindered the advertising and the use of a board. Therefore, he said that unless there D was a change of attitude towards advertising and the use of a board, it would be very difficult to find an owner/occupier to buy at a higher price. It was decided at the preliminary meeting that the strategy to be adopted was to tell Mr Tapping and Mr Payne that they were one of several agents interested in the property and to hint strongly that the only way to get it would be to increase the offer.

There then followed the meeting with Mr Tapping and Mr Payne. They were told that if the offer was increased the defendants might be tempted, and it was suggested that an increase of £100,000 might be acceptable. Mr Tapping said that they were not interested in that. After that meeting there was a further discussion, briefly, about E remarketing the property and again Mr Kingston said £265,000 was not acceptable.

Mr Goddard's account of the meeting was different. He said that after he had received the offer of £265,000 he was advised that it might go to £365,000. He contacted Mr Shepherd, of Shepherd Dear & Co, because he thought that it was desirable to obtain another opinion, and the opinion of Mr Shepherd was that £365,000 was acceptable. Mr Shepherd was not retained, but at the meeting Mr Goddard told Mr Kingston the substance of the advice given by Mr Shepherd, namely that it could be marketed over the next six months, a further offer might be forthcoming, and that a price of £300,000 to F £350,000 might be forthcoming, and then it was for Mr Goddard to make up his mind what to do. He said that, in effect, he was being advised at the meeting that, if an offer of £300,000 were made, it was acceptable and should be accepted.

On January 6 Mr Tapping did increase the offer to £300,000, with the proviso that there should be a contract within seven days. The offer was discussed on the telephone between Mr Goddard and Mr Kingston. Mr Kingston said that he had put the pros and cons of the offer, leaving it to his client to decide whether to accept it, saying that it was on the light side and that Mr Goddard might get more.

According to the judge, Mr Goddard's recollection of that telephone conversation was not clear. He could not remember discussing it with Mr Kingston, but accepted that he must have done. G Eventually, after discussing it with Mr Broomfield, they decided to accept the offer, because Mr Kingston advised that there was no guarantee that £350,000 would be forthcoming and that the offer of £300,000 was available only for a short time. Shortly after that, the offer was accepted.

After those negotiations were complete and a binding contract was entered into, Ideal (the firm I have already mentioned) became interested and made an offer of £555,000 to Mr Payne. That was in the form of an option which was to be exercised by them if they wished to do so, and in fact they exercised the option in June 1990.

There was evidence from a Mr Robert Meacock [FRICS FSVA], who H was an estate agent, who was called on behalf of the defendants, that in April 1989, when he was looking for premises for another client of his, Houlihan & Co, the planning officer at Spelthorne suggested that he might be interested in the defendants' premises. He got in touch with Mr Goddard, but by that time they were committed to Mr Payne. Mr Meacock in fact proposed that his client should buy out Mr Payne, as I understand it, at a premium, paying £500,000 for the site, but, although negotiations took place with Mr Payne, they fell through, not surprisingly perhaps because Mr Payne was going to get £555,000 from Ideal.

That is a brief summary of the facts of the case. The judge accepted J the evidence of Mr Kingston where it differed from that of Mr Goddard. At p20 of the judgment he said:

> It is also convenient at this stage to state my impression of Mr Kingston and Mr Goddard as witnesses. Mr Kingston often found it very difficult to answer, even a simple question, in short, simple and unconvoluted language. But I am satisfied, having seen and heard him for a substantial period of time, that he was not fencing with any of the questions, that he was trying to answer them accurately even though he used circumlocution. I find him to have been a truthful and reliable witness. I very much regret to say that that was not my impression of Mr Goddard. He frequently fenced with and was evasive in answering questions put to him. As I have already found, he attempted to shift to Mr Broomfield alone the desire to avoid Messrs Little and Dean learning K about the sale.

I should point out that Mr Broomfield, although he was an important witness in relation to the meeting of November 1 and indeed of January 4, was not called to give evidence. The judge then went on to deal with the evidence of Mr Denham, and again he said that he preferred Mr Kingston's account of the meeting of January 4.

With regard to the breaches of contract pleaded in para 14 of the defence and counterclaim, he held, first, in relation to the alleged failure to erect a "for sale" board, that the plaintiffs had received specific instructions on November 1 not to use a for sale board. In L relation to the allegation that the plaintiffs failed to circularise other estate agents with particulars of the sale, he accepted Mr Kingston's evidence that he circulated 65 to 70 local agents, and that was a sufficient discharge of the duty. With regard to para (c) that they failed to market the property properly, he held that the marketing strategy carried out was perfectly proper, having regard to the constraints put upon the plaintiffs by the requirement that there was to be no local publicity so that Mr Little should not get wind of the matter. With regard to the allegation in para (d) that the plaintiffs failed to market, or attempt to market, the property to persons or organisations desirous of purchasing the premises for business use (ie as owner/occupiers), he M held that there was no breach of that. It was agreed that developers would be targeted in the first place, but, when there was no offer, the question of targeting the owner/occupiers was raised in the letter of December 19. Then they received the offer of £265,000, which eventually became one of £300,000, which, as it were, superseded the alternative of further marketing the property. With regard to the final allegation that the plaintiffs advised the defendants to accept offers which were substantially below the true market value of the property, the judge held that Mr Kingston did not advise acceptance. He put the pros and cons of accepting the offer of £300,000 and left it to the clients to accept or not as the case may be. In fact, as the case was developed, the real allegation was that the plaintiffs should have

A advised the defendants not to accept the offer of £300,000. The judge also rejected that contention.

The criticisms which are made in the grounds of appeal and developed by Mr Augustus Ullstein QC [counsel for the appellants] are entirely upon the judge's findings of fact. That presents the plaintiffs with a very difficult task in this court. The principles to be applied are set out in the White Book at p940 and they are well known. It is necessary only for me to refer briefly to two well-known passages in the speech of Lord Wright in *Powell* v *Streatham Manor Nursing Home* [1935] AC 243, first at p265 where Lord Wright is dealing with the speech of Lord Sumner in *The Hontestroom* v *The*
B *Sagaporack* [1927] AC 37 at p40. Lord Wright in citing the speech of Lord Sumner, after referring to the fact that the Court of Appeal can interfere with a judge's findings of fact because it is a rehearing, said:

> None the less, not to have seen the witnesses puts appellate judges in a permanent position of disadvantage as against the trial judge, and, unless it can be shown that he has failed to use or has palpably misused his advantage, the higher Court ought not to take the responsibility of reversing conclusions so arrived at, merely on the result of their own comparisons and criticisms of the witnesses and of their own view of the Probabilities of the case.

Then he said:

> C If his [the trial judge's] estimate of the man [the witness] forms any substantial part of his reasons for his judgment the trial judge's conclusion of fact should, as I understand the decisions, be let alone.

Then at p267 Lord Wright said:

> In truth, it is not desirable, in my opinion, to do more than state, as I think Lord Sumner was stating, principles which will guide the appellate Court in the majority of such cases. The problem in truth only arises in cases where the judge has found crucial facts on his impression of the witnesses: many, perhaps most cases, turn on inferences from facts which are not in doubt, or on documents: in all such cases the appellate Court is in as good a position to decide as the trial judge. But where the evidence is conflicting and the issue
> D is one of fact depending on evidence, any judge who has had experience of trying cases with witnesses cannot fail to realize the truth of what Lord Sumner says: as the evidence proceeds through examination, cross-examination and re-examination the judge is gradually imbibing almost instinctively, but in fact as a result of close attention and of long experience, an impression of the personality of the witness and of his trustworthiness and of the accuracy of his observation and memory or the reverse. He will not necessarily distrust a witness simply because he finds him inaccurate in some details: he can give such inaccuracy its proper place, particularly if he sees that the witness is tired, or antagonised, or confused, or perhaps impatient, and especially if the matter of the inaccuracy is of minor or collateral importance. But such inaccuracies may appear in a very different light when pointed to as
> E isolated passages in the shorthand notes and abstracted from the human atmosphere of the trial and from the totality of the evidence. The judge will form his impression from the whole personality of the witness; he can allow for the nervous witness, standing up in a crowded Court or worried by the strain of cross-examination. The judge may be deceived by an adroit and plausible knave or by apparent innocence: for no man is infallible; but in the main a careful and conscientious judge with his experience of Courts is as likely to be correct in his impressions as any tribunal, unless perhaps, as some would say, a jury of twelve members is preferable.

I do not think I need recite any more from that passage. It shows that the defendants in appealing these findings of the very experienced judge have an uphill task.

F Mr Ullstein criticises, first, the judge's finding as to the date and time the contract was completed, namely November 1, and the judge's decision that there was no breach of contract in failing to display a board. Mr Ullstein, in effect, takes three points. First, he submits that the judge's conclusion is inconsistent with the board minute of October 25. That, so far as is relevant, simply reads:

> It was agreed that the [managing director, ie Mr Goddard] should instruct Russell Cash & Co to dispose of the company's freehold property.

Mr Ullstein submitted to this court, as he did to the judge, that that meant that he was instructed to complete the agreement on the terms of the letter of September 26. In my judgment, that is not so. First, it is common ground that it was arranged that Mr Broomfield would be G present at the meeting, and it is clear there was a need to see Mr Broomfield. Moreover there is no reference in the minute to the letter of September 26, still less the fact that it could be accepted only on those terms. Second, Mr Ullstein points to the fact that Mr Kingston's notes of the meeting of November 1, although, according to him, there were two important changes in the strategy as set out in the letter of September 26, contain no express reference to those changes, namely that there was to be no board and no advertising in the *Estates Gazette*. What Mr Kingston did record in his very sketchy notes of that meeting was that the property was to be quietly marketed. It is common ground that Mr Kingston was anxious about this and, in my judgment, there was nothing improbable about those instructions being altered in view H of the reservations that both directors of the defendant company had about the secrecy of the operation and the fact that Mr Little was not to know about it.

Then it is said that it is surprising that these two important changes were not recorded in writing. It is correct to say that there was no further letter from either side confirming the terms of the agreement. It is common ground that the acceptance of the instructions was oral. In my judgment, this is merely a factor to be taken into account by the judge, as he did in fact take it into account, in reaching a conclusion. It seems to me that the probability is that, if Mr Kingston had suggested both these two things, ie the board and the advertising in the *Estates* J *Gazette*, especially since the client would be charged separately for it and in his opinion it would be an aid to the marketing, it is likely, it seems to me, that would have been done unless the clients changed their instructions. In my opinion, the probability supports the plaintiffs' account of the matter.

In addition Mr Broomfield was not called (as I have already indicated) to give a different account of that meeting, and I agree with the point made by Mr Simon Brown [counsel for the respondents] in his skeleton argument that there does not appear to be any really convincing explanation why Mr Broomfield should have been brought into the matter at all if in fact it had already been finalised by the K telephone conversation accepting the contract in the terms of the letter.

The second point which Mr Ullstein criticises really goes to the nub of the case and that is to the judge's finding that Mr Kingston circulated 65 to 70 of the local agents. Mr Kingston said that that is what he did. He said that he took the agents from the *Estates Gazette* list. He took them from the list for those in the locality. He gave the names to the secretary, and she typed out the addresses to be put on the envelopes. The judge accepted that there was a paucity of documentary evidence supporting the plaintiffs' account of this matter. He dealt with it at p24A-E in relation to such documentary evidence as there was. The first was a fax, which appears in the bundle, which L appears to be a response to an inquiry by Messrs Stimpsons, and the judge says:

> According to Mr Kingston, he received a telephone call from Messrs Stimpsons to whom he had sent the particulars and as a result of the enquiry, he faxed the site and location plans to them. I accept that evidence. That is in fact not the only documentary evidence showing that local agents were circulated. At page 84 section II is a letter from the Cardiff Property plc of Egham dated 23rd November 1988 and written to a Mr Richards of Bryant & Partners under the heading "Spelthorne Lane Ashford . . ." in which the writer states:
>
> "I refer to your faxed letter of 22nd November and would confirm that we M have already received particulars of the same direct from Russell Cash."
>
> The Plaintiffs also called Mr Tapping of Bryant & Partners, the agents who were acting for Mr Payne. In evidence-in-chief, Mr Tapping said, although he was not sure, that he thought he had received the particulars from the Plaintiffs.

However, in cross-examination he plainly became less certain of that and the judge did not rely on Mr Tapping's evidence in support of the plaintiffs' case in that matter. Mr Ullstein submits that the judge's finding about this matter cannot stand. First, he submits that Mr Kingston's evidence about it was contradictory. He refers to a passage at day one p59 in his evidence-in-chief where he said:

A There is a system, my Lord, for mailing London agents utilising the estate agent's services. This is a service which we're registered with and mails something in the region of 400 to 500 other practices in the London and the surrounding area. In addition to that, clearly in this instance I felt the local agents should receive the information also, and I subsequently sent information on to them. This was done by use of an *Estates Gazette* agents' directory, which I might explain, my Lord, is something which comes out once a month which is like a bible of agents across the country.

Mr Ullstein says that that has to be compared with what he said at p70F where he is again being asked in chief:

> Q. So those are the particulars you prepared, and those particulars you sent
> B out, did you, to London agents?
> A. To all the developers on our list and investors, to all the Central London agents and to the local agents.
> MR JUSTICE TUDOR EVANS: Just say that again, would you?
> A. I do apologise.
> Q. It is my fault. You sent it to?
> A. The developers, 300 plus, as I discussed earlier; the Central London agents which, if you recall, was by way of the estate agent's services, and the appropriate local agents.

Mr Ullstein's point appears to be that in that passage Mr Kingston is saying that he got the estate agent's service to circularise the local
C agents. I do not read it in that way at all, and I think there is nothing in the point that Mr Kingston's evidence was contradictory.

Second, Mr Ullstein submits that there was no clear documentary evidence that the circulation was done. He submits that Mr Tapping and Stimpsons may have been contacted by developers who were circulated directly by the plaintiffs and that they in turn contacted their own property agents who took the matter up with the plaintiffs. That is a possible explanation, but, in my judgment, it seems no more likely than that they were circulated direct by the plaintiffs.

Third, Mr Ullstein says that it is very surprising that there were no responses direct from those agents whereas there were from the
D developers. On the other hand, as I understand it these particulars were sent out on a speculative basis to the agents concerned. They were then probably filed away for future reference or thrown in the waste paper basket, as the case may be, and looked at as the occasion demanded. So although I think perhaps it is a little surprising that there was no reaction other than the three possible ones to which I have referred, in my judgment, it is not wholly inconsistent with the plaintiffs' case.

Fourth, the defendants called a Mr Robert Meacock [FRICS FSVA], who was a local agent who had three offices in the area. Mr Kingston said each of those offices would have received the particulars. Mr
E Meacock said that at the material time he had some 40 clients on his books who were interested in this kind of property, and in particular that he had two firms (Ideal and one called Houlihan), both of whom later showed very considerable interest in the property, who at all material times were interested. So it is said that, if Mr Meacock's offices received these particulars, it is most improbable that Mr Meacock, who had these clients available, would not instantly have got on to the plaintiffs to say, "We have some clients who are interested". The judge's assessment of Mr Meacock's evidence is to be found at p25F-G. He said:

> . . . There are reasons why I do not find it possible to accept Mr Meacock as a
> F reliable witness. In my view of him, he was a very partisan witness, always anxious to make every point he could in favour of the defendants. Quite apart from this impression derived from seeing him in the witness box, I note that in a letter of the 24th June 1989, he was offering himself to Mr Goddard to assist any claim which he might have against the plaintiffs.

It seems to me that it is impossible for this court to overcome that assessment by the learned judge. It is true that it is perhaps surprising, if Mr Meacock was a reliable witness, that the particulars did not come to his attention in one of the three offices, but that is the way the judge dealt with it. He was by no means satisfied that they would have come to Mr Meacock's attention personally, and that is the way it was left. The judge's conclusion on this point can be summarised at p27 A-D.

He said: G

> Quite apart from from my impression of him

— ie Mr Kingston —

> I can find no reason in the evidence or in common sense why he should not circulate the particulars to an appropriate number of local agents.

That is a clear finding that on the probabilities Mr Kingston would have done exactly what he said. Mr Ullstein submits that the judge was wrong in that because he failed to appreciate that Mr Kingston's strategy was to target the developers and not local owner/occupiers,
and therefore there was no real need to circulate local agents, and that H
Mr Kingston might be reluctant to share his commission with some other local agent. But it seems to me that the answer to this point was provided by my lord, Kennedy LJ, in argument, namely that it was part of the original strategy by Mr Kingston, even if only as a fall-back position, that the local agents should be circulated and there was no reasonable explanation as to why he should have changed that strategy. It seems to me that it is improbable that he would have changed the strategy, particularly since he was as anxious as anybody to get the best price for the property. In my judgment, even if we thought on the balance of probabilities that Mr Meacock's evidence was right, it would not, in my view, entitle us to reverse the judge's
acceptance of the truth of Mr Kingston's evidence to the contrary. In J
fact, I can see nothing improbable about the fact that Mr Kingston did circulate those local agents. It seems to me that it is likely that he would have done so for the reasons I have given.

The next criticism is that the plaintiffs advised the defendants to accept an offer below the market value. As I have already pointed out, the real case here is not that which was pleaded, but that the plaintiffs (Mr Kingston) ought to have advised the defendants not to accept the offer, whereas in fact Mr Kingston says that he did not give any positive advice; he put the pros and cons of accepting it and left it to the clients to make up their minds.

The judge had a great deal of evidence about the value of the K
property. His conclusion was expressed at p36C in these terms:

> I am satisfied on the evidence that, in the circumstances which governed the marketing of this property in the autumn of 1988, the acceptable range of values was £300 to £350,000.

That is a clear reference to the inhibiting circumstances of the marketing, namely that it had to be done quietly without Mr Little getting wind of it.

Mr Ullstein relies particularly on the evidence of Mr Guy Egerton-Smith [FRICS], who was the expert witness called on behalf of
the plaintiff. At day three p49, there are some answers which L
culminate in a passage in cross-examination of that witness. The judge says to Mr Egerton-Smith:

> Just pause, I would like to get this down. You agree that at £300 [that is, £300,000] the advice should be: "You can't do worse; unless you've got special reasons, go on"?
> A. Exactly.
> Q That's it?
> A. Yes, my lord.

Mr Ullstein fastens on that passage and says: "There is the
plaintiffs' own expert witness saying that that is what ought to have M
happened". But the judge had a good deal of evidence about the value of this property, including evidence from Mr Tapping, who was the agent for Mr Payne who bought the property, and the judge records his evidence at p34F as follows:

> Mr Tapping said in evidence that at the time of the meeting on the 4th January, in his view the value of £300,000 was about right. In giving this evidence, Mr Tapping may, of course, have been justifying his deal for Mr Payne at that figure, but having seen him, I do not think that that is so. I find him to have been a reliable witness.

I am bound to say that, for my part, I would have thought that Mr Tapping might have taken the opposite view that, if he thought it was

A an undervalue, he had done extremely well for his clients by getting substantially less than the market value. However, the judge accepted Mr Tapping as a reliable witness and that is what his estimate was at that time.

There was evidence from a Mr John Gay [FRICS], who inspected the property in July 1989 for the purposes of a building society. The judge said at p35:

> His valuation was £350,000. He accepted that the prime purpose of his valuation was to be satisfied that the property provided good security and that therefore a mortgage valuation would be on the conservative side. But in his view, and his opinion is quite independent of the parties in this case, the
> B mortgage value was not far below the full market value.

That, of course, was six months later on a rising market.

The evidence about Mr Shepherd is also of particular importance. The judge refers to it at p36E:

> . . . Mr Goddard's evidence about his conversation with Mr Shepherd after the offer had been made is, in my judgment, good evidence that the advice given by Mr Kingston was competent advice. It shows that another surveyor practising in Ashford, who knew the local market and who knew the property having inspected it in November 1987 gave similar advice. I do not consider that Mr Kingston ought to have told Mr Goddard that his advice would be not to accept the offer.

C

In my judgment, that evidence in relation to Mr Shepherd is of very great importance for the reason that the judge gives. It seems to me that with valuations at and around that figure, particularly having regard to the instructions given by the client which circumscribed the way in which the property should be marketed, it was not possible to say that Mr Kingston was in breach of his duty in leaving the matter for the client to make up his mind one way or another at a figure of £300,000.

The final point which is raised by Mr Ullstein in his notice of appeal and argument relates to the judge's finding about who was D present at the meeting on January 4. Was it, as Mr Kingston said, a representative from Shepherd Dear & Co or was it Mr Goddard's nephew, Mr Denham? In their reply the plaintiffs' at p71 of the bundle had pleaded this in relation to that meeting:

> Mr Goddard advised Mr Kingston that a firm of Local Estate Agents, namely Shepherd & Dear, had independently advised the Defendants that:
> (A) £300,000 would be a reasonable price for the freehold of the Site in view of the planning difficulties; and
> (B) the Defendants should not "Re-Market" the Site if an offer of £300,000 was forthcoming.

There is no reference in the reply to the presence, at the meeting, of E Mr Shepherd.

On July 10 — ie a few days before the trial started — Mr Kingston's proof of evidence was submitted, in which it was said for the first time that Mr Shepherd or a representative of the firm was present at the meeting. It was put in cross-examination to him that he was wrong about that, that in fact there was no one present at the meeting, but it appears that Mr Goddard thought that the possible explanation may have been that it was his nephew Mr Denham, who was a student at the time and who was working at the firm with his uncle, who may have been present. Mr Denham was called. He said that he recognised Mr Kingston in court, but he had no recollection of F the meeting, still less of what was discussed at it. Finally, when Mr Kingston was asked about this matter again, he accepted, as I understand it, that he might have been mistaken and that the person present might have been somebody like Mr Denham, but at any rate he was told about the information that Mr Shepherd had given.

The judge on this point preferred the evidence in chief of Mr Kingston. It may be that the judge was mistaken on that point. It may be that in fact it was Mr Denham who was present and Mr Goddard related to Mr Kingston at the meeting what he had been told by Mr Shepherd, but, in my judgment, it really makes no difference. Mr Kingston may have been mistaken about that matter. The real point is that Mr Shepherd's advice was substantially the same as that of Mr Kingston, and to my mind it does not matter whether that advice was G given before the meeting or at the meeting. It does not, in my judgment, seriously affect the credibility of Mr Kingston. As I have said, in the original reply it does not appear to have been the case for the plaintiffs that Mr Shepherd was present at the meeting. It may be that Mr Kingston thought that that was so because he had a very clear and detailed recollection of what the advice was, which of course could have come only either from Mr Shepherd, or from a representative of that firm or from Mr Goddard having related it to him, and, in my opinion, that matter, which really was conceded by Mr Ullstein, cannot affect the judge's conclusion.

I have come to the clear conclusion that we cannot disturb the H judge's careful findings of fact here, which were based on ample evidence, that, although Mr Ullstein has pointed to certain matters which affect questions of probability, they do not in any way, in my judgment, undermine the judge's findings of fact, and I would dismiss the appeal.

Kennedy and **Butler-Sloss LJJ** agreed and did not add anything.

Appeal dismissed with costs.

J

Queen's Bench Division

January 14 1994

(Before Mr Justice COLMAN)

McCULLAGH v LANE FOX & PARTNERS LTD

Estates Gazette February 26 1994
[1994] 08 EG 118

K

Estate agents — Whether vendor's estate agent owes duty of care to purchaser in respect of negligent misstatement

On May 11 1990 the plaintiff and his wife saw an advertisement offering for sale a house fronting the River Thames, with gardens of nearly one acre. During a visit to the property at 61 Hartington Road, Chiswick, London W4, on May 12, the plaintiff was given a copy of the sale particulars prepared by the defendants, which referred to the area of the gardens as nearly 1 acre. The plaintiff alleged that this area was confirmed by S, of the defendant firm, during the visit. The plaintiff informed S that he was prepared to exchange contracts on May 14 and would not be relying on a survey report. On May 14 1990 L the plaintiff exchanged contracts to purchase the property for a purchase price of £875,000. On May 24 the plaintiff was informed by his architect that the true area of the gardens was 0.48 acres. The plaintiff claimed that as a result of the oral negligent misstatements of the defendants' representative he had suffered loss and damage, because had he known that the true area of the gardens was just under half an acre either he would not have purchased the property, or he would have made a lower offer. He alleged that the property was worth only £550,000.

***Held*: The plaintiff's claim was dismissed. S, on behalf of the defendants, made an oral misrepresentation of the area of the M gardens and there was a casual link between that misstatement and the plaintiff's alleged loss. The necessary ingredients of a duty of care owed by the defendants to the plaintiff existed: (1) S was aware that the plaintiff was dispensing with a survey and therefore the facts of the area would affect his decision to purchase; (2) accordingly, there existed a neighbour or proximity relationship; (3) there was a professional estate agent knowing that what he said about an important characteristic of the property was material to its value and to a decision to purchase and then knowing that the purchase was to be without independent investigation; (4) it was just and reasonable to impose**

A **a duty of care; and, (5) there was foreseeability that the plaintiff might offer more than the market value of the property. The possible availability of an alternative remedy against the vendor would not make it less just and reasonable that a duty should be imposed. However, on the evidence, the purchase of the property was not at a price in excess of its market value and therefore the plaintiff had suffered no loss.**

The following cases are referred to in this report.

Amalgamated Investment & Property Co Ltd (in liquidation) v *Texas Commerce International Bank Ltd* [1982] QB 84; [1981] 3 WLR 565; [1981] 3 All ER 577; [1981] Com LR 236, CA

B *Bango* v *Holt* (1971) 21 DLR (3d) 66

Caparo Industries plc v *Dickman* [1990] 2 AC 605; [1990] 2 WLR 358; [1990] 1 All ER 568, HL

Computastaff Ltd v *Ingledew Brown Bennison & Garrett* (1983) 268 EG 906

Edgington v *Fitzmaurice* (1885) 29 ChD 459

Gran Gelato Ltd v *Richcliff (Group) Ltd* [1992] Ch 560; [1992] 1 EGLR 297

Harris v *Wyre Forest District Council* [1990] 1 AC 831; [1989] 2 WLR 790; [1989] 2 All ER 514; (1989) 87 LGR 685; [1989] 1 EGLR 169; [1989] 18 EG 89; 21 HLR 424, HL

C *Hedley Byrne & Co Ltd* v *Heller & Partners Ltd* [1964] AC 465; [1963] 3 WLR 101; [1963] 2 All ER 575; [1963] 1 Lloyd's Rep 485, HL

Komarniski v *Marien* (1979) 100 DLR 81

McNaughton (James) Paper Group Ltd v *Hicks Anderson & Co* [1991] 2 QB 113; [1991] 2 WLR 641; [1991] 1 All ER 134, CA

Mutual Life & Citizens' Assurance Co Ltd v *Evatt* [1971] AC 793; [1971] 2 WLR 23; [1971] 1 ALL ER 150; [1970] 2 Lloyd's Rep 441, PC

Richardson v *Norris Smith Real Estate Ltd* [1977] 1 NZLR 152

Smith v *Chadwick* (1884) 9 App Case 187

Smith v *Eric S Bush (a firm)* [1990] 1 AC 831; [1989] 2 WLR 790; [1989] 2 All ER 514; (1989) 87 LGR 685; [1989] 1 EGLR 169; [1989] 17 EG 68 & 18 EG 99, HL

D *Sutherland Shire Council* v *Heyman* (1985) 60 ALR 1; [1985] 59 ALJR 564; 2 Const LJ 150; (1985) 157 CLR 424

This was a claim by the plaintiff, Edward McKim Lyell McCullagh, for damages for negligent misstatement against the defendants, Lane Fox & Partners Ltd.

Philip Havers (instructed by Myshcon de Reya) appeared for the plaintiff; Richard Lynagh (instructed by Cameron Markby Hewitt) represented the defendants.

Giving judgment, COLMAN J said: This is a claim in tort for negligent misstatement by the defendant estate agents. It raises an issue of fundamental importance to estate agents, namely whether the
E vendor's agent owes a duty of care to the purchaser of property in respect of negligent misstatements upon which the purchaser relies in entering into the purchase contract. On this point there appears to be only one previous decision of the English courts.

The plaintiff purchased the residential property at 61 Hartington Road, Chiswick, London W4. Exchange of contracts took place on May 14 1990. The defendants were the vendors' estate agents. Andrew Scott, then an associate director and now a full director of the defendants, was the person who was handling the property and who dealt with the plaintiff. The claim for damages for negligent misstatement is founded on the plaintiff's contention that he was
F induced to enter into the contract by the defendants negligently misrepresenting that the house stood in gardens of 0.91 acres, whereas in reality the area of the entire plot including the space occupied by the house was only 0.48 acres. The plaintiff says that he would never have been prepared to pay what he did for the property — £875,000 — if he had known the truth about the garden area. He contends — in reliance on expert evidence — that the true value of the property with the area accurately described would not have been more than £550,000 and that he has therefore suffered a loss of £325,000, which he claims as damages.

It is admitted by the defendants that if, which they deny, there was a representation that the area of the gardens was 0.91 acres that representation was (i) untrue, and (ii) negligently made, the true area G
being 0.48 acres.

Plaintiff's case

On Friday May 11 1990 the plaintiff and his wife saw in the current issue of *Portrait Magazine* an advertisement by the defendant estate agents of the house at 61 Hartington Road. Alongside a photograph taken of the rear garden looking towards the house was the following description.

> A quite exceptional family house occupying a unique position on the River Thames, offering extensive accommodation, private parking, gardens of nearly one acre and a private 90' river frontage. H
>
> Front and rear gardens of nearly one acre, 90' river frontage with potential private moorings, 12m swimming pool.

The asking price was stated to be £850,000. The plaintiff and his wife, who lived in Kensington, were looking for a house on the Thames and, being particularly attracted by the riverside position and area of 1 acre, they made an appointment with the defendants to view no 61 the following morning, Saturday May 12, at 11.30am.

Mr McCullagh was a relatively wealthy man, the owner of a group of companies, including a substantial engineering company, having its head office in west London. He had been building up his company for most of 30 years and was in every sense an experienced businessman. J
He had experienced the selling and buying of houses on three occasions since 1972.

The plaintiff and his wife and two young children drove over to no 61 and arrived there at 11.30am on the Saturday. There they met Andrew Scott of the defendants. According to the plaintiff, Mr Scott met them and conducted them quickly through the house without showing them over it and out into the gardens at the rear. The plaintiff and his wife were greatly attracted by the gardens, which ran down to the river. There was no intervening tow path. The lower rear gardens contained a swimming pool and were attractively laid out. The upper part of the rear gardens included a most unusual cedar extending K
horizontally across a substantial proportion of the width of the lawn and supported on struts. They found the house unattractive. It was built in the 1920s with late additions to both sides. Almost immediately the plaintiff and his wife came to the conclusion that they would like to acquire the property for the purpose of demolishing the house and building what the plaintiff described as their dream house on the site together with a tennis court.

According to the plaintiff, having walked to the end of the garden with Andrew Scott, they were openly enthusiastic about the property. The plaintiff told Mr Scott that a 1-acre site on the river was unique and this was a golden opportunity to acquire such a site. This was in L
the context of a conversation in which the plaintiff mentioned his company had acquired a half- to three-quarter-acre site on the Isle of Dogs, which they had bought for about £390,000 and was then valued at £5m. He recalled making some flippant remark about commuting to work down the river. When he mentioned that no 61 was a 1-acre site Mr Scott corrected him, saying that it was in fact only 0.91 acres. The plaintiff then told Mr Scott that they would want to demolish the house and replace it and build a tennis court alongside the swimming pool at the lower end of the garden. They spent what the plaintiff said was about half an hour to three-quarters of an hour in the garden. They then went with Mr Scott to look at the house itself, although they M
made it plain to him that they were not really interested in it because they envisaged demolishing it. They found what the plaintiff described as a rather tired 1930s house, in poor condition with insufficient bathrooms.

According to the plaintiff, while they were at the property, Mr Scott told them that an offer had been made by some friends of the owners, Mr and Mrs Bonser, which was near to the asking price. Mr Scott told them that if they wanted the property they would therefore have to move quickly. Mrs McCullagh, the plaintiff's wife, substantially confirmed his account of what happened during their visit on that Saturday morning. In the context of her saying that they would demolish the house, she asked Mr Scott if it was listed and he

A said that it was not. She also said that Mr Scott had stated that the property was 0.91 acres. She also recalled her husband's reference to his company's property on the Isle of Dogs.

The plaintiff and his wife left the house after that visit, determined to acquire it. What made them so enthusiastic was that the site was nearly 1 acre on the Thames. They believed that there was room for them to build their dream house and to put in a tennis court, while retaining the existing swimming pool. Just as they were leaving the house they met Mrs Bonser who asked if the defendants had provided them with the particulars of sale. They had not been given those particulars and she then fetched a copy and passed it over
B to them.

Having discussed the matter with his wife and children, the plaintiff then moved with great determination. That same evening he telephoned Mr Scott at home and said that his family loved the property and wished to buy it. He asked what was the exact position about the existing offer. Mr Scott said that the Bonsers had agreed terms and the purchasers were going round to have tea with them on the following day. Later that evening the plaintiff again telephoned Mr Scott and made an offer of £850,000, the asking price. Mr Scott remarked that he did not want his clients to fall between two stools. The plaintiff felt that the implication of those remarks was that, in
C view of the fact that the vendors had already accepted an offer, he would have to be prepared to exchange contracts very quickly. For that reason he thereupon said that he was prepare to exchange contracts on the Monday.

On the Sunday morning the plaintiff contacted the vendors direct and asked if he could again visit the house. He and his family spent another half-hour there and in the course of that visit, feeling that he wanted to be sure of beating off competitive bids from those whose offer the vendors had already accepted, which he believed to be near the asking price, he then increased his offer to £875,000. There was a further conversation with Mr Scott on that Sunday evening in which the plaintiff repeated his offer of £875,000 and confirmed that he was
D prepared to exchange contracts the next day. In the event the vendors accepted that offer and contracts were exchanged on Monday May 14 1990.

It was the plaintiff's evidence that what attracted him and his family to the property was the riverside location and the area of the ground, as confirmed by Andrew Scott, of 0.91 acres, which the plaintiff believed to be: (i) almost unique on the river; and (ii) spacious enough to contain his dream home, a tennis court and the existing swimming pool. He believed that a tennis court would fit between the swimming pool and boundary, but he did not know the dimension of a court or the dimensions of that space. At the time when the offer was
E made he did not know which area would be occupied by his dream home.

The particulars of sale provided to the plaintiff in the course of the Saturday visit included the following:

An exceptional family house occupying a unique position on the River Thames, with gardens of nearly one acre and a 90' river frontage
Front and rear gardens of nearly 1 acre.

Description

The property, approached by a 90' driveway, is a detached freehold house built in 1931 and set in 0.92 of an acre of its own private gardens. The location on the River Thames offers a large river frontage which could easily provide
F extensive private moorings.

At the end of the particulars of sale there appear in small but legible print the following provisions:

1. These particulars do not constitute, nor constitute any part of, an offer or contract.

2. All statements contained in these particulars as to this property are made without responsibility on the part of Lane Fox or the vendors or lessors.

3. None of the statements contained in these particulars as to this property are to be relied on as statements of representations of fact.

4. Any intending purchasers must satisfy themselves by inspection or otherwise as to the correctness of each of the statements contained in these particulars.

5. The vendors do not make or give and neither Lane Fox nor any person G in their employment has any authority to make or give any representation or warranty whatever in relation to this property.

The plaintiff's evidence was that because by the time he left the property on the Saturday morning they envisaged demolishing the house, he did not do more than glance at the photographs and contents of the particulars of sale, but, having some experience of moving house, he knew that estate agents' particulars invariably include disclaimers of liability of the kind in the defendants' particulars.

In the course of his evidence the plaintiff said that if, having decided to acquire the property but before exchanging contracts, he had been told that the area was, after all, not 0.91 of an acre but about H half an acre, he would have reduced his offer, but more probably not continued with the purchase.

Defendants' case

Mr Scott said in evidence that an appointment had been made on Friday May 11 for the McCullaghs to view the property the following morning, he subsequently tried to cancel the appointment because an offer of £810,000 made by a Mr and Mrs Land had already been accepted by the vendors. Unfortunately, he used the wrong telephone number and left a message for the McCullaghs on somebody else's answering machine. In the event he was most surprised when they J arrived at no 61 the next day at 11.30am. He was just leaving. He had an appointment with another client in Chelsea at 12.30 and he therefore aimed to take the McCullaghs over the property as quickly as possible. In his mind he had already accomplished a sale the previous day.

He said that he showed them over the house before they went into the rear garden. They spent about 10 minutes in the house. They were "quite positive" in their response to the house and "generally enthusiastic".

After looking round the house he showed them out across the terrace into the rear garden. They went off down the garden, but he did K not accompany them. He heard the plaintiff say "we could fit a tennis court into there". They spent about five minutes in the rear garden while he stayed on or near the terrace by the house, part of this time in conversation with Mrs Bonser. He remembered Mrs Bonser giving them a copy of the particulars of sale on their return from the garden. They then left. They were at the property for 20 minutes at the most. There was no mention of any plans for demolishing the house. There was no reference to the plaintiff's company having a site on the Isle of Dogs. There was no discussion of the area of the property and in particular he had no recollection of saying that its area was 0.91 acres. He had no conversation with Mrs McCullagh. If he had mentioned the L area at all, he would have given it as 0.92 acres, which was the area that he personally had calculated (unknown to him inaccurately) in 1989 and which was reproduced in the particulars of sale.

Mr Scott said that the earliest references to the prospect of the house being demolished was made by the plaintiff in the course of a telephone conversation which took place on the morning of Monday May 14, when Mr Scott told the plaintiff for the first time that his offer was accepted by the vendors. On that occasion the plaintiff said to him: "We're not quite sure what we're going to do with the place. We might even knock it down."

Was there an oral misrepresentation by Mr Scott? M

In considering what was said by Mr Scott to the plaintiff in the course of the visit, I have considered not only the evidence and demeanour of Mr Scott, the plaintiff and Mrs McCullagh, but also the contents of the transcripts of two telephone conversations between the plaintiff and Mr Scott which the plaintiff recorded on May 24 1990, the day the plaintiff had been told by his architect that the site was only 0.48 acres, as well as the contents of a very detailed letter written by the plaintiff to Mr Scott dated June 5 1990, setting out the events from May 10 onwards and the contents of a further transcript of a telephone conversation on June 15.

I accept the evidence of the plaintiff and Mrs McCullagh as to Mr

A Scott's confirmation of the area as being less than an acre. I consider it more likely that he said in the course of their Saturday visit that the true area was 0.92 acres and that Mr McCullagh wrongly recollected the figure as 0.91 acres when he discussed the error with Mr Scott on May 26. That is not to say that I find that Mr Scott has knowingly given untruthful evidence. I take the view that he has simply forgotten the exact circumstances and details of the conversations that he had with the plaintiff and his wife in the course of their visit to no 61. The recollections of the plaintiff and his wife are, in my view, more reliable. They over-estimated the amount of time they spent in the garden, but they have certainly not, in my judgment, concocted their B evidence that Mr Scott said that the area was 0.91 acres, although 0.92 was probably what he said.

The plaintiff's reliance on the misrepresentation

If the plaintiff is to recover damages for negligent misrepresentation he must establish that the misrepresentation caused his damage, that is to say in the present case he must show that the misrepresentation caused him to enter into the contract to purchase no 61. In the context of a claim in respect of fraudulent misrepresentation it was stated by Lord Selborne LC in *Smith* v *Chadwick* (1884) 9 App Case 187 at p190 that, in order to make good a claim in deceit, the plaintiff must establish that the "fraud was an inducing cause to the C contract; for which purpose it must be material, and it must have produced in his mind an erroneous belief influencing his conduct". It is, however, clearly unnecessary for the purposes of the tort of deceit that the misrepresentation should be the only consideration which caused the representee to act as he did. It is enough if his judgment was *influenced* by the misrepresentation in addition to other considerations. This is clear from the judgments of the Court of Appeal in *Edgington* v *Fitzmaurice* (1885) 29 ChD 459, the famous case of the misrepresentation in a company prospectus inviting subscriptions for debentures of the purpose to which it was intended that the company would put the money so subscribed. There the D representee mistakenly believed that the prospectus offered a charge on the property of the company and, but for that belief, would not have subscribed for the debentures, but he was also shown to have relied on the misrepresentation as to the purpose to which the money was to be put. His claim succeeded before Denman J, whose judgment was affirmed by the Court of Appeal. Cotton LJ stated at p480:

> But it was urged by the counsel for the appellants that the plaintiff himself stated that he would not have taken the debentures unless he had thought they were a charge upon the property, and that it was this mistaken notion which really induced the plaintiff to advance his money. In my opinion this argument does not assist the defendants if the plaintiff really acted on the statement in E the prospectus. It is true that if he had not supposed he would have a charge he would not have taken the debentures; but if he also relied on the misstatement in the prospectus, his loss none the less resulted from that misstatement. It is not necessary to show that the misstatement was the sole cause of his acting as he did. If he acted on that misstatement, though he was also influenced by an erroneous supposition, the defendants will be still liable.

Then at p483 Bowen LJ analysed the position thus:

> But such misstatement was material if it was actively present to his mind when he decided to advance his money. The real question is, what was the state of the plaintiff's mind, and if his mind was disturbed by the misstatement of the defendants, and such disturbance was in part the cause of what he did, F the mere fact of his also making a mistake himself could make no difference.

At p485 Fry LJ said:

> . . . in my opinion if the false statement of fact actually influenced the plaintiff, the defendants are liable, even though the plaintiff may have been also influenced by other motives.

In my judgment, the same analysis of the required causal linkage between the misrepresentation and the plaintiff representee's loss must apply equally in cases of negligent misrepresentation like the present case. The same test of causation has been applied in relation to the question whether a representation gives rise to an estoppel: see *Amalgamated Investment & Property Co Ltd (in liquidation)* v *Texas* G *Commerce International Bank Ltd* [1982] QB 84 *per* Robert Goff J at pp104-105. There is, therefore, good reason in the interests of consistency why the test in cases of negligent misrepresentation should be the same.

Whereas it is unquestionably true that the plaintiff's decision to purchase the property was influenced by his having inspected the site, seen the layout and its relationship to the river, as well as the immediate environment of Hartington Road, his decision was also influenced by the statement in the advertisement in *Portrait Magazine* that the area was nearly an acre and further by the oral statement, which I have found to have been made by Andrew Scott, that the true H area was 0.92 of an acre. I accept the plaintiff's evidence that he would have made no offer at the level at which he did, if he had discovered on that Saturday morning that the true area was only about half an acre. He had a certain notion of the desirability of the very special combination of the approximately 1-acre area and the river frontage. If the area were substantially less than he had been told, he would not have purchased merely on the appearance of the site. Whereas his understanding of the area originated with the advertisement, that belief was clearly confirmed and fortified by Andrew Scott's oral statement of the precise measurement. It follows that the oral misrepresentation thus made was material and influenced J the plaintiff's decision to purchase the property. Accordingly, the plaintiff has established the necessary causal connection between the oral misrepresentation and the decision to purchase.

Did the defendants owe a duty of care to the plaintiff?

Mr Philip Havers, on behalf of the plaintiff, contends that an estate agent does owe a duty of care to a potential purchaser notwithstanding that he is the agent exclusively employed by the vendor. He relies in particular on a passage in the speech of Lord Bridge in *Caparo Industries plc* v *Dickman* [1990] 2 AC 605 in which, having reviewed the leading authorities, including *Smith* v *Eric S Bush (a firm)* [1990] 1 AC 831*, he identified the essential factual basis for liability in tort for K negligent misstatement in the following words [at p629]:

> The salient features of all these cases is that the defendant giving advice or information was fully aware of the nature of the transaction which the plaintiff had in contemplation, knew that the advice or information would be communicated to him directly or indirectly and knew that it was very likely that the plaintiff would rely on that advice or information in deciding whether or not to engage in the transaction in contemplation. In these circumstances the defendant could clearly be expected, subject always to the effect of any disclaimer of responsibility, specifically to anticipate that the plaintiff would rely on the advice or information given by the defendant for the very purpose for which he did in the event rely on it. So also the plaintiff, subject again to the effect of any disclaimer, would in that situation reasonably suppose that he L was entitled to rely on the advice or information communicated to him for the very purpose for which he required it.

He further relies on the passage from Lord Oliver's speech at p638:

> What can be deduced from the *Hedley Byrne* case, therefore, is that the necessary relationship between the maker of a statement or giver of advice ("the adviser") and the recipient who acts in reliance upon it ("the advisee") may typically be held to exist where (1) the advice is required for a purpose, whether particularly specified or generally described, which is made known, either actually or inferentially, to the adviser at the time when the advice is given; (2) the adviser knows, either actually or inferentially, that his advice will be communicated to the advisee, either specifically or as a member of an M ascertainable class, in order that it should be used by the advisee for that purpose; (3) it is known either actually or inferentially, that the advice so communicated is likely to be acted upon by the advisee for that purpose without independent enquiry, and (4) it is so acted upon by the advisee to his detriment.

The only English authority on the point is the decision of McNeill J in *Computastaff Ltd* v *Ingledew Brown Bennison & Garrett* (1983) 268 EG 906. The material issue in that case was whether, for the purposes of the Civil Liability (Contribution) Act 1978, the third party estate agents acting for the lessor would, if sued, have been liable to

*Editor's note: Also reported at [1989] 1 EGLR 169.

A the plaintiff tenants by reason of the third party having negligently misrepresented to the plaintiff's agent the rateable value of the property. Such liability could arise only under the tort of negligent misstatement. In the course of his judgment McNeill J reached the conclusion that the third party estate agent would, if sued, have been liable in negligence. His reasoning was as follows at p911:

> It is, I think, without setting out in any citation the *rationes decidendi* of *Hedley Byrne*, correct to say that it must be shown that the plaintiff was reasonable in seeking the information from the third party and, if so reasonable, reasonable in relying upon it. So much, I think, can be summarised from the speech of Lord Reid in the report of *Hedley Byrne & Co Ltd* v *Heller* B *& Partners Ltd* [1964] AC 465 at p486, and the speech of Lord Morris of Borth-y-Gest at p502 and of Lord Hodson at p514.
>
> This is a case in which lines of inquiry were reasonable down both channels, but neither was of itself sufficient: neither of itself, so far as the solicitors were concerned, discharged their obligation, for reasons that I have already said: that is to say, in going back to Hewitt, Woollacott & Chown to question their figure. It was also reasonable to inquire down the estate agents' channel. The fault there, as I have indicated, is in the posing and answering of the questions down that channel. But as between plaintiff and third party, it is to my mind reasonable for the plaintiff to inquire, as one of the sources of information, of the third party and reasonable to rely upon the information that is given.
>
> C It may be, at the end of the day, that the information achieved from source down both channels differs, in which case the professional advisers have responsibilities. The plaintiff is entitled reasonably to investigate down both channels and to rely, for what it is worth, on the information which is provided. In my view, the third party, if sued by the plaintiff, could have been held liable under what I have, for convenience only, called the *Hedley Byrne* line of authority.

Having regard to the very substantial developments in the law of negligent misstatement since that case was decided it is right that I should consider afresh the existence of a duty of care and, in particular, whether those developments point towards the imposition D of such a duty on a vendor's estate agent.

The substance of the passages in *Hedley Byrne* v *Heller* relied upon by McNeill J is sufficiently expressed in the speech of Lord Reid at p486:

> A reasonable man, knowing that he was being trusted or that his skill and judgment were being relied on, would, I think, have three courses open to him. He could keep silent or decline to answer with a clear qualification that he accepted no responsibility for it or that it was given without that reflection or enquiry which a careful answer would require: or he could simply answer without any such qualification. If he chooses to adopt the last course he must, I think, be held to have accepted some responsibility for his answer being E given carefully, or to have accepted a relationship with the enquirer which requires him to exercise such care as the circumstances require.

Mr Havers has also helpfully drawn my attention to three decisions of Commonwealth courts at first instance, which had also concluded that an estate agent does owe a duty of care to a potential purchaser, notably of the British Columbia Supreme Court in *Bango* v *Holt* (1971) 21 DLR(3d) 66, a decision applying *Hedley Byrne* v *Heller*; of the New Zealand Supreme Court in *Richardson* v *Norris Smith Real Estate Ltd* [1977] 1 NZLR 152, applying the decision of the Privy Council in *Mutual Life & Citizens' Assurance Co Ltd* v *Evatt* [1971] AC 793, and of the Saskatchewan Court of Queen's Bench in F *Komarniski* v *Marien* (1979) 100 DLR 81.

Mr Havers contends that there exists here a clear factual basis for a duty of care inasmuch as:

(i) the defendants held themselves out as a reputable firm of competent surveyors and estate agents who could be expected to have accurately and carefully measured any site of which they stated the dimensions;

(ii) the purpose of the provision of information as to the site by Andrew Scott was to encourage potential purchasers such as the plaintiff to offer to purchase the property;

(iii) it was obvious to Andrew Scott that the plaintiff would rely on his statement as to the area when deciding whether to buy the G property;

(iv) the information given by Andrew Scott that the site had a particular area was presented as having been carefully considered;

(v) accordingly, the plaintiff was reasonable in seeking the information from the defendant and reasonable in acting upon it.

On the evidence in this case the plaintiff has, in my judgment, established each of (i) to (v) above. As to (iii) Mr Richard Lynagh [for the defendants] contended that on Mr Scott's evidence in his experience 95% of purchasers of property had their own survey conducted before exchanging contracts and that accordingly there was no "overwhelming probability" or "high degree of probability" (*per* H Lord Griffiths in *Smith* v *Bush (supra)*, at p864) that the plaintiff would conduct a site survey to satisfy himself of the area. I cannot accept this submission. Whatever might have been the position on the Saturday on the first viewing when Andrew Scott first misrepresented the area, by the time of the telephone conversation that very evening it was apparent to him that the plaintiff was prepared to exchange contracts on the Monday without a survey. It must have been perfectly obvious that there would not be time for such a survey and that in the absence of any request from the plaintiff for access to a surveyor, none was intended. There was, therefore, a continuing representation in circumstances, where to the knowledge of the representor, it was obviously being relied upon as true by the plaintiff and, unless J corrected prior to exchange of contracts, which it was not, it would obviously be one of the matters in reliance upon which the contract would be entered into.

But do these facts, as Mr Havers contends, give rise to a duty of care and, if loss be sustained in consequence of acting in reliance on the misrepresentation, liability in damages for negligence?

The law of negligent misstatement, no less than the general law of negligence, as it has now developed, requires that three elements should be present before liability can arise, namely foreseeability of damage, proximity or neighbourhood and that the situation is one where it is fair, just and reasonable that the law should impose a duty K of the scope contended for upon one party for the benefit of the other. Thus in *Caparo Industries* v *Dickman (supra)* Lord Bridge analysed the requirements at pp617-618:

> What emerges is that, in addition to the foreseeability of damage, necessary ingredients in any situation giving rise to a duty of care are that there should exist between the party owing the duty and the party to whom it is owed a relationship characterised by the law as one of "proximity" or "neighbourhood" and that the situation should be one in which the Court considers it fair, just and reasonable that the law should impose a duty of a given scope upon the one party for the benefit of the other. But it is implicit in the passages referred to that the concepts of proximity and fairness embodied L in these additional ingredients are not susceptible of any such precise definition as would be necessary to give them utility as practical tests, but amount in effect to little more than convenient labels to attach to the features of different specific situations which, on a detailed examination of all the circumstances, the law recognises pragmatically as giving rise to a duty of care of a given scope. Whilst recognising, of course, the importance of the underlying general principles common to the whole field of negligence, I think the law has now moved in the direction of attaching greater significance to the more traditional categorisation of distinct and recognisable situations as guides to the existence, the scope and the limits of the varied duties of care which the law imposes.

I do not read that passage from the speech of Lord Bridge as M suggesting that in deciding whether in any particular factual situation a duty of care to avoid making negligent misrepresentations was imposed, it was necessary only to consider foreseeability of loss and proximity without considering whether it would be just and reasonable that the law should impose such a duty. On the contrary, the last sentence of the passage cited above strongly implies that when it is necessary to ask whether there is to be a duty of care in a hitherto *unexplored* factual situation the presence of all three ingredients must be investigated. Certainly in *Smith* v *Bush (supra)* Lord Griffiths at pp864-865 applied all three ingredients to test the existence of a duty of care on the building society and local authority surveyors in the

A cases before him; and Lord Keith and Lord Brandon both agreed with his speech. I do not read the omission of both Lord Templeman and Lord Jauncey expressly to refer to all three ingredients as indicating any difference in approach in principle on their part from that of Lord Griffiths. Accordingly, I reject the submission advanced by Mr Havers, on behalf of the plaintiff, that in cases of negligent misstatement the third ingredient that the imposition should be fair and reasonable is unnecessary.

In my judgment, where the court has to consider whether a duty of care not to cause loss by negligent misstatement exists in the context of a relationship the facts of which have not previously been analysed
B by the courts for this purpose, the three ingredients of foreseeability of loss, proximity and the justice and reasonableness of the imposition of the duty are indeed necessary. However, the closer the factual situation of the relationship in question to that found in other relationships which the courts have held to give rise to a duty of care, the readier will be the courts to recognise in a case where the requirements of foreseeability of loss and proximity have been established, that it is just and reasonable that a duty should be imposed. It is in this way that the law develops incrementally, as envisaged by Brennan J in the frequently cited passage from his judgment in *Sutherland Shire Council* v *Heyman* (1985) 157 CLR 424
C at p481.

On behalf of the plaintiff, Mr Havers relies heavily on the decision of the House of Lords in *Smith* v *Bush* and *Harris* v *Wyre Forest District Council* [1990] 1 AC 831*. He argues that, if the building and local authority valuers were held liable to the mortgagor/purchaser in those cases, so also must the estate agents in this case. In both cases the representor was the servant or independent adviser of the other contracting party; the mortgagee in that case and the vendors in this case. In both cases a representation as to the characteristics of the property was made which it was highly probable could be relied upon by the particular representee in deciding whether to enter into the particular transaction with the employer or principal of the representor,
D which must have been envisaged by the representor. Moreover, just as in the *Smith* v *Bush* cases, the representee paid the employer/principal a fee to cover the cost of the survey of the property, so in this case the purchaser's payment of the purchase price to the vendor's solicitors would partly be utilised for the purpose of payment of the estate agents' commission on the transaction.

I am unable to accept that there is any feature of the present case analogous to the payments by the representees made in the *Smith* v *Bush* cases by way of valuation fee (Mr and Mrs Harris) or survey fee (Mrs Smith). Those payments were made pursuant to transactions between the representee and the local authority mortgagee and the
E building society mortgagee respectively, which were free-standing engagements whereby those bodies would entertain the mortgage applications and in the course of so doing have the properties surveyed. The fees were not paid pursuant to or in performance of the transaction induced by the representation, but pursuant to separate transactions entered into in order to obtain information on the very matter which became the subject of the representation. In the present case the vendors engaged the estate agents to market the property on their behalf in consideration of a commission payable by reference to the sale price. The purchaser paid nothing in addition to the consideration payable to the vendors under the transaction induced by
F the representation. In particular there was no separate transaction to which the representee was a party under which the vendors would employ the estate agents to provide information which was then passed to the representee. The subsequent utilisation of part of the consideration to discharge the vendor's obligation to the estate agents in respect of commission does not create a factual situation materially similar to that in the *Smith* v *Bush* cases.

Accordingly, at least one of the factual consideration, which led the House of Lords to conclude in these cases that there was sufficient proximity and that it was just and reasonable to impose a duty of care on the surveyors is absent from the present case. In this connection it G
is helpful to consider the following analysis of the ingredients of proximity and what was just and reasonable on the facts in the *Smith* v *Bush* cases as identified by Lord Griffiths in a passage from his speech at pp864-865.

> I therefore return to the question in what circumstances should the law deem those who give advice to have assumed responsibility to the person who acts upon the advice or, in other words, in what circumstances should a duty of care be owed by the adviser to those who act upon the advice? I would answer — only if it is foreseeable that if the advice is negligent the recipient is likely to suffer damage, that there is a sufficiently proximate relationship between the parties and that it is just and reasonable to impose liability. In the case of a H
> surveyor valuing a small house for a building society or local authority the application of these three criteria leads to the conclusion that he owes a duty of care to the purchaser. If the valuation is negligent and is relied upon damage in the form of economic loss to the purchaser is obviously foreseeable. The necessary proximity arises from the surveyor's knowledge that the overwhelming probability is that the purchaser will rely upon his valuation, the evidence was that surveyors knew that approximately 90% of purchasers did so, and the fact that the surveyor only obtains the work because the purchaser is willing to pay his fee. It is just and reasonable that the duty should be imposed for advice given in a professional as opposed to a social context and liability for breach of the duty will be limited both as to its extent and amount. The extent of the liability is limited to the purchaser of the house — I J
> would not extend it to subsequent purchasers. The amount of the liability cannot be very great because it relates to a modest house. There is no question here of creating liability of indeterminate amount to an indeterminate class. I would certainly wish to stress that in cases where the advice has not been given for the specific purpose of the recipient acting upon it, it should only be in cases where the adviser knows that there is a high degree of probability that some other identifiable person will act upon the advice that a duty of care should be imposed. It would impose an intolerable burden upon those who give advice in a professional or commercial context if they were to owe a duty not only to those to whom they give the advice, but to any other person who might choose to act upon it.

Here the ingredient of proximity and the ingredient of the K
imposition of the duty being just and reasonable must rest primarily, if not exclusively, on the purpose for which the representation was made, the fact that it was made in a professional capacity to a particular potential purchaser in relation to a particular and identifiable transaction and the fact that to the knowledge of the representor it was highly probable that it would be relied upon by that representee in entering into that transaction without making his own investigation.

A further factor of considerable weight which is, in my judgment, material to the question of whether it is in the present case fair and reasonable to impose a duty of care is the circumstance present in this case of the plaintiff proceeding to exchange contracts for a property L
for which the consideration was £875,000, no small amount, without independent investigation of the facts represented by the estate agents. Mr Scott's evidence, which I accept, was that in his experience in 95% of purchases of expensive property the purchaser procured an independent survey before exchange of contracts. The course adopted by the plaintiff of exchanging contracts before survey, although not unknown, was therefore unusual. This was not, therefore, a case in which the representee proceeded to finalise the transaction, like most representees in a similar commercial setting, without an independent survey. He did so in the knowledge of clause 13 of the contract which provided: M

> 13. The Purchaser hereby acknowledges that the Purchaser has not entered into this Contract in reliance wholly or partly upon any statement or representation whether made orally or in writing or otherwise howsoever by or on behalf of the Vendor save for the written replies of the Vendor's Solicitors to enquiries made by the Purchaser's Solicitors.

The plaintiff therefore knowingly took the risk that if the estate agents had miscalculated the area he would have no recourse against the vendors. It was therefore his decision to take that risk which put the defendant in the relatively unusual position of being relied upon as the party solely responsible for the accuracy of the information upon the basis of which the purchaser chose to commit himself to the

*Editor's note: Also reported at [1989] 1 EGLR 169.

contract. Even leaving aside, as I must for the purpose of this analysis, the existence of the defendants' disclaimer in their particulars of sale, the plaintiff chose to adopt a course which as a matter of common prudence the great majority of purchasers would not have adopted.

In this connection it is right to have regard to a passage in the judgment of Neill LJ in *James McNaughton Paper Group Ltd* v *Hicks Anderson & Co* [1991] 2 QB 113 at p126 where, having listed five matters which were likely to be of importance in most cases, in reaching a decision as to whether a duty existed, he added the following sixth matter:

> (6) *Reliance by the advisee*
> In cases where the existence of a duty of care is in issue it is always useful to examine the matter from the point of view of the plaintiff. As I have ventured to say elsewhere the question "Who is my neighbour?" prompts the response "Consider first those who would consider you to be their neighbour." One should therefore consider whether and to what extent the advisee was entitled to rely on the statement to take the action that he did take. It is also necessary to consider whether he did in fact rely on the statement, whether he did use or should have used his own judgment and whether he did seek or should have sought independent advice. In business transactions conducted at arms' length it may sometimes be difficult for an advisee to prove that he was entitled to act on a statement without taking any independent advice or to prove that the adviser knew, actually or inferentially, that he would act without taking such advice.

Whereas Mr Scott certainly knew prior to exchange of contacts that there would be no independent survey, it is, in my judgment, material to what is just and reasonable that it was the plaintiff's decision to adopt this unusual course, which invested the representation by Mr Scott with a causal effect which it would not normally have had.

Although the decision of the plaintiff to dispense with a survey prior to exchanging contracts was unusual, it was a course known to Mr Scott. His statement confirming the area of the plot then became a matter which it would be obvious to anybody would be one of the facts which, without further investigation, a purchaser would have in mind so that it influenced his decision when he entered into the contract. It is at the time when Mr Scott appreciated that there was to be no independent survey that the question of duty and, in particular proximity and what is just and reasonable, has to be tested. Having full regard to the approach of Neill LJ in the passage from his judgment which I have cited above, I have come to the conclusion that at that time there did exist the neighbour or proximity relationship between Mr Scott and the plaintiff, which is a necessary ingredient of duty of care. Here was a professional estate agent knowing that what he had said about an important characteristic of the property was material to its value and to a decision to purchase and then knowing that the purchase was to be without independent investigation. In my judgment, that in the circumstances of this case was sufficient to create proximity.

Mr Lynagh, in the course of his submissions for the defendants, placed great weight on the recent decision of Sir Donald Nicholls V-C in *Gran Gelato Ltd* v *Richcliff (Group) Ltd* [1992] Ch 560*. In that case the defendant solicitors had acted for the vendors of an underlease on the sale of an underlease to the plaintiff and had made a negligent misrepresentation in response to inquiries by the plaintiff's solicitors about the superior leases. They failed to disclose the break clause in the headlease. The plaintiff, being unable to sell the underlease due to the presence of the break clause, sued the defendant solicitors for negligent misrepresentation at common law and under the Misrepresentation Act 1967. Sir Donald Nicholls V-C concluded at pp569-571 that the defendant solicitors owed no duty of care on the following grounds.

(i) The purchaser received the solicitor's representations as made on behalf of the *vendor* and it was not fair and reasonable that the representee purchaser should have a remedy against the vendor's solicitor as well as against the vendor;

(ii) The vendor would be liable in damages for negligent misrepresentations made by his solicitor in the course of his employment and it was not fair and reasonable that the solicitor should also be liable to the purchaser. In *Smith* v *Bush* the mortgagors would not have been able to sue anyone if they had had no remedy against the valuers personally or their employers.

(iii) There was no good reason why in addition to the duty which the solicitor/agent owed to his client/principal he should also owe a duty to the third party who had already had a perfectly viable alternative remedial course. At p571 he put his approach in these words:

> . . . it does seem to me that in the field of negligent misrepresentation caution should be exercised before the law takes the step of concluding, in any particular context, that an agent acting within the scope of his authority on behalf of a known principal, himself owes to third parties a duty of care independent of the duty of care he owes to his principal. There will be cases where it is fair, just and reasonable that there should be such a duty. But, in general, in a case where the principal himself owes a duty of care to the third party, the existence of a further duty of care, owed by the agent to the third party, is not necessary for the reasonable protection of the latter. Good reason, therefore, should exist before the law imposes a duty when the agent already owes to his principal a duty which covers the same ground and the principal is responsible to the third party for his agent's shortcomings. I do not think there is good reason for such a duty in normal conveyancing transactions.

There is no doubt that, in general, an estate agent when selling a property on behalf of his client acts in the course of that authority when he makes a representation about some feature of the property in question. If that representation is inaccurate due to his negligence, then, apart from any exclusion clause in the sale contract, the vendor will be liable for damages for such misrepresentation. The vendor would then have a right of indemnity against the agent. If, in such a case, the estate agent owed an independent duty of care to the third party the latter would have a duplicate remedy to that which the law already gave him.

Whereas I accept that the duplication of remedy may in some cases be a consideration material to the question whether it is just and reasonable to impose on an agent a duty of care for negligent misrepresentation, I am unable to accept that in general the consequence that there will be a duplication of remedy should point away from the imposition of a duty of care on the agent.

In the first place, the English courts have not hitherto regarded the existence of a remedy against the employer or principal as a reason for excepting the negligent employee or agent from liability in tort in cases not involving negligent misstatements. Indeed, in cases outside that field, the whole doctrine of vicarious liability depends upon proof that the servant or agent has committed an *independent* tort in the course of his employment. No doubt in most cases the injured third party sues the employer or principal because the latter has greater substance than the servant or agent, but there may be many commercial situations, such as that of the sale of an expensive house or of commercial property, where the party of substance will be the agent and not the principal. To introduce a principle into the laws of negligent misstatement which inhibited the imposition on a servant or agent of a duty of care *because* the existence of a duplicate remedy against the employer or principal did not render it just and reasonable that such a duty should be imposed would be to introduce into the law of negligent misstatement an anomalous policy consideration not reflected in any other branch of the law of tort. If an injured third party can recover damages for deceit from an employee acting in the course of his employment who makes a fraudulent misrepresentation which induces a contract with his employer, why as a matter of policy should not an injured third party be entitled to recover damages for negligence from a negligent employee or agent acting in the course of his employment who makes a negligent misrepresentation which induces a contract?

Second, in my judgment, there are very strong policy considerations why the law should provide a remedy directly against a party whose personal act or default has brought about the injury which is the basis of the third party's remedy. If it is the fault of the estate

*Editor's note: Also reported at [1992] 1 EGLR 297.

agent that a misrepresentation has been made, what consideration of policy should confine the third party to a remedy against the perfectly innocent principal? There is no good reason why he should be involved in litigation in order to attach to the party at fault a remedial route which could sensibly and effectively and consistently with legal principle in other areas of the law of negligence and of tort in general be attached directly to him.

For these reasons, with great respect to the more general remarks by Sir Donald Nicholls V-C relating to negligent misstatement and without wishing in any way to question the conclusion reached by him in relation to a solicitor acting for his client in a conveyancing matter, I am not persuaded that even if there normally would exist or did exist in the present case an alternative remedy against the vendors, the availability of that remedy would make it any less just and reasonable that a duty of care should be imposed on the defendant estate agents.

Having regard to the facts of this case, including the circumstances in which the information was originally imparted by Mr Scott to the plaintiff, and the circumstances in which it became known to Mr Scott that the purchase was to be concluded without a survey, I have come to the conclusion that it would indeed be just and reasonable to impose a duty of care on the defendants. I do not consider that the plaintiff's determination to exchange contracts without a survey, unusual though it was, renders it in any way unfair to impose on the defendants, who were fully aware of this course, a duty of care in respect of the accuracy of what they had told him about the area of the property.

Having regard to what I have said about the duplication of remedy argument, it is therefore unnecessary to consider whether clause 5 of the conditions of sale by expressly removing from the ambit of the agent's authority any representation relating to the property would have made it more just and reasonable that a duty of care should be imposed on the agents. However, had I taken another view on the duplication of remedy argument, I would have held that, in as much as this clause was sufficiently brought to the attention of the plaintiff so as to shut him out from any remedy against the vendors, it would have been just and reasonable to impose a duty of care on the defendants.

Thus far I have considered the ingredients of proximity and whether it would be just and reasonable that a duty of care should be imposed. But before there can be a duty of care there must be present all three ingredients, including foreseeability of loss. In the present case, it was certainly reasonably foreseeable to Andrew Scott that the plaintiff would rely on his representation as one of the matters that induced him to enter into the contract. Loss would be sustained only in consequence of reliance if the plaintiff were induced to purchase the property for more than its true value. If all that was reasonably foreseeable was that in reliance on the representation the plaintiff might purchase at no more than the true value, the necessary ingredient of foreseeability of loss would not be established. In deciding whether the foreseeability of loss ingredient has been established in cases of negligent misstatement, the correct approach is, in my view, to inquire whether the representor ought to have foreseen that the representee would be induced to act in reliance on the representation by entering into the transaction in question and that one consequence of his so doing might well be that he sustained economic loss. It is enough that it is reasonably foreseeable that the representee would act on the representation and that if he were so to act there would be a real and substantial risk of his suffering economic loss. Tested in this way, there was, in my judgment, reasonable foreseeability at the time when the representation was made that the plaintiff might thereafter offer more than the true value and thereby suffer loss because he relied upon the area of the site being greater than it truly was and the property being more valuable than in truth it was.

Accordingly, I conclude that the defendants did owe a duty of care to the plaintiff to avoid making the negligent misstatement which Mr Scott did indeed make and that they were therefore in breach of that duty by his making it.

I am unable to accept Mr Lynagh's submission that the disclaimer of liability in the particulars of sale protects the defendants in the circumstances of an oral misrepresentation made by Mr Scott. On the proper construction of clauses 1-4 of that disclaimer, they relate exclusively to the contents of the particulars of sale. It is therefore unnecessary to consider the application of the Unfair Contract Terms Act 1977.

Plaintiff's loss

The plaintiff contends that the true value of no 61 with its half-acre site and in the tired condition in which it was did not exceed £550,000. He relied on the evidence of Mrs Land whose offer of £810,000 had been accepted by the vendors on the day before the plaintiff and his family visited the property on May 12 1990. He also relied on the expert evidence of Mr Anthony Copping Joyce, an estate agent of 30 years' experience and a Fellow of The Royal Institution of Chartered Surveyors, for which body he was national spokesman on the housing market.

Mrs Land's recollection of the sequence of two offers which she and her husband made to the vendor in the week before May 14 1990 differed from that of Andrew Scott, but the differences are immaterial. Her evidence was that she and her husband first offered £775,000 which was accepted and that subsequently, when informed that another party had offered £875,000, she indicated that they were prepared to increase their offer to £810,000. According to Mr Scott, it was the £810,000 offer which was accepted by the Bonsers. On this point I accept his evidence. However, the important part of Mrs Land's evidence is that they were prepared to pay that price on the assumptions that: (i) the house stood in about 1 acre; and (ii) they would have to spend about £100,000 on renovations. What led them to offer as much as £810,000 was the area of 1 acre alongside the Thames. Notwithstanding the fact that it was her evidence that over the years she and her husband had looked at many houses and that she was chief executive of Collins Publishing and her husband managing partner in Ernst & Young and that accordingly they could both be expected, very broadly, to appreciate the relative values of different residential property, I do not find persuasive her evidence that had they been told that the site area was only about half an acre, they would have been prepared to pay only £500,000. Whereas I have no reason to doubt that this is what she genuinely believed, I do not derive much assistance from her judgment in arriving at the true value of the property given that it was only half an acre in area. I am not persuaded that she knew enough about relative property values to be able to provide firm evidence of the lower value. Her evidence merely supports the proposition that the house was perceived by the public to be less valuable if set in half an acre than if set in 1 acre. That is hardly a contentious issue. The crucial point is whether, being set in half an acre, it was worth less than the plaintiff paid for it. In my view, the evidence of the expert witnesses was more to the point and carried greater weight.

However, before coming to the experts, it is necessary to mention a witness called by the defendant albeit not as an expert, whose evidence I found particularly helpful. That was Mr William Hagon [FSVA]. He was a qualified valuer who had worked as such in offices in Chiswick High Road for 17 years. Not only did he work in Chiswick but he had lived in Chiswick for 18 years just about half a mile away from no 61. He was familiar with the property, having often seen it from the block of flats next door known as Hartington Court. He described no 61 as in one of the best locations in Chiswick and just the sort of property he would like to own. He knew that in May 1990 the local property market was "very full", that is, prices were high, but when no 61 came on the market at £850,000 he felt that it was very reasonably priced having regard to the fact that the "very ordinary" houses with which he had recently been professionally concerned were then fetching about half a million pounds and "this was a very special house". He inspected it. He realised that a great deal would have to be spent to refurbish and modernise it. He arranged for his builder to give a preliminary costing. The figure was about £100,000. He was well aware, as a valuer, that there was nothing like an acre of ground and indeed he was attracted to the house itself, if it could be refurbished, and to the very good proportion and layout of the gardens running down to the river. He offered £775,000. That represented what, having regard to his own available resources, he was prepared to spend. He did not increase his offer when he came to know from the defendants that others had offered more.

In my judgment, great weight must be attached to Mr Hagon's evidence. He was not just an ordinary house-hunting member of the public. He knew the property in the neighbourhood very well and he knew the relative market prices at the time. Moreover, he was proposing to buy no 61 out of his own pocket. His aim was to acquire it below the asking price of £850,000. It is very strongly to be inferred that he would never have offered a price which he had believed to be above the market value for this "very special house". Indeed, his professional background and familiarity with the neighbourhood strongly suggests, significantly, that even with knowledge that the site area was substantially less than an acre, the true value of no 61 in May 1990 was much more in the region of £850,000 than £550,000.

I now turn to the expert evidence. Much of this was, not surprisingly, directed to an investigation of the current prices of a number of "comparables", that is, houses considered by the experts to be indicators of local market values at the relevant time. But, before considering that evidence, it is necessary to consider the evidence as to what was the general state of the market for upper-priced residential property in Chiswick in May 1990. Having carefully considered the evidence of Mr Copping Joyce, I am prepared to accept that in those parts of London with which he was familiar, notably Islington and Bloomsbury, there was a distinct slackening in prices at the top end of the market from a peak in 1988. That said, however, I am unable to accept that this extended throughout Greater London and in particular to Chiswick. Mr Keith Pierson [ARICS], a chartered surveyor with Tyser Greenwood & Co, who was closely familiar with property in Chiswick, called on behalf of the defendants, said that, far from prices peaking in 1988, they continued to rise until 1990. What happened was that the number of sales began to decline.

Mr Graham Marks [FSVA], another expert valuer called on behalf of the defendants, who was more familiar with the central London property market rather than with Chiswick, did not agree that from the end of 1988 the market went into "severe decline" as claimed by Mr Copping Joyce. Although newly constructed residential developments such as that at Chelsea Harbour had dropped steeply, the prices of substantial residential properties at the top end of the market, such as no 61, had not. Having considered the whole span of expert evidence on the general state of the market, I find that in the Chiswick area prices of high-quality properties, that is in the £500,000-£1m bracket had not begun to fall to any discernible extent by May 1990, although the number of sales was tending to slow down. Thereafter, they did begin to fall and by a year later they had dropped by perhaps 7%-8%.

The evidence of the expert witnesses on the comparables was extremely detailed and considered the relative merits and demerits of each house put forward by comparison with no 61. It would not be helpful to rehearse in this judgment the often very detailed comparative analyses that were conducted by Mr Copping Joyce, Mr Pierson and Mr Marks. There are, however, certain identifiable principles of comparative valuation that emerged from the evidence of all three experts. These are the following:

(i) Within any particular residential area, such as Chiswick, there is likely to be a broad range of house prices which reflect the characteristics of the area as a whole and the perceived desirability of the area in the eye of the public. Thus, houses of a roughly similar style, size and plot area in one residential area are likely to fetch roughly the same price at any given time which may differ substantially from the price attainable for houses of similar characteristics in a different residential area. Chiswick was at all material times a distinctly high-priced area and Hartington Road, Chiswick, was, if not quite as expensive as Strand on the Green or Chiswick Mall, certainly at the upper end of the Chiswick price range.

(ii) Although in a given residential area there may be many houses of closely similar characteristics which may be relatively easy to value, there may be other houses with unusual or unique characteristics which are very much less easy to value because particular purchasers may attach a premium value to the unusual or unique characteristics in question. The extent to which the price at any time will be affected by such characteristics will depend on the requirements and taste of the purchasers in the market at that time. One group of purchasers might be particularly attracted by the chance of acquiring an elegant 18th-century residence with handsome well-proportioned rooms, fit to receive their antique furniture, and be inclined to attach relatively little importance to the availability of space for a tennis court or the existence of a heated swimming pool, whereas others might attach little or no importance to whether a house was specially noted in Pevsner and might instead, above all else, require substantial space for living accommodation and children's outdoor sporting activities. That said, houses having similar *special* characteristics in the same residential area are likely to attract roughly similar prices, all other things being equal.

(iii) The closer to the top end of the market a particular house may be and the more unusual its characteristics, the more difficult it is likely to be to predict the price it will fetch. This may be particularly so in the case of a house which combines highly unusual characteristics with unattractive characteristics because at a particular time a particular buyer might be prepared to pay a high premium to acquire the unusual characteristics and be prepared to overlook or attach little importance to the unattractive characteristics.

Against this background the expert evidence established the following in relation to no 61.

(i) Its 90-ft of river frontage was almost unique, a 30ft-40 ft frontage being more usual in Chiswick.

(ii) It was set well back from Hartington Road, behind a wall, much further back than other houses in the road, and therefore well away from the traffic flow.

(iii) It had a particularly attractive and well-laid out mature rear garden of substantial size, which was not separated from the river by a tow path or traffic route.

(iv) The house was detached and well-positioned in the overall plot area so that front and rear gardens were of pleasing proportions.

(v) The overall size of the plot represented one of the largest gardens in Chiswick, where many of the houses, particularly in Hartington Road, are large, Victorian semi-detached houses with very narrow rear gardens leading down to the river.

(vi) The house was in need of substantial refurbishment and modernisation of the kitchen and bathrooms. That would, I find, have cost roughly £100,000 or more depending on the lavishness of the modernisation.

(vii) Some people might regard the side wings that had been added later as aesthetically unattractive features.

(viii) Many potential buyers would regard it as a serious inconvenience that the only access from the front to the rear gardens was through the house or garage, since the building spanned the entire width of the plot.

(ix) It was overlooked on one side by a very large four-storey block of low-value, brick and concrete flats, known as Hartington Court, which extended from alongside about one-third of the depth of the front garden to about 80% of the depth of the rear garden. The nearest point of this block was only 41ft from the house. From various positions on the balconies of the flats it was possible to look into the windows of no 61. This characteristic seriously detracted from the degree of privacy available to occupants of no 61. The problem could have been completely solved only by planting a line of fully grown trees along the northern boundary of the front and rear gardens of no 61. In so far as they were planted in the rear garden, they would have to be lifted into it by crane over the top of no 61. The problem of privacy in the gardens might partially have been solved by erecting a very high trellis along the northern boundary and growing creepers up it.

Of the above characteristics those numbered (i) to (v) were strong or very strong selling points, whereas those numbered (vi) to (ix) were disadvantages to selling. It has to be observed that in the sales literature prepared by the defendants none of the photographs shows any part of Hartington Court.

Although I accept Mr Copping Joyce's point that properties of totally dissimilar characteristics to no 61 can be used to, as he put it, set the "tone" of prices in the area, this is a more helpful indicator of the relationship of prices in the given area with prices of similar property in other areas. It is of much less assistance in arriving at the value differential between the properties in question and totally dissimilar properties in the same area. For this reason I am unable to derive much help in relation to no 61 from Mr Copping Joyce's comparables at Chiswick Staithe, Ibis Lane or Chiswick Quay. As for Zoffany House, the elegant Queen Anne three-storey house on Strand on the Green, which sold for £685,000 in June 1988, some two years before the sale of no 61, I find this to be a poor comparable. Apart from its number of rooms and the fact that it overlooked the river, albeit separated by a tow path, which in the summer, on the evidence was heavily populated by tourists, it had nothing in common with no 61. For one thing it was attached at both sides and its main garden was only about 1,000 sq ft, a minute fraction of the size even of the rear garden alone at no 61. Again 62 Hartington Road is a very different house from no 61, although it is in close proximity. It had a small garden and was close to a road junction with two fairly busy roads alongside it. It was a six-bedroomed house with four bathrooms and therefore had plenty of living accommodation, which had been recently refurbished, but it lacked both the vitally significant characteristics of no 61 — river frontage and considerable plot-size with well-laid out gardens. It fetched £465,000 on a dull market in August 1990, having been put on the market at £610,000 at almost the same time as no 61. It is a property which would clearly be attractive to purchasers looking for substantial accommodation who were largely disinterested in plot size, architectural appearance or being adjacent to the river. At least it is an indicator of what that quantity of refurbished accommodation in Hartington Road would fetch at the relevant time.

Comparables which I find somewhat more helpful are the two Victorian semi-detached houses at 29 and 35 Hartington Road, relied on by Mr Marks. Both have rear gardens with river frontages about one-third that of no 61. The plot size of both is about 8,000 sq ft, again little more than one-third of that of no 61. Of these no 29 fetched £575,000 on a falling market in December 1991 and no 35 fetched £540,000 also on a falling market in March 1991. Both would have fetched more when the market was more buoyant in early 1990 and I find that they probably would then have sold at slightly higher prices in the range £550,000 to £600,000. Among Mr Copping Joyce's working papers produced in preparing his report is a reference to the adjacent no 59 Hartington Road. This was put on the market early in 1991 at £690,000, the agent's view being that it might make £650,000 and the vendor having rejected an offer of £600,000 in March 1991.

A comparable heavily relied upon by Mr Pierson and Mr Marks was Staveley House, which was sold by the defendants for £835,000 in April 1993. It stood in grounds of about three-quarters of an acre, had been built in the 1920s, was in good condition and above all had magnificent and very secluded gardens, beautifully laid out with lawns, pathways, mature shrubs and trees. It also had a substantial heated conservatory and had been sold primarily on the size and quality of its gardens. The defendants' expert, Mr Marks, believed this to be the only true comparable. Mr Copping Joyce did not agree. No 61 was, in his view, far less desirable than Staveley House, notwithstanding its river frontage: it had far less seclusion, the gardens were less attractive and it was in a far less satisfactory state of repair and much less well fitted out. I am satisfied on the whole of the evidence having regard to the fall in the market between then and 1993 that in May 1990 Staveley House could have been sold for rather more than £1m. I am also satisfied that at any given time it would have sold for more than no 61. The differential at any one time is difficult to arrive at except on the basis of a range. Of one thing there can be no doubt at all and that is that at no time would it have been anything like £500,000. I would have expected no 61 to sell for very roughly 75%-80% of the price of Staveley House at any given time.

Taking the evidence as a whole, I am not persuaded that Mr Copping Joyce's valuation of no 61 in May 1990 at £550,000 can be sustained. A Victorian semi-detached house in Hartington Road with a garden and river frontage far smaller than no 61 would have fetched at least that. In my judgment, the value of no 61 in May 1990 lay in the range £800,000 to £875,000, but the precise selling price within that range would be a matter of chance depending on who was in the market at the time and the relative importance attached by them to the attractive characteristics of substantial, well-laid-out gardens and a big river frontage on the one hand and the proximity of Hartington Court and the relatively tired condition of the interior on the other hand.

Accordingly, I conclude that the plaintiff's purchase of no 61 at £875,000 was not at a price in excess of its market value. It follows that he has suffered no loss by reason of his reliance on Mr Scott's misrepresentation as to the total area. He paid no more than it was worth with the area that in truth it had. It is to be observed that it is common ground that he cannot recover damages for loss of a bargain, that is to say he cannot recover damages calculated by reference to the difference between its value if the representation had been true and its real value.

Conclusion

Breach of a duty of care to avoid making negligent misstatements can give rise to a cause of action for damages only if the representee plaintiff has suffered loss by reason of reliance on the misstatement. That he has acted in reliance upon it without suffering loss gives him no right action for damages. In the present case, I have found that it has not been established that the price at which the plaintiff purchased 61 Hartington Road was more than the true value of that property. Accordingly, the plaintiff has failed to prove that the defendants' breach of duty caused him any loss and it follows that he can have no cause of action in the tort of negligence against the defendants.

I therefore conclude that the plaintiff's claim must be dismissed.

HOUSING ACTS

A **Court of Appeal**

October 11 1993

(Before Lord Justice DILLON and Mr Justice HOLLIS)

MASON v NWOKORIE

Estates Gazette February 5 1994
[1994] 05 EG 155

Housing Act 1988 **—** ***Damages for unlawful eviction* — *Whether***
B ***common law damages fell to be set off against statutory damages — Whether exemplary or aggravated damages — Whether damages excessive***

On November 6 1988 the plaintiff was granted a six-month tenancy of a bed-sitting-room in a house in which the defendant landlord and other tenants also resided. Cooking and washing facilities were shared. In April 1989 the defendant gave the plaintiff 14 days' notice to quit (instead of the 28 days to which he was entitled) and then enforced possession without a court order. In the court below the judge awarded the plaintiff £6,000 damages, of which £500 was general damages, £1,000 exemplary damages and £4,500 damages under sections 27 and
C 28 of the Housing Act 1988. The defendant appealed, contending that: (1) the general and exemplary damages fell to be set off against the 1988 Act damages; and (2) the 1988 Act damages were excessive.

Held: **The appeal was allowed in part. Although it was clear that the written evidence put in by the defendant's valuer failed to allow for the possible six months it would take to get possession when making a valuation on the assumption in section 28(1)(a) of the 1988 Act that the plaintiff continued to have the same rights of occupation, the assessment of damages could not be interfered with. Because the award of general damages covered the same loss as that awarded under the 1988 Act, the £500 fell to be set off**
D **against the award of £4,500. The defendant did not evict the plaintiff for financial motives; the award of £1,000 could not have been an award of exemplary damages and must be treated as aggravated damages. Because it as an award for the same loss as that awarded under the 1988 Act it too fell to be set off against the £4,500.**

The following cases are referred to in this report.

Drane v *Evangelou* [1978] 1 WLR 455; [1978] 2 All ER 437; (1977) 36 P&CR 270; 246 EG 137

Jones v *Miah* (1992) 24 HLR 578; [1992] 2 EGLR 50; [1992] 33 EG 59

E *Mafo* v *Adams* [1970] 1 QB 548; [1970] 2 WLR 72; [1969] 3 All ER 1404, CA

Ramdath v *Daley & Daley* (1993) 25 HLR 273

Rookes v *Barnard* [1964] AC 1129; [1964] 2 WLR 269; [1964] 1 All ER 367; [1964] 1 Lloyd's Rep 28, HL

Tagro v *Cafane* [1991] 1 WLR 379; [1991] 2 All ER 235; [1991] 1 EGLR 279, CA

This was an appeal against an order of Judge Goldstein in Bow County Court, who on October 23 1992 awarded damages to the plaintiff, Mr Nwokorie, against the defendant, Mr Mason, in respect of the unlawful eviction of the plaintiff.

William Geldart (instructed by Aleem Khan & Co) appeared for F
the appellant; David Watkinson and Stephanie Harrison (instructed by Edwards Son & Noice) represented the respondent.

Giving the first judgment, **DILLON LJ** said: This is an appeal by the defendant, Mr Nwokorie, against a judgment against him awarded by Judge Goldstein after the trial of this action on October 23 1992. The judgment was in sums amounting to £6,000 with costs.

The matter arose out of the eviction by the defendant of the plaintiff (Mr Mason) from residential accommodation in a property belonging to the defendant, 8 Upper Road, Plaistow, as long ago as April 22 1989. The plaintiff had a bed-sitting-room in the property. G
The defendant was a resident landlord with a sitting-room on the ground floor and a bedroom on the upper floor. Cooking, washing and lavatory facilities were shared. Therefore, it is common ground that the plaintiff did not have a protected tenancy or security of tenure under the Rent Acts, but he was entitled to 28 days' notice of termination of his tenancy. In addition, he had the right not to be evicted without an order of the court.

The defendant, however, gave him a mere 14 days' notice to quit and proceeded to enforce that without any order of the court.

In the notice of appeal it is put as the first ground of appeal that the judge's decision that there was an eviction was against the weight of the evidence, and the judge should have held that what happened H
amounted to a surrender of his tenancy by the plaintiff. Furthermore, it is put that the judge misdirected himself in holding that the Housing Act 1988 applied to the plaintiff's tenancy, because it was a tenancy not protected by the Rent Act. It is therefore said that there was no right to damages under that section.

Both these points, however, have been abandoned by the defendant in this court as unarguable in the light of the evidence and the judge's findings. The appeal is therefore limited to the *quantum* of damages.

The sum of £6,000, which the judge awarded the plaintiff, is broken up, under his judgment, as: £500 general damages, £1,000 exemplary damages and £4,500 Housing Act damages; total: £6,000. J

"Housing Act damages" mean damages under sections 27 and 28 of the Housing Act 1988.

Section 27(1), as it was in force at the relevant time, provides that:

> . . . if at any time after 9th June 1988, a landlord (in this section referred to as "the landlord in default") or any person acting on behalf of the landlord in default unlawfully deprives the residential occupier of any premises of his occupation of the whole or part of the premises.

It is then provided by subsection (3) of section 27:

> Subject to the following provisions of this section, where this section applies the landlord in default shall, by virtue of this section, be liable to pay to the K
> former residential occupier, in respect of his loss of the right to occupy the premises in question as his residence, damages assessed on the basis set out in section 28 below.

That is subject to the provisions of subsection (4) and (5) of section 27. Subsection (4) provides that:

> Any liability arising by virtue of subsection (3) above —
> (a) shall be in the nature of a liability in tort; and
> (b) subject to subsection (5) below, shall be in addition to any liability arising apart from this section (whether in tort, contract or otherwise)

Subsection (5) provides that:

> Nothing in this section affects the right of a residential occupier to enforce any liability which arises apart from this section in respect of his loss of the right to occupy premises as his residence; but damages shall not be awarded both in respect of such a liability and in respect of a liability arising by virtue of this section on account of the same loss.

There can therefore be no duplication of damages.

It is submitted for the appellant landlord that the result of subsection (5) is that the £500 which the judge awarded as general damages, and also the £1,000 which the judge awarded under the head of "exemplary" damages, but which the plaintiff seeks to justify under the alternative head of aggravated damages, both fall to be set off against the damages awarded under section 27. In other words, the plaintiff has an option. He can claim the damages under section 27, or he can claim the general damages and the exemplary or aggravated damages, but he cannot have both.

The landlord also asserts that the judge's award of £4,500 under section 27 is excessive and cannot be maintained.

Subsection (3) provided that the damages under section 27 were to be assessed on the basis set out in section 28. Section 28 provides by subsections (1) and (2):

> (1) The basis for the assessment of damages referred to in section 27(3) above is the difference in value, determined as at the time immediately before the residential occupier ceased to occupy the premises in question as his residence

— That of course is the date of eviction in April 1989 —

> . . . between
>
> (a) the value of the interest of the landlord in default determined on the assumption that the residential occupier continues to have the same right to occupy the premises as before that time; and
>
> (b) the value of that interest determined on the assumption that the residential occupier has ceased to have that right.
>
> (2) In relation to any premises, any reference in this section to the interest of the landlord in default is a reference to his interest in the building in which the premises in question are comprised (whether or not that building contains any other premises) together with its curtilage.

Thus it is necessary to look not just at the room which the plaintiff occupied but also at the whole of the building, 8 Upper Road, Plaistow.

I will deal, first, with the landlord's attack on the award under sections 27 and 28 of the Housing Act. What had to be valued was the difference between the value of the landlord's interest in the building, on the assumption that the plaintiff had ceased to have the right of occupation — the landlord's interest was a freehold interest in the building — and the value of that interest determined on the assumption that the plaintiff continued to have the same right to occupy the premises as before. That was a right which could have been the subject of a 28 days' notice to quit at any time. If the plaintiff did not quit on the expiration of that notice the landlord could at once have applied to the court for a possession order. The plaintiff would have had no defence to the claim for a possession order and the maximum time which the court could have allowed for possession to be given under a possession order was three months.

There is, therefore, 28 days for notice to quit, a period for proceedings to be instituted and to come on for hearing and a period of up to three months before possession would have to be given under a possession order obtained at the hearing. Both sides have agreed that that means that possession could have been obtained against the plaintiff in up to six months.

No valuation evidence was called at the trial, but each side instructed a valuer to put in a written valuation. The plaintiff had the valuation of a Mr Bernard Living [FRICS], of Charles Living & Son.

His valuation was as follows:

> Having regard to the foregoing,

— his description of the property —

> we are of the opinion that the following values are fairly stated as at April 1989.
>
> (a) £45,000 on a full vacant possession basis;
>
> (b) £40,500 with possession being obtainable in one to six months.
>
> (c) £22,500 subject to a protected tenancy of one room let at £40 per week inclusive of rates and with vacant possession of the remainder.

£40 a week was the rent payable by the plaintiff.

The defendant landlord's surveyor was Cornerstone. A Mr Phillip Bellamy [ARICS], a chartered surveyor, was acting on behalf of them. It seems that he would have had Mr Living's valuation before him. He agreed that, as at April 1989:

> (a) freehold vacant possession value £45,000 and . . .
>
> (c) valuation of house, subject to statutory tenancy,

— where the entire property was subject to a statutory tenancy and the landlord is unable to gain possession —

> . . . we would consider the house to have an open market value, subject to that tenancy of £22,500.

In relation to (b), however, he said:

> (b) Value subject to occupation of one room. It is understood that the tenant took up occupation of one room under the terms of the old Rent Act on the basis of a resident landlord and would therefore be unable to oppose the landlord obtaining possession and selling the property in the usual way. On this basis and bearing in mind that during the normal course of events the sale and conveyancing of the property is likely to take at least two to three months, it is considered that the average landlord would serve notice upon the tenant requiring his room to be vacated which would then enable the landlord to sell the property with full vacant possession. Therefore it is considered highly unlikely that there would be any effect upon the freehold value of the property and we would still consider this to be £45,000. This is on the assumption that the tenant has no legal right to oppose the landlord's possession.

It is common ground that there is no question on the facts of this case of a valuation under (c), subject to a statutory tenancy.

One of the complicating factors in the case is that the plaintiff was not the only occupier of the property as well as the defendant landlord.

In his evidence, the defendant said:

> I gave him a six months tenancy from 9th November 1988, £40 per week, five of them in the house, one my brother.

"Five of them" seems to indicate that the defendant himself was not to be counted among the five. The five would thus have included the plaintiff, the defendant's brother and three other people. How they all fitted in is unclear. There were two living rooms on the ground floor, one occupied by the plaintiff and one by the defendant. There were three bedrooms on the upper floor, one occupied by the defendant.

Mr Bellamy's valuation for the defendant states:

> No access was possible to the small front bedroom or rear bedroom as these were locked during the time of our inspection.

There is no evidence to show on what terms the defendant's brother, and the other three people, were in occupation of their rooms. It seems clear, however, that none of them could have had security of tenure under the Rent Act because of the shared kitchen and washing facilities, and because also of the fact that the defendant was a resident landlord.

Consequently, it would seem that none of them could have had greater security than the plaintiff. There would be the same right, at most, to have not less than 28 days' notice and not to be evicted without a court order.

Therefore, as it seems to me, the other occupants can be ignored, because they could be got out as easily as the plaintiff could be got out if the same lawful steps were taken.

The fact remains, however, that while Mr Living, in his valuation, assumes possession being obtainable in one to six months, which necessarily means, for valuation purposes, up to six months, Mr Bellamy has assumed that the landlord, if he is proposing to sell the

A house and the conveyancing is likely to take at least two to three months, can proceed to sell, serving notice upon the tenant at the time he sells. He does not envisage a period of more than three months being required to obtain vacant possession.

It seems startling to me that the market value of the property would be depreciated by as much as £4,500 if it would take six months to get vacant possession, but would not be depreciated at all if vacant possession could be obtained in two to three months, the normal period for a conveyancing transaction. The two to three months must include the time before exchange of contracts. I do not regard three months after exchange of contracts as the normal course of events.

B It seems, therefore, that Mr Bellamy has overlooked the three elements that go to make up the time factor for obtaining vacant possession against the plaintiff.

1. The 28 days' notice.
2. The period thereafter until a possession order can be obtained.
3. The period allowed under the possession order.

If the period is six months, it could have the effect of discouraging would-be purchasers, who are eager to move in, possibly for reasons of change of job, and do not want to have to face a prolonged completion. I would regard a six-month completion from the time that the purchaser comes on the scene, which might be immediately after

C the property is put on the market, as an unusually deferred completion, which might well defer some purchasers. Whether that would have the effect of reducing the freehold value by £4,500 is another matter.

The judge, however, dealt with this by saying:

> Without in any way seeking to distinguish the two reports, it is right to say that the report of Charles Living & Son, which is that produced by the plaintiff, addresses the very problem of damages that the Act sets out in section 28, and the report from Cornerstone does not. It therefore follows that I should accept, and do accept, the plaintiff's report in preference to that of the defendant.

We were referred by Mr David Watkinson, for the plaintiff, to the decision of this court in *Tagro* v *Cafane* [1991] 1 WLR 379*. In that

D case a question had arisen on a claim for damages for unlawful eviction under sections 27(3) and 28 of the Housing Act 1988. The plaintiff called valuation evidence putting the amount of the plaintiff's loss at £31,000. The defendant called no valuation evidence.

Apart from deciding certain other points this court held that, having regard to the defendant's failure to adduce expert valuation evidence, the judge had been entitled to accept that proffered by the plaintiff and, accordingly, although the amount awarded was high, the court would not interfere either by remitting the matter or by varying the judge's order.

The leading judgment was given by Lord Donaldson of Lymington

E MR. Russell and Nolan LJJ agreed. Lord Donaldson said at p386C:

> Mr Cafane called no evidence to dispute this figure. Mr Carnwath is therefore reduced to saying that the judge can be faulted on the footing that no reasonable judge could have accepted that evidence. That submission he made simply and with force and with slightly more difficulty when Russell LJ said, "Well, what should he have done?". It is true, I suppose, that he could have adjourned the hearing and urged Mr Cafane to call expert evidence. But it is difficult to see how he could have said, "I am not satisfied that any damages are due on this evidence"and, if he is not to say that, then he would have in some way to reduce the surveyor's figure on the basis, I suppose, of what he (the judge) thought were proper values. Had he done so, this court would probably have quashed his decision on the grounds that he was acting
> F not upon evidence and not upon something of which a judge should take judicial knowledge but upon some extraneous view as to a matter of fact.

Then it was urged that the damages seemed to be high. Lord Donaldson said, at p387E:

> I accept that the damages do seem to be high, but I have to warn myself against using any knowledge that I may have gained in other ways to support that view, and I am quite unable to say that the judge was at fault. If we were to interfere on this ground, it could only be on the basis of sending it back for a rehearing designed to enable Mr Cafane to call valuation evidence. It is not clear to me why he should have a second opportunity to call valuation evidence when he had the opportunity originally and did not choose to avail G
> himself of it.

The only difference is that in the present case the defendant's valuer failed to appreciate the legal position and so, in relation to (b), asked the wrong question, although it would seem he had the plaintiff's valuation before him. He gave no reasons for adopting a different basis in his calculation. I do not believe that makes any difference.

Accordingly, I reach the conclusion that it is not open to this court to interfere with the judge's award of £4,500 under the 1988 Act.

Then there is the award of £500 general damages. That is not disputed by the defendant, but it is said that it falls to be set-off against H
the £4,500 awarded under sections 27 and 28.

Mr Watkinson says that there should be no set-off. As I understand his submissions, they are that the award of general damages was indeed an award of general damages, and an award of general damages is not necessarily limited to damages in respect of the loss of the right to occupy the premises as the plaintiff's residence.

That may be so, but it is difficult to see what, in the context of this case, could conceivably have been covered by the award of £500 except that right. The wording used is "loss of the right to occupy the premises as his residence", not "loss of his tenancy or estate or interest in the premises". It is essentially damages for wrongful eviction J
which deprived him of his right to occupy the premises as his residence.

I take the view, therefore, that this is clearly a case where the general damages fall to be set off against the award under the 1988 Act.

I come then to the £1,000 which was awarded as exemplary damages. The judge, on the first page of his judgment, referred to the claim as a claim for damages for unlawful eviction including: (a) a claim for exemplary damages or aggravated or punitive damages; and, (b) damages under sections 27 and 28 of the Housing Act 1988. He does not appear, in the course of his judgment, to devote much K
attention, if any, to the distinction between exemplary and aggravated damages.

On p9 of his judgment, starting at E, he sets out that from the breach of legal obligations the plaintiff seeks to recover damages under four quite separate and distinct headings. Then he sets out, first of all, any special damages. Second, general damages for the inconvenience and hardship resulting from the eviction. Third, exemplary damages to point out to any person minded to do this the financial consequences that might follow from it. Last, damages under the 1988 Housing Act.

He seems there to use the term "exemplary damages" as possibly L
covering aggravated or punitive damages. He ends up by saying:

> Without being the worst case of its kind, it is a case of unlawful eviction, deliberate eviction, and there will be an award of £1,000 exemplary damages.

During the course of the argument it has been pointed out to us that in his speech in *Rookes* v *Barnard* [1964] AC 1129 Lord Devlin sets out the distinction between aggravated and exemplary damages.

Thus, at p1221 he said:

> It must be remembered that in many cases of tort damages are at large, that is to say, the award is not limited to the pecuniary loss that can be specifically M
> proved. In the present case, for example, and leaving aside any question of exemplary or aggravated damages, the appellant's damages would not necessarily be confined to those which he would obtain in an action for wrongful dismissal.

He continued a bit further down that page:

> Moreover, it is very well-established in cases where the damages are at large the jury (or the judge if the award is left to him) can take into account the motives and conduct of the defendant where they aggravate the injury done to the plaintiff. There may be malevolence or spite or the manner of committing the wrong may be such as to injure the plaintiff's proper feelings of dignity and pride. These are matters which the jury can take into account in assessing the appropriate compensation.

*Editor's note: Also reported at [1991] 1 EGLR 279.

Then he continues, at the foot of p1221:

> But there are also cases in the books where the awards given cannot be explained as compensatory, and I propose therefore to begin by examining the authorities in order to see how far and in what sort of cases the exemplary principle has been recognised.

He then goes through various cases and he reaches the conclusion that there are only a limited number of categories where exemplary damages can be awarded. The first category is oppressive, arbitrary or unconstitutional action by the servants of the Government.

That is not this case. There is then a second category, namely those cases in which the defendant's conduct has been calculated by him to make a profit for himself which may well exceed the compensation payable to the plaintiff.

In addition, there must be added any category in which exemplary damages are expressly authorised by statute, but that does not cover the present case. From then onwards the courts have recognised that, in considering exemplary damages, the position is different from a claim for general damages or, indeed, aggravated damages. That is echoed in the recent judgment of Nourse LJ in *Ramdath* v *Daley & Daley*, decided on January 18 1993, and reported in (1993) 25 HLR 273. At p279 he said:

> Aggravated damages and exemplary damages are not only different in themselves, they are awarded for two different purposes,

which he proceeded to set out.

On the authorities there has been, as it seems to me, a measure of confusion, particularly in certain county court cases to which we were referred, where aggravated damages and exemplary damages seem to have been regarded as substantially the same thing.

There is also the complication that, in the case of *Drane* v *Evangelou* [1978] 1 WLR 455*, the majority of the court, Lawton and Goff LJJ, regarded the damages in question as aggravated damages, whereas Lord Denning regarded them as exemplary damages. That, it would seem on Lord Devlin's test, would have been appropriate if the object of the exercise in evicting or endeavouring to get out a protected tenant was to achieve, as it very often was, a financial advantage for the landlord.

Where, however, there was no evidence whatsoever of a financial motive, it was held in *Mafo* v *Adams* [1970] 1 QB 548 that there was no case for exemplary damages: see the judgments of Sachs LJ at pp555 and 556 and Widgery LJ at p559. Sachs LJ pointed out that the onus to establish a financial motive was squarely on the plaintiff.

I do not propose to go through a number of the other authorities and decisions of this court which were drawn to our attention.

In the present case it is accepted that there was no financial motive in the defendant's course to get rid of the plaintiff from the property. The difficulty was that the two did not get on and the defendant wanted the plaintiff out of the property. It seems to me that the award cannot be an award of exemplary damages, but the plaintiff, by a respondent's notice, seeks to support it as a justifiable award of aggravated damages.

In *Ramdath* v *Daley & Daley* there were special considerations which precluded what had purported to be an award of exemplary damages from being regarded as an award of aggravated damages. In that case the judge had awarded general damages in a different figure which he had said included aggravated damages.

In the present case, there is no such difficulty and, as I have said, there are indications in the judgment that the judge was not drawing any distinction between exemplary and aggravated damages. This was indeed a case where the conduct of the defendant in evicting the plaintiff, who was a man who had a particular need through certain temperamental difficulties, was calculated to cause him humiliation and shame and was calculated to be thoroughly offensive to his feelings. He was refused to be let in when he came back to the property. He had to go away. When he went to the police and brought them to the house the defendant denied knowing of the plaintiff's existence.

*Editor's note: Also reported at (1977) 246 EG 137.

Therefore, I would treat the £1,000 award as an award of aggravated rather than exemplary damages.

But it is still, in my judgment, an award of damages in respect of his loss of the right to occupy premises as his residence, damages aggravated by the manner in which he lost that right, but still aggravated damages for the same loss, the loss of the right to occupy the premises as his residence.

Therefore, in my judgment, the £1,000 ought also to be set-off against the £4,500 damages under sections 27 and 28.

Our attention was drawn to a decision of this court in *Jones* v *Miah* (1992) 24 HLR 578*, in which I gave the leading judgment, sitting with Nourse and Leggatt LJJ. In that case there had been an award of £3,000 for aggravated damages, £1,500 for loss and inconvenience, personal discomfort and distress and a considerably higher sum for the tort of unlawful eviction under sections 27 and 28 of the Housing Act 1988. The parties there agreed that the £1,500 fell to be set off against any damages awarded under the Housing Act, but they also agreed that the £3,000 for aggravated damages did not fall to be set off. There was no issue as to the set off of the £3,000 in this court. The court simply accepted what the parties had agreed on that point. The case is not authoritative on that point, although it is an authority on quite a lot of points which are irrelevant to the present case.

It follows that, in my judgment, the award of damages should be reduced from £6,000 to £4,500. To that extent I would allow this appeal and vary the order of the judge.

Agreeing, **Hollis J** said: The most important factor in this case is the meaning of section 27 of the Housing Act 1988. So far as material, subsection (1) reads:

> (1) This section applies if at any time after 9th June 1988, a landlord (in this section referred to as "the landlord in default") or any person acting on behalf of the landlord in default unlawfully deprives the residential occupier of any premises of his occupation of the whole or part of the premises.

That, in fact, is just what happened in this case. Subsection (3):

> Subject to the following provisions of this section, where this section applies the landlord in default shall, by virtue of this section, be liable to pay to the former residential occupier, in respect of his loss of the right to occupy the premises in question as his residence, damages assessed on the basis set out in section 28 below.

Subsection (4):

> Any liability arising by virtue of subsection (3) above —
> (a) shall be in the nature of a liability in tort; and
> (b) subject to subsection (5) below, shall be in addition to any liability arising apart from this section (whether in tort, contract or otherwise).

Subsection (5):

> Nothing in this section affects the right of a residential occupier to enforce any liability which arises apart from this section in respect of his loss of the right to occupy premises as his residence; but damages shall not be awarded both in respect of such a liability and in respect of a liability arising by virtue of this section on account of the same loss.

I do not consider that, at any stage during the course of the proceedings, the plaintiff has to elect formally under which branch he seeks damages, whether under the Housing Act or under the common law. But whatever he is awarded, it seems to me that the smaller amount must be deducted from the larger in reaching the net amount of damages that he should receive. In other words, a set-off. This, in my view, not only applies to general damages but also to exemplary or aggravated damages.

It will be noted that the wording in subsection (5) of the Housing Act is "any liability". Therefore, the plaintiff can obtain damages either under sections 27 and 28 of the Housing Act or at common law, but he cannot have both.

*Editor's note: Also reported at [1992] 2 EGLR 50.

A I entirely agree that, on the facts of this case, the plaintiff would have obtained aggravated damages and not exemplary damages for the reasons given by Dillon LJ.

I also would allow the appeal, in part, and reduce the damages awarded to the plaintiff to the sum of £4,500.

Appeal allowed in part.

Court of Appeal

B May 28 1993

(Before Lord Justice NEILL, Lord Justice STEYN and Lord Justice ROSE)

COVENTRY CITY COUNCIL v COLE AND OTHERS

Estates Gazette February 12 1994

[1994] 06 EG 133

Housing — Lease granted pursuant to the right to buy provisions in Housing Act 1985 — Whether covenant to pay service charge void C ***and unenforceable by reason of para 18 (alternatively para 11) of Schedule 6 to Housing Act 1985***

By a lease dated July 4 1988, granted pursuant to the right to buy provisions of the Housing Act 1985, the respondent council granted the appellants a lease for the term of 125 years of a flat for a premium and subject to a ground rent of £10 pa. By clause 5(2) of the lease the appellants covenanted to pay a service charge at a rate of £208 a year subject to an increase to take account of any escalation in building costs established by the RICS Index. The council made a claim in accordance with clause 5(2) for the year from October 1 1988 in the sum of £244. In the court below the judge held that clause 5(2) of the lease was valid and enforceable. The lessees appealed contending that D clause 5(2) was void and unenforceable as it fell within para 18(*a*) of Schedule 6 to the 1985 Act, alternatively, the sum claimed would fall to be treated as part of the rent and would be irrecoverable by reason of para 11 of the same Schedule.

Held**: The appeal was dismissed. Para 16A of Schedule 6 to the 1985 Act entitles a local authority landlord to include in a lease a provision that the tenant should bear a reasonable part of the costs incurred by the landlord in discharging the obligations imposed by the covenants for repair implied by virtue of para 14(2). The sum claimed is not "a reasonable part of the costs incurred by the landlord" within the meaning of para 16A, but a** E **fixed charge payable without any reference to the actual costs incurred or to be incurred in the future. Schedule 6 and the protection which it confers on a tenant have to be looked at in the context of the statutory control relating to variable charges. The reasonableness of a fixed charge can be examined at the time when the long lease is being negotiated. Assuming the fixed charge is reasonable the tenant is protected over the whole period of the lease from fluctuating and unpredictable costs. His only exposure to risk is in the risk attendant on a clause which depends on inflation. Clause 5(2) therefore fell outside para 16A and accordingly outside the scope of para 18 of Schedule 6. Further, it** F **did not infringe para 11 of Schedule 6 because the word "rent" in that paragraph cannot embrace all sums payable from time to time as consideration under the lease.**

No cases were referred to in this report.

This was an appeal by Michael David Cole and Patricia Eleanor Cole from an order of Judge Harrison-Hall QC of June 2 1992 setting aside an arbitration award of Deputy District Judge Sharpe of November 15 1991 in proceedings brought against the appellants by the respondents, Coventry City Council.

Lance Ashworth (instructed by Field Overell, of Coventry) appeared for the appellants; Michael Briggs (instructed by the solicitor to Coventry City Council) represented the respondents; the third G respondents did not appear and were not represented.

Giving judgment, NEILL LJ said: This is an appeal by Mr Michael David Cole and Mrs Patricia Eleanor Cole (whom I shall call "the lessees"), from the order of Judge Harrison-Hall QC dated June 2 1992 setting aside an arbitration award of Deputy District Judge Sharpe dated November 15 1991 in proceedings brought against the lessees by Coventry City Council (the council). The appeal is brought by leave of the judge.

The appeal raises an important question relating to the right to buy provisions contained in the Housing Act 1985 (as amended).

For the purpose of reaching our decision on this appeal it will be H necessary to consider not only the right to buy provisions themselves but also the history of the statutory control of service charges. The control of service charges was first introduced in 1972 in relation to leases on property granted to tenants by private landlords. First, however, I propose to set out the relevant facts.

In 1988 the lessees were the tenants of a first-floor flat at 2 Deal House, Dunhill Avenue, Coventry. The landlords were the council from whom the lessees held the flat as secure tenants. The lessees decided to exercise their right to take a long lease of the flat in accordance with the provisions of the Housing Act 1985. By a lease dated July 4 1988 the council demised the flat to the lessees for the J term of 125 years in consideration of the sum of £5,175. The lease was at a fixed rent payable throughout the term at the rate of £10 pa, payable annually in advance on October 1.

By clause 4 of the lease the lessees covenanted, *inter alia*, to put keep and maintain all parts of the interior of the flat in good and tenantable repair order and condition, and to permit the council and the other specified covenantees and their agents to enter the flat for the purposes of repairing any other part of the building of which the flat formed part, and for other specified purposes. In addition by clause 4(4) the lessees covenanted as follows:

K

Structural defects

[To] bear a reasonable part of the cost of (a) carrying out repairs not amounting to the making good of any structural defects, (b) the cost of making good any structural defects of which the council have notified the lessee before the date of this lease or of which the council do not become aware of earlier than ten years after the date of this lease and (c) insuring against risks involving such repairs or the making good of such defects.

By clause 5 of the lease the lessees entered into a number of specific covenants with the council. The relevant part of clause 5 was in these terms:

5. The lessee hereby covenants with the council as follows: L

(1)...

To pay service charge

(2) To pay sum (hereinafter called "the Service Charge") as is specified in the fifth schedule hereto subject to and in accordance with the provisions therein mentioned in arrears on the 1st day of October in every year.

The fifth Schedule to the lease contained provisions relating to the service charge. It is sufficient to refer to only part of the fifth Schedule as follows:

The Service Charge to be paid by the lessee to the council in accordance with M clause 5(2) thereof shall be:—

(a) The sum of £208 per annum and

(b) By way of additional charge a sum bearing the same proportion to the said sum as shall be borne by any increase in the Building Cost Information Service Tender Price Index published by the Royal Institution of Chartered Surveyors to the figure shown therein for May... and...

It is not necessary to set out the terms of the proviso. It will be seen that the service charge was to be at the rate of £208 a year, but was subject to an increase to take account of any escalation in building costs as established by the RICS index.

Clause 6 of the lease contained a covenant by the council in these terms:

> The council hereby covenants with the lessee in addition to the covenants implied by virtue of Part III of the sixth schedule to the Housing Act 1985 or any statutory amendment or re-enactment thereof (subject to the lessee paying the rents hereby reserved and performing and observing the several covenants conditions and agreements herein contained and on the lessee's part to be performed and observed) but not so as to be personally liable after it shall have transferred all its estate and interest in the said property that the lessee shall and may peaceably and quietly hold and enjoy the said property during the said term without any lawful interruption or disturbance from or by the council or any person or persons rightfully claiming under or in trust for the council.

In due course the council made a claim against the lessees for the service charge expressed to be payable in accordance with clause 5(2) of the lease for the period from October 1 1988 to September 30 1989. The sum claimed was £244 calculated, it seems, in accordance with the fifth Schedule to the lease. The lessees failed to pay the sum claimed or any part of it and on April 1990 the council issued a summons in Coventry County Court. The case came before Deputy District Judge Sharpe in November 1991 under the small claims procedure. By a judgment dated November 15 1991 Deputy District Judge Sharpe dismissed the claim on the basis that the council were not authorised to recover the sum claimed by reason of the provisions of the Housing Act 1985 relating to service charges. The council then appealed.

On June 2 1992 Judge Harrison-Hall QC allowed the council's appeal and set aside the order of the deputy district judge. He granted leave to appeal to this court. Judge Harrison-Hall held that clause 5(2) of the lease was valid and enforceable.

Statutory right to buy

The right to buy houses and flats was first conferred on council tenants by the Housing Act 1980. The right could be exercised either by taking a conveyance of the freehold or by obtaining the grant of a long lease of not less than 125 years. By section 17 of the 1980 Act, it was provided that a grant of a lease should conform with Parts I and III of Schedule 2 to the Act.

The 1980 Act was amended by the Housing and Building Control Act 1984, which introduced new paragraphs into Part III of Schedule 2 and a requirement that the landlord should provide estimates of costs of making good structural defects.

The lease with which this appeal is concerned was granted pursuant to the right to buy provisions contained in Part V of the Housing Act 1985 (the 1985 Act), which replaced the 1980 Act. It is therefore necessary to consider in some detail the provisions of the 1985 Act and to take account of the amendments to the 1985 Act introduced by the Housing and Planning Act 1986. By section 139(1) of the 1985 Act (replacing section 17 of the 1980 Act) it was provided that the grant of a lease should conform with Parts I and III of Schedule 6.

Before I come to Schedule 6, however, I should refer first to the definition of "service charge" contained in section 621A of the 1985 Act and to the provisions relating to estimates and information about service charges contained in section 125A.

Section 621A of the 1985 Act was inserted by section 24(2) of the Housing and Planning Act 1986. So far as is material it is in these terms:

> (1) In this Act "service charge", means an amount payable by a purchaser or lessee of premises —
>
> (*a*) which is payable, directly or indirectly, for services, repairs, maintenance or insurance or the vendor's or lessor's costs of management, and
>
> (*b*) the whole or part of which varies or may vary according to the relevant costs.
>
> (2) The relevant costs are the costs or estimated costs incurred or to be incurred by or on behalf of the payee, or (in the case of a lease) a superior landlord, in connection with the matters for which the service charge is payable.
>
> (3) For this purpose —
>
> (*a*) "costs" includes overheads, and
>
> (*b*) costs are relevant costs in relation to a service charge whether they are incurred, or to be incurred, in the period for which the service charge is payable or in an earlier or later period.

It is provided in section 62lA(4) that the "payee" means the person entitled to enforce payment of the charge.

Section 125 of the 1985 Act provides that where a secure tenant has claimed to exercise the right to buy the landlord should serve on the tenant a notice containing the information prescribed in the section. Section 125 was amended by the Housing and Planning Act 1986 and it is now provided by section 125(4) that where the notice contains provisions which would enable the landlord to recover service charges from the tenant the notice should also contain the estimates and other information required by section 125A. Though it is not necessary to set out the terms of section 125A at length, it is to be noted that the broad effect of this section (which was inserted by the Housing and Planning Act 1986) is to require a landlord to set out in the notice under section 125 estimates of the likely cost of, and of the tenant's likely contribution in respect of, each item of work which the landlord considers may have to be carried out by way of repair in the first five years after the grant of the lease. Where works are so itemised the notice has to contain an estimate of the average annual amount which the landlord considers is likely to be payable by the tenant.

I come now to Schedule 6. As to Part I of Schedule 6 it is sufficient to refer only to para 5. It is in these terms:

> Subject to paragraph 6 [which has no relevance in this case], and to Parts II and III of this Schedule, the . . . grant may include such covenants and conditions as are reasonable in the circumstances.

Part III of Schedule 6, however, requires fuller treatment. It is provided by paras 11 and 12 of Part III that where the landlord holds the freehold interest the lease should be for a term of not less than 125 years and at a rent not exceeding £10 pa.

Para 14 contains statutory implied covenants by the landlord where the dwelling-house is a flat. Para 14(2) is in these terms:

> There are implied covenants by the landlord —
>
> (*a*) to keep in repair the structure and exterior of the dwelling house and of the building in which it is situated (including drains, gutters and external pipes) and to make good any defect affecting that structure;
>
> (*b*) to keep in repair any other property over or in respect of which the tenant has rights by virtue of this schedule;
>
> (*c*) to ensure, so far as practicable, that services which are to be provided by the landlord and to which the tenant is entitled (whether by himself or in common with others) are maintained at a reasonable level and to keep in repair any installation connected with the provision of those services;

Para 14(3) contains an additional implied covenant by the landlord that he should "rebuild or reinstate the dwelling-house and the building in which it is situated in the case of destruction or damage by fire, tempest, flood or any other cause against the risk of which it is normal practice to insure".

It is further provided by para 14, however, as follows:

> (4) The county court may, by order made with the consent of the parties, authorise the inclusion in the lease or in an agreement collateral to it of provisions excluding or modifying the obligations of the landlord under the covenants implied by this paragraph, if it appears to the court that it is reasonable to do so.

Para 16 of Schedule 6 sets out the covenants to be implied on the part of the tenant. It is in these terms:

> Unless otherwise agreed between the landlord and the tenant, there is implied a covenant by the tenant —
>
> (*a*) where the dwelling-house is a house, to keep the dwelling-house in good repair (including decorative repair);
>
> (*b*) where the dwelling-house is a flat, to keep the interior of the dwelling house in such repair.

A I come now to paras 16A to 19 of Schedule 6. Paras 16A to 16D and para 18 were introduced by the Housing and Planning Act 1986. It is necessary to set some of these paragraphs out *in extenso*. It is not necessary, however, to refer to paras 16C or 17.

Service charges and other contributions payable by the tenant.

16A.—(1)The lease may require the tenant to bear a reasonable part of the costs incurred by the landlord —

(*a*) in discharging or insuring against the obligations imposed by the covenants implied by virtue of paragraph 14(2) (repairs, making good structural defects, provision of services, etc.), or

B (*b*) in insuring against the obligations imposed by the covenant implied by virtue of paragraph 14(3) (rebuilding or reinstatement, etc.),

and to the extent that by virtue of paragraph 15(3) (effect of provision of superior lease) such obligations are not imposed on the landlord, to bear a reasonable part of the costs incurred by the landlord in contributing to costs incurred by a superior landlord or other person in discharging or, as the case may be, insuring against obligations to the like effect.

(2) Where the lease requires the tenant to contribute to the costs of insurance, it shall provide that the tenant is entitled to inspect the relevant policy at such reasonable times as may be specified in the lease.

(3) Where the landlord does not insure against the obligations imposed by the covenant implied by virtue of paragraph 14(3), or, as the case may be, the C superior landlord or other person does not insure against his obligations to the like effect, the lease may require the tenant to pay a reasonable sum in place of the contribution he could be required to make if there were insurance.

(4) Where in any case the obligations imposed by the covenants implied by virtue of paragraph 14(2) or (3) are modified in accordance with paragraph 14(4) (power of county court to authorise modification), the references in this paragraph are the obligations as so modified.

(5) This paragraph has effect subject to paragraph 16B restrictions in certain cases as regards costs incurred in the initial period of lease.

16B.—(1) Where a lease of a flat requires the tenant to pay service charges in respect of repairs (including works for the making good of structural defects), his liability in respect of costs incurred in the initial period of the D lease is restricted as follows.

(2) He is not required to pay in respect of works itemised in the estimates contained in the landlord's notice under section 125 any more than the amount shown as his estimated contribution in respect of that item, together with an inflation allowance.

(3) He is not required to pay in respect of works not so itemised at a rate exceeding —

(*a*) as regards parts of the initial period falling within the reference period for the purposes of the estimates contained in the landlord's notice under section 125, the estimated annual average amount shown in the estimates;

E (*b*) as regards parts of the initial period not falling within that reference period, the average rate produced by averaging over the reference period all works for which estimates are contained in the notice;

together, in each case, with an inflation allowance.

(4) The initial period of the lease for the purposes of this paragraph begins with the grant of the lease and ends five years after the grant, except that —

(*a*) if the lease includes provision for service charges to be payable in respect of costs incurred in a period before the grant of the lease, the initial period begins with the beginning of that period;

(*b*) if the lease provides for service charges to be calculated by reference to a specified annual period, the initial period continues until the end of the fifth such period beginning after the grant of the lease; and

F (*c*) if the tenant served notice under section 142 deferring completion, the initial period ends on the date on which it would have ended if the lease had been granted on the date on which the notice was served . . .

16D.—(1)The Secretary of State may by order prescribe —

(*a*) the method by which inflation allowances for the purposes of paragraph 16B or 16C are to be calculated by reference to published statistics; and

(*b*) the information to be given to a tenant when he is asked to pay a service charge or improvement contribution to which the provisions of paragraph 16B or 16C are or may be relevant.

(2) An order —

(a) may make different provision for different cases or descriptions of case, including different provisions for different areas;

(*b*) may contain such incidental, supplementary or transitional provisions as the Secretary of State thinks appropriate; and

(*c*) shall be made by statutory instrument which shall be subject to annulment in pursuance of a resolution of either House of Parliament . . . G

18. Where the dwelling-house is a flat, a provision of the lease or of an agreement collateral to it is void in so far as it purports —

(*a*) to authorise the recovery of such a charge as is mentioned in paragraph 16A (contributions in respect of repairs, etc.) otherwise than in accordance with that paragraph and paragraph 16B (restrictions in initial period of lease); or

(*b*) to authorise the recovery of any charge in respect of costs incurred by the landlord —

(i) in discharging the obligations imposed by the covenant implied by paragraph 14(3) (rebuilding or reinstatement, etc.), or those obligations as modified in accordance with paragraph 14(4), or H

(ii) in contributing to costs incurred by a superior landlord or other person in discharging obligations to the like effect; or

(*c*) to authorise the recovery of an improvement contribution otherwise than in accordance with paragraph 16C (restrictions in initial period of lease).

19. A provision of the lease, or of an agreement collateral to it, is void in so far as it purports to authorise a forfeiture, or to impose on the tenant a penalty or disability, in the event of his enforcing or relying on the preceding provisions of this Schedule.

Statutory control of service charges J

Since 1972 Parliament has recognised the need to protect tenants from excessive demands made in respect of service charges. The control was introduced by section 90 of the Housing Finance Act 1972, which provided as follows:

(1) Where the service charges which are payable by the tenant of a flat in any calendar year, or which are demanded from the tenant as being so payable, exceed [£80] . . ., the tenant shall, in accordance with this section, be entitled to obtain a summary in writing of the relevant costs in the accounting year ending in or with that year, certified by a qualified accountant as being in his opinion —

(*a*) a fair summary of those costs, set out in a way which shows how they are or will be reflected in demands for service charges, and K

(*b*) sufficiently supported by accounts, receipts and other documents which have been produced to the accountant,

and the certificate shall identify the accounting year to which the summary relates.

It was made plain, however, by the definition of "service charge", in section 90(12) that the statutory control was restricted to any charge for "service, repairs, maintenance or insurance" which varied or might vary according to any costs (including charges for overheads) incurred from time to time by or on behalf of the landlord. It was further provided that the amount of £80 could be varied from time to L time by the Secretary of State. Local authorities and certain other bodies were excluded from the statutory control by section 91.

By section 124 of the Housing Act 1974 a right to challenge service charges was given to tenants by the insertion of a new section 91A into the Housing Finance Act 1972. In 1980 sections 90 to 91A of the Housing Finance Act 1972 were replaced by the provisions contained in Schedule 19 to the Housing Act 1980. A service charge was defined in para 1 of Schedule 19 as meaning:

an amount payable by the tenant of a flat as part of or in addition to the rent

(*a*)which is payable, directly or indirectly, for services, repairs, maintenance or insurance or the landlord's costs of management; and M

(*b*) the whole or part of which varies or may vary according to the relevant costs.

The relevant costs were defined as the costs or estimated costs incurred or to be incurred in any period by or on behalf of the landlord.

By para 14 of Schedule 19 an exception from the statutory control was made in relation to service charges payable by a tenant of the local authority and certain other bodies.

It is to be noted that the definitions of "service charge" contained in section 46(1) of the 1985 Act (before it was amended), in section 18(1) of the Landlord and Tenant Act 1985 and in section 621A of the

1985 Act (as amended by the Housing and Planning Act 1986) include as part of the definition the words "the whole or part of which varies or may vary according to the relevant costs".

Case for the lessees

Counsel for the lessees drew attention to the right to buy legislation and to the provisions relating to long leases of flats imposed first by Schedule 2 to the 1980 Act and later by Schedule 6 to the 1985 Act (as amended in 1986). He pointed out that clause 4(4) of the lease, though apparently included in the lease by an oversight, would have been apt as a tenant's covenant at the time when paras 16 and 17 of Schedule 2 to the Housing Act 1980 were in force in their original form. He submitted, however, that the charge mentioned in clause 5(2) of the lease (whether properly described as a service charge or by some other name) was void and unenforceable as it fell within para 18(*a*) of Schedule 6 to the 1985 Act. Alternatively, the sum mentioned in clause 5(2) would fall to be treated as part of the rent and would then be irrecoverable because of the provision in para 11 of Schedule 6 that the rent could not exceed £10 pa.

Case for the council

It was submitted on behalf of the council that clause 5(2) of the lease was unaffected by para 18(*a*) of Schedule 6. That paragraph, in common with the statutory control relating to private landlords, was concerned with service charges as defined in the various Acts since 1972, or at any rate was concerned with variable charges. Neither para 18(*a*) nor paras 16A and 16B were concerned with fixed charges of the kind specified in clause 5(2). Furthermore, it was clear from the context that the word "rent" in para 11 was not used in the wide sense as meaning any periodic consideration payable under a lease, but meant a ground rent. It was to be noted that the words "the ground rent" were used in section 127(3)(*c*) with reference to the assumptions to be made for calculating the value of a dwelling-house or the purposes of the statutory notice to be provided by the landlord in accordance with section 125. It was argued that the statutory control of service charges which were variable was fully justified but the need for such control had no relevance to a fixed charge. A variable charge was beneficial to the landlord under long leases because he was thereby insured against financial risks. On the other hand, if the service charge was not controlled, the tenant was exposed to the unpredictability of charges for repairs which could be undertaken at times decided by the landlord. In addition, the right to recover costs incurred provided the landlord with no incentive to make sure that any repairs and maintenance were carried out as economically as possible. In the case of a fixed charge it was the landlord who was at risk of having to incur substantial expenditure at short notice. A fixed charge (subject only to an indexed escalation clause) enabled a tenant to budget for the future. Furthermore, a tenant exercising his right to buy had the added security of being able to refer the terms of his long lease to the county court before any contract was concluded.

Validity of clause 5(2) of the lease

It was accepted on behalf of the council that if clause 5 (2) was held to be valid they could not rely in addition on clause 4(4). Counsel stated that the council are willing to give an undertaking on the matter which can be recorded in any order of the court.

The central question for decision is whether clause 5 (2) is avoided by the following words in para 18 of Schedule 6:

> Where the dwelling-house is a flat, a provision of the lease or of an agreement collateral to it is void in so far as it purports —
>
> (*a*) to authorise the recovery of such a charge as is mentioned in paragraph 16A (contributions in respect of repairs, etc.) otherwise than in accordance with that paragraph and paragraph 16B (restrictions in initial period of lease);

Certain matters appear to be clear:

(a) Although the sum referred to in clause 5(2) is described as "the service charge", both in clause 5(2) and in the fifth Schedule to the lease, the description is of no particular significance in this case. The reference to the price index published by the RICS makes it clear, however, that the sum is payable as a contribution towards the cost of repairs and maintenance.

(b) The "service charge" in clause 5(2) is not a "service charge" within the meaning of the 1985 Act or of any of the relevant provisions of the 1985 Act such as sections 125A and para 16B of Schedule 6.

(c) Para 16B of Schedule 6 has no application because the restrictions on liability imposed by that paragraph during the initial period of a lease apply only "where a lease of a flat requires the tenant to pay service charges in respect of repairs". The sum payable under clause 5(2) is not a "service charge" within the meaning of the 1985 Act.

One therefore turns to consider the words "such a charge as is mentioned in para 16A (contributions in respect of repairs etc) . . ." in para 18.

The implied covenant contained in para 14(2) of Schedule 6 obliges the landlord, *inter alia*, to keep in repair the structure and exterior of the dwelling-house and of the building in which it is situated and to ensure, so far as practicable, that services which are to be provided by the landlord are maintained at a reasonable level. Para 16A entitles the landlord to include in the lease a provision that the tenant should bear a reasonable part of the costs incurred by the landlord in discharging the obligations imposed by the covenants implied by virtue of para 14(2).

In the present case the council are bound by the implied covenants in para 14(2): see clause 6 of the lease. They do not seek, however, to recover any contribution from the lessees by way of a charge authorised by para 16A. The sum claimed is not "a reasonable part of the costs incurred by the landlord" within the meaning of para 16A, but a fixed charge payable without any reference to the actual costs incurred or to be incurred in the future. Clause 5(2), it is said, is a term which is permissible under para 5 of Schedule 6, which authorises the inclusion in the grant of a lease of "such covenants and conditions as are reasonable in the circumstances".

After some hesitation I have come to the conclusion that the submissions advanced on behalf of the council are correct. Schedule 6 and the protection which it confers on a tenant have to be looked at in the context of the statutory control relating to variable charges. The reasonableness of a fixed charge can be examined at the time when the long lease is being negotiated. Assuming that the fixed charge is reasonable, the tenant is protected over the whole period of the lease from fluctuating and unpredictable costs. His only exposure to risk is in the risk attendant on a clause which depends on inflation. I consider that the judge was right to uphold the clause and to treat clause 5(2) as falling outside para 16A and therefore outside the scope of para 18 of Schedule 6.

I am also satisfied that clause 5(2) does not infringe para 11 of Schedule 6. The rent "not exceeding £10" stipulated in that paragraph clearly means "the ground rent". The provisions set out in paras 16A and 16B make it clear that the word "rent" in para 11 cannot embrace all sums payable from time to time as consideration under the lease.

Subject to the inclusion in the order of the court of a term that will record the undertaking given by the council that they will not hereafter seek to enforce or rely upon clause 4(4) of the lease, I would dismiss the appeal.

STEYN and **ROSE LJJ** agreed and did not add anything.

Appeal dismissed.

LANDLORD AND TENANT
GENERAL

A **Official Referees' Business**

May 13 1993

(Before Judge FOX-ANDREWS QC)

HAFTON PROPERTIES LTD v CAMP AND ANOTHER

Estates Gazette January 22 1994

[1994] 03 EG 129

Landlord and tenant — Repairs — Covenants to be performed by
B ***management company — Whether covenant can be implied to make lessor liable for breach of management company's covenants***

By a lease dated July 28 1978 the defendants held a term for 99 years at an initial rent of £10 of a flat in a block of flats the reversion to which was owned by the plaintiffs. The defendants were additionally plaintiffs in their counterclaim against S Ltd, a company incorporated to manage the building and provide services for the lessees. By the lease the defendants covenanted with the lessor and S Ltd to pay to S Ltd a maintenance charge being a specified proportion of S Ltd's expenditure on specified matters and S Ltd covenanted to carry out repairs and maintenance: the lease contained a right of
C re-entry upon failure of the lessee, inter alia, to pay the maintenance charge to S Ltd. By their claim the plaintiffs alleged that the defendants had failed to pay £9,890.14 in arrears of rent and service charges and sought possession of the flat following forfeiture. In their defence and counterclaim the defendants alleged that S Ltd had allowed the building to fall into a stage of disrepair and that the plaintiffs were in breach of an implied duty to perform the covenants of repair and maintenance on S Ltd's behalf. The plaintiffs sought a determination of a question of law under Ord 14A of the RSC.

Held: **A term could not be implied into the lease that the plaintiffs would observe the covenants for repair and maintenance**
D **on behalf of S Ltd. The more comprehensive a code in a lease the less room there is for the implication of a term. The defendants have a right to apply for the appointment of a receiver in respect of any failure of S Ltd to perform its covenants under the lease.**

The following cases are referred to in this report.

Barrett v *Lounova (1982) Ltd* [1990] 1 QB 348; [1989] 2 WLR 137; [1989] 1 All ER 351; [1988] 2 EGLR 54; [1988] 36 EG 184, CA

Blawdziewicz v *Diadon Establishment* [1988] 2 EGLR 52; [1988] 35 EG 83

Clayhope Properties Ltd v *Evans* [1986] 1 WLR 1223; [1986] 2 All ER 795; [1986] 2 EGLR 34; (1986) 279 EG 855, CA

Daiches v *Bluelake Investments Ltd* [1985] 2 EGLR 67; (1985) 275 EG 462

E *Duke of Westminster* v *Guild* [1985] QB 688; [1984] 3 WLR 630; [1984] 3 All ER 144; (1983) 48 P&CR 42; [1983] EGD 541; 267 EG 762, CA

Gordon v *Selico Co Ltd* [1986] 1 EGLR 71; (1986) 278 EG 53, CA

Halsall v *Brizell* [1957] Ch 169; [1957] 2 WLR 123; [1957] 1 All ER 371

Hart v *Emelkirk Ltd* [1983] 1 WLR 1289; [1983] 3 All ER 15; (1982) 267 EG 946

Liverpool City Council v *Irwin* [1977] AC 239; [1976] 2 WLR 562; [1976] 2 All ER 39; (1976) 74 LGR 392; [1976] EGD 282; 238 EG 879, HL

This was a hearing of a summons issued by the plaintiffs, Hafton Properties Ltd, under RSC Ord 14A for the determination of a question of law in proceedings against the defendants, Mr and Mrs M F
D Camp, for forfeiture.

Michael Daiches (instructed by Rumke Joseph & Rabin) appeared for the plaintiffs; Stephen Hockman QC and Richard Barraclough (instructed by Amphlett Lissimore) represented the defendants.

Giving judgment, **JUDGE FOX-ANDREWS QC** said: The plaintiffs are the two owners of five-storey blocks of flats known as Silchester Court, London Road, Thornton Heath, Croydon, Surrey. It appears that the blocks consisted of 58 flats of which approximately 15 are let out on rack-rents with the rest let out on ground rents.

In this action the plaintiffs claim, *inter alia*, possession of no 22 of G
which the defendants are the present occupiers. The action is one of some 28 similar actions, but this judgment is only as between the plaintiffs and these defendants.

The lease under which the defendants held their interest was made on July 28 1978 between the plaintiffs as lessor, Silchester Court (Croydon) Management Co Ltd (the company) as the company and Brian Godfrey Johnson as the lessee. It appears that on July 2 1979 Mr Johnson transferred his interest in the lease to a Mr Timothy Sedgwick. I have no information as to whether the defendants acquired their interest from Mr Sedgwick, but nothing turns on this.

The term was for 99 years from January 1 1976 at a premium of
£11,500 at an initial yearly rent of £10. H

I should read some terms of the lease. Recital (2):

> The Lessor intends when the occasion arises to demise flats comprised in the Building in accordance with a general scheme and then in every such lease (other than leases at rack rents) of an individual flat the lessee thereof shall enter into obligations and restrictions on or substantially as stated in clause 3 of this lease to the intent that the lessee for the time being of any flat (other than lessees of rack rents) may be able to enforce in equity the performance and observance thereof by the lessee for the time being of other flats as demised.

Recital (3): J

> The Company has been incorporated with the object (*inter alia*) of providing services to and for lessees of the flats comprised in the Building and otherwise managing the same as hereinafter appears.

Clause 1 provided, *inter alia*:

> The lessor HEREBY DEMISES unto the lessee all that flat the details whereof are set forth in Part 3 of the said First Schedule . . . TOGETHER with (to the exclusion of all others) the easements, rights and privileges specified in the Second Schedule hereto subject as therein mentioned EXCEPT AND RESERVING as specified in the Third Schedule . . .

It is not necessary to consider the second and third schedules. K

By clause 2 the lessees entered into the usual covenants. Clause 3 provided:

> In accordance with the said general scheme and for the benefit of the lessees of the other flats comprised in the Building (other than lessees at rack rents) the lessee hereby covenants with the lessor and the Company and the lessees for the time being of the other flats comprised in the Building (other than the lessees at rack rent) and with each of them that the lessee will from time to time and at all times hereinafter during the said term
>
> (A) Keep the said flat (other than the parts thereof to be maintained by the Company pursuant to the Fifth Schedule hereto) . . . in good substantial and

A tenantable repair and condition and in particular so as to support, shelter and protect the parts of the Building other than the said flat . . .

(D)(i) Pay to the Company in respect of the year ending on the 31st December 1979 and each year thereafter [2%] of the costs, charges, expenses and management fees from time to time incurred by the Company in carrying out its obligations under the Fifth Schedule hereto (hereinafter called "The Maintenance Charges") such amount to be paid as hereinafter provided.

I should read part of the fifth schedule:

Subject to the due performance by the lessee of his obligations to contribute to the maintenance charges as herein provided, the Company will at all times during the said term

B

(1) Whenever reasonably necessary maintain, repair, redecorate and renew

(a) The external walls and structures and in particular the roof foundations chimney stacks gutters and rainwater pipes of the building

(b) The gas and water pipe sewers drains vent pipes and electric cables and wires in under and upon the Building or enjoyed or used by the lessee in common with the lessees of other flats in the Building

(c) The main entrances common lifts passages landings and staircases and other parts of the Building so enjoyed or used by the lessee in common as aforesaid

(d) The boundary walls and fences of the lessor's land

(e) The flat or flat spaces and accommodation (if any) for the time being
C occupied or used by the caretaker and other persons employed by the Company as hereinafter mentioned.

Reverting to clause 3 by (D)(ii) it was provided that the lessee would:

Pay to the Company in advance on the first day of January and the first day of July in each year (commencing on the 1st day of January 1979) on account of the maintenance charges payable by the lessee for such year one half of such amount as the Company or its Managing Agents shall certify to be the estimated amount of the maintenance charges attributable to the said flat for that year (such certificate to be final or binding on the lessee).

(iii) Within 28 days after the accounts for the Company of the year have
D been audited and a certificate signed by the auditor stating the amount of the maintenance charges attributable to the said flat for that year (or a certified copy thereof) has been served on the lessee (such certificate to be binding on the lessee) to pay to the Company the amount (if any) by which the maintenance charges payable in respect of the said flat for such year exceed the amount paid on account in respect of such year PROVIDED that if the amount of such maintenance charges payable for the year in respect of the said flat is less than the amount paid in advance on account the excess shall at the discretion of the Company either be repaid to the lessee or retained by the Company on account of payment due from the lessee in future years.

Clause 3(F) provided:

E Upon any transaction or disposition to which the lessee is a party or over which he has control involving a change or a contract for a change in the ownership of the said flat to procure that the person becoming or contracting to become as a result of such transaction or disposition the owner of the said flat (which expression shall be deemed to include any assignee of this lease and any person holding as under-tenant for substantially the whole of the unexpired part of the term hereby granted but excluding any mortgagee) (a) becomes (in the case of an assignee of this lease) a member of the Company and (b) at the expense of the lessee enters into a Deed of covenant with the Company to observe and perform all the covenants by the lessee with the Company contained in this lease in the terms specified in the Sixth Schedule hereto and pays the Company's solicitors reasonable charges in
F connection with the said Deed.

The sixth schedule somewhat surprisingly related not only to the service charge but to the rent as well.

Clause 3(H) contained provisions whereby the lessee had to permit the company and its respective surveyors and agents access to the flat and made provision as to what had to be done by the lessee in the event of a consequent notice specifying any repairs necessary to be done and for which the lessee was liable.

By clause 4 the plaintiffs covenanted with the lessee:

(A) That the lessee paying the rents hereby reserved and performing and observing the several covenants conditions and agreements herein contained and on the lessor's part to be performed and observed shall and may peaceably and quietly hold and enjoy the said flat during the said term without any lawful interruption or disturbance from or by the lessor or any person or G
persons rightfully claiming under or in trust for the lessor.

(B) That the lessor will require every person to whom they shall hereafter grant a lease of any flat comprised in the Building (other than leases at rack rents) to covenant to perform and observe obligations as or substantially as stated in Clause 3 of this lease (with such modifications as regards contribution to the maintenance charges as the lessor may deem appropriate to the flat concerned) and that if so required by the lessee will take all reasonable steps to enforce the same PROVIDED . . .

(C) That the lessor will allow the Company and persons authorised — by the Company to have such access to the Building and other parts of the lessor's land as may be necessary and proper for enabling the Company to carry out its obligations hereunder. H

2. The Lessor HEREBY FURTHER COVENANTS with the lessee . . . that the lessor will while any flat comprised in the Building is let to a tenant at a rack rent make such payments to the Company in respect of such flat as are provided in Clause 3(D)(i) and (iii) hereof.

Clause 5 contained rights of re-entry, *inter alia*, if there should be any breach of any covenant or agreement on the part of the lessee.

By clause 6 it was thereby agreed and declared that:

(A) The lessor shall not be liable or responsible for any damage suffered by the lessee or any visitor employee invitee or licensee of the lessee or any other person to their person or goods by reason of any act neglect default or omission of any other lessee or occupier of the Building and of any J
contractor employee visitor invitee or licensee of such other lessee or occupier . . . or by reason of any default in any fixture pipe wire staircase or thing or the absence of lighting in or upon the Building or any part thereof including the said flat.

(B) If during the term hereby granted the Company shall go into liquidation the lessor shall be entitled (but not bound) upon giving notice to the lessee to undertake the obligations hereby undertaken by the Company and if the lessor elects so to do the lessor shall be entitled to the benefit of all provisions herein contained concerning the Company and to recover from the lessee all monies hereby covenanted to be paid by the lessee to the Company.

By clause 7 the company covenanted with the lessor and the lessee K
to perform and observe the obligations set out in the fifth schedule (to which I have made partial reference).

I should read clause 8:

NOTWITHSTANDING anything herein contained the Company shall not be liable to the lessee nor shall the lessee have any claim against the Company in respect of

(A) Any loss or inconvenience occasioned by any interruption of any of the services mentioned in the Fifth Schedule from whatever cause.

(B) Any act neglect default or omission of any caretaker porter or other of its staff servants or any person acting under any such caretaker porter staff or servants. L

(C) Any loss or damage or interference or annoyance suffered by the lessee during the carrying out by the Company of repairs decorations additions alterations or other work which may appear to the Company to be necessary or desirable in the said flat or the Building.

Clause 9 made provision for the supply or non-supply of central heating and hot water.

In these proceedings, commenced on April 18 1991, the plaintiffs alleged that the residue of the term created by the lease granted to Mr Johnson was vested in the defendants. The statement of claim set out the covenant in the lease in clause 2(1) relating to the payment of rent, the provisions of clause 3(D) and the proviso for re-entry. M

It was alleged that on June 5 1990 the plaintiffs had served a section 146 notice. Para 5 reads:

The defendants have failed to pay the rent and service charge in accordance with the terms of the lease and the arrears amount to £9,890.14.

The plaintiffs asserted that the lease had been forfeited on April 18 1991. The plaintiffs sought to recover, *inter alia*, possession and £9,890.14 arrears of rent and service charges.

During the course of submissions my attention was drawn to the statement of claim. It did not appear that the plaintiffs were entitled to seek judgment for the amount claimed in respect of the service charge.

A It may be, but I do not have to make a determination thereon, that the plaintiffs could seek a declaration that the defendants were obliged to pay to the company the service charge. But that is not what is sought. Mr Michael Daiches for the plaintiffs, while not making a formal concession, was disposed to think that the plaintiffs were not entitled to recover the costs of the service charge.

The defendants served their defence and counterclaim on July 8 1991. Although the heading of the document does not make this apparent, the plaintiffs were also defendants to the defendants' counterclaim. By their defence the defendants did not expressly raise the point that under the terms of the lease the service charge was
B payable to the company and not to the plaintiffs. By para 7 the defendants pleaded the plaintiffs' covenant for quiet enjoyment. By para 8 was set out the company's covenant in clause 7 to perform and observe the obligation set out in the fifth schedule, which they then pleaded in para 9.

Para 10 set out alleged implied terms so far as the company was concerned.

I should read paras 11, 12 and 13:

> 11. The Plaintiffs owed the Defendants a fiduciary duty to ensure that the Company did not breach the covenants and terms referred to in paragraphs 8, 9 and 10 hereof.
>
> C 12. It was an implied term of the lease that in the event of the Company failing to observe the covenants and terms referred to in paragraph 8, 9 and 10 hereof, the Plaintiffs would perform them for the benefit of the tenant.
>
> 13. In breach of the covenant for quiet enjoyment referred to in paragraph 7 hereof and the duty imposed on the Plaintiffs by reason of the matters pleaded in paragraphs 11 and 12 hereof the Plaintiffs by themselves or their agents and in particular the Company in breach of the covenants pleaded in paragraphs 8 and 9 hereof, have allowed the Building to fall into a state of disrepair and condition . . .

Particulars of such disruption were then set out. I do not at this stage refer to other matters pleaded in the defence and counterclaim.

D By a summons dated January 19 1993 the plaintiffs made the following application:

> (a) An application under Order 14A of the RSC for the determination by the Court of the following questions of law, namely whether it was an implied term of the lease dated 20th December 1976 between (1) the Plaintiff and (2) the defendant by Counterclaim (hereinafter referred to as "the Company" and (3) Brian Godfrey Johnson, that in the event of the Company failing to observe the covenants in terms referred to in paras 8, 9 and 10 of the defence and counterclaim, the Plaintiff would perform them for the benefits of the tenants (as alleged in paragraph 12 of the Defence and Counterclaim).
>
> (b) An application under Order 18 r19(l)(a) of the RSC that paragraph 13
> E of the defence and counterclaim be struck out as against the plaintiff as disclosing no reasonable defence to the plaintiffs' claim and/or no reasonable cause of action against the plaintiffs.
>
> (c) That, in the event that the answer to the question of law posed at (a) is "no", the defendants' counterclaim against the plaintiff be struck out under Ord 18 r19(1)(a) of the RSC as disclosing no reasonable cause of action against the plaintiff.
>
> (d) That the applications set out in paras (a) to (c) above also apply to actions number 1992-ORB-1025 to 1051 inclusive and the parties in relation to which are set out in the Schedule below.

Application (d) failed because the defendants in the other actions
F were not present.

As regards Ord 14A I should set out the provisions of that order:

> *Determination of questions of law or construction (O14A r1)*
>
> 1.—(1) The Court may upon the application of a party or of its own motion determine any question of law or construction of any document arising in any cause of matter at any stage of the proceedings where it appears to the Court that —
>
> (*a*) such question is suitable for determination without a full trial of the action, and
>
> (*b*) such determination will finally determine (subject only to any possible appeal) the entire clause or matter or any claim or issue therein.
>
> (2) Upon such determination the Court may dismiss the cause or matter or make such order or judgment as it thinks just.
>
> (3) The Court shall not determine any question under this Order unless the G
> parties have either —
>
> (*a*) had an opportunity of being heard on the question, or
>
> (*b*) consented to an order or judgment on such determination . . .
>
> (5) Nothing in this Order shall limit the powers of the Court under Order 18, rule 19 or any other provision of these rules.
>
> *Manner in which application under rule 1 may be made (O14A, r2)*
>
> 2. An application under rule 1 may be made by summons or motion or (notwithstanding Order 32, rule 1) may be made orally in the course of any interlocutory application to the Court.

I was not satisfied that it was appropriate to deal with the matter under Ord 14A. In the event, however, it may make little difference H
because if the plaintiffs are successful under (*b*) then the claim against them will in large measure be struck out.

It is well-settled law that a breach of covenant for quiet enjoyment requires a finding that the landlord has breached some duty to the tenant. An omission which was a breach of duty could constitute a breach of a covenant for quiet enjoyment. But an omission causing a tenant damage which is not a breach of any duty owed by the landlord to the tenant cannot amount to a breach of the covenant of quiet enjoyment.

As has been seen the only duties allegedly owed by the plaintiffs to the defendants are those set out in paras 11 and 12 of the defence. As J
to the fiduciary duty alleged in para 11 Mr Stephen Hockman QC frankly and fairly conceded that he was unable to support the existence of such a duty. He did contend that there was an implied term of a similar nature, but this was never clearly formulated and I consider that the plea if made would not be sustainable.

As to the implied term alleged in para 12 my attention was drawn particularly to the decisions of *Halsall* v *Brizell* [1957] Ch 169; *Liverpool City Council* v *Irwin* [1977] AC 239*; *Duke of Westminster* v *Guild* [1985] QB 688†; *Gordon* v *Selico Co Ltd* [1986] 1 EGLR 71; *Barrett* v *Lounova (1982) Ltd* [1990] 1QB 348 and to various paragraphs in *Woodfall* 28th ed. K

A copy of the memorandum and articles of association of the company was proffered in the course of the hearing. As Mr Heckman was concerned to consider these before agreeing to my considering them, in the event I have not considered them at all.

I have therefore the position that the lease contemplated that a lessee would be a member of the company, but I have no indication as to how control of the company was to be exercised. The implication of a term in a lease is dependant upon the same considerations that apply in any other contract. The question is whether the insertion of the term is necessary to give business efficacy to the contract between landlord and tenant. L

An obligation placed upon the tenant may require the imposition of a correlative or corresponding obligation on the landlord. If it can be shown that a landlord is taking the benefit of a covenant he may be under an obligation to accept a corresponding burden. If a lease is silent as to an obligation which necessarily falls on either the landlord or the tenant this may give rise to an implication that the obligation is that of the landlord on the one hand or of the tenant on the other.

Where it is sought to raise an implied term one matter to which the court should have regard is whether the lease provides a comprehensive code for the carrying out of repairs and the payment of them. The more comprehensive the code the less room there is for the M
implication of a term.

The defendants' case is that the lease not only imposed a duty on the lessee to pay the company the maintenance charge but further, by clause 3, the lessee covenanted with the lessor to make such payment and the lease provided that the landlord had a right of re-entry in the event of noncompliance. Yet, say the defendants, the obligation to comply with the fifth schedule was that solely of the company unless on the liquidation of the company the lessor elected to undertake those obligations. The lease was silent as to what was to happen in the event

---

*Editor's note: Also reported at [1976] 238 EG 879.

†Editor's note: Also reported at (1983) 267 EG 762.

of the company failing to discharge its obligations or as to what was to happen in the event of the company going into liquidation and the landlord not making an election.

Yet, say the defendants, these obligations had to be discharged by someone if the company failed to do so.

A question of implication ultimately turns on a detailed examination of the terms of the lease. *Prima facie* the terms of the lease and in particular clause 6(B) militate against the existence of such a term. Even assuming that the lessees could not control the company they had undoubted rights to enforce the compliance by the company of its obligations. Of course these might prove nugatory. The company might go into liquidation and the lessor might not make an election. But that is not the end of the matter.

Where a landlord has an obligation to carry out repairs but fails to do so a court frequently appoints a receiver and manager: see, for example, *Hart* v *Emelkirk Ltd* [1983] 1 WLR 1289; *Daiches* v *Bluelake Investments Ltd* [1985] 2 EGLR 67; *Clayhope Properties Ltd* v *Evans* [1986] 2 EGLR 34 and *Blawdziewicz* v *Diadon Establishment* [1988] 2 EGLR 52.

It is not contended that the court lacked power to appoint a receiver and manager where it is not the landlord but the management company which fails to discharge its obligations. Among the relief sought in the defendants' counterclaim was an order for the appointment of a receiver and manager of the building under Part II of the Landlord and Tenant Act 1987, alternatively under the Supreme Court Act 1981, section 37.

Interesting points were canvassed as to whether such relief could be claimed in the action, whether an official referee, as distinct from a judge of the Chancery Division, had jurisdiction to hear the application and as to whether the application had to be made by originating application rather than by the pleading itself. My attention was drawn, *inter alia*, to section 37 of the 1981 Act, sections 21 and 24 of the 1987 Act and RSC Ord 2 r1, Ord 3 r5, Ord 4 r3, Ord 5 r3, and Ord 97.

But these matters are academic so far as the striking out application is concerned. I do not consider it appropriate to determine in this application this matter, not least because there is no application to do so. The relevant fact is that the defendants are entitled to apply for a receiver if the company fails to discharge its obligations. How this should be done is a matter of no relevance to this application.

The contention that the lease contained the term alleged in para 12 is, I find, unsustainable.

I have reached the conclusion that para 13 of the defence in so far as it raises allegations against plaintiffs should be struck out as therefore should paras 11 and 12. Obviously the defendants' claim for relief from forfeiture stands. As indicated I make no finding as to their right in these proceedings to claim for the appointment of a receiver. Until such time as the plaintiffs amend their statement of claim in respect of the service charge, para 15 of the defence should stand. Nothing of course in this judgment in any way touches on the right of the defendants to pursue their counterclaim against the company.

Because this judgment indirectly affects others this judgment is given in open court.

Chancery Division

March 4 1993

(Before Mr Justice ALDOUS)

HANDEL AND OTHERS v ST STEPHENS CLOSE LTD

Estates Gazette February 5 1994
[1994] 05 EG 159

Landlord and tenant — Parking rights on private areas of block of flats —Whether interlocutory injunction should be granted restraining obstruction of parking — Whether serious issue to be tried

The 42 named plaintiffs are tenants of a block of flats known as St Stephens Close, London NW8; the defendant is the landlord. The premises were built in about 1936 as a purpose-built block of 50 flats. The block is approached down two private roads; one off Avenue Road and another known as Broxwood Way. The block forms part of the Avenue Close Estate and within that estate are private roads and other areas where cars can be parked. Long leases were granted in the early 1970s. The tenants contended that they have the right to park in the areas around the block by reason of the terms of their leases, properly construed, by reason of section 62 of the Law of Property Act 1925 or as a matter of implication.

Held: **On an application by motion for an interlocutory injunction to restrain the landlord from carrying out works to the private ways, there was a serious issue to be tried as to whether the landlord was estopped from denying the tenants a right to park. Although the evidence established a perceived requirement of the tenants to be able to park, the point did not have to be decided. There was evidence that the tenants had for many years parked in the private ways and that raised a serious issue to be tried as to whether section 62(2) of the 1925 Act applied. On the affidavit evidence the tenants had parked in the private areas for years because they thought they were allowed by the landlord. An injunction was granted as damages would not have been an adequate remedy for the plaintiffs.**

The following cases are referred to in this report.

Bulstrode v *Lambeth* [1953] 1 WLR 1064; [1953] 2 All ER 728
Gregg v *Richards* [1926] Ch 521
Habib Bank Ltd v *Habib Bank AG Zurich* [1981] 1 WLR 1265; [1981] 2 All ER 650; [1982] RPC 19, CA
Lokumal (K) & Sons (London) v *Lotte Shipping Co Pt (The August Leonhardt)* [1985] 2 Lloyd's Rep 28, CA
Newman v *Jones* unreported
Pwllbach Colliery Co Ltd v *Woodman* [1915] AC 634

This was an application by notice of motion by the plaintiff tenants, James Handel and 41 others, for an injunction to restrain the defendant landlord, St Stephens Close Ltd, from obstructing parking in certain private roads in St Stephens Close, London NW8.

Simon Berry QC and Phillip Kremen (instructed by Rochman Landau) appeared for the plaintiffs; Kim Lewison QC and Mark Warwick (instructed by Edge & Ellison Calow Easton) represented the defendant.

Giving judgment, **ALDOUS J** said: The 42 named plaintiffs sue on behalf of themselves and on behalf of the St Stephens Close Residents Association. They are the tenants of a block of flats known as St Stephens Close, London NW8. They claim a right to park in the private areas around St Stephens Close.

The defendant is the freehold owner of the flats and the private areas. It claims that the plaintiffs do not have the right to park in those areas. By a letter dated November 9 1992, the defendant informed the plaintiffs that work to provide secure parking bays in the private areas around St Stephens Close would begin on November 16. The erection of such bays would enable the defendant to let them to the plaintiffs and to others. The plaintiffs sought assurances that the work would not take place, but these were not forthcoming. Thus, the plaintiffs issued a notice of motion on November 30 1992, and obtained an *ex parte* order from Knox J restraining the defendant from obstructing or otherwise interfering with the free parking of the plaintiffs. The motion was stood over to come on as a motion by order and directions were given to enable the evidence to be completed.

St Stephens Close was built in about 1936 and is a purpose-built block of 50 flats. They are expensive flats costing a few hundred thousand pounds. They are approached down two private roads. The first is off Avenue Road and the other is a private road known as Broxwood Way, which exits into St Edmund's Terrace. St Stephens Close forms part of the Avenue Close Estate and within that estate are the private roads and other areas where cars can be parked and over the years have been parked.

The defendant was formed in 1973 with Mr Reginald Shaw and Mr Bob Holston as shareholders. The intention was that the defendant would purchase the freehold of St Stephens Close from Regalion Securities Ltd and thereafter the shares of the defendant would be sold to residents of St Stephens Close. The defendant did purchase the freehold, but there was insufficient interest in the purchase of the shares. Thus, the shares were retained by Mr Shaw and Mr Holston. In 1984 those shares were sold to a company controlled by Mr Cook.

In the 1970s long leases were offered to the tenants. These were in most cases purchased in the early 1970s and what was said to be an example of a standard form of lease was exhibited as LH 2 to the first affidavit of Mrs Holston.

After the hearing had been completed, counsel for the plaintiffs became aware that exhibit LH 2 might not be representative of all the leases in St Stephens Close. An adjournment was requested and granted. At the restored hearing, fresh evidence established that the leases of the 10th, 16th, 36th and 38th plaintiffs were different from the leases of the other plaintiffs. Those plaintiffs' leases are in the form exhibited as DRL-1 to the first affidavit of David Liebeck. I will refer to that form of lease as the modified lease so as to differentiate it from the form of lease that was acquired by the other tenants.

The plaintiffs' case is that they have the right to park in the areas around St Stephens Close. They contend that their right comes from: (a) the terms of their leases when properly construed; alternatively, (b) from the terms of their leases which include the right to park by reason of section 62 of the Law of Property Act 1925; and alternatively, (c) because such a right should be implied into their leases. The plaintiffs also submitted that if a right to park was not given by the leases, the defendant was estopped from contending that the plaintiffs did not have such a right and that their leases should be rectified to include such a right.

The defendant disputes the plaintiffs' case. The defendant went further and submitted that there was no serious issue to be tried between the parties based on the LH 2 lease and, on that ground, all but four of the plaintiffs' requests for interlocutory relief should be dismissed. The defendant did accept that there was a serious issue to be tried between the four plaintiffs with modified leases, but submitted that no injunction should be granted having regard to the balance of justice.

I have come to the conclusion that there is a serious issue to be tried between all the plaintiffs and the defendant. As will appear later in my judgment, it is my belief that there is a serious issue to be tried as to whether the defendant is estopped from denying that the plaintiffs have the right to park. It is therefore unnecessary for me to decide the other issues raised, but in deference to the full and interesting submissions of counsel I will shortly set them out.

Construction of the lease

The plaintiffs submitted that there was a serious issue to be tried as to whether the right given in the second schedule of the lease exhibited as LH 1, which gave the tenants the right to pass and re-pass, included the right to park. Counsel for the plaintiffs accepted that, although the grant of a right to pass and re-pass was found in other leases, there was no case in which it had been held that such words granted a right to park. Further, in *Newman* v *Jones* (unreported), Sir Robert Megarry V-C concluded that the right "to go, pass and re-pass" did not *per se* include a right to park. However, in *Bulstrode* v *Lambert* [1953] 1 WLR 1064, Upjohn J when considering a right to pass and re-pass held, at p1071, that it included the right to stop, load and unload. That, he held, was "... only an incident of the right of way expressly granted and may be described as ancillary to that easement, because without that right he cannot substantially enjoy that which has been reserved to him".

It was submitted that the right to park, for example overnight, was an incident of the right granted only in connection with the permitted use. In essence the tenants could not substantially enjoy their tenancies without the right to park.

The evidence does establish a perceived requirement of the tenants to be able to park. However, I doubt whether that evidence is sufficient to establish that the flats, being flats in London, could not be substantially enjoyed without the right to park, even though the value of the flats would be reduced if the leases did not include that right. The plaintiffs' submission that the tenants could not substantially enjoy their tenancies without the right to park appears to me to be somewhat speculative. However, I do not have to decide whether the plaintiffs' case raises a serious issue to be tried and I decline to do so.

Section 62

The plaintiffs submitted that the right to park was imported into their leases by reason of section 62 of the Law of Property Act 1925. The relevant parts of that section are:

> (2) A conveyance of land, having houses or other buildings thereon, shall be deemed to include and shall by virtue of this Act operate to convey, with the land, houses, or other buildings, all outhouses, erections, fixtures, cellars, areas, courts, courtyards, cisterns, sewers, gutters, drains, ways, passages, lights, watercourses, liberties, privileges, easements, rights, and advantages whatsoever, appertaining or reputed to appertain to the land, houses, or other buildings conveyed, or any of them, or any part thereof, or, at the time of conveyance, demised, occupied, or enjoyed with, or reputed or known as part or parcel of or appurtenant to, the land, houses, or other buildings conveyed, or any of them, or any part thereof . . .
>
> (4) This section applies only if and as far as a contrary intention is not expressed in the conveyance, and has effect subject to the terms of the conveyance and to the provisions therein contained.

On the plaintiffs' evidence, both the landlord and the tenants believed that the tenants had and would have the right to park when the long leases were agreed in the 1970s. Further, the plaintiffs' evidence was to the effect that parking had been carried out for many years as if that right existed.

In *Newman* v *Jones*, the tenants had the right "with or without motorcars and other vehicles at all times and for all purposes in connection with the permitted user of the flats to go pass and re-pass over and along the road or driveway leading or adjacent to the block of flats". That is similar to the right granted in this case "to pass and repass . . . over and along all roads and footpaths serving the demised premises". Sir Robert Megarry V-C, held:

> In my opinion, where there is a block of flats, and the tenants in general regularly park their cars within the curtilage of the block, the liberty, privilege, easement, right or advantage of being allowed to do this will rapidly become regarded as being something which appertains or is reputed to appertain to each of the flats in the block, and as being reputed appurtenant to each of those flats. Accordingly, on the grant of a lease of one of the flats, I think that section 62(2) of the Law of Property Act 1925 will operate to give the lessee an easement of car parking appurtenant to his leasehold. I do not think that it matters whether the previous occupant of the particular flat did or did not park their car within the curtilage of the block, or, indeed, whether they had any car. In all ordinary cases the reputation will be that of a right of parking which goes with each of the flats, for there will be no reason for one lessee to have greater rights than another in this respect. The question, "can the tenants park their cars round the block?" would receive a simple yes, and not an answer which distinguished between one flat and another on the basis of whether previous occupants of the flat in question had been accustomed to park their cars round the block.

I do not of course say that such a rule applies to every block of flats. Thus no reputed right to park will arise if the lessor operates some scheme whereby tenants are required to obtain some licence or parking permit from the lessor, or pay for a specified parking space. Again, in the present case, if the yellow parking space had actually been constructed, it would have been at least arguable that the grant of a right to park there was intended when operative, to take the place of any general right to park on the forecourt. Nor can any rights arise under section 62(2) if the contrary intention has been expressed in the lease, thereby bringing section 62(4) into play. But none of these points applies to the present case. In view of *Wright* v *Macadam* [1949] 2 KB 744 (which was not cited in *Copeland* v *Greenhalf* [1952] Ch 488) I feel no hesitation in holding that a right for a landowner to park a car anywhere in a defined area nearby is capable of existing as an easement.

At trial it will be necessary to analyse the evidence as to the perceived right of the tenants in the 1970s. However, there is evidence that the tenants had for many years parked in the private areas around St Stephens Close and perceived that they had a right to park in those areas. Leaving aside section 62(4), that evidence is sufficient, in my view, to raise a serious issue to be tried as to whether section 62(2) applies. If so, the conclusion of the Vice-Chancellor in *Newman* v *Jones* appears apt.

The defendant did not seriously dispute that the wording of the grant and the evidence filed raised a serious issue to be tried as to whether a grant should be implied having regard to section 62(2). The defendant's submissions started from the assumption that such a conclusion could be reached. On behalf of the defendant, it was submitted that subsection (4) of section 62 applied because of the words "but to the exclusion of all others" in clause 1 of the exhibit LH 2 lease. That clause reads:

THE Lessor hereby demises unto the Lessee ALL THAT the flat details of which are specified in Part 3 of the First Schedule hereto and which is for the purposes of identification shown edged red and coloured red on the plans attached hereto (hereinafter called "the demised premises" which flat forms part of the building (hereinafter called "the building") known as St Stephens close Avenue Road Regents Park TOGETHER with (but to the exclusion of all others) the rights set out in the Second Schedule hereto but EXCEPT AND RESERVED the matters set out in the Third Schedule hereto TO HOLD the same unto the Lessee for the term commencing on the date hereof and terminating on the 25th March 2069 PAYING therefor during the said term the yearly rents specified in Part 2 of the First Schedule hereto all such rents being payable by quarterly instalments in advance on the usual.

On behalf of the defendant, it was submitted that pursuant to clause 1 all rights were expressly excluded, including the right given by section 62(2) of the 1925 Act, except those actually set out in the second Schedule.

In *Gregg* v *Richards* [1926] Ch 521, the plaintiff claimed the right to use the whole width of a road for the purpose of access of vehicles to the back of her premises. The Court of Appeal held that such a right passed to her by the predecessor to section 62 of the 1925 Act. At p534 Sargant LJ said:

I think it is important to point out that because the words are in the statute and do not appear in the conveyance they must not be looked upon as mere implied words or anything of the sort. Lindley LJ said in terms that it must be looked on as being an express grant, though no doubt an express grant which may fail altogether if an intention is expressed to exclude it, or may be limited if an intention is shown to limit it. It is a little difficult to see how that is to be dealt with in practice, but I think perhaps the way to regard it may be this, to consider that in such a conveyance there is added to the parcels a printed form of words such as you find in the common form printed forms of clause, and that then if the conveyance expresses an intention to exclude, you may consider that printed form of words struck out in ink and removed from the conveyance in that way; while, if an intention is shown to limit or alter the form, the limitation or alteration being introduced in the particular conveyance would of course prevail, according to the ordinary rules of construction of such documents, over the common form which you find in the print. It may be that that would practically give effect to the provisions of the statute, but it is to be noticed that it is by way of express grant that it operates and not by way of implied grant, and that is for the grantor who seeks to show that that express grant is limited to prove affirmatively that there is some limitation of that express grant.

On behalf of the plaintiffs, it was submitted that I should notionally do that which was suggested by Sargant LJ, namely write into the lease the right to park and thereafter consider whether the lease contained an express intention to exclude it. If so, I should write into the second Schedule the right to park and not cross it out as the words in clause 1 did not express an intention to have it excluded. In essence, it was submitted that the words "but to the exclusion of all others" excluded implied rights, but not expressed rights whether they were actually expressed in the second Schedule or were deemed to be so expressed by the statute.

The plaintiffs also submitted that the words "to the exclusion of all others" referred to other persons not other rights or only excluded rights of a nature similar to those expressed. If so, section 62(4) did not apply. In any case, the correct construction and effect of those words should not be decided on motion.

During the hearing, I favoured the submissions of the defendant. However, as I have said, I have come to the conclusion that there is a serious issue to be tried and therefore there is no need for me to express a concluded opinion. I decline to do so.

Implied term

The plaintiffs submitted that if the right to pass and re-pass did not include the right to park, then such a right should be included by implication: see *Pwllbach Colliery Co Ltd* v *Woodman* [1915] AC 634. The defendant did not submit that such a right could not be implied in appropriate circumstances, but submitted that this was not such a case. The defendant drew attention to the words "but to the exclusion of all others" in clause 1 of the lease and submitted that no rights other than those referred to in the second Schedule could exist.

If the words "but to the exclusion of all others" in clause 1 exclude the effect of section 62(2) of the 1925 Act, I believe they must also exclude any implied right. However, as I have said, I have come to the conclusion that there is a serious issue to be tried and therefore need not decide this issue either.

Estoppel

I cannot decide issues of fact at this stage of the proceedings. Thus, to decide whether there is a serious issue to be tried, I will assume that the facts set out in the affidavits filed on behalf of the plaintiffs will be proved at trial.

Mr Kim Lewison QC, for the defendant, referred to the statement of Kerr LJ of the law of estoppel in *The August Leonhardt* [1985] 2 Lloyd's Rep 28, at p35.

The applicability of the doctrine of estoppel in any given case can also be tested in another way. There cannot be any estoppel unless the alleged representor has said or done something, or failed to do something, with the result that — across the line between the parties — his action or inaction has produced some belief or expectation in the mind of the alleged representee, so that, depending on the circumstances, it would thereafter no longer be right to allow the alleged representor to resile by challenging the belief or expectation which he has engendered.

That statement of the law differs in words from that set out in *Taylor Fashions Ltd* v *Liverpool Victoria Friendly Society*, which was approved by the Court of Appeal in *Habib Bank Ltd* v *Habib Bank AG Zurich* [1982] RPC 19, at p36.

Furthermore the more recent cases indicate, in my judgment, that the application of the *Ramsden* v *Dyson* principle — whether you call it proprietary estoppel, estoppel by acquiescence or estoppel by encouragement is really immaterial — requires a very much broader approach which is directed rather at ascertaining whether, in particular individual circumstances, it would be unconscionable for a party to be permitted to deny that which, knowingly or unknowingly, he has allowed or encouraged another to assume to his detriment than to inquiring whether the circumstances can be fitted within the confines of some preconceived formula serving as a universal yardstick for every form of unconscionable behaviour.

It is not my task, on motion, to decide whether a difference in wording between those two statements of the law is material and, if so, which is right. It is sufficient for me to decide whether there is a

A serious issue to be tried as to whether the defendant or its predecessors, knowingly or unknowingly, allowed or encouraged the plaintiffs to assume to their detriment that they had the right to park and, if so, whether it would be now unconscionable to allow the defendant to deny that which it has allowed or encouraged the plaintiffs to assume.

Estoppel is, of course, a personal defence and therefore at trial it may be necessary to look at the position of a number of the plaintiffs; but at this stage of the proceedings, I believe I can assume that if one of them has a good defence based on estoppel, then the others have also.

B On behalf of the defendant, it was submitted that it was not suggested in the evidence that the landlord had ever expressly represented to the tenants that they had the right to park. Further, nothing was said about parking when the 1970 leases were granted. Upon the plaintiffs' evidence, the tenants believed that they had the right to park and none of the matters relied on, such as parking discs and road repairs, influenced their minds. According to the defendant, the evidence was to the effect that the tenants parked their cars and believed they had the right, but no action or inaction of the defendant influenced that belief.

I have come to the conclusion that the evidence does establish a

C serious issue to be tried. The flats were built in about 1936 and up to about 1970 they were let on leases which were bought and sold. Mrs Holston has lived in her flat since 1954. She said in para 8 of her first affidavit:

> 8. To my knowledge successive landlords of the Estate including the present landlords who are the Defendants to this Action and their respective managing agents have throughout treated the residents on the Estate as entitled to free and uninterrupted parking along the Road. When flats are sold by local estate agents they are always advertised as having private parking rights and this would be known by St Stephen's Close Limited and their agents, Flintcote Management Limited ("Flintcote"). Evidence of this is in part provided by a
>
> D bundle of correspondence now produced and shown to me marked "LH3".

In para 10 of her second affidavit she said:

> 10. Had my husband or I been told then that the Landlord could interfere with our free and uninterrupted parking rights which had been exercised by us since 1954 we would have insisted on that right being expressly granted or provided for in the Lease so as to remove any measure of doubt. The reason why this question was (quite simply) never addressed nor raised was, in my view, because the right existed, had then been exercised for over twenty years by some of the tenants of the block and for about sixteen years by ourselves. Neither Hipparchus nor the defendant (until October 1992) ever asserted otherwise.

E

It is pertinent to ask why Mrs Holston believed the right to park existed. The answer, as she says, was that the landlord allowed parking and never suggested that the right did not exist. Thus, the landlord stood back and by inaction encouraged the residents to believe that they had the right. That was their belief. They acted upon that belief when they purchased their leases in 1970.

Mr Astor purchased the end of a short lease from a German diplomat in about 1959 and has lived in that flat ever since. He was informed by that diplomat that the flat had the benefit of free and unrestricted parking for residents and their visitors. He renewed the lease and subsequently purchased a long lease in 1971. He said in

F para 9 of his affidavit:

> In para 5 of his Affidavit, Mr Pickering states that it is not correct to say that the Managing Agents have treated the residents as if they were "entitled" to free and uninterrupted parking. Since I first moved into the block in 1959 I have always considered this to be an established right, as do my fellow residents, and this is clearly reflected in the prices paid for the original long Leases. That this is the clearly established view is evidenced by the fact that whenever a flat is placed on the market, it is stated in the appropriate estate agents particulars that the block has the benefit of off-street parking ("OSP"). At pages 9 to 14 there are examples of various estate agents particulars (recent and old). For Mr Pickering to suggest that he has never been aware of this is, to say the least, most surprising.

It is a reasonable inference that the diplomat had the belief that free G parking was available because of the action or inaction of the landlord. As Mr Astor said, that belief was communicated to him and thereafter everything that happened appears to have been consistent with the belief being correct. Thus the inaction of the landlord encouraged the belief that there was a right to park, resulting in Mr Astor paying a price consistent with that right existing.

Mr Silvert's parents moved into a flat in St Stephens Close in 1948. He said parking was not interfered with at that time. He was instrumental in the refurbishment of the block on behalf of the landlord and negotiated the sale of long leases with the tenants in the 1970s. He said that the question of parking never arose because he H believed and everyone considered that there was an existing right to park. That, he said, was one of the attractions of the flats.

The belief that there was an existing right to park appears universal and seems to have arisen from custom and the inaction of the landlord. His evidence supports the plaintiffs' case that the long leases were entered into upon the basis that such a right existed. Such evidence is consistent with the plaintiffs' case that the landlords have, over the years, encouraged the tenants to believe that they had the right to park and that it would now be unconscionable for the defendant to resile from that position.

Mr Shaw said in para 6 of his affidavit: J

> 6. I bought the property in the belief that the residents had the right to park on the roadway day and night which I would have confirmed to any purchaser or lessee had I been asked. I do not remember whether I was asked about this in terms but if I was not it can only have been because residents' parking rights were obvious and apparent to everyone concerned. From time to time discussions took place about preventing unauthorised non-residents from using the roadway for the purpose of parking. At some time during my ownership of the landlord company we introduced stickers to identify the cars of residents. We also considered granting specific bays to residents but we came to the conclusion that this would be impracticable and cause difficulties in the longer term, especially as by then many of the residents had more than one car and, of course, one also had to allow for visitors parking. Overall the K general consensus was that the only way to operate the roadway in a fair fashion was to allow free parking on the roadway as and where spaces became available (on a first come first serve basis) but taking such steps as were deemed to be practical to prevent unauthorised parking by strangers. As far as I can recollect this unauthorised parking was irregular and casual.

In essence his evidence was that residents' parking rights obviously existed. That view is likely to have been due to the inaction of the landlords, which has continued until 1992.

The exchange of correspondence between the solicitors acting for the lessees of penthouse no 2 and the defendant's solicitors is also consistent with the plaintiffs' evidence that they were encouraged to L believe they had the right to park. The defendant's solicitors were asked: "What are the provisions for parking along St Stephens Close?" The reply was: "Resident parking is on a first come first served basis. There is no specific allocation between flats".

According to the plaintiffs' evidence, the tenants have exercised a right to park since 1936 and they have for nearly 50 years believed that they had that right. It is, in my view, reasonably arguable that both the landlord and the tenants have over many years acted upon the assumption that the tenants had a right to park and that it would now be inequitable for the present landlord to be allowed to resile from that position. M

There is a dispute between the parties as to whether the tenants' belief was encouraged by window stickers and the erection of a hut and gates. There is also a dispute as to whether the tenants paid for improvements to enable parking, such as work to strengthen the road. The plaintiffs also believe that the defendant had knowledge that the leases were being advertised and sold upon representations by estate agents that off-street parking was available and have taken no action. Those are matters that will have to be gone into in detail at the trial. However, I believe that on the evidence there is a serious issue to be tried as to whether the conduct of the landlords from the 1940s up to 1992 encouraged the belief that a right to park existed and, if so,

A whether that belief was acted upon by the tenants to their detriment. I conclude that I should go on to consider whether damages would be an adequate remedy for the plaintiffs and, if not, where the balance of justice lies.

Pending trial, the defendant wishes to carry out work to provide secure parking bays. This will involve erecting posts in the private areas surrounding St Stephens Close. Thereafter, the defendant will issue keys to the tenants which will enable them to park in one space and also to park in six spaces close to the flats if they are vacant. The defendant wishes to let out the other spaces and I understand that there is a market for them at a price of about £2,500 pa each. That is B unacceptable to the plaintiffs. Their evidence points to certain tenants having two or more cars and some being disabled. In any case, the plaintiffs believe that the work will cause severe disruption which they contend will amount to a substantial and continuing interference with their right to park. The plaintiffs realise that any money that they will have to pay to park will be held in a joint account in the names of solicitors and therefore it will be returned to them if they succeed at trial. However, they wish to be able to park anywhere in the private areas where there is a space and submit that damages will not be an adequate remedy for the disturbance caused by the works and the inconvenience of having to park in one allocated bay.

C The proposed work to provide secure parking bays will substantially interfere with the plaintiffs' claimed right to park and I believe that damages will not adequately compensate them for that interference nor the disturbance created by the work. Not all the plaintiffs will have parking bays close to the entrance to St Stephens Close and in any case no provision will be made for two cars, let alone visitors'.

As to the defendant's damage, it was not disputed that the plaintiffs could pay any award of damages ordered pursuant to the cross-undertaking as to damages. That being so, it is likely that any loss suffered by reason of the grant of an injunction could be reasonably accurately calculated.

D I believe that I should, pending trial, maintain the status quo. I do not believe that it would be right to allow the defendant to inflict upon the plaintiffs the disturbance that will occur if the security bays are erected, particularly as damages would be a reasonably adequate remedy for the defendant under the cross-undertaking as to damages.

The defendant submitted that if I came to the conclusion that an injunction pending judgment in the action should be granted, it should be limited to the areas adjacent to St Stephens Close and exclude the access road known as Broxwood Way. I realise that the plaintiffs park in Broxwood Way only if there is no parking available adjacent to the flats, but the compromise suggested does not appeal to me. If E the plaintiffs have the right to park in Broxwood Way, that right will be disturbed by the erection of security parking bays. Damages would be unlikely to be adequate compensation for the plaintiffs and I conclude that it would be best to maintain the present position, a position that has existed since 1936, until after trial. That means that the injunction will prevent erection of bays in Broxwood Way.

Interlocutory injunction granted.

F

Chancery Division G

September 22 1993

(Before Mr Gavin LIGHTMAN QC, sitting as a deputy judge of the division)

IN RE VEDMAY LTD

Estates Gazette March 12 1994

[1994] 10 EG 108

H ***Insolvency — Landlord and tenant — Statutory tenancy — Terms of discontinuance of application by statutory tenants for a vesting order — Whether statutory tenants had locus standi to apply for vesting orders — Whether statutory tenants should be excluded from all interest in the property — Costs***

By a lease ("the headlease") dated May 26 1934, 31-41 Sussex Place, London W2, which consists of five flats and four shops, was demised for a term of 60 years from September 29 1933. The lease was assigned to Vedmay Ltd on September 2 1983. On June 24 1990 the underleases of two flats held by the applicants expired and thereafter the applicants remained in possession as statutory tenants. On July 24 1992 the liquidator of Vedmay Ltd disclaimed the J headlease. On October 29 1992 the applicants applied for vesting orders under the Insolvency Act 1986. In the present application the statutory tenants applied for orders of discontinuance. Although the substantive application was not opposed by the respondent landlord, it was contended on its behalf that such orders should only be made on terms for the statutory tenants should be excluded from all interest in the flats and pay the costs of the respondents on the applications for vesting orders.

Held: **The application was allowed. Statutory tenants have sufficient interest in the affected property to apply for vesting orders. However, where a statutory tenant declines to accept a** K **tenancy upon such an application, he cannot be excluded from all interest in the property under section 182(4) of the Insolvency Act 1986; his statutory tenancy cannot be so determined and such exclusion therefore cannot be a term of discontinuing such applications. The costs of the application for the orders should be paid by the tenants with no order as to the costs of the present application.**

The following cases are referred to in this report.

A E Realisations (1985) Ltd, In re [1988] 1 WLR 200; [1987] 3 All ER 83 L
Finlay, In Re, ex parte Clothworkers' Co (1888) 21 QBD 475

This was an application, by Margaret Ritson and Alexander Begg, to discontinue applications for vesting orders in respect of their interests as statutory tenants in premises the freehold of which was owned by the respondents, the Church Commissioners for England.

Kim Lewison QC (instructed by Denton Hall Burgin & Warrens) appeared for the applicants; Robert Reid QC and Timothy Harry (instructed by Radcliffes & Co) represented the respondents.

Giving judgment, **Mr Gavin Lightman QC** said: I have before me an application by Margaret Ritson ("Miss Ritson") and Alexander M Begg ("Mr Begg"), whom I shall call the statutory tenants, for orders for discontinuance of applications made by them on October 29 1992 for vesting orders in respect of the leasehold premises at flats C and E, 35 Sussex Place, London W2, occupied by them as statutory subtenants.

The occasion for this application was the disclaimer of the headlease of the property, 31-41 Sussex Place, by the liquidator of the then lessee, Vedmay Ltd. The respondent, the Church Commissioners for England, do not oppose the making of such orders, but contend that such orders should only be made on terms that the statutory tenants, one, be excluded from all interest in the flats and, two, pay the

A costs of the Church Commissioners of the applications for the vesting orders.

Mr Robert Reid QC, for the Church Commissioners, contends that the condition for payment of costs was the only condition applicable on a discontinuance where there are no special factors justifying any departure from the ordinary rule. As regards the condition for surrendering the statutory tenancy, he rests his case on two sections of the Insolvency Act 1986. He contends, first, that the statutory tenants were indeed entitled to apply for vesting orders as persons interested within the meaning of section 181 of the Insolvency Act 1986, and that the proceedings were accordingly properly constituted. He
B contends, second, that on that application it was open to the court to order the vesting of the leases in the statutory tenants as sublessees within the meaning of section 182 and that if they had declined to accept such order their interest would have been automatically determined under section 182 (4).

Accordingly, he submits that on a discontinuance the court should insist that as a condition they submit to an order for determination of their interests, a determination which would automatically have had effect if they had proceeded with their application, but declined to accept the order for a vesting.

Mr Kim Lewison QC, for the statutory subtenants, challenges each
C of these contentions unabashed by the necessity for this purpose of contending that his clients' applications for a vesting order, which initiated these proceedings, was totally misconceived.

The relevant facts and dates are as follows. These proceedings relate to a property which I shall refer to as "the property", namely 31-41 Sussex Place, which consists of five flats and four shops. By a lease ("the lease") dated May 26 1934 made between: (1) the trustees of Paddington Estate (who were the freeholders), (2) the Church Commissioners, and (3) Eric C Lyne Tutt, the property was let for 60 years from September 29 1933. By January 1 1953 the freehold had become vested in the Church Commissioners. On October 16 1981 the
D then lessees underlet flat C, 35 Sussex Place, to Mr Begg for eight years from June 24 1982. On September 6 1982 a similar underlease was granted of flat E, 35 Sussex Place, to A H Begg, again for eight years from June 24 1982. I shall refer to these as "the flats".

On September 2 1983 the lease was assigned to Vedmay Ltd. On June 18 1990 Mr Begg assigned the underlease of flat E to Miss Ritson. On June 24 1990 the underleases of flats C and E expired and it is common ground that the subtenants became statutory tenants. On February 19 1992 Vedmay Ltd was wound up and on May 14 1992 a Mr Wesley was appointed liquidator. On July 24 1992 a notice of disclaimer was served by Mr Wesley on Mr Begg and Miss Ritson. On
E October 29 1992 application for a vesting order of flats C and E were made by the statutory tenants. At some date in August or September 1993, those statutory tenants made application for discontinuance of that application for a vesting order.

Working ahead, on September 29 1993 the headlease will expire and it is common ground that on its expiration, in view of the mixed business and residential users of the property, in accordance with two recently decided decisions of the Court of Appeal*, at any rate unless the House of Lords should decide otherwise, the statutory tenancies will determine.

I turn now to the relevant legislation. Section 178 of the Insolvency
F Act 1986 empowers a liquidator to give notice of disclaimer of a lease and the liquidator duly gave such notice in this case on July 24 1992. Section 179, however, provides that such disclaimer shall not take effect if an application is made to the court under section 181 until the court so directs.

Under section 181 an application may be made to the court by any person who claims an interest in the disclaimed property or any person who is under any liability in respect of the disclaimed property not being a liability discharged by the disclaimer.

*Editor's note: See *Pittalis* v *Grant* [1989] 2 EGLR 90; *Bromley Park Garden Estates Ltd* v *George* [1991] 2 EGLR 95.

The first question I have to consider is whether the statutory tenant G
had *locus standi* to make the application under section 181, ie did they have an interest in the disclaimed property. The disclaimed property for this purpose is the lease. In my judgment, the answer is in the affirmative. The term "interest" is not, in my view, confined to a proprietary interest. It extends to any financial interest in the subsistence or otherwise of the lease and includes, in particular, any interest that would be adversely affected by the disclaimer.

It is for this reason and on this ground that a landlord himself may be an interested party entitled to make an application under this section: see *In Re Finlay, ex parte Clothworkers' Co* (1888) 21 QBD 475 at pp484-485 and *In Re A E Realisations (1985) Ltd* [1988] 1 H
WLR 200. It seems to me that there is no reason in principle why a statutory tenant, if he may be adversely affected by the disclaimer, should not be qualified to apply notwithstanding that he has no lease himself or other proprietary interest, but enjoys merely a personal right to retain possession or, as it has been called, a mere status of irremovability: see *Megarry on The Rent Acts*, vol I, p252. It is common ground that on the facts of this case the statutory tenants may be so affected.

The second question accordingly arises, whether section 182(4) requires such applicants in default of acceptance of a vesting order to be excluded from all interest in the property and accordingly to be J
deprived of their statutory tenancy.

Section 182(4) provides that, where upon an application under section 181 the court orders the vesting of a disclaimed lease in an underlessee and the underlessee declines to accept the order, that person is excluded from all interest in the property.

In my view, section 182(4) has no possible application. It is limited in its application to cases where there is an underlessee with a proprietary interest in the land. In that case, on an application to the court by an interested party, for example a landlord, if the underlessee declines to accept a vesting order, that proprietary interest is statutorily extinguished. But a statutory tenant is on no basis an underlessee and on no basis has any proprietary interest in the property to be K
extinguished. There can be no determination of a statutory tenancy under section 182(4) whether the application is made for a vesting order by the landlord, tenant or statutory tenant.

Accordingly, it seems to me, the statutory tenants were in principle qualified to make the application and, further, since in the proceedings whatever the outcome they were not at risk of losing their statutory protection, there is no statutory or other justification for requiring as a condition of discontinuance the surrender or loss of their status.

As regards the second condition, I can see no reason why in respect of the costs, other than the costs of the application for discontinuance, the statutory tenants should not be ordered to pay the cost of a L
discontinued application for the vesting orders. It is common ground that my discretion as to costs is an exercise on ordinary principles and on these principles I can reach no other conclusion. Mr Lewison seeks a special order, namely no order as to costs, on the grounds that the action has now become academic with the lease now practically expired and it is in view of this fact, he says, that the action is discontinued and that in these circumstances a fair and proper course is to make no order.

I do not agree. The length of the headlease was apparent from the beginning of the action and no suggestion has been made that any blame for any delay in obtaining a trial date is attributable to the M
Church Commissioners. I, accordingly, order that the cost of the application for a vesting order, other than the cost of the application for discontinuance, should be paid by the statutory tenants. As regards the cost of the application for discontinuance, it seems to me that the proper order is no order as to costs. Both parties have succeeded and lost on one issue. But I have not heard argument on this question and if either or both of the parties wish to seek to persuade me that some other order ought to be made, I shall of course listen to what they have to say.

Application allowed.

Chancery Division

September 2 1993

(Before Mr Justice LINDSAY)

CREDIT SUISSE v BEEGAS NOMINEES LTD

Estates Gazette March 19 1994 and March 26 1994

[1994] 11 EG 151 and [1994] 12 EG 189

Landlord and tenant — Recently erected building — Covenant to repair — Cladding joints allowed ingress of water — Whether breach of covenant — Assessment of damages

By an agreement dated December 14 1977 D, a developer, contracted with a trustee for the defendant landlord that, in the events which happened, to erect a building to consist mainly of offices at 66-67 St James's Street, London SW1. The building was erected and clad by G, a subcontractor, in aluminium panels each of which interlocked both vertically and horizontally with its neighbours. The windows consisted of double-glazed units inserted into the panels. On January 28 1985 D, who had been granted a headlease of the building, and who had found R, who had taken possession under an agreement to take an underlease in July 1982, sold the headlease to the trustee; on the same date the underlease was granted to R for a term of 25 years from July 25 1983. On June 26 1985 the plaintiff tenant took an assignment of the underlease and at all material times held the term directly from the defendant landlord. Under the terms of the underlease the landlord covenanted by clause 5(c) (subject to payment of a service charge representing 75% of the costs) "to maintain repair amend renew . . . and otherwise keep in good and tenantable condition . . . the structure . . . roof . . . and walls . . .". From October 1982 water leaked through the panels into the building and both D and G were informed; by reason of the failure of D or G to cure the cause of those leaks no final certificate was issued in respect of the 1977 agreement. Water continued to penetrate the panels and in January 1988 the plaintiff resolved to move its West End office elsewhere. In January 1989 the rent was agreed at review at £365,000 (on the basis of £45 per sq ft). Thereafter the plaintiff sought to market the underlease but, because of the leaking panels, one prospective assignee withdrew and no further prospective assignee could be found. Meanwhile, in proceedings by the landlord against D and G, G consented to a *Tomlin* order by which, in the events which happened, it agreed to pay the defendant £2.3m as compensation for the failed cladding. The tenant brought the present proceedings against the landlord, claiming damages for breach of clause 5(c). At the end of the trial, during closing speeches, the landlord applied to amend its pleading to enlarge its claim to seek recovery of the appropriate proportion of the costs of replacing the cladding under the service charge provisions of the underlease.

Held: **1. The landlord was in breach of clause 5(c). In order to put and keep premises in a specified condition of good and tenantable repair the work needed may not be a repair strictly so called. In such cases the required condition is not by reference to the state of the actual building at the date of the demise, but the requirements as to condition of reasonably minded tenants of the class likely to take the premises at that date. In the present case that meant a condition that the building would be substantially watertight. The cladding was not put into that condition. Having regard to "good and tenantable condition" the replacement of the cladding was within clause 5(c), it was not a substantial rebuilding and the cause of the water leaks was irrelevant. Further, having regard to the obligation "to repair, amend and renew" in the clause, the evidence showed that the water leaks had become worse owing to the movement of the panels and the failure of the seals at their joints. Accordingly, there was a state of disrepair. Although the replacement of the cladding was too extensive to amount to a "repair" it fell within the obligation to "renew and amend".**

2. The following damages were recoverable: (1) general damages for inconvenience suffered by staff and customers £40,000; loss of premium expected on sale of underlease and abortive costs in attempts to dispose of the same £260,173; rent, insurance and service charges payable after the date when the underlease would have been disposed to trial of £1.642m, and continuing liability to pay rent to the term date of 2008 in the sum of £2.298m after deducting the landlord's counterclaim for arrears of rent.

3. The landlord's application to amend the pleadings was disallowed on the ground, *inter alia*, that by reason of the recovery of £2.3m from G there was no injustice if the landlord could not recover the costs of replacing the cladding under the service charge provisions from the tenant.

The following cases are referred to in this report.

A/S Awilco of Oslo v *Fulvia SpA di Navigazione of Cagliari* [1981] 1 WLR 314; [1981] 1 All ER 652, HL

Amalgamated Investment & Property Co Ltd (in liquidation) v *Texas Commerce International Bank Ltd* [1982] QB 84; [1981] 3 WLR 565; [1981] 3 All ER 577; [1981] Com LR 236, CA

Anstruther-Gough-Calthorpe v *McOscar* [1924] 1 KB 716, CA

Bailey (CH) Ltd v *Memorial Enterprises Ltd* [1974] 1 WLR 728; [1974] 1 All ER 1003; [1974] EGD 211; (1973) 229 EG 613, CA

Barrett v *Lounova (1982) Ltd* [1990] 1 QB 348; [1989] 2 WLR 137; [1989] 1 All ER 351; [1988] 2 EGLR 54; [1988] 36 EG 184, CA

Battishill v *Read* (1856) 18 CP 696

Calabar Properties Ltd v *Stitcher* [1984] 1 WLR 287; [1983] 3 All ER 759; [1983] EGD 578; (1983) 268 EG 697, CA

Collins v *Flynn* [1963] 2 All ER 1068

Lurcott v *Wakely and Wheeler* [1911] 1 KB 905, CA

McDougall v *Easington District Council* (1989) 87 LGR 527; (1989) 58 P&CR 201; [1989] 1 EGLR 93, CA

Mira v *Aylmer Square Investments Ltd* [1990] 1 EGLR 45; [1990] 22 EG 61, CA

Norwegian American Cruises A/S v *Paul Mundy Ltd; Vistafjord, The* [1988] 2 Lloyd's Rep 343, CA

Norwich Union Life Insurance Society v *British Railways Board* [1987] 2 EGLR 137, (1987) 283 EG 846

Plough Investments Ltd v *Manchester City Council* [1989] 1 EGLR 244

Post Office v *Aquarius Properties Ltd* [1987] 1 All ER 1055; [1987] 1 EGLR 40; (1987) 281 EG 798, CA

Proudfoot v *Hart* (1890) 25 QBD 42

Ravenseft Properties Ltd v *Davstone (Holdings) Ltd* [1980] QB 12; [1979] 2 WLR 897; [1979] 1 All ER 929; (1978) 37 P&CR 502; [1979] EGD 316; 249 EG 51, DC

Smedley v *Chumley & Hawke Ltd* (1981) 44 P&CR 50; [1982] EGD 473; 261 EG 775, CA

Stent v *Monmouth District Council* (1987) 19 HLR 269; [1987] 1 EGLR 59; 282 EG 705, CA

Torrens v *Walker* [1906] 2 Ch 166

This is a claim for damages by the plaintiff tenant, Credit Suisse, for breach of repairing obligations under an underlease dated January 28 1985 against the defendant landlord, Beegas Nominees Ltd.

Jonathan Gaunt QC and Richard Gray QC (instructed by Fladgate Fielder) appeared for the plaintiff; Brian Knight QC and Martin Rodger (instructed by Braby & Waller) represented the defendant.

Giving judgment, LINDSAY J said: This case is between tenant and landlord and concerns water leaking from the exterior into a very striking modern office and residential building known as 66 St James's Street, London SW1. The building is sometimes referred to as "Map House" but, to some, a little unfairly, it is also known (by reason of its almost circular towers or turrets and metallic appearance) as "The Rocket Ship". The freehold of the building is owned by the defendant landlord and it is still let, as to parts up to and including the fourth floor, to, and was earlier occupied by, the plaintiff tenant. Above the fourth floor (which is residential) there is a residential penthouse which is not let to the plaintiff and which plays little part in the events. I shall be principally concerned with the ground to fourth floors. The ground floor and the first to third floors are offices and the

A ground floor was used by the plaintiff as its West End banking hall. The total approximate area, let to and previously occupied by the plaintiff, is some 8,120 sq ft.

No 66, as my reference to "The Rocket Ship" may have suggested, presents an unusual appearance. It is on a site at the corner of St James's Place and St James's Street; it is roughly "L" shaped with the left and lower faces of the L facing those respective streets. At the corner of the L there is a tower describing a good portion of a circle and capped at the top by a sloping and very tall elliptical window some two storeys or so high. Less emphatic turrets with similar sloping windows mark the other ends of the "L". On the outside of the B building there are a number of circular or rectangular vertical ducts which are made of, or are clad in, metal to match in appearance the dark bronze anodised aluminium which clads the rest of the building. Each of the three turrets has set in it on each floor, with which I am concerned, curved double-glazing units. In between those curved units, on each of the flat lengths of the "L", are flat double-glazing units on each floor. The penthouse's outer walls slope sharply away from the street rather as would a mansard roof, and almost all the ducts bend in conjunction with those outer walls of the penthouse. No 66 was completed in the very early 1980s and, like much of modern architecture, it has both adherents and opponents, but all, I apprehend, would agree that it is in a prominent position on a major thoroughfare C in the capital city, an area (and here I quote from a brochure produced by the plaintiff's agents) "now recognised as a premier office location by many well-established and international organisations".

The questions which arise for decision relate principally to a covenant in the plaintiff's underlease which is of a kind usually (but without necessarily thereby completely indicating its true nature) called a repairing covenant. Water leaks from the exterior to the interior, says the plaintiff tenant, because there are now and have been breaches of the repairing covenant in relation to the aluminium cladding of the building.

As it may become material to consider the class of tenant which at D and around the time of the relevant demise would be likely to take the underlease of no 66, I should mention that the first underlessee was J Rothschild & Co Ltd and the surety Charterhouse J Rothschild plc. Rothschild's assignee was the plaintiff, Credit Suisse, a bank with a rating of Aaa at Moody's for its long-term deposits, one of the world's strongest banks and the third largest and most profitable of Swiss banks. The landlord-defendant is custodian trustee of the British Gas Corporation Pension Scheme. The scheme's property manager has the supervision, through agents, of literally hundreds of tenancies and the scheme has immense resources. I am entitled to regard the defendant as a respectable landlord, well able, out of its own resources, to E finance any of the works to which I will later need to turn in more detail.

The convenient course will be for me to divide this judgment into a number of broad headings as follows:

(i) A description of the cladding and the technical terms which are used in relation to it;

(ii) The conveyancing title with which I shall be concerned, giving, in particular, the relevant terms of the plaintiff's underlease;

(iii) A chronology of the relevant events;

(iv) Breach or nay: whether on the true construction of the underlease and in the events which have happened the defendant has F breached the terms of the repairing covenant in the underlease?

(v) If the answer to breach or nay is breach, then the next subject will be damages: were any suffered and, if they were, how should they be computed?

(vi) Estoppel;

(vii) Counterclaim;

(viii) Finally, I shall mention a point on the pleadings and then attempt to summarise my conclusions.

Cladding

No 66's two faces to the streets are clad in bronze-coloured anodised-aluminium panels each of which interlocks both vertically and horizontally with its neighbours. On the turrets or towers at the G three extremities of the L-shape of no 66 there are curved panels, but the rest of the panels are flat. Each separate flat panel is 3.3 m high and 1.2 m wide. It is convenient to have the outline of a ladder in mind; the 3.3 m lengths are, so to speak, the vertical sides of the ladder and the 1.2 m pieces are as if the rungs, but, unlike a ladder, the rungs are infrequent and not at regular intervals. The sides of the ladder are called "mullions", which are basically hollow vertical rectangular tubes, an inner and an outer, with additions and plates designed to link into the neighbouring mullions in such a way as also to accommodate horizontal expansion and contraction between one panel and the next. The mullions also have horizontal divisions to accommodate H expansion and contraction in the vertical plane as between one panel and the next. As for the "rungs", the two most important are called "transoms". The gap between the head transom and the sill transom (one above and the other below a window) is a space designed to be fitted with one of the double-glazing units ("DGUs") used at no 66. Each transom is, in effect, a rectangular hollow tube with additions designed to link to the adjoining mullions and to accommodate the DGU intended for that cladding unit. These additions create the double-glazing rebate. Each transom in a flat panel is fixed to its neighbouring mullions by, as it transpires, screws from the mullions passing into receptacles called "screw ports" in the transom. A typical J DGU is 1.9 m tall in the flat cladding panels; the rest of the 3.3 m height of each flat cladding unit consists not only of the head and sill transoms (one immediately above and one immediately below the window, as I have described), but also of horizontal flat plates of aluminium filling in the rest of the ladder and covered in part by other plates in a rather angular "C" shape with the open end of the letter C pointing in towards the building. The C-section plates are largely or perhaps entirely cosmetic. These horizontal plates running above the head transom and beneath the sill transom are sometimes referred to as "spandrels", a term used in particular to distinguish between these horizontal pieces and the horizontal transoms. The cladding panels I K have described are commonly fixed to stanchions set in the concrete floors of no 66, stanchions which are themselves embedded in and covered by the concrete of the flooring. Apart from the DGUS, the cladding units, which are something of the order of 9 ins in depth inside to out, represent generally the totality of what lies between the fresh air outside and a person standing in no 66. In large part the anodised-bronze finish of the cladding units framing the windows is what one would see looking out from no 66 to the outside, but parts, even of the internal faces of the cladding units, are invisible from within no 66 by reason of the concrete floors of the building and also by reason of suspended ceilings added beneath the upper floors.

There will be reference in the course of my judgment to the so- L called "mullion transom" joint. This term refers, plainly enough, to any joint between a vertical mullion and an adjoining horizontal transom. There are four such joints on each panel (ie at each of two ends of two separate head and sill transoms) but, depending on its context, the term can mean the mullion transom joint visible on the outer face of the building, the upper or lower joint running, inside to outside, along the upper or lower junction between each mullion and each transom or the vertical joint between mullion and transom as visible from inside no 66.

The detailed design of the transoms and hence of the mullion M transom joints differs as between the curved cladding units and the flat cladding units, but the broad description I have given suffices for the moment. The manner in which the DGUs are installed also differs between curved and flat. In flat units DGUs are held in the glazing rebate by neoprene seals, which are held compressed against the glass by way of glazing clips which have been hammered home and then covered by snap-on glazing beads (which beads serve merely to cover and protect the otherwise unsightly and vulnerable glazing clips). The curved DGUs are retained in their glazing rebates by "rubboseal" glazing compound and by an aluminium bead of box section which screws either into the adjoining transom or mullion or both. "Rubboseal" is a proprietary compound which, when its two principal ingredients are mixed together, becomes a rubbery substance which can then be used for bedding glass within a rebate.

The transoms in the flat panels have aluminium vertical "upstands" which are of a piece with the adjoining transom. The outcome is that there is an area, described in the case as "the bath", as to which, if water is allowed in, it may, if not evaporating at a sufficiently fast rate, collect, build up and either escape over the edges of the upstands downwards into the transom below or downwards on to the faces or "reveals" of the mullion below.

The cladding panels or units interlock, as I have mentioned, both horizontally and vertically and it is far from easy to arrange for the inspection or removal of any one panel; indeed, without damaging one or more adjoining panels it may very well be impossible fully to inspect a panel or remove it. DGUs by comparison, though not easy to remove, may be removed, albeit at some risk, especially in relation to curved units, of breakage either of the glass or of the bonding of the sealant between the two sheets of glass.

I have mentioned that there is provision for vertical and horizontal expansion and contraction as between one unit and the next; there is not, it would seem, provision, at all events by way of movement joints, to deal with movement within the confines of each separate panel. Of course, there may be scope for expansion and contraction of each individual transom in its relation with its adjoining mullion by way of the broadening or narrowing of the relevant mullion transom joint, but movement in such joints is a subject to which I shall need to return in more detail below.

I shall need to go in more detail below into some aspects of the cladding and in particular its drainage but, for the moment, that suffices and I turn to my next main heading.

Conveyancing title

On December 14 1977 an agreement was made between Samuel Properties (Developments) Ltd ("the developer") and Williams & Glynn's Trust Co Ltd ("the fund") in its capacity as custodian trustee of the central fund of the Gasworkers Pension Scheme. The fund was to buy the corner site known as 66-67 St James's Street and 1 St James's Place. I shall call the site and the premises eventually built there "no 66". If the fund was able to buy no 66 then the developer was to proceed to erect a building there for shopping, office and residential use. The fund was to finance the development. On completion of the building the fund was to grant the developer a 99-year lease with an option for further renewal. This development agreement made provision for a "certificate of practical completion" and "certificate of final completion". The developer was required to submit plans for the building to be erected. The fund was not unreasonably or unnecessarily to withhold its consent to the developer's plans, but withholding of approval on the ground, *inter alia*, that the building would, in the opinion of the fund, be unlikely to be readily lettable was not to be deemed unreasonable. The fund's surveyors were to have liberty to inspect the site to view the state and progress of the building works. The developer bound itself to endeavour to negotiate the first underlettings of no 66 to a tenant approved by the fund. The 99-year term was to begin on the date of the certificate of practical completion. There were later supplemental agreements which I need not mention. The fund did acquire the freehold of no 66 and the development thus proceeded.

The approved plans provided for a building readily separable into a residential penthouse on its fifth and sixth floors and the rest, which I shall call "the offices". I shall need to say something later about the class of building which emerged or was intended to emerge as no 66 but, to continue with the conveyancing side of the matter, the developer found an underlessee for the offices, J Rothschild & Co Ltd, who, on July 7 1981, agreed to take an underlease. I shall refer to this agreement more fully below. The building was, at that date, still far from completion. On July 16 1982 a certificate of partial practical completion was signed (as to the offices) and on November 10 1982 a further one was signed as to common parts servicing the offices. Rothschild cannot have been in occupation long. It is unclear precisely when they moved in or out; they began paying rent from July or August 1982, presumably under the agreement for an underlease, but by August 1983 they were minded to move out. As yet no headlease from the fund to the developer had been granted, but that emerged on June 25 1984. Then, on January 28 1985, the developer granted J Rothschild & Co a 25-year term from July 25 1983 (which had been the date of practical completion of the penthouse). This is the underlease under which the plaintiff holds and to which I shall need to return later in greater detail. On the same day, January 28 1985, the developer sold its interest in the headlease of June 25 1984 to the fund, subject to and with the benefit of the Rothschild's agreement for an underlease. The intention and the result was that the developer's headlease would be surrendered by operation of law leaving J Rothschild & Co Ltd holding under the terms of their underlease, but directly from the fund. On June 26 1985 J Rothschild & Co Ltd assigned their underlease of the offices at no 66 to the plaintiff for the unexpired residue of their term. The fund had consented to the assignment. The fund and the plaintiff agreed some minor amendments to the underlease to reflect that the plaintiff would wish to use the ground floor of no 66 for retail banking purposes, a change of use for which planning permission was thought to be necessary. The changes included that when the rent review came round if, by then, planning permission had been granted for this intended use then the review would look toward whichever would yield the highest rack-rent of office, showroom and, as was now to be added, banking use. At the time of the assignment to the plaintiff the rent payable under the underlease was £200,000 pa, subject to review.

By virtue of the Royal Bank of Scotland Act 1985 the undertaking of the fund was vested in the Royal Bank of Scotland plc. Presumably there was included the property held by the fund as custodian trustee; certainly the case has proceeded on that footing. In that way (as would it seem to be) the Royal Bank of Scotland became custodian trustee of the British Gas Corporation Pension Scheme, which I take it was a new name for or included the central fund of the erstwhile Gasworkers Pension Scheme. On May 15 1980 the Royal Bank of Scotland transferred the freehold of no 66 to Beegas Nominees Ltd, the defendant, to hold the same as trustee of the British Gas Corporation Pension Scheme. Thus it is that the plaintiff holds no 66 for the residue of term of the underlease of January 28 1985 — until 2008 — of the defendant upon the terms, with slight modification, of the underlease. That suffices to deal with title. I must now look at the underlease in more detail.

In the underlease the terms "tenant" and "landlord" were expanded in conventional ways to include the respective successors in title of J Rothschild & Co and the developer and hence the plaintiff and the defendant fall within those respective definitions. Rent is payable quarterly in advance on the usual quarter days. In addition to "ordinary" rent a service charge described as "further and additional rent" is payable, as well as a contribution to insurance. The tenant's annual service charge was to equal 75% of certain specified outgoings incurred by the landlord performing his, the landlord's, repairing covenant in the year in question, as to the ascertainment of which outgoings there are detailed provisions. It is plain from those quantification provisions that the landlord was to be able recoup outgoings other than only those that were of a periodically recurring nature. The remaining 25% of the specified outgoings were, in effect, attributed to the penthouse and to be recoverable from its tenant. If all went well the landlord would thus recover all its outgoings of a repairing kind.

The tenant covenants to repair and keep the interior and all landlord's fixtures and fittings in good and substantial repair, to decorate and to yield up the premises in such state and condition as should in all respects be consistent with a full and due performance by the tenant of the tenant's covenants. The landlord is entitled to enter no 66 to inspect the state of repair and condition and may by notice require the tenant to make good his shortcomings in that respect. The landlord is expressly entitled to enter no 66 to do work which is its responsibility within its covenants. There is a user clause which (in the

A event which, I apprehend, happened, of planning permission being granted for retail banking on the ground floor) reads as follows:

> 16a. Not to occupy and use the demised premises or any part thereof or permit the same to be used except as to the ground floor and the basement thereof for a shop for retail sales (or upon planning permission being granted as offices or showrooms or for banking purposes), as to the first second and third floors as offices and as to the fourth floor for residential purposes.

I understand that the fourth floor was used for residential purposes during the plaintiff's occupancy and that all other user was within the prescribed terms. The user clause and other clauses expressly forbad a number of uses and practices in a manner consistent with an intention
B of the landlord's part to preserve the value and tone of the premises and there is a covenant against alienation in a full form, including a provision that the landlord should be able to withhold consent to subunderletting where the rent to be paid by the subunderlessee:

> Is less than the full rent and is not reviewable upwards only.

Turning to the landlord's covenant, the crucial one for the purposes of the case is found at clause 5(c) which, with the omission of parts which need not now be read (but with my emphasis added) is as follows:

C (c). (Subject to the payment by the tenant of the rents and service charge) to maintain repair amend renew cleanse repaint and redecorate and *otherwise* keep in good and tenantable condition:

> 1. The structure of the building and in particular the roof foundations and walls thereof but excluding nevertheless therefrom;—
>
> . . .
>
> (ii) the internal faces of boundary walls that enclose the demised premises
>
> 2. The boundary walls . . . of and in the curtilage of the Building . . .
>
> Provided that the landlord shall not be liable to the tenant for any *defect or* want of repair hereinbefore mentioned unless the landlord has had notice thereof nor in respect of any obligation hereunder that is to be constructed [*sic*] as falling within the ambit of any of the tenant's covenants hereinbefore
D contained.

I shall call the passage I have read "the principal covenant". In clause 5(d) the landlord goes on (using the same phrases as in the principal covenant) to covenant to maintain, repair, amend, renew, cleanse, repaint and redecorate and otherwise keep in good and tenantable repair the common parts and service facilities. Different verbs of obligation are used in different parts; for example, clause 5(e)(i) has "to supply maintain repair and renew", clause 5(e)(ii) has "supply provide purchase maintain renew replace repair and keep in good and serviceable order and condition", clause 5 (g)(i) reverts, in relation to the heating and air-conditioning and lighting, to the phrase
E "maintain repair amend renew cleanse and otherwise keep in good and tenantable condition" as used in the principal covenant, as does clause 5(h) in relation to the lifts.

Clause 6 represents several pages of provisions which include, at clause 6(12)(b), that the landlord should not be liable for any breach of the express or implied covenants for renewals repair and maintenance until it shall have written notice of such repairs being necessary and then only after a lapse of a reasonable time for the carrying out of the same. There is also an express provision that the landlord should not be liable to make any refund to the tenant in the event of any breach of the landlord's express or implied covenants. No particular written
F notice is pleaded as complying with clause 6(12)(b), but it has been acknowledged by the defendant that it takes no point as to that absence. There are a number of communications which could amount to notice within clause 6(12)(b), but in any event the case has proceeded as if, so long as notice shall have been clear, it matters not whether it shall have been oral or written.

Clause 7 of the underlease dealt with review of the rent at the expiration of the fifth year and thereafter at five-yearly intervals. The rent can never move downwards; it will always be either as it shall have been in the preceding five years or higher. There is no express assumption of a basis that the landlord shall have performed his covenants.

The third schedule, in addition to passing on to the tenant, as I have G mentioned, 75% of the costs incurred by the landlord in discharging the obligations under, *inter alia*, the principal covenant, provides also for the landlord similarly to pass on:

> 2 The cost of periodically inspecting examining maintaining overhauling and where necessary replacing any and every part of the building and the appurtenances thereof.

referred to in, *inter alia*, the principal covenant.

While I shall need at some point to refer back to the clauses in the underlease which I have already cited and indeed, to some other clauses, that, for the moment, suffices to describe the conveyancing H background of the case before me.

Relevant events; a chronology

In 1979, as the design of the proposed new building on the site of no 66 took shape in the hands of its architects, a decision was made to invite Messrs Josef Gartner ("Gartners") to supply and fit the curtain-wall cladding to the building. Gartner's reputation in relation to cladding and windows was, at all events at that time, a very high one. Because of the nature of the building's design the curtain walling was going to be complex. Terms were agreed and the building was erected with Gartners' cladding forming a high proportion of its external surfaces and in particular the totality of the important J frontages to St James's and St James's Place.

On July 7 1981 an agreement for an underlease of what was to become no 66 was, as I have mentioned, made between the developer, Samuel Properties (Developments) Ltd ("the developer") and J Rothschild & Co Ltd. Annexed to that agreement, as a draft, was a form of underlease in the form which was later granted. The developer was there described as landlord, a term which was expressed to include the estate owner for the time being of the reversion. The term tenant extended to persons deriving title under J Rothschild & Co Ltd. Rothschild's surveyors had previously approved specific plans and a specification which had been produced to them and which described K the building works which were to be done at the site. There was a covenant with Rothschilds that the described works would be done in good and substantial and workmanlike manner and in accordance with the approved plans. Prior to the certificate of practical completion a "snagging list" was to be drawn up in consultation with the tenant's surveyors who could hold up the issue of the certificate of practical completion. Clause 19 of the agreement for an underlease provided that the developer as landlord should, to the reasonable satisfaction of the tenant, make good as soon as practicable all defects *and* wants of repair which might appear within two years after the date of practical completion, all those in the "snagging list" or otherwise which were L such that the contractor was liable to remedy them within the defects' liability period there referred to. I have no doubt but that the parties to the underlease contemplated that the building demised was to be a practical building of a high quality, a quality consistent with its important site and such as would attract tenants of considerable substance.

On July 16 1982 the certificate of partial practical completion relating to basement, ground floor and the first to fourth floors was signed by the architects. That certificate had annexed to it a schedule of outstanding works and defects that went into some detail, but the only leak reported was described as a slight one, on the third floor, and M was required to be investigated by Gartners.

There was a fire at the premises before they were completed but it was in or around the penthouse and, although there were arguments as to its consequences, it did not affect the letting of the lower floors which were to be Rothschilds, who began to pay rent as from July or August 1982. A formal underlease was not, though, for the moment granted.

A little later, in late October 1982, the architects were receiving complaints that in heavy rainfalls the building leaked. Rothschild had expressed concern that their carpets and electronic equipment might be damaged as a result. The architects instructed Gartners, whose men were still then on site, that it was imperative that those men should

A remain until the facade was made completely watertight. The architects prepared a schedule of leaks, being leaks through Gartners facade; Rothschild's surveyors did much the same. As early as late October 1982 Gartners were blaming blocked drainage holes within the cladding and glazing as a cause of the leaks and Gartners were then suggesting a cleaning of the drainage holes as a cure. If a cladding system is a drained system then if its drainage holes become blocked water will collect and may then run into the interior. This is what Gartners asserted was the case.

On November 10 1982 a further certificate of partial practical completion was signed by the architects, but all was not well. Gartners
B were informed of continuing leaks and were asked to give their immediate attention to the position. Gartners blamed fire damage to the silicone seals to deflector spigots and water tests with a hosepipe were arranged. Rothschild were concerned that they should not proceed further with installation of carpets and decorations because of the risk of water damage. However, by March 1983 the contractors understood from Gartners that the remedial work had been attended to. On March 10 1983, it seems, Rothschild confirmed to the architect that no further ingress of water had occurred.

By early August 1983 Rothschild had decided that the premises were unsuitable for their requirements and that they would vacate and
C they told the developer of their intention. It was not said to the developer that leaks were still occurring or that Rothschild were moving out because of them.

On November 9 1983 the architects drew up a schedule of defects at the end of the defects' liability period. No leaks are mentioned in it as such although some discolouration is mentioned in relation to a sill.

On February 9 1984 the architects wrote as follows:

> I have checked with Samuel Properties who report that they believe Gartners facade is watertight. RIT [in effect, Rothschilds] have not complained of leaks recently; I consider it safe to conclude that all is well. I intend to visit the building tomorrow to inspect the repainted Gartner windowsills, which are the
> D only remaining defects in the work included in the first practical completion certificate. If all is well I hope to be able to issue the certificate of making good defects next week.

However, the correspondence by March 26 1984 contains a reference to one leak on the third floor. Gartners arranged to visit while it was raining to trace the leak and put it right.

From March 1983 to March 1984 there is no record of any complaint as to leaks in the evidence in this litigation. Given the very full availability of documents and the interest in the subject on the part of a number of professionals and others, I see the proper inference as being that there was a year — March 1983 to March 1984 — during
E which leaks were either non-existent or had become so trivial as not to be worthy of complaint. The architects did not issue a certificate as to the making good of defects because later on, in April 1984, a typist at no 66 told them of water penetrating the cladding to her office. The architects received other complaints as to the ground, first, second and third floors and Gartners were again called in.

Gartners blamed the blockage of drainage slots or holes as to the upper floors and said that that was found to be the only problem. The cladding was therefore cleaned, but in September 1984 it was found that it still leaked. Gartners were called back in early October to the site, but could not find the cause. They decided to obtain from
F Germany the drawings in accordance with which the cladding was thought to have been constructed.

By now Rothschild had begun a marketing of the underlease. Credit Suisse were interested and on November 5 1984 Rothschild sent them a brochure (which has not survived to form part of the evidence). The drift of the brochure is not hard to guess as Rothschild's agents wrote to the plaintiff's as follows:

> We have pleasure in enclosing herewith a copy of our brochure in relation to this unique office building which stands as a prominent landmark in St James's at the corner of St James's Place. The entire building has been fitted out to an extremely high standard providing a number of executive offices and open-plan areas.

As for leaks, although Gartners indicated that they would take further steps to investigate and remedy, the architects felt it right to G
withhold a final certificate from the contractors because of the problems with the cladding. The architects indicated that they would issue a certificate when the defects had been made good. When Mr Richard Bennett [ARICS] of the landlord's agents first went round the interior of no 66, in late 1984 or early 1985, with a representative of the developer he was told that the leakage problem had been solved and he saw no signs of water penetration (although it is fair to add he was not then looking for them).

In the meantime the plaintiff, Credit Suisse, was studying the building and its suitability for its wish to have a private banking H
presence in the West End. The plaintiff was acting through agents, Dron & Wright, whose Mr John Gordon [FRICS] advised them on December 18 1984, after a cursory inspection of the building, that the costs of maintenance and repair of such a building would likely to be higher, because of its use of untried systems, than the corresponding costs of a building of conventional construction. He was, in general, unimpressed by the building and found serious condensation running down the windows and causing damage to carpets and staining the ceilings below. He recommended attempts being made to cure the condensation. He suggested a budget of £200,000 for the fitting out of no 66 for banking accommodation.

The landlord's agents, then called Harold Williams Bennett & J
Partners, took the view that it would be much to the landlord's advantage in valuation terms to have a tenant of the plaintiff's stature in occupation and so advised the landlord.

The plaintiff, perhaps because of Dron & Wright's unenthusiasm, approved its agents bringing in consulting engineers, the ZMG Partnership, to advise as to the building's structural suitability. They were asked to include the cladding within the subject-matter of their report. In turn ZMG brought in Bickerdike Allan Partners, specialist engineers with particular experience in relation to cladding. ZMG had been told by Gartners that the drawings for the cladding would not be K
available for something like six weeks, although ZMG had already obtained less detailed layout drawings from Pell Frischmann, the developer's engineers who had been instructed while no 66 was being built. At a client-to-client level the plaintiff was informed directly by Rothschilds that the problems as to leakage through the cladding had been solved.

On March 4 1985 Dron & Wright reported by way of a preliminary report to the plaintiff. As to the cladding, they reported as follows:

> The cladding of the building on the two outer elevations is constructed from anodised-aluminium and has generally been well designed and installed but leakage into the hollow section reduces the value of the insulation installed L
> there and, in the long term, may give rise to gradual deterioration. There are certain areas where a cold bridge exists which may give rise to internal condensation, and heat losses through the external envelope may be rather higher than with conventional construction. These features are unlikely to cause serious problems of occupation in the foreseeable future but merely somewhat raised running costs.
>
> The most serious feature of the external cladding is a problem with the seals of the double glazing. It has been found that some 30% of the windows have faulty seals which admit moisture to the metal section into which the glazing is fitted. Moisture in this position will, in the longer term, give rise to damage to the double glazing seals resulting in condensation between the units. In certain instances the leakage is sufficient to allow water penetration M
> to the interior of the building.
>
> The seals may be improved by applying an additional bead of mastic along the top of the existing mastic seal. Such temporary work would need to be renewed from time to time but should, whilst it remains effective, prevent further deterioration. You should anticipate the renewal of the temporary beads every 2/3 years.
>
> The only alternative to this temporary repair is to take out and reglaze the windows, and at the same time, incorporate a modification to prevent a recurrence of the problem. The costs of these two alternative solutions are set out in the schedule below.

Later in the same report they advised as follows:

A At first the condition of the external cladding had given us grave cause for concern but, more recently, we have examined this in more detail with Bickerdike Allan Partnership and now understand the problems are associated largely with the glazing rather than with the general cladding system. Because of the nature of the cladding and difficulty of access maintenance costs will inevitably be high. Whilst substantial sums of money may be needed to remedy the faults that exists with the glazing, they are, nevertheless, capable of repair and provided you are prepared to budget for expenditure involved and regard it as part of the occupation costs of the building, these need not deter you from proceeding to take up the lease. The most serious consideration is the lack of direct control over the carrying out of these works and this is something you may choose to take up with the lessors prior to your taking the assignment of the lease.

B

Dron & Wright went on to give an estimate of the cost of the temporary making good of windowsills as £4,000 and for the refurbishment of windows over a period of between five and 15 years as to £50,000 to £70,000.

On March 8 1985 Dron & Wright sent the plaintiff the final report of ZMG into which Bickerdike Allan's report was incorporated. Looking to the structure generally, ZMG reported that no 66 was generally well designed and should not require more than the normal routine maintenance and expense to be expected for a structure of its kind, but Bickerdike Allan had real doubts. Neoprene seals might need C to be renewed during the life of the building; water could collect in the cladding and overflow if the drainage holes understood to exist became blocked. Indeed, there were some signs of seepage. Leakage through windows was a serious problem, they thought, probably by reason of the blockage of drainage holes. About a third of the windows were affected. This threatened not only the cosmetic appearance of things but also the failure of DGUs. Making good the windowsills would cost about £4,000 they thought; replacement of the DGUs would cost some £50,000 to £70,000 over a span of five to 15 years, although the DGUs could not be expected to fail simultaneously and it would be difficult to determine the point at which replacement of all D would become necessary as opposed to replacement window by window as they might fail. Bickerdike Allan warned that no 66 could involve a lessee in substantial maintenance and remedial costs and disruption. ZMG and Bickerdike Allan had had produced to them some Gartner drawings or extracts therefrom, some of which drawings were marked (either by Gartners or by those to whom the drawings were produced) to show drainage holes which had not been shown on the original Gartner drawings, but which were now marked in positions said by Gartners to be the appropriate positions for those drainage holes.

Dron & Wright, I apprehend, had also been told by Gartners that the cladding had drainage holes.

E The building was leaking and the plaintiff, who had not as yet committed itself to an assignment of the underlease, sent a copy of the ZMG report to the defendant's agents. There had been a serious ingress of water during the then recent heavy rain. The defendant's agents passed the report to Gartners, describing the problem as one which required urgent rectification. The contractors, too, were reminding Gartners that they had said they would attend to the leaks.

On June 26 1985 Rothschild assigned the underlease to the plaintiff, although the plaintiff did not for the time being move in. Mr Murray, the plaintiff's premises maintenance manager, had complained to the defendant's agent of water ingress and on July 4 F 1985 the defendant's agents wrote to him as follows:

Meanwhile I would confirm that we are taking most seriously the question of repairs required to the building, particularly in relation to the external cladding. My clients' solicitors are considering the most appropriate course of action to pursue as I understand that the developers are still bound by their agreement in relation to the construction of the building.

Gartners inspected the building in May and June 1985 and continued principally to blame the blocked weepholes which had caused water to stagnate, they said, within the cladding. They said they had taken steps such that the weepholes had been unblocked. However, the building continued to leak and the plaintiff, fearing that the carpets that it was about to have laid would be damaged, drew the G attention of the landlord's agents to the continuing problem. A meeting was attempted to be arranged. The plaintiff's agents wrote to the defendant's saying, as to the ingress of water:

I would ask you to do all in your power to ensure that remedial works are put in hand urgently as we aim to complete the ground floor fitting out by the 23rd September. My clients will take an extremely dim view of their new furnishings being damaged and I trust you will do all you can to remove this risk.

The landlord's agents replied on August 28 1985, saying that they were:

H

. . . raising the matter with Joseph [*sic*] Gartner who were the manufacturers of the cladding and glazing units and also the structural engineers responsible for the original design of the building. It seems to us important to try and identify the problems and get those put right as soon as possible so as to satisfy your clients without waiting for a lengthy battle or litigation as to responsibility — which my clients will have to pursue in due course. I am therefore awaiting a report from the engineers and from Gartners and we will get in touch with you as soon as that is received.

The meeting that was arranged took place on September 6 1985. Gartners' representative claimed, rightly or not I cannot say, to have prior engagements and failed to attend, but the plaintiff, ZMG, J Bickerdike Allan, Pell Frischmann and the architects did attend. Bickerdike Allan reported that about 30% of the windows leaked. It was recognised that the weepholes which Gartners had mentioned were inaccessible. It was decided to do nothing without prior notification to both the developer and to Gartners. The meeting on September 6 had begun with the landlord's agents by their Mr Richard Bennett [ARICS], referring in these terms to the underlease and responsibility under it:

Richard Bennett explained that in accordance with the terms of the leases of the offices and of the penthouse the lessors, The British Gas Corporation Pension Scheme, retained responsibility for the maintenance of the main K structure of the building, including the glazing and external cladding, the costs thereof being recovered from the tenants by way of the service charge.

On September 11 1985 the landlord's agents told the plaintiff that they had in mind beginning investigation and remedial works as a matter of urgency. They would be looking for an indemnity from the developer. The developer did not accept liability to the landlord and, other arguments apart, cited Gartners' belief that the blocked weepholes caused the leakage and that their blockage was caused by the landlord's poor maintenance and cleaning. The landlord, faced with this argument, therefore saw fit to instruct its own engineers, Mott Hay & Anderson, who were handed a copy of the earlier ZMG L report.

Mott Hay & Anderson sent a site visit report to their client, the landlord, on November 15 1985 and then later an interim report dated November 19 1985. They had not been able to obtain a copy of the drawings from Gartners. They identified three problems, but felt it necessary to inspect the building during a rainy period.

By January 1986 Mott Hay had received Gartners' drawings which, it was found, did not show drainage holes of the kind which Gartners had been asserting to exist. Mott Hay took a note, window by window, of precisely where leaks were still being encountered. It was M not based on direct personal observation as there had not been any continuously wet period, but was compiled from discussion with the plaintiff's staff.

By late January Mott Hay had been able to visit no 66 during rainy conditions. They, who it will be remembered are the landlord's engineers, reported:

A large number of leaks were observed, some of which are of quite intolerable intensity and will cause damage to finishes and fittings in their vicinity.

Taking into account the comparatively low rate of precipitation, the wind direction at the time of inspection and the fact that there has been relatively low rainfall for a considerable period, it is considered that the number of leaks observed is only an indication of the possible total number of leaks that can occur.

We would strongly recommend remedial action be taken as soon as possible to eliminate the leaks which are occurring at the windowsills by the external application of black silicon mastic. This remedial action should, however, not prejudice any further investigation work and remedial action which is still considered necessary to locate and remedy other leaks which are occurring or may occur due to water penetration at points other than the window glazing gaskets.

Mott Hay sent a detailed record of the leaks they had observed. The landlord's agents studied this report and their building surveyor, Mr Richard Bennett, considered that it was becoming increasingly apparent that the problem of water leaks was becoming widespread and that the application of mastic could be regarded only as a temporary repair. None the less, on March 4 1986, the landlord's solicitors indicated to the developers that unless they heard that the developers were themselves arranging the mastic application, as recommended by Mott Hay, then the landlord would itself go ahead with the application of mastic and with further investigations.

On March 10 1986 Mott Hay sent their final report to their client's agents. They understood that the cladding included drainage holes not shown on Gartners' drawings. As a priority they still advised the application of silicon mastic and they still needed to make further investigations as to the causes of leaks other than those at windowsills. They had not found any cause of leaks other than such as would, in their view, be held to be the responsibility of the installer of the cladding. They followed up their report on March 19 1986 with a specification of the immediate remedial works considered necessary to overcome the majority of the leaks. They thought these works would cost some £9,000 to £11,000. The work consisted of the application by gun of a high-modulus black-silicon-based joint sealant to all sides and to the bottom edges of all glazed panels. A life expectancy of 20 to 25 years was anticipated for the sealant.

The developer's reaction to the landlord's indication that the landlord itself would do the remedial works if the developer did not was to warn that so doing might release other parties from claims that might otherwise be made against them, presumably thus referring to Gartners, who, in correspondence between themselves and the developer, were declining to accept any responsibility, especially referring to the failure, as they said it was, of the owners or occupiers to keep the drainage holes properly cleaned. The outcome was that the landlord did not proceed with the works which its own engineers had suggested — the mastic sealant; this work, as was said, was put "on hold". Gartners were still insisting that blocked drainage holes were the root cause. Although they had put the matter "on hold", in correspondence the landlord's solicitors continued to threaten that the remedial steps recommended by Mott Hay would be taken in order to meet what it indicated was its duty to mitigate, presumably referring to mitigation of such damage as might be claimed by the plaintiff. Something like an impasse now existed. Without a full inspection involving an opening up of the cladding, it was unlikely that the root cause of the leaks would be traced, but Gartners, insisting that poor maintenance and blocked drainage holes were the causes and were not their responsibility, would not open up the cladding until they had been paid in full; the developer's architects and contractors would not have been keen to see Gartner paid in full unless and until their cladding system was verified to be sound. The landlord wished to press ahead with an investigation into causes, but the developer was fearful that opening up might involving losing Gartners' guarantee and in the meantime the tenant's premises leaked, although the tenant had understood that the landlord would see the problem cured.

On July 14 1986 the plaintiff wrote to the landlord's agents a letter which included the following:

Your company and ours, together with other interested parties, have met and discussed this situation in the past with the result that your responsibility to undertake to find a method of resolving this problem was recognised.

By July 1986 the dampness and water was causing what the tenant described as "problems of an operational nature". Water penetrating the ground floor caused a glass tungsten electric uplighter lamp to explode in an area used for client consultations. With some justice the tenant wrote:

. . . this is not a tolerable situation for an organisation in our position. Interviewing and counselling of clients is a major part of the function performed by personnel employed at this location and we feel that enough embarrassment and hazard has already been suffered since our occupation. We would like the situation resolved as quickly as possible and we look to you and your company as managing agents to elicit a solution from the contractors involved.

The landlord's agents answered that they fully appreciated that the matters complained of were by no means acceptable. They indicated they were pressing the developers.

In September 1986 Mott Hay added further to their detailed window-by-window notation of leaks that they had produced in January. The tenant continued to complain that the cladding was leaking to an intolerable extent. Mr Murray's evidence was that every time it rained, water leaked into no 66. Promotional material on the ground floor was destroyed or rendered useless. Water ran down a cable into an electricity distribution box. The staff laid towels and cloths on windowsills in an attempt to absorb water and had from time to time to stop work to mop up. Carpets became wet and were stained. Baskets and bins were set to catch drips. Machinery was moved in, away from windows. The facade of the building was cleaned by specialist industrial cleaners, who had been directed to pay special attention to weepholes. Yet still the tenant was experiencing serious water penetration and on March 25 1987 the landlord's agents reported to the architects:

It is quite clear that the cleaning out of the weepholes which has been the mainstay of Gartner's argument in this matter is no longer applicable.

and a little later they added:

May I please hear from you as a matter of urgency with a set of positive proposals [from] yourselves and Gartners on this subject.

By June 9 1987 the developers saw fit to serve writs against the architects, Gartners and others; the writs had been issued almost a year earlier. This development led to Gartners declining to attend further site meetings which had been arranged in the hope of a more thorough investigation of the causes of the problems. However, on August 4 1987 Gartners made proposals for remedial works which involved sealing all expansion joints to mullions, considering the use of silicon in place of mastic seals, sealing expansion joints and so on. The developers introduced yet further experts into the equation and in the meantime the tenant was suffering water ingress of a kind which damaged carpets and left water on windowsills.

Even Gartners were by now beginning to be less certain that the blockage of drainage holes was a material factor. Indeed, the very existence of the drainage holes of the kind to the blocking of which Gartners had attributed the problem began to be questioned by some. Gartners' drawings showed only a few drainage holes, but they had been representing to others that there were drainage holes not shown on the drawings and to those who had questioned this Gartners had replied that the drawings did not show the holes because they were incorporated into their work as a matter of standard practice.

On January 7 1988 the tenant's executive board met. The board decided in principle to move the West End office elsewhere. The decision was not dictated solely or, I would think, even principally by the state of the building but the fact, as the board recorded it, that during the frequent rainy periods parts of the building became uninhabitable was plainly *a* contributory factor feeding into the decision.

In February 1988 Mott Hay, under the heading "Suggested remedial works", advised their clients, the landlord, of alternative remedial measures. The first was removal of the entire cladding system and its replacement by a redesigned system. This, I think, was the first appearance of so radical a solution being proposed as a

A remedy. The alternative they suggested included removal of the existing system, replacement of mullions and transoms by redesigned ones, the drilling of drainage holes and the reassembly of the units with fresh seals.

On March 1 1988 two cladding panels were, at last, removed from the building, in Gartners' presence, but if it had been hoped that this inspection would point to a clear cause of the leakage that hope would have been disappointed. By now the tenant was in touch with surveyors and estate agents, Conway Relf, with a view to considering a disposal of the underlease at no 66. It was recognised that the then current rent payable of £200,000 pa would shortly come up for review.
B The first rent review date would July 25 1988. The market was buoyant; the likelihood was that the landlord would wait until as late as possible to begin the review negotiations so as best to take advantage of the rising market. Conway Relf, by their Mr Nicholas Berrill [ASVA], advised the tenant to settle the review before disposing of the underlease. Any incomer, he said, would want to know for certain the rent he would need to pay.

On May 25 1988 the landlord's agents gave notice to the tenant of an intention to increase the rent with effect from July 25 1998. By July the landlord's agents were proposing an increase to £385,000.

The tenant's consideration of moving out of no 66 was,
C unsurprisingly, sharpened by the prospect of so great an increase. It felt that it would be possible to move its private banking operations back to its main Bishopsgate premises. It made sense for the tenant to assign the underlease while there seemed to be sufficient demand, the tenant adding, with a thoughtful caution which may be typical of Swiss bankers:

> We do not know whether the climate will change in two or three years.

The decision was taken to sell the lease and to return private banking into Bishopsgate.

If the move was to be completed before the underlease was
D disposed of then the tactics to be adopted in the review would require some thought. The building still leaked; on July 21 1988 there was a heavy downpour and within minutes water started dropping on to the carpet, on to desks and on to leather chairs. The bankers at Credit Suisse felt that in the negotiations for the review the tenant should rely on an argument that the building was not tenantable and was unsuitable for a private banking operation. If a low rent were thus to be obtained, it would, thought the tenant, be easier to find a new tenant. By August 31 1988 the decision had been taken to vacate no 66 as from November 1 1988. Among the reasons for the move was dissatisfaction with no 66 as being technically deficient.

By October 1988 Conway Relf had been given instructions by the
E tenant in relation to the rent review. Their Mr Berrill knew of the problem of water penetration. He recognised that it could affect the disposability of the underlease. His view was that it might be wise to agree a slightly higher rent than would otherwise be the case in order to obtain early agreement to the new rent so as to be able to proceed with the disposition without delay. The tenant confirmed that such a strategy should be adopted. The tenant and its agents had in mind not for the time being telling the landlord of the intention to vacate; it is apparently not uncommon for tenants to feel, as a rent review looms, that knowledge on the landlord's part that the tenant wishes to assign gives the landlord a stronger bargaining position in the review. He can,
F for example, embarrass the tenant who wishes to leave by delaying the conclusion of the review. So, as yet, the landlord was not told of the tenant's wish to leave no 66, but in late October that wish came indirectly to the landlord's notice. In November 1988, perhaps on the 25th, the tenant moved out of no 66. Its private banking was now conducted once again from its Bishopsgate office.

In December 1988 the landlord's agents, who had only a few months before proposed a new rent of £385,000 pa on the review, were now pressing for £430,000 pa, presumably taking the view that the buoyant market had justified such a further increase. Conway Relf had not played the "water penetration" card in their dealings with the landlord's agents, nor had they been instructed to do so.

In January 1989 the landlord's and tenant's agents agreed a new G
rent for £365,000, roughly £45 per sq ft. The tenant confirmed the acceptability of this new rent and the tenant's marketing of the underlease was now set to begin. Letting boards were put up at the premises. Conway Relf felt that a premium of £125,000 would be appropriate, although they warned that survey by interested parties might disclose the defects in the building and that the result of a survey would be used in negotiation against the tenant to reduce the size of the premium otherwise recommended to be asked. A handsome brochure was prepared and circulated, *inter alia*, to major banks and building societies. Advertisements were placed in the *Financial Times* and the *Estates Gazette*. There is nothing to suggest that the marketing H
was not appropriate and thoroughly competent.

The marketing campaign soon bore fruit. Interest was widely shown and offers began to come in. Knight Frank & Rutley, as agents acting for Coast Securities, offered a premium of £50,000. They indicated that their client was in a position to move very quickly. Coast Securities was a substantial company as was confirmed by reference to its referees. As the jargon has it, Coast Securities would be a very good "covenant". Other serious offers were received, but the offerors were told that the tenant was proceeding with another party but that they would be informed should negotiations with that other party prove abortive. There is and could be no criticism of this J
business-like decision. Conway Relf were holding out for a larger premium than the £50,000 offered by Coast Securities; they were pressing for £85,000. They confirmed to Coast that there would be no further marketing while Coast's offer was being processed, which is what Coast had wished. Knight Frank & Rutley trusted that the matter would proceed to a speed and satisfactory conclusion. Solicitors were instructed. A deadline of March 9 1989 for exchange of contracts was being worked towards.

On March 6 1989 Knight Frank & Rutley, for Coast, indicated that they had been instructed to carry out a building survey. At much the same time the tenant's solicitors answered the inquiries before contract K
of Coast's solicitors. It was quite properly felt that the problem of water penetration could not be left quite unmentioned. The tenant's solicitors answered one question that "the purchaser must rely on its own investigations and survey and no warranty is given in this respect. However, our clients in the past have complained to the lessor of water penetration and the purchaser must rely on its own survey". The inquiries also raised the question of whether there was any breach of covenant by either landlord or tenant; the answer was given that the tenant was aware of none, but it was added that the purchaser was to take subject to any breach that there might be. On March 7 1989 Coast's solicitors specifically asked for information as to the dispute L
between the landlord and the suppliers of the cladding. Knight Frank & Rutley had made inquiries directly of the landlord's agents who had answered, inevitably, that there was water penetration into the building. That reference to the dispute between the landlord and the supplier of the cladding was perhaps a reference to the fact that on March 3 1989 the landlord had issued a writ in the Queen's Bench Division, headed "Official Referee's Business", against the developer for breach of the development agreement of December 14 1977. There had, I think, already begun other proceedings in Queen's Bench between the developer and six defendants, including Gartners, and at some date arrangements were made for those two sets of proceedings to proceed together. M

On March 9 1989, the deadline date for exchange, the tenant's solicitors faxed the landlord's indicating that the proposed assignees had asked for a letter to be produced from the landlord acknowledging that the cost of remedying any defect in the building in respect of water penetration would be borne by the landlord and would not be passed on as part of the service charge. The landlord's solicitors were warned that the prospective assignee was threatening to withdraw if such a letter was not produced. The tenant's solicitors warned that failure to supply such a letter, "a comfort letter", could cause the tenant yet further damage. Nothing was heard from the landlord's solicitors, who were again pressed, on March 13 1989, for such a "comfort letter". On March 15 1989 the landlord's solicitors indicated that they were unable to give the tenant the acknowledgment it sought.

Conway Relf, attempting as best they could to keep the deal with Coast alive, suggested a reduction of the asked-for premium and also the giving of an indemnity to Coast. They told the tenant that Coast's agents, as was the case, were pressing for more information as to the extent of the defects in the cladding, as to the time likely to elapse before the problems in the cladding were made good and as to the disruption that any making good would be likely to cause.

At the end of March 1989 the tenant, recognising that it would need to have a thorough and professional answer to such questions, again instructed Bickerdike Allan & Partners to return to the problem. Their Mr Barry Josey visited no 66. He had ZMG's report of March 1985 with him, which incorporated his firm's earlier report. He had a number of Gartner drawings. His report of April 5 1989 would have made depressing reading to the tenant; the disruption to an occupier inevitably to be associated with the making good of such defects as he found was described by him as "likely to be substantial".

On April 7 1989 the landlord's agents informed Knight Frank & Rutley that they were unable to give any assurance as to whether the tenant would be liable for any of the costs of rectification of the water penetration and that they could not give any indication as to how the remedial works would be undertaken. Bickerdike Allan's report was sent to Knight Frank & Rutley; it had been one of the intended purposes of the report that it should be.

On April 11 1989 Knight Frank & Rutley, for Coast, wrote to confirm that Coast had instructed them to withdraw from negotiations; they added:

> This decision has been taken in the light of the findings in both Bickerdike Allan & Partners' report together with my firm's building survey. Coast are not only concerned about the immediate problems arising from the cladding/windows but also the long term effect on the property as they would be taking a full repairing and insuring lease.

Conway Relf accordingly revived contact with the other offerors who had been put on one side while negotiations with Coast had been in progress, only to find that their first question was "Why did your first proposed assignee withdraw?". Conway Relf had honestly to say that the first offeror had withdrawn because of the problem of water penetration. The premises, which were, of course, by now empty, were littered with moisture-monitoring devices and smelt of damp. Conway Relf advised the tenant that:

> . . . due to the problems associated with the cladding/windows I do not believe that we can seriously expect to secure a tenant in the circumstances. I therefore reluctantly, would strongly recommend that the building is withdrawn from the market and that your position is carefully reviewed.

On April 21 1989 the tenant instructed Conway Relf to withdraw no 66 from the market.

On June 14 1989 the tenant's solicitors wrote a letter before action to the landlord's, asserting that had the premises only been in repair the tenant could have assigned them and for a premium. The tenant had been advised, they said, that they would set-off against future liability for rent any claim for damages that they might have. They indicated that rent would not be paid from and including June 24 1989, although they indicated also that they were still willing to assist with the landlord's proceedings against the contractors and others.

Attempts to ascertain the causes of the leaks were proceeding in the context of the landlord's action against the developers (by now called "Clayform" rather than by their earlier name, Samuel Developments).

On August 17 1989, 12 individuals met, representing eight parties or firms. Glazing was removed and metal was cut away and the majority agreed that the cladding was not a fully drained system and differed from Gartners' drawings. The landlord was represented at this meeting by Crocker Tooth Associates, cladding experts, who had come on the scene in connection, principally I think, with preparation of the landlord's statement of claim against the developers and contractors. It was agreed that further tests were necessary. An opening up of further cladding in the presence of Miss Elizabeth Tooth [RIBA], of Crocker Tooth, and Mr Josey, of Bickerdike Allan, (for the tenant) was arranged for late November 1989.

On December 1 1989 the landlord's solicitors understood that the landlord (without admitting that the tenant was entitled to withhold the same) was not going to proceed for the unpaid rent and service charges and was going to include such losses in its claim against the developers, which by then had been fixed to come on before the official referee in February 1991. However, by March 1990 the landlord had been advised not to delay seeking rent and service charges from the tenant and the landlord then demanded some £333,000 or so in March, a claim which had grown to £447,000 odd by April. Perhaps prompted by this, the tenant issued the writ in the action now before me on April 4 1990.

On March 8 1991 heads of agreement were signed to settle the actions between the landlord and its developer and the architect and between the developer and the architect and others including Gartners. The tenant had assisted the plaintiff in pressing those claims. The landlord was to appoint an architect to devise a new cladding scheme for no 66 which Gartners, after removing the existing cladding, were to design, manufacture and erect to the reasonable satisfaction of the landlord's architects and at Gartners' cost. The new system was to include a thermal break. Gartners were, to an agreed extent, to indemnify the landlord against the tenant's claims against it in the present action. If a building contract for the new cladding was not to be entered into within a specified period then, unless an extension to the specified period were to be agreed, Gartners would be obliged to pay the landlord £2,300,000 by a date in March 1993. By a consent order in *Tomlin* form before Judge Newey QC on official referee's business on March 8 1991 all proceedings as to cladding in that court were stayed. Gartners paid the landlord's costs. The arrangements I have just described were incorporated in the schedule to the *Tomlin* order in the usual way.

The architects chosen by the landlord for devising the new cladding were Broadway Malyan, who soon set about their task. One of their objects was to produce a design that dispelled the stigma, as it has been called, which by now was attached to no 66 and to make it readily marketable and capable of reoccupation. Planning permission would not be necessary for a renewal of the existing cladding with matching materials, but as time passed the design proposals began to contemplate other materials, such as masonry, being incorporated in the new cladding. The landlord was by now on the horns of a dilemma; if the recladding was in something other than matching materials, such as masonry, planning permission would be necessary and would delay matters, perhaps substantially. But, on the other hand, the Royal Fine Arts Commission in July 1991 commended the masonry solution and indicated that other proposals would be wholly unacceptable. Broadway Malyan were by now producing proposals for cladding which bore scarcely any resemblance to the Rocket Ship. Proposals for the inclusion of masonry in the cladding led to consideration of whether the structure could reasonably be expected to take the weight and consideration, also of how much further cost masonry would add to the recladding that had earlier been in mind. The landlord instructed quantity surveyors, who costed various alternatives. The cost of replicating the existing cladding but to an improved specification and with thermal breaks was in August 1982 estimated to be £940,000, although a simplified proposal could be achieved for about £685,000. Gartners foresaw difficulties and devised an alternative over-cladding scheme. The landlord had instructed new engineers by now and they, Emmer Pfennigger Partners, went round the building in January 1993 with Miss Tooth, the landlord's cladding expert. Emmer Pfennigger reported that Gartners' overcladding proposal was unworkable. By now it was February 1993. The deadline, if £2.3m was not to become payable by Gartners, approached. In March 1993 Emmer Pfennigger costed a satisfactory proposal for recladding at £1,154,516.

The landlord's agents, of course, by now recognised that under the heads of agreement and the *Tomlin* order they were due £2.3m from Gartners but they did not, for the moment, press for payment. By May

12 1993 the landlord, having given Gartners more than time enough, thought it unlikely that Gartners would be used in any new scheme. I have not been told that Gartners cannot or will not pay the £2.3m nor that the agreement that was made in that behalf has been varied or supplanted by any other agreement. So far as it is relevant to any question before me, I should, in my judgment, proceed on the basis that the landlord will recover £2.3m or so (interest may be payable) from Gartners; I have been given no reason to think that that is not a proper assumption for me to make.

To this day no final certificate as to the completion of the whole of no 66 has been granted.

That it is all I need to say by way of a chronicle of the relevant events and so I turn next to the subject of which I have called "Breach or nay".

Breach or nay

I shall deal first with a number of minor arguments on the construction of the underlease.

I am invited by Mr Brian Knight QC, for the landlord, to proceed on the basis of giving commercial effect to the principal covenant. He cites *CH Bailey* v *Memorial Enterprises Ltd* [1974] 1 WLR 728* *per* Sir Eric Sachs at p735B-C. I thus approach the underlease having in mind that the arrangements likely to have been intended to have been made by business people will be business-like arrangements. But I do not see the notion as going further than that; it does not enable me to give other than a natural meaning to the words used, although, doubtless, where two natural meanings equally beckon I should, in this context, prefer the more business-like.

Mr Knight invites me, in connection with terms commonly referred to in the authorities, to give a construction consistent with that derivable from those authorities because inconsistency is undesirable, he says, in standard forms of commercial contracts. Plainly, I should accept guidance from authorities, but I do not regard this lease as a standard form of contract. Although certain words recur in the authorities dealing with different repairing covenants and although some guidance can thus be derived from other cases, there is a world of difference between the case before me and cases where the courts properly strive for consistency as to the construction of, say, the printed standard form terms of the New York Produce Exchange: see eg *A/S Awilco of Oslo* v *Fulvia SpA di Navigazione of Cagliari* [1981] 1 WLR 314 HL to which Mr Knight referred.

It is argued that the fact that this is a landlord's covenant (but with recovery under the service charge) should not be a factor pressing me towards some construction other than that which would obtain had it been an identical covenant on the tenant's part. My attention was drawn to *Torrens* v *Walker* [1906] 2 Ch 166 at p174. There Warrington J held that the principle (applicable to a tenant's repairing covenant) that a lessee was not bound to give back to the lessor at the end of the term a different thing from that which had been demised to him, was matched, in a landlord's covenant, by one under which the lessor would not be bound by a similar covenant to give to the lessee during the term a different thing from that which the lessee had taken at the beginning of the tenancy. There was, he held, no difference in principle. The case does not, so far as material, go beyond that, but it does seem correct to me that unless other circumstances compel otherwise, similar words must have similar effect whether they be found in a landlord's or in a tenant's covenant.

It is an irony in this case that the landlord has been saying that it would be wrong to give a wide effect to the landlord's covenant for repair because, as the expense of compliance is passed on to the tenant, so to do would, says the landlord, unreasonably burden the tenant. The tenant, by contrast, hopeful that even if given works are within the principal covenant and hence ordinarily within the service charge, it none the less has a prospect of escaping the need to pay for extraneous reasons, presses for a wide construction of the landlord's covenant. I am also invited to recognise, as did Scott J in *Plough Investments Ltd* v *Manchester City Council* [1989] 1 EGLR 244 that, when the covenant is cast in the way it is in the underlease here, the landlord's obligation to repair, although nominally an obligation, is in a sense a right. It enables the landlord to keep its building in repair at the tenant's expense and, as to works within the covenant, it is the landlord rather than the tenant who determines how the work is to be done. A tenant who might, had the choice been his, have got away, so to speak, with a cheap solution to a given problem as to repair might find, under this form of a landlord's covenant, that a more expensive solution is foist upon him by way of the landlord's choosing to employ it. I bear this in mind, but I do not see it as having a force sufficient to drive me away from the ordinary meaning of words.

The tenant argues that it is not to be expected that the parties left a gap in their obligations such that work needing to be done should transpire to be an obligation neither of the tenant nor of the landlord. That is far from obvious to me; parties can overlook some needs or even deliberately shy away from some problems. Mr Jonathan Gaunt QC, for the tenant, argued that if the necessary works were not within the landlord's obligations then, as neither were they within the tenant's, the situation would be that no one was responsible for them which, he said, would be uncommercial. I was taken to *Barrett* v *Lounova (1982) Ltd* [1990] 1 QB 348* CA at p358. In that case the Court of Appeal upheld the implication, in a demise of a house which contained an express covenant that the tenant should keep the interior in good order, of an obligation on the landlord to repair the outside. However, what persuaded Kerr LJ to uphold the implication was not so much that if there was no such implication no one would be responsible for the upkeep of the outside as the fact that if the outside was not kept up then in time the express covenant by the tenant as to the interior could no longer be complied with and yet it was plain that the tenant did have to comply with it throughout the whole term. It was that, rather than a gap such that otherwise no one would have had to keep up the exterior, which enabled the court to imply an obligation to keep the outside in repair. *Barrett* might assist a court in coming to a business-like conclusion and in filling a gap where the language of a lease can fairly be construed either to leave a gap or to fill it, but it cannot be authority for a proposition that whenever one encounters what seems to be an unbusiness-like gap then the court is able or obliged to fill it.

The landlord refers to the ability of the landlord to enter during the term and argues that such entry is limited to entry for works of repair strictly so called, which thus points to the principal covenant also being limited only to works of repair. However, clause 4(11)(d) permits entry "for the purpose of carrying out the landlord's obligations hereunder" and Part 3 of the second schedule reserves to the landlord the right to enter not merely for the purposes of "repairing" but also "for the purpose of carrying out any work or doing anything whatsoever comprised within the landlord's obligations herein contained . . . for which the tenant is liable hereunder to make a contribution". The argument thus rebounds on the landlord; if the draftsman has seen fit to give entry for works under the landlord's obligation, but which are other than of repair strictly so called, then surely the principal covenant must correspondingly go beyond works only of repair? The lease is over 45 pages of single-spaced typescript and I am far from confident that its draftsmanship is of a quality such that one can derive very much from this sort of exercise, which requires a contrast to be made between references, for example, on p12, with those some 30 pages or so later. However, what little weight this point does have ends up, as I see it, as being in the tenant's favour.

The landlord argues that it is wrong to look at the eight verbs in the principal covenant and to see each one as intended to add to the meaning ascribable to each of the others. In *Post Office* v *Aquarius Properties Ltd* [1987] 1 All ER 1055† the covenant included "well and substantially to repair . . . amend . . . renew and keep in good and substantial repair and condition" yet the Court of Appeal and Hoffmann J below ruled on the case on the basis that the words "keep

*Editor's note: Also reported at (1973) 229 EG 613.

*Editor's note: Also reported at [1988] 2 EGLR 54.

†Editor's note: Also reported at [1987] 1 EGLR 40.

A in good and substantial repair" were the critical words. However, it is trite law, as *Aquarius* acknowledges at p1064B, that the parties to a lease may agree such provisions as they choose and *Aquarius* cannot be authority for a proposition that separate words are not to be given separate meanings where that is possible, if only because the Court of Appeal was careful to indicate that no reliance had been placed in the argument before them on any contention that the words "amend", "renew" or "condition" added anything to the words which Hoffmann J had taken to be the critical words: see, in the Court of Appeal at p1059E *per* Ralph Gibson LJ, *per* Slade LJ at p1065B-C and see also p1066G.

B The landlord sought to advance the same argument by reference to *Norwich Union Life Insurance Society* v *British Railways Board* [1987] 2 EGLR 137, a rent review case in which the arbitrator had adjusted the new rent downwards to reflect that the covenant to repair was especially onerous upon the tenant. The covenant for repair included not only "to keep the demised premises in good and substantial repair and condition" but also "where necessary to rebuild reconstruct or replace the same". The 1968 lease was for a term of 150 years from 1964. With an irony corresponding to that which I have mentioned as present in this case, the landlord was there arguing that the covenant was to be construed so as to lighten the apparent burden
C on the tenant; the added words "rebuild", "reconstruct" and "replace" should not, he urged, be taken as imposing obligations substantially beyond those in a conventional covenant for repair and hence should be limited to the rebuilding etc of subsidiary parts only of the building. In rejecting that, Hoffmann J concluded that the covenant had two distinct parts: one involved the concept of repair; the other, distinctly and separately, of rebuilding reconstructing and replacing, and that both were imposed on the tenant. However, and this is the point most relied upon by Mr Knight, Hoffmann J said:

> Now I accept that in the construction of covenants such as this one cannot, for the reasons I have already given, insist upon giving each word in a series a
> D distinct meaning. Draughtsmen frequently use many words either because it is traditional to do so or out of a sense of caution so that nothing which could conceivably fall within the general concept which they have in mind should be left out. I also accept that if the language is not entirely clear the covenant should not readily be assumed to impose unusual obligations. In the ordinary way a covenant in a lease to rebuild the entire premises would be unusual. — see *Norwich Union supra* at p138.

The "reasons already given" included that the lease in that case employed what Hoffmann J described as "a torrential style of drafting". As Hoffmann J held that the obligation upon the tenant did, on the language of that covenant, extend to a rebuilding of the
E whole building he was plainly not of the view that added words can never add to the meaning. The argument that added words should lead to added meaning had not needed to be pressed; *Lurcott* v *Wakely and Wheeler* [1911] 1 KB 905, for example, was not, it seems, cited to the learned judge, who in any event accepted, of course, that, according to normal rules of construction, additional words should be given some additional meaning. *Norwich Union* is not and could not be authority for a proposition that that normal rule does not apply to repairing covenants. Even where there is a torrent, each stream of which it is comprised can be expected to have added to the flow.

F Mr Knight refers to Slade LJ's remarks at p1066A-B in *Aquarius* that clear words are needed to impose a contractual obligation on a tenant to remedy defects in the original construction of the building "at least at a time before these have caused any damage. This is not an obligation which tenants under a commercial lease might reasonably be expected readily to undertake". Slade LJ does not say, of course, that if clear words to that effect are found they can be overridden. Moreover, as will later appear, I shall hold there to have been damage to the subject-matter of the principal covenant.

I now turn to more central issues of construction. As Fletcher Moulton LJ emphasises in *Lurcott (supra)* at p95, the duty of the court is to give a proper and full effect to each word used in repairing covenants. Atkin LJ endorsed this approach in *Anstruther-Gough-Calthorpe* v *McOscar* [1924] 1 KB 716 at p731. I G
see the verbs "amend" and "renew" as capable of going outside the verb "repair", especially where, as here, the use of the phrase "defects *or* wants of repair" in the proviso to the principal covenant shows that some meaning beyond repair is contemplated. Moreover, and again having regard to that phrase in the proviso, I see the words "*otherwise* keep in good and tenantable condition" as having a potential going beyond repairs strictly so-called (by which I mean works of a kind which would fall within a covenant merely "to repair"). Just as did Fletcher Moulton LJ in *Lurcott* at p916 ("keep in good condition"), I shall first take the phrase as to tenantable condition in the principal covenant and look at it as if it were a separate covenant.

A number of points emerge. First, although *Proudfoot* v *Hart* H
(1890) 25 QBD 42 has lost ground as to some parts of its reasoning and has been said often to have been misunderstood, the notion illustrated by it that a covenant to keep in a specified state includes an obligation to put into that state has been acted upon, by now, in probably hundreds of cases and still survives. The obligation here is thus not merely to keep in good and tenantable condition, but that the landlord should put the building into that condition.

Next, to transpose *dicta* directed to "tenantable repair" to the expression "good and tenantable condition" then the expression refers to such condition as, having regard to the age, character and locality of the property, would make it reasonably fit for the occupation of a J
reasonably minded tenant of the class who would be likely to take it — see *Proudfoot (supra)* at p55; *Lurcott (supra)* at p921 — though one cannot lower the nature of that condition by reference to deterioration in the expectations of the hypothetical tenant consequential upon a deterioration in the class of tenants after date of the lease — *Anstruther-Gough-Calthorpe (supra)*.

Next, while I accept the inevitability of the conclusion of the Court of Appeal in *Aquarius (supra)* that one cannot have an existing obligation to repair unless and until there is a disrepair, that reasoning does not apply to a covenant to keep (and put) into good and tenantable condition. One cannot sensibly proceed from "No disrepair, K
ergo no need to repair" to "No disrepair, ergo no need to put or keep in the required condition". Leaving aside cases, such as this, where there is special provision for there to have been prior knowledge or notice in the covenantor, all that is needed, in general terms, to trigger a need for activity under an obligation to keep in (and put into) a given condition is that the subject-matter is out of that condition.

Next, where, as here, one has not only the verbs other than "repair" but also a context showing that "defects" in express contradistinction to "repairs" are within the covenant, the great body of law debating whether a given shortcoming is or is not a repair or is or is not an inherent defect such that remedying it goes beyond repair does not, in L
my judgment, apply. Of course, as with covenants to repair, questions as to the age and nature and class of the subject matter to be put and kept in the specified condition have to be borne in mind — see Fletcher Moulton LJ's examples of the difference between keeping in condition (in 1911) the Mauretania and a "tramp that has been at sea for 15 years" — see *Lurcott (supra)* at p916. So cases on "repair" cannot be wholly discarded but, as I see it, it is no necessary escape from liability under a covenant to put and keep in a specified condition that the work needed in order that that condition should be achieved is not a repair strictly so called. Nor do I see as important on this part of the landlord's covenant that the work required to achieve the given M
condition is of a particular great cost. I can see that that may well affect and has been frequently taken to affect whether the work is one of "repair", but I do not see it as material to an obligation to put and keep in a given condition. I have not understood any authority cited to me as disabling me from giving to the words of the covenant "*otherwise* keep in good and tenantable condition" the meaning which *Proudfoot* and *Lurcott* (*supra*) ascribe to similar phrases, nor as obliging me to attach to these words all the qualifications which have been held to circumscribe or affect an obligation to "repair".

It may perhaps be (but it is, as I mention below, unnecessary for me to decide) that a proper limitation on the obligation to put and keep

A and tenantable condition is that the landlord cannot be required to provide the tenant with something "of a wholly different character from that which had been let" see *McDougall* v *Easington District Council* [1989] 1 EGLR 93 CA at p96. I thus say nothing as to that.

Next, if the meaning I have attributed to "good and tenantable condition" is right, then the required condition in which the premises are to be kept and into which they are to be put falls to be judged not (if they differ) by reference to the actual state of the building as at the date of the demise but rather to what at that time were the requirements as to condition of the hypothetical reasonably minded tenant of the class likely to take that building. There is warrant in
B *Smedley* v *Chumley & Hawke Ltd* (1981) 44 P&CR 50* CA at pp56 and 57 for the court to have regard, in relation to a new building, as the starting point in judging whether the effect of the words in question would be to yield something different to that which had been demised, not to what was the state at the demise but what was *contemplated* as then being the state. The extensive foundation work done in that case — providing a concrete raft, bracing it with concrete piles driven in, placing joists on the raft and then lowering the building on to that new foundation — was held to be such:

> . . . that after the work done the difference to the premises was that the walls and roof were in the condition that both parties contemplated as their
> C necessary condition at the date of the lease

— *per* Cumming-Bruce LJ at p57.

If that is appropriate in a case where reference to a hypothetical reasonably minded tenant was not in play, *a fortiori* it is relevant when the requirements as to condition of such a tenant are introduced. *Anstruther-Gough-Calthorpe (supra)*, by its emphasis that the later lowered expectations of the subsequent hypothetical tenants are not to lower the standard beneath that applicable at the time of the demise, itself thus makes the point that it is expectation or contemplation which is important. Of course, in most circumstances the hypothetical
D required condition and the actual condition as at the date of the lease will coincide. If there is no reason to regard the actual tenant as other than a reasonably minded member of the class likely to take the building of the kind in question and if he has inspected before he takes the lease it can reasonably be inferred, if the parties are silent on the point and if the surrounding circumstances add nothing, that the building reached the standard that the hypothetical tenant would have thought fit for his occupation. But so far as concerns good and tenantable condition I cannot see it as relevant minutely to examine into whether, on the very day of the lease, no 66 was or was not watertight and if it was not, why it was not. If, as I think it is, the
E relevant standard of condition is what I shall call the putative condition, namely the condition that the tenant of the class I have described would require, then what is important is not so much the state of fact as at the date of the demise, but what the tenant in particular then thought was the fact or, in the absence of direct evidence as to that, what it can be inferred the tenant is likely then to have thought in relation to the condition of the building. I cannot see it ought to be effective to lower the standard of condition in which the landlord has to keep the building beneath the putative condition were it to be proved, for example, that the building leaked on January 28 1985 or July 25 1983 (the date from which the term ran) if it could be claimed that the tenant reasonably believed that the new building was
F then good or would have had no reason to think that it was poor or if he understood or expected that the building was in the course of being, or was to be, made good by the landlord. The yardstick by which the necessary condition is, on this basis, to be ascertained is derived from the putative requirements of the reasonably minded tenant and so it is not to be lowered either by proof that the actual tenant was deceived or delayed in his receipt of that which he could be inferred to have expected.

Turning to what that hypothetical equivalent might have expected, I cannot think that any tenant of the class of "undoubted covenant" that the landlord would have then been looking for, taking, in 1985, a G
wholly new and expensive architect-designed office and residential building in such a prime site as that occupied by no 66 would have required of the cladding of no 66 anything less than that it should be very substantially watertight. Short of the cladding being put into and kept in that state, it seems to me that the building would not be reasonably fit for the occupation of such a tenant.

If I am right so far, then the chief question becomes a relatively simple one. Beyond any doubt the cladding at no 66 (which is conceded to be within the subject-matter of the principal covenant) has during the plaintiff's occupancy not been put into, nor has it been kept in, the good and tenantable condition which I have supposed to be the H
putative necessary condition. The failed DGUs, the water stains on carpets, the buckets and bins needing to be placed to collect drips, the wet and stained windowsills, the stained and salt-encrusted mullions, the spoiled display material and the bursting tungsten lamp all thus testify. Moreover, the chronology shows that the landlord was sufficiently put on notice from 1984 on.

Nor, moving on to another point, can I regard no 66, even with a wholly new and redesigned cladding system, as being of a wholly different kind from that which had been let. Although, as will have been seen from the chronology, the landlord's engineers, Mott Hay, advised the application of black mastic that seems since to have been J
regarded as only a temporary solution at best and probably not comprehensive even so long as it lasted. The landlord has not pressed that it would suffice to apply and reapply mastic from time to time. It may be a further factor that the building has acquired a "stigma", as it has been described, the consequence of which is that short of comprehensive works being *seen* to have been done it will be poorly regarded by the agents letting in the area. Rather, the experts agreeing that the dismantling of the whole existing cladding and its entire replacement by a newly designed system is the only practical solution. Where what a sensible man would do would be thus comprehensively to eliminate the shortcomings in a way which would avoid repetition K
of temporary works, that factor, so long as the making good of the defect does not involve a substantial rebuilding of the whole subject-matter of the covenant, tends to put the works into the category of "repair" — *Stent* v *Monmouth District Council* [1987] 1 EGLR 59 CA at p65. I would not regard the replacement of the cladding as amounting to a substantial rebuilding of the whole subject-matter of the covenant, which extended to the whole structure of the building, including its foundations. The cladding is but the exterior of two faces of the "L" shape of the building. If it were wholly to be replaced with watertight cladding, the tenant would merely be getting what the hypothetical reasonably minded tenant would always have expected and what, I have no doubt, Rothschild also thought they L
were contractually entitled to and would be getting and what the landlord hoped to be demising, namely a modern building completed to a good specification with anodised-aluminium cladding faces which were watertight. It is for this reason that I do not need to decide whether the "wholly different character" argument — see *McDougall (supra)* — applies to "good and tenantable condition"; even if I assumed it did apply, the point would fail on the facts. Nor, if I am wrong in disregarding cost, would I think that the cost of dismantling the present cladding and replacing it with wholly new and redesigned cladding, a cost of some £1.2m, would be such to put the works (if this be relevant) beyond the description of "repair", given the value of M
no 66 as demonstrated by a rent which has varied between £200,000 to £365,000 and in the light of no 66 having had a capital value (on an assumed basis of being in repair) of as much as £7,000,000.

Accordingly, in my judgment, and looking for the moment just to the obligation comprised within the words "and otherwise keep in good and tenantable condition" the landlord is in breach, and has long been in breach, of the principal covenant. I should add that if, contrary to my views as to the proper yardstick being what I have called the putative condition at or about the time of the lease, the true measure were instead to be the actual condition then or thereabouts, then even so I should hold the landlord to be in breach. I shall need to go further

*Editor's note: Also reported at (1982) 261 EG 775.

into this subject below, but I shall be holding that an earlier and better condition of the building has not been maintained. Leaks have got worse, DGU failures have increased, encrustations have grown worse and, of course (though this relates to a period before the demise), the leak-free year between March 1983 and March 1984 represents a state that had not yet been regained.

If I am right to have regarded the words "keep in good and tenantable condition" in the way that I have then there is no need to ask in more detail why the building leaks. It is enough on that basis that it has leaked to the extent that it has. But if the verbs "repair and renew" in the principal covenant are to be fastened upon, then, as to the verb "repair", as *Aquarius* shows, if the landlord is to be liable to "repair" there must first be a disrepair. It is thus relevant, lest I am wrong on "tenantable condition" and if "renew" adds nothing material to "repair", to find out why it is that the building leaks in order to ascertain whether there is some disrepair. Equally, it would be appropriate to see if what needs to be done to make the building appropriately watertight can fairly be described as work of "repair" or of "amendment" or of "renewal" and whether the work falls outside all or some of such descriptions as being, for example, the remedying of some inherent design or other defect which has afflicted the building from its outset.

To assist on such questions as these I received evidence by way of both written reports and oral evidence from two experts as well as such indications as one can glean from the correspondence and earlier reports which I have described in the chronology.

Turning to the experts' evidence, Miss Elizabeth Tooth RIBA, instructed by the landlord, a chartered architect and principal of Crocker Tooth Associates, gave evidence in her written report:

(i) That water passing down the exterior faces of the mullions of the flat panels could and did penetrate the inner chambers of the mullions and thereafter penetrated to the interior of the building at the gaps of the movement joints just above windows and into glazing rebates;

(ii) Direct rainwater penetration occurred in flat panels at external panel/mullion, panel/transom joints and at mullion/transom joints;

(iii) The neoprene glazing seals were not watertight in bad weather;

(iv) As for curved panels, they were more vulnerable to water penetration than the flat ones and there was, during hose testing, severe water penetration at panel joints;

(v) It could not be excluded that there was some water penetration into glazing rebates in bad weather;

(vi) Entry at transom/mullion butt joints in the curved units was a major source of water penetration into the glazing rebates. A significant quantity of water penetrated the glazing rebates of the curved DGUs and the DGUs themselves were threatened with failure or failed;

(vii) Most water which entered the cladding system drained away by way of inadequately sealed joints rather than accumulated within the system, but, while most drained to the exterior, some drained to the interior of the building;

(viii) That there was no reason to suppose other than that this water penetration had occurred since the cladding had been completed in July 1982. In other words, the building had leaked from new.

(ix) To judge from observations made when the cladding was opened up, metal-to-metal interfaces at joints had either no sealant or insufficient sealant in that it had not been applied to the full width of the interface or was just a small fillet on the edge;

(x) Where sealant was found it was still well bonded. The sealant itself had not deteriorated.

(xi) If, as the design here contemplated, there should be voids around the DGUs in the glazing rebates, then such voids should be drained, but these were not;

(xii) Water had penetrated the glazing rebates of the flat units by way of penetration of the neoprene gaskets. Premature seal failure was a possibility.

(xiii) Some rubboseal around curved DGUs had become "crumbly" and had lost its bond. At first Miss Tooth attributed this either to poor workmanship when the units were put in or deterioration caused by exposure to the water which had penetrated and accumulated in the glazing rebates or to a combination of those two factors.

(xiv) The design of the cladding was such that it would not be watertight. It had been intended to be a drained system, but it had transpired that it had no provision for drainage of such water as penetrated the external skin.

(xv) Neither the cladding itself nor the insulation within it had deteriorated as a result of water penetration.

Her theme, if I may so describe it, was that the building had not got worse over the years, but that it had leaked from the beginning and now leaked not significantly more than it had earlier done. It was, as she put it, in a phrase not altogether unknown to lawyers, "doomed from the start", although that might suggest, contrary to her theme, that there was a time before its doom overtook it.

In oral supplement to her written report Miss Tooth introduced further plans showing the construction of the transoms and transom/mullion joints in the curved units to be other than she had previously had in mind and to be better than she had previously had in mind. She now played down the transom/mullion butt joints in the curved units as having been a major source of penetration as her written report described them to be. In cross-examination she was taxed with her written report, as I have summarised it in (vi) above and also with the similar views which she had expressed in written evidence prepared for the proceedings in the Official Referee's Court. She explained her change of view by reference to her having since had a sight of the later plans, but her explanation of the later plans was itself unconvincing as she needed to *assume* the existence of the spigots in order to make the transom/mullion joints watertight or less leaky than she had previously advised, yet the new drawings themselves showed no spigots. When taxed with her reports having said, as I have summarised at (xiii) above, that the rubboseal may have deteriorated by reason of exposure to water entering into and accumulating in the glazing rebates, she was constrained to say that she could not remember having written that and that, on thinking about it, it had not happened. That was an unimpressive part of her evidence, as was also her inability satisfactorily to explain how it could be that, although in her view the building had leaked and had always leaked, much as it now does, from day one, the architects responsible for inspecting and who plainly had inspected in relation to certificates and snagging lists had not found the major leaks which on her evidence must have always existed. While I have no doubt but that in the vast majority of her evidence Miss Tooth was technically sound and gave evidence that can very readily be accepted, there were some areas in which, as I shall explain, I prefer the evidence of the plaintiff's expert, Mr Josey.

Mr Barry Josey is an associate of Bickerdike Allan Partners, architects. For the past 15 years he has specialised in design and construction of curtain walls and cladding systems. In his written evidence he referred back to his firm's 1985 report (for which he had not been responsible) and now found a marked deterioration by comparison with the 1985 state of things. He found mullion/transom joints which had failed. Movement between one panel and its neighbours was accommodated in the Gartner design, but not expansion and contraction within the panel itself. This could, he thought, have caused panel joints to open over a period in the flat panels and hence for the joints to allow the passage of water. Lack of sealant or of adequate sealant in the spandrel areas permitted the entry of water. It could then pass through screw and bolt holes and through the joints. The rubboseal had been poorly applied and the DGUs were threatened by reason of the voids in the glazing rebates being penetrated by water. Water entering by way of gaps in the copings at the top of no 66 then getting into the mullions beneath was another form of entry of water and occurred either because seals had deteriorated or failed or because they had not been applied in the first place. Loose glazing clips were found and where the clips were loose the gaskets to the DGUs were no longer compressed tight and could allow water either to stand in the glazing rebate (and thus threaten the

DGU) or leak into the interior of the building. The failure rate being encountered with DGUs was far higher (some five or more times higher) than one would ordinarily expect. When I went for a view of the premises, on July 1, close to the beginning of the trial, it was during a spell of what, by the standards of an English summer, was a prolonged period of fine weather. Even so there was water in the bins placed to catch drips and the curved DGU on the 3rd floor had, at its base, some 2ins-3ins of water standing within it.

In oral evidence, Mr Josey spoke of the joints within individual panels. Joints generally were found to be sealed, but it had been found that seals had failed. On this subject he differed to some extent from Miss Tooth. The internal faces of panels would be in relatively static condition with no wind and an air-conditioned constancy of temperature. The exterior of the same panel would be exposed to the variability of the English weather. The outside of a panel would be seeking to move in a way which the inside was not. One thus had "restless" joints and in small joints such as the ones in issue the sealant would be caused over a period to deteriorate and fail. This it "restless" joint theory explained why, after a period, the mullion/transom joints had become a major source of water penetration. It was, I think, to avoid this conclusion that Miss Tooth had been constrained to abandon her earlier conclusion that at any rate in the curved units the mullion/transom joints were a major source of water penetration. I accept Mr Josey's evidence that it was not so much that there were no seals in the system that contributed to leaks, but rather that such seals as there were have not uncommonly deteriorated over a period and have thus failed and permitted the entry of water. They failed because they were unable to cope with the "restlessness" of the materials surrounding the seals. I accept Mr Josey's evidence that things have been getting worse, in particular in relation to the number of window units being found to have something wrong with them, a subject on which he drew up a table of other persons' conclusions as between October 1982 and June 1993. It was unavoidably a rather rough and ready exercise because, on subjects such as whether a window is leaking and how seriously it is leaking, different people are likely to have different views and, of course, the weather in which the windows observed has an effect on the conclusion. Even so, his table does seem to me to be *some* evidence of a worsening of the problem of leakage of the building. Quite apart from the technical credibility of the "restless" joint theory, Mr Josey's evidence was consistent, in a way which Miss Tooth's was not, with the inspection by architects in relation to certificates and snagging lists having found no or only minor leaks and with the period of about a year to May 22 1984 during which, although no 66 continued to be occupied by a tenant who had previously seen fit to complain of leakage, there is no record of any complaint as to leaks. It would not be appropriate to rely too heavily on the inspection by architects and on the lack of complaint by Rothschild as I have not heard from any of those sources, but it is of some weight in favour of Mr Josey's evidence that his technical explanation is consistent as I have explained.

As for making the building watertight, it is common ground between Miss Tooth and Mr Josey that nothing short of the removal of the whole of the existing cladding and its total replacement with a new and redesigned system will suffice. The cost of such operation is estimated at about £1.2m.

What, then, are the conclusions which I should draw from all this evidence? I accept the situation as to leaks has got progressively worse in recent years, but I am not convinced that proving that the leaks have got worse necessarily shows that the problem is one of repair strictly so called or that it is not one of inherent bad design or of initial bad construction. It does not seem to me to follow, for example, that an initial shortcoming in design or construction should reveal itself by symptoms which are observably of a steady frequency and of unincreasing severity, especially where variable factors such as wind and weather are material. Nor does it seem to me to follow from the starting point that one cannot be obliged to repair that which is not in disrepair, that where there is disrepair the work that is needed to make it good must necessarily be described as work of repair. Still less does it necessarily prove that what is needed is work of repair for it to be shown that "deterioration" has taken place. The cladding is not designed for ease of dismantling, for reduction to its component parts or for the replacement of the thin seals, which I accept have existed between many of its material parts. Failed seals cannot be made good by a smear of sealant on their faces nor can sealant to any useful degree be forced in between the assembled components. On the technical and other evidence I have heard I conclude as follows:

(i) There has been actual damage caused to the subject-matter of the principal covenant in the sense that DGUs have failed and there have been encrustations and stains on the faces of the internal mullions and seals. The landlord itself alleged this and other damage to no 66 in its proceedings against others in the Official Referee's Court.

(ii) The damage has been caused by water penetration into the cladding and its passage from the cladding into the interior and, in the cases of DGUs, by penetration into the glazing rebates and from there into the DGUs themselves.

(iii) There were, from the outset, grave shortcomings in design and, to a lesser extent, of construction in the cladding and glazing systems, which had a propensity to leak.

(iv) Seals have failed and amongst the causes of the penetration of water an important one is the failure of seals. In large part their failure has occurred after, and has grown worse over the time since, the building was demised.

(v) In the sense that the failure of seals is the failure of subsidiary parts it can be described as a disrepair and some at least of the damage to the subject-matter of the covenant which has occurred can be attributed to the failure of the seals.

(vi) However, even if I should thus describe the cladding as being in a state of disrepair I would none the less not think it right to speak of the total removal of the existing system and its entire replacement with a new and redesigned system as work of mere "repair". The question is established to be one of degree and, in my judgment, the whole cladding system, while well short of being the whole subject-matter of the covenant, is too substantial and important a portion of the whole subject-matter of the principal covenant for its complete removal, redesign and replacement to be regarded as an operation in respect only of a subordinate or subsidiary part — see Buckley LJ in *Lurcott (supra)* at p924.

(vii) However, I would think it right to describe such remedial work as a "renewal" and within the obligation to "repair amend renew" in the principal covenant, again bearing in mind that the proviso, by its reference to "defects *or* want of repair", indicates a possibility of liability outside mere repair. In a covenant which appears to go beyond mere repair and which, in so doing, refers to renewal, "renewal" can be taken to embrace reconstruction even of the entirety of the subject-matter of the covenant — see Buckley LJ in *Lurcott (supra)* at p924. I do not see it as a necessary part of a renewal (in contrast with a restoration) that the new should be exactly as was the old, particularly where the old had proved defective. The fact that the new cladding will be to a different design does not, in my view, deny it the description of being a renewal. On this basis, whether I am right or wrong as to the total recladding being outside "repair", I find it within the obligation to "amend, renew". I am conscious of the argument that if "renew" is to be given a separate meaning beyond "repair" then there is probably no stopping point short of its ability to require the rebuilding of the whole subject-matter of the covenant: see *Collins* v *Flynn* [1963] 2 All ER 1068. However, the landlord in a case such as the present would be likely to have wished that and the tenant, while recognising the size of the potential burden on him, could well have thought, in relation to a new architect-designed building thought to have been completed to a high specification, that the likelihood of that burden actually turning into a present liability during the 25 years of the underlease was so remote as to be discounted. Thus if the word if "renew", in a context plainly going beyond repair, can properly extend, as I believe it can, as far as a total replacement of the subject-matter to which it relates, then I see no good reason why it should not here do so. If it is asserted that clear words would be

A needed for such a conclusion I would say that in a context, as here, where "defects" are within the covenant as alternatives to "repairs", the words "amend" and "renew" are clear enough.

It is not as if the work proposed will extend the practical life of the building. The new cladding, said Miss Tooth, would last 60 years, the normal life of the building. I note that in *Ravenseft Properties Ltd* v *Davstone (Holdings) Ltd* [1980] QB 12* the covenant included an obligation to "renew", but, as appears at p22G-H, Forbes J found it unnecessary to give the word any separate consideration.

Accordingly, even if I am wrong in seeing the landlord as being in breach of its covenant as to good and tenantable condition, I would B find it to have been in breach of its obligations to "renew" and to "amend". It would only be if I were constrained by authority to regard the clear apparent extension of the principal covenant beyond a liability merely to "repair" that I would conclude the landlord not to be in breach, but I find no such constraint. Indeed, so far from finding such constraint, I have already cited high authority for the propriety of giving a proper and full effect to each word used. On the basis that the landlord is in breach, I turn next to the subject of damages.

Damages

The plaintiff argues that if no 66 had begun to be put into C tenantable condition (or amended, or renewed, or as the plaintiff would have it also, repaired) within a reasonable time after the landlord had received notice of "such renewal, repair and maintenance" (clause 6(12) of the underlease) then that work would have been completed long before the tenant came to market its underlease of no 66 in early 1989 and, in the buoyant conditions of the time, the likelihood is, says the plaintiff, that the underlease would have been promptly assigned for a premium of £85,000 or thereabouts. It is not said that, in a lease of this kind, it need not have been foreseen that the tenant might wish to assign: see *Mira* v *Aylmer Square Investments Ltd* [1990] 1 EGLR 45.

D I hold, as Mr Berrill of Conway Relf and Mr Geoffrey Castle [FRICS] of Dron & Wright deposed, that the problems encountered on the attempt at disposition were entirely due to shortcomings in the state of no 66, which I hold would not have existed had the principal covenant been duly performed. It cannot be doubted but that the landlord had notice of severe leaks from 1985 at the latest. Its own engineers, Mott Hay, reported to it in 1985 and 1986. Mott Hay's recommendations for further investigations were not fully pursued. Had they been and had the cladding been investigated in detail its lack of drainage would have come to light literally years before that was discovered and the need wholly to renew the cladding would similarly have been revealed as the only real solution years before that E eventually emerged. I hold that had the landlord taken reasonable steps to cure the breaches of its obligations under the principal covenant within the time allowed by the underlease then no 66 would have been assigned by the tenant in or about the first week of March 1989 and for a premium of £85,000 or so, the figure which Conway Relf, on the tenant's behalf, were pressing to obtain from Knight Frank & Rutley's client, Coast Securities.

Mr Gaunt cited in this context *Calabar Properties Ltd* v *Stitcher* [1984] 1 WLR 287†, CA at pp295G to 296 where Stephenson LJ says:

> In measuring and assessing any tenant's damages for breach of a landlords' F repairing covenant the Court must, I think, always start with the fundamental principle that they are
>
> "so far as is possible by means of a monetary award, to place the Plaintiff in the position which he would have occupied if he had not suffered the wrong complained of, be that wrong a tort or a breach of contract".
>
> I take that statement from the judgment of Donaldson LJ in *Dodd Properties (Kent) Ltd* v *Canterbury City Council* [1980] 1 WLR 433 . . . So the true measure of damages for persons owning or occupying land, whether in tort or contract, depends on the position of the Plaintiffs and all the circumstances in which they have suffered loss and damage, in the light of the fundamental principle to which I have referred.

*Editor's note: Also reported at (1979) 249 EG 51.
†Editor's note: Also reported at (1983) 268 EG 697.

I willingly accept, as that citation suggests, that there is no one G inflexible form of computation to be applied to all breaches, but that the court is to look as best it can to the particular position which the plaintiff in the case before the court would have occupied if he had not suffered the wrong in question.

Building on that citation from *Calabar (supra)*, the plaintiff seeks damages under three heads. First, there is general damages for inconvenience. As to this, the plaintiff points to the inconvenience suffered by customers and, more particularly, by staff of the bank. The plaintiff argues that the breach was complete by 1986 and that compensation under this head is payable in respect of the period from then until the plaintiff moved out of the premises in late 1988. I would H not put the breach quite so early, but I would see it as occurring by or at the beginning of 1987. I am thus asked to make an award in respect of nearly two years' inconvenience. Mr Gaunt seeks to use the landlord's own expert's view as a guide, that expert, Mr Ian Hayward [FRICS], having suggested that the premises were halved in rental value by reason of their state or, upon some assumptions little related to reality, that the premises were reduced by some 20% in rental value. Mr Gaunt modestly claimed only the lesser basis of 20%. However his modesty, I apprehend, was not true modesty, but was arrived at from an awareness that the drop of 50% in value would produce a figure that would likely be seen to be hopelessly high for general damages of J the kind in question. In my judgment, the 20% basis is also unacceptable; I am unconvinced that one can evaluate inconvenience to a tenant in occupation by reference simply to the diminution in prospective letting value to some hypothetical other tenant. General damages are notoriously at large, but doing the best I can to have regard to the inconvenience suffered by staff and customers in these expensive and prestige premises over the period I have described I fix general damages at £40,000.

The plaintiff's second head of damage includes the lost premium and costs incurred in the abortive attempts to dispose of the underlease in 1989. I have already indicated that I would expect a premium of K £85,000 to have been achieved and I find this to be within the plaintiff's special damage suffered. As for the costs incurred in the abortive attempt to dispose of the underlease, the landlord says that such costs would have been incurred even had the premises been in the best order and condition and even had the assignment gone ahead. However, it is to be borne in mind that those costs were incurred, but that they profited the tenant not one iota and that the tenant will incur a further set of costs when it does come, albeit belatedly, to underlet the premises once they are put into good order. (It is accepted that, because of the fall in rents, an assignment, at all events without a "negative premium", cannot realistically be supposed to be possible). I L accept that these wasted costs are to be recovered by the tenant. The landlord must therefore pay a series of sums (the precise sums, I understand, are or will be agreed) in respect of fees of Dron & Wright, Bickerdike Allan, Conway Relf and Fladgate Fielder. Also within this second head is a whole range of outgoings — cleaning, storage, security, rates, fire protection, electricity, water and heating services — which the tenant has borne since early March 1989 which would not have been borne, at any rate by him, had the underlease been assigned. I am by no means confident of the individual figures agreed as they have changed from time to time in the course of argument and I have not received the final schedule from the defendants which I had M understood I was to receive. If there are difficulties on these figures then they can be mentioned to me, but I apprehend the current figure down to September 1993 and, including the abortive costs I have mentioned, is £260,173.37 plus £6,305.12 VAT.

To such, and subject to the effect of the counterclaim, which I deal with below, must be added rent, insurance and service charges payable by the tenant after the date at which, had the principal covenant been honoured, the tenant would have disposed of the property. The relevant figure is of the order of £1,551,250. Again, if a precise figure is not agreed then I will hear counsel on the point. Insurance required by the underlease in the period which is to be added to the special damage to be payable by the landlord is of or about £17,130.02 and service charges add a further £170,529.07.

Still within the tenant's second head are two forms of claims as to interest. They are, I think, affected by my finding the first week of March 1989 as the date of the putative disposal of the underlease, but rather than my attempting to put a value on these claims or fully describing them, the simpler course is for me merely to indicate that if interest is not agreed then I will hear counsel further on the subject.

The third head of loss claimed by the plaintiff is occasioned by its remaining liable under the underlease for the rest of the term to 2008. This liability is to be computed on the basis referred to during the hearing as option 2B, under which (the landlord having given evidence on this subject, which I accept) it is assumed that irrespective of recovery from Gartners or others the landlord will now very shortly set about a total recladding of no 66.

Option 2B is an elaborate table (happily largely agreed as to the figures) which assumes the tenant remains tenant until 2008, but assumes also that soon as is practicable after no 66 is put into good order the tenant will subunderlet at the best rent obtainable to a description of tenant of a kind to whom the landlord would be likely to permit a subunderletting. The present tenant is to bear rates, service charges and insurance until the putative new tenant comes in. In the fullness of time the putative incoming tenant will suffer upwards-only rent reviews. The plaintiff, under option 2B, never pays a greater rent to the landlord than the £365,000 pa which it now pays, as both experts agree that 1983 rent levels are unlikely to be reached again throughout the whole term until 2008. The plaintiff-tenant bears service charges until the putative new tenant takes them over. These putative incomings and outgoings of the tenant are, in option 2B, converted to a net cash flow in or out, quarter by quarter, down to 2008 and that net cash flow is discounted back to today's date at an agreed rate of 8% pa. Today's pound is thus worth 100p whereas that to be paid in 2008, for example, is now worth 32p.

Option 2B assumes that the specifying of the necessary recladding works and their execution will be followed first by a period for the marketing of the underlease and then (in the conditions foreseen as likely) by the incoming tenant needing to be given an initial "rent holiday". The experts differed as to the precise length of "rent holiday" likely and of the marketing period needed. The plaintiff's computation assumes no 66 does not come on stream as a yielder of rent till the quarter day March 25 1997. I accept on the evidence that that is the probability, as seen from today. Although, looking to the rest of the period from March 25 1997 to mid-2008, the experts differed in only one part of their respective computations, because one is looking to expenses likely over a relatively long period, even a small difference can yield a large difference in the resulting cash figure. The one difference is this: both experts expect the first rent, when, after the necessary works and the "rental holiday", no 66 will first yield rent, to be £170,000 pa, which is also what they both see as the rent which would be the value of the premises today were they in good condition. Both experts are content to accept that as a matter of general experience over a long period rents can be expected to increase by 7% pa in respect of premises such as no 66 in a location such as no 66's. There will be two reviews (upwards only) of the rent that the putative incoming subtenant will need to pay, in 1998 and in 2003. The difference between the experts in relation to the rent increases likely on those rent reviews is as to from what date the expected 7% pa increase should be taken to operate. The landlord says that that 7% should be computed to run as from today whereas the tenant urges that it should run from the date of the supposed letting to the putative tenant in 1995. I prefer the landlord's view; if the 7% figure is a figure generally observable over a long term I see no good reason to delay the start of that long term. The fact that this approach leads to a conflict between the view that the premises, if in good order, would be worth £170,000 pa today, but only that same figure in 1995 when they are relet, is simply a conflict between a generally observable pattern and what is specifically expected in the relatively short term in this particular case. If, as seems accepted and indeed inevitable, the rent in the longer term is to be computed by reference to what is *generally* observed, then the adoption of that general position seems to me to lead to a computation increasing from today at 7% pa from today's figure of £170,000 pa.

Option 2B is computed on the basis (which I accept to be correct) that the tenant's tenancy of no 66 since it left the premises in November 1988 has been of no value whatsoever to it and, moreover, incapable of being turned to value. Any subunderlessee of the kind who might have been willing to pay *something* to be in possession in the short term of premises in such disorder would, on the evidence, have been a tenant an underletting to whom the landlord would not for a moment have countenanced.

Mr Knight, while agreeing many of the figures in option 2B, resisted it in principle and submitted that any award of damages in respect of the liability of the tenant to pay rent, insurance and service charges in respect of no 66 after the date of my judgment would be wrong. The principal covenant, he urges, is a continuing covenant under which damages may be awarded from time to time as they accrue. The tenant cannot, he says, sue for future unascertained damages. He relies on para 979 in the 15th ed of *McGregor on Damages* on this point. The tenant can come back and sue as often as it likes, he says, so long as any breach of the landlord's covenant (which is denied) continues. Mr Gaunt's answer is that what the tenant here invites to be assessed is not further losses flowing from future breaches, but future losses from a past breach, the breach from 1987 or so to March 1989 which had the consequence that the tenant, instead of being able to dispose of the premises, was unable to rid itself of them, with the consequences, on the evidence, that it will be unable to shift liability for rent and so on during the rest of the term to 2008. Mr Gaunt argues that if the tenant were to fail now fully to claim for these losses it would be in difficulties were it to attempt to recover them later; the landlord could in such a case reply *res judicata*. I accept Mr Gaunt's argument and I do not see *Battishill* v *Read* (1856) 18 CP 696 as barring that conclusion. In that case the plaintiff, who held the reversion in some premises, complained that some tiles in the eaves of his property had been removed by the defendant and that a taller building had been erected so close by on the immediately adjoining land by the defendant that water no longer fell as it had on to and from his eaves. Although the cost of replacing the tiles was only 30s the plaintiff claimed that the 40s paid into court was not enough. The jury found it was sufficient. On appeal it was argued that the plaintiff ought to have been allowed to give evidence to show that the saleable value of his premises had been materially depreciated by the defendant's wrongful acts. The plaintiff had suffered a permanent injury, it was argued, in being prevented from raising the level of his property, if so minded, and had suffered a diminution in its market value, evidence of which should have been admitted. The crucial difference between that case and this (although it is a difference only one of the four judges expressly brought out) is that the plaintiff there did not aver that he had wished to sell his reversion. Cresswell J, in finding that the evidence of diminution in value was properly excluded, said:

> So, here, the consideration of the subject might have been very different if the plaintiff had alleged and proved an attempt to sell the premises, which had failed in consequence of the nuisance.

Williams J, next following, began by saying that he was entirely of the same opinion, as did Willes J. By contrast, the evidence before me did show an attempt to dispose which failed in consequence of the landlord's breach of covenant. In the light of that crucial difference I do not see the tenant as properly to be confined, as was the plaintiff in *Battishill*, to initial nominal damages coupled with the ability to bring second and subsequent actions, perhaps for increasing damages, so long as the defendant's default should continue.

I am thus in a position to award general and special damages to the tenant broadly as I have mentioned them, although I recognise that I may need to be addressed further as to the final detail if figures are not agreed in the light of my findings. More important, as I shall mention below, there is a counterclaim by the landlord to be borne in mind and this will have an effect on the total damages to be awarded because of the indemnity or set-off I shall there discuss.

Estoppel

The tenant submits that the landlord is estopped, by way of an estoppel by convention, from denying that it has been and still is liable to remedy such defects as have caused the ingress of water. Mr Gaunt relies on *Amalgamated Investment & Property Co Ltd* (*in liquidation*) v *Texas Commerce International Bank Ltd* [1982] QB 84 at p121C *per* Lord Denning MR and *"The Vistafjord"* [1988] 2 Lloyd's Rep 343 CA *per* Bingham LJ at pp 351-2, where he said:

> I do not propose to essay a definition of estoppel by convention. It is sufficient for the purposes of this case to adopt the description given by the Court of Appeal in *Keen* v *Holland* [[1984] 1 WLR 251] *supra* at p261, of the essential nature of the *Amalgamated* case, and to say that it applies where (1) parties have established by their construction of their agreement or their apprehension of its legal effect a conventional basis, (2) on that basis they have regulated their subsequent dealings, to which I would add (3) it would be unjust or unconscionable if one of the parties resiled from that convention.

In the light of my earlier conclusions I may take this further argument in the tenant's favour fairly shortly. I am willing to treat the underlease as if it was an agreement between the landlord and the tenant, which it was not, but even so I do not find the facts necessary to establish a convention as to its construction or effect. In the context that a landlord may indicate a willingness to do works outside his strict obligation and in a context, as here, in which the landlord believes he may have a good case to get others to do works or to recover the expense of works from others, merely to point to a landlord's apparent willingness to do works or have them done is not, of itself, material sufficient to show a convention as to the effect of the underlease. The landlord by its agents used expressions such as that it was taking most seriously the question of repairs, that it would draw up a scheme for investigation and remedy and that it seemed important to try to identify the problems with the cladding and get them put right without waiting for a lengthy battle or litigation as to responsibility (which the landlord's agents, be it noted, indicated *they would have to pursue in due course*). The landlord's agents said they were in contact with their client and its solicitors with a view to initiating the necessary further investigation and remedial works as a matter of urgency, but, as the landlord also made clear, it was pursuing the developer to have the defects remedied and was maintaining a pressure on the developer to that end. Nowhere do I find that the landlord was indicating that it would see the defects made good by reason of its liabilities under the principal covenant or under the underlease generally. Nowhere is the meaning and effect of the principal covenant discussed or assumed. Nowhere at any time relevant to estoppel is the ordinary consequence of the landlord's liability under the principal covenant — namely the tenant's liability under the service charge — discussed or assumed. Mr Murray, the tenant's premises maintenance manager, undoubtedly and rightly believed that the landlord was a substantial pension fund which had accepted that it would get the necessary work done, but he did not assert that allied to that was that it had accepted that the work had to be done by reason of the principal covenant. The tenant asserts that not once in the correspondence did the landlord ever deny its responsibility to do the necessary works or have them done, but not once, either, was it ever accepted in the correspondence that the landlord was so responsible by way of the principal covenant or otherwise under the underlease. Had it been material for me to deal with estoppel as a matter of decision I would have held the estoppel to fail because the first of Bingham LJ's three ingredients is missing. I would not have been satisfied, either, that the second was present and in the absence of those first two the third would not have arisen.

Counterclaim

The current rent is £365,000 pa. Option 2B is computed on the basis that the tenant will be liable for and will pay that rent punctually quarter by quarter till the expiry of the underlease in 2008. There is and can be no counterclaim as to future rent. As to past rent after the date, early March 1989 (which I have accepted as the date at which the tenant would have assigned if the premises had been in the proper order), down to the September quarter day in 1993, this has not been paid, after June 1989, by the tenant. I have mentioned, as a head of loss by the tenant, that it was liable for rent over this period, a liability which amounts to some £1,642,500 when added together with the further sums for service charges and insurance. If that total were to be allowed as a head of recoverable damage by the tenant then the landlord would need to be put in a position in which he was to be paid that money. However, the tenant's claim was expressed not only as damages in respect of that global liability but, alternatively, for indemnity against that liability. Although any potential difference in the net effects between, on the one hand, my giving judgment on damages and ordering payment to the landlord on his counterclaim and, on the other, my simply setting one against the other, has not been the subject of any argument, I take it to be tacitly accepted that either way the overall effect is zero and that the more convenient course is not to award a sum and then a countersum but to proceed on the basis that no damages are to be awarded to the tenant in respect of rents, insurance and service charges payable prior to the September 1993 quarter day and that no rent etc is to be ordered payable to the landlord for that period either. On that basis I need grant no separate relief on the landlord's counterclaim, but rather adjust the tenant's damages downwards by deleting the part otherwise attributable to rents, insurance and service charges payable prior to September 1993 and not paid.

Pleadings and conclusion

At the end of Mr Castle's oral expert evidence on behalf of the plaintiff I asked whether his schedule of the computation of future losses suffered by the tenant reflected that if the landlord had been liable under the principal covenant to do the various works and if they had been done then their cost would have been passed on to the tenant under the service charge. I was told that neither Mr Castle's figures nor those of Mr Hayward (the landlord's expert) reflected that possibility. This troubled me because, if the measure of damages was to include reference to what the tenant's position would have been if only the landlord had not breached the covenant, it seemed to me that that position would include that the tenant would become liable by way of service charge. I later raised the point with Mr Gaunt who was, I think, entitled to answer, as he did, that it was not clear that the landlord would even attempt so to recover and that it was for the landlord to plead factors which reduced the damages (if any) payable by him and that nothing of the kind was pleaded. Mr Gaunt also then outlined some other reactions he would need to enlarge upon were the landlord to seek to have his prospective recovery by way of service charge brought into account. Some while later, after Mr Knight had completed his speech and when Mr Gaunt had nearly finished his reply at the very end of the case, Mr Knight applied further to amend the defence and counterclaim on the basis that the landlord would indeed seek to recover under the service charge and that it wished to deduct from the tenant's computation known as option 2B that part (said to be 75%, under the service charge) of the cost of the recladding works (£1.2m) which the landlord had sworn would now be put in hand. The prospective future deduction would need to be discounted back to give its present value. No expert or other evidence had been directed to these ideas and I was not confident that there could have been no relevant evidence directed to them. Moreover, the tenant, were the point to be raised, would have wished to argue that for the landlord to recover from the tenant for such works would have led to an unjust entitlement, a double recovery, because it had already recovered therefor from Gartners, or was about to do so. Questions of subrogation were said to arise and the landlord was set to argue that recovery from Gartners was *res inter alios acta*. None of these questions had been explored earlier. Had I felt that any true hardship or material injustice would have been visited on the landlord if I

A refused it leave to amend I would have given leave even though an adjournment for amendment of pleadings, for further argument and, not improbably, for further evidence would have ensued. However, if I make no deduction from the damages to reflect prospective recovery under the service charge then the landlord will be entirely free to seek to recover from the tenant the 75% (if that is appropriate) of whatever the true cost of the works within the covenant proves to be when it is eventually done. Even if there were to be an injustice to the landlord here (and that is far from clear, given the tenant's arguments against double recovery and other arguments, perhaps including subrogation and estoppel, against its liability under the service charge) it would be B short-lived. In all the circumstances I refused the landlord leave further to amend. There was, at the same time, another subject as to which the landlord sought to amend, but in the event, which happened, that Mr Gaunt opposed it, it was not pressed and I need say no more about it.

So, by way of conclusion, I hold the landlord to be, and long to have been, in breach of the principal covenant, that substantial damages are to be awarded to the tenant and that, because of the landlord's counterclaim, the sum finally to be awarded is to take into account that the tenant, while liable under the underlease for rent, service charges and insurance, has in fact paid no rent since June 1989 and other sums, if at all, only sporadically. As I hand this judgment C down while sitting as vacation *judge the convenient course will be for further discussion as to the precise awards in the light of this judgment and as to other questions, such as to costs, to be the subject of a restored short hearing in the new term.

Court of Appeal

November 19 1993

D (Before Lord Justice Ralph GIBSON, Lord Justice HIRST and Lord Justice Peter GIBSON)

DALLHOLD ESTATES (UK) PTY LTD v LINDSEY TRADING PROPERTIES INC

Estates Gazette April 30 1994

[1994] 17 EG 148

Landlord and tenant — Rent — Whether section 48 of the Landlord and Tenant Act 1987 applies to an agricultural holding — Whether notice given pursuant to that section

E ***Agricultural holdings — Notice to pay rent — Whether effective notice to pay served pursuant to Case D under the Agricultural Holdings Act 1986***

The plaintiff was the tenant of an agricultural holding under a five-year term from July 30 1987 at a yearly rent of £50,000 payable on the usual quarter days. The holding included a manor house and several cottages. The tenant failed to pay the rent in respect of the quarter days June 24 1990 to and including June 24 1991. By notice dated August 7 1991 the defendant landlord gave notice pursuant to Case D, Part I of Schedule 3 to the Agricultural Holdings Act 1986 requiring the tenant to pay the arrears of rent within two months. On October 9 1991 the F landlord gave the tenant notice to quit the holding under Case D on the ground that the notice to pay had not been complied with. A provisional liquidator of the tenant company challenged the service of the notice to pay rent and requested the landlord's agents to state whether any notice under section 48 of the Landlord and Tenant Act 1987 had been served. On December 2 1991 the landlord's agents sent a letter to the tenant's agents referring to various notices served by them establishing their address as a service address. By a further notice dated December 3 1991 the landlord's agents confirmed for the purposes of section 48 their client's address for service of all notices, including notices in proceedings, and enclosed a further notice to pay rent.

G The tenant's application for declarations that by reason of a failure to serve a notice under section 48 of the 1987 Act no rent was due on August 7 1991, and that the purported notices to pay rent of August 7 and December 3 1991 were invalid, was allowed by Chadwick J: see [1992] 1 EGLR 88; [1992] 23 EG 112. The landlord appealed, relying on the contentions advanced before Chadwick J, and the tenant cross-appealed on the grounds that the failure to serve a notice under section 48 of the 1987 Act before December 3 1991 had in law the effect that rent which would otherwise be due before that date was irrecoverable for all time, alternatively that such rent did not become due until some later date, but was not due on December 3.

H ***Held*****: The landlord's appeal was allowed in part and the tenant's cross-appeal dismissed in part on the following grounds: (1) "Premises" in section 46(1) of the 1987 Act means the subject-matter of the letting and nothing in that section provides any guide which enables the court to discern any limit to the meaning of "premises" narrower than the subject-matter of the letting; accordingly, the section applies to an agricultural holding which included a dwelling-house. (2) No notice satisfying section 48 was given by the landlord prior to December 3 1991 and it was not necessary to decide whether the letter of December 2 1991 was such a notice, although in the context of the correspondence it was J so capable. (3) Section 48 does not provide that a failure to serve a notice destroys the right to claim the rent due before such service. All the rent in respect of which the notice of December 3 1991 was served was "otherwise due from the tenant" on that date; the notice was a valid section 48 notice and upon its service the rent then became due. There was no justification for any extension of time. (4) The notice to pay rent of December 3 was valid notwithstanding that it stated by reference to a schedule the quarter days from which the arrears of rent were alleged to be due and unpaid. In so far as the notice inaccurately asserted that the rent "otherwise due" had been due on and from the quarter K days listed, it did not mislead and could not reasonably have misled the tenant in any way; the statutory purpose of the notice was fully satisfied.**

The following cases are referred to in this report.

Afovos Shipping Co SA v *Pagnan (R) & F.lli* [1983] 1 WLR 195; [1983] 1 All ER 449; [1983] 1 Lloyd's Rep 335, HL

Bank of Baroda v *Panessar* [1987] Ch 335; [1987] 2 WLR 208; [1986] 3 All ER 751

Bracey v *Read* [1963] Ch 88; [1962] 3 WLR 1194; [1962] 3 All ER 472

Dallhold Estates (UK) Pty Ltd v *Lindsey Trading Properties Inc* [1992] 1 EGLR 88; [1992] 23 EG 112

L *Dickinson* v *Boucher* [1984] EGD 19; (1983) 269 EG 1159, CA

Horford Investments Ltd v *Lambert* [1976] Ch 39; [1973] 3 WLR 872; [1974] 1 All ER 131; (1973) 27 P&CR 88, CA

London & Westminster Loan & Discount Co v *London & North Western Railway* [1893] 2 QB 49

Maunsell v *Olins* [1975] AC 373; [1974] 3 WLR 835; [1975] 1 All ER 16, HL

Official Solicitor v *Thomas* [1986] 2 EGLR 1; (1986) 279 EG 407

Pickard v *Bishop* (1975) 31 P&CR 108; [1975] EGD 24; 235 EG 133, CA

South Tottenham Land Securities Ltd v *R&A Millett (Shops) Ltd* [1984] 1 WLR 710; [1984] 2 All ER 614; [1984] EGD 96; (1983) 269 EG 630, CA

Urwick v *Taylor* (1969) EGD 1106; 212 EG 1275

M *Whitley* v *Stumbles* [1930] AC 544; 46 TLR 555

This was an appeal and cross-appeal by respectively the landlord, Lyndsey Trading Properties Inc, and the tenant, Dallhold Estates (UK) Pty Ltd, from the judgment of Chadwick J who on March 2 1992 had made declarations on an application by the tenant.

David Neuberger QC and Joanne Moss (instructed by Franks Charlesly & Co) appeared for the appellants; Edward Bannister QC and Martin Rodger (instructed by Mills & Reeve, of Cambridge) represented the respondents.

Giving the first judgment, **Ralph Gibson LJ** said: This is an appeal by Lindsey Trading Properties Inc, defendants in the

A proceedings, from the order of Chadwick J made on March 2 1992 on the trial of the action. The judgment of Chadwick J is reported at [1992] 1 EGLR 88. I shall refer to Lindsey Trading Properties Inc as the landlord, and to Dallhold Estates (UK) Pty Ltd, who are the plaintiffs in the proceedings and respondents to this appeal, as the tenant.

The appeal raises questions as to whether section 48 of the Landlord and Tenant Act 1987 applies to tenancies of an agricultural holding; by what date any notice was given by the landlord under that section; and whether an effective notice to pay was served by the landlord under Case D of Part 1 of Schedule 3 to the Agricultural B Holdings Act 1986 (AHA).

The facts are set out in the judgment of Chadwick J. I shall repeat a brief summary:

The landlord, a company incorporated and having its registered office in Panama, is owned and controlled by Mr Jorg Bolag, who is resident in Switzerland. The landlord is the freehold owner of an agricultural and sporting estate of 940 acres in Hertfordshire, which was let to the tenant by a lease dated July 30 1987. The estate includes a manor house, Upp Hall, and a number of cottages, some of which are occupied by estate workers. The estate is an agricultural holding for the purposes of the AHA.

C The tenant is a company incorporated in Western Australia. On August 9 1991, a provisional liquidator of the tenant company was appointed by an order of the Federal Court of Australia and on August 14 1991 a provisional liquidator was appointed in England. On January 24 1992 an administration order was made by Chadwick J to achieve a more advantageous realisation of the tenant company's assets.

The lease is for a term of five years from July 30 1987 at a yearly rent of £50,000, payable by equal quarterly payments in advance on the usual quarter days and provides for payment to Franks Charlesly & Co, solicitors, of Hulton House, London, or to such other agent as the landlord should from time to time notify to the tenant. At the time of D the grant of the lease the landlord was Wydgee Pastoral Co Ltd of Perth, Western Australia; the tenant was Teilas Pty Ltd also of Perth and the present landlord was party to the lease as surety. The circumstances in which the tenant acquired the leasehold interest and the landlord acquired the freehold reversion are not relevant.

The tenant did not pay the rent provided to be paid by the terms of the lease on June 24, September 29 and December 25 1990, or on March 25 and June 24 1991: a total of £62,500. On August 7 1991 a notice was served on the tenant in Perth requiring the tenant to pay within two months the rent due in respect of the estate, said to be £62,500. The notice was said to be given "in accordance with Case D E in Part 1 of Schedule 3 to the AHA" by Franks Charlesly & Co, solicitors and authorised agents for the landlord of Hulton House, 161-166 Fleet Street, London EC4.

No rent was paid in response to that notice. No rent was paid on the next rent day, September 29 1991. On October 9 1991 Frank Charlesly & Co, as agents for the landlord, gave notice to the tenant to quit the estate on July 30 1993 for the ground and reasons set out in Case D, namely failure to comply with the written notice of August 7 1991. By section 26(1) of AHA, upon service by the tenant of a counternotice, it is provided that a notice to quit shall not have effect unless, on an application by the landlord, the tribunal consent to its operation; but in F certain cases, including Case D in Schedule 3, consent of the tribunal to operation of the notice to quit is not required. On failure to pay rent within two months of service of a valid notice to pay, the tenant loses the protection of the AHA: see *Pickard* v *Bishop* (1975) 31 P&CR 108*.

(vi) By letter of October 31 1991 the tenant's solicitors asked Franks Charlesly whether any notice had been served under section 48 of the Landlord and Tenant Act 1987. That section, as material to this case, provides as follows:

> (1) A landlord of premises to which this Part applies shall by notice furnish tenant with an address in England and Wales at which notices (including notices in proceedings) may be served on him by the tenant. G
>
> (2) Where a landlord of any such premises fails to comply with sub-section (1), any rent or service charge otherwise due from the tenant to the landlord shall (subject to subsection (3)) be treated for all purposes as not being due from the tenant to the landlord at any time before the landlord does comply with that subsection.

Subsection (3) has no application to this case. The application of section 48 is defined by section 46(1):

> This part applies to premises which consist of or include a dwelling and are not held under a tenancy to which Part II of the Landlord and Tenant Act 1954 applies. H

The word "dwelling" in that phrase is defined by section 60(1) as:

> . . . a building or part of a building occupied or intended to be occupied as a separate dwelling, together with any yard, garden, outhouses and appurtenances belonging to it or usually enjoyed with it;

The letter of October 31 1991 also enclosed a copy of the notice requiring arbitration of the notice to quit dated October 9 1991. The landlord's solicitors were asked to state if they were not in fact instructed to accept service on behalf of the landlord. By letter of November 4 the tenant's solicitor requested confirmation that the J landlord's solicitors were authorised to accept the notice requiring arbitration and by further letter of the same date it was asserted, on behalf of the tenant, that there were:

> reasonable grounds to expect that the notice to quit may be defeated on the grounds . . . that the landlord does not appear to have complied with the requirements of section 48 of the 1987 Act.

By letter of November 8 1991 the tenant's solicitors wrote:

> to confirm that the landlord's solicitors are both instructed to accept service and acknowledge receipt of the notice requiring arbitration. K

By a letter of November 12 1991 the landlord's solicitors confirmed receipt of the notice requiring arbitration and expressed surprise at the tenant's request in relation to section 48 since "you and your clients have known where to serve notices on the landlord and indeed have served counternotices for the landlord here". By letter of November 28 1991 the tenant's solicitors asked whether it was admitted that no section 48 notice had been served and in reply, on December 2 1991, the landlord's solicitors, with reference to the matter of the section 48 notice, referred the tenant's solicitors to a number of matters including "various notices served by this firm on behalf of [the landlord] establishing us as a service address and, of L course, you have served purported counternotices to Lindsey upon ourselves". By amendment of the notice of appeal in this court the landlord has contended that the passage cited from that letter of December 2 1991 was, in context, an effective notice under section 48(1) of the 1987 Act.

The landlord then delivered to the tenant in Perth on December 3 1991 a letter of that date, which reads:

> We enclose by way of service on behalf of our client Lindsey Trading Properties Inc, Notice to Pay Rent under the Agricultural Holdings Act 1986.
>
> This is without prejudice to any other Notices served or demands made in M respect of the outstanding rent and also without prejudice to our contention that section 48 of the Landlord and Tenant Act 1987 has been complied with.
>
> We would confirm on behalf of our client and for the purposes of Section 48 aforesaid that our client's address for service of all Notices, including Notices in proceedings, is care of this firm.

The notice enclosed, in form 1 in the Agricultural Holdings (Forms of Notice to pay Rent or to Remedy) Regulations 1987, said "that I require you to pay within two months from the date of service of this Notice the rent due in respect of the above holding (Upp Hall estate) as set out below". Under the "particulars of rent not paid" there was, in effect, set out in an appendix the following:

*Editor's note: Also reported at (1975) 235 EG 133.

A

Date when due	*Amount due*
24th June 1990	£12,500
29th September 1990	£12,500
25th December 1990	£12,500
25th March 1991	£12,500
24th June 1991	£12,500
29th September 1991	£12,500
Total	£75,000

It has been common ground that the letter of December 3 1991 was in itself an effective notice for the purposes of section 48 of the 1987 B Act. The contention for the tenant has been that, nevertheless, the notice to pay was defective because of defects in form and by reference to the time at which unpaid rent could properly be held to have become payable: in particular, even if the letter removed the legal requirement created by section 48(2), that "rent otherwise due from the tenant to the landlord shall be treated for all purposes as not being due from the tenant to the landlord at any time before the landlord does comply with subsection (1)", nevertheless this notice to pay was fatally defective in form because it asserted that the unpaid rent was due on the quarter days listed whereas in truth the arrears became due only on delivery to the tenant of the letter of December 3 C 1991.

On January 27 1992 the tenant commenced these proceedings by originating summons. The declarations sought at the trial were, in summary, that: (i) on the true construction of section 48 of the 1987 Act no rent was due from the plaintiff to the defendant under the lease of July 30 1987 on August 7 1991 and December 3 1991; and (ii) that the notices of August 7 and December 3 1991 purporting to be notices to pay rent were invalid and of no effect. The primary contention of the landlord was that Part VI of the 1987 Act had no application to land held under a tenancy of an agricultural holding.

Before Chadwick J in March 1992 undertakings were given by the D landlord to extend time for compliance with the notice to pay rent of December 3 1991 until, at latest, the seventh day after the final determination of the action, including any appeal.

Grounds of decision of Chadwick J

The judgment of the learned judge is reported* and I shall state his grounds in brief summary.

The primary question was whether the word "premises" in section 46(1) of the 1987 Act is to be given the meaning which it bears in ordinary legal parlance, namely the subject-matter of a letting, or some more limited meaning which would not extend to the agricultural E holding let under the lease of July 30 1987.

The word "premises" is capable of a wide range of meanings: reference was made to *Maunsell* v *Olins* [1975] AC 373; to *Whitley* v *Stumbles* [1930] AC 544; and to *Bracey* v *Read* [1963] Ch 88. Within that wide range the meaning which it was intended to have for the purposes of the 1987 Act must be determined by construing the words used in the Act.

To that end it was necessary, first, to consider the 1987 Act as a whole in order to see whether some "overall statutory purpose" could be identified from which the word "premises" must take its meaning. After reference to the long title of the 1987 Act; to the terms of Parts I, II and III of the Act which apply only to flats; to Parts IV and V which F apply both to flats and to dwellings other than flats; and to the various sections in Part VI, in which section 48 is included, Chadwick J concluded that it was impossible to regard the 1987 Act as having a single theme or purpose which should colour the court's approach to the meaning of the word "premises" in section 46(1).

Further, it was not possible to identify any context in Part VI of the Act, or in the Act as a whole, which points to a particular meaning of the word "premises" other than the definition in section 46(1).

Since the requirement there is simply that the premises "consist of or include a dwelling", the word "premises" must be taken to have a wider meaning than dwelling, which word is defined in section 60(1) G of the Act as "a building, or part of a building occupied or intended to be occupied as a separate dwelling, together with any yard, garden, outhouses and appurtenances belonging to it or usually enjoyed with it". Thus, the word "premises" can include land which is the subject-matter of the same letting: but how much land?

The express exclusion in section 46(1) of business tenancies to which Part II of the 1954 Act applies shows that the word "premises" could include premises held under such a tenancy. Part II of the 1954 Act does not apply to tenancies of an agricultural holding: section 43(1) of that Act. It therefore appeared that Parliament understood that, if tenancies of an agricultural holding were to be H excluded from the definition in section 46(1), express provision was necessary.

Nothing in section 46(1) provides any guide which enables the court to discern any limit to the meaning of the word "premises" narrower than the subject-matter of the letting.

The second question was stated thus: was any notice under section 48(1) of the 1987 Act given prior to December 3 1991? For the reasons set out in his judgment Chadwick J rejected each of the candidates proposed on behalf of the landlord.

As to the third question, whether there had been an effective notice to pay under Case D, after reference to the regulations, SI 1987 J No711, and to para 10 of Part II of Schedule 3 to the AHA, Chadwick J held that a notice would not comply with the regulations if the information which is required to be included in the prescribed form is inaccurate. The notice served on August 7 1991 was invalid because at that date, by reason of section 48(2), no rent was due from the tenant. As to the notice to pay rent served with the letter of December 3 1991, since the rent had not become due until receipt of the letter, the information required to be included in the prescribed form was inaccurate since it was there asserted that the rent was due on the quarter days from June 24 1990 to September 29 1991. Reference was made to *Pickard* v *Bishop* (1975) 31 P&CR 108, to *Dickinson* v K *Boucher* (1983) 269 EG 1159 and to *Official Solicitor* v *Thomas* [1986] 2 EGLR 1.

Appeal and cross-appeal

The contentions of the landlord are those advanced before the judge. The tenant has submitted that Chadwick J was right for the reasons given by him and has contended that the decision of the judge should be affirmed or varied on additional or alternative grounds, namely that:

(i) failure to serve notice under section 48 of the 1987 Act before December 3 1991 had in law the effect that rent which would otherwise have been due before that date was irrecoverable for all L time;

(ii) or, such rent did not become due until the contractual rent day following service, and thus no rent was due on December 3 1991;

(iii) or, such rent did not become due until a reasonable time had elapsed after service of the notice and therefore, again, no rent was due at any time on December 3 1991;

(iv) finally, such rent was not due at earliest until midnight on that day, and therefore a notice to pay rent served on the same day was premature.

The submissions of Mr David Neuberger QC, on the primary question, included the following: M

The word "premises" in its natural and ordinary meaning in section 46(1) does not extend to include a farm: see *Maunsell* v *Olins* at p386G-H *per* Lord Wilberforce and *per* Viscount Dilhorne pp383H to 384A. Lord Reid expressly agreed with the speech of Lord Wilberforce. These passages indicate that the natural meaning of "premises" is buildings of some sort.

The 1987 Act is housing legislation only and, in the context of such legislation, that natural meaning of the word "premises" is the more clear: in *Maunsell's* case at p389G Lord Wilberforce said:

Nobody can be unaware that the law relating to tenancies of agricultural land is one of considerable political import and delicacy and I am very reluctant to

*Editor's note: See [1992] 1 EGLR 88.

A believe that the particular aspect of this law which relates to subtenancies, treated with evident circumspection in 1948, would have been dealt with in a general provision such as appears in section 41 of the Act of 1954.

Further, at p384E, Viscount Dilhorne said:

The use of the word "premises" is not, in my opinion, sufficient to justify the conclusion that Parliament intended to restrict the common law rights of the owner of the freehold of a farm.

To exclude business premises under Part II of the 1954 Act (section 46(1) of the 1987 Act), but not to exclude agricultural holdings, would be extraordinary because business tenancies have the
B same sort of protection in relation to notices (see section 66 of the 1954 Act and section 23 of the Landlord and Tenant Act 1927) as agricultural tenancies have under section 93 of AHA. There can be no logic in, and there should not be attributed to Parliament assent to, the giving of protection under section 48 to a tenant of a 1,000-acre agricultural estate with several houses on it, but not giving the same protection to a tenant of a house who happens to use one room as his professional surgery.

The law relating to agricultural holdings was consolidated in the AHA: it would be surprising if by "a side wind", and without express reference to agricultural holdings, the 1987 Act impinged upon the
C relationship between the landlord and tenant of an agricultural holding.

Upon the tenant's contentions, section 3 of the Landlord and Tenant Act 1985 would apply to any agricultural holding with a dwelling or dwellings on it; but section 3 should not be so widely construed not least because by section 3(3) a criminal offence is created.

The word "premises" in section 46(1) of the 1987 Act and in section 3(1) of the 1985 Act should be held to mean either: (a) a building being a single unit of habitation or a building or buildings containing one or more units of habitation in either case with ancillary
D land or other buildings (eg garage, paddock or orchard); or (b) a building or buildings containing one or more units of habitation which would not be excluded from being "premises" simply because it included yards, gardens, outhouses or appurtenances belonging to or enjoyed with the unit of habitation, but it would be excluded if it contained more land.

Further, "a dwelling" in the phrase "premises which consist of or include a dwelling", in section 46 (1), should not be read as meaning "dwellings": see *Horford Investments Ltd* v *Lambert* [1976] Ch 39 at pp46D to 47D and 52C-G. Such a restrictive reading is consistent with the purposes of the 1987 Act which, when read as a whole, should be
E taken as being to protect tenants in their homes.

The scope of the 1987 Act can be seen from the long title: the reference to "connected purposes" should not be seen as sufficient to permit the Act to be treated as extending to agricultural holdings.

Nothing indicates that the mischief at which section 48(1) was directed in the case of housing accommodation was seen by Parliament as requiring attention in the case of agricultural holdings, but not in the case of lettings of business premises under Part II of the 1954 Act.

The provisions of section 49 (extension of circumstances in which notices are sufficiently served) are unlikely to have been intended by Parliament to extend to agricultural holdings, to which a separate set
F of provisions are applied by section 93 of the AHA.

The provisions of section 52 in Part VII of the 1987 Act, which provide that the county court shall have jurisdiction to hear and determine any question arising under any provision of section 46 to section 48, do not accord with, and should be taken as showing that they were not intended to apply to, the provisions for arbitration under the AHA.

The reference to the "Secretary of State" in section 53, presumably the Secretary of State for the Environment, as contrasted with the minister for agriculture or the Lord Chancellor, supports the contention that the powers there given were not intended to apply to agricultural holdings.

As to the 1985 Act, sections 1 to 3 (of which section 3 contains the G
reference to "a tenancy of premises which consist of or include a dwelling") derive from sections 121 and 122 of the Housing Act 1974 and it is therefore unlikely that they could have been intended to apply to agricultural holdings.

Section 17 of the 1985 Act derives from section 125 of the Housing Act 1974 and provides for specific performance of a landlord's repairing obligations. It would be surprising, although not logically inconsistent with the AHA, if an agricultural tenant could use this provision to obtain specific performance as a basis of claim additional to the code provided by section 7 of the AHA and the regulations (SI 1973 No 1473 and SI 1988 No 291) made thereunder which provide H
for arbitration.

In answer to these submissions Mr Edward Bannister QC, on behalf of the tenant, submitted, in effect, that they left the reasoning of Chadwick J unimpaired.

Conclusion on the primary question

Mr Neuberger demonstrated, in my view, that it is fairly to be described as surprising that Parliament should exclude tenancies within Part II of the 1954 Act from the effects of Part VI of the 1987 Act, but include therein agricultural tenancies; and he demonstrated that the application of Part VI to agricultural tenancies was left to take effect without any such amendment of or addition to the statutory code J
applicable to agricultural tenancies as might well have been devised. Mr Neuberger, however, acknowledged in the course of his careful and detailed examination of the provisions of the 1985 Act and of the 1987 Act that the construction of section 46(1) adopted by Chadwick J could not be shown to produce any clear conflict between the statutory provisions separately or specially applicable to agricultural tenancies, on the one hand, and the additional or cumulative provisions thus made applicable by section 46(1) of the 1987 Act and section 3 of the 1985 Act, on the other hand. His list of provisions which, it was said, were "unlikely" to have been intended to apply to agricultural tenancies was advanced as providing added force to the contention, K
based upon the statements in *Maunsell* v *Olins*, that the word "premises" in section 46(1) should be given a restricted meaning, such as those contended for by Mr Neuberger. For my part, I cannot accept the submission as to the meaning of "premises" in the context of the 1985 and 1987 Acts; and the asserted improbability that Parliament can have intended to apply provisions, such as section 48, to agricultural holdings by way of addition to the existing and unmodified statutory code is, in my judgment, incapable of bearing the weight placed upon it for the purposes of Mr Neuberger's argument. As Mr Bannister, for the tenant, pointed out, it is not required of Parliament that it should explain its decision as a matter of policy to L
exclude business tenancies from the operation of section 46(1), but to include agricultural tenancies.

The long title of the 1987 Act may be looked at for the purpose of interpreting the statute as a whole and of ascertaining its scope, although not for the purpose of contradicting the clear and unambiguous language of particular provisions: see *Halsbury's Laws* vol 44, 1983, para 812. This long title shows, as is confirmed by examination of the provisions in Parts I to V of the 1987 Act, that the statute was primarily concerned with housing and the benefit of its provisions would be enjoyed mainly by the tenants of flats, but, in my judgment, it is impossible to say that the application of section 48 to M
agricultural tenancies is not a connected purpose.

Mr Bannister, for the tenant, in summary submitted that the landlord's contention on this part of the case:

(a) provided no explanation why Parliament excluded business tenancies under Part II of the 1954 Act, but failed by express words to exclude agricultural holdings;

(b) equated "premises" in section 46(1) with "dwelling", which is defined in section 60(1) of the 1987 Act, and thus deprived the phrase "premises which consist of or include a dwelling" of any acceptable meaning; and

(c) ignored the fact that in other parts of the 1987 Act Parliament

A enacted express words to restrict the meaning of the word "premises": see section 1(2); section 2(1); section 3(2); section 21(2); and section 25(2).

For my part, I agree with Chadwick J, substantially for the reasons set out in his judgment, that the word "premises" in section 46(1) means the subject-matter of the letting.

As to the second question, whether a notice under section 48(1) of the 1987 Act was given prior to December 3 1991, I agree with Chadwick J that no such notice was given. The requirement is that the landlord "by notice furnish the tenant with an address in England and Wales at which notices (including notices in proceedings) may be
B served on him by the tenant". By section 54(1) such a notice shall be in writing and may be sent by post. There is no prescribed form for such a notice. Chadwick J proceeded upon the basis that such a notice must state that the address given is the address at which notices, including notices in proceedings, may be served on the landlord by the tenant; and that it would not be sufficient to state an address which is shown to be such that, if notice in proceedings were served on the landlord at that address, it would in any particular circumstances be held to be effective service. In short, the tenant is to be told at what address notices, including notices in proceedings, may be served. In my judgment, Chadwick J was right to proceed on that basis.

C In this court, by amendment of the notice of appeal, Mr Neuberger submitted that the letter of December 2 should be accepted as an effective notice in the context of the correspondence conducted between the parties. For my part, I accept that a letter may be properly construed for this purpose in the light of the preceding correspondence and might be held to satisfy the requirement of section 48(1) although the terms of the letter itself, without the help of that context, could not be so construed. The final paragraph of the letter of December 2 could so qualify and the context in the correspondence might also be held to show that the tenant's solicitors were impliedly authorised to accept service of such a notice. It is not, in my judgment, necessary to decide these questions for the purposes of this part of this case. The letter of
D December 3 is and has throughout been accepted as a valid notice for section 48(1) and, on the point of law which would be relevant to any difference in effect between a notice of December 2 and a notice of December 3, the tenant cannot, for the reasons which follow hereafter, in my judgment, succeed.

The third question is whether there was service of an effective notice to pay under Case D in Part I of Schedule 3 to the AHA. Chadwick J held that the notice of August 7 1991 was defective because at that date no rent was due by reason of section 48(2), and in that holding he was clearly right. As to the notice of December 3 1991, he held that the information as to the date upon which the unpaid rent
E was due was inaccurate and therefore the notice was bad because of failure to comply with the regulations. The unpaid rent was not due on the quarter days stated, but became due only on receipt of the letter of December 3 1991. It mattered not that the tenant was not in fact misled. There was need for strict compliance with the statutory requirements in respect of notices to pay which had been affirmed in *Official Solicitor* v *Thomas*. The defect in this notice to pay was not in the nature of a misdescription, but was simply an inaccurate statement as to the date when rent was due.

Mr Bannister submitted that the judge's conclusion should be upheld for the reasons which he gave and, in addition, he contended
F that his conclusion should be upheld on the additional grounds stated in the respondent's notice. In summary he submitted:

(i) the use of a Case D notice by a landlord towards an exclusion of the section 26 protection is in the nature of a forfeiture proceeding and requires there to have been some default on the part of the tenant before it can be invoked and, accordingly, a Case D notice served at the same time as a notice complying with section 48 must be ineffective;

(ii) that the use of Case D is in the nature of a forfeiture proceeding is established by the decisions of this court in *Pickard* v *Bishop*; and the requirement of default for Case D is established by *Urwick* v *Taylor* (1969) EGD 1106*; Bridge J;

(iii) at the very lowest, the landlord must allow the minimum time G
required for the process of effecting payment: see *Bank of Baroda* v *Panesser* [1987] Ch 335;

(iv) the better view is that the "arrears", made due by compliance with section 48, do not become due and payable until the next contractual rent day: see the reasoning in *South Tottenham Land Securities Ltd* v *R&A Millett (Shops) Ltd* [1984] 1 WLR 710† at pp714H to 715B;

(v) alternatively, the "arrears" should be held not to become due and payable until a reasonable time after compliance by the landlord with section 48: see *London & Westminster Loan & Discount Co* v *London & North Western Railway* [1893] 2 QB 49, at p54; H

(vi) and, last, if all these submissions be rejected, the tenant should be allowed until the end of the day, upon which the landlord complied with section 48, to make payment of what then became due: *Afovos Shipping Co SA* v *Pagnan (R) & F.lli* [1983] 1 WLR 195 at p201B:

> I take it to be a general principle of law not requiring authority that where a person under an obligation to do a particular act has to do it on or before a particular date he has the whole of that day to perform his duty.

For my part, I cannot accept the submissions on this part of the case put forward for the tenant. The contention that failure to serve a section 48(1) destroys the right to claim the rent otherwise due before J
service of a notice was not formally abandoned. It was not pursued in argument. It is, I think, unarguable for the simple reason that the section does not so provide.

All the rent in respect of which the notice of December 3 1991 was served was "otherwise due from the tenant" on that date, ie it was due but for the effect of section 48(2). The letter of December 3 was a valid notice under section 48(1). This provision can be given no effect in derogation of the landlord's legal rights beyond that required by the terms of the enactment. The rent "otherwise due", therefore, is to be treated as not due from the tenant "at any time *before* the landlord does comply with" section 48(1); but such rent becomes due at the K
time when the landlord so complies and continues due thereafter. There is no justification for any extension of the period of time over which the rent is treated as not due whether until the end of that day, or for a reasonable time, or until the next rent day. The cases cited for this purpose are, in my judgment, of no relevance. No question of construction of contractual obligations arises as to when the rent was "otherwise due from the tenant". The rent in respect of which the notice was served was due from the tenant when he received the notice.

As to the ground upon which the judge held the notice to pay rent to be invalid, I have, after some hesitation, found myself unable to L
accept it.

The restriction on operation of notices to quit, contained in section 26 of AHA, "shall not apply in any of the Cases set out in Part I of Schedule 3 to this Act": see section 26(2). It is provided by Case D:

> At the date of the giving of the notice to quit the tenant had failed to comply with a notice in writing served on him by the landlord, being either —
>
> (*a*) a notice requiring him within two months from the service of the notice to pay any rent due in respect of the agricultural holding to which the notice to quit relates, or
>
> (*b*) a notice requiring him within a reasonable period specified in the notice to remedy any breach by the tenant that was capable of being remedied of any term or condition of his tenancy which was not inconsistent with M
> his responsibilities to farm in accordance with the rules of good husbandry,
>
> and it is stated in the notice to quit that it is given by reason of the said matter.

The requirement in Case D is that the notice requires the tenant within two months from the service of the notice to pay any rent due.

The Agricultural Holdings (Forms of Notice to Pay Rent or to Remedy) Regulations 1987 (SI 1987 No 711) requires, by regulation 3, that "a notice to pay rent shall be in form 1". By regulation 2 (2)

*Editor's note: Also reported at (1969) 212 EG 1275.

†Editor's note: Also reported at (1983) 269 EG 630.

A form 1 "means the form so numbered in the schedule to these regulations or a form substantially to the same effect".

Para 1 of the form reads as follows:

> I hereby give you notice that I require you to pay within two months from the date of service of this notice (note: this notice may not be served before the rent is due) the rent due in respect of the above holding as set out below:
>
> Particulars of rent not paid:
>
> Date when due Amount due

The form contained information to the tenant as to the effect of the B notice and the steps open to the tenant in response to it.

The statutory purpose of the requirement of the giving of the notice in the prescribed form is to identify to the tenant the payments of rent which the landlord alleges to be due and to give to the tenant the information which the tenant requires in order to protect his rights. If the rent alleged to be due is not in fact due the notice may be held to be invalid: see *Dickinson* v *Boucher* (1983) 269 EG 1159 where Oliver LJ (as he then was) said at p1162:

> The court decided in [*Pickard* v *Bishop*] . . . that the notice to pay, although it did not have to comply with any prescribed form under the 1948 Act, must comply strictly with the requirements of the section; that the notice must get C the landlord right; that the notice in that case had not purported to be served by the landlords . . .; and that, accordingly, it was a bad notice. The judgment of the Master of the Rolls . . . makes it clear what the ratio of the decision was. He refers to the meaning in the Act of the word "landlords" . . . in reference to the notice: "It requires payment of the rent due to Mr Arkwright, the previous owner, who was not the "landlord" . . .
>
> Then he goes on:
>
> In construing section 24(2)(*d*) it must be recognised that it is a provision which enables the landlord to forfeit the tenant's interest and obtain possession, to the great advantage of the landlord and great loss of the tenant. Now it is a settled rule of construction that forfeiture clauses are to be construed strictly. The party who seeks to enforce a forfeiture must comply D strictly with every requirement leading to it. So I would construe section 24(2)(*d*) strictly so as to require the landlords to fulfil it to the letter.

Later, in his judgment, Oliver LJ continued:

> The difficulty that arises, as it seems to me, in this case, is that where the notice does go on to specify what the rent is and gets it wrong, it is then difficult to say that the tenant has failed to comply, if he does not pay what is demanded, with a proper notice under the Act. It is not simply that the notice to quit is allowed to operate if the tenant does not, within two months of the service of the notice, pay the rent due at the date of service to the landlord. The notice to quit is to operate if, at the date of the giving of the notice to quit, the tenant had failed to comply with a notice in writing served on him by the E landlord. So one is to ask, did the tenant fail to comply with a proper notice, ie one with which he was bound to comply? Of course, he failed to comply here with the obligation which was cast on him by the lease to pay his rent, but what was required by the notice was that he should pay a sum of £650, that being erroneously described as the rent due, and though no doubt he did not comply with that, he was not, of course, bound to do so.

Finally, Oliver LJ said at p1164:

> It seems to me that the approach of this court in *Pickard* v *Bishop* is one which, albeit in that case applicable to a misdescription of the landlord, must equally be applicable to the misstatement of the requirement with which the tenant is notified that he must comply.

F In this case, the rent which was said in the notice to be due from the tenant was in fact due when the notice was served, for the reasons stated above, and accordingly the requirement with which the tenant was notified that he must comply was accurately stated.

Does it matter that the notice referred to the quarter days, upon which the arrears of rent were unquestionably "otherwise due" as "the date when due" under the particulars of rent not paid? Chadwick J held that to be a fatal inaccuracy. In my judgment, it was not.

In *Official Solicitor* v *Thomas* (1986) 279 EG 407, CA, Nicholls LJ, in a judgment with which Slade LJ agreed, considered the validity of a Case D notice in which criticism was directed at the description of the "term or condition" of the tenancy which, it was said, the tenant had broken. The inaccuracy was that the arrangements whereby the G tenant agreed to pay the interest charges consisted of two supplemental agreements and not one only so that the particulars should have read "supplemental agreements" in the plural and not "supplemental agreement"; and it was said that the location of the agreement should have been identified by a reference to the memorandum and the correspondence. Nicholls LJ said:

> I cannot accept the tenant's submission to the effect that any inaccuracy in this description, however trivial or immaterial or however obviously a slip, and regardless of whether the tenant may be misled by it or not, is necessarily fatal. So to decide, with regard to the part of the notice with which this appeal is concerned, would be to carry the need for strict compliance with the H statutory requirements to an absurd length, and for no apparent purpose. In the present case the description was terse, but it did identify the source as a supplemental agreement (which, plainly, was being contrasted with the tenancy agreement itself) and the essential subject matter of that agreement (payment of interest on the landlord's improvements). The only inaccuracy relied on is the omission of letter "s" after the phrase "supplemental agreement". Given that it is accepted for the tenant that, despite the inaccuracy and the terseness, this tenant could not reasonably have been misled, in my view the particulars given were good enough.

In this case the reference to the dates when due in the schedule to the notice was intended to identify the payments which the landlord asserted had not been made: that, in my judgment, is plain on the face J of the document and is put beyond question in the context of the correspondence between the parties in which the tenant had raised the question of compliance with section 48 and the landlord had responded by serving a sufficient notice to satisfy the requirement of that section.

In so far as the notice inaccurately asserted that the rent "otherwise due" had been due on and from the quarter days listed, it did not mislead and could not reasonably have misled the tenant in any way. Furthermore, it did not and could not affect the clarity of the notice as to what the tenant was required to do or what the effect would be if the tenant did not comply with it. To treat this notice as invalid, therefore, K would be to carry the need for strict compliance with the statutory requirement to a length beyond any useful purpose. The statutory purpose of the notice was fully satisfied. On this ground alone I would allow this appeal.

Agreeing, **Hirst LJ** said: On the point of construction, I entertained the gravest doubt during the course of the argument whether Parliament can have intended by section 48 in Part IV of the Landlord and Tenant Act 1987 to make provisions affecting notices in the case of agricultural tenancies, seeing that they had already enacted a detailed code for such notices in the previous year in section 93 of the Agriculture Holdings Act 1986, which consolidated the L pre-existing statutes relating to agricultural holdings, subject to certain amendments.

It also seemed to me that the House of Lords decision in *Maunsell* v *Olins* [1975] AC 373, while not binding authority in Mr Neuberger's favour, provided very strong support for his argument that the section should not be so construed.

However, after very considerable hesitation, I have reached the conclusion that, for the reasons given by Ralph Gibson LJ, it is not open to the court as a matter of strict construction to construe the word "premises" in section 48 otherwise than as meaning the subject-matter of the letting, without any exception other than that expressly provided M for in relation to business tenancies.

So far as *Maunsell*'s case is concerned, I have come to the conclusion that on careful examination there are a number of important distinguishing features which render that case much less favourable to Mr Neuberger than appears at first sight. These are as follows:

The question at issue was whether section 18(5) of the Rent Act 1968 applied to agricultural holdings, with a result which would have greatly enlarged the rights of subtenants of such holdings. It was in this context that Lord Wilberforce, in the leading judgment with which Lords Reid and Dilhorne agreed, referred to the political delicacy of

A the law relating to agricultural land at p389G, in the passage quoted by Ralph Gibson LJ. The requirements of section 48 of the 1987 Act in relation to notices are in no way comparable in their impact or sensitivity.

Lord Wilberforce stated at p389C that:

> One suggested alternative is to give to "premises" a general meaning but to exclude from it any property forming part of an agricultural holding. The basis for this would be that agricultural holdings have a code of their own . . . and should not be brought under the general provisions of the Rent Acts. I see the force of this, but I think that a particular exception limited in this way would have received specific legislative mention and this is not to be found.

B The same applies in the present case.

Lord Reid, while agreeing with Lord Wilberforce's speech, also cited at p383A-C a special feature of the section in question which he considered to be "a fairly clear indication that the draftsman did not have in mind any 'superior letting other than a letting for residential purposes'." There is no similar indication in the section presently under consideration.

I am therefore in agreement with Ralph Gibson LJ as to the construction of section 48.

I also agree with Ralph Gibson LJ in his reasoning and conclusions on the issues relating to the form and efficacy of the notice which was C served, and would also, on this ground only, allow the appeal.

Also agreeing, **Peter Gibson LJ** said: My first impression, on reading the Landlord and Tenant Act 1987, was to doubt whether Parliament could have intended to include agricultural holdings in such an Act without express reference to such holdings. As a practical matter, the draft of the Bill and its carriage through Parliament would no doubt have been the responsibility of the Department of the Environment, not the Ministry of Agriculture, and it was not inconceivable that the applicability of the provisions of Part VI of the 1987 Act to agricultural holdings could have escaped attention. The word "premises" in ordinary parlance means buildings of some sort D and *Maunsell* v *Olins* [1975] AC 373 shows that, in an appropriate context in housing legislation, it can properly be interpreted in a way that limits its meaning (in that case to what are dwelling-houses for the purposes of the Rent Acts) so as to avoid encroaching on what is properly the preserve of the agricultural holdings regime.

Further, I found it surprising that Parliament should be enacting provisions relating to notices (in sections 48 and 49) and giving the county court jurisdiction (in section 52) without express reference to the separate regimes for notices and jurisdiction in the Agricultural Holdings Act 1986. I was also impressed by the correspondence between the provisions to which Part VI applied and those to which E section 3 of the Landlord and Tenant Act 1985 applied. I was not impressed by the point to which the judge drew attention that in the 1987 Act there are references to the Rent (Agriculture) Act 1976: provisions to which that Act applies are, in effect, outside the agricultural holdings regime and within the Rent Acts regime.

However, I have been persuaded by the careful judgment of the judge and the cogent arguments of Mr Bannister that it is simply not permissible to cut down the meaning of the words in section 46(1) so as to exclude agricultural holdings. The words "premises which consist of or include a dwelling" and the definition of "dwelling" in section 6(1) ("a building or part of a building occupied or intended to F be occupied as a separate dwelling, together with any yard, garden, outhouses and appurtenances belonging to it or usually enjoyed with it") make it plain that "premises" means more than a dwelling (with its extended meaning) and, like the judge, I find it impossible to find any satisfactory stopping place short of giving the term its ordinary legal meaning of all that is the subject-matter of a letting. The difficulty which Mr Neuberger had in formulating alternative definitions of "premises" seemed to me to underline this point.

Maunsell v *Olins* with its different statutory context is plainly distinguishable from the present case. Lord Wilberforce, after referring to the argument that from "premises" should be excluded property forming part of an agricultural holding because such holdings have a code of their own, observed ([1975] AC 373 at p389): "I think G that a particular exception limited in this way would have received specific legislative mention and this is not found". Further, as Mr Neuberger, with his characteristic and commendable candour, acknowledged, there is nothing in the 1987 Act which can be said to conflict with the agricultural holdings regime. It cannot be said to have been impossible for Parliament to have decided as a matter of policy that tenants of agricultural holdings which include a dwelling-house should have the benefits now afforded by Part VI of the 1987 Act. As for the 1985 Act it is, in my view, significant that this Act contains an express exclusion of agricultural holdings, albeit in a different context: see section 14(3). H

Nor can I accept Mr Neuberger's argument that a "dwelling" should not be read as extending to dwellings. Premises which include two dwellings logically must include one dwelling. In any event, I can see no contrary intention appearing from section 46 to the application of the ordinary rule that words in the singular are to be read as including the plural: section 6(c) Interpretation Act 1978.

Accordingly, for these and the other reasons given by Ralph Gibson LJ, I have concluded that section 48 of the 1987 Act does apply to a tenancy of an agricultural holding which includes a dwelling and so is applicable to the tenancy in question.

On what Ralph Gibson LJ has called the second and third questions J I am in entire agreement with him and like him I would differ from the judge only on the notice of December 3 1991. Because of the serious consequences for the tenant of an agricultural holding, it is quite right, as Mr Bannister submitted, that strict compliance with the statutory provisions should be required. However, I would be sorry to see the law in this field develop the unattractively rigid features that disfigured at least one other area of the law where strict compliance was necessary (*viz* the law of bankruptcy prior to the Insolvency Act 1986), and, for my part, I welcome the approach adopted in *Official Solicitor* v *Thomas* [1986] 2 EGLR 1, whereby errors in completing a statutory notice which could not reasonably have misled the tenant to K whom it was addressed may be held in appropriate cases not to invalidate the notice. For the reasons given by Ralph Gibson LJ I too would allow the appeal on this ground alone.

Plaintiff to have half of the costs of appeal; no order as to costs in the court below; application for leave to appeal refused. Money paid into court by defendant to be paid out.

Chancery Division L

May 28 1993

(Before Mr Roger KAYE QC, sitting as a deputy judge of the division at Luton County Court)

McMULLEN & SONS LTD v CERRONE AND ANOTHER

Estates Gazette May 14 1994
[1994] 19 EG 134

Landlord and tenant — Insolvency — Whether landlord entitled to distrain for arrears of rent when interim order under section 252(2) of the Insolvency Act 1986 in force in respect of tenant's affairs M

By a lease dated November 27 1981 the respondent tenants held the term of retail premises owned by the applicant landlords at a passing rent of £15,000. On November 26 1992, and following a voluntary arrangement proposed by the respondent and his wife, Mr and Mrs C, Luton County Court made an interim order under section 252(1) of the Insolvency Act 1986; this was subsequently extended. On January 14 1993 bailiffs acting for the landlords entered the premises and distrained for arrears of £15,000 by taking possession of a quantity of jewellery. Following an order of the district judge of

Luton County Court restraining the sale of the jewellery, the applicants applied to discharge that order on the ground that leave of court is not necessary to levy a distress when an interim order is in force.

***Held*: The application was allowed. The limitation in section 252(2)(b) of the Insolvency Act 1986, that no proceedings and no execution or other legal process may be commenced or continued against a debtor without leave of the court, do not cover the common law remedy of distress. The distress was therefore lawful.**

The following cases are referred to in this report.

Bellaglade Ltd, In re [1977] 1 All ER 319
Bishopsgate Investment Management Ltd v *Maxwell* [1993] Ch 1; [1992] 2 WLR 991; [1992] 2 All ER 856, CA
Bristol Airport plc v *Powdrill* [1990] Ch 744; [1990] 2 WLR 1362; [1990] 2 All ER 483, CA
Debtor, A, Re (No 1 of 1987) [1989] 1 WLR 271, CA
Exchange Travel Agency Ltd v *Triton Property Trust plc* [1991] 2 EGLR 50; [1991] 35 EG 120; [1991] BCLC 396
Exhall Coal Mining Co Ltd, Re (1862) 4 DJ&S 377
Fanshaw & Forston, In re, ex parte Birmingham & Staffordshire Gaslight Co (1871) LR 11 Eq 615
Herbert Berry Associates Ltd v *IRC* [1977] 1 WLR 1437
International Tin Council, In re [1987] Ch 419; [1987] 2 WLR 1229; [1987] 1 All ER 890
Lancashire Cotton Spinning Co, Re, ex parte Carnelly (1887) 35 ChD 656
Memco Engineering Ltd, In re [1986] Ch 86; [1985] 3 WLR 875; [1985] 3 All ER 267
Naeem (a bankrupt), In re [1990] 1 WLR 48
Olympia & York Canary Wharf Ltd, Re [1993] BCC 154
Roundwood Colliery, Re [1897] 1 Ch 373
Seagull Manufacturing Co Ltd, Re [1993] BCC 241, CA
Smith (a bankrupt) v *Braintree District Council* [1990] 2 AC 215; [1989] 3 WLR 1317; [1989] 3 All ER 897, HL
Winterbottom (G) (Leeds) Ltd, Re [1937] 2 All ER 232

This was an application by motion by the applicant landlords, McMullen & Sons Ltd, to determine the validity of a distress levied on property of the respondents, Mr and Mrs Cerrone.

Nicholas Peacock (instructed by McLellans, of Hertford) appeared for the applicants; Martin Collier (instructed by Pellys, of Bishop's Stortford) represented the respondents.

Giving judgment, **Mr Roger Kaye QC** said: The matter before me raises a question of the true construction of the provisions of section 252(2) of the Insolvency Act 1986. The question, which can be shortly stated, is can a landlord exercise the ancient self-help remedy of distress for arrears of rent when an interim order is in force under that section?

Section 252 provides as follows:

> (1) In the circumstances specified below [that is a reference to section 253] the court may in the case of a debtor (being an individual) make an interim order under this section.
> (2) An interim order has the effect that, during the period for which it is in force —
> (*a*) no bankruptcy petition relating to the debtor may be presented or proceeded with, and
> (*b*) no other proceedings, and no execution or other legal process, may be commenced or continued against the debtor or his property except with the leave of the court.

The facts giving rise to this question are not in dispute. The applicants are the freehold owners and landlords of retail premises at 36 Bucklersbury, Hitchin. They demised those premises to the respondents, Mr and Mrs Cerrone, under a lease dated November 27 1981 and a supplemental lease dated November 15 1985. The current passing annual rent under both leases is £15,000. The premises were used by Mr and Mrs Cerrone for the purposes of running a jewellery business, trading as "Hamilton Cerrone".

The business, although initially successful, did not prosper. It declined through 1990 and 1991 and by November 1992 Mr and Mrs Cerrone found themselves in real financial difficulties. They proposed an individual voluntary arrangement under the provisions of Part VIII of the Insolvency Act 1986. On November 26 1992 Luton County Court made an interim order under section 252(1) of that Act. That, of course, would have lasted only 14 days (see section 255(6)), but the order was subsequently extended to January 14 1993 and on that date extended still further (no doubt under section 256(4) or section 256(5)). There seems some doubt as to the exact date to which it was ultimately extended and, moreover, it seems now to have expired without being further extended, but it is common ground that at all material dates for the purposes of the matter before me the order was in force. Moreover, what the fate was of the proposed voluntary arrangement itself is unclear.

By the end of November 1992 the arrears of rent owing by Mr and Mrs Cerrone to their landlords had risen to £16,500 odd. At that stage the landlords appear to have been ignorant of the interim order. In view of the size of the amount owing they resolved to distrain for the arrears by seizing goods on the demised premises. On November 30 bailiffs were instructed and they visited the premises on December 4 1992. During that visit they discovered the existence of the interim order, but in view of certain proposals made by Mr and Mrs Cerrone for payment of the current rent over the Christmas period the landlords instructed the bailiffs to withdraw. Following that withdrawal the landlords received a cheque from Mr and Mrs Cerrone for some £1,600 odd which they appropriated in reduction of the arrears, leaving some £15,000 odd still owing.

Unfortunately, the Christmas period did not prove to be profitable for Mr and Mrs Cerrone. On January 8 1993 the nominee under the proposed voluntary arrangement wrote to the landlords' solicitors informing them that the business had not done as well as expected and that it was proposed to close down the business on January 16 and sell the stock.

The landlords thereupon decided to act quickly and to distrain for the arrears. On January 13 they reinstructed the bailiffs to distrain for the arrears of £15,000. On January 14 the bailiffs entered the demised premises and took possession of a quantity of jewellery.

Mr and Mrs Cerrone immediately protested about the distress in view of the interim order. The landlords disagreed, taking the view that, notwithstanding the order, they were entitled to act as they did. They made plans to sell the goods at auction on January 21 1993. Mr and Mrs Cerrone accordingly attended before the district judge on January 20 and obtained an order restraining the continuation or completion of the distress and an order that the goods be restored to them. The district judge granted liberty to apply, but directed that any application should be made to a judge of the High Court. It is not entirely clear to me what jurisdiction he had to direct that unless it be under the provisions of section 42(2) or (3) of the County Courts Act 1984.

On January 21 1993 the landlords applied to Ferris J to suspend the order of the district judge. On the landlords' giving undertakings in damages, to keep the seized goods in safe custody and not to dispose of them, the judge suspended the district judge's order until January 28. On that day those undertakings were continued before Mummery J pending the effective hearing of the landlords' motion to discharge the order of the district judge. It is that motion which is now before me.

I now turn to the consideration of section 252. The starting point is the nature of the court's approach to construction of the Insolvency Act 1986. There are three general points to make at the outset.

First, the Insolvency Act 1986 is a consolidating Act. It consolidated the reforms of the law of both individual and company insolvency reflected in the Insolvency Act 1985. It also consolidated, in the same Act, the retained provisions relating to company insolvency found in the Companies Act 1985. It thus includes in one single Act the principal statutory provisions relating both to individual as well as company insolvency. It is well known that many of the reforms in the 1985 Act were passed in the light of the *Report of the Review Committee on Insolvency Law and Practice* (1982, Cmnd 8558), under the chairmanship of Sir Kenneth Cork (the *Cork Report*).

A It is, however, also well known that not all the recommendations of that report were accepted by Parliament. Two of the reforms recommended were that both a company and an individual should be able to promote a voluntary arrangement with their respective creditors under the supervision of a provisional trustee: see chapter 7 of the *Cork Report*. Here, again, much of the substance of the proposal, but not all the detail, was accepted by Parliament. The reforms, as enacted by Parliament, permit both companies and individuals to propose a voluntary arrangement with their respective creditors under the supervision of a suitably qualified insolvency practitioner called a supervisor. The scheme or code applicable to B company voluntary arrangements is to be found in Part I of the 1986 Act. That applicable to individual voluntary arrangements is in Part VIII. There are many common features to both codes: both involve supervision of the arrangement by a supervisor, both require the convening of a statutory meeting of creditors, both bind the creditors and the company to the arrangement (with or without modifications) if approved by the creditors at the statutory meeting, both entitle a challenge to be made if the arrangement is unfairly prejudicial to the creditors or if there is some material irregularity at the meeting, and both entitle applications to be made to the court for directions in certain cases.

The section that falls for construction in this case, section 252, falls C within Part VIII of the 1986 Act. Part VIII was first enacted in Chapter 1 of Part III of the 1985 Act. It is thus part of the new provisions and does not derive from any previous legislation even if the language of certain of the sections, including section 252, echoes words or phrases featuring in earlier insolvency legislation. Accordingly, in my judgment, the task of the court must be to approach the construction of the section afresh, unfettered by previous legislation or decisions under earlier Acts. I am fortified in the view that this is the correct approach in that this has been held to be the correct approach on previous occasions in the construction of different parts of the 1986 Act, different that is to Part VIII: see, for example, *Re A Debtor (No 1* D *of 1987)* [1989] 1 WLR 271 CA at p276G-H; *Bishopsgate Investment Management Ltd* v *Maxwell* [1993] Ch 1 CA at p21A-C; and *Smith (a bankrupt)* v *Braintree District Council* [1990] 2 AC 215 HL.

Second, it is apparent from the Act itself that the intention of Parliament was, so far as possible, to harmonise the two systems of company and individual insolvency: see, in this respect, *Bishopsgate Investment Management Ltd* v *Maxwell* [1993] Ch 1 CA, at p21D; and *Re Seagull Manufacturing Co Ltd* [1993] BCC 241 CA at p242H. That, however, must not be taken too far, for the statutory provisions may indicate from their very nature that Parliament intended that the two systems, although to be similar in many respects, should differ in E some material respects. In the field of voluntary arrangements one noticeable difference is the scope for interim orders. An interim order may be obtained in connection with an individual voluntary arrangement, but not in the case of a company. In the latter case a company voluntary arrangement may be entered into where a winding-up order has been made or where an administration order is in force: see section 1(3). But it may also be entered into outside these circumstances: see section 1(1). Where a winding-up petition has been presented the court has power to stay pending actions or proceedings: see section 126(1). After the winding-up order no action or proceeding can be commenced or proceeded with against the company without F leave: section 130(2) and (3). In the case where an administration order is in force the Act contains wide provisions inhibiting the rights of creditors against the company: see section 11(3). But outside these circumstances, and in the absence of any agreement binding the creditors to the voluntary arrangement, no power is conferred by the Insolvency Act 1986 on the court to prohibit creditors from enforcing their rights in whatever manner they may legitimately do so against the company or its property. In the case of a proposal for an individual voluntary arrangement the position is different. Here the court has power, pending approval or rejection of the proposal, to make an interim order under section 252 prohibiting creditors from acting in the manner prescribed by the section. Thus the two systems of company and individual voluntary arrangements, while adopting a G common form or approach, are not entirely uniform.

Third, it is axiomatic that in construing an Act of Parliament regard must be had to the scope and purpose of the Act and of the part of the Act within which the provision in question falls.

With these preliminary points in mind I must now deal with Part VIII of the 1986 Act. The scheme is clear. An individual who is a debtor may, as an alternative to bankruptcy, propose an arrangement with his creditors. If he does, he may seek and obtain an interim order under section 252: see section 253(1) and section 255. Even before obtaining that order he may, so long as he has applied for an interim order, seek interlocutory relief, whereupon the court "may stay any H action, execution or other legal process against the property or person of the debtor": see section 254. The interim order, once obtained, lasts for 14 days: section 255(6). The court may grant the order only if it thinks that it is appropriate for the purpose of facilitating the consideration and implementation of the debtor's proposal: section 255(2). The 14-day period is to enable the nominee, the qualified insolvency practitioner, to consider the proposal and prepare an independent report to the court stating whether the prescribed statutory meeting of creditors should be convened to consider it: section 256. If the nominee requires more time, the court may extend the 14-day period: section 256(4). If the nominee recommends a meeting be J convened, the court is to extend the interim order to enable that to happen: see section 256(5) and section 257. If the debtor shows no real intention of carrying through with his proposal, or if the nominee does not recommend a meeting, the interim order may be discharged: section 256(6). Once the meeting has been held, the chairman of the meeting (usually the nominee) must report the result to the court: section 259(1). If the meeting approves the proposal, it becomes binding on the creditors and the debtor (section 260(1),(2)) and the interim order ceases to have effect on the expiration of a 28-day period running from the date the chairman presents his report to the court: section 260(4). If the meeting does not approve the arrangement, the K interim order may be discharged immediately: section 259(2). The approved arrangement, of course, binds only those creditors who, in accordance with the rules (this is the Insolvency Rules 1986), had notice of, and were entitled to vote at, the meeting: section 260(2)(*b*). Other creditors are unaffected and can, subject to the limits of section 252 while any interim order is in force, enforce their rights.

It is thus clear that the purpose of the interim order is to protect and preserve the debtor's estate from claims of creditors for a relatively short period of time to enable the creditors to consider the debtor's proposals. Once they have had that opportunity, if they approve the proposal there is no longer any need for the interim order, for they will be bound by the arrangement and will all share in the arrangement in a L just and fair manner or at least in a manner which is not unfairly prejudicial: see section 262(l)(*a*). In such circumstances no creditor could present a bankruptcy petition in respect of any existing debt, ie one which is subject to the terms of the arrangement. If, on the other hand, the creditors decline to approve the arrangement, then there is likewise no need for the order to continue, for the creditors, in those circumstances, should be, and are, left free to pursue whatever remedy each considers appropriate, including the presentation of a bankruptcy petition. Thus, the voluntary arrangement is a true alternative to bankruptcy. That being so, it is, I think, legitimate to see how the 1986 Act deals with the rights of creditors in the interim period between M presentation and disposal of the bankruptcy petition which is, I think, the closest analogy to the period between the initiation of the proposed voluntary arrangement and its disposal by the creditors' meeting. In this connection section 285(1) and (2) is relevant. The remaining provisions of section 285 are not relevant for these purposes because they deal solely with the situation after a bankruptcy order has been made.

The relevant parts of section 285 provide as follows:

> (1) At any time when proceedings on a bankruptcy petition are pending . . . the court may *stay any action execution or other legal process against the property or person of the debtor* . . .

A (2) Any court in which *proceedings* are pending against any individual may, on proof that a bankruptcy petition has been presented in respect of that individual . . . either *stay the proceedings* or allow them to continue on such terms as it thinks fit.

Emphasis supplied.

Thus, an application can be made either to the bankruptcy court to stay *any* action, execution or other legal process or to any court in which proceedings are pending to stay *those proceedings*. Again, however, what is contemplated by the legislation, at least in the usual case, is that for the relatively short period between presentation and hearing the rights of creditors should be restricted. This is obvious for
B the scheme is again to protect and preserve the estate for the benefit of all creditors should a bankruptcy order be made: see, for example, section 284 restricting the debtor's right to dispose of his estate during the period between the presentation of the petition and the vesting of the estate in a trustee in bankruptcy.

Now it is true that the wording of section 285 differs slightly from section 252. The former refers to "action" in subsection (1), whereas in section 252(2)(*b*) this word is avoided, preferring instead the wider expression "other proceedings". In the latter case there can, I think, be no doubt to what "other proceedings" refer. The word "other" must be read in conjunction with what has immediately preceded it, namely the
C reference to "bankruptcy petition" in section 252(2)(*a*). In my judgment, therefore, the words "other proceedings" in section 252(2)(*a*) are intended to relate solely to proceedings before a court, ie of a judicial or adjudicative nature. They are not apt to cover the ancient self-help remedy of distress for rent. This is entirely consistent with section 285(2), which plainly refers only to such proceedings.

The words "execution or other legal process" appear in both section 252 and section 285. The word "execution" is, I think, a term of art, meaning execution or enforcement of some judicial process. It, too, is not apt to describe the process of distress.

I thus turn to the crucial words in this case: "or other legal
D process". In both sections what must be considered is "other legal process" "against the property or person of the debtor" (section 285(1)) or "against the debtor or his property" (section 252(2)), which is the same thing. A "process" can be described as some course of action, but it must be "legal". That, I do not think, for one moment, is to be used in contradistinction to the word "illegal" or criminal for it is self-evident that criminal acts could not be contemplated on any footing. Additionally, some regard must be had to the word "other" before the words "legal process". I therefore reach the conclusion that it was plainly intended that the word "legal" in the context of para (*a*) of section 252(2) was intended to colour the word "process" and that
E the word "other" was intended to tie it to "proceedings" and "execution" giving the whole of the paragraph a distinct flavour of judicial or enforcement proceedings or process. I do not think that the word "legal" is to be construed as wide as contended on behalf of Mr and Mrs Cerrone, namely as meaning any lawful process or any process recognised by the law or having a legal consequence. That, in my judgment, is not justified and goes too far.

In my judgment, therefore, section 252(2)(*b*) was intended to restrict only those rights which a creditor could enforce by "legal process" against the debtor or his property, that is process of a judicial or adjudicative or quasi-adjudicative nature and not rights which might be enforced without recourse to such process, ie the distress in this
F case. Some support for this comes from the language of section 254, dealing with the court's powers to make an interlocutory order on a pending application for an interim order, where the wording is clearly very similar to section 285(1) and (2) but, at the same time, intended to complement section 252 since it deals with what is to happen up to the effective hearing of the application for an interim order. It would be surprising if the jurisdiction under one section (section 252) was very different from the other (section 254).

I must now consider whether this view is affected by other analogous provisions or decided cases to which I have been most helpfully referred by the researches of counsel.

I must start, I think, with *Smith (a bankrupt)* v *Braintree District Council* [1990] 2 AC 215 HL for that was a decision on section G
285(1). There the House of Lords held that the words "or other legal process" in section 285(1) were apt to cover the process in Part VI of the General Rate Act 1967 for the recovery of rates, including proceedings for committal under that Act. Under section 96(1) of that Act the payment of unpaid rates may be enforced by distress and sale under warrant issued by the magistrates' court. If there is insufficient distress the individual ratepayer may be liable to imprisonment. Section 102 enables the court to issue a warrant for commitment where the distress is insufficient. At p230F-G Lord Jauncey of Tullichettle (with whom the rest of the House agreed) said:

My Lords, the words "or other legal process" must be construed in the context H
of the underlying purpose of section 285, namely, the protection of the bankrupt's estate for all his creditors. It follows that proceedings by one creditor to enforce payment to himself are the sort of proceedings contemplated by the section. It cannot be in doubt that the issue of a warrant of distress would fall within the description "or other legal process" . . . It would be both strange and illogical if the bankruptcy court could stay such proceedings but had no power to stay the next stage of the proceedings when distress had not been wholly successful. In my view, as a matter of pure construction, the words "or other legal process" in section 285(1) are quite wide enough to comprehend all the machinery provided by Part VI of the Act of 1967 for the recovery of unpaid rates, including proceedings for the issue of a warrant of commitment. J

Mr Martin Collier, who appeared for Mr and Mrs Cerrone, relied heavily on that passage. He submits that that is apt to describe the warrant, or authority, the landlords issued to the bailiffs authorising them to enter the demised premises to carry out the distress. But, as was pointed out by Mr Nicholas Peacock for the applicant landlords, in that passage Lord Jauncey is referring to a warrant for distress issued under section 96. The "next stage" to which he is referring is likewise the process of application to a judicial tribunal for a warrant of commitment. In my view, therefore, this passage does not assist Mr and Mrs Cerrone.

In *ex parte Birmingham and Staffordshire Gaslight Co* (1871) LR K
11 Eq 615 Sir James Bacon, CJ, had to consider the words "execution or other legal process" in section 13(1), Bankruptcy Act 1869. That section, and its successor section 9(1), Bankruptcy Act 1914 were, in some respects, the precursors to section 285(1) of the 1986 Act. The question was whether the gas company could distrain for rent under special powers conferred on them by the relevant Act of Parliament enabling them to recover arrears for gas "by the same means as landlords may recover rent in arrear". The judge held:

I think it quite plain that a distress for rent cannot be considered to be included in the expression "legal process", inasmuch as no legal process L
whatever is necessary; and the landlord may, if he thinks proper, distrain with his own hands. As little can it be called an execution, for an execution is the result of a judgment which has been recovered in some Court of Law, but to a distress no legal proceedings whatever are necessary.

Now that decision is not, of course, binding on me since, as I have previously mentioned, the provisions in the 1986 Act are to be construed afresh. It does not, however, assist Mr Collier.

Finally, in the realm of individual insolvency, Mr Collier relied on a decision of Hoffmann J (as he then was) in *In re Naeem (a bankrupt)* [1990] 1 WLR 48. In that case Hoffmann J allowed an appeal against the decision of the registrar to revoke an individual voluntary M
arrangement entered into by a bankrupt. The landlord had sought to revoke the arrangement on the basis that it was unfairly prejudiced in that it could not forfeit the lease for arrears of rent. It was held that the arrangement did not interfere with its proprietary right to forfeit, but that it could forfeit only for the amount due to it under the arrangement. At p5OD-H the judge said:

Of course the effect of the interim order was to hold up the landlord's right to forfeit, but under section 260(4) of the Act of 1986 the interim order ceases to have effect 28 days after the date on which the report on the outcome of the creditors' meeting has been given to the court under section 259. Accordingly the interim order no longer applies and the landlord is free to commence such

A proceedings for forfeiture as it may be advised.

That the interim order held up the landlord's right to forfeit seems to have been an assumption of the judge and formed no part of the ratio of his decision. Moreover, it is not entirely clear from the report precisely what type of forfeiture was contemplated, by proceedings in court or by exercising a right of re-entry without recourse to legal proceedings. If anything, the last sentence of the passage quoted points to the former rather than the latter. If so, then the passage quoted is entirely consistent with my construction of the section. But, if I am right in my construction of section 252 so far, it is also right to point out, as does Mr Collier, that this produces something of an anomaly
B in that a landlord could not enforce his right to forfeit by court action where an interim order is in force under section 252, but could by right of re-entry. But I see nothing in the judgment of Hoffmann J compelling me to reach a different conclusion to that reached so far.

I must now consider the provisions applicable to the scheme of company insolvency. I have already pointed out that there is no directly comparable provision in this sphere for obtaining an interim order. But, again, as I have already pointed out, a company voluntary arrangement may be proposed where an administration order is in force. In such a case section 11(3), Insolvency Act 1986 applies. This
C subsection provides as follows:

> During the period for which an administration order is in force —
>
> (*a*) no resolution may be passed or order made for the winding up of the company,
>
> (*b*) no administrative receiver of the company may be appointed,
>
> (*c*) no other steps may be taken to enforce any security over the company's property, or to repossess goods in the company's possession under any hire-purchase agreement, except with the consent of the administrator or the leave of the court and subject (where the court gives leave) to such terms as the court may impose; and
>
> (*d*) no other proceedings and no execution or other legal process may be
> D commenced or continued, and no distress may be levied, against the company or its property except with the consent of the administrator or the leave of the court and subject (where the court gives leave) to such terms as aforesaid.

It is to be noted that that subsection is in much wider terms than section 252. I must also refer here to section 126 of the InsolvencyAct 1986 which I mentioned above. That section derived from section 251 Companies Act 1985 and in turn from section 226, Companies Act 1948 and from earlier legislation. In many decisions under that section (and under the predecessor to section 130(2) of the 1986 Act) it has been held that where a landlord has commenced his distress before winding up, but not by the time of the order completed the distress by
E sale, the court may restrain completion of the sale as a "proceeding". It is clear, however, from those decisions, that this construction reached in the course of the last century has not been greeted with judicial favour but has, until the 1986 Act at least, been followed in view of the passage of time that the construction stood: see *Re Exhall Coal Mining Co Ltd* (1862) 4 DJ&S 377; *Re Lancashire Cotton Spinning Co, ex parte Carnelly* (1887) 35 ChD 656; *Re Roundwood Colliery Co*[1897] 1 Ch 373; *Re G Winterbottom (Leeds) Ltd* [1937] 2 All ER 232; *In re Bellaglade Ltd* [1977] 1 All ER 319; *Herbert Berry Associates Ltd* v *IRC* [1977] 1 WLR 1437 HL and *In re Memco Engineering Ltd* [1986] Ch 86. Thus, as it seems to me, the intention
F of Parliament was quite plainly to leave no doubt, at least in relation to administration orders (and, thereby, where relevant, any period prior to approval of a proposal for a voluntary arrangement where an administration order is in force), that distress was to be covered also in section 11(3)(*d*). Nevertheless I was urged to follow this line of cases.

But they do not, I think, assist Mr Collier. I have already considered that the word "proceedings" in section 252(2)(*b*) does not cover distress and I see nothing in these cases that affects that view. If anything it tends to show that Parliament was alive to the point and intended, at least so far as companies were concerned, specifically to deal with a point that had previously given rise to judicial disfavour by express words. Since the opening words of section 11(3)(*d*) so closely resemble that of section 252(2) it would have been a relatively simple G
matter to have included in section 252(2) the words "and no distress may be levied". It is to be noted, too, that nothing resembling section 11(3)(*c*) is to be found in section 252.

I was referred to three cases on section 11(3)(*d*). The first is *Bristol Airport plc* v *Powdrill* [1990] Ch 744. There the Court of Appeal held that the detention of an aircraft under the provisions of the Civil Aviation Act 1982 was caught by section 11(3)(*c*). Sir Nicolas Browne-Wilkinson V-C added that he did not consider that the right of retention fell within the concept of distress in section 11(3)(*d*) nor did it fall within the words "other proceedings" which he considered referred to judicial proceedings: see pp765-766. H

Next I was referred to *Exchange Travel Agency Ltd* v *Triton Property Trust plc* [1991] BCLC 396* where Harman J, *obiter*, decided that a landlord's right of re-entry could be covered by the words "other legal process" in section 11(3)(*d*). At p401 he said:

> It is clear that the terms of s130(3) provision, which I have already referred to as providing for "proceedings . . . against the company or its property", cover the peaceable exercise of a right of re-entry pursuant to a proviso for re-entry in a lease, and it seems to me astonishing if "other legal process" in s11(3)(*d*) do not also cover that step. It seems to me "legal process" is exactly how one would describe the exercise of a right by a person not in contractual relationship with his tenant but holding a relationship by privity of estate J
> between them. In my judgment it is a correct use of words to describe a landlord who has legal rights arising out of the privity of estate which he exercises by the process of peaceable re-entry as exercising "legal process". It seems to me therefore that if this exercise of the right of re-entry be not a security, it is quite certainly the commencement of "other legal process", and as I see it that must have been one of the purposes for which Parliament plainly intended the Act to operate.
>
> It would be astonishing, in my mind, if Parliament has plainly prevented a landlord from issuing a writ for forfeiture or taking steps in the courts to effect a forfeiture and a re-entry, but has left entirely open and untouched the right to effect a peaceable re-entry.

On that passage, too, Mr Collier relied. He argues, with some force, K
that the landlord's right of distress may be regarded in the same light. It would be anomalous, he submits, for a landlord enforcing a right of re-entry by action to be restrained, but one enforcing a right by re-entry not to be. So, he argues, by parity of reasoning, a landlord exercising his self-help remedy of distress. He relies on the plain purpose of Part VIII of the 1986 Act, which is to effect a complete moratorium on the rights of creditors pending the outcome of the proposal while an interim order is in force.

This passage of Harman J's judgment has, however, been subjected to judicial scrutiny in *Re Olympia & York Canary Wharf Ltd* [1993] BCC 154. There the question arose whether the existence of an L
administration order, and the operation of section 11(3)(*d*) of the 1986 Act in particular, prevented the service by the applicant upon the respondent company (which was in administration) of a notice to complete making time of the essence, or prevented the applicant accepting a repudiatory breach of contract. Millett J held that the section did not. Millett J entirely agreed with Harman J's actual decision (that the right of re-entry was covered by section 11(3)(*c*), but disagreed with his approach on section 11(3)(*d*). After referring to section 9(1), Bankruptcy Act 1914 (which, it will be remembered, is the precursor to section 285(1) and (2) and resembles the language of section 254) he said at pp156-157: M

> In sec 10 and sec 11 of the 1986 Act, Parliament has effected certain modifications to sec 9 of the *Bankruptcy Act* 1914. The word "action" has been replaced by the more compendious expression "proceedings" which is apt to denote any legal or quasi-legal proceedings whether commenced by writ or otherwise; and a prohibition on the levying of distress has been expressly included, thereby dealing with a question which had previously caused difficulty.
>
> In *Stroud's Judicial Dictionary* (5th ed, 1986) it is said that the word "process" is:
>
> "the doing of something in a proceeding in a civil or criminal court, and that

*Editor's note: Also reported at [1991] 2 EGLR 50.

A which may be done without the aid of a court is not a 'process'."

Therefore a distraint, whether for rent or any other payment, and whether the right of distress be given by common law, statute or, as it would seem, by any other authority was not a "process" nor was it an "execution, or other legal process" within the *Bankruptcy Act* 1869.

Then, after quoting the passage in *In re Fanshaw & Forston, ex parte Birmingham and Staffordshire Gaslight Co* (1871) LR 11 Eq 615 I have set out above, he continued:

> The presence of the word "stay" in sec 9 of the *Bankruptcy Act* 1914, and the insertion of the word "legal" before "process" in sec 11(3)(d) of the 1986 Act support the impressions that the word "process" in each of the Acts means a
> B process which requires the assistance of the court, and does not extend to the service of a contractual notice, whether or not the service of such a notice is a precondition to the bringing of legal proceedings.

He also referred to his own judgment, in *In re International Tin Council* [1987] Ch 419 at p453H, where he had to consider the meaning of the phrase "immunity from suit and other legal process" and considered "suit" extended to all forms of the adjudicative and the phrase "legal process" as apt to describe all forms of the enforcement, jurisdiction.

With this reasoning I respectfully agree. It seems to be that the cases on the construction of the language used in section 252,
C particularly on the phrase "other legal process", if anything, supports the view that I have taken as to the true construction of section 252(2). I must regretfully decline to adopt Harman J's approach. I can find, in particular, no good reason for distinguishing the approach under section 285 to that under section 252. When a comparison is made with section 11(3) it is noteworthy that Parliament sought fit expressly to include provision for distress. Like both counsel before me, however, I can discern no good reason for this distinction between the scheme applicable to companies under the Act and that applicable to individuals. Given, however, there is an express distinction on the wording and despite the clear legislative purpose I have considered
D above I do not feel able to decide that the words in section 252(2)(*b*) are apt to cover the common law remedy of distress. That this produces anomalies I have no doubt, but to yield to the temptation to include distress within the embrace of section 252 would, I think, create greater uncertainties for there would then, I anticipate, be great difficulty in drawing the line as to what was or was not included within the phrase "legal process". I can express only the hope that, at some stage, Parliament recognises the anomaly of including express provision for distress in section 11, but not in section 252. There seems no good reason why it should not be expressly included in the latter.

E I would finally add that I have not found much assistance from the words "commenced or continued". Those words were thought to be of some influence in both *Bristol Airport plc* v *Powdrill* [1990] Ch 744 and in *Re Olympia & York Canary Wharf Ltd* [1993] BCC 154 as being indicative of the adjudicative process. That I accept, but they are not entirely inapt to describe the commencement of the process of distress (eg by entry) and its continuation (eg by making arrangements for the sale of the goods seized).

In the result, therefore, the judicial and legislative authorities cited to me have not caused me to depart from my initial view that the distress was not prevented by section 252, rather they have tended to
F reinforce this view. I therefore accede to the application of the landlords.

Chancery Division G

February 16 1994

(Before Mr Justice KNOX)

PELLICANO AND ANOTHER v MEPC PLC AND ANOTHER

Estates Gazette May 14 1994

[1994] 19 EG 138

Landlord and tenant — Forfeiture — Disclaimer of lease — Application by underlessee for vesting order — Forfeiture of headlease by peaceable re-entry — Whether underlessee entitled to be restored to possession pending outcome of application for vesting order under section 146(4) of the Law of Property Act 1925 H

On October 6 1970 Metropolitan Estates Property Co Ltd granted David Lavender a 50-year term of 63 Molton Street, London W1, from December 25 1969 ("the lease"). By 1986 the lease had been assigned to L Ltd. On July 28 1989 L Ltd granted a 16-year term of the shop and basement part of the building to the first plaintiff at a rent of £75,000 pa ("the underlease"). A receiver was appointed for L Ltd in March 1992 and the first plaintiff thereafter paid the receiver a reduced rent of £800 per week previously agreed with L Ltd. On J August 10 1992 L Ltd went into voluntary liquidation owing rent arrears of £11,932.33 under the lease. In October 1992 the first plaintiff ceased paying rent. In early January 1993 the first plaintiff received a notice under section 6 of the Law of Distress (Amendment) Act 1906 requiring him to pay direct all future payments of rent until L Ltd's arrears were duly paid. On January 21 1993 the liquidator filed notice of disclaimer of the lease. On June 23 1993 Mr Lavender sought a vesting order in respect of the lease under section 181(2) of the Insolvency Act 1986; on July 18 the first plaintiff made a similar application in respect of his part of the premises to which the defendant indicated that it would not play any role. On February 11 K 1994, when the arrears under the lease were £90,945.47, the defendant's bailiffs re-entered the building. The plaintiffs challenged the validity of the defendant's forfeiture and sought an order to be restored to possession claiming that by reason of the application by the first plaintiff for a vesting order under the Insolvency Act 1986 the right of forfeiture was suspended pending the outcome of that application.

***Held*: The defendant's right to forfeit the lease was not affected by the existence of an application for a vesting order, the defendant's indication that it would not oppose it, or the steps taken under the 1906 Act. The forfeiture gave the first plaintiff an L independent right to apply for relief under section 146(4) of the Law of Property Act 1925. If a vesting order were to be made, it was not retrospective; accordingly, until it was made an applicant who remained in possession was a trespasser liable to pay mesne profits. The first plaintiff had been removed from possession and there was no arguable defence to the right of the defendant to retain possession until the making of a vesting order.**

The following cases are referred to in this report.

A E Realisations (1985) Ltd, In re [1988] 1 WLR 200; [1987] 3 All ER 83
Finlay, In Re, ex parte Clothworkers' Co (1888) 21 QBD 475 M
Hill v *East & West India Dock Co* (1884) 9 App Cas 448
Official Custodian for Charities v *Mackey* [1985] Ch 168; [1984] 3 WLR 915; [1984] 3 All ER 689

This was an *inter partes* application by motion by the plaintiffs, Luciano Pellicano and Recordlook Trading Ltd, to be restored to possession of premises following peaceable re-entry by the defendant, MEPC plc.

Edward Cousins (instructed by Robert Gore & Co) appeared for the plaintiffs; Guy Fetherstonhaugh (instructed by Reynolds Porter Chamberlain) represented the first defendant; the second defendant, the Sherriff of Greater London did not appear and was not represented.

A Giving judgment, **Knox J** said: These proceedings are brought by Mr Luciano Pellicano, who is underlessee of the shop and basement ("the shop and basement") at 63 South Molton Street, London W1, and Recordlook Trading Ltd, which is a company controlled by Mr Pellicano and his wife which trades from that shop and basement in ladies' fashion wear, against, first, MEPC plc, the landlord of a superior lease and, second, the Sheriff of Greater London, who has taken no part in the argument before me and I need not deal further with his position.

There is before me a motion *inter partes*, which seeks the continuation of *ex parte* relief granted by Evans-Lombe J on February
B 12 last, whereby he restrained MEPC from entering upon the shop and basement, save pursuant to an order or with the leave of the court.

The chronology of the matter is as follows. There is a headlease to MEPC or its predecessor, which is not affected by these proceedings and I propose to ignore its existence in the course of this judgment. At the top of the chain of leases and underleases comes the lease ("the lease"), which was granted by Metropolitan Estates Property Co Ltd on October 6 1970 to a Mr David Lavender of the whole of the premises at 63 South Molton Street, both the shop and basement and the other floors. The term was one of 50 years from Christmas 1969. There was provision for an upwards-only rent review every seven years and the evidence is that there was in 1986 a review to £43,000
C pa rent and that the rent was varied by a deed of variation on April 8 1988 to a figure of £48,000 pa. In 1986 that lease was vested in Linewool Ltd ("Linewool") (pursuant to a licence in that behalf granted by MEPC) by an assign of Mr Lavender. I need not trace the devolution of the lease at earlier stages from Mr Lavender to Linewool. On February 6 1989 Linewool underlet the first, second and third floors to a company called Periplan Ltd. More significantly, on July 28 1989 by an underlease ("the underlease") Linewool underlet the shop and basement to Mr Pellicano for a term of 16 years from Lady Day 1988 at a rent of £75,000 pa, with an upwards-only rent review provision at four-yearly intervals. Mr Pellicano, in his affidavit
D in this matter, says about that:

> There should have been a rent review in March 1992 but nothing was formally agreed. I went to see Linewool Ltd because I could not afford to continue paying £75,000 pa and they agreed to accept £800 pw plus VAT.

That works out at £41,600 pa plus VAT.

In March 1992 a receiver was appointed of Linewool. The evidence is that Mr Pellicano paid him that rent of £800 pw plus VAT. Mr Pellicano says:

> In about March 1992 a receiver, Healey & Baker, was appointed by Barclays Bank plc in respect of the underlease and so far as I recall I was asked to and
> E did pay rent of £800 pw plus VAT to the receiver from approximately March to October 1992, but in about October 1992 I was informed by somebody in the offices of Healey & Baker that there was little point in continuing to pay rent to the receiver, so I stopped.

That, so far as the evidence goes, is the present position with regard to the payment of rent by Mr Pellicano.

On August 10 1992 Linewool went into creditors' voluntary liquidation and at that point of time the arrears of rent under the lease were some £11,932.33.

In October 1992, as I have mentioned, Mr Pellicano accepts that he ceased paying rent and in early January 1993 notices under section 6
F of the Law of Distress (Amendment) Act 1906 were served both on Mr Pellicano and on Periplan Ltd. They asked this. First of all there was a statement that there was due to MEPC, as superior landlord of 63 South Molton Street, from Linewool the sum of £33,382.18 due and owing in respect of arrears of rent for those premises and the requirement both to Periplan Ltd and to Mr Pellicano was that as undertenant of the premises, or some part of them,

> to make to it direct at MEPC's office all future payments of rent whether now accrued or to accrue during the future from you to the said Linewool Ltd until such arrears have been duly paid.

The arrears referred to are obviously the arrears under the lease.

On January 21 1993 the liquidator of Linewool filed a notice of G disclaimer of the lease. It is not clear exactly when or if formal notice of disclaimer was given to Mr Pellicano. He clearly knew about it at the latest in April of that year, because he says in his affidavit:

> It was only when the first defendant commenced proceedings against me in about March or April

— and the context shows that that meant 1993 —

> that I was given notice that the liquidator had disclaimed the under-lease.

The disclaimer was of course in fact of the lease, not of what I am calling the underlease. H

On March 25 1993 MEPC sued Mr Lavender for arrears of rent under the lease in the sum of £33,480 and obtained summary judgment, but it remains unsatisfied. Similarly on April 21 1993 MEPC sued Periplan Ltd for sums due under the lease, by then some £47,500. There was a defence served by Periplan Ltd on June 21 1993 that the disclaimer made MEPC unable to recover under the covenants in the underlease as against Periplan Ltd and a similar defence was advanced by Mr Pellicano against whom proceedings were started a month later by MEPC on May 21 1993, Mr Pellicano's defence being in the same terms as Periplan Ltd's, namely that the disclaimer rendered the covenants no longer directly enforceable against him. J

On June 23 1993 Mr Lavender, by originating application in the Companies Court, applied for a vesting order pursuant to section 181(2) of the Insolvency Act 1986 in respect of the whole premises at 63 South Molton Street and on July 18 Mr Pellicano in turn issued an originating application seeking a vesting order pursuant to section 181(2)(*b*) of the Insolvency Act 1986 in relation to the shop and basement.

On July 14 1993 there was written by the solicitors, then acting for MEPC, to the solicitors then acting for Mr Pellicano, a letter in response to a letter from Mr Pellicano's solicitors, who were also actually acting for Periplan Ltd and had indicated their intention to K seek a vesting order. In the solicitors for MEPC's letter of July 14 the following passage is included:

> We note your intention is to seek a vesting order in favour of your client [Mr Pellicano]. We would stress that provided we receive an assurance from yourselves that it is not your intention to seek any form of relief against MEPC plc, that we will not be playing any active role in the question of a vesting order, it would not be our intention to attend upon the application of David Lavender returnable on the 19th July, nor for that matter would it be our intention to attend on any application made on your client's behalf. Therefore we await your confirmation by return that no relief is sought against MEPC plc. With regards to the question of an adjournment of Mr Lavender's application for a vesting order, we confirm that we have no objection to your L seeking an adjournment.

On February 4 1994 the rent arrears under the lease had risen to £90,945.47. On the 11th of this month MEPC's bailiffs claim peaceably to have re-entered the premises. The solicitors then acting for both Periplan Ltd and Mr Pellicano negotiated a reoccupation by Periplan Ltd of the upper floors, but there was no such negotiation in respect of the shops and premises and Mr Pellicano has changed his solicitors and is now represented by a different firm.

The following day (Saturday) there was made by Evans-Lombe J the order to which I have already referred. That was a short-term order M to hold the position until the matter was argued before me *inter partes* earlier this week. There was a waiver of any question of lack of full notice, as required by the rules, for an *inter partes* hearing and I treated the matter as fully before me *inter partes*.

There is one further collateral matter that I should mention regarding Mr Pellicano. That is that there is evidence from MEPC that in other premises in South Molton Street there was a lease vested in Mr Pellicano between July 1990 and August 1992, when he surrendered it to MEPC, during which period arrears of rent and service charge in the sum of £92,094.60 accrued as at March 16 1993. That figure is not as such challenged by Mr Pellicano's evidence. The only materiality of this claim, which I emphasise does not affect the

A premises at 63 South Molton Street, is that it may well be thought to be relevant to Mr Pellicano's financial position.

Finally, so far as the factual background is concerned, there is evidence before me that the chattels, mainly stock, on the shop and basement premises do not belong to Mr Pellicano, but belong to the second plaintiff, Recordlook Trading Ltd, which is the entity that trades from those premises.

On these facts the question arises — what is the *locus standi* of Mr Pellicano as regards the shop and basement at 63 South Molton Street?

Assuming the disclaimer of the lease by the liquidator of Linewool is effective, and it is not clear how Mr Pellicano is in a position to B challenge this given, first, that he has relied upon it in his defence to the action brought by MEPC to recover arrears of rent consequent upon the failure to comply with the notice under section 6 of the Distress for Rent (Amendment) Act 1906 and, second, that he has on foot an originating application for a vesting order in his favour, then Mr Pellicano was entitled to possession under the underlease in his favour, notwithstanding the disclaimer of the lease, but subject to the right of MEPC to re-enter for non-payment of rent. That this is the effect of a disclaimer by a liquidator or a trustee in bankruptcy of a headlease is clear from, *inter alia*, *In re A E Realisations (1985) Ltd* [1988] 1 WLR 200 where at p211B Mr Vinelott J described the C position of an underlessee where the headlessee's trustee in bankruptcy had disclaimed the lease in the following terms:

> The underlessee did not become bound by any privity of contract or of estate to the lessor but was entitled to remain in possession of the property during the term of the underlease and so long as the lease would, apart from the forfeiture, have remained in existence.

That was said in terms of the provisions of section 55(2) and (6) of the Bankruptcy Act 1883, but those provisions are substantially re-enacted by section 178(4) and sections 181 and 182 of the Insolvency Act 1986 and the principles applicable remain as stated in the authorities referred to by Vinelott J in *In re A E Realisations (1985)* D *Ltd*, notably *Hill* v *East & West India Dock Co* (1884) 9 App Cas 448 and *In re Finlay, ex parte Clothworkers' Co* (1888) 21 QBD 475. In the latter case Lindley LJ, at p486, said this of the situation where there was a lease and a sublease and the lessee/underlessor became bankrupt:

> We must next consider the case as between the original lessor and the sub-lessee . . . the sub-lessee, although freed from his covenants to his own immediate lessor, must perform the covenants of the original lease or he will be liable to be distrained upon and to be ejected by the original lessor. That is obviously his position.

E That, in my judgment, is applicable here. Mr Pellicano's rights under his underlease remain subject to MEPC's right to distrain and forfeit for non-payment of rent.

Second, Mr Pellicano also had the benefit of his application for a vesting order. Mr Edward Cousins, on his behalf, argued that the existence of this application, in respect of which it is claimed MEPC were out of time in filing their evidence, should freeze the operation of any proceeding for forfeiture for non-payment of rent. Mr Guy Fetherstonhaugh, for MEPC, submitted that the two processes: (1) application for a vesting order under section 181 of the Insolvency Act 1986; and (2) forfeiture of a lease for non-payment of rent are independent of each other and that there is no provision requiring the F latter to be stayed or otherwise held up because of the pendency of the former. I prefer this analysis, given that Mr Cousins was not able to point to any provision in the Insolvency Act 1986 or authorities upon it, or indeed elsewhere, in support of his submission regarding the freezing of forfeiture provisions during the pendency of an application for a vesting order. I am also unpersuaded that there was anything in the nature of a waiver of MEPC's rights in the letter from which I have read an extract, written by MEPC's solicitors on July 14 1993. At most that letter stated MEPC's position at that stage in relation to the application for a vesting order. It did not, in my view, affect or prejudice their right some seven months later, in the absence of any payments of rent under the headlease, to take steps to recover possession. If the disclaimer was for some reason not effective, it is G not easy to see how that could improve Mr Pellicano's position, since it must on that view remain possible for MEPC to take the normal steps by way of forfeiture for non-payment of rent. Equally, I do not regard the steps taken under section 6 of the Distress for Rent (Amendment) Act 1906 as affecting MEPC's rights to forfeit the lease. That seems to me to be a quite separate matter, which has in fact run into the sand at least for the time being, but it does not affect, in my view, the position as between MEPC and Mr Pellicano in relation to the exercise by MEPC of its rights under the lease.

The situation regarding MEPC's right to re-enter for non-payment of rent is that Mr Pellicano has an independent right to apply for relief H against forfeiture under section 146(4) of the Law of Property Act 1925. In *Woodfall Landlord and Tenant* at p17/77, para 17.163 one finds regarding subsection (4);

> This subsection is an independent enactment which is not dependent on relief being available to the lessee. Accordingly it extends to cases where the headlease has been forfeited for bankruptcy or for non-payment of rent, which are excluded from the general operation of section 146.

Mr Pellicano is seeking to exercise that right and has included a prayer for such relief in the writ which has now, pursuant to the undertaking given to Evans-Lombe J, been issued.

J The practice of the court on such applications for relief by an underlessee is stated in *Woodfall* at p17/82, para 17.170, in terms which include the following:

> *Discretion to grant relief to sub-tenant*
>
> Whether relief should be granted to a sub-lessee, and on what terms, are matters for the discretion of the court. The jurisdiction to grant relief should be exercised sparingly, because its effect is to thrust upon a landlord a person whom he has never accepted as tenant and creates privity of contract between them. The factors to be taken into account will include those applicable to a case between landlord and tenant. In addition the following factors to be considered are . . .

K Then there is a long list from (*a*) to (*i*) of factors of which I need only refer to two:

> (*b*) whether the property sub-let is capable of convenient occupation under a lease divorced from the remainder of the property comprised in the head-lease;
>
> . . .
>
> (*g*) whether the breach can be remedied and if there are breaches of the head-lease, whether the sub-tenant is willing to remedy them.

The relevant relief under section 146(4) of the Law of Property Act L 1925 is, on the authorities, not retrospective. Until an order is made an undertenant, where the headlease has been forfeited is, strictly speaking, a trespasser. That appears most clearly from the decision in *Official Custodian for Charities* v *Mackey* [1985] Ch 168 at p181. That decision was concerned with the position of mortgagees of a lease or leases, but the same principle applies to underleases. The plaintiffs were the lessors and the relevant defendants were mortgagees of a forfeited lease. Scott J said at p181:

> I now turn to the second point in this case, namely the question whether an order can be made under section 146(4) with retrospective effect.
>
> The mortgagees contend that they will in due course, if the lease stays M forfeited, obtain relief from forfeiture under section 146(4). Such relief, they submit, can, and, in the circumstances of this case, would, be given retrospective effect so as to restore them to the position in which they would have been had the lease not been forfeited.
>
> Mr Mowbray, for the plaintiffs, concedes that the defendant mortgagees will, if the lease remains forfeited, all succeed in obtaining relief under section 146(4). He concedes that, if the court has power under subsection (4) to make an order with the retrospective effect to which I have referred, it would make such order. But he contends that on the true construction of subsection (4), the court has no such power. He submits that the relief contemplated by subsection (4) must take the form of the grant of a lease for an appropriate term and on appropriate conditions to the applicant and that such lease cannot do other than take effect from the date of the order.

If it is correct that the power of the court under subsection (4) is thus limited, the result is, as it seems to me, unfortunate and may well work injustice to under-lessees and mortgagees in some cases. Under-lessees and mortgagees cannot apply for relief from forfeiture of the head-lease until the head-lease has been forfeited. There is, therefore, bound to be a lapse of some time before the applications for relief can be dealt with. In a case like the present, where the forfeiture of the lease is hotly contested by the lessee up to House of Lords level, the lapse of time is likely to extend over a period of years. An under-lessee can obtain relief under section 146(4) against forfeiture of the lease. But, if Mr Mowbray's point is right, an under-lessee who remains in actual occupation is necessarily a trespasser vis-à-vis the freeholder and liable to pay mesne profits accordingly from the date of forfeiture down to the date on which the subsection (4) relief is granted. This may not matter in a case where the under-lessee is a tenant at a rack rent, but in a case where the original lease and the under-lease have been granted at substantial premiums and nominal rents, the windfall to the freeholder and the detriment to the under-lessee may be very considerable. In the present case, at the date of the Court of Appeal order, the rent reserved by the lease was £27,500 per annum while the rents being received from the various occupiers and sub-lessees by the receivers were not far short of £700,000 per annum. As I have said, Mr Mowbray accepts that the court would, if it had power to do so under section 146(4), restore the mortgagees to their security as if the lease had not been forfeited. The question is whether the court has such power under that statutory provision.

He went on to hold that the court had in fact no such power as is there mentioned.

There is in the case before me no specific challenge to the efficacy of the physical re-entry by MEPC, although it is referred to by Mr Pellicano in his affidavit in the phrase "the first defendant purportedly peaceably re-entered upon the premises, which were then secured by means of locks and shutters". No specific ground was advanced in argument for holding that the re-entry was unlawful or otherwise ineffective.

Mr Cousins relied on the balance of convenience and the maintenance of the status quo. But, as Scott J in *Official Custodian for Charities* v *Mackey* at p187 observed, those considerations arise only where there is an arguable defence to the plaintiff's claim. He said at p186:

> The court has, in my judgment, no power under section 146(4) to attach to such vesting order an order that the plaintiffs pay over to the mortgagees mesne profits received by the plaintiffs since the forfeiture but before the date of the vesting order, nor to relieve the recipient of any such mesne profits from the obligation to account therefor to the plaintiffs, nor to relieve the actual occupants of the property from their obligation to pay mesne profits to the plaintiffs in respect of their occupation.

And later:

> Accordingly, a defence to the plaintiffs' claim in this action based on the proposition that the plaintiffs will, by an order made in due course on the section 146(4) applications, be deprived of their proprietary rights in respect of the period before such order is made is, in my view, misconceived.
>
> It follows, further, that in my opinion the only obstacle the defendants can present to the success of the plaintiffs' claim in the present action is the possibility that the House of Lords may, in the forfeiture action, reverse the Court of Appeal or by interlocutory order suspend the operation of the Court of Appeal order.
>
> In these circumstances, I was invited by Mr Nugee (counsel for some of the mortgagees) "to apply the principles of *American Cyanamid Co* v *Ethicon Ltd* [1975] AC 396. He submitted that I should consider the balance of convenience and that that balance came down strongly in favour of maintaining the status quo under which the first and second defendants are collecting the rents of the property and managing it. I do not, however, think that this is a case to which the *Cyanamid* principles can be applied. Those principles are not, in my view, applicable to a case where there is no arguable defence to the plaintiffs' claim.

The position regarding possession of the shop and basement seems to me somewhat stronger in favour of MEPC than it was in favour of the plaintiffs in *Official Custodian for Charities* v *Mackey*, because there is nowhere near the certainty which was conceded in *Mackey*'s case that relief would be granted under section 146(4). Mr Pellicano, on the evidence before me, seems to me a somewhat dubious asset as a tenant at a substantial rent. He has paid no rent at all since October 1992 and, although it was accepted on his behalf that he might be required to pay the appropriate fraction of the rent under the lease of £48,000 pa pending the hearing of his application for relief under section 146(4), no offer with regard to the arrears since October 1992, when Mr Pellicano had been enjoying the premises, has been forthcoming. I appreciate, of course, that there was no receiving hand available to Mr Pellicano in the period between October 1992 and the present. But, in my judgment, there is no arguable defence to MEPC's claim to possession pending the hearing of an application under section 146(4). Even if, contrary to my view, there was such an arguable defence, I should be disposed to allow Mr Pellicano to continue in possession of the basement and ground floor only on terms that he pay within 14 days £20,000 towards the arrears of rent since October 1992 and that he continue to pay during his occupation, pending the hearing of the application under section 146(4), the appropriate fraction of the rent reserved by the lease attributable to the basement and ground floor. If necessary I would have directed an inquiry or further evidence to ascertain what that appropriate fraction would be. That should not prove a particularly difficult problem, since there were underleases granted at closely connected dates of the two parts in which the premises have in the past been underlet.

However, on the view that I take, that does not arise and I decline, therefore, to make an order on this application.

Court of Appeal

December 1 1993

(Before Lord Justice NOURSE and Mr Justice THORPE)

WILLOWGREEN LTD v SMITHERS

Estates Gazette May 21 1994

[1994] 20 EG 147

Landlord and tenant — Address for service of county court summons seeking forfeiture — Whether service at demised premises at which defendant tenant had never lived good service

By a lease dated August 30 1978 the defendant and his mother were demised a 99-year term of a flat at an initial yearly ground rent of £25. The lease contained provisions for the payment of a service charge and for forfeiture of the lease for breach of covenant. Although the flat was acquired for the defendant's mother, neither the defendant nor his mother ever lived at the premises. The defendant's mother died in 1979 and the defendant allowed his stepfather to live at the flat. The last payment of rent and service charges was made in 1987 since when the stepfather disappeared. On July 9 1990 the plaintiff landlord sent a notice under section 146 of the Law of Property Act 1925 concerning arrears of service charges addressed to the defendant at the flat. On September 24 1990, pursuant to Ord 3, r3(1) of the County Court Rules 1981, the plaintiff requested the county court to issue a summons seeking possession of the flat. In accordance with Ord 7, r10(1)(b) the summons was sent by first-class post to the defendant at the flat. It never came to his notice. On December 6 1990 judgment for the plaintiff was entered for recovery of possession and arrears of rent, service charges and mesne profits. A warrant for possession was executed on February 5 1991. Although there has been dealings with the flat thereafter, the defendant remained the registered proprietor of it. The defendant appealed the decision of Judge Quentin Edwards QC, who decided that the summons had been properly served and dismissed the defendant's application to set aside the judgment.

***Held*: The appeal was allowed. Ord 7, r10(1)(b), of the County Court Rules 1981 requires the plaintiff to state the defendant's address for the purpose of service. A person's address is a place where he can receive communications. A place where a person is**

A **never present at all cannot be his address. The word "address" in the rule could be construed according to its ordinary meaning. Thus it does not include a place at which the defendant is never present and at which the process does not come to his notice. The summons was therefore not properly served. The judgment of December 6 1990 must therefore be set aside.**

The following cases are referred to in this report.

Cooper v *Scott-Farnell* [1969] 1 WLR 120; [1969] 1 All ER 178, CA

R v *London County Quarter Sessions Appeals Committee, ex parte Rossi* [1956] 1 QB 682; [1956] 2 WLR 800; [1956] 1 All ER 670, CA

B *Rolph* v *Zolan* [1993] 1 WLR 1305, CA

White v *Weston* [1968] 2 QB 647; [1968] 2 WLR 1459; [1968] 2 All ER 842, CA

This was an appeal by the defendant, Douglas Smithers, from the decision of Judge Quentin Edwards QC, who on April 14 1993 refused to set aside a judgment of Judge Martin QC, who on December 6 1990 had, *inter alia*, made an order for possession of premises on the application of the respondent, Willowgreen Ltd.

Beverley Lang and Javan Herberg (instructed by Parfitt Cresswell Carnt & Mudie) appeared for the appellant; Anthony Radevsky (instructed by Finers) represented the respondent.

C Giving the first judgment, **NOURSE LJ** said: Ord 3, r3(1) of the County Court Rules 1981 provides that a plaintiff desiring to commence a default or fixed date action shall file a request for the issue of a summons. Ord 7, r10(1), provides that service of the summons shall usually be effected (a) by the plaintiff delivering the summons to the defendant personally; or (b):

> by an officer of the court sending it by first-class post to the defendant at the address stated in the request for the summons.

Here, in a landlord's action for forfeiture and possession of a flat, D the address stated in the request for the summons was that of the flat, where the tenant had never lived or worked. The question, which may be one of some general importance, is whether, on its delivery there pursuant to Ord 7, r10(1)(b), the summons was properly served on the tenant.

By a lease dated August 30 1978 and made between the then landlord of the first part, a management company of the second part and the defendant Douglas Smithers and his mother of the third part, a flat known as 135 Peters Court, Porchester Road, London W2, was demised to the defendant and his mother for a term of 99 years from 25 December 1971 at a premium of £17,500 and an initial yearly E ground rent of £25. The lease contained provisions for payment of service charge and a provision for re-entry in standard form. By clause 8(f) it was agreed and declared that:

> Any demand for payment notice or other document required or authorised to be given to the Tenant shall be well and sufficiently given if sent by the Lessor or the Company or its agent for the time being through the post addressed to the Tenant by name or by the general description of "the Tenant" or it may be left for the Tenant at the Demised Premises . . .

The flat was owned by the defendant and his mother as joint tenants at law and in equity. They bought it as a home for the defendant's mother and stepfather, but she never lived there. She died F on September 14 1979, while still living at 85 Princess Court, Queensway, London W2, the address stated in the lease to have been that of the defendant as well as of herself. However, since 1971 the defendant has lived at an address in another part of London: 4 Brunel House, 105 Cheyne Walk, SW10.

On his mother's death, the defendant became the sole legal and beneficial owner of the flat. He offered it to his stepfather and allowed him to live there rent free on the understanding that he would pay all the bills, including the demands for ground rent and service charge. This arrangement was not notified to the landlord or the management company. Nor did the defendant notify them or their successors of any other address to which communications should be sent.

The last time that the defendant saw his stepfather was in 1986. He G did not visit the flat between then and December 1991, when he called there and found someone else in occupation. The porter told him that his stepfather had died. The defendant's solicitors have been unable to find any recorded death between 1986 and 1990 for someone of the stepfather's name and age. But whether he be dead or alive, it is certain that there has been no trace of him since.

It appears that the last payment of rent and service charge was made in 1987, presumably by the stepfather. By that time the plaintiff, Willowgreen Ltd, had acquired the reversion immediately expectant on the determination of the lease. A letter addressed to the defendant at the flat dated September 15 1987 and written by the plaintiff's H managing agents shows that arrears of service charge and ground rent amounting to £567.08 and £12.50 respectively were then claimed. On July 9 1990 a notice under section 146 of the Law of Property Act 1925 addressed to the defendant at the flat was served under cover of a letter from the plaintiff's solicitors similarly addressed.

On September 24 1990, pursuant to Ord 3, r3(1), the plaintiff filed a request for the issue of a summons against the defendant in Bloomsbury County Court. The request was made on form N204, which is headed, "Request for Issue of Summons for Possession of Land". It is not a form prescribed by the County Court Rules, but a practice form approved by the Lord Chancellor. At the top on the left are four boxes: the first for "the plaintiff's full name and address"; the J second for "the name and address for service and payment (if different from the above)"; the third for the "Defendant's name and address"; beside the fourth appeared the words "The claim is for possession of — give the full address of the land claimed".

In both the third and fourth boxes the address stated was that of the flat. Curiously, the defendant's name was omitted from the third, although nothing turns on that. The plaintiff also filed particulars of claim claiming possession of the flat, together with arrears of rent and service charge, interest payable pursuant to the terms of the lease and mesne profits at the daily rate of seven pence.

In purported pursuance of Ord 7, r10(1)(b), an officer of the court K sent the summons by first-class post to the defendant at the flat. It never came to his notice. There having been no reply or response on the part of the defendant, on December 6 1990 Judge Martin QC, in the absence of the defendant, entered judgment for the plaintiff for recovery of possession of the flat on January 3 1991, and for the sum of £4,464.89, together with mesne profits at the rate claimed and costs on scale 3. A warrant for possession was duly issued. It was executed on February 5 1991. Meanwhile, on July 30 1991, the plaintiff had sold the freehold of Peters Court to Etablissement Croy ("Croy") and on the same day Croy granted the plaintiff a 125-year leaseback of the flat. However, on May 9 1991 the plaintiff surrendered that lease to L Croy, since when the plaintiff has had no interest in the flat. It appears that there have since been other dealings with it, although the defendant remains the registered proprietor of the lease at the Land Registry.

Soon after he discovered the position in December 1991, the defendant instructed solicitors to act on his behalf. In due course he applied to set the judgment of December 6 1990 aside. The application came before Judge Quentin Edwards QC on April 14 1993, when he held, first, that the summons had been properly served on the defendant and, second, that in the circumstances, especially since rights of innocent third parties had arisen, he ought not in his M discretion to set the judgment aside. The defendant now appeals to this court.

On the question of service the judge thought that the only obligation of the plaintiff under the rules was to state in the request an address which had "a direct and immediate connection" with the defendant. If that was correct, I would be inclined to think that the plaintiff did what it had to do in this case. In view of the terms of clause 8(f) of the lease, the fact that demands and notices under it were always sent to or left for the defendant at the flat and the further fact that he never notified the landlord or the managing agents of any other address to which communications should be sent, I would be inclined

A to think that the address of the flat did have a direct and immediate connection with him. However, as Mr Anthony Radevsky, for the plaintiff, has recognised, that interpretation of the rules might, in certain circumstances, include a number of addresses, perhaps up to four or five residential addresses alone. For my part, I do not think that the judge's interpretation can be correct.

Since form N204 is not a prescribed form, neither its front nor the notes on its back can be resorted to in order to construe the rules. Thus we are left with Ord 7, r10(1)(b), in which it is implicit that the plaintiff is obliged to state the defendant's address in the request for the summons. So far I am in agreement with the judge. But what is the
B defendant's "address" for this purpose?

It is remarkable that a question seemingly so fundamental should seemingly be so free from authority. We have been referred to only four decided cases. Beverley Lang, for the defendant, has referred us to *R* v *London County Quarter Sessions Appeal Committee, ex parte Rossi* [1956] 1 QB 682, *White* v *Weston* [1968] 2 QB 647 and *Cooper* v *Scott-Farnell* [1969] 1 WLR 120. Mr Radevsky has also referred us to *Rolph* v *Zolan* [1993] 1 WLR 1305. I will say at once that I am unable to see in what way that last decision assists either side in the present dispute.

I start by considering the question apart from authority. In ordinary
C parlance a person's address is a place at which written communications can be delivered to him. In order that they can be delivered to him, he must, to a greater or lesser extent, be present to receive them. The extent to which his presence is necessary to make it his address may vary, and vary significantly, with the circumstances. But if he is never there at all, it cannot properly be called his address. It can, if communications will be sent on to him from there, be called a forwarding address. But that is not the same thing as an address.

In construing the word "address" in a rule which permits service of proceedings to be made by post, it is important to remember that that mode of service involves a significant departure from the rule, which prevailed until well into this century, that an originating process
D should, in the absence of agreement, authority or special order, be served on the defendant or the respondent personally. As Denning LJ (as he then was) put it in *ex parte Rossi*, at p691:

> When construing this section, it is to be remembered that it is a fundamental principle of our law that no one is to be found guilty or made liable by an order of any tribunal unless he has been given fair notice of the proceedings so as to enable him to appear and defend them. The common law has always been very careful to see that the defendant is fully apprised of the proceedings before it makes any order against him.

Having referred to the common law writ of *capias*, which required
E the sheriff to bring the defendant to court, Denning LJ continued:

> That has all been done away with, but the law still insists in most cases that the defendant shall be served personally so as to be sure that he knows of the proceedings against him. In modern times there have been a few statutes and rules which allow service by registered post, and this is one of them. The merit of registered post in this regard is that the postman will only deliver the letter to the person to whom it is addressed or to someone who will take responsibility for seeing that he gets it. Otherwise he will return it to the sender, who will thus get to know, sooner or later, if the letter is not received.

What we have here is a rule which does not even require service to be made by registered post. The postman who delivers the first-class
F post is expected to do no more than put it through the letter-box or leave it on the premises, a process fraught with the risk that the addressee will not receive it. The rule-making authority could hardly have intended that the risk should be increased still further by permitting first-class postal service at a place where the addressee is never present and where it does not come to his notice.

Further support for this view can be derived from *White* v *Weston*, where it was held by this court that a summons had not, for the purposes of Ord 8, r8(3) of the County Court Rules 1936, been sent to the defendant by ordinary post because the letter containing it had been sent to a place at which the defendant had ceased to reside more than five months earlier; with the result that it was not "properly addressed", as required by section 26 of the Interpretation Act 1889. G
At p658A, Russell LJ said:

> But a summons addressed to an address with which the defendant has had no connection for five months or more cannot be said to be properly addressed.

At p658C, he said:

> The function of service is primarily to bring to the attention of the person to be served the fact that he is being sued, and particular language is in my judgment required if something short of that is to constitute service.

Sachs LJ said, at p661D:

> I can find no warrant in any of the County Court Rules or in section 26 of the H
> Interpretation Act 1889, for holding that service in purported pursuance of Ord 8, r3, at an address which at the relevant time was not the abode, residence, or place of business of a defendant is good service should the relevant document not in fact reach him.

Mr Radevsky submits that the first of those observations of Russell LJ supports the judge's "direct and immediate connection" view of the question. I reject that submission. I think that Russell LJ, like Sachs LJ, regarded it as essential that the defendant should have had some continuing presence there before the residence in question could properly have been called his address. It seems that that was the view
of Russell LJ's observation taken by Willmer LJ in *Cooper* v J
Scott-Farnell at p127B-C.

The position, therefore, is that the three relevant authorities to which we have been referred support the view that the word "address" in Ord 7, r10(l)(b), should be construed in accordance with its ordinary meaning. Thus, it does not include a place at which the defendant is never present and at which the process does not come to his notice, albeit that it is a place which, in the circumstances of this case, may well have had a direct and immediate connection with him. It follows that the summons here was not properly served on the defendant.

What then should be the fate of the judgment entered by Judge
Martin QC on December 6 1990? The judgment having been obtained K
in proceedings initiated by a summons which was not properly served on the defendant, it seems plain that it must be set aside *ex debito justitiae*. Mr Radevsky has submitted that there is some sort of discretion in the matter, but it is clear both on principle and from *White* v *Weston* that that is not the case.

I would therefore allow the appeal, discharge the order of Judge Quentin Edwards and set aside the judgment of Judge Martin QC entered on December 6 1990.

Agreeing, **THORPE J** said: The facts and circumstances which give rise to this appeal are no doubt highly unusual. However, it is agreed L
between counsel that the county court rule that governed the service of the respondent's default summons on the appellant was Ord 7, r10(1)(b). That subrule provides that service shall be effected:

> by an officer of the court sending it by first-class post to the defendant at the address stated in the request for the summons.

The request for issue of summons for possession of land is a practice form N204. On its face are seven numbered boxes for completion. Box 3 requires the insertion of "Defendant's name and address". Beside box 4 appears "the claim is for possession of — give the full address of the land claimed". In this case box 4 was accurately
completed "Flat 135 Peters Court, Porchester Road, London W2". M
Precisely the same words were inappropriately typed into box 3. The defendant's name was not stated and the premises in dispute had never been his address. Had due care been given to the completion of box 3 it might have read "Douglas Smithers whose present address is unknown to the plaintiff", for the request for summons was dated September 24 1990, by which date the premises had been empty for three years.

However, the question of law in the appeal is whether the judge was right to construe the words "the address" in Ord 7, r10(1)(b), as "an address which has direct and immediate connection with the defendant". In my judgment, the judge was plainly wrong in that

A construction. There is no authority for its adoption and despite the submissions of Mr Radevsky it is contrary to the principles stated in the judgments in *White* v *Weston* [1968] 2 QB 647. The principle is stated by Sachs LJ in these terms:

> It follows that only an explicit and clear provision in a statute, or in rules having statutory force, can operate to deprive a citizen of his right to receive notice of the commencement of process against him; and to permit service other than personal at an address which is not in fact his abode (to use the word employed, for instance, in rule 19(*b*) of the Magistrates' Courts Rules, 1952) nor his residence, nor his business address would be a provision clearly calculated to deprive him of that right.

B Russell LJ in his judgment had said:

> But a summons addressed to an address with which the defendant has had no connection for five months or more cannot be said to be properly addressed.

That sentence seems to me to support the appellant's argument. Mr Radevsky claims it for his case by advancing it by corollary as the foundation for the judge's test, namely did the address have a direct and immediate connection with the defendant? It is plain to me that far from formulating a test so wide and potentially productive of injustice, Russell LJ was upholding the principle that the function of service is to bring to the attention of the person to be served the
C fact he is being sued and particular language is required if anything short of that is to constitute service. I do not, myself, see any useful distinction between the proposed defendant's abode and the proposed defendant's residence. In most cases "the address" in Ord 7, r10(l)(b), will be the address at which the defendant ordinarily resides or works.

I therefore conclude that the summons was not served in accordance with the rules, and the judgment should be set aside as of right *ex debito justitiae*. Mr Radevsky's argument, that even if the appellant was not served in accordance with the rules he is not entitled to have the order set aside *ex debito justitiae*, is also precluded by the
D judgment of Russell LJ in *White* v *Weston*.

I, too, would allow this appeal.

Appeal allowed.

Chancery Division

February 17 1994

E (Before Mr Justice ALDOUS)

ROYTON INDUSTRIES LTD v LAWRENCE

Estates Gazette May 21 1994

[1994] 20 EG 151

Landlord and tenant — Assignment — Whether obligation in reddedum clause to pay service charges bound original tenants after assignment

On June 24 1986 the plaintiff granted the defendants a 110-year lease of commercial premises. On July 27 1987 the defendants
F assigned the lease to a third party, who in turn assigned on. The plaintiff sought to recover from the defendants service charges of just over £15,000 in accordance with the provisions of the lease. The defendants contended that, having regard to the construction of the lease, their liability was limited to their time of occupation as tenants; the lease imposed a covenant on the tenant and assigns to pay the rent during the term, but there was no similar obligation by covenant on the tenant to pay service charges after an assignment. The only obligation relating to service charges appeared in the *reddendum*, and before the tenant's express covenants, "YIELDING and PAYING . . . the Tenant's proportion of the maintenance costs".

***Held*: The defendants were liable for the service charges. The expression "YIELDING and PAYING" amounted to a covenant by the tenant to pay rent and service charges during the term of the lease. The defendants were bound by privity of contract with the plaintiff and therefore all express obligations remained binding after assignment.** G

The following cases are referred to in this report.

Allied London Investments Ltd v *Hambro Life Assurance Ltd (No 2)* (1984) 270 EG 948
Newton v *Osborn* (1653) Sty 387
Steward v *Wolveridge* (1832) 9 Bing 60
Walkers Case 1587 3 Co Rep 22a
Youngmin v *Heath* [1974] 1 WLR 135; [1974] 1 All ER 461, CA H

This was the hearing of a preliminary issue as to whether the defendants, Paul Gerald Lawrence and Maureen Lawrence, were liable under a lease dated June 24 1986 to pay service charges to the plaintiff, Royton Industries Ltd.

Jonathan Gaunt QC (instructed by Bury & Walkers, of Leeds) appeared for the plaintiff; Vivian Chapman (instructed by Fox & Co, of Taunton) represented the defendants.

Giving judgment, **ALDOUS J** said: By a lease dated June 26 1981, Bartlett Construction Group Ltd granted and demised to Coin Controls Ltd a 125-year lease over units 1-5 Bartlett Mews, Priorswood J Industrial Estate, Venture Way, Taunton. Coin Controls Ltd changed its name to Royton Industries Ltd and on June 24 1986 granted and demised a 110-year lease over the units to Paul and Maureen Lawrence. I will refer to that lease as the "intervening lease". On September 15 1987 Paul and Maureen Lawrence gave notice to Royton that on July 27 1987 they had assigned the intervening lease to Dornar Ltd. Since then, the intervening lease has been assigned on more than one occasion and I was told that Royton no longer know the identity of the person in whom the lease is vested.

The intervening lease provides for payment by the tenant of a K proportion of maintenance costs expended by the landlord. Such costs arose and were paid to the landlord by Royton pursuant to the landlord's demands. Royton sought to recover them from Paul and Maureen Lawrence. That was resisted. Therefore, Royton, in November 1991, issued the writ in this action. In the amended statement of claim, Royton refer to the intervening lease and allege that the defendants, Paul and Maureen Lawrence, are liable to repay to them the maintenance costs they had paid, amounting to just over £15,000 and interest. In the defence the lease is admitted, but liability to pay is denied because of the assignment by the defendants of the lease. The defendants also put in issue the reasonableness of the L maintenance costs and the amount claimed.

By agreement between the parties it was ordered on February 10 1993 that the following issue be tried as a preliminary point:

> Whether on the true construction of the lease dated 26th June 1986 between the Plaintiff and the Defendants of premises known as Units 1-5 Bartlett Mews, Priorswood Industrial Estate, Venture Way, Taunton, Somerset ("the intervening lease") there is any liability imposed on the Defendants to make the relevant service charge payment in respect of sums accruing due after they had assigned the intervening lease.

The intervening lease is stated to be between Royton, called the landlord, and Paul and Maureen Lawrence, called the tenant, "which M expression shall where the context so admits include his successors in title".

The lease states that in consideration of £150,000 "and in consideration of the rent and covenants hereinafter reserved and contained the landlord" grants to the tenant the rights of the landlord over units 1-5. The grant was:

> TO HOLD the same except and reserved as aforesaid unto the tenant from and including the 1st January 1986 for the term of 110 years . . . YIELDING AND PAYING therefor unto the landlord on the 31st day of December in each year the yearly rent of One Pound AND ALSO YIELDING AND PAYING therefor unto the Landlord the Tenant's proportion of the maintenance costs as defined in the

> Second Schedule hereto once a year within 14 days after the payment thereof having been demanded by the landlord of the tenant.

Thereafter the underlease contains this clause:

> 2. The Tenant for itself and its assigns and to the intent that the obligations may continue throughout the term hereby granted hereby covenants with the landlord as follows that is to say:
>
> (1) To pay the reserved rent as herein provided at the times and in the manner aforesaid without any deduction.

Clause 2 continues with 24 further subclauses dealing with such matters as payment of rates, repair of the premises and painting it.

Clause 4 provided:

> 4. Provided always and it is hereby agreed and declared
>
> (i) that these presents are made upon the express condition that if the said rent or any part thereof shall be unpaid for twenty one days after any of the days hereinbefore appointed for payment thereof whether the same shall have been lawfully demanded or not or if any covenant on the tenant's part herein contained shall not be performed or observed then and in any of the said cases and thenceforth it shall be lawful for the Landlord or any person or persons duly authorised by the Landlord in that behalf into or upon the demised premises or any part thereof in the name of the whole to re-enter and the same to repossess and enjoy as if these presents had not been made without prejudice to any right of action or remedy of the Landlord in respect of any antecedent breach of any of the covenants by the tenant herein contained
>
> (ii) In the interpretation of this Lease the masculine where the context so admits shall include the feminine and any corporate body and the singular shall include the plural and vice versa and all covenants entered into by more than a single person shall be deemed to be entered into by all persons and Corporations entering into the same jointly and severally.

The second schedule, which refers to the maintenance costs, stated:

> The Maintenance Costs payable by the tenant in accordance with the provisions of Clause 1 of this Lease shall be 50.75 per cent of the cost to the Superior Landlord of providing the following services and carrying out the following maintenance and repairs and otherwise in fulfilling its obligations under the Head Lease including the provision of a suitable sinking fund or reserve namely . . .

Thereafter there are set out obligations to maintain the access roads, the walls, fences and the like.

It is settled law that a lessee remains liable upon express covenants in the lease for the whole term of the lease despite any assignment. As long ago as 1587 that was stated to be the law in *Walkers Case* 1587 3 Co Rep 22a. As Walton J said in *Allied London Investments Ltd* v *Hambro Life Assurance Ltd (No 2)* (1984) 270 EG 948 at p950:

> I do not think that Mr Gaunt's attempt to bring this case under that umbrella will really stand up, because here we have an original covenant by the defendant to pay the rent. That is an original covenant which binds him by virtue of privity of contract throughout the term of lease, and indeed, as has been pointed out in a number of cases dealing with the question of what happens when the lease is assigned, the covenant is economically to the effect that: "Either I will pay you or the assignee will pay you". But it is not something which only comes into effect when the assignee fails to pay. Of course the expectation, commercially speaking, is that the assignee will pay, but the assignor does not by assignment get rid of one jot or tittle of his original liability.

The defendants did not disagree with that proposition of law as a general proposition of law. They submitted that the intervening lease did not contain an express obligation that the defendants would pay the maintenance costs during the whole term of the lease. They submitted that, when clauses 1 and 2 were read together, the implication was that the obligation on the defendants to pay the maintenance costs was limited to their time of occupation as lessees.

The plaintiff submitted that the word "rent" in clause 2(1) of the underlease referred to all the sums reserved by the *reddendum*. If so, the obligation to pay both the yearly rent and the tenant's share of the maintenance costs remained after assignment. The defendants accepted that if the word "rent" was to have such a broad meaning then the conclusion advanced by the plaintiff followed. However, they submitted that the word "rent" in clause 2(1) clearly referred back to the word "rent" in clause 1 and did not include the maintenance costs.

Clause 1 states that the grant is in consideration of £150,000, "and in consideration of the rent and covenants hereinafter reserved and contained". Thereafter the tenant holds the land "Yielding and paying therefor unto the landlord on the 31st day of December in each year the yearly rent of One Pound and also yielding and paying therefor" the tenant's portion of the maintenance costs. As will appear later in this judgment, I believe that those words which I have quoted place an obligation on the tenant amounting to a covenant to pay rent and also to pay the tenant's proportion of the maintenance costs. Thus both the grant and the *reddendum* point to the obligation to pay rent being separate from the other obligations under the lease. That is consistent with what I believe to be the natural way to construe the word "rent" in both clauses 1 and 2, namely as referring to the sum which is to be paid annually under the lease, but not to sums for maintenance which would vary in amount and need not necessarily arise in any particular year.

I accept that pursuant to clause 4(ii) the word "rent" may be construed as meaning rents, but, even so, I believe that the construction sought to be placed by the plaintiffs upon the word "rent" in clause 2 is not correct. The word "rent" refers to the annual rent in clause 1 and not to the maintenance costs. I therefore reject this part of the plaintiff's submissions. Thus the dispute between the parties must be decided upon the basis contained in clause 1, and clause 2 contains a covenant in respect of only one of them, namely the obligation to pay rent.

The plaintiff submitted that even if the word "rent" in clause 2 should be construed as I have construed it, the obligation in clause 1 was not avoided by assignment. That submission is based upon a number of propositions.

First, the plaintiff submitted that the *reddendum* imposed express obligations upon the defendants and that, in the case of rent, it was an obligation to pay the rent during the term of the lease. I believe that that submission is well founded and in any case it was not disputed by the defendants. There is ample authority to establish that the words "yielding and paying" impose an express obligation: see, for example, *Newton* v *Osborn* (1653) Sty 387 and *Steward* v *Wolveridge* (1832) 9 Bing 60 in particular at p67. Further, it is of the essence of the agreement that rent will be paid during the term of the lease: see *Youngmin* v *Heath* [1974] 1 WLR 135.

Second, the plaintiff submitted that the words "and also yielding and paying therefor" imposed an express obligation to pay the tenant's proportion of the maintenance costs for the full term of the lease. That submission is also correct and was also not disputed by the defendants. If the words "yielding and paying therefor", when referring to rent, impose an express obligation to pay for the full term of the lease, then those words must impose a similar obligation when referring to the maintenance costs.

Third, the plaintiff submitted that, in the absence of indications to the contrary, the express obligation to pay the rent imposed in clause 1 bound the persons in privity of contract with the lessor even though such persons had assigned the lease. That submission is also correct and was not disputed by the defendants.

Fourth, in the absence of the indications to the contrary, the express obligation to pay the tenant's proportion of the maintenance costs imposed in clause 1 bound the persons in privity of contract with the lessor even though such persons had assigned the lease. That submission was disputed by the defendants. They submitted that it did not follow that, because the obligation to pay rent remained after assignment, the obligation to pay the maintenance costs also remained.

I believe that the submission of the plaintiff is correct. There being privity of contract between the plaintiff and the defendants it follows that all express obligations will continue to be binding between them even after assignment unless there be agreement to the contrary.

To accept the submission of the defendants, it is necessary to read into clause 1 a limitation on the obligation to pay the tenant's share of the maintenance costs, namely that the tenant, the defendants, would pay only during the time that the lease was vested in them. I see no

A reason for implying such a limitation from the words in clause 1.

The primary submission of the defendants was that there were clear indications in the lease that the obligation to pay the tenant's share of the maintenance costs should be qualified so that the defendants were liable only while the lease was vested in them. They drew to my attention the opening words of clause 2, namely "The Tenant for itself and its assigns and to the intent that the obligations may continue throughout the term hereby granted hereby covenants . . .", and the fact that clause 2 contained an obligation to pay the rent, but did not contain an obligation to pay the maintenance costs. The conclusion suggested to be drawn was that the draftsman had made it clear that
B there was an obligation to pay rent imposed upon the tenant and its assigns for the term of the lease, and by implication there was no similar obligation in respect of the duty to pay the tenant's share of the maintenance costs. It followed that the obligation to pay the maintenance costs did not remain with the tenant, the defendants, after assignment.

I cannot accept the submission of the defendants. The fact that the obligation on the tenant in respect of rent is spelt out in clause 2(1) does not mean that the obligation to pay the tenant's share of the maintenance costs in clause 1 should be altered by implication. There is an express obligation to yield and pay the tenant's share of the
C maintenance costs. I can see no reason for cutting down that obligation by reason of the wording of clause 2. There is privity of contract between the plaintiff and the defendants and therefore the express obligation to pay the tenant's share of the maintenance costs remained even though there was assignment by the defendants.

I conclude that the question to be tried as a preliminary point should be answered in the affirmative. There is a liability imposed upon the defendants to make the relevant service charge payment in respect of sums accruing due after they had assigned the intervening lease.

For these reasons I answer the question posed in the affirmative.

D *Declaration accordingly.*

Chancery Division

January 14 1994

(Before SIR DONALD NICHOLLS V-C)

E ESTATES GOVERNORS OF ALLEYN'S COLLEGE OF GOD'S GIFT AT DULWICH v WILLIAMS AND ANOTHER

Estates Gazette June 11 1994

[1994] 23 EG 127

Leasehold Reform Act 1967 — Management scheme — Whether arbitrator applied correct test in deciding that managers of scheme unreasonably withheld consent to proposed building

In January 1974 the High Court made an order under section 19 of the Leasehold Reform Act 1967 approving a scheme in respect of the Dulwich Estate, in south-east London, conferring powers of
F management on the applicant estate governors. The management scheme, *inter alia*, prohibited the erection of any new structure which is visible at ground level from beyond the boundaries of an enfranchised property without the prior approval of the applicants. In May 1992 the respondents, the owners of an enfranchised detached house, sought the estate governors' consent to the construction of a two-car detached garage, for which they had planning permission, at the front of their house. On the refusal of the applicants to give their approval, and pursuant to the provisions for arbitration in the scheme, the appointed arbitrator held that the proposed garage would be visible from beyond the boundaries of the property, so that consent was needed, but that the applicants' decision was unreasonable. The applicants contended that the arbitrator applied the wrong test in G
making his award.

Held: **The application was allowed and the award was remitted to the arbitrator. Under the scheme the applicants must not withhold their consent arbitrarily; they may withhold consent if reasonable managers of the estate might do so. Applying principles derived from the authorities relating to the withholding of consent in landlord and tenant for assignments and subletting to the applicants' original decision, the question is whether the conclusion of the applicants might have been reached by a reasonable man in the circumstances. If the decision were reviewable by a court, the court cannot consider the matter afresh H
and only if the applicants' decision was shown to be one which no reasonable body of estate governors could reach would the disgruntled property owner succeed. The arbitration clause does not enable a decision of the applicants which a reasonable estate manager might reach to be overturned by an arbitrator. Accordingly, the arbitration clause does not give the arbitrator a wider role than the court would have. The arbitrator's conclusion, that the applicant's decision was unreasonable, could not be read as a conclusion that no reasonable estate managers could, in all the circumstances, have refused the respondents' application. It was impossible to be sure what would have been the arbitrator's conclusion if he had correctly directed himself on how he should J
approach the issue before him.**

The following cases are referred to in this report.

F&G Sykes (Wessex) Ltd v *Fine Fare Ltd* [1967] 1 Lloyd's Rep 53
International Drilling Fluids Ltd v *Louisville Investments (Uxbridge) Ltd* [1986] Ch 513; [1986] 2 WLR 581; [1986] 1 All ER 321; (1985) 51 P&CR 187; [1986] 1 EGLR 39; 277 EG 62, CA
Pimms Ltd v *Tallow Chandlers Co* [1964] 2 QB 547; [1964] 2 WLR 1129; [1964] 2 All ER 145, CA
Shanly v *Ward* (1913) 29 TLR 714

This was an appeal by consent under section 1 of the Arbitration K
Act 1979 by the applicants, the Estates Governors of Alleyn's College of God's Gift at Dulwich, on a question of law arising out of an award by an arbitrator, Mr C R Martin FRICS, appointed to determine differences arising between the applicants and the respondents, Mr and Mrs John Williams, following the refusal by the applicants of approval under a management scheme approved under section 19 of the Leasehold Reform Act 1967 to a garage.

Terence Etherton QC and Timothy Dutton (instructed by Lovell White Durrant) appeared for the applicants; John McDonnell QC (instructed by Compton Carr) represented the respondents. L

Giving judgment, **SIR DONALD NICHOLLS V-C** said: On January 21 1974 Walton J made an order under section 19 of the Leasehold Reform Act 1967 approving a scheme in respect of the well-known Dulwich College Estate in south-east London. The scheme conferred on the Estates Governors of Alleyn's College of God's Gift at Dulwich, to give them their full title, powers of management over the estate. A statutory prerequisite to the making of that scheme was a certificate from the minister to the effect that, in order to maintain adequate standards of appearance and amenity and regulate redevelopment in the area in the event of tenants acquiring the landlord's interest in their house and premises under the Act, it was in M
the minister's opinion likely to be in the general interest that the landlord should retain powers of management in respect of enfranchised property. A further prerequisite was that the scheme appeared to the court to be fair and practicable and not to give the landlord a degree of control out of proportion to that previously exercised by him or to that required for the purposes of the scheme. Both those prerequisites were duly satisfied in the case of the Dulwich Estate.

The estate comprises some 1,500 acres, on which there are about 4,500 privately owned houses. Of these, some 3,000 have now been enfranchised so that the freeholds belong to the respective

A householders and not the estates governors. One of the enfranchised properties is a detached house, 33A Alleyn Park, belonging to Mr and Mrs John Williams. They wish to build a two-car detached garage at the front of their house, for which they have obtained the appropriate planning permission. Under the scheme of management, however, they also need the approval of the estates governors. The effect of clauses 3(a) and 15 of the scheme is to prohibit the erection of any new structure which is visible at ground level from beyond the boundaries of an enfranchised property without the prior written approval of the estates governors, such approval not to be unreasonably withheld.

B The Williams applied for consent on October 21 1990, but that was refused by the estates governors. They made a further application on May 29 1992, but that also met with a refusal. The estates governors' reasons were that the proposed garage would be in front of the main building line and that to permit this garage would establish an undesirable precedent. Mr and Mrs Williams were unhappy at this outcome. They considered that the estates governors were acting unreasonably. The scheme contains an arbitration clause, clause 17. Pursuant to this clause Mr C R Martin FRICS was appointed as arbitrator to determine the dispute between the parties. By a reasoned award dated August 23 1993 the arbitrator held that the proposed garage would be visible from beyond the boundaries of the property, C so consent was needed. He held further that the estates governors' decision was unreasonable.

The estates governors, in turn, were unhappy with this adverse decision. They contend that the arbitrator applied the wrong test in deciding the issue before him, that is whether they had unreasonably refused consent. The matter is of some general importance to them because this is not the first time this point has arisen. So they have appealed to this court on this question of law, pursuant to section 1 of the Arbitration Act 1979. The estates governors have agreed that in any event they will pay the costs of the claimants, Mr and Mrs Williams, and, for their part, the claimants have consented to this D appeal being brought.

Estates governors' decision and reasonableness

The first point to be addressed is the scope of the prohibition in clause 3(a). Clause 3 is one of several clauses which inhibit, in one or other respect, a property owner's freedom to exercise the ordinary rights of property owners. For example, clause 6(a) confines the permitted use to a use which could lawfully be carried out at the date of enfranchisement. Under clause 6(b) trees may not be cut down or lopped, caravans or temporary buildings may not be kept on the property if visible beyond the property's boundaries, advertisements E (other than "For sale" notices or temporary notices of a forthcoming function) may not be displayed. In each of these cases the prohibition is against these acts being done without the written consent of the estates governors.

The scheme applies to each enfranchised property from the date of enfranchisement and is enforceable by the estates governors against the owner of the property as though he had covenanted with the estates governors to be bound by the scheme (see clause 2(a)). Likewise, the Act gives statutory force to the provisions of the scheme by making them enforceable by the estates governors against the property owners and occupants as though they had entered into covenants with the estates governors to be bound thereby: see section 9(10).

F Clause 3(a) provides:

> No material alteration shall be made to the external appearance of any building or structure for the time being on an enfranchised property and no new or additional building or structure shall be built or erected thereon so as to be visible at ground level from beyond the boundaries of the same without (in any such case) the prior written approval of the Managers . . .

The managers are the estates governors or their successors in title.

Standing alone, subclause 3(a) imposes no express fetters on the circumstances in which, or the reasons for which, the governors may refuse their consent. That subclause, however, must be read with clause 15, which provides:

> Any requirement in the Scheme of any approval or consent to be given by the Managers shall be subject to the proviso that the same shall not be unreasonably withheld and the Managers shall not serve any notice under the Scheme except such as shall be reasonable in the circumstances and no premium or fine should be required for any consent. G

The effect of clause 15 is to set bounds to the estates governors' freedom to give or withhold their consent. They must not act arbitrarily. They must not withhold consent arbitrarily. They may withhold consent only if to do so would, in all the circumstances, be reasonable. In other words, they may withhold consent if reasonable managers of the estate might do so, but not otherwise.

This strikes a fair balance for the benefit of all those who have H interests in the estate, including the estates governors, their tenants, and enfranchised property owners. Restrictions against building are desirable for the general benefit of the estate and all who are interested in it. It is for this reason they exist. Clause 16(a) underlines this by providing that the rights and powers conferred on the estates governors by the scheme are conferred on them for the purpose of enabling them to preserve the amenities of the estate for the common benefit. But these building restrictions, in all fairness, should be confined to circumstances where, acting reasonably, the estates governors consider there should be no new building. The governors do not have unfettered control, exercisable arbitrarily or capriciously or J oppressively.

The most cursory consideration of these provisions reminds one immediately of similar, commonplace provisions in leases precluding assignment, subletting or the making of alterations without the lessor's consent. Such provisions normally go on to state that such consent shall not be unreasonably withheld, but even if they do not include these words such words are written in by statute, by section 19 of the Landlord and Tenant Act 1927.

Thus, what this scheme has done is to incorporate a control regime regarding new building and user, and in certain other respects, corresponding to that commonly found in many leases. Provisions to K this effect in leases have been before the courts on many occasions. It is now well established that, in relation to covenants against assignment, the effect of the common-form provision regarding consent not being unreasonably withheld is the same as I have already stated regarding clauses 3(a), and 15 of this scheme. Before me it was common ground, and in my view rightly so, that in this respect no distinction is to be drawn between covenants against assignment and covenants against alterations or improvements. I need refer only to two cases. First, *Shanly* v *Ward* (1913) 29 TLR 714, which was a decision of the Court of Appeal. There the plaintiff tenant was challenging the defendant landlord's refusal to permit assignment of a L lease. The Master of the Rolls said, at p715:

> The first point was that the plaintiff coming to the Court and asking for declaration that the defendants had unreasonably withheld their consent, the burden was thrown on him to prove his case. It was not for the defendants to prove that they were justified in withholding their consent but for the plaintiff to prove that it was unreasonably withheld. What did unreasonableness mean? It was not enough to show that other lessors might have accepted the proposed assignees: the lessors were not to be held to have withheld the licence unreasonably if in the action they took they acted as a reasonable man might have done in the circumstances.

The second authority I mention is also a decision of the Court of M Appeal *Pimms Ltd* v *Tallow Chandlers Co* [1964] 2 QB 547. At p564, delivering the judgment of the court, Danckwerts LJ referred to *Shanly* v *Ward* and other authorities and said:

> There is no doubt that the onus of proving that consent has been unreasonably withheld is on the tenant . . . Further, it is not necessary for the landlords to prove that the conclusions which led them to refuse consent were justified, if they were conclusions which might be reached by a reasonable man in the circumstances . . .

Mr John McDonnell QC, submitted that a gloss has now been placed on these authorities by observations made by Balcombe LJ in the decision of *International Drilling Fluids Ltd* v *Louisville*

A *Investments (Uxbridge) Ltd* [1986] Ch 513*. I do not agree. I do not read Balcombe LJ's summary of the state of the law in his seven propositions, and in particular his proposition (7) at p521, as departing in any way from the authorities I have mentioned. His proposition (7) was expressly subject to his earlier propositions, and in his propositions (3) and (4) he set out the effect of the two cases I have already mentioned. In his proposition (7) Balcombe LJ stated that the issue whether consent was being unreasonably withheld was one of fact. That is so, but the touchstone to be applied when seeking an answer to that question is whether the conclusion might have been reached by a reasonable man in the circumstances.

B There is nothing surprising or exceptional about that result in this case. Under the scheme decisions on the future development of the estate remain with the estates governors. This conforms to the statutory prerequisites for a scheme of management under section 19(1). But the powers of management retained by the estates governors are circumscribed by an overriding requirement of reasonableness. There will be instances where different minds, quite reasonably, may reach different conclusions on whether a particular type of development is or is not desirable, or on whether it would or would not be likely to damage desirable amenities. On these matters the decision is entrusted to the estates governors, so long as they keep C within the bounds of reasonableness. That gives them a degree of control over the estate not out of proportion to that required for the purpose of preserving the estate amenities for the common benefit, in accordance with section 19(5).

Thus far there was a large measure of agreement between the parties. At this point I pause to observe that if a householder and the estates governors unfortunately are at loggerheads over whether consent for a new building should or should not be given, and if proceedings are instituted in court to resolve this dispute, the issue before the court is not an issue at large as to whether consent should or should not be given. The issue before the court is whether the estates D governors acted unreasonably in refusing consent. If, having heard the parties, the court finds that the estates governors' decision falls within the band of possible decisions a reasonable body of estate governors could reach, then the estates governors' decision stands. The judge would not consider the matter afresh as though the estates governors had not reached a decision. He would not stand in the shoes of the estates governors for all purposes. He would stand in their shoes only in the sense that he would be concerned to decide whether they, acting reasonably, could have refused the application. Only if the decision by the estates governors was shown to be one which no reasonable body of estates governors could reach would the disgruntled property owner succeed.

E

Clause 17

That is the position if an application were made to the court. As already noted, this scheme contains an arbitration clause. Clause 17 provides that:

Any dispute or difference between the Managers and any owner

(a) under clauses 3(a) and (b) . . .

— or under various other specified clauses or regarding certain other specified matters —

F shall be referred to a single arbitrator to be appointed in default of agreement by the President for the time being of the Royal Institution of Chartered Surveyors . . .

Mr McDonnell submitted that the combined effect of clauses 3(a) and 15 and this clause is to confer on the arbitrator much wider powers than the court would have in a case such as the present. He submitted that while that might not be so if the arbitration clause applied only to any "dispute", the additional words "or difference" went wider than this and went beyond justifiable disputes. Those words were apt to include a difference of opinion which might exist between reasonable men over whether consent should be given. The arbitrator would be an G experienced surveyor, and the intention was that he should step into the shoes of the estates governors for all purposes when deciding such a case and not be confined to deciding whether a reasonable estate manager could have withheld consent.

I do not accept this. I am unable to attach to the word "difference" the far-reaching effect for which Mr McDonnell contended. His contention would effectually rewrite clauses 3(a) and 15 so as to provide that a householder who wished to build needs either the consent of the estates governors, such consent not to be unreasonably withheld, or the approval of the arbitrator. It would mean that, even if the estates governors had reached a conclusion which a reasonable H estate manager might reach, their conclusion could nevertheless still be overturned by an arbitrator. I can see no justification for reading clause 17 in this way.

Mr McDonnell prayed in aid the observations of Danckwerts LJ in *F&G Sykes (Wessex) Ltd* v *Fine Fare Ltd* [1967] 1 Lloyd's Rep 53. In that case there was a lacuna in an agreement concerning the supply of chickens for processing. Danckwerts LJ, at p60, noted that the word "differences" in that arbitration clause was particularly apt for a case where the parties had not agreed. In other words, submitted Mr McDonnell, Danckwerts LJ considered that the arbitrator could fill a gap in the contractual provisions, even though the court could not. In J my view, the observation of Danckwerts LJ in that case must be read in the context of that particular agreement and the nature of the issue which had arisen. In the present case there is no question of a lacuna. I do not think the arbitration provision in this agreement can be taken to have been intended to affect the substance of the parties' rights and obligations under the scheme provisions to which it relates. Clause 17 provided a method for resolving disputes or differences about those rights and obligations.

Mr McDonnell sought to place reliance on the fact that the arbitration provision does not apply to every dispute, or even to every dispute in which the reasonableness of the estates governors' decision K was in issue. I consider that this fact is neutral. Consideration of which matters fall within the arbitration provision, such as clause 3(a), and those which do not, such as clause 7 relating to causing an obstruction by parking a car on the road, or clause 8 relating to the estates governors giving notice of entry to inspect and carry out works, does not point to the arbitration clause being given a wider or a narrower interpretation.

Likewise neutral is the fact that the arbitrator will normally be a surveyor. Even with what I may call the narrower interpretation of clause 17 there is need, and ample scope, for the expertise of a surveyor. Whether the estates governors have acted in a way no reasonable body of estates managers could act is an issue calling for a L surveyor's professional expertise just as much as would be the case if the issue were whether the arbitrator himself would, in all the circumstances, consider it desirable for the relevant permission to be granted.

Nor am I impressed by the argument that fairness requires or suggests that the arbitrator should have the wider role. True it is that the scheme makes no provision for the estates governors to follow any particular procedure when considering applications for consent. There is no express requirement, for instance, that any representations made to the estates governors by the advisory committee should be shown to the applicant for his comment. However, I cannot see that that M provides any support for a construction of the scheme which would give the arbitrator, under this unexceptionally worded arbitration clause, a wider role than the court. As to the procedure to be followed by the estates governors, that must be a matter for them. I have heard no argument on what they ought to do, nor have I seen any evidence on what in practice they do. So I note only that it cannot be in the interests of the estate as a whole that the estates governors should adopt any procedure which is open to criticism as being unfair or unreasonable.

I recognise that the consequence of rejecting Mr McDonnell's arguments is to draw a distinction between instances where the

*Editor's note: Also reported at [1986] EGLR 39.

arbitrator's conclusion replaces the views of the estates governors on a particular point and instances where his decision does not. An example of the former category will be a decision on whether a proposed structure would be visible from outside the property boundary. On that the arbitrator would reach his own conclusion. In the latter category are cases where the issue is whether the estates governors' view was reasonable or whether they have acted reasonably. In these cases it is only if the estates governors' view is unreasonable or if they acted unreasonably that the arbitrator's intervention would avail the dissatisfied householder. This distinction, however, follows from an essential difference between the two categories. In the first category the relevant issue is the existence or not of a stated fact: whether the proposed house extension would be visible outside the boundaries. The second category stands differently. In the second category are cases where a decision, on a point on which minds may differ, has been entrusted to the managers so long as they act reasonably. It is only if they acted unreasonably that they have erred and thus it is only in that circumstance that the independent arbitrator can step in and correct their error.

Arbitrator's approach

Mr McDonnell's alternative argument was that, even so, this award should not be disturbed because, fairly read, the award shows that the arbitrator did approach the issue before him correctly. He did apply the test enunciated in the landlord and tenant cases which, for convenience, I shall call "the reasonable estate manager approach".

The arbitrator made his award on the basis of written representations and a site visit. In their written representations the parties' solicitors disagreed on how the arbitrator should approach his task. In para 9.1 of their initial submissions the estates governors contended:

> By clause 3 of the Scheme, it is the managers' consent that is required. Similarly, by clause 16(a) the powers conferred by the Scheme are conferred on the Managers. The only question to be decided on arbitration is therefore whether their consent was reasonably withheld. In this context, it is important to remember that people acting reasonably may still disagree. The question is therefore not whether the arbitrator would have come to the same view as the Estates Governors, but whether he is satisfied that they acted unreasonably in reaching their decision. The test that has been applied in other cases where bodies have been exercising powers derived from statute is whether it is correct to say that no reasonable person, properly instructed in the relevant law, could have come to the decision under review. The burden is on the Claimants to show that the answer to that question is "yes".

In their response the solicitors acting for the claimants, Mr and Mrs Williams, submitted:

> Your functions as arbitrator are first to test the reasonableness of the decision of the Estates Governors, and secondly to consider the rival merits of the propositions put before you and arrive at a balanced judgment . . . The issue in this case is not, as suggested by the solicitors acting for the Estates Governors, whether you are satisfied that the Estates Governors have acted unreasonably in reaching their decision. The solicitors state that that test has been applied "in other cases where bodies have been exercising powers derived from statute". That is correct but the facts of those cases were different . . .

The solicitors then submitted that the *Wednesbury* judicial review test alluded to, or hinted at, by the estates governors' solicitors was not in point because here, unlike in those cases, the scheme provided for a right of appeal.

In para 15.1 of their further submissions, the estates governors' solicitors returned to the point. Omitting immaterial words they said:

> Paragraph 5 of the claimants' written submissions suggests that the arbitrator should put himself in the position of the respondents [that is, the estates governors] and decide whether consent should have been granted. That is not the correct way for the arbitration to be conducted. There is only one question for the arbitrator and that is . . . whether the respondents acted unreasonably when considering the claimants' application for consent. The question is therefore not whether the arbitrator is of the view that the claimants' submissions are reasonable, nor whether they are correct, but whether the views of the respondents were unreasonable, ie whether they were "so wrong that no reasonable person could sensibly take that view" . . . Accordingly, while it is of course hoped that the arbitrator will take the view that the respondents were not only "not unreasonable" but also right, it is necessary only for the arbitrator to make a finding on whether the respondent was unreasonable or not.

In his award the arbitrator referred to the rival submissions. Then, in para 36, he expressed his conclusion succinctly:

> On the matter of whether the refusal is unreasonable I have considered the comments made by the parties. The President of the Royal Institution of Chartered Surveyors appoints arbitrators in property disputes having regard to their background experience of property matters. It is my conclusion that the interpretation put upon the matter by the [estates governors] in paragraph 9.1 of their submissions and 15.1 of the counter-submissions is too exacting and from my experience and the submissions I have considered I conclude that in this case the decision was an unreasonable one.

Thus the arbitrator rejected the estates governors' contention as too exacting, without going on to state expressly what was the approach he himself was adopting. In my view, the fair and natural reading of para 36 is that the arbitrator, rejecting the estates governors' approach, by implication must have accepted the claimants' rival submission. He is to be taken to have done that, because he does not state or suggest otherwise. Thus his conclusion, that the estates governors' decision was unreasonable, cannot be read as a conclusion by him that no reasonable estate manager could, in all the circumstances, have refused the claimants' application.

Mr McDonnell submitted that what the arbitrator rejected was the estates governors' invocation of the *Wednesbury* test and that the test requires more than mere unreasonableness before the court will intervene. I need not examine whether that analysis of the *Wednesbury* test is correct. What matters in this case is what was submitted to the arbitrator regarding the alternative approaches. Here, in the rival submissions, the reference to the circumstances in which the court exercises its supervisory powers over bodies exercising discretions conferred by statute was unhelpful. It was unhelpful, because it was irrelevant. In my view, however, this does not assist Mr McDonnell because, as adumbrated in the rival submissions, the test said to be derived from the judicial review line of cases did not differ materially from the reasonable estate manager approach. In para 15.1 the estates governors did interpret "unreasonable" as meaning "so wrong that no reasonable person could sensibly take that view", but that is to be read with para 9.1 of the initial submission. In his conclusion in para 36 of the award the arbitrator put those two submissions together and rejected this interpretation of his role as "too exacting", by which I understand him to have meant that it placed upon Mr and Mrs Williams too heavy a burden.

Mr McDonnell finally submitted that even if the approach adopted by the arbitrator when making his award is not clear, still the matter should not be remitted to him for further consideration, because his conclusions on the underlying facts, about the building line and so forth, would mean that, inevitably, whatever approach is adopted, the claimants would succeed. I do not agree. It is impossible to be sure what would be the arbitrator's conclusion if he had correctly directed himself on how he should approach the issue before him. Accordingly, the matter must be remitted to him for further consideration. In so directing I ought, in fairness to him, to note that he, a surveyor, was confronted with a point of law on which he had to make a decision on the basis only of written submissions. Those written submissions did not develop the rival arguments as fully as they have been developed before me, nor did they refer the arbitrator to any of the most relevant authorities, namely those relating to the landlord and tenant cases.

A

Court of Appeal

October 20 1993

(Before Lord Justice RUSSELL, Lord Justice STAUGHTON and Lord Justice STEYN)

R v WORTHING BOROUGH COUNCIL, EX PARTE BRUCE

Estates Gazette June 18 1994

[1994] 24 EG 149

B ***Landlord and tenant — Secure tenancy — Order by consent dismissing tenant's claim to secure tenancy and right to buy — Whether tenant entitled to judicial review of order — Whether order contained admission by tenant that he was not a secure tenant***

In 1983 the appellant was granted a tenancy of a flat by the respondent council. At all material times he also owned a bungalow in Kent. In August 1988 the council served a notice to quit, contending that the appellant was not a secure tenant. The appellant brought proceedings in which he claimed that the council wrongly denied that he was a secure tenant and wrongfully refused to enable him to C exercise the right to buy under the Housing Act 1985. The proceedings were compromised by a consent order in which the appellant's claim was dismissed. The appellant thereafter sought judicial review of the order, contending that it should be set aside because his agreement to it was reached in circumstances where he was subjected to undue pressure by counsel and solicitor. The appellant appealed Otton J's rejection of his application, contending that it was not possible to contract out of a secure tenancy and that the judge making the consent had failed to obtained the appellant's express admission that he had no protection.

***Held*: The appeal was dismissed. It was not necessary for there to be express admission that the appellant did not have a secure D tenancy. In any event the consent order, if it was not an express admission that the appellant did not enjoy protection, was undoubtedly an implied admission.**

The following cases are referred to in this report.

Barton v *Fincham* [1921] 2 KB 291

Middleton v *Baldock* [1950] 1 KB 657; [1950] 1 All ER 708, CA

Morris v *Barnet London Borough Council* Unreported May 26 1989

R v *Bloomsbury and Marylebone County Court, ex parte Blackburne* [1985] 2 EGLR 157; (1985) 275 EG 1273, CA

Wandsworth London Borough Council v *Fadayomi* [1987] 1 WLR 1473; E [1987] 3 All ER 474; (1987) 86 LGR 176, CA

This was an appeal from the decision of Otton J who had dismissed the application by the appellant, Vaughan Morris Synott Bruce, seeking judicial review of a consent order made by Judge Hammerton in Worthing County Court. The respondents were Worthing Borough Council.

Robin Allen (instructed by Tarleton Davies, of Ashford) appeared for the appellant; Malcom Gibney (instructed by Thomas Eggar Verrall Bowles, of Worthing, agents for the solicitor to Worthing Borough Council) represented the respondents.

F

Giving judgment, **RUSSELL LJ** said: This is an appeal from an order of Otton J who, on July 5 1991, refused the appellant's application for judicial review of an order made by consent in Worthing County Court on April 4 1990, before Judge Hammerton. The case has generated a vast amount of paper. Solicitors delivered no less than nine arch files before the hearing in this court. Fortunately it has not been necessary to refer to more than a very few documents for the factual issues which it is necessary to examine, and the law falls within a comparatively small compass. We have to say we strongly deprecate the unnecessary compilation of files in a case of this kind, particularly at public expense. We hope that when the costs are taxed our views will be borne in mind in respect not only of photocopying charges but also professional fees. That said, I proceed to consider the G questions for our determination.

The appellant is a gentleman who is affected by polio. To his credit, he has made great efforts, despite his disabilities, to earn his own living. To this end, in March 1983 he obtained from the local authority, namely Worthing Borough Council, the tenancy of a flat at 105 Canterbury Road, Worthing. Originally the tenancy was for a period of 11 months, but that period was extended at the expiration of the term. At all material times the appellant also owned a bungalow in Kent, which had been adapted to his disability. The litigation which subsequently ensued between the appellant and the local authority related to the status of the appellant in the flat in Worthing and whether he was a secure tenant there enjoying the right to buy the H freehold pursuant to the provisions of the Housing Act 1985 to which I now turn. Sections 79, 80 and 81 define a secure tenancy. For the purposes of this appeal it is sufficient to observe that such a tenancy is created when a local authority let a dwelling-house to a tenant who, in accordance with section 81:

> . . . occupies the dwelling-house as his only or principal home.

Section 82, in general terms, provides that such a tenancy cannot be brought to an end by the landlord except by obtaining an order of the court for possession, after service of an appropriate notice required by section 83. The provisions of section 84 are important. They read, J so far as material to this appeal, as follows:

> (1) The court shall not make an order for the possession of a dwelling-house let under a secure tenancy except on one or more of the grounds set out in Schedule 2.
>
> (2) The court shall not make an order for possession —
>
> (*a*) on the grounds set out in Part I of that Schedule (grounds 1 to 8), unless it considers it reasonable to make the order.

One of the grounds under Schedule 2 is non-payment of rent lawfully due. The tenant's right to buy is contained in Part V of the Act. No more need be said about it, except it is the essence of Part V K that that right can be exercised only by a secure tenant, as earlier defined in the statute, and in particular by one who occupies the property as his sole or principal home.

There is no doubt that during 1985 and 1986 the local authority in this case entertained the appellant's right to buy. But by July 1988, the respondents took the view that the appellant had not occupied the flat as his sole residence. On August 15 1988, the authorities served a notice to quit, it being their case that no protection under the Act was available to this appellant. The notice to quit was followed by the proceedings under review in Worthing County Court. They were launched, however, by the appellant, whose particulars of claim alleged as follows: Para 2: L

> The plaintiff entered into possession of the said property in March 1983 pursuant to a Tenancy Agreement with the defendant, and has occupied the said property as his principal residence at all material times. . . .
>
> 3. In the premises the plaintiff is a secure tenant within the meaning of the Housing Act 1985.
>
> 4. Further by reason of the matters aforesaid, the plaintiff has the "right to buy" the said property pursuant to section 118 of the Housing Act 1985. The said right is vested in the plaintiff himself.
>
> 5. By a notice (Form RTB2) dated 13th September 1985 the defendant admitted the plaintiff's right to buy the said property. Thereafter the plaintiff elected to defer purchase until 16th August 1988 by completing form TRB13 M and making the payment of £150.
>
> 6. Thereafter the defendant has wrongly denied that the plaintiff is a secure tenant as aforesaid and wrongfully refused to take the requisite statutory steps to enable the plaintiff to exercise his right to buy the said property, is therefore in breach of statutory duty under section 138 of the Housing Act 1985.

Paras 7 and 8 sought damages and a refund of rent paid, together with costs. There was a claim for a declaration that the plaintiff was a secure tenant within the Housing Act 1985.

By their defence and counterclaim the respondents admitted para 5 of the particulars of claim, but contended that their employees had completed form RTB2 in the mistaken belief that the plaintiff then

A occupied the property as his only or principal home. The pleading went on to allege that the plaintiff had not occupied the property as his only or principal home and had not been a secure tenant thereafter. The respondent sought possession of the flat, together with mesne profits.

There were further pleadings to which it is unnecessary to refer. Thus the battle lines were drawn for the hearing in Worthing County Court. The appellant contended that he had always been a secure tenant because he had, and did, occupy the flat in Worthing as his sole or principal home; the respondents contending that the appellant was not entitled to the protection of the 1985 Act, or any rights under Part
B V, because he was not a secure tenant and never had been as defined in the Act.

The case came on for hearing, both sides being represented by counsel. The appellant gave evidence, as did his wife and two other witnesses. Their testimony occupied more than two days. For my part, I do not think it necessary to rehearse the content of the evidence. To some extent, Otton J did so in his judgment and it is reported in (1991) 24 HLR 261. On the third day, counsel indicated to the judge that the parties had come to terms. They had drafted and produced to the court a consent order. We have no note of precisely what was said, but I am prepared to assume that the judge, without further inquiry, initialled
C the consent order and the proceedings thereupon came to an end. The terms of the consent order are as follows:

> By consent it is ordered:
> 1. The plaintiff makes no claim against the defendant in respect of rebates for any rent or rates paid or payable by him in respect of 105, Canterbury Road, Worthing.
> 2. The defendant makes no claim against the plaintiff for arrears of rent or mesne profits . . .
> 3. The plaintiff's claim be dismissed with no order as to costs.
> 4. On the defendant's counterclaim for possession there be an order for possession . . . in 28 days from today with no order as to costs.

D Thereafter, the appellant clearly had second thoughts. He obtained legal aid to enable him to seek judicial review of the order. He contended in those proceedings that the order ought to be set aside, because his agreement to it was reached in circumstances where he was subjected to undue pressure by counsel and solicitor. This ground for relief Otton J rejected, saying [at p278]:

> I am satisfied that he was a full and willing party to the agreement in the way that it was negotiated and finally drawn in the consent order.

Sensibly, in my view, that complaint has not been pursued before this court. I deal now with what is at the heart of this appeal.

E Mr Robin Allen, who did not appear either in the county court or before Otton J and who, it is right to say, disclaims any responsibility for the preparation of the files in this case, correctly reminds us of the fundamental principle that parties cannot contract out of the Rent Acts or, by parity of reasoning, out of the Housing Act 1985: see, for example, *Barton* v *Fincham* [1921] 2 KB 291. A party who is entitled to protection cannot lawfully relinquish that protection for a consideration or otherwise.

Mr Allen contends that that is what the appellant did in the instant case. Developing the submission, Mr Allen contends that only an express admission to the effect that no protection under the Housing Act 1985 exists enables the court to make a possession order in a case
F such as the instant one where the tenant is contending that he is a secure tenant with a statutory right to buy. Counsel relies in particular on the mandatory terms of section 84(1):

> The court shall not make an order for possession of a dwelling-house let under a secure tenancy.

The judge should have made inquiries of the appellant and obtained from him an express admission that he had no protection before making the order. He did not do so and counsel submits that failure therefore vitiated the order.

For my part, I am satisfied that the submission that there has to be an express admission is contrary to authority and principle: see the judgment of Glidewell LJ in *Morris* v *Barnet London Borough* G
Council, decided in this court on May 26 1989 (unreported), where Glidewell LJ said:

> It is also clear, as Hutchinson J concluded, that such an admission need not be express and can be implied.

Alternatively, Mr Allen submits there was nothing in the circumstances I have described which could be construed as an implied admission by the appellant that he did not have the protection of the Act. Mr Allen contends that there may have been, and on his client's instructions there were, many reasons why he settled his case, not least among them being the escalation of costs. In my judgment, H
this is the wrong approach. The county court judge was confronted with a consent order drafted by counsel. Its terms were unequivocal. In particular, I repeat para 3 of the order:

> The plaintiff's claim be dismissed with no order as to costs.

What was the plaintiff's claim? It was, *inter alia*, a claim for a declaration that he was and remained a secure tenant within the meaning of the Housing Act 1985. In my opinion, para 3 of the consent order, if it was not an express admission that the appellant did not enjoy protection — and I am inclined to think that it was — was undoubtedly an implied admission that did not call for any further J
inquiry from the judge. This being my interpretation of what transpired at Worthing County Court, it suffices for the purposes of resolving this appeal.

Otton J, however, took a different route. He reviewed the evidence up to the stage when the settlement was reached. He said [at p277]:

> In reaching my conclusions I am satisfied that this is not a case where there was no investigation, or insufficient investigation, on the security point. There was a sufficient investigation to entitle the judge to conclude that he had jurisdiction to make the order on the basis that the applicant was not a secure tenant. I decline to draw the inference that the learned judge overlooked section 84(1) of the Housing Act 1985. I note, without paying undue weight to K
> it, that he has been on the bench for 18 years and is clearly a most experienced judge.

Otton J then went on to make what we are told was a mistaken reference to the judge's editorship of a legal volume of *Halsbury's Laws of England*. He continued:

> I am satisfied on the balance of probabilities that he,

— meaning the judge —

> would have been fully aware of the implications of this case and, in particular, the provision of section 84(1). It would have been desirable if he had in fact L
> passed some comment or given some express approval with a direct or passing reference to section 84(1) but I am satisfied that the fact that he did not make any such comment does not invalidate his decision.
>
> It is clear from the evidence that he heard, the documents which were presented to him, the discussions which took place in his chambers and, in particular, the terms of the consent order, that it was open to the judge to draw the inference that an implied admission was being made by counsel on Mr Bruce's behalf that: (1) he never was a secure tenant; (2) or if he was at the time that he made his application in September 1985, he ceased to be a secure tenant between that time and the time when the point was taken in the pleadings of the defence and reamended defence; (3) or that the plaintiff could not discharge the burden of proof that he was a secure tenant, as he was M
> obliged to do in order to establish his claim to his right to buy; (4) or that Mr Bruce was not entitled to the protection of the Housing Act and; (5) that he was not entitled to exercise his right to purchase as he had claimed in his particulars of claim.
>
> It is implicit in what occurred, in my judgment, that if he had not been satisfied that he could draw an inference which was adverse to Mr Bruce he would have without hesitation have said so and would not have approved the consent order which had been placed before him.
>
> I am satisfied that this agreement was arrived at not on a compromise without prejudice basis, as was suggested in the *Hussain* case, but at arm's length and after full consultation and advice from his counsel. I am satisfied that the evidence and the documents unequivocally pointed to an admission along the lines that I have set out.

A Furthermore, I am satisfied that if there had been any doubt about it, Mr Bruce would have made it clear both to his own counsel and to the other side when he was present at the discussion which took place in his presence. I draw no inference either in favour of the council or adverse to Mr Bruce from the terms of the order itself.

For my part, I am satisfied that Otton J was entitled to make all these findings, although I part company with him in observing, as he did, that he drew no inference either in favour of the council or adverse to Mr Bruce from the terms of the order itself. Contrary to Otton J's observation, I see the terms of the order itself, fully consented to by the appellant, as crucial.

B Only one other authority need be mentioned, for it was relied on heavily by Mr Allen. *R* v *Bloomsbury and Marylebone County Court, ex parte Blackburne* [1985] 2 EGLR 157 was an appeal from Glidewell J (as he then was). In that case a tenant was entitled to protection under the Rent Act 1987, which by section 98(1) also contained the words:

The court shall not make an order for possession.

Glidewell J quashed a consent order for possession and his order was upheld in the Court of Appeal. In my judgment, the facts in *Blackburne*'s case are plainly distinguishable from the instant C case. In particular, the court was entitled to draw the inference, as it did, that settlement in *Blackburne*'s case was achieved by the payment of money to the tenant, effectively to relinquish his tenancy. There was no admission of a lack of statutory protection express or to be implied from the terms of the consent order, nor from anything that Mr Blackburne had told the court. In the context of this appeal, *Blackburne*'s case does not, in my view, assist the appellant.

Although, therefore, by a somewhat different route from that adopted by Otton J, I, too, would uphold the order made in the county court and would dismiss this appeal.

D Agreeing, **Staughton LJ** said: In our adversarial system a judge who is asked to make a consent order should do so, provided that the parties are of full age and understanding and that the order is not illegal, immoral or so equivocal as likely to give rise to further dispute. But sometimes the inquisitorial system takes over. In the present case that happened, because the particular wording of the statute requires it.

So far as we have heard, this doctrine originated with section 5(1) of the Increase of Rent and Mortgage Interest (Restriction) Act 1920. That said:

No order or judgment for the recovery of possession of any dwelling-house to which this Act applies, or for the ejectment of a tenant therefrom, shall be E made or given unless . . .

It was held in *Barton* v *Fincham* [1921] 2 KB 291 that an order could not be made unless one of the requirements of the section was proved. That was a case where the landlord and tenant had made a written agreement to end the tenancy, but before the action was brought the tenant refused to keep his part of the bargain. The issue was whether the parties could contract out of the Act, rather than the power of the court to make a consent order.

Middleton v *Baldock* [1950] 1 KB 657 was concerned with section 3(1) of the Rent and Mortgage Interest Restrictions (Amendment) Act F 1933, which was in the same terms so far as relevant. There it was held that the court could make an order for possession only on proof of one of the grounds stated in the section, unless the tenant in court admitted facts which would give rise to one of those grounds. Jenkins LJ at p669 said:

The court cannot go further than that and exercise a general jurisdiction to make a consent order without inquiry or investigation simply because the tenant appears in court and says: "I consent to an order", or says in the witness box that he does not contest the landlord's right.

R v *Bloomsbury and Marylebone County Court, ex parte Blackburne* [1985] 2 EGLR 157 turned on section 98(1) of the Rent Act 1987:

A court shall not make an order for possession of a dwelling-house which is G for the time being let on a protected tenancy or subject to a statutory tenancy unless . . .

In that case there was a consent order, but the Court of Appeal held it to be invalid. Sir John Donaldson MR said at p158:

The question which then arises is whether the learned county court judge could have been satisfied that the requirements of section 98(1) were met. There was no express admission to this effect. The appellants, however, submit that such an admission was implicit in the consent to judgment. Mr Bartlett, on their behalf, submits that this consent is explicable only on the basis that Mr Blackburne recognised that (a) he was unable to claim the protection of the 1977 Act because he had never been or had ceased to be a H statutory tenant, or (b) although he was a statutory tenant, he fell within Case 1 of Schedule 15 and it was reasonable to make an order for possession. I disagree. There is a third possibility, namely (c) Mr Blackburne, although a statutory tenant whose arrears of rent, if any, were extinguished by his counterclaim, nevertheless thought that the sum offered by the landlords was so attractive that he should surrender his rights or at least thought that any doubts which he had upon this score should be resolved by a compromise involving his acceptance of a large lump sum. Inference (c) would not have entitled the learned judge to make the order.

It is to be noticed that the consent order in that case provided for a substantial payment to the tenant, who was claiming damages for failure to maintain the premises in good repair. This may have been J thought difficult to reconcile with an admission by the tenant through the consent order that he was not a statutory tenant. If that be right, one can understand why there was held to be no implied admission that the landlord was entitled to possession under the section. I do not regard that case as having decided that one can never imply such an admission from the terms of an order construed in the light of the issues in the case.

However, in *Wandsworth London Borough Council* v *Fadayomi* [1987] 1 WLR 1473 at p1479 Parker LJ said:

. . . in order to justify an order being made without an investigation of the K facts, what is required is not a mere consent to the making of the order but an express admission of facts which, if they had been proved, would give the court jurisdiction to make the order.

That was a case under section 84 of the Housing Act 1985, the section with which we are presently concerned.

A different view was expressed by Glidewell and Parker LJJ in *Morris* v *Barnet London Borough Council* May 26 1989 (Lexis). Glidewell LJ said that an admission need not be expressed and can be implied, but he added at p2:

If a court does not have before it any material at all and merely has counsel or L solicitors appearing and saying that they have agreed an order, then it may be difficult to imply an admission.

It is thus apparent that the views of judges have fluctuated in recent times. I take the law now to be that there can, in a draft consent order, be found an implied admission of the facts which justify the judge in making it. Whether there is depends on the terms of the order construed in the light of the surrounding circumstances, including the issues in the case. For my part, I would not be at all reluctant to imply such an admission. I would certainly do so in this case. If a tenant has been persuaded by duress to agree to an order, he has other remedies. If, on the other hand, he has voluntarily agreed M to it, perhaps with legal advice, consumer protection does not require us to ignore an implied admission if common sense suggests that there is one.

I mentioned at the beginning of this judgment that in cases such as this, the inquisitorial system takes over. That appears to me to result from the language of the statute in some cases, but by no means in all. Section 4 of the Statute of Frauds 1677, for example, provided:

. . . noe action shall be brought . . . whereby to charge the defendant upon any special, promise to answere for the debt default or miscarriages of another person . . . unlesse the agreement upon which such action shall be brought or some memorandum or note thereof shall be in writeing, and signed by the partie to be charged therewith or some other person thereunto by him lawfully authorized.

It has never been held that a court may not make a consent order where the requirement of that section is not shown to be satisfied. Indeed, the court is not required to give effect to the section at all unless it is pleaded. Why then are there some statutes which exercise a more powerful restraint upon judges? Bankes LJ noticed the problem in *Barton*'s case. He said at p296:

In my opinion s 5 is an instance of a case where the Legislature has in clear and unmistakable language restricted the jurisdiction of the Court, and where no agreement between parties can give the Court a jurisdiction which the Legislature has said it is not to exercise. This is by no means the only statute in which the Legislature has secured the object it had in view by limiting the exercise by the Courts of their full jurisdiction. It is true that in some cases the strictness of this limitation has been relaxed by the pleading rule which, while rendering the pleading of a statutory defence optional, declares that unless pleaded it cannot be relied on. No point of this kind arises in the present case, as I assume that all necessary conditions had been fulfilled entitling the appellant to raise the statutory defence in the county court.

So it would seem to depend upon the language of the statute.

I would dismiss this appeal. I agree with what Russell LJ has said about there being far too many documents before the court. For my part, I have found it necessary to look only at the pleadings in the county court, the consent order, the judgment, the notice of appeal and the outline arguments of counsel. We were told that the costs in the county court, as they might eventually turn out to be, were estimated at various times at £10,000, £20,000 or £50,000. That is all too likely to be true. It is appalling. What is the point of having a county court if that is what it costs to go there? I do not know why so much was being spent or expected to be spent, but I am quite sure unnecessary proliferation of documents is in some cases a major cause of undue expense.

Steyn LJ agreed and did not add anything.

Appeal dismissed.

Chancery Division

February 7 1994

(Before Mr Timothy LLOYD QC, sitting as a deputy judge of the division)

GRAY AND OTHERS v STANDARD HOME & COUNTIES PROPERTIES LTD AND ANOTHER

Estates Gazette June 18 1994
[1994] 24 EG 152

Landlord and Tenant Act 1987 — Acquisition order — Whether appointment of a manager would be an adequate remedy — Terms of acquisition order

The applicants for an acquisition order under Part III of the Landlord and Tenant Act 1987 were the tenants holding long leases of three of the five flats in a converted Georgian listed building. In January 1980 the defendant sold the freehold to V Ltd; since that transfer there were further transfers, the last being to Linway Ltd. None of the transfers was registered and, accordingly, the defendant remained the registered proprietor of the freehold. Since 1980 there had been no active management on behalf of the freeholder. In 1984 a receiver was appointed by the court. Linway Ltd applied for an adjournment so that it could become a party and contest the proceedings. The plaintiffs served a notice under section 27 of the 1987 Act in January 1993 stating, *inter alia*, the allegations of breach of covenants and that the appointment of a manager would not be an adequate remedy.

***Held*: An acquisition order was granted. The application by Linway Ltd for an adjournment to be joined as third defendant was refused. The landlord was in substantial breach of various obligations relating to the repair, maintenance, insurance or management of the premises, which had been outstanding for a long time. Those circumstances were likely to continue. The fact that the receiver, although not appointed under the 1987 Act, had been able to keep the problem in check, but not to remedy the situation altogether, demonstrated that the appointment of a manager would not solve the problem. It was directed that, if and so far as the consideration fixed by the rent assessment committee or by agreement is more than adequate to cover and secure the discharge of all charges existing and registered, the amount of the plaintiffs' costs of the application is to be deducted from that consideration before payment to the persons entitled to it.**

No cases were referred to in this report.

This was an application by originating summons under Part III of the Landlord and Tenant Act 1987 by Stuart Gray, Shirley Gray, Michael Colwell and Angela Gill for an acquisition order in respect of a building of which the first defendant was the registered proprietor.

Simon Redmayne (instructed by Parlett Kent & Co) appeared for the plaintiffs; Jonathan Small (instructed by Guy Clapham & Co) represented the proposed third defendant, Linway Ltd; the first and second defendants did not appear and were not represented.

Giving judgment, **Mr Timothy Lloyd QC** said: This is an application by originating summons under Part III of the Landlord and Tenant Act 1987 on the part of the tenants of three out of the five flats in a converted Georgian listed building called Orchard House, Thames Street, Sunbury, Surrey. The defendants are the registered proprietors of the freehold, Standard Home & Counties Properties Ltd, and the nominated person to take the transfer of the title, if so ordered, Thames Street Orchard House Management Ltd. The freehold is vested in the first defendant and has been since it was registered as proprietor in August 1976. It performed its duties and functions as freeholder for some time, but in January 1980 it sold the freehold to a company called Vervale Ltd for £8,150 and since then has taken no part in the management of the property. It would seem at first sight strange that it is still the registered proprietor. What has happened is that the transfer to Vervale Ltd was never registered and that in March 1985 a further transfer was executed by Vervale in favour of a company called Tuppercrest Ltd for the same price and then in July 1982 a yet further transfer was executed by Tuppercrest to a company called Linway Ltd. None of those transfers has been registered. During that time since 1980, no one has taken any active part in the management of the property on behalf of the freeholder.

In 1984, in a situation in which there were serious breaches of the landlord's repairing covenants, the then long leaseholders applied to the court for the appointment of a receiver and a Mr Ronald Garner was appointed in this division. He is still acting as the receiver of the property and is coping as best he can with the repairing obligations.

I should mention the leases. Four leases on 99-year terms were granted between 1973 and 1977 of flats 1, 2, 4 and 5. They were on standard terms and I have seen as a specimen the lease of flat 1. They require the lessee to deal with the internal repairs and to contribute to the landlord's maintenance costs, but in clause 5 they impose covenants on the landlord, which include a covenant to insure and a covenant that, subject to contribution and payment, the lessors will throughout the term and from time to time wherever necessary maintain, repair, redecorate and renew first the main structure of the block, in particular the external walls, foundations, roof, roof tank, roof gutters and rainwater pipes and the external parts of the block, but not to include glass and sash-cords of windows; second, sewers, water courses, pipes, drains etc; third, the main entrances passages, landings

A and staircases used in common; and then as a further covenant that the lessor will in 1977 and thereafter every fourth year decorate the exterior of the block in a good and workmanlike manner; and there is a separate covenant that the lessor will keep the garden properly cultivated and the grass mowed, and so on.

As I mentioned, there are five flats. In 1980, flat 3 was, I am told, subject to a Rent Act protected tenancy, which came to an end in about 1984 and that flat has since then been empty. It appears from an affidavit sworn by a Mr Frederick Madden on behalf of Linway Ltd that in 1985 Tuppercrest Ltd granted him a long lease of that flat for a premium of £8,500. I have not seen that lease and it has not been B registered. He has never taken up occupation of the flat. The flat is vacant, and, as is shown by photographs which are in evidence before me, it is in a particularly terrible state with a serious outbreak of dry rot.

In those circumstances, such repair and maintenance as has been done to the property as a whole since 1980 has been done either by the receiver, financed by the long leaseholders, or by the local district council, which I am told entered flat 3 under statutory powers and carried out emergency repairs and I am also told that they have a statutory charge which is protected by the registration of a local land charge for the recovery of that expense.

C The plaintiffs, as I say, are the leaseholders of three of the four occupied flats. The evidence shows that the fourth, who was the first person to take the initiative in 1984 for getting the appointment of a receiver, is still entirely happy with the proposition that the freehold should be acquired, but has had enough of litigation and for that reason only has not joined in the present exercise.

The first defendant (the registered freeholder) is not represented and, in the circumstances which I have mentioned, regards itself as having no commercial interest in the premises. Linway Ltd, a company which does regard itself as having a commercial interest in the premises, is not a party to these proceedings and indeed made an D application this morning to me, which I rejected, for the case to be adjourned so that it could become a party and contest the proceedings. Thus, there is no representation hostile to the plaintiffs. The application is made under Part III of the Act, and I will now proceed by steps through the Act to see whether I can be satisfied that the order should be made.

The first section in Part III is section 25, which provides that Part III

> has effect for the purpose of enabling qualifying tenants of flats contained in any premises to which this Part applies to make an application to the court for an order providing for a person nominated by them to acquire their landlord's E interest in the premises without his consent.

By subsection (2):

> Subject to subsections (4) and (5),

— which do not apply —

> this Part applies to premises if —
> (a) they consist of the whole or part of a building; and
> (b) they contain two or more flats held by tenants of the landlord who are qualifying tenants; and
> (c)

F — a requirement set out in subsection (3) is satisfied. These premises consist of the whole of a building and because section 26 is satisfied I can accept that they do contain two or more flats held by tenants of the landlord, who are qualifying tenants.

The appropriate requirement in subsection (3) is that, where the premises contain more than three but less than 10 flats, all but one of the flats are let by the landlord on long leases. So far as that is concerned, the evidence appears to indicate that four out of the five are so let. If Mr Madden's evidence is correct and, subject to any questions that arise from non-registration of his lease, it may be the case that all the flats are so let. But at any rate all but one of them are so let, so that that requirement is satisfied.

By section 26 a person is a qualifying tenant of a flat — subject to G various provisos that do not apply — if he is the tenant of a flat under a long lease other than one constituting a tenancy to which Part II of the Landlord and Tenant Act 1954 applies. That is the case.

Then by section 27 provision is made whereby, before the tenants can apply for an acquisition order, a notice must be served on the landlord by qualifying tenants of the flats contained in the premises, who at the date when it is served constitute the requisite majority of such tenants. The requisite majority is more than 50% of the tenants by virtue of subsection (4) and the notice in the present case was served by the plaintiffs, who hold three votes out of either four, if one is to disregard flat 3, or five if one is not to disregard flat 3. Either way H the requisite majority is satisfied.

The notice has to specify the names of the qualifying tenants by whom it is served and other information relating to them. That is done by the actual notice, which was served on the first defendant on January 11 1993.

It must state the tenants' intention to apply for an acquisition order, but that they will not do so if the landlord complies with the requirement specified in pursuance of para (d). It must specify the grounds on which the court would be asked to make such an order, the matters that would be relied on by the tenants for the purpose of establishing those grounds; and it must, where those matters are capable of being remedied by the landlord, require the landlord within J such reasonable period as is specified to take such steps for the purpose of remedying them as are so specified.

A notice was served, as I say, in January 1993, addressed to the first defendant. I note that the first defendant was misdescribed as Standard Home & Counties Ltd, but its registered address was correctly given, and I proceed on the basis that that was an inadvertent error, which would have been obvious to the recipient and is therefore not a ground for holding that the notice is invalid. It states the names of the relevant qualifying tenants; it gives an address for service; it states their intention to apply for an acquisition order in respect of Orchard House registered at the Land Registry under a given title K number, but that they would not do so if the landlord complied with the requirement set out in 5. At para 4 it states the grounds on which the court would be asked to make the order and the matters that would be relied on for the purpose of establishing those grounds. The grounds are that the landlord is in breach of its obligations to the qualifying tenants under their leases relating to the repair, maintenance, insurance and management of Orchard House. The circumstances relating to the breach are that:

> you, the first defendant, and the company which purports to be your successor, Vervale Limited

L — I interpose that at the stage this notice was given it seems that nothing had been heard by the tenants of the two other companies that had successively taken the place of Vervale —

> have taken no part in the management, maintenance or repair or upkeep of the property since about 1981. Those circumstances are likely to continue and since January 30 1984 pursuant to the order of that date Mr Ronald Garner has acted as receiver and manager of Orchard House, the appointment of a manager under Part II of the 1987 Act would not be an adequate remedy for the matter set out above.

Those matters look ahead to section 29(2), which is one of the two alternative conditions which must be fulfilled if the court is to make an M acquisition order.

The matters then specified in the notice as being relied on in support of those grounds are the failure to keep the building insured and the failure to maintain, repair, redecorate and renew the main structure, walls, foundation, roof, gutters and pipes of Orchard House, the pipes and drains of Orchard House, the main entrance, passages, landings and staircases of Orchard House enjoyed in common by the tenants; and the failure to decorate the exterior of Orchard House in 1981, 1985 or 1989 and the failure to keep the garden of Orchard House properly cultivated and the grass mowed when necessary and so on.

A Then the notice sets out a long list of matters, which are required in order to remedy the various matters complained of which the landlord is required to take within three and a half months. These list obligations of maintenance and repair, maintenance of the rainwater goods, the brickwork, the external decorations and then it goes on; there are up to 40 items under those categories. Then it proceeds with certain internal matters in flats 2 and 5, which are said to be the direct consequence of breaches of the covenants in respect of the main structure and also a long list of wants of repair in respect of flat 3, extending from items 46 through to 68. Then there are some external matters, items 69 through to 79.

B That is the notice. Section 28 permits the making of the application which is now made by qualifying tenants of the flats contained in the premises, who at the date when it is made constitute the requisite majority of such tenants. That condition is satisfied.

> No application shall be made unless the time specified in the notice under section 27 has expired.

That was the case; this application was made in June 1993.

Then one comes to section 29, which sets out the conditions on which the court may make an acquisition order. The court must be satisfied that the premises were and are premises to which the Part C applies and that they have not ceased to be such premises. I am satisfied of those matters. It must be satisfied that the conditions specified in either subsection (2) or subsection (3) are fulfilled — and here it is subsection (2) that is relevant — and the court must consider it appropriate to make the order in the circumstances of the case. The conditions in subsection (2) are, first, a breach or breaches of the obligations owed by the landlord to the applicants under their leases relating to the repair, maintenance, insurance and management of the premises or any part of them; second, that the circumstances by virtue of which the landlord is in breach of any such obligation are likely to continue; and, third, that the appointment of a manager under D Part II to act in relation to those premises would not be an adequate remedy.

Looking at those matters, it is clear that the landlord is in substantial breach of various obligations owed by him to the applicants relating to the repair, maintenance, insurance or management of the premises; many of the matters specified in the section 27 notice are breaches of that nature. It seems to me that the internal repairs to flats 2 and 5 are capable of being relied on by the tenants in that respect only if and to the extent that they are directly consequential on the landlord's breaches of its obligations in respect of the structure. It appears to me that at least some of those matters are either, truly considered, external work or directly consequential on E external defaults. Others may not be.

So far as flat 3 is concerned, it may well be that some of these wants of repair, although badly needing to be done, are in fact not within the scope of the landlord's obligations to the other tenants and therefore I do not take those into account in the process of considering whether the statutory conditions have been fulfilled. However, I proceed on the basis that there are substantial breaches, that they have been outstanding for a very long time, and in the present situation it seems clear that those circumstances are likely to continue. I also consider that the appointment of a manager under Part II of the Act would not be an adequate remedy. Although the receiver was not F appointed under Part II of the Act, since he was appointed in 1984, the fact that he has been managing the property by virtue of the appointment as receiver and has been able to keep the problem in check but not to remedy the situation altogether demonstrates, as it seems to me, that the appointment of a manager would not solve the problem.

In those circumstances it seems to me that it is a case in which it is entirely appropriate to make the order sought and the conditions having been fulfilled I will make such an order.

Section 30 prescribes the content of acquisition orders so far as relevant. The order is to provide for the nominated person, that is to say the second defendant, to be entitled to acquire the landlord's interest in the premises specified in the order on such terms as may be G determined (a) by agreement between the landlord and the qualifying tenants in whose favour the order is made, or (b) in default of agreement by a rent assessment committee under section 31; and an acquisition order may be granted subject to such conditions as the court thinks fit and in particular its operation may be suspended on terms fixed by the court.

Section 31 deals with the determination of terms by rent assessment committees in the absence of agreement. Subsection (1) deals with the terms on which the landlord's interest is to be acquired, and those terms are to be such as appear to the committee to be fair and reasonable, subject to the question of consideration. This is dealt H with by subsection (2), under which:

> Where an application is made under this section for such a committee to determine the consideration payable for the acquisition of a landlord's interest in any premises, the committee shall do so by determining an amount equal to the amount which, in their opinion, that interest might be expected to realise if sold on the open market by a willing seller on the appropriate terms. . .

that is to say all those except as to price which have been determined by the rent assessment committee, and on the assumption that none of the tenants of the landlord of any premises comprised in those premises was buying or seeking to buy that interest.

I should note also section 32, under which: J

> Where the landlord's interest in any premises is acquired in pursuance of an acquisition order, the instrument by virtue of which it is so acquired shall

— subject to other provisions —

> operate to discharge the premises from any charge on that interest to secure the payment of money or the performance of any other obligation by the landlord or any other person

— that is to say it operates as if it were a sale by a prior mortgagee. In particular, that subsection does not apply:

> if . . . the court is satisfied, whether on the application for the order or an K application made by the person entitled to the benefit of the charge, that in the exceptional circumstances of the case it would be fair and reasonable that the landlord's interest should be . . . acquired [subject to the charge], and orders accordingly.

In my judgment, it is appropriate that an order should be made for the acquisition by the nominated person of the whole of the first defendant's interest in the property as registered under title no SY374031. The premises will be acquired on such terms as may be determined, in practice, by a rent assessment committee under section 31. There may be some theoretical possibility of an attempt to agree the terms, but in practice in the present situation I think that is L unlikely. Nevertheless, the attempt should be made.

Mr Simon Redmayne, for the plaintiffs, submits to me that the only terms that need to be settled by the rent assessment committee are the consideration and I think that is probably right, but I should not prejudge what the committee may have to decide. All I can do is order that the nominated person be entitled to acquire the landlord's interest in the premises to which I have referred on the terms to be determined in accordance with sections 30 and 31.

So far as mortgages are concerned, the property is subject to a number of cautions: one in favour of the receiver, the others in favour of one or more of the plaintiffs. As I have mentioned, I have been told M that the local authority have a charging order, which would be registered as a local land charge, not at the Land Registry and which entitles them to the recovery of expenditure incurred under statutory powers.

I am also invited to order — this would be as a condition of the acquisition order — that, if and to the extent that the purchase price as fixed by the rent assessment committee is sufficient to cover all prior charges, the amount of the plaintiffs' costs should also be deducted from that amount before anything is paid over to the landlord or into court. I think that is an appropriate order to make.

In those circumstances I will make an order as provided for by section 30(1). I will not make any special order as to the non-discharge

A of any particular charges under section 32(2), and I will direct that, if and in so far as the consideration fixed by the rent assessment committee or by agreement is more than adequate to cover and secure the discharge of all charges existing and registered (whether at the Land Registry or at the local Land Charges Register), as at today's date, the amount of the plaintiffs' costs of these proceedings is to be deducted from that consideration before payment to the persons entitled to it. I will make that order and I will order that the plaintiffs' costs be taxed on the standard basis if not agreed with a representative of the landlord.

B

Queen's Bench Division

January 28 1994

(Before Mr Justice GARLAND)

ALLIED DUNBAR ASSURANCE PLC v FOWLE AND OTHERS

C Estates Gazette June 25 1994
[1994] 25 EG 149

Landlord and tenant — Assignment of lease — Surety covenants — Whether sureties' liability discharged by Crown disclaimer of lease — Whether sureties' liability revived on restoration of dissolved company

By an assignment dated September 1 1980 Chicago Pneumatic Tool Co Ltd ("Pneumatic") acquired the term of a lease dated September 30 1977. by the licence to assign Pneumatic covenanted with the lessor to observe and perform the lessee's obligations under
D the lease. On December 22 1987 the landlord, by then the plaintiff, gave licence to Pneumatic to assign the lease to a company, which became known as KGP Technical Services Ltd ("KGP"). By the terms of that licence the first and second defendants, directors of KGP, ("the sureties") covenanted as sureties that KGP would comply with the lessee's covenants, and in default the sureties would comply. By the assignment the sureties covenanted with Pneumatic that they would observe and perform the lessee's covenants and indemnify Pneumatic for any breach. On April 7 1992 KGP was struck off the register and dissolved pursuant to section 652 of the Companies Act 1985 and the lease vested in the Crown as *bona vacantia*. On April 23 1993 the Crown, on the application of the sureties, disclaimed the lease
E pursuant to section 656 of the 1985 Act. On September 30 1993 KGP was restored to register upon the application of Pneumatic. In the first of two actions the landlord obtained summary judgment on October 29 1993 for arrears of rent, service charges and insurance premiums against the sureties and Pneumatic for the period December 25 1991 to December 26 1992; Pneumatic was given judgment under its contribution notice against the sureties for the same principal sum. In the second action the landlord obtained summary judgment for arrears of rent, service charges and insurance against Pneumatic, and Pneumatic against the sureties on the third party notice, for the period March 25 to June 30 1993. The sureties appealed the orders giving
F summary judgment, contending: (1) that they were discharged from liability by the Crown disclaimer, which is deemed to take effect from the moment the lease was vested in the Crown; and (2) their liability did not revive on September 30 1993 because the disclaimer on April 22 1993 constituted a "disposition" under section 655 of the 1985 Act and that the disclaimer therefore remained unaffected by the restoration (entitling KGP to compensation) and relieving them of their liabilities under their covenants both to the landlord and Pneumatic.

Held: **The appeals were dismissed. The sureties were released from their covenants to the landlord and to Pneumatic on the crown disclaimer: see *Stacey* v *Hill* [1901] 1 KB 660 as applied in *Re No 1 London* [1991] BCC 118 and *Re Yarmarine (IW) Ltd* G [1992] BCC 28. On the restoration of KGP to the register on September 30 1993, the sureties' covenants were revived with retrospective effect. The Crown's disclaimer is not a "disposition" for the purposes of section 655 of the Companies Act 1985. KGP was therefore deemed to have continued in existence as if its name had not been struck off. If the lease is deemed never to have vested in the Crown, the Crown cannot have disclaimed.**

The following cases are referred to in this report.

A E Realisations (1985) Ltd, In re [1988] 1 WLR 200; [1987] 3 All ER 83
British & Commonwealth Holdings plc v *Quadrex Holdings Inc* [1989] QB H 842; [1989] 3 WLR 723; [1989] 3 All ER 492, CA
Carter (RG) Ltd v *Clarke* [1990] 1 WLR 578; [1990] 2 All ER 209, CA
Dixon (CW), Re [1947] Ch 251; [1947] 1 All ER 279; [1947] 63 TLR 155
Finlay, In Re, ex parte Clothworkers' Co (1888) 21 QBD 475
Lister, In re, ex parte Bradford Overseers & Bradford Corporation [1926] 1 Ch 149
Murphy v *Sawyer-Hoare* [1993] 2 EGLR 61; [1993] 27 EG 127
No 1 London, Re [1991] BCC 118
Nottingham General Cemetery Co, Re The [1955] Ch 683; [1955] 3 WLR 61; [1955] 2 All ER 504; 53 LGR 519
Paradise Motor Co Ltd, In re [1968] 1 WLR 1125; [1968] 2 All ER 625, CA
Stacey v *Hill* [1901] 1 KB 660 J
Stratton's Disclaimer, In re, Stratton v *Inland Revenue Commrs* [1958] Ch 42; [1957] 3 WLR 199; [1957] 2 All ER 594, CA
Thompson & Riches Ltd, In re [1981] 1 WLR 682; [1981] 2 All ER 477
Tymans v *Craven* [1952] 2 QB 100; [1952] 1 All ER 613; [1952] 1 TLR 601, CA
Warnford Investments Ltd v *Duckworth* [1979] Ch 127; [1978] 2 WLR 741; [1978] 2 All ER 517; (1977) 36 P&CR 295
Yarmarine (IW) Ltd, Re [1992] BCC 28

These were appeals from orders of District Judge Lichfield, sitting as a deputy master, who in two actions gave summary judgment to the plaintiff, Allied Dunbar Assurance plc, for sums representing arrears of rent against the first to third defendants, Kevin Fowle, Patrick Hicks K and Chicago Pneumatic Tool Co Ltd, and in each action gave summary judgment to the third defendant under contribution and third party notices against the first and second defendants.

Timothy Fancourt (instructed by Nabarro Nathanson) appeared for the plaintiff; Gavin Lightman QC and Elizabeth Jones (instructed by Clifford Harris & Co) represented the first and second defendants; David Hodge (instructed by Reynolds Porter Chamberlain) represented the third defendant.

Giving judgment, GARLAND J said: These are appeals by Mr Fowle and Mr Hicks, to whom I shall refer as "the sureties", against orders of L District Judge Litchfield giving summary judgment in action 914 for the plaintiffs against the sureties and Chicago Pneumatic Tool Co Ltd, to which I shall refer as "Pneumatic", the third defendant, and for Pneumatic on its contribution notice against the sureties.

In action 2319, judgment was given for the plaintiffs against Pneumatic and for Pneumatic on its third party notice against the sureties. Since there were no issues of fact between the parties and the only matters to be determined were of law, I was invited by Mr David Hodge for Pneumatic to decide all matters of law and give judgment in open court. Mr Gavin Lightman QC, for the sureties, submitted that I should not accept this invitation and should not give final judgment, M but instead give leave to defend and transfer the action to the Chancery Division so that the issues could be fully argued before being taken to the Court of Appeal as important matters of principle requiring determination by a higher court, the more so if the applicability of *Stacey* v *Hill* [1901] 1 KB 660 to a Crown disclaimer is to be challenged.

I was referred to the note 13/7/5 at p153 of the annual practice and to the passages in *R G Carter Ltd* v *Clarke* [1990] 2 All ER 209 (also reported at [1990] 1 WLR 578) and *British & Commonwealth Holdings plc* v *Quadrex Holdings Inc* [1989] QB 842. It is to be noted that both these cases were decided before Ord 14A came into force, but, with all due respect to Mr Lightman's submissions, I find myself in complete agreement with that part of the note which reads:

> Where the court is satisfied that there are no issues of fact between the parties, it would be pointless to give leave to defend on the basis that there is a triable issue of law, and this is so even if the issue of law is complex and highly arguable, but it is otherwise if the issue of law is not decisive of all the issues between the parties or if the issue is of such a character as would not justify its being determined as a preliminary issue or if the answer to the question of law is in any way dependant on undecided issues of fact.

The extent of the facts necessary to define the issues of law do not take up a very significant part of this judgment and, as has already been said, are not in dispute at all. The points of law are wholly determinative of the rights of the parties. I was told quite frankly that these proceedings were merely a step on the way to the Court of Appeal and it would be quite wrong, in my view, to put the parties to the delay and expense of a further hearing in the High Court. I therefore propose to deal with the issues before me.

Genesis of the actions

By a lease dated September 30 1977, Hambro Life Assurance Ltd granted a 20-year term of part of CP House, 97-107 Uxbridge Road, Ealing, to Consolidated Pneumatic Tool Co Ltd, which subsequently changed its name to Chicago Pneumatic Holdings Ltd, and to which I shall refer as "Holdings". On August 26 1980, the lessor granted Holdings a licence to assign to a company which became Chicago Pneumatic Tool Co Ltd to which I have already referred as "Pneumatic", the third defendants in action 914, and the sole defendant in action 2319. Pneumatic covenanted with the lessor to observe and perform the lessee's obligations under the lease and to assume liability for any existing breaches of covenant or condition. The assignment was executed on September 1 1980.

On December 22 1987, the lessor, which by then had become Allied Dunbar Assurance plc, the plaintiffs, granted Pneumatic a licence to assign to Millrow Ltd, which subsequently became KGP Technical Services Ltd and to which I shall refer as "KGP". The assignment was executed on the same day.

The sureties, Mr Fowle and Mr Hicks, were directors of KGP and covenanted by clause 9 of the general conditions of the licence that the assignee, KGP, would comply with the lessee's covenants and conditions and that in default of compliance, notwithstanding any time or indulgence given to the defaulting person, the sureties would comply with those obligations. By the assignment, the sureties further covenanted with Pneumatic to procure and further guarantee to Pneumatic that at all times KGP would observe and perform the covenants under clause 2 of the assignment (KGP's covenants to Pneumatic) and, in the event of any default, the sureties would observe and perform the same and indemnify Pneumatic for any breach. The covenants in clause 2 by KGP to Pneumatic were, in summary, to pay the rent, observe and perform the lessee's covenants and to indemnify Pneumatic against any claims on account of any omission to pay the rent or observe the covenants.

On April 7 1992 KGP was dissolved having earlier been struck off the register, pursuant to section 652(5) of the Companies Act 1985. The interest under the lease became vested in the Crown as *bona vacantia* pursuant to section 654. On April 22 1993, the Crown disclaimed the lease, at the instance of the sureties, pursuant to section 656. On September 10 1993 KGP was ordered to be restored by the registrar and on September 30 was restored pursuant to section 653. The issues in these appeals concern the effect of the disclaimer and restoration on the liabilities of the sureties to the plaintiffs and to Pneumatic, having particular regard to the effect of sections 655 and 657, and the application by subsection (2) of section 657, of section 178(4) and of sections 179 to 182 of the Insolvency Act 1986.

Statutes

It is therefore necessary to turn to the statutes. Section 652 of the Companies Act 1985 empowers the registrar to strike off the register a company which he has reasonable cause to believe is not carrying on business or is not in operation. On publication of a notice of the striking off in the *London Gazette*, the company is dissolved. On dissolution, property vested in the company immediately before its dissolution, including leasehold property, is deemed to be *bona vacantia* and to belong to the Crown. Section 656 allows the Crown to disclaim property vested in it by virtue of section 654, subject to time-limits which are set out in subsection (3). On disclaimer, section 657 provides by subsection (1):

> Where notice of disclaimer is executed under section 656 as respects any property, that property is deemed not to have vested in the Crown under section 654.

Subsection (2) provides:

> As regards any property in England and Wales, section 178(4) and sections 179 to 182 of the Insolvency Act shall apply as if the property had been disclaimed by the liquidator under the said section 91 immediately before the dissolution of the company.

Section 653 permits the revival or reanimation of the company in the following terms:

> (1) The following applies if a company or any member or creditor of it feels aggrieved by the company having been struck off the register.
>
> (2) The court, on an application by the company or the member or creditor made before the expiration of 20 years from publication in the Gazette of notice under section 652, may, if satisfied that the company was at the time of the striking off carrying on business or in operation, or otherwise that it is just that the company be restored to the register, order the company's name to be restored.
>
> (3) On an office copy of the order being delivered to the registrar of companies for registration the company is deemed to have continued in existence as if its name had not been struck off; and the court may by the order give such directions and make such provisions as seem just for placing the company and all other persons in the same position (as nearly as may be) as if the company's name had not been struck off.

Section 655 deals with the effect of such a revival on the vesting of property in the Crown under section 654 as follows:

> (1) The person in whom any property or right is vested by section 654 may dispose of, or of an interest in, that property or right notwithstanding that an order may be made under section 651 or 653.
>
> (2) Where such an order is made
>
> (*a*) it does not affect the disposition (but without prejudice to the order so far as it relates to any other property or right previously vested in or held on trust for the company), and
>
> (*b*) the Crown, or, as the case may be, the Duke of Cornwall shall pay to the company an amount equal to —
>
> (i) the amount of any consideration received for the property or right, or interest therein, or
>
> (ii) the value of any such consideration at the time of the disposition,
>
> or, if no consideration was received, an amount equal to the value of the property, right or interest disposed of, as at the date of the disposition.

Section 651 empowers the court to declare a dissolution of a company void:

> (1) Where a company has been dissolved, the court may . . . , on an application made for the purpose by the liquidator of the company or by any other person appearing to the court to be interested, make an order, on such terms as the court thinks fit, declaring the dissolution to have been void.
>
> (2) Thereupon such proceedings may be taken as might have been taken if the company had not been dissolved.

Although this section is *prima facie* concerned with companies in liquidation, it can provide an alternative route to reanimation or revival for a company struck off under section 652: see *In re Thompson & Riches Ltd* [1981] 1 WLR 682.

Finally, I turn to the Insolvency Act 1986 as applied by section 657(2). Section 178 deals with a liquidator's power to disclaim onerous property. Subsection (4) reads as follows:

> A disclaimer under this section —
>
> (*a*) operates so as to determine, as from the date of the disclaimer, the

> A rights, interests and liabilities of the company in or in respect of the property disclaimed; but
>
> (*b*) does not, except so far as is necessary for the purpose of releasing the company from any liability, affect the rights or liabilities of any other person.

Section 181 defines the court's powers to make orders in respect of disclaimed property: sections 179 and 182 deal specifically with leaseholds. In summary, any person claiming an interest in a disclaimed leasehold, or who is under any liability in respect of it other than a liability discharged by the disclaimer, may apply to the court, which has very wide powers, for an order vesting the property in him.

B Section 181 provides as follows:

> (1) This section and the next apply where the liquidator has disclaimed property under section 178.
>
> (2) An application under this section may be made to the court by —
>
> (*a*) any person who claims an interest in the disclaimed property, or
>
> (*b*) any person who is under any liability in respect of the disclaimed property, not being a liability discharged by the disclaimer.
>
> (3) Subject as follows, the court may on the application make an order, on such terms as it thinks fit, for the vesting of the disclaimed property in, or for its delivery to —
>
> C (*a*) a person entitled to it or a trustee for such a person, or
>
> (*b*) a person subject to such liability as is mentioned in subsection (2)(b) or a trustee for such a person.

Reference should also be made to subsection (4) of section 182 which provides:

> Where subsection (1) applies and a person claiming under the company as underlessee or mortgagee declines to accept an order under section 181, that person is excluded from all interest in the property.

In the course of argument, counsel drew attention to differences between a crown disclaimer and liquidator's disclaimer. First, a Crown D disclaimer under section 657(1) results in the property being deemed not to have vested in the Crown under section 654 whereas a disclaimer under section 178(4) operates as from the date of the disclaimer. The Crown's right to execute a notice of disclaimer may be waived by taking possession of the property (section 656(2)) whereas a liquidator may disclaim notwithstanding that he has taken possession of the property, endeavoured to sell it or otherwise executed rights of ownership in relation to it: section 178(2).

Counsel examined the historical evolution of the present legislation both as a guide to construction and also to put in context authorities decided under, in particular, the Bankruptcy Act 1883 and the E Companies Act 1948. It is necessary to refer briefly to the 1948 Act and to the Companies Act 1981. Sections 352 to 355 of the 1948 Act contain parallel provisions to sections 651, 652, 653, 654, 656 and 657(1) of the 1985 Act, but the equivalent of section 655 was introduced by section 108 of the Act of 1981; and, as has already been noted, section 657(2) was added, by amendment, by the Insolvency Act 1986. Accordingly, before the 1981 Act the Crown could not safely dispose of *bona vacantia*, and before the 1986 Act an interested party could not apply for a vesting order following a Crown disclaimer as he could in the case of liquidator's disclaimer. Subsections (2) and (3) of section 178 of the Insolvency Act are in almost the same terms F as section 55(1) of the Bankruptcy Act 1883, subsection (4) as section 55(2) and subsections 181 and 182 as section 55(6).

Actions

In action 914, the plaintiffs recovered judgment on October 29 1993 against all three defendants in the principal sum of £158,999.51 together with interest, in respect of rent, service charges and insurance premiums due under the lease from December 25 1991 to December 26 1992. If this principal sum falls to be apportioned as at April 7 1992, the amount due up to that date would be £63,898.79 and the amount due thereafter would be £95,100.72. Pneumatic were given judgment on their contribution notice against the sureties for the same principal sum.

In action 2319, judgment was given for the plaintiffs against G Pneumatic in the principal sum of £59,214.04 and for Pneumatic against the sureties on the third party notice in the same amount. This represented rent, service charges and insurance falling due between March 25 and June 30 1993, that is before KGP were restored to the register on September 30. The sureties now appeal against the judgments based on liabilities said to have arisen between April 7 1992 and June 30 1993 both to Allied and Pneumatic.

Issues

These can be very simply stated. The sureties' argument is:

(1) That they were discharged from liability by the disclaimer which is deemed to take effect from the moment that the leasehold H interest was vested in the Crown as *bona vacantia* on April 7 1992: section 657(1).

(2) That their liability did not revive on September 30 1993 because the disclaimer on April 22 1993 constituted a "disposition" under section 655 entitling KGP to compensation under subsection (2), and relieving them of their liabilities under their covenants both to Allied and Pneumatic.

Allied and Pneumatic submit:

(1) Since on disclaimer the property is deemed never to have vested in the Crown and on restoration to the register the company is deemed to have continued in existence as if its name had not been J struck off, the rights and liabilities of the sureties are revived with retroactive effect.

(2) That in any event the sureties' liability was not extinguished by the dissolution. They concede that there is cogent authority against them but wish to keep the point open on the basis that there is a valid distinction to be drawn between the effect of a Crown disclaimer and that of a liquidator's disclaimer.

(3) Mr Timothy Fancourt, for the plaintiffs, submitted that on the construction of the covenant between the sureties and the plaintiffs it was possible to find an intention that the sureties should continue to be bound notwithstanding disclaimer. K

A number of subsidiary issues emerged, including arguments *in terrorem* based on the possible consequences of the acceptance of other parties' submissions.

1. *Release of surety on disclaimer.*

It is necessary to distinguish two situations:

(i) the simple lease, with no underletting or assignment;

(ii) where there is an underlease or assignment: see *Warnford Investments Ltd* v *Duckworth* [1979] Ch 127.

In the case of a simple lease, *Stacey* v *Hill* [1901] 1 KB 660 decided that section 55(2) of the Bankruptcy Act 1883, which is not L materially different from section 178(4) of the Insolvency Act 1986, had the effect of determining the lease as between lessor and lessee from the date of the disclaimer so that the surety's liability for future arrears of rent was also determined. If it were not so, a surety liable to the lessor would be entitled to an indemnity from the lessee company and would require to be released under what is now subsection (4)(b) "for the purpose of relieving the company".

Mr Timothy Fancourt for the plaintiffs and Mr Hodge for Pneumatic submitted that in the second situation, where there had been an assignment or where there was a subsisting underlease, the lease continues in existence and the company, on disclaimer, is M relieved only of its rights and liabilities under the lease so that the first reason for the decision in *Stacey* v *Hill* does not arise; nor does the second, because the company has ceased to exist and cannot be liable to indemnify the surety. There may well be considerable force in these arguments, but the trend of authority is against them. In chronological order, the following decisions must be considered.

Re No 1 London [1991] BCC 118 (Hoffmann J as he then was). A company was dissolved under section 652. The Crown disclaimed the lease guaranteed by the applicant who applied for a vesting order under section 181(2)(b) of the Insolvency Act 1986 as a person "under any liability in respect of the disclaimed property, not being a liability discharged by the disclaimer". The disclaimed property was, of course, the unexpired term and Hoffmann J said at p119F:

> Any liabilities in respect of the disclaimed property were, by virtue of the law as laid down, for example, in *Stacey* v *Hill* [1901] 1 KB 660, discharged by the disclaimer.

Mr Fancourt and Mr Hodge pointed out that, although this was under sections 652, 656 and 657, Hoffmann J, at the foot of p118 and top of p119, said:

> . . . the effect of the disclaimer is the same as if the property had been disclaimed by a liquidator under the Insolvency Act immediately before the dissolution of the company. The effect of a disclaimer by a liquidator is, by virtue of sec 178(4) of the Insolvency Act, to determine as from the date of the disclaimer the rights, interests and liabilities of the company in or in respect of the property disclaimed.

Had he, they ask, lost sight of the difference between a liquidator's disclaimer, which must of necessity precede dissolution, and a Crown disclaimer, which is deemed to take effect immediately before dissolution although in fact dissolution precedes disclaimer?

Re Yarmarine (IW) Ltd [1992] BCC 28 (Millett J). The essential facts were the same as in *Re No 1 London*. Dealing with the surety's future liabilities, Millett J said at p30G:

> But sec 178(4)(b) is, to all intents and purposes, in the same terms as its predecessors going right back to sec 55(2) of the Bankruptcy Act 1883, and it was decided in *Stacey* v *Hill* [1901] 1 KB 660 that that section required the release of the surety in order to release the tenant, since otherwise the continuing liability of the surety and the surety's continuing right of indemnity would not bring about the release of the tenant. Accordingly a disclaimer by a liquidator under sec 178(4) of a lease brings to an end the surety's obligation in respect of future liabilities under the lease. Consequently the effect of Crown disclaimer under sec 656 of the Companies Act 1985 has the same effect by virtue of sec 657(2).

At p31C, Millett J added:

> The language of sec 657(2) is very narrow. It does no more than apply certain specified sections of the Insolvency Act as if there had been a liquidation. It does not deem there to have been a liquidation or provide that leases and other contractual arrangements should have effect as if the Crown disclaimer were a disclaimer by a liquidator.

Mr Fancourt and Mr Hodge drew attention to this passage in support of the point they take that a distinction can and should be drawn between the different forms of disclaimer.

Finally, *Murphy* v *Sawyer-Hoare* [1993] 27 EG 127*, a decision of Mr Grabiner QC, sitting as a deputy judge of the Chancery Division. The second defendant was the guarantor of the assignee of a lease. The assignee went into liquidation and the liquidator disclaimed the lease. The deputy judge, following *Warnford Investments Ltd* v *Duckworth*, held that the lease continued to exist and that the position as between lessor and lessee was unaffected. However, he took the view that *Stacey* v *Hill* applied as between assignee and surety:

> If, on the true construction of the covenant given by the surety, he would be discharged if the principal debtor ceases to have any liability then, in my view, it is necessary for the purpose of releasing the company from any liability that the liability of the surety is affected.

He also expressed the view that Parliament, by continuing to use the wording of section 55(2) of the 1883 Act, was impliedly adopting the decision in *Stacey* v *Hill* in section 178(4).

Were the matter *res integra*, I would have regarded the arguments advanced by Mr Fancourt and Mr Hodge as very persuasive. However, while acknowledging that there is a fundamental distinction between a liquidator's disclaimer taking effect from the date of the disclaimer and a Crown disclaimer deeming the property never to have vested in the Crown at all; and between a disclaimer taking effect some time before dissolution and one that takes place after dissolution though deemed to take place immediately before it, I feel constrained by the decisions of Hoffmann J (as he then was) and Millett J to accept that *Stacey* v *Hill*, a decision of the Court of Appeal, is binding on me both as to covenants to Allied and the covenants to Pneumatic. Counsel took the points that *Murphy* v *Sawyer-Hoare* and, more important, *Stacey* v *Hill*, were wrongly decided.

Mr Fancourt picked up a point made by Mr Grabiner QC at p129 of *Murphy* v *Sawyer-Hoare* that a covenant could be so drafted as to ensure that a surety's liability continued notwithstanding the liquidation of the obligor, whether original tenant or assignee, and submitted, perhaps in a somewhat muted way, that the sureties' covenants to the plaintiffs did so continue. Their covenants are contained in the licence of December 22 1987 which is in what can best be described as a law stationer's common form designed for use for an underlease as well as an assignment. The relevant words under the headings "Appendix, General Conditions, are:

> Where a surety joins this Licence it covenants that in consideration of this Licence being issued, then throughout the term of the Lease (including any continuation of that term) the Assignee . . . will comply with the lessee's covenants and the conditions in the Documents [the Lease and any earlier Licences, Assignments, Guarantees or Indemnities] . . . and that in default of compliance (notwithstanding any time or indulgence given to the defaulting person) the Surety will comply with those obligations . . .

In my view, this covenant falls a long way short of the sort of provision to be found in chattel leasing agreements, contracts for the sale and delivery of goods, or in construction contracts providing for retention of property in materials delivered to the site but not incorporated in the works or, if incorporated, not yet paid for.

2. *Is the sureties' liability revived?*

I therefore pass to the larger question: what happens when the company is revived under section 653(3)? Is it an "as you were" situation with the company liable from the moment of dissolution and the sureties likewise; or, as Mr Lightman submits, are the sureties at any rate excused all liability from April 7 1992 to September 30 1993?

Arguments

The argument for Allied and Pneumatic can be summarised in the following way.

The effect of section 653(3) is that on September 30 1993 KGP was "deemed to have continued in existence as if its name had not been struck off". The lease therefore is deemed never to have vested in the Crown so that no question of *bona vacantia* ever arose.

If the lease is deemed never to have vested in the Crown, the Crown cannot have disclaimed. However, if there had in fact been a disposition while the company's property was vested in the Crown, section 655 provides for the necessary "tidying up", but a disclaimer is not a disposition. The construction of the section requires a disposition to a third party. A mere divesting from the Crown does not fall within the section. If it did, two undesirable consequences would follow:

(a) The Crown's right to disclaim is to protect the Crown from having to assume burdens: if, in the event, the lease is shown to have had some value on assignment, the Crown would have to pay compensation under section 655(2)(b)(ii).

(b) A company (or interested surety, eg a director) could obtain relief from liability by the company failing to file returns, being struck off, persuading the Crown to disclaim, being revived (at very modest cost) and then continuing to trade, but without liability for the period from dissolution to revival.

Mr Lightman QC accepted that prior to the enactment of section 108 of the Companies Act 1981 the effect of subsequent restoration to the register would have been to revest the lease in KGP as was the case in *Re Dixon (CW)* [1947] Ch 251 and *Tymans* v *Craven* [1952] 2 QB 100. However, he submitted that the legislative scheme introduced by section 108 and now found in sections 654(2) and 655 is that restoration of the company cannot effect any "disposition" by the Crown of the lease during the period before KGP was restored to the register. On restoration, the Crown would be liable to account for any consideration received for it, or if none was received, an amount equal

*Editor's note: Also reported at [1993] 2 EGLR 61.

A to the value of the lease at the date of disposition: see section 655 (1) and (2). If, therefore, a Crown disclaimer is a "disposition", it would have the effect of releasing the sureties from liability and any subsequent restoration could not revive such liability. The central issue is accordingly whether a Crown disclaimer constitutes a disposition. While acknowledging that the word has to be construed in its context, Mr Lightman looked to other statutes and situations in support of his submission that it was sufficient that there should be a disposition *from* (emphasis supplied) without there being a disposition *to* (emphasis supplied).

Meaning of disclaimer

B It was submitted that an appropriate definition is "a renunciation of all rights and obligations in respect of the property disclaimed". Reference was made to *In re Finley, ex parte Clothworkers' Company* (1888) 21 QBD 475. The decision concerned the construction and effect of section 55 of the Bankruptcy Act 1883 where a trustee in bankruptcy had disclaimed a lease which the bankrupt had mortgaged by subdemise and the original lessor applied for an order under subsection (6) vesting the property in the mortgagee. At p482, Lindley LJ said:

> The effect of the disclaimer, speaking roughly and generally, is plain enough;
> C it is to put an end to both the rights and the liabilities of the bankrupt and his trustee in respect of the disclaimed property. That is the primary object. By the disclaimer the trustee elects to have nothing to do with the property, which is onerous, and from which nothing can be got for the benefit of the creditors. He is at liberty to disclaim it, and the effect of the disclaimer is to extinguish all the rights and liabilities, both of the bankrupt and of the trustee himself.

The Court of Appeal again considered the meaning of the section in *In Re Lister, ex parte Bradford Overseers & Bradford Corporation* [1926] 1 Ch 149. The bankrupt occupied a small house under an oral tenancy. The official receiver, as trustee, entered the premises and took possession of the furniture, remaining in occupation from
D February 8 until May 15 when he sold the furniture. On June 1 he disclaimed the tenancy. The local authority claimed rates during the period of the trustee's occupation. The Court of Appeal held that, although he had incurred liability by his voluntary occupation of the premises, that liability was discharged by section 54 as one flowing from the vesting in him of the property disclaimed. At p165, Sargent LJ said:

> The use throughout s 54 of the words "disclaim" and "disclaimer" is in itself a strong indication of the scope of the section. The words, in my view, connote a renunciation of, or refusal to claim, rights or property which would automatically devolve on or accrue to the person making the disclaimer. They
> E are precisely applicable to a renunciation of a property which with all its rights, interests and liabilities has vested in the trustee under s 53, Subs 2, of the Act. They are not appropriate to any interests, rights or liabilities that have been acquired or undertaken by the voluntary act of the trustee himself, and independently of devolution of the bankrupt. "Disclaime, disclamare . . . signifieth utterly to renounce . . .:" Co Litt 102*a* . . . And the word is habitually used in this sense at the present day with reference to the non-acceptance or renunciation of the office and estate of a trustee under a deed or of an executor and trustee under a will.

Mr Lightman pointed to the decisions in *Re No 1 London Ltd* and *Re Yarmarine (IW) Ltd* treating a disclaimer by a liquidator on the
F same footing as a disclaimer by the Crown.

Meaning of disposition

It was submitted that a disclaimer by a liquidator is a disposal or disposition on the authority of *Re The Nottingham General Cemetery Co* [1955] Ch 683. The liquidator of a company incorporated by special Act of Parliament and which prohibited it from selling or disposing of any of its land which had been consecrated, set apart or used for the burial of the dead sought an order that he be at liberty to disclaim the land constituting the company's cemetery together with all implied contracts with holders of grave certificates and all contracts for the upkeep of the graves. The actual decision, summarised in the headnote, was: (i) the land was "burdened with onerous covenants" within the meaning of section 323 of the Companies Act 1948; (ii) the G
subject-matter of the disclaimer was not necessary for the carrying on of the company's business because that business had come to an end; (iii) the right to disclaim the property was exercisable personally by the liquidator so that no question as to whether such a disclaimer was *ultra vires* the company could arise.

Mr Lightman directed my attention to a passage in the judgment of Wynn-Parry J at p694 where he said:

> In the present case the act for which authority is sought is the disposal of land because the undertaking has by force of circumstances come to an end, or is about to come to an end. There can be no reason for retaining land which was used for the purpose of carrying on the undertaking when the undertaking has H
> come to an end. Further, unlike the case of a sale, the liquidator in making a disclaimer does not have to use the name of the company. The right to disclaim is a right conferred on the liquidator, as such, by section 323 of the Companies Act, 1948 — a right which did not exist prior to 1929.

The disclaimed land would have vested in the Crown as *bona vacantia* because there was no other person entitled to the freehold. There would accordingly in a broad sense be both a disposition *from* the company, in whom the land remained vested until disclaimer, *to* the Crown in which it would vest as *bona vacantia*.

For my own part, I take the view that Wynn-Parry J was using the word "disposal" in the widest and most informal sense and I do J
not regard it as an indication of the meaning of the word in section 655.

Mr Lightman passed to gifts *inter vivos* and by will conceding that a disclaimer operates by way of avoidance or extinguishment not disposition, although an oral direction to hold property on trust was a purported "disposition" within the Law of Property Act section 53(1)(*c*) and ineffective because it was not in writing. Reference was made to *In re Stratton's Disclaimer* [1958] 1 Ch 42 where, at p55, Jenkins LJ said:

> As to the second question, I entirely agree with the judge that the disclaimer must be held to have operated as an extinguishment of Mrs Stratton's right in K
> respect of the specific bequest and devise. It seems to me that a disclaimer such as this is a typical example of the extinguishment of a right.

In Re Paradise Motor Co Ltd [1968] 1 WLR 1125 is to the same effect. A donee disclaimed a gift of sharesection. The point was taken that, because the disclaimer would by retransfer be a disposition of an equitable interest in the property, it was required to be evidenced in writing signed by the donee by virtue of section 53(2) of the Law of Property Act 1925. Danckwerts LJ, delivering the judgment of the court, dealt with the point very shortly (p1143):

> We think that the short answer to this is that a disclaimer operates by way of L
> avoidance and not by way of disposition. For the general characteristics of disclaimer we refer briefly to the discussion in *In Re Stratton's Disclaimer* . . . .

Mr Lightman submits that that part of the decision was *per incuriam* and referred to the note in *Wolstenholme and Cherry* vol 1 p131:

> The Court of Appeal has held that the disclaimer of a beneficial interest in shares is not within para (c), *supra*, [section 52(2)(*c*)] for the short reason that a disclaimer operates by way of avoidance and not by way of disposition . . . . But no reference was made to the fact that the Act defines "disposition" as M
> including a conveyance and defines "conveyance" as including a "disclaimer": s 205(1)(ii), *infra*.

Section 52(1) provides:

> All conveyances of land or of any interest therein are void for the purpose of conveying or creating a legal estate unless made by deed.

Subsection (2):

> This section does not apply to —
>
> . . .
>
> (*b*) disclaimers in accordance with [sections 178 to 180 or sections 315 to 319 of the Insolvency Act 1986], or not required to be evidenced in writing;

A (*c*) surrenders by operation of law, including surrenders which may, by law, be effected without writing.

Section 205(1)(ii) defines "conveyance" as including:

. . . a mortgage, charge, lease, assent, vesting declaration, vesting instrument, disclaimer, release and every other assurance of property or of an interest therein by any instrument, except a will.

Mr Hodge pointed out that section 205 is prefaced by the phrase: "In this Act unless the context otherwise requires" and that, notwithstanding the amendment to section 52 to include reference to the Insolvency Act, no reference has been made to Crown disclaimers
B under section 656 of the Companies Act 1985 which provides that a disclaimer is by notice signed by the Crown representative, whereas section 52 of the Law of Property Act 1925 requires a deed subject to the exceptions set out in subsection (2). Mr Hodge therefore submits that Crown disclaimers stand alone outside the provisions of the Law of Property Act. He also submits that *In Re Paradise Motor Co Ltd* was correctly decided, pointing out that Russell LJ (as he then was) had been counsel for the plaintiffs in *Re Stratton* and that Danckwerts LJ had been the judge of first instance whose decision was upheld by the Court of Appeal. The court cannot, he submits, have simply ignored section 201: subsection (2)(*b*) of section 52 then read:

C disclaimers made in accordance with section 54 of the Bankruptcy Act, 1914, or not required to be evidenced in writing.

Allied and Pneumatic rely on *In Re Paradise Motor Co Ltd*; a disclaimer, they submit, operates by way of avoidance, section 655 does not include a disclaimer under section 656.

Policy

Mr Lightman answered Mr Fancourt's points to which reference has already been made, first by accepting that, if a Crown disclaimer is a disposition, the Crown may have to pay for the consequences.
D Probably most property disclaimed will be of no or negative value, but, if it is not, and the Crown has to make up its mind within 12 months overall or three months on application, compensation becomes payable. Second, Mr Lightman asks how far should a landlord be left at risk of a company being restored to the register and the lease being revived after disclaimer? Counsel for Allied and Pneumatic made the E
following points:

(1) A lessor can under section 656(3)(b) force the Crown to make up its mind rather than wait for the 12-month period to expire.

(2) Until disclaimer, the lessor can forfeit as against the Crown: reference was made to *In Re A E Realisations (1985) Ltd* [1988] 1 WLR 200 at pp206H-207A.

(3) An alternative to (2) above would be peaceful re-entry.

(4) The lessor is entitled to apply under sections 181 and 182 of the Insolvency Act for a vesting order or exclusion order, but it was conceded that the wording of subsections (2) and (3) of section 181 (and of the corresponding sections of the Bankruptcy Act 1883) were F
"somewhat opaque". This is because if the landlord is the "person who claims an interest in the disclaimed property" the court may make a vesting order in favour of "the person entitled to it". To what is the lessor entitled? He may in fact recover his reversion by the *exclusion* of someone claiming under the company or seek an order vesting the lease in a surety who had covenanted to accept a lease should the tenant or assignee default, but he is not on the face of it entitled to the unexpired term.

(5) The court would have power, on the restoration of the company to the register, to grant relief against forfeiture in any event.

Policy arguments based on *lacunae* in detailed legislation are a G
legitimate guide to construction, but should be used with caution. There are bound to be anomalies and uncertainties where matters are deemed to have taken place quite contrary to what happened in actual fact: the deeming required by sections 653(3) and 657(1) is bound to create some anomalies.

Conclusion

The answer to the central issue: "Is a Crown disclaimer a disposition for the purposes of section 655?" must, in my judgment, be "No". I accept the arguments advanced by Allied and Pneumatic which I have summarised earlier. In my view, both the natural meaning of the words used and the policy of Chapter VI of Part XXI H
of the 1985 Act point to an "as you were" situation once the company is restored to the register, subject to the court's power to "tidy up" under sections 653(3) and 655.

Accordingly, I dismiss the appeals.

For further cases on this subject see p 243

LANDLORD AND TENANT

BUSINESS TENANCIES

A **Chancery Division**

November 8 1993

(Before Judge RICH QC, sitting as a judge of the High Court)

M&P ENTERPRISES (LONDON) LTD v NORFOLK SQUARE HOTELS LTD AND OTHERS

Estates Gazette April 9 1994

[1994] 14 EG 128

B ***Landlord and Tenant Act 1954 — Whether where reversion severed between four landlords, four separate section 25 notices relating to separate properties within single tenancy valid — Whether copies of the same notices relying on para (g) served on tenant's solicitors effective counternotices to section 26 request***

By an agreement dated March 10 1988 T Ltd made block bookings of two hotels owned by the four defendant companies, one of which consisted of four adjoining buildings; by a judgment in the High Court made on December 20 1990 the agreement was held to be a tenancy, accepted on behalf of the landlords as a single tenancy of the five buildings. The tenancy was assigned to the plaintiff on February 11
C 1991. The landlords served four separate notices, each relating to a separate building, dated March 6 1991 under section 25 of the Landlord and Tenant Act 1954 on the tenant, accompanied by a single letter referring to the five properties, determining the tenancy in each case on September 8 1991 and relying on section 30(1) ground (*g*). Copies of these notices were subsequently served on the tenant's solicitors. The tenant served four notices, also dated March 6 1991, on the landlords under section 26 of the 1954 Act requesting a new tenancy; these notices were received by the landlords after the service on the tenant of the section 25 notices, but prior to the receipt of the copies of the section 25 notices by the tenant's solicitors. The
D landlord contended that: (1) the section 25 notices were effective notwithstanding that each separate notice purported to terminate the tenancy only of one building demised by the tenancy; and (2) if the section 25 notices were ineffective, the copy section 25 notices, which relied on para (*g*) were effective counternotices to the tenant's section 26 request.

***Held*: (1) A valid section 25 notice must be a notice in respect of the whole of the premises the subject of the demise. Although the four notices might be construed as one notice contained within four, or together with the covering letter five, documents, they were capable of prejudicing a tenant because they did not make clear whether the tenancy which was bound to be treated as a**
E **single tenancy was in fact being so treated and accordingly were invalid. (2) The copies of the section 25 notices received by the tenant's solicitors after the landlords received the section 26 request were effective counternotices allowing the landlords to rely on ground (*g*) in the tenant's application for a new tenancy.**

The following cases are referred to in this report.

Long (Nevill) & Co (Boards) Ltd v *Firmenich & Co* (1983) 47 P&CR 59; (1983) 268 EG 572, CA

Pearson v *Alyo* [1990] 1 EGLR 114; [1990] 25 EG 69, CA

Southport Old Links Ltd v *Naylor* [1985] 1 EGLR 66; (1984) 273 EG 767, CA F

This was an application by the defendants, Norfolk Square Hotels Ltd and three other companies, to have determined two preliminary questions in an application by the plaintiff, M&P Enterprises (London) Ltd, for a new tenancy of premises owned by the defendants.

Kenneth Munro (instructed by Graham Harvey & Co) appeared for the plaintiff (the respondent to the preliminary issue); Philip Kremen (instructed by Georgiou Nicholas) represented the defendants (the applicants to the preliminary issue).

Giving judgment, **JUDGE RICH QC** said: This is a preliminary issue G
to be determined under an order made by Master Dyson on October 25 1991 as to the validity of certain notices under section 25 of the Landlord and Tenant Act 1954, served by the defendants in this case, and as to the effect of a notice served under section 26 and the response thereto served by the plaintiff upon the defendants. It is convenient to refer in cases of this sort to the plaintiff as "the tenant" and to the defendants as "the landlords", although it is necessary for me to explain at the outset what it is that the defendants are in fact landlords of.

The tenancy, which is the subject-matter of the proceedings in respect of which the preliminary issue arises, arose out of an H
agreement made between the four defendant companies and a company called Tours & Travel UK Ltd, which agreement was made on March 10 1988, for what were called block bookings in the hotels owned by the defendants. Each of the four defendants owned one building, or in the case of the defendant called Norfolk Square Hotels No 2 Ltd two buildings. Four of the buildings were contiguous in Norfolk Square, London W2, and together were called the Continental Hotel. The fifth building at 92 Sussex Gardens, London W2, owned by Sussex Gardens Hotels Ltd, was known as the Classic Hotel.

The agreement that was made for the block bookings at these hotels was claimed by Tours & Travel UK Ltd to constitute a tenancy. J
That was disputed by the present defendants, but the matter was determined by the judgment of Judge Sumner, sitting as a High Court judge on December 20 1990, whereby he adjudged and declared that the plaintiff did have a tenancy from the defendants of the premises which he there described.

That decision I construe as meaning that there was a single tenancy of all five buildings from the four landlords to Tours & Travel UK Ltd and it has been accepted by Mr Philip Kremen, who appears for the landlords who are before me, that that is the effect of the judgment, which has not been appealed. That appears to me to be binding as between the landlords and the present plaintiff, to whom these K
premises and that tenancy was assigned on February 11 1991.

Immediately following that assignment, the landlords were minded to bring that tenancy to an end under the terms of the Landlord and Tenant Act 1954. For that purpose, they served four separate notices upon the defendant. Each notice followed a similar form, in that each referred to one — or in the case of the landlord who owned two buildings two — of the buildings the subject of the agreement and gave notice terminating the tenancy in each case on the same date more than six months ahead from the date of service, namely September 8 1991. Each gave as the address of its agent the name of

A its solicitors, Georgiou Nicholas, and gave the same address for the service of all correspondence in regard to the notice.

These notices, when served on the defendant, were accompanied by a single letter from the landlords' solicitors, addressed to the defendant tenant, referring to the premises by title, 36 to 42 Norfolk Square and 92 Sussex Gardens, and saying "We refer to the above matter and enclose herewith notices under section 25 of the Landlord and Tenant Act 1954". The notice was said to be served without prejudice to an appeal in respect of the order of December 20 of the previous year (to which I have referred), but nothing, I think, turns upon that qualification upon the service of the notices for the purposes
B of this present issue. It is, I think, probably helpful before turning to the effect of those notices, to take the history forward a little so that the nature of the issues between the parties can be more fully understood. Each of those notices recited the intention of the landlord, if the recipient of the notice applied for a new tenancy under the Landlord and Tenant Act 1954, to oppose such application on the grounds mentioned in para (*g*) of section 30(1) of the Act.

The service of these notices, as it were, crossed an attempt to serve a request under section 26 of the 1954 Act on behalf of the tenant for a new tenancy. The service of such request was, I think, complicated by the desire of those who were wishing to serve it to serve each one of
C the four landlords separately, but they were each served with a copy of a request in identical form, which referred to all five of the properties and requested all four of the landlords to grant a new tenancy under the Act.

That form of notice, dated likewise March 6 1991, is agreed to have been served on the landlords only later than the service of the section 25 notice, but prior to the receipt by the tenant's solicitors of copies of the section 25 notices. The section 26 request, in the same way as the section 25 notices, stated the name and address of the tenant's agent — namely their solicitors, Graham Harvey & Co — and invited the recipient to address all correspondence about the request to
D that agent.There was no response to that notice of request specifically relying upon a ground of opposition to the grant of a new tenancy under the Act, save the effect (if any) of the service upon the tenant's solicitors of copies of the notices served already on the tenant himself, subsequent to the receipt by the landlords of the section 26 request.

The section 26 request, to which I have referred, contained a typographical error in the reference to one of the properties, in that the Sussex Gardens property, instead of being referred to as no 92 as was its true address, was by typographical error referred to as a half-bracket and 2 Sussex Gardens. It has been accepted on both sides that such error in no way misled the recipients of the document and therefore in no way affected its validity, and so, although as a
E precaution a yet further version of the section 26 request was served on the landlords, it is agreed that that is unnecessary for me to consider, because, if the section 25 notices had been valid, both section 26 requests would be invalid; and, if the section 26 request is not invalidated by the typographical error to which I have referred, the second section 26 request was unnecessary and therefore of no further effect.

The first question which I am called upon to determine is whether or not notices in this form on behalf of a group of landlords in respect of a group of properties is an effective form of notice to comply with the terms of section 25 of the Act.

F On the face of it, each separate notice could not be a valid notice under the Act, because each separate notice purports to determine the tenancy — which is a single tenancy of the five hotels — in respect of only one, or in one case two, of the buildings, and it is common ground that the Act does not permit the determination of part only of a tenancy or of a tenancy in respect of part only of the demise. Authority for that proposition can be found, if nowhere else, in the decision of the Court of Appeal in *Southport Old Links Ltd* v *Naylor* [1985] 1 EGLR 66. But it is said on behalf of the landlords that it is wrong to consider each notice separately and the combined effect of the four notices was to determine the tenancy in respect of all five buildings, and therefore service in that manner constituted a valid notice. Indeed it is said, persuasively, that service in that form was contemplated by G the Court of Appeal in what is accepted to be really a parenthetic observation in the course of the judgment in *Nevill Long & Co (Boards) Ltd* v *Firmenich & Co* (1983) 47 P&CR 59*. In that case Fox LJ gave the judgment of the Court of Appeal in respect of the application of the 1954 Act to a situation where there was a severed reversion such that the property held by the tenant was in part held from a landlord called Meek and in part — the relevant part being in fact the incorporeal hereditament of a right of way — from the defendant in that case. Fox LJ, having decided that in that case the right of way was held upon terms which were protected under the 1954 Act, went on to consider how the tenancy could be determined. H
He added, at the bottom of p66:

> We should add that it seems to us that the defendants and Meek together could constitute "the landlord" for the purposes of section 25 (either by serving a single notice or separate notices operating at the same time) since together they are entitled to the entirety of the land comprised in the relevant reversion.

Mr Philip Kremen, for the landlords in the present case, draws my attention to the proposition that the Court of Appeal were contemplating in such circumstances the serving of separate notices operating at the same time. He points out that the four notices served in this case were indeed operating at the same time, in the sense that they all bear the same date and all gave notice for termination of the J respective parts of the demise to which they referred on the same date, namely September 8 1991.

I am not sure whether that is what was in the mind of the Court of Appeal in making this parenthetic observation. It may well be, with the greatest of respect to the court, that they had not actually considered the mechanics of notice in the way that it has been necessary for me to do after hearing argument on this present issue. But I think that it is at least possible that in the circumstances of that case the problem that here arises would not have arisen, because, in that case, the description of the property, which was the subject of the demise, would have been effectively the same whether as described by K the landlord of the buildings, which were held by the tenant in that case, or by the defendant in that case who held only the reversion upon a right of way attached to those same buildings. Whether or not the Court of Appeal had in mind the possibility that the description of the premises to be the subject of the notice was to be the same or different, I have come to the conclusion that a valid notice must be a notice in respect of the whole of the premises the subject of the demise and one can construe the four notices, which are before me, as being *a* notice in respect of all five buildings only if one reads them as if they were in fact a single notice, that is to say not treating them as, in the Court of Appeal's words, separate notices operating at the same time, but as a L single notice contained within four, or maybe with the combination of the covering letter five, documents.

In order to discover whether that is the way in which these particular notices should be construed, I think, I am entitled to look at the letter under cover of which they were in fact served upon the tenant. That refers to "the notices" not as a single notice but as notices, and that is indeed what they purported to be, each and every one of them separately operating in respect of part only of the demise under the single tenancy. That leaves, of course, the possibility that cumulatively the four invalid notices might effect a valid termination of the tenancy under section 25. If that is the only way in which those M four, or with the letter five, documents could be construed by a reasonable tenant receiving them, that is a result which would follow from the service of such documents. That it must, however, be the necessary conclusion does, I think, follow, as Mr Kenneth Munro [who appears for the tenant] has submitted, from the decision of the Court of Appeal in *Pearson* v *Alyo* [1990] 1 EGLR 114. At p116, Nourse LJ in the leading judgment, with which Bingham LJ agreed, said:

> it must be emphasised that the validity of a section 25 notice is to be judged,

*Editor's note: Also reported at (1983) 268 EG 572.

A and judged objectively, at the date at which it is given. The question is not whether the inaccuracy actually prejudices the particular person to whom the notice is given, but whether it is capable of prejudicing a reasonable tenant in the position of that person.

I think that this combination of notices was capable of prejudicing a reasonable tenant in the position of the tenant in this case, because it did not make clear whether or not the tenancy which was bound to be treated as a single tenancy was in fact being so treated. It would have been a perfectly logical conclusion to draw from the notices in fact served that the tenant was being invited to consider the situation in regard to each of the four lots of property separately, and to apply B perhaps in respect of one, but not in respect of another of those properties, for a new tenancy. That was not, if it were a single tenancy, an application that was open to him to make and in that sense the notices as served were misleading or capable of being misleading. For that reason the defect of serving four notices without sufficient further explanation to convert them into a single notice was, in my judgment, an invalidating defect.

Mr Munro has sought to rely additionally upon another defect to which he has drawn attention, that in respect of that notice served on behalf of Norfolk Square Hotels No 2 Ltd, the name of the landlord is given merely as Norfolk Square No 2 Ltd, omitting the word "hotels". C In the absence of any evidence that Norfolk Square No 2 Ltd is the name of an actual company, rather than what it appears to be and has been treated as by the parties — merely a misdescription of the company with its full name — I do not think that such misdescription is in any sense more invalidating of such a notice than is the mistyping of the "9" in 92 Sussex Gardens by its replacement by a bracket in the section 26 request served on behalf of the tenant.

It is right to say that, although I hold that this was a potentially misleading series of documents, it does not appear in actuality to have misled — at least if one judges the matter by the actions of the solicitors for the tenant, who did issue within the statutory period an D application for a single new tenancy on behalf of the tenant and served the single application upon all four of the landlords. But, as was decided in the *Pearson* v *Alyo* case, to which I have already referred, I must judge the matter as at the date of the service of the document, and as viewed then it appears to me to contain sufficient of a potential misleading nature as to invalidate the section 25 application.

I have invited, in the course of argument, the canvassing of the question as to whether or not these notices may not be further misleading in their reference merely to para (*g*) of section 30(1), which I would construe in the context of this situation as being properly satisfied only if the landlords together form an intention that each of the properties the subject of the demise should be occupied by E at least one of them on behalf of all in the furtherance of the business, being a single business, of all. It may well be that the form of separate notices served under section 25, referring merely by reference to the paragraph, would not make clear to the recipient whether that is the allegation upon which the separate landlords were proposing to rely should an application be made under the Act. That is a potentiality certainly of confusion, but I am not satisfied — nor does it arise so far as the consideration of the section 25 notice — that that would of itself be a sufficient reason for holding these notices to be invalid. Rather, it is an example of the difficulty of construing the four documents, as completed, as together forming a single notice and thus being in effect F a single document sufficient to operate section 25 in respect of this particular tenancy.

I have indicated already that there was a request made under section 26 of the Act immediately after the effective service of these four notices; indeed so immediately that the copies of the notices together with a covering letter addressed to the tenant's solicitors, arrived after the service of the section 26 notice, although it was despatched, as appears from the covering letter, before the effective service of the section 26 request. The covering letter indeed from the landlords' solicitor to the tenant's solicitor, which is dated March 7, refers to a letter which was apparently concerned with the proposal on behalf of the tenant to serve such a request, but went on to say, "As at the time of writing this letter we have been advised by your client that G they have not yet been served with your letters" — that is to say the letters containing the section 26 request. "In any event your clients were served by hand" — I must insert "at" — "their registered office with section 25 notice this morning". The letter goes on to make clear that the copies of the letter addressed to the tenant together with the section 25 notices enclosed are being sent only for information and not for the purpose of operating as a counternotice to the request made under section 26 for a new tenancy.

But I have to consider whether, under section 30 of the 1954 Act, the ground of opposition referred to in those notices is a ground of opposition open to the landlord to rely upon in an application for a H new tenancy made on behalf of the tenant following the section 26 request. Section 30 provides that the grounds on which a landlord may oppose an application are such of the following grounds as may be stated either in the landlord's notice under section 25 where that is applicable or as the case may be under subsection (6) of section 26. Subsection (6) of section 26 provides that within two months of the making of a tenant's request for a new tenancy a landlord may give notice to the tenant that he will oppose an application to the court for the grant of a new tenancy, and any such notice shall state on which of the grounds mentioned in section 30 of the 1954 Act the landlord will oppose the application. It was after the making of the tenant's request J and within the period of two months thereafter that there was received by the agent referred to in the notice containing the section 26 request the series of notices, which stated in terms that, if the tenant applied to the court for the grant of a new tenancy, the individual landlords each would separately oppose such application on the grounds mentioned in para (*g*) of section 30(1) of the Act.

That appears to me to be clearly a notice which satisfies the terms of section 26(6). It follows therefore that the ground so mentioned — namely para (*g*) of section 30(1) — is a ground upon which the landlord may in the proceedings begun following the service of such section 26 request rely upon by virtue of section 30(1) and thus I K would answer the first of the two preliminary issues that are ordered to be tried by me in favour of the tenant, but the second in favour of the landlords.

Court of Appeal

January 25 1994

(Before Lord Justice MANN, Lord Justice HIRST and Sir Michael KERR) L

NICHOLLS v KINSEY

Estates Gazette April 23 1994
[1994] 16 EG 145

Landlord and Tenant Act 1954 — Order under section 38(4) purporting to exclude security of tenure — Lease for one year then year to year containing agreement excluding sections 24 to 28 — Where tenancy not a term certain whether tenant bound by agreement to exclude Act M

On November 19 1979, and following a joint application by both parties, an order was made in the county court under section 38(4) of the Landlord and Tenant Act 1954 authorising the exclusion of sections 24 to 28 in relation to a proposed tenancy for a term of one year and to continue thereafter from year to year. On the same date the respondent landlord granted a tenancy to the appellant tenant; the tenancy recited the court order and was endorsed with the parties' agreement excluding sections 24 to 28. On December 10 1992 the tenant issued proceedings in the country court seeking the grant of a new tenancy under section 26 of the Act. On the hearing of a preliminary issue, the learned assistant recorder held that the tenancy was not for a term certain within the meaning of section 38(4)(*a*) of the Act, but rejected the tenant's contention that the agreement excluding the protection of the Act was a nullity and of no effect. The tenant appealed.

***Held*: The appeal was allowed. Section 38(1) is a declaratory anti-avoidance provision rendering all agreements as there defined void unless they escape through section 38(4). Section 38(4) is itself carefully and narrowly prescribed and renders such agreements valid provided only that there is a prospective tenancy for a term of years certain, the court authorises an agreement excluding the relevant provisions of the Act in relation to that tenancy and the agreement is made pursuant to such authorisation. The agreement fell outside section 38(4) because the tenancy was not for a term certain and consequently there was no authorisation; section 38(1) therefore applied and the agreement excluding the Act was void.**

The following cases are referred to in this report.

Barton v *Fincham* [1921] 2 KB 291

Inland Revenue Commissioners v *Hoogstraten* [1985] QB 1077; [1984] 3 WLR 377; [1984] 3 All ER 25, CA

Isaacs v *Robertson* [1985] AC 97; [1984] 3 WLR 705; [1984] 3 All ER 140, PC

Joseph v *Joseph* [1967] Ch 78; [1966] 3 WLR 631; [1966] 3 All ER 486, CA

Smith v *East Elloe Rural District Council* [1956] AC 736; [1956] 2 WLR 888; [1956] 1 All ER 855; (1956) 54 LGR 233; 6 P&CR 102, HL

This was an appeal by the tenant, William John Nicholls, from an order of Mr Assistant Recorder Bush, sitting at Macclesfield County Court, who held on the hearing of a preliminary issue that a tenancy granted by the landlord, William Nield Kinsey, was excluded from the protection of the Landlord and Tenant Act 1954.

Martin Rodger (instructed by Burges Salmon, of Bristol) appeared for the appellant; Paul Lowenstein (instructed by Rigby Stringer & Holmes, of Sandbach) represented the respondent.

Giving the first judgment at the invitation of Mann LJ, **Hirst LJ** said: This is an appeal against the order of Mr Assistant Recorder Bush made in Macclesfield County Court on July 26 1993, whereby it was determined that a tenancy agreement dated December 19 1979 made between the appellant tenant, Mr William John Nicholls, and the respondent landlord, Mr William Nield Kinsey, is not an agreement to which sections 24 to 28 in Part II of the Landlord and Tenant Act 1954 ("the Act") apply.

The appeal raises a short, but important, point as to the construction and application of section 38 of the Act.

The background facts are that on December 19 1979 the respondent landlord granted to the appellant tenant a business tenancy of a field at Windy Ridge, London Road, Cranage, Chester. The tenancy was expressed to be for a period of 12 months from December 19 1979, and to continue thereafter from year to year, but determinable by the landlord giving to the tenant 12 months' notice in writing of his intention to terminate the agreement.

As is very well known, the purpose of Part II was to provide security of tenure for business tenancies. Section 24 lays down the provisions for the continuation of such tenancies; section 25 lays down a procedure for the termination of such tenancies by the landlord; section 26 prescribes the procedure for a business tenant's request for a new tenancy and sections 29 to 37 lay down the procedures for court applications within the ambit of the Act.

Section 38, which is the critical section for the purposes of this case, provides so far as relevant as follows:

(1) Any agreement relating to a tenancy to which this Part of this Act applies (whether contained in the instrument creating the tenancy or not) shall be void (except as provided by subsection (4) of this section) in so far as it purports to preclude the tenant from making an application or request under this Part of this Act or provides for the termination or the surrender of the tenancy in the event of his making such an application or request or for the imposition of any penalty or disability on the tenant in that event.

. . .

(4) The Court may—

(*a*) on the joint application of the persons who will be the landlord and the tenant in relation to a tenancy to be granted for a term of years certain which will be a tenancy to which this Part of this Act applies, authorise an agreement excluding in relation to that tenancy the provisions of sections 24 to 28 of this Act; and

(*b*) on the joint application of the persons who are the landlord and the tenant in relation to a tenancy to which this Part of this Act applies, authorise an agreement for the surrender of the tenancy on such date or in such circumstances as may be specified in the agreement and on such terms (if any) as may be so specified;

if the agreement is contained in or endorsed on the instrument creating the tenancy or such other instrument as the court may specify; and an agreement contained in or endorsed on an instrument in pursuance of an authorisation given under this subsection shall be valid notwithstanding anything in the preceding provisions of this section.

Subsection (4) was not contained in the statute as originally enacted, but was added by section 5 of the Law of Property Act 1969.

On December 10 1992 the tenant issued proceedings in Macclesfield County Court seeking the grant of a new tenancy under section 26 of the Act. That claim was met by the landlord's answer dated March 4 1993, which contended that there was no jurisdiction under section 26 to grant a new tenancy, because, by an order made in the same county court on December 19 1979 under the provisions of section 38 of the Act, it had been ordered that the tenancy in question should exclude certain provisions of the Act, including sections 24 to 28 inclusive.

As a result, on March 25 1993 Mr District Registrar Tynas ordered the trial of the following preliminary issue:

Whether or not by virtue of the order of this court dated 19th December 1979 the tenancy dated 19th December 1979 made between the parties is a tenancy agreement which falls inside the statutory protection of Part II of the Landlord and Tenant Act 1954.

The course of events in December 1979 was that, on a joint application of the two parties to the proposed lease, the district registrar (Mr Tynas) made the following order dated December 19 1979:

Upon the joint application of the applicants, it is ordered pursuant to Section 38 (4) of the Landlord and Tenant Act 1954 (as amended by Section 5 of the Law of Property Act 1969) that the applicants be at liberty to enter into an agreement in the form of the draft annexed to the Originating Application herein excluding the provisions of sections 24 to 28 of the Landlord and Tenant Act 1954 in relation to a tenancy of the premises intended to be created by a Lease to be granted by William Nield Kinsey to William John Nicholls in the draft form annexed to the Originating Application herein and that such agreement be endorsed on the Lease.

Following upon this authorisation, on the same day, the parties entered into the following agreement which was endorsed on the tenancy agreement:

Whereas:

By an Order of the Macclesfield County Court dated the nineteenth day of December One thousand nine hundred and seventy nine in proceedings in which the Lessor and the Tenant were Joint Applicants this Agreement was duly authorised pursuant to Section 38(4) of the Landlord and Tenant Act 1954 (as amended by section 5 of the Law of Property Act 1969) and it was ordered that this Agreement should be endorsed on the attached Lease

Now this agreement witnesseth:

That the Landlord and Tenant hereby agree and declare that the provisions of section 4 to 28 of the said Act of 1954 (as amended) shall not apply to the attached Lease

The first issue before the assistant recorder was whether the prospective tenancy to which the district registrar's order and the above quoted agreement applied was for "a term of years certain", within the ambit of section 38(4)(*a*), seeing that the term, as already noted, was for a period of 12 months and to continue thereafter from year to year until determined by 12 months' notice in writing. This issue was resolved in the tenant's favour, the assistant recorder holding that it was not for a term of years certain, and this conclusion is not now challenged by Mr Paul Lowenstein on behalf of the landlord. It follows, to put it for the time being as neutrally as possible, that the district registrar was plainly not entitled to make the order he did under section 38, though I stress that he is open to no criticism on this count, since the point was never drawn to his attention. The learned recorder then proceeded to consider the submission from the tenant that the order was a nullity and the agreement consequently void; he held that it was not a nullity, but a valid and effective order under section 38(1) which, unless and until set aside in the proper procedural fashion, remained in force. Mr Martin Rodger, on behalf of the appellant, submits that the case turns entirely on a question of law, namely the proper construction of section 38, which he submits is clear and unambiguous and renders the agreement in the present case void. His chain of reasoning is as follows:

(1) Section 38(1) is a declaratory anti-avoidance provision rendering all agreements as there defined void unless they can escape through the loophole provided by section 38(4). This, he submits, is apparent not only from the clear wording of section 38(1) itself but also having regard to the decision of the Court of Appeal in *Joseph* v *Joseph* [1967] Ch 78 (Lord Denning MR, Diplock and Russell LJJ) where the Court of Appeal unanimously held that the word "purported" in section 38(1) means "has the effect of"; and that, in the words of Diplock LJ (at p90) section 38(1):

> . . . does render void any provision of an agreement between landlord and tenant whereby the tenant undertakes to do in the future any act which will have the effect under the statute of disqualifying him from applying for a new tenancy.

Joseph's case was decided at a time prior to the insertion of section 38(4) by the Law of Property Act 1969.

(2) The section 38(4) loophole is itself carefully and narrowly prescribed and renders such an agreement valid provided only that:

(a) there is a prospective tenancy for a term of years certain; and

(b) the court authorises an agreement excluding the relevant provisions of the Act *in relation to that tenancy*; and

(c) the agreement is made *in pursuance of such authorisation.*

(3) In the present case the prospective tenancy was not for a term of years certain; consequently there was no authorisation by the court in relation to such a tenancy, and the agreement was not made in pursuance of an authorisation given under the subsection.

(4) The agreement therefore fell outside the ambit of section 38(4) and was not valid under the terms of that subsection.

(5) The loophole thus having closed, section 38(1) applied with its full rigour, and the agreement was void.

Mr Rodger further submitted that since the jurisdiction of the county court has been expressly defined by section 38 the general principles as laid down by the Court of Appeal (Bankes, Scrutton and Atkin LJJ) in *Barton* v *Fincham* [1921] 2 KB 291 applied; Atkin LJ stated as follows in relation to a similar provision in the Increase Rent and Mortgage Interest (Restrictions) Act 1920 (p299):

> The section appears to me to limit definitely the jurisdiction of the Courts in making ejectment orders in the case of premises to which the Act applies. Parties cannot by agreement give the Courts jurisdiction which the Legislature has enacted they are not to have . . . A Court of limited jurisdiction cannot increase its jurisdiction by a pleading rule, and is bound itself to take notice of a defect of jurisdiction appearing on the face of the proceedings.

Mr Lowenstein's principal submission is that unless and until the registrar's 1979 order is set aside, it remains in force and the court has no jurisdiction to entertain the tenant's application under section 26 of the Act. Consequently, the judge was correct when he held that the order remained effective until such time as it was set aside.

In support of that argument he cited two leading authorities in very different fields. In *Isaacs* v *Robertson* [1985] AC 97 the Privy Council had to consider a case where a court granted an injunction notwithstanding the fact that under the RSC of the West Indies Associated States the action had been deemed "altogether abandoned" as a result of delay beyond the prescribed period. Giving the judgment of the board, Lord Diplock held that the order of the court granting the injunction was not a nullity (p101):

> upon the short and well-established ground that an order made by a court of unlimited jurisdiction, such as the High Court of St Vincent, must be obeyed unless and until it has been set aside by the court.

A similar conclusion was reached by the Court of Appeal in the case of *Inland Revenue Commissioners* v *Hoogstraten* [1985] QB 1077 in relation to a sequestration order imposed by a High Court judge which should not have been made, but none the less remained valid unless and until set aside (at p1089 *per* Dillon LJ with whom Stephenson LJ and Sir Roger Ormrod agreed).

With all respect to Mr Lowenstein's able argument, these two cases, which are fully authoritative in their own sphere, have, in my judgment, no bearing on the present case. The jurisdiction of the county court is entirely and exclusively statutory, for present purposes in the terms of section 38, and it follows that any departure from the criteria there laid down renders the authorisation of the court as given by the registrar in the present case inherently invalid.

As a subsidiary argument, Mr Lowenstein submitted that, even if the 1979 order were invalid, the agreement itself was nevertheless valid because it was entered into by the parties at a time when the 1979 order was a subsisting order of the court. This, in my judgment, is a hopeless argument, since, under the express terms of the subsection itself, the agreement can only be valid if it is made pursuant to an authorisation given under subsection 38(4), which this one was not; in any event, this approach would fatally undermine the whole purpose of section 38 as an anti-avoidance provision and render section 38(1) a virtual dead letter, since the signing of an agreement will, as in the present case, almost certainly take place immediately following the grant of an authorisation.

Returning to Mr Rodger's admirable argument, I am satisfied and I hold that every link in his chain of reasoning, which I do not need to repeat, is sound and is furthermore fully in accordance with the principles laid down in *Barton*'s case.

In a nutshell, the registrar's order was invalid and as a result the agreement was not made pursuant to an order under section 38(4) and therefore falls within and is declared void by section 38(1).

For these reasons I would allow this appeal, and would answer the question posed by the preliminary issue in the affirmative, and declare that the tenancy agreement dated December 19 1979 is a tenancy agreement which falls within the statutory protection of Part II of the Landlord and Tenant Act 1954.

Agreeing, **Sir Michael Kerr** said: This appeal should be allowed for the reasons stated in the judgment of Hirst LJ. The only question before this court, which at first sight appeared to present difficulty, was whether the order made on December 19 1979, although now seen to have been made without jurisdiction, was valid unless or until set aside, or whether it was always a nullity. The assistant recorder, Mr Bush, took the former view, but after hearing the argument before us I have no doubt that the latter is correct. We have not had to grapple, as he did, with the issue whether a tenancy for a period of 12 months and thereafter from year to year is one "granted for a term of years certain". The appellants now accept that this was correctly answered in the negative. The arguments and citations before us have therefore concentrated on what was a secondary issue below.

One of the cases not cited below which we have seen and to which Hirst LJ has already referred is *Barton* v *Fincham* [1921] 2 KB 291 (CA). This shows that Mr District Registrar Tynas, who also formulated the present preliminary issue, is not to be criticised in any way for the order which he had made in 1979. In that case a landlord and tenant purported to conclude an agreement which was in contravention of a precursor of the Rent Acts and which was then unwittingly enforced by a county court. In agreeing with the

A unanimous decision of the Court of Appeal that the order must be set aside, Atkin LJ said at p299:

> If the parties before the Court admit that one of the events has happened which give the Court jurisdiction, and there is no reason to doubt the bona fides of the admission, the Court is under no obligation to make further inquiry as to the question of fact; but apart from an admission the Court cannot give effect to an agreement, whether by way of compromise or otherwise, inconsistent with the provisions of the Act.

However, in that case no question arose about the validity of the order between the time when it purported to be made and when it was
B set aside. In dealing with that issue, the learned assistant recorder said in the present case:

> The question of whether or not the court had jurisdiction is a matter which requires careful consideration of law and careful investigation and examination of the terms of the tenancy agreement. In my judgment, to characterise the order which Mr Registrar Tynas made on the material placed before him, about which I make clear I know nothing, as being a nullity and as having no effect or being void, would be quite wrong. In my judgment, that order is a perfectly valid order of the court under section 38(4) which, unless and until it is set aside in the proper procedural fashion, remains in force. I cannot embark upon a rehearing on the basis of different evidence and on the basis of different, or it may be additional, submissions of law, and set myself
> C up as a Court of Appeal against the decision of Mr Registrar Tynas, that is, unless I were invited to do so by a proper procedural device.

I am bound to say that, in my view, this approach is untenable in the circumstances of the present case. The jurisdiction of Mr District Registrar Tynas was restricted by statute in a double sense. First, the jurisdiction of all county courts is limited by statute, though this plays no direct part in the present case. Second, however, no court could make such an order unless it complied with section 38(4) of the Landlord and Tenant Act 1954. But the order in question (I omit the use of words such as "purported" for brevity) was on its face in
D conflict with subsection (4). It stated that it was made in relation to "an agreement in the form of the draft annexed to the Originating Application herein". But that draft agreement referred on its face to a tenancy for 12 months and thereafter from year to year, and therefore not to a tenancy "for a term of years certain" as required by the subsection. No question of "evidence" arises, as mentioned by Mr Bush. In effect, therefore, the order expressly incorporated its own death sentence. Or, to adapt the colourful language of Lord Radcliffe in *Smith* v *East Elloe Rural District Council* [1956] AC 736 at p769, it bore the brand of invalidity upon its forehead.

This, I think, is the fundamental distinction between the present case and cases such as *Isaacs* v *Robertson* [1985] AC 97 and *Inland*
E *Revenue Commissioners* v *Hoogstraten* [1985] QB 1077, to which Hirst LJ has also already referred. In such cases the invalidity of the order in question is not intrinsic, as here, but dependent upon the establishment of extrinsic facts, which may then reveal that the order had been made without jurisdiction. Thus, in the former case the determination of the status of the order required the investigation and ascertainment of facts showing that the action in question was deemed to have been already "altogether abandoned". Similarly, in the latter case, the invalidity of the sequestration order in question could be determined only after an investigation which showed that the defendant had ceased to be in contempt of court by the time when it
F was made: see *per* Dillon LJ at p1086-87.

The present case was treated by the assistant recorder as falling into the same category, because the validity or invalidity of the 1979 order required the determination of the legal issue which he had to resolve: is a tenancy for a period of 12 months and thereafter from year to year, a tenancy for a term of years certain? That, however, is an issue of law, not of fact, and one which arose on the face of the order. In my view, that is a crucial distinction. The fact that it may have been a difficult issue of law makes no difference. In law, and on its face, the order was there and then either good or bad, without the need for any further investigation. It was intrinsically bad. Therefore it was always a nullity.

What we do not have to decide on this appeal is whether the G
position would be any different if the 1979 order had been made in the High Court, being a court of unlimited jurisdiction, to use the phrase of Lord Diplock in *Isaacs* v *Robertson* (*supra*). The position in that case was, of course, not the same as here, since the validity or invalidity of the order did not depend, as here, upon an intrinsic issue of law which was apparent on the face of the order itself, without recourse to any extraneous factors. However, the generality of the passage from p102G to 103E of the judgment purports to cover all High Court cases. But, since cases such as the present raise no issue which is relevant to the limited statutory jurisdiction of the county courts, but fall to be determined solely by reference to the intrinsic H
statutory jurisdiction relating to the particular order, it is difficult to see any logical difference between the two jurisdictions, assuming that we are right in our conclusion on the present appeal. However, that is an issue which I gladly leave for argument and decision on another occasion.

Mann LJ agreed and did not add anything.

Appeal allowed with costs; declaration that the tenancy agreement is a tenancy agreement which falls within the statutory protection of Part II of the Landlord and Tenant Act 1954; leave to appeal to the House of Lords refused.

J

Court of Appeal

February 15 1994

(Before Lord Justice STAUGHTON and Sir Francis PURCHAS)

TURNER AND OTHERS v WANDSWORTH LONDON BOROUGH COUNCIL

K

Estates Gazette June 25 1994

[1994] 25 EG 148

Landlord and Tenant Act 1954 — New tenancy opposed under section 30(1)(f) — Whether landlords had established intention under ground (f) by agreement with adjoining owner to remove buildings and occupy site under four-year lease — Whether landlord's real motive to sell irrelevant

The appellant landlords are the owners of a holding occupied by the respondent trustees as the City Farm, a strip of land upon which stand a number of buildings housing sheep and a few chickens. The L
landlords opposed the respondents' application for a new tenancy on the ground set out in section 30(1)(*f*) of the Landlord and Tenant Act 1954. In April 1993 the landlords entered into an agreement with Newton Prep Ltd, a company running an adjoining private school, by which that company would remove the old hardstanding of the site, demolish the buildings, put down some tarmac for a car park on part of the site and landscape the rest. On completion of these works the company would be given a four-year lease. There was evidence that the landlords intended eventually to sell the site when the market was more favourable. The landlords appealed the decision of the trial judge that they had failed to establish ground (*f*). M

***Held*: The appeal was allowed. The motive of the landlords in ultimately wishing to sell was irrelevant if, as was accepted, it had a genuine intention to carry out ground (*f*) works. The landlords intended that the demolition work be carried out by the company, that counted as intention within the meaning of section 30(1)(*f*). The real motive of the landlord, which was to sell the freehold, was irrelevant and the landlord was entitled to oppose a new tenancy.**

The following cases are referred to in this report.

Atkinson v *Bettison* [1955] 1 WLR 1127; [1955] 3 All ER 340, CA

Betty's Cafés Ltd v *Phillips Furnishing Stores Ltd* [1959] AC 20; [1958] 2 WLR 513; [1958] 1 All ER 607; [1958] EGD 92; (1958) 171 EG 319, HL; affirming [1957] Ch 67; [1956] 3 WLR 1134; [1957] 1 All ER 1, CA

Fisher v *Taylors Furnishing Stores Ltd* [1956] 2 QB 78; [1956] 2 WLR 985; [1956] 2 All ER 78, CA

Gilmour Caterers Ltd v *St Bartholomew's Hospital Governors* [1956] 1 QB 387; [1956] 2 WLR 419; [1956] 1 All ER 314, CA

Housleys Ltd v *Bloomer-Holt Ltd* [1966] 1 WLR 1244; [1966] 2 All ER 966, CA

Spook Erection Ltd v *British Railways Board* [1988] 1 EGLR 76; [1988] 21 EG 73, CA

This is an appeal by the appellants, Wandsworth London Borough Council, from a decision of Judge Hunter, sitting in Wandsworth County Court, who had decided a preliminary issue in an application for a new tenancy under the Landlord and Tenant Act 1954 Part II in favour of the respondent tenants, Messrs Turner, Speigal, Plant and Smith.

Guy Fetherstonhaugh (instructed by the solicitor to Wandsworth London Borough Council) appeared for the appellants; Nicholas Bowen (instructed by Howard Kennedy) represented the respondents.

Giving judgment, **STAUGHTON LJ** said: The plaintiffs in this action, Mr Turner, Mr Speigal, Mr Plant and Mr Smith, are the trustees of a property in Wandsworth which is the City Farm; or rather, they were the trustees when the action was commenced, but now they have been replaced by others. It has been agreed that the new trustees will be substituted as the plaintiffs upon proof that their predecessors have retired and that they have been appointed. No objection has been taken to the hearing of the appeal on that ground.

The City Farm occupies a strip of land in the London Borough of Wandsworth, apparently near to some railway lines. It is a long and narrow strip. It has some buildings; there is some old hardstanding, which looks as though it has been there for some time; the buildings are either prefabricated or of concrete blocks, no more than one storey high. The rest of the strip is rough vegetation, which might perhaps be used for grazing. We have some photographs.

The present use of the land is, we were told, only for one sheep and a few chickens, but it would seem that there have been more animals at other times. Mr Nicholas Bowen, for the plaintiffs, tells us that several thousand children in the London Borough of Wandsworth visit this farm and derive enjoyment, and indeed education, from seeing what animals get up to.

The trustees have occupied this strip for quite a long time (since the early 1980s) as tenants of Wandsworth London Borough Council. They were first granted an actual lease in December 1985. In 1987 there was an application to the court for a new tenancy under the Landlord and Tenant Act 1954. That was opposed. But in May 1991 there was an order for a new tenancy of 20 months from July 24 1990. Almost at once that tenancy had expired. In October 1991, the tenants sought another new tenancy under the 1954 Act. The landlords opposed that on the grounds set out in section 30(1)(*f*) of the 1954 Act. That provides that it is a ground for refusing a new tenancy:

> that on the termination of the current tenancy the landlord intends to demolish or reconstruct the premises comprised in the holding or a substantial part of those premises or to carry out substantial work of construction on the holding or part thereof, and that he could not reasonably do so without obtaining possession of the holding.

Certain points are common ground. First, the two alternatives "demolish" or "reconstruct" are disjunctive; the landlord can succeed if he shows an intention to do either. Second, that which is to be demolished or reconstructed has to be a substantial part of the premises, but not necessarily a substantial part of the holding as a whole.

In April 1993, the landlords entered into an agreement with a company called Newton Prep Ltd, which runs a neighbouring private school. The terms of the agreement were, in brief, that Newton Prep would, once the tenancy of City Farm had been terminated, dig up the old hardstanding on the site, demolish the buildings, put down some tarmacadam or concrete for a car park on part of the site, and arrange for the rest to be planted with grass and trees so that there would be an attractive, levelled site. On doing that, Newton Prep would be entitled to a lease for four years. There was also agreement about two other sites belonging to the council nearby.

The hearing of the dispute came on before Judge Hunter in Wandsworth County Court on July 16 1993. It was in the form of a preliminary issue: whether the landlord council could make out ground 30(1)(*f*) of the 1954 Act. If they could not do so, of course there was the remaining issue of the terms of the new tenancy for the tenant.

There was evidence at the trial of the council's intention to go ahead with the Newton Prep agreement; but there was also the minute of a committee of the council meeting on June 23 1992. The council's property committee considered a report on the Elm Farm site and neighbouring land and decided that, although it was intended eventually to sell the freehold of the three sites, the current market was not favourable for an early disposal. The committee resolved that the borough valuer and estate surveyor should enter into negotiation for the grant of a building agreement and a lease to Newton Prep School for the demolition and development of the three sites on terms to be agreed by the committee.

The judge, in the light of that minute, notwithstanding the evidence that he heard, was sceptical about the intention and motive of the council. The note we have of his reasoning is by no means complete. We fully understand that in the county court counsel have to do the best they can to take a note of the judgment, and so we have to do the best we can in reading it.

The judge said that the council's intention was colourable, not genuine, a ploy and not long term. Judging from the evidence of the minute, what he meant was that the council did indeed intend to have some demolition done, but that was because they wanted eventually to sell the freehold, and not for demolition's sake. In the result, the judge decided the preliminary issue in favour of the tenants, the trustees of City Farm. The council appeal against that decision.

We were told that the evidence of intention of the council was not challenged at the hearing and indeed that is accepted by Mr Bowen today. This experienced judge was fully justified in regarding the evidence with caution. But what we have to consider is the effect of his findings in relation to that evidence. Mr Bowen, it would seem, accepts that the motive of the council is neither here nor there. If they had an intention to demolish, it did not matter why they intended to demolish. At all events, if Mr Bowen does not accept that, it is plain law all the same.

We have been referred to a number of cases, first of all, *Atkinson* v *Bettison* [1955] 1 WLR 1127, where the Court of Appeal dismissed an appeal of the landlord on two grounds. The first was that demolition or reconstruction was an ancillary purpose and not the main or primary purpose of the landlord. In other words, motive was relevant. The second was that it was not a substantial part of the premises which was going to be demolished or reconstructed. Denning, Hodson and Morris LJJ on both those grounds rejected the landlord's appeal. That did not last long. Indeed, it has been referred to by Diplock LJ in a later case as a "short-lived fallacy". In *Fisher* v *Taylors Furnishing Stores Ltd* [1956] 2 QB 78, the Court of Appeal resiled from that. Sitting this time were Denning, Morris and Parker LJJ. It was there held that (p84):

> . . . the court must be satisfied that the intention to reconstruct is genuine and not colourable; that it is a firm and settled intention, not likely to be changed; that the reconstruction is of a substantial part of the premises, indeed so substantial that it cannot be thought to be a device to get possession; that the work is so extensive that it is necessary to get possession of the holding in order to do it; and that it is intended to do the work at once and not after a time.

That view, as we have said, was accepted by Diplock LJ in the case of *Housleys Ltd* v *Bloomer-Holt Ltd* [1966] 1 WLR 1244 at p1251, and also by the House of Lords in *Betty's Cafés Ltd* v *Phillips*

Furnishing Stores Ltd [1959] AC 20. Mr Bowen has to face the problem that, in general, motive is irrelevant, provided there is a genuine intention to demolish or reconstruct.

He meets that problem in this way. In terms of section 30(1)(*f*), it must be the landlord who intends to demolish or reconstruct and not somebody else. There is a definition of "landlord" in section 44, but, with respect to Mr Bowen and indeed the Parliamentary draftsman, that does nothing but confuse for the purpose of solving the present problem.

It is accepted that the landlord, as Sir Francis Purchas put it, is not obliged to do the work with his own hands or to intend to do it with his own hands; he can do it by his employees or by a contractor whom he engages on his own behalf. There is authority that the law goes further than that. The case of *Gilmour Caterers Ltd* v *St Bartholomew's Hospital Governors* [1956] 1 QB 387, shows that it is sufficient if the landlord arranges by means of a building lease for a new tenant to do the work. In that case the lease was for 48 years. Nevertheless, it was held to be, in effect, a landlord intending to do the work by his agent, the new tenant.

There then came the case of *Spook Erection Ltd* v *British Railways Board* [1988] 1 EGLR 76. There it was a case of a building lease for 99 years. Still it was held that this was the landlord intending to do the work by a tenant/agent on his behalf under a building lease.

Faced with those two authorities, Mr Bowen says: but the contemplated lease in this case was for only four years and was not really a building lease at all. That, to my mind, does not improve his position. If it be right, as was held in *Spook*'s case, that the landlord who arranges a building lease for 99 years, under which the new tenant would do the work, is still himself intending to do the work by his agent, how much the more so when the agreement is for only four years? It is not a stronger case for saying that he does not intend to do the work; it is a stronger case for saying that he does intend to do it. I would not accept the distinction, the ground upon which Mr Bowen says that the *Gilmour* case and the *Spook* case does not apply here.

It seems to me that, first of all, as is accepted, the landlords intended that demolition work should be carried out by Newton Prep; second, that that counts as the intention of the landlords within the meaning of the section; third, notwithstanding the judge's justified suspicion that the real motive was eventually to sell the freehold, still motive is irrelevant and the landlords are entitled to refuse a new tenancy.

I would allow the appeal and decide the preliminary issue in favour of the landlords, which results in the refusal of a new tenancy.

SIR FRANCIS PURCHAS agreed and did not add anything.

Appeal allowed.

County Court

September 20 1993

(Before Judge O'BRIEN at Brentwood County Court)

BUSBY AND ANOTHER v CO-OPERATIVE INSURANCE SOCIETY LTD

Estates Gazette February 12 1994
[1994] 06 EG 141

Landlord and Tenant Act 1954 — Compensation — Whether quitting tenant of business premises without domestic element entitled to compensation determined by reference to rateable value of premises on March 31 1990 — Whether para 4 of Schedule 7 to the Local Government and Housing Act 1989 so provides

By a lease dated March 20 1964 the applicants held a 21-year term of purely business premises. By notice dated September 22 1992 the respondent landlord determined the tenancy continued by the Landlord and Tenant Act 1954. In due course the applicants abandoned their application for a new tenancy, but applied to the court to have determined that they were entitled to compensation under section 37 of the 1954 Act, subject to the transitional provisions of para 4 of Schedule 7 to the Local Government and Housing Act 1989: compensation on this basis would be 16 times the rateable value on March 31 1990 as they had been in possession more than 14 years. The respondent landlord contended that: (1) the transitional provisions applied only where the premises were partly or wholly domestic because, having regard to the scheme in section 37 for compensation, the transitional provisions were designed to deal with the abolition of domestic rating; (2) the application of the rule in *Pepper* v *Hart* [1992] 3 WLR 1032 showed that the intended purpose of para 4 was to limit the transitional provisions to partly or wholly domestic properties; and (3) the applicants failed to serve an effective notice within the meaning of para 4(1)(c) of Schedule 7 claiming the higher basis of compensation.

***Held*: The transitional provisions in para 4 of Schedule 7 to the Local Government and Housing Act 1989 does not apply to purely business premises. All that para 4 does is to put a gloss on the preceding para 2, which introduced amendments to section 37 of the 1954 Act concerned with domestic property. Its provisions apply "instead" of those introduced into section 37 by para 2 of the Schedule. There was no ambiguity and the principle in *Pepper* v *Hart* therefore did not apply: however, if it did apply *Hansard* makes clear that the transitional provisions were concerned with partly or wholly domestic properties. In the event the applicants has served a notice which satisfied para 4(1)(c) of the Schedule.**

The following case is referred to in this report.

Pepper v *Hart* [1992] 3 WLR 1032

Edmund Cullen (instructed by AG Freeman & Son, of Benfleet) appeared for the applicants; Guy Fetherstonhaugh (instructed by the solicitor to the Co-operative Insurance Society Ltd) represented the respondent.

Giving judgment, JUDGE O'BRIEN said: In this case the applicants Alan Busby and Malcolm Roy Stockwell practised for many years as accountants in the premises, the subject-matter of the application, and seek the construction by this court of the compensation provisions in section 37 of the Landlord and Tenant Act 1954 (as amended). Against them is the Co-operative Insurance Society Ltd.

The applicants were the tenants of the property in question and the respondent the landlord. The facts of this matter are agreed between the parties. There is one quasi-factual dispute which turns on the admissibility of a letter headed "without prejudice". If inadmissible, there is one particular fact which is not before the court.

A The tenants were granted a business lease of the relevant premises on March 20 1964; it was a lease of 21 years less three days because, as I understand it, the landlords were themselves lessees, granting in effect a sublease.

On September 22 1992 the respondent landlord served a notice of termination of the lease under section 25 of the Landlord and Tenant Act 1954. The applicant tenants replied by solicitor's letter of December 14 1992 headed "without prejudice". They refer to the fact that they have until January 23 1993 to issue proceedings for a new lease. The letter then went on to consider various possibilities, indicating that the tenants might be prepared to go quietly without a
B dispute if they got a covenant from the landlord indicating that the premises would not be let to other accountants; and if they got a favourable answer, they might be prepared to deliver up possession, in which case:

> There will, of course, be the question of compensation and in this respect it seems that our clients have the alternative of claiming twice the current rateable value, which is £6,825

— I interpose: on that basis, compensation would be in the region of £13,650. I go back to the text —

> or alternatively, 16 times the old rateable value which is £1,113, making a
> C total of £17,808. Clearly, it will be in our client's interest to adopt the latter basis. Perhaps you would confirm that if our clients did give up possession on 25th March, this is the figure that would be payable. We look forward to hearing from you.

The respondent's solicitors replied to that letter on January 22 1993. It was an open letter. In the second paragraph they said:

> It is their intention to occupy these premises for the purposes of their own business. Accordingly, it is not my clients, intention to relet these premises. They are not prepared to enter into any covenant such as your clients have requested. What is your authority for saying that your clients have the option of seeking compensation of twice the current rateable value or 16 times the old
> D rateable value? My clients are concerned that your clients have not in fact been in occupation of the whole of the premises they currently occupy for in excess of 14 years and I should therefore be pleased to know what evidence is available to substantiate that claim.

I note that before me today it is not suggested that the applicants have not been in occupation for 14 years. Certainly the respondent's solicitors have taken note of what had been said regarding compensation and were disputing the basis that the applicants had indicated would be more advantageous. The applicants did in due course serve a counternotice dated January 19 1993. By that counternotice they sought a new lease.

E The respondent, on February 9 1993, formally opposed the grant of a new lease and on July 30 1993 the applicants formally and effectively abandoned their application for a new lease and indicated that they were quitting the premises in question. On June 8 1993 the applicants had given notice to the court, and to the landlord, that they would apply to the court today for a termination of the amount of compensation payable following the termination of the new lease on the ground set out in para (*g*) of section 30(1) of the Landlord and Tenant Act 1954.

The dispute is whether the basis of compensation amounting to £13,650 should be adopted or one which would give them £17,808.
F Counsel for the respondent raises three arguments why the lower basis of compensation is applicable. He argues that it is based on the scheme set out in section 37 of the Landlord and Tenant Act 1954. He argues that that provision so far as these premises are concerned, which are purely business, is unaffected by the legislation passed to amend the 1954 Act in the Local Government and Housing Act 1989, which he says was designed to provide for a new state of affairs when domestic rates were being abolished and the community charge was introduced. Those amendments were required for premises containing a domestic element because the old rating basis of compensation was no longer appropriate. So he says the whole point of this amending legislation was to deal with domestic or partly domestic property only.

Counsel for the applicants argues that Schedule 7 to the Local G
Government and Housing Act 1989 absolutely and clearly, by its terms, unambiguously and literally, has amended the situation in respect not only of domestic but also of premises with no domestic element. There is no authority that either party has been able to find after what I am quite sure, having seen the other fruits, were the most diligent researches. I have been helpfully referred to a number of standard texts and legal articles. None of them is binding on me. I approach this material, therefore, merely as an example of how the arguments have been set out. But it is clear that opinion in the profession is divided.

Woodfall says that a literal reading of para 4 of Schedule 7 to the H
Local Government and Housing Act 1989 suggests that para 4 applies to any tenancy and not merely tenancies of wholly or partly domestic premises, thereby assisting the applicants.

Delyth Williams in *Handbook of Business Tenancies* [published by Sweet & Maxwell] is silent on the matter, without suggesting that para 4 is limited to domestic premises and therefore the applicants say that it assists them. *Aldridge on Leasehold Law* expresses the matter in such a way that both parties say he assists them. *Hill and Redman's* suggests in terms that para 4 is limited to domestic premises and therefore assists the respondent. Legal journals: Howel Lewis in two articles dated July 1 1990 and January 29 1993 in *New Law Journal* J
has argued that the amending provisions relied upon by the applicants are limited to premises with a domestic element. On the other hand, it has been argued in *Estates Gazette* by Patrick McLoughlin (a solicitor with Theodore Goddard) on August 18 1990 and two solicitors in Finers [Michael Bibring and Katherine Miller] on July 28 1990, and in the *Law Society's Gazette* on September 12 1990 by two solicitors in Titmuss Sainer & Webb [Margaret Spencer and Steven Fogel], that these provisions apply as well to purely business premises as they do to domestic premises. These are the arguments as set out in those articles to which I have been referred.

The basic scheme of compensation as set out in the 1954 Act is K
based upon the rateable value. Because these premises had been occupied for more than 14 years compensation would have been twice the product of the appropriate multiplier and the rateable value and would have yielded compensation in the sum of £13,650. The alternative basis contended for is introduced by the Local Government and Housing Act 1989, and in particular by section 149, which in turn brings in Schedule 7 headed "Compensation provisions of Landlord and Tenant Act 1954 Part II". The first paragraph of that schedule says that:

> Any reference in this Schedule to a section which is not otherwise identified is a reference to that section of the Landlord and Tenant Act 1954, Part II of L
> which relates to security of tenure for business, professional and other tenants.

Para 2, it seems, is the effective amending provision of that Schedule. It has four subparagraphs, the first of which says:

> Subject to the following provisions of this Schedule, section 37 . . . shall have effect with the amendments set out below.

It then sets out the amendments in the following three subparas (2), (3) and (4). The effect, principally, is to add four subsections, 5(A), 5(B), 5(C) and 5(D) to section 37 of the principal Act. They are then set out as subpara (3) of para 2, Schedule 7. There is a fourth subparagraph which alters the way a multiplier is set out. My M
understanding of para 2(1) is that when it says "Subject to the following provisions of this Schedule", it means the provisions after para 2; and when it says "the amendments set out below", it means those numbered subparagraphs; and, in my judgment, it is only by approaching para 2 in that way that one understands the whole system and it all makes sense. After para 2 there are three more paragraphs, all of which I understand to be "the following provisions of this Schedule" and all of which I understand to be qualifying in some way para 2. For example para 3 says:

> The amendments made by paragraph 2 above do not have effect unless the date which, apart from paragraph 4 below, is relevant for determining the

A rateable value of the holding under sub-section (5) of section 37 is on or after 1st April 1990

That is a provision perfectly sensibly pointing the ambit or effect of the preceding paragraph. Then we get to para 4, which says:

(1) Subject to paragraph 3 above and paragraph 5 below, in any case

— words relied upon by the applicant —

where —

(a) the tenancy concerned was entered into before 1st April 1990 or was entered into on or after that date in pursuance of the contract made B before that date, and

(b) the landlord's notice under section 25 or, as the case may be, section 26(6) is given before 1st April 2000, and

(c) within the period referred to in section 29(3) for the making of an application under section 24(1), the tenant gives notice to the landlord that he wants the special basis of compensation provided for by this paragraph,

the amendments made by paragraph 2 above shall not have effect and section 37 shall, instead effect with the modification specified in sub-paragraph (2) below.

Subpara (2) provides what has been called the special basis of C compensation which in this case provides the greater figure, and reads:

The modification referred to in sub-paragraph (1) above is that the date which is relevant for the purposes of determining the rateable value of the holding under section 37(5) shall be 31st March 1990 instead of the date upon which the landlord's notice is given.

For the sake of completeness, the statutory instrument which deals with the multipliers — ie the Landlord and Tenant Act 1954 (Appropriate Multiplier) Order 1990 (SI 1990 No 36) — provides that the multiplier in the normal case will be one, but in the special basis will be a multiplier of eight.

D If my reading of para 2 is correct then all that para 4 is doing is putting on a gloss upon para 2. It is providing this special basis if three certain conditions can be met. And if that approach to Schedule 7 is correct, then the word "instead" in para 4 makes perfect sense. The contrary argument is that right at the beginning of para 4 it says: "in any case" which means "in any case within Part II of the Landlord and Tenant Act 1954". That is not the correct interpretation of this Schedule of this statute because, as I have already said, the real amendment is made by para 2. "In any case" in para 4 must be subject to the main amendment and "instead" is not surplusage and makes sense whereas on the applicants' argument "instead" is pure E surplusage. So I come to the conclusion that the construction contended for by the applicants is wrong. On that ground alone the applicants' contention fails.

I think it right in deference to the arguments which have been addressed to me to deal with the other arguments put forward. It was also argued for the respondent that if, contrary to its arguments, the provisions of Schedule 7 were ambiguous, obscure, or led to an absurdity, the recently enunciated doctrine in *Pepper* v *Hart* [1992] 3 WLR 1032 would come into effect and this court would be entitled to look at material which was a statement by a minister or a promoter of the Bill provided that the same was clear. With all due deference to F the applicants' argument, once I had grasped the matter and considered it as carefully as I am able to do, I concluded that the matter is clear perhaps not beyond a peradventure but clear. I do not find that there is any ambiguity or obscurity in the matter. Certainly it would not lead to any absurdity. But in case I am wrong on this point then the respondent would, in my judgment, be entitled to rely upon *Pepper* v *Hart* and to pray in aid what one of the ministers moving the Bill said about the matter. That is to be found in *Hansard* for October 11 1989 beginning at col 305 when Lord Hesketh moved that the House do resolve itself into committee on the Local Government and Housing Bill and when the Earl of Caithness (the Government's Paymaster General at the time), at col 338, moved the provision. At col 339 the Earl of Caithness said:

The abolition of the present rating system does not alter the basis of G compensation under the 1954 Act for purely non-domestic leasehold premises.

— I should have thought that one could hardly wish for a clearer statement —

The amendments in this schedule are particularly

— Again I regard that as clear; he could have said "particularly but not exclusively" but he did not. He said "particularly" —

directed towards business tenants' holdings which contain a domestic element, for example, shops with flats above. Existing tenants of such mixed hereditaments will have a choice for a transitional period from 1st April 1990 as to the basis of compensation. H

So if the respondent was to have failed on his first argument, he would have succeeded on this. He (counsel for the respondent) goes further in any event and says that the applicants failed to get over the third condition in para 4 of Schedule 7 (para 4(1)(c)) regarding the notice to be given in that they did not give notice to the landlord requiring the special basis of compensation. That turns on the construction of the applicants' solicitor's letter dated December 14 1992 in which they said:

There will, of course, be the question of compensation and in this respect it seems that our clients have the alternative of claiming twice the current J rateable value . . . or alternatively, 16 times the old rateable value . . . Clearly it will be in our client's interest to adopt the latter basis.

Well, in my judgment, that did just enough to give notice that they wanted the special basis of compensation. It would not be the form of notice that I would recommend to practitioners, but it just passes muster. There is a second objection: that the letter was "without prejudice". It was answered by an open letter, but it is said that the applicants cannot waive privilege unilaterally. I do not think that either party came to court expecting to argue this point, because if they did they would have come armed with authority as they have done on the first two points. In those circumstances I do not propose to decide K that point. I have not had enough assistance on it. I determine that the amount which was due to the applicants by way of compensation was £13,650.

Declaration accordingly.

County Court

October 21 1993 L

(Before Judge Glyn MORGAN, sitting in Newport (Gwent) County Court)

FORBOUYS PLC v NEWPORT BOROUGH COUNCIL

Estates Gazette June 18 1994
[1994] 24 EG 156

Landlord and tenant — Landlord and Tenant Act 1954 — Terms of new tenancy — Whether rent review clause to be upwards only or upwards or downwards M

The plaintiff applied by originating application for the grant of a new tenancy under Part II of the Landlord and Tenant Act 1954 following the expiration of a 21-year term that it held granted from July 1 1970. The parties agree all the terms of the new lease save for certain provisions of the rent review clause. The defendants agreed to grant the plaintiff a nine-year term of a shop in the Bettws Shopping Centre, Newport, at a rent of £4,600 pa subject to a rent review every three years. The defendants contended that the rent at review should be upwards only, whereas the plaintiff sought an upwards or downwards clause.

***Held*: A rent review clause upwards or downwards would not be unfair on the landlords and would be fair and reasonable in all the circumstances. Such a clause should therefore be included in the new lease.**

The following cases are referred to in this report.

Amarjee v *Barrowfen Properties Ltd* [1993] 2 EGLR 133; [1993] 30 EG 98
Blythewood Plant Hire Ltd v *Spiers Ltd* [1992] 2 EGLR 103; [1992] 48 EG 117
Boots the Chemists Ltd v *Pinkland Ltd* [1992] 2 EGLR 176; 28 EG 118
Janes (Gowns) Ltd v *Harlow Development Corporation* [1980] EGD 110; (1979) 253 EG 799
O'May v *City of London Real Property Co Ltd* [1983] 2 AC 726; [1982] 2 WLR 407; [1982] 1 All ER 660; (1982) 43 P&CR 351; 261 EG 1185, HL
Stylo Shoes Ltd v *Manchester Royal Exchange Ltd* (1967) 204 EG 803

Jonathan Brock (instructed by Donne Mileham & Maddock, of Brighton) appeared for the plaintiff tenant; Caroline Stent (instructed by the solicitor to Newport Borough Council) represented the defendant landlords.

Giving judgment, **JUDGE GLYNN MORGAN** said: By originating application dated July 4 1991, the plaintiff applies to the court for the grant of a new tenancy pursuant to Part II of the Landlord and Tenant Act 1954. The premises to which the application relates are the Corner Shop, 6-8 Bettws Shopping Centre, Newport. The defendants had granted a 21-year lease to the plaintiff from July 1 1970 at a rent of £2,200 pa. The lease expired by effluxion of time on June 30 1991. The plaintiff is still in occupation of the demised premises, paying an interim rent.

It has been agreed between the parties that the defendants will grant a new lease of the premises. All the terms of the lease have been agreed save for one, upon which the decision of the court has been sought. The lease will be for a nine-year term at a rent of £4,600 pa with a rent review every three years. The sole issue for the decision of the court is whether, as the defendants contend, at each review date the rent should be equal to the rent previously payable or such revised rent as may be ascertained, whichever be the greater, or, as the plaintiff contends, the rent should be such revised rent as may be ascertained. In more simple terms, the defendants contend that the rent review should be upwards only and the plaintiff contends that the rent review should be upwards or downwards.

Part II of the Landlord and Tenant Act 1954 which deals with business tenancies gives wide powers to the court to grant and settle the term on which the business tenant is to have a new lease. Section 33 relates to the duration of the tenancy. Section 34 relates to the rent under a new tenancy. Section 34(3), as amended by the Law of Property Act 1969, provides:

> where the rent is determined by the Court, the Court may, if it thinks fit, further determine that the terms of the tenancy shall include such provisions for varying the rent as may be specified in the determination.

Section 35 relates to other terms of a new tenancy.

It was decided by Cross J in *Stylo Shoes Ltd* v *Manchester Royal Exchange Ltd* (1967) 204 EG 803 that the court, in its discretion, could provide for a rent review clause which provided for review of the rent downwards as well as upwards.

I was referred to a number of cases where an upwards or downwards rent review has been considered: in *Janes (Gowns) Ltd* v *Harlow Development Corporation* (1979) 253 EG 799 Judge Finlay QC, sitting as a judge of the High Court, ordered a rent review clause, five-yearly upwards or downwards. In *Boots the Chemists Ltd* v *Pinkland Ltd* [1992] 28 EG 118* Judge Thompson ordered a rent review clause, five-yearly, upwards and downwards. In *Blythewood Plant Hire Ltd* v *Spiers Ltd* [1992] 48 EG 117† Judge Diamond QC declined to order a rent review clause upwards or downwards and ordered that the clause to be inserted in this lease should be an upwards-only review clause. In *Amarjee* v *Barrowfen Properties Ltd* [1993] 30 EG 98* Judge Beddard ordered a rent review clause, five-yearly, upwards or downwards.

I heard the expert evidence of Mr Jerry A Burton FSVA ACIArb, on behalf of the plaintiff. His report dated September 27 1993 with appendices was put in evidence. I heard the expert evidence of Mr Wayne F Morgan ARICS, on behalf of the defendants. His report dated October 7 1993 with appendices was put in evidence.

On behalf of the landlords, Miss Caroline Stent submitted that the purpose of rent review clauses in business leases was to protect the long-term interest of the landlord and to protect the value of the landlord's capital investment and the yield to be obtained from such capital investment. Where there is a rent review clause upwards only in a business lease, the tenant is able to procure better terms of the lease than he otherwise might get, for example in the length of the lease, the basic rent, repairing and other conditions. Miss Stent submitted that market conditions in Newport are such that any further decline in rental values is unlikely and that the demised premises at Bettws were in any event at the bottom end of the market in Newport, so that there was no justification for a downwards rent review. Miss Stent further submitted, that market conditions in Newport were quite different from those which have prevailed and which are likely to be in future in London and South East England. Hence, the views which prevailed in all the cases, except that decided by Judge Diamond QC, were not relevant in consideration of the present case. She submitted that the correct approach of the court should be to consider the concept of these fairness of this landlords and to this tenant in the prevailing market conditions and that such approach must lead to the conclusion that an upwards-only rent review clause was the fair and appropriate form of order.

Mr Jonathan Brock submitted, on behalf of the tenant, that except for the decision of Judge Diamond QC, who decided as he did without hearing any evidence on the point, that all the recent authorities shewed that an upwards and downwards rent review clause was appropriate in modern conditions. Only a landlord could benefit from an upwards-only rent review clause, whereas there were benefits to landlord and tenant from an upwards and downwards rent review clause. An upwards and downwards rent review clause necessarily produces a fair market rent. Why, Mr Brock asked, should a landlord obtain or preserve institutional benefits at the expense of a tenant? Market conditions should be considered. Rents have dropped in Newport over the past few years. Neither of the expert witnesses could guarantee that rents at Bettws would not go down in three or six years' time. Where rents are going down, it would be grossly unfair for rents to go up pursuant to a review. The facts of this case justify the ratio of all the recent authorities, except for the decision of Judge Diamond QC. Applying the concept of overall fairness to both sides, the court should hold in favour of the tenant and to decide that an upwards and downwards rent review clause was appropriate in this case.

The shopping centre at Bettws was built in the 1960s to service a large council housing estate. It comprised a development of 24 trading units, together with a public library, health centre and a supermarket. Mr Burton annexed a plan dated September 27 1993 to his report, which set out the occupancy of the trading units. I am satisfied as to the accuracy of the information shewn in the plan. Of the original 24 shops, 11 now trade as retail units, four are occupied by other occupants such as a dentist, five are occupied by the defendant as a housing office, two are occupied by the police and two are vacant.

In October 1991, William Hill Bookmakers and Gateway Supermarket renewed the leases of the premises occupied by them for terms of 15 and 20 years respectively subject to five-yearly upwards- only rent reviews. I accept the view, put forward by Mr Burton, that at the time when the renewal of these leases was being negotiated the effects of the recession were not being felt to the same extent as at the present time and that the prevailing market conditions were different. I also accept that a three-year rent view, albeit in a

*Editor's note: Also reported at [1992] 2 EGLR 176.
†Editor's note: Also reported at [1992] 2 EGLR 103.
*Editor's note: Also reported at [1993] 2 EGLR 133.

A shorter lease, which has been agreed in this case, is more onerous to a tenant than a five-year review.

In the 1980s there was a significant growth in business rents in London and South East England. The content of that growth was not so pronounced in South Wales and in particular in Newport, though growth there undoubtedly was. I accept the evidence of Mr Burton that, in general terms, units at the shopping centre at Ringland, another large council estate in a different part of Newport, are over-rented, while rents in the centre of Newport at Commercial Street have dropped. Mr Morgan accepted that rents in the prime areas of Newport have not risen, and that there were places in Newport where retail
B tenants were paying too much rent. He made the point, which I accept, that secondary and tertiary trading areas do not experience the same peaks and troughs of the local market as the primary trading areas. I accept that Bettws Shopping Centre is at the bottom end of the local market.

The effects of the recession over the past year or two have been more exaggerated in London and South East England than in South Wales. Neither of the expert witnesses was confident that the end of the recession is in sight, though Mr Morgan was more optimistic than Mr Burton. Where there is a conflict between them, I prefer the view of Mr Burton that "the recession is still bumping along the bottom"
C and that Newport is below average in recovering from the effects of the recession.

In *O'May* v *City of London Real Property Co Ltd* [1982] 1 All ER 660* the House of Lords held that since, under section 35 of the Landlord and Tenant Act 1954, the court was required to have regard to the terms of the current lease, the burden of persuading the court to change those terms rested on the party proposing the change and, although the court had wide discretionary powers to change the terms and impose new terms, the party proposing the change had to show that the change was fair and reasonable in all the circumstances. At p671F-G, Lord Wilberforce said:

> D There is no obligation under section 35 of this Act, to make the new terms conform with market practice if to do so would be unfair to the tenant. And there is no inherent necessity why the terms on which existing leases are to be renewed should be dictated by those of fresh bargains which tenants may feel themselves obliged to accept. The court has to compare the advantage desired by the landlord with the detriment to be suffered by the tenants, and to E consider whether any monetary compensation offered against that detriment ought fairly to be imposed on the tenants in exchange for the acceptance of that detriment.

It is the case that, until about 30 years ago, rent review provisions were not usual in leases for 21 years or less. But inflation completely changed the position. In the case of leases which contained no rent review provisions, landlords saw the return on their investment dwindle to relatively insignificant levels towards the end of the life of the lease. So the practice grew of incorporating provisions for rent reviews at various intervals to enable landlords to maintain the level of their rental return during the period of high inflation. F

It is true that the particular facts of this case can be distinguished in various ways from the facts of the cases cited to me and to which I have referred. I bear in mind that I have to have regard to the provisions of the old lease, while at the same time bearing in mind that all the terms of the new lease, save for the rent review case, have been agreed. I bear in mind that the order which I make must be fair and reasonable in all the circumstances.

On the facts of this case, I have come to the conclusion that to order a rent review clause upwards only would be unfair to the tenant and would not be fair and reasonable in all the circumstances. On the other hand, a rent review clause upwards or downwards would not be G unfair to the landlords and would be fair and reasonable in all the circumstances. It is not inevitable that the rent would be adjusted downwards on a review. It seems to me that the rent is as likely to be adjusted upwards as downwards. Like Judge Thompson in *Boots the Chemists* v *Pinkland Ltd* I do not have a crystal ball. It seems to me that an upwards and downwards rent review clause is fair to both parties; at reviews, the rent to be determined will largely depend on prevailing market conditions.

Accordingly, I rule that clause 2 B(b) of the draft lease should be in these terms:

> During each successive review period a rent equal to such revised rent as may H be ascertained as herein provided.

It having been agreed between the parties that the costs of this application should follow the event, I order that the defendants pay costs of the plaintiff on County Court Scale 2.

*Editor's note: Also reported at (1982) 261 EG 1185.

For a further case on this subject see p 249

LANDLORD AND TENANT

RENT ACTS

Court of Appeal

June 16 1993

(Before Lord Justice LLOYD and Lord Justice ROCH)

DAWNCAR INVESTMENTS LTD v PLEWS

Estates Gazette April 2 1994

[1994] 13 EG 110

Landlord and tenant — Rent Act 1977 — Suitable alternative accommodation — Reasonableness — Weight to be placed on environmental considerations.

Since 1975 the defendant has been the protected tenant of a first-floor flat at 50 Pilgrims Lane, London NW3, owned by the plaintiffs. The defendant, who was aged 42 and had one child, worked as a self-employed textile designer in Covent Garden, London. The plaintiffs sought possession of the flat on the ground that they had offered as suitable alternative accommodation a first-floor flat at 4 Iverson Road, London NW6. In the county court the judge, in dismissing the claim for possession, considered it would not be reasonable to make such an order as it would be entirely wrong to expect the defendant to live at Iverson Road, an area of noise, traffic, heavy lorries, railway lines and rough inhabitants, notwithstanding that the alternative accommodation, as accommodation, was in many ways superior. The plaintiffs appealed, contending that where environmental considerations are relevant they should be given less weight where a landlord puts its case under the Rent Act 1977, Schedule 15, Part IV, para 5(1)(*a*) rather than para 5(1)(*b*).

***Held*: The appeal was dismissed. The court is not obliged to give less weight to environmental considerations in assessing overall reasonableness because of the way the plaintiffs choose to put their case. The weight to be put to factors which are relevant are essentially a question for the trial judge, especially where the judge has had the opportunity of seeing both flats.**

The following case is referred to in this report.

Redspring Ltd v *Francis* [1973] 1 WLR 134; [1963] 1 All ER 640, CA

This was an appeal from the decision of Judge Martin QC (sitting in Bloomsbury County Court), who on April 16 1992 dismissed a claim by the plaintiffs, Dawncar Investments Ltd, for possession of premises occupied by the defendant Miss Vanessa Plews.

Romie Tager (instructed by JJ Goldstein & Co) appeared for the appellants; Paul Staddon (instructed by Oliver O Fisher & Co) represented the respondent.

Giving judgment, **LLOYD LJ** said: This is an appeal from a judgment of Judge Martin QC, given in Bloomsbury County Court on April 16 1992, whereby he held that the landlords, Dawncar Investments Ltd, had failed to make out a claim for possession of a dwelling-house on the first floor of 50 Pilgrims Lane, London NW3.

The defendant, Miss Vanessa Plews, is a protected tenant for the purposes of the Rent Act 1977. She is a self-employed textile designer, working in Covent Garden. She has been in occupation of the flat since 1975. She is now aged 42 and has one child.

The plaintiffs' case is that there is suitable alternative accommodation available for her on the first floor of 4 Iverson Road, London NW6. The learned judge, who is very experienced in cases of this kind, visited both flats. He formed the view, which he expressed in forthright language, that it would not be reasonable to make an order in this case. He said it would be entirely wrong to expect the defendant to live in Iverson Road and that he had reached that view without any hesitation. It is obvious, therefore, that the plaintiffs must face a difficult task in persuading this court to take a different view on the facts, unless the judge has misdirected himself in law, or taken account of some wholly irrelevant factor.

The description and character of the two flats are well set out in the judgment. Briefly, Pilgrims Lane is a quiet road in a pleasant part of Hempstead. It is on the fringes of the Heath, with good-quality shops nearby. Iverson Road, on the other hand, is a busy commercial road, running between Westend Lane and Kilburn High Road. There is a lot of traffic, including heavy lorries turning into a timber yard immediately opposite the house. The noise of the bandsaw in the timber yard is described as being very intrusive. There is a railway at the front of the house and another at the back, two public houses nearby and a lot of drunkenness.

In the course of her evidence the defendant said that she would not feel safe in the area. The judge accepted that evidence. He found that she would be afraid to live there and with good reason. A woman police officer gave evidence as to the crime rate in the locality. The judge summed it all up when he said in two sentences:

> I would be very unhappy for a woman like Miss Plews, having to live in 42 Iverson Road because of the noise, traffic, heavy lorries, proximity of railway lines, general roughness of the area and of the inhabitants. I find as a fact that it is not suitable as regards character.

If one could ignore the neighbourhood all together, then the judge found that the alternative accommodation, as accommodation, was in many ways superior. But I need not go into details.

I turn then to the relevant statutory provisions, which are contained in section 98 of the Rent Act 1977, and Part IV of Schedule 15. These provisions are so well known that I need not take time to read them, save only para 5 of Part IV of Schedule 15, which sets out the conditions which have to be fulfilled under para 4. Para 5 reads:

> (1) For the purposes of paragraph 4 above, the relevant conditions are that the accommodation is reasonably suitable to the needs of the tenant and his family as regards proximity to place of work, and either —
>
> (*a*) similar as regards rental and extent to the accommodation afforded by dwelling-houses provided in the neighbourhood by any authority [local housing authority] for persons whose needs as regards extent are, in the opinion of the court, similar to those of the tenant and of his family; or
>
> (*b*) reasonably suitable to the means of the tenant and to the needs of the tenant and his family as regards extent and character;

No question arises in the present case with regard to proximity to the tenant's place of work. But Mr Tager draws our attention to the inclusion under para 5(1)(*b*) of a reference to the character of the alternative accommodation, as well as to its extent; whereas in para 5(1)(*a*) there is no reference to the character of the accommodation. Yet, it was by reason of the character of 42 Iverson Road, as affected

A by the character of the neighbourhood, that the judge found that it would not be reasonable to make an order for possession in this case.

If Mr Tager had felt able to submit that the character of the house was irrelevant, when considering the question of reasonableness under section 98 of the Act, then he would have been on strong ground in arguing that we ought to disturb the judge's decision, for the judge would, on that view, have taken account of an irrelevant consideration. But Mr Tager rightly did not feel able to press his argument that far. He accepted that reasonableness under section 98 is an overriding consideration and that in assessing reasonableness all relevant consideration had to be taken into account, including the character of

B the accommodation. Moreover, in the case of *Redspring Ltd* v *Francis* [1973] 1 WLR 134, this court held that what Buckley LJ called "environmental considerations" are relevant under what is now section 98, as well as under para 5(1)(*b*) of Schedule 15. I quote from Buckley LJ's judgment at p138E:

> Under section 10 of the Act of 1968 it is possible that this sort of consideration might be regarded as coming into consideration under one of two heads: either as a consideration which affects the suitability of the accommodation as regards its character, or as being a matter which affects the question of whether it is proper for the court to regard it as reasonable to make a possession order under the section. For my part, I do not think that it
> C very much matters under which head one regards it as coming. I am, for myself, satisfied that such environmental questions are proper matters to be taken into consideration in deciding whether or not a possession ought to be made.

In other words, environmental considerations may do double service, both under para 5(l)(*b*) and also in assessing reasonableness under section 98.

Mr Tager was, for those reasons, right to accept that the environmental considerations, which impressed the judge so much, D were relevant to his decision under section 98.

How then does Mr Tager put the appeal? He says, in effect, that environmental considerations, although relevant, should be given less weight when the landlords put their case under para 5(1)(*a*) (as they did in this case) than when they put their case under para 5(1)(*b*), as they did in *Redspring Ltd* v *Francis*. He says that the judge's failure to appreciate that distinction was an error or, alternatively, that the judge erred in giving too much weight to the environmental considerations on the facts of this case.

I cannot accept either branch of that argument. I do not agree that the judge was obliged to give less weight to environmental E considerations in assessing overall reasonableness, because of the way the plaintiffs chose to put their case. But, in any event, the weight to be given to factors which are, admittedly, relevant is essentially a question for the trial judge. I find it difficult to imagine a case in which this court would interfere because the judge had given, in this court's view, too much weight to one factor, and not enough weight to another factor, unless it could be said that the judge was plainly wrong in his conclusion. This is especially so where the judge has had the opportunity of seeing both flats and the environment in which they are set, which we have not. Mr Tager has thus failed to satisfy me that the judge was plainly wrong to form the view which he did.

On that short ground I would dismiss this appeal. In those F circumstances, I need not consider whether the judge was right to hold, as he did, that the plaintiffs had not succeeded in bringing their case within para 5(l)(*a*) at all, for reasons which do not matter. I would dismiss the appeal.

ROCH LJ agreed and did not add anything.

Appeal dismissed.

LANDLORD AND TENANT

RENT REVIEW

A # Chancery Division

November 29 1993

(Before Mr Justice HARMAN)

BROADGATE SQUARE PLC v LEHMAN BROTHERS LTD

Estates Gazette January 29 1994

[1994] 04 EG 135

B ***Landlord and tenant — Rent review — Arbitration — Open market rent defined as best yearly rent after expiry of a rent-free period — Whether deduction should be made for incentive period from rent payable after expiration of rent-free period to arrive at open market rent***

An arbitrator made an award pursuant to the provisions of a rent review clause in a lease of office premises at 1 and 2 Broadgate, London EC2, owned by the applicant landlord and let to the respondent tenant. For the purposes of the rent review, the open market rent was defined as "the best yearly rent which would reasonably be expected to become payable after the expiry of a rent-free period of such length as would be negotiated in the open C market between a willing landlord and a willing tenant upon a letting of the Premises . . .". The arbitrator concluded that, when analysing comparable transactions, the "additional" or "inducement" rent-free period (additional to any period for fitting out) may be amortised and a suitable deduction made from the "face" rent payable at the expiration of the rent-free period. He applied a factor on a discounted cash-flow basis to reflect the "incentive" rent-free period of 24 months amortised over 10 years at 6%. That was done on the basis that the fitting-out period was to be ignored, but the inducement rent-free period should be taken into account. He determined the open market rent of the premises at £5,281,365 pa for 1 Broadgate and £3,931,795 pa for 2 D Broadgate; alternatively, if he was held to be incorrect and no regard should be had to inducement rent-free periods, £7,033,316 pa for no 1 and £5,236,046 pa for no 2. The landlord appealed the award by notice of motion under section 1 of the Arbitration Act 1979, contending that the arbitrator erred in law in making deductions from the "face" rent in arriving at the open market; the tenant cross-appealed the arbitrator's decision that the rent review clause was not onerous.

Held: **The landlord's appeal was allowed and the tenant's dismissed. The definition of "open market rent" in the lease requires the arbitrator to find the rent which will actually be payable when such rent-free period as would have been E negotiated shall have expired. In looking at comparables, the arbitrator does not take the rent reserved as being the true rent of the comparable premises, but is entitled to adjust such rents to arrive at the true rent. This the arbitrator appears to have done in arriving at the "headline" or "face" rent. However, there was no warrant in terms of the rent review clause for the arbitrator applying some further adjustment to the rent. The tenant's contention that the arbitrator wrongly failed to make a deduction to reflect the onerous nature of the rent review clause was dismissed on the ground that it was a question of fact and not law, and therefore was not appealable. The judge gave a certificate under section 1(7)(b) of the Arbitration Act 1979, but refused leave to appeal.** F

The following cases are referred to in this report.

British Gas Corporation v *Universities Superannuation Scheme Ltd* [1986] 1 WLR 398; [1986] 1 All ER 978; (1986) 52 P&CR 111; [1986] 1 EGLR 120; 277 EG 980

British Gas plc v *Dollar Land Holdings plc* [1992] 1 EGLR 135; [1992] 12 EG 141

This was the hearing of notices of motion under section 1 of the Arbitration Act 1979 seeking to appeal an award of an arbitrator in rent review proceedings between the applicant landlord, Broadgate Square plc, and the respondent tenant, Lehman Brothers Ltd. G

Kim Lewison QC (instructed by Herbert Smith) appeared for the landlord; David Neuberger QC (instructed by Freshfields) represented the tenant.

Giving judgment, **Harman J** said: I have before me four notices of originating motion, two on behalf of a landlord, Broadgate Square plc, and two on behalf of a tenant, Lehman Brothers Ltd. All arise out of an award by an arbitrator on a rent review clause. They were entered pursuant to a consent of the parties by notices by the landlord of April 28 1993 and by the tenant on May 4. The matter involves extremely H large sums of money and obviously is, to the parties, of considerable commercial importance. The clause is a specific one, I think frequently called a one-off clause, which was before the arbitrator on the rent review hearing. The award contains extracts from the lease. The lease itself is not appended to or incorporated expressly by reference in the award. Accordingly, when it was submitted to me that I should look at the terms of the lease itself, I held that under Ord 73 and the practice note set out at p1266 of the current White Book, only documents incorporated in or expressly appended to the award are proper for the court to look at, and I ruled that I should not look further at the lease than the extracts of it in the award.

The rent review clause itself, as with so many such clauses, is one J where the longer you look at it the less clear becomes its meaning. I will read it, quoting from p2 of the award under the rubric 5:

> "Open Market Rent" means the best yearly rent which would reasonably be expected to become payable in respect of the premises after the expiry of a rent free period of such length as would be negotiated in the open market between a willing landlord and a willing tenant upon a letting of the Premises as a whole by a willing lessor to a willing lessee in the open market at the relevant Review Date for a term of ten years or a term equal to the residue then unexpired of the Term (whichever is the longer) with vacant possession without fine or premium . . .

K I think that both the extremely able and learned counsel who have argued this matter before me agreed in the end that the clause contains tautologies and unhappinesses. The arbitrator himself plainly found the matter difficult and he asked for a legal assessor, who sat with him on the first day, having read the skeleton arguments presented by leading counsel to the arbitrator and heard that argument, and on the last day also, when points which the assessor wanted dealt with were put to counsel for both sides. As a result the assessor, Mr Joseph Harper QC, one of Her Majesty's counsel, gave an opinion which, although it runs to 12 pages, is comparatively short and clear to read. I

A say "comparatively" considering other such documents that the court has seen.

He sets out the questions of law as to the true construction of the clause, as to the meaning of the rent-free period point, as it was defined. He refers to the current market conditions, which I think must be a reference to the market conditions as at December 1991, which is the relevant date for the rent-review terms, and he refers to various authorities, notably the decision of Sir Nicolas Browne-Wilkinson V-C in *British Gas Corporation* v *Universities Superannuation Scheme Ltd* [1986] 1 WLR 398* at p403 A-C, and finds (as I consider entirely correctly) that clear words in a rent review
B should be given effect to, but that subject to special circumstances it is proper to give effect to the underlying commercial purposes of a rent review clause and to construe the words so as to give effect to that purpose.

Mr Harper concludes, and it was accepted by both sides as correct, at para 9:

> In my opinion, an arbitrator should if possible give effect to the literal meaning of the words used in the rent review clause.

He then went on to say that he was of the view that it is possible to do so in the present case without arriving at an absurd result. He then
C considers the terms of the clause which I have read. He observes at para 11 that the formula, that is the terms of the rent review clause, says nothing whatever about comparables. It relates to "the Premises" which are, as he says, the demised premises, and it is said that the formula is telling the arbitrator that he must direct his attention to achieving such a rent as would be payable for these, that is the demised premises, after the expiry of the hypothetically market negotiated rent-free period. That would be the new rent which will be payable from day one of the review period.

It was pointed out in argument, in considering this opinion, that the new rent as defined by the clause as an open market rent is not in fact the rent payable passing between these two parties, because the rent
D passing between them is, in the case of one of the two buildings in Broadgate, 90.275% of the open market rent, and the other 90 point (some other slight fractional difference)% of the open market rent. That seems to me an entirely passing point which in no way affects the soundness of the legal advice tendered.

He went on in para 12 to say:

> Accordingly, what an arbitrator does when evaluating comparable transactions on the present review is a matter for the valuer's art. It is not a matter touched by this formula. In particular therefore the arbitrator is not compelled, *in respect of such comparables*, to "look only at the rent payable after the expiry
> E of any rent free period". He has a free hand.

That advice being tendered, the arbitrator had to conclude what he believed was the true effect of this clause. He held at p5, para 11(a) of the award:

> When analysing comparable transactions, regard may be had to rent free periods granted in the market over and above rent free periods granted for fitting out that, when analysing such transactions the "additional" or "inducement" rent free period may be amortised and a suitable deduction be made from the "face" rent. The amortised deduction shall only relate to rent free periods granted in excess of rent free periods for fitting out.

F I can pass over the next point. He then held at (c):

> That the "open market rent" as defined in the Third Schedule is £5,281,365 per annum
>
> and
>
> (d) if the award at (a) is held to be incorrect no regard should be had to any rent free periods or inducements, then I would have awarded an open market rent of £7,033,300 per annum.

Finally, it was held:

> That if it is held that the Respondent's secondary contention is correct and regard should be had to *all* rent free periods and inducements, including those granted for fitting out then I would have awarded an "open market rent" of G
> £4,874,323 per annum.

Mr Kim Lewison QC, for the landlords, in attacking this award, went on to look particularly at the reasoning of the arbitrator set out at pp 15 and 16 of the award, notably the phrases below line 20, that:

> It has been customary for many years for landlords to grant rent-free periods to assist tenants against the cost of fitting-out work. However, the situation has now changed and landlords are granting relief for periods in excess of the norm and that extra relief is seen to take the form of an inducement.

He then refers to any "normal period". He referred further to p22 of the award. He accepts Mr Harper's opinion at line 13 on p22: H

> As Mr Harper's Opinion clearly illustrates, the word "premises" as used in the clause is defined on page 4 of the lease as the demised premises. There is nothing in the clause to direct me on the treatment of comparables and, indeed, I would not expect there to be anything. I therefore reject Mr Cullen's

— who was leading counsel for the tenant below —

> submission in paragraph 11(c) of his Statement of Claim. The review clause is merely making clear that the hypothetical tenant of the premises would not be able to claim a rent-free period at the review date.

and that sentence was attacked by Mr Lewison as unfounded. He goes on to say in the next paragraph: J

> ... Mr Harper's Opinion hits the point squarely on the head, and in answer to Mr Harper's first question, both counsel agreed that the word "premises" ... meant "the demised premises" and nothing else. In my view, the correct way to treat the clause is to ignore rent-free periods granted for fitting out but to make a deduction for rent-free periods that can be looked upon as taking the form of an inducement.

That was attacked as wholly unsupported by anything in the clause. The clause made no distinction between rent-free periods granted for fitting out and rent-free periods granted as an inducement. He went on to record that: K

> Counsel agreed that the hypothetical tenant did not have the benefit of a rent free period and that the rent would become payable from day 1, the review date. Any rent free period that might have run would have been deemed to have expired immediately before the grant of the hypothetical lease.

He then goes on at considerable, though not unnecessary, length, through the matter and comes to his conclusions where he sets out a section beginning "14 CONCLUSIONS AND VALUATION". He refers to his conclusions on the legal issues, accepts the respondent's secondary contention that allowance should be made for rent-free periods over and above that achieved in the market for fitting out. He takes the
appropriate period for fitting out to be six months and an additional L rent-free period to have been amortised on a discounted cash-flow basis. I can pass over the next paragraph.

He refers to three transactions which he says he relied upon and he then comes to what he states as:

> my valuation of the "headline" or "face" rent of each building is as follows: ...

and he reaches the conclusion that 1 Broadgate, the prime office amount of the space, by far the larger amount, is to be valued at £38 per sq ft. Various other rents for ancillary smaller parts of the building are added in, producing a total figure of £7,033,316 pa. For M 2 Broadgate, a somewhat similar process is gone through to produce a figure of £5,236,046 pa.

The arbitrator, having thus said that he has found the headline or face rent, then goes on to a further series of steps and these are the steps that are said by Mr Lewison to be illegitimate within the rent review process as determined by this clause. The arbitrator says that he has:

> applied a factor on a discounted cash flow basis to reflect an "incentive" rent free period of twenty-four months amortised over ten years at 6%

and he arrives at figures which I have already said are the figures that

*Editor's note: Also reported at [1986] 1 EGLR 120.

he stated in his award as the amounts to be fixed of £5,281,365 pa for no 1 and £3,931,795 pa for 2 Broadgate. He does that on the basis that the respondent's secondary contention, that is that the fitting-out rent-free period should be ignored but the inducement rent-free period should be taken into account, was correct. He then goes on to find a yet further variation on the respondent's primary contention.

Having said all that, I go back to the open market rent definition in the third schedule to the lease itself.

— The "Open Market rent" should be the best yearly rent which would reasonably be expected to become payable in respect of the premises. That rent is to be fixed after the expiry of a rent free period of such length as would be negotiated in the open market between a willing landlord and a willing tenant upon a letting of the Premises as a whole by a willing lessor to a willing lessee in the open market at the relevant review date . . .

So that in finding the best yearly rent reasonably expected to become payable the arbitrator is to find the rent which will actually be payable when such rent-free period as would have been negotiated shall have expired. That rent he can reach only by looking at comparables which show other premises let in adjacent or similar areas of the City of London, in similar conditions to these.

In looking at the comparables the arbitrator obviously does not have to take the rent reserved in the allegedly comparable lease as being the true rent of the comparable premises. As Mr Harper said, he had to adjust that according to the art or magic of valuers in reaching the true rent which is agreed by that lease. If it were the fact that someone had signed a lease agreeing to pay a rent of £100 per sq ft for some floors of an office block for a term of five years, but made that agreement and became the tenant for that term only in consideration of a payment of £50m down as an inducement to the tenant to take that lease, naturally the arbitrator in looking at these premises, which might be nearby and might be very similar physically to the demised premises which he is required to value, would say that the reserved rent of £100 per sq ft is not the true rent for these premises at all. The arbitrator, in exercising his own skill and judgment, would be entitled to hold that the comparable premises are a good physical comparable, but that the rent gave him no guide to the open market rent which the premises would fetch were the rent, that I ventured to call in argument "a straight rent", that is a rent payable from day one throughout the term without regard to fancy inducements by way of large payments, to take on a lease. Quite clearly on the hypothesis I have given the reserved rent of £100 per sq ft is no guide to the letting value. That is what the arbitrator was told he was entitled to consider by Mr Harper and as part of the art of valuation he must take the various alleged comparables and find what they truly represent as straight rents, in order to reach what is the true open market rent as defined here. That is the valuation exercise which the arbitrator had to perform.

It follows that upon that having been done he reaches a figure, and that figure he appears to have reached when he says he has found what he has defined earlier as the rent reserved by the lease by the words "the 'headline' or 'face' rent of each building is as follows". It seems to me that Mr Lewison is right when he says that there is no warrant in the terms of this clause for the arbitrator applying some further adjustment to the rent which he has found, is in fact the best yearly rent reasonably expected to become payable for a term of 10 years or the surviving part of the lease, because once he has found that figure by considering and adjusting the comparables, bearing in mind that this is to be the amount payable after the expiry of such rent-free period as may be negotiated, that is indeed the open market rent. It seems to me that the arbitrator, in his attempt to reach some figure between the top and the bottom, has taken the finding of the true rent to be reserved and adjusted it down. I am well aware that the true rent to be reserved is not the rent to be payable, but that is, in my view, entirely of assistance to Mr Lewison, who observed to me, and in my view correctly, that it is a *non sequitur* to say that because the actual tenant has no rent-free period the hypothetical tenant has no rent-free period. There are many other differences and no exact correspondence between reality and hypothesis upon this set of terms. One such difference is the reduction of the open market rent to about 10% below the figure to reach the rent, which is actually payable and reserved, as opposed to that which is to be assessed as the open market rent. Another is the fact that the premises are described as being in a particular state up to a particular specification, although that will not be in fact the state of specification in which the premises are because considerable work has been done to them, so that the arbitrator is not assessing the premises as they stand, but assessing them upon an assumed basis. Both those points seem to me to justify Mr Lewison's submission that there is no exact correspondence between reality and hypothesis in this case and the fact that there is no rent-free period here, in actuality, does not require the hypothesis to have a rent-free period inserted in for the purpose of discounting the rent which would otherwise be reached.

The rent, as Mr Lewison pointed out, is payable in respect of the premises after the expiry of a rent-free period. The formula must be there pointing to the rent which is that actually payable and not then reduced by some discounted cash-flow calculation to a discounted figure. It seems to me that there is an error of law disclosed by the arbitrator's award, although, as it seems to me, he was advised by his legal assessor correctly in what he should have done. Upon that basis, it seems to me that the right answer is that I should allow the landlord's appeal against the arbitrator's award and dismiss the tenant's, as it were (although not technically), cross-appeal contending that the rent should be fixed in accordance with the tenant's primary submission at the lowest of the three amounts fixed by the arbitrator as possible choices.

I should also consider the point raised by Mr David Neuberger QC, for the tenant, that in any event the arbitrator has made an error of law in that at p58 of the award the arbitrator dealt with a point raised before him under the title "The 'onerous' rent review clause". He referred there to a deduction of 10% made by a valuer from his valuation to reflect the fact that the hypothetical rent review clause was drawn in the same terms as the actual clause and is on an onerous basis. The arbitrator held:

I do not regard the clause as onerous and reject this point.

He went on:

If my legal interpretation is held to have been incorrect, [as I have held] and no allowance or deduction should be made for rent free periods, I would still reject Mr Hyder's submission for a deduction

— he then adds the words:

for it might introduce an element of double counting.

Mr Neuberger observed that the phrase "it might introduce an element of double counting" is a thoroughly unsatisfactory one. First, to say that it might do so leaves one entirely uncertain whether it in fact does, or in fact does not, and one is left wondering what the effect is. Second, the reference to double counting is entirely obscure. What could be double counting is, to my mind, very difficult to perceive. None the less, Mr Lewison said that it was not his business to try to deal with every step in the reasoning. I was confined by the Act to considering whether there was a question of law arising out of the award on which I could confirm, set aside or vary the award: see section 1(2) of the 1979 Act. In particular, he observed, whether a term is onerous must be a question of fact. It is a question of comparative exercise in judgment, is this term inflicting an excessive burden, because almost every term inflicts a burden and that cannot be the sense of "onerous" in this phrase. It must be that the term is more burdensome than normal similar terms. It must be a question of comparisons, and that plainly must be a question of fact, said Mr Lewison, and I agree. Mr Lewison submitted further that whether, if the term were onerous, it would in fact produce a discount on the rent is also a question of fact. There are many onerous covenants, that is covenants which are more restrictive than usual covenants, which, although more restrictive, will not in practice, because they are bearable in the market, produce any reduction in rent. Whether they will or not is a matter of evidence and a matter of assessment by an expert valuer and must be a question of fact. The arbitrator rejected the proposition that there should be a discount for the onerous, as it was called, clause on both views of the law and on that footing no question of law is raised by this part of the award.

Mr Neuberger submitted that the extremely unsatisfactory nature of the last phrase in the last sentence of that paragraph ought to raise such doubts in my mind that it would be proper to exercise the jurisdiction under section 1(5) where the court may order an arbitrator to state the reasons for his award in sufficient detail to enable the court to deal with it. I follow that there are doubts about the meaning of that phrase, but I remain convinced that the arbitrator, who is the sole judge of fact and who should not be interfered with by the court on questions of fact, has on matters which are primarily factual on both bases of legal view held that the clause is not onerous, and upon that basis I do not think that I ought to intervene and send the award back for further findings in that way.

I have therefore come to the conclusion that on all the questions raised before me the landlords succeed and obviously, consequentially, that the tenant's claim fails. On that basis I will make such precise order as counsel agree.

Following submissions by counsel on certificate and leave to appeal, **Harman J** said: It seems to me that the sums of money in this case, which when capitalised on any reasonable application of years' purchase must run into tens of millions, are such that it is impossible not to accept that there are special reasons for allowing an appeal by certificate under section 1(7)(*b*) of the Arbitration Act 1979. It does not seem to me that there is any general public importance in the question of construction of this somewhat obscure clause. No doubt some of the phrases in it are used in other clauses, but that is not a sufficient general interest, in my view. If it were a clause from a precedent book, which it could be said had been used in dozens of leases up and down the City, there might be a general public importance, but that is not the case.

I therefore propose to grant a certificate that there is a question of law arising upon the first part, the landlord's appeal in this matter. It seems to me that the second part of this matter, as to whether the clause was "an onerous clause", is purely raising questions of fact, it does not raise any question of law and no certificate from me ought to be granted upon that question.

I then have to turn to whether I should give leave to appeal on this matter, and it seems to me I am, perhaps not very surprisingly, clear that the decision to which I have come is correct and does not really warrant going to the Court of Appeal, although it does raise questions of special interest to the parties. I therefore think the right thing to do is to follow my brother Knox J in *British Gas plc* v *Dollar Land Holdings plc**, grant a certificate on the point I have identified, but refuse leave to appeal.

*Editor's note: Reported at [1992] 1 EGLR 135.

Court of Appeal

July 5 1993

(Before Sir Thomas BINGHAM MR, Lord Justice STEYN and Lord Justice HOFFMAN)

BRITISH RAILWAYS BOARD v MOBIL OIL CO LTD

Estates Gazette March 5 1994

[1994] 09 EG 181

Landlord and tenant — Rent review — Construction — Base date for calculation of increase — Whether direction to apply rise in values of comparable properties merely arbitrary trigger to rent review or provides index to increase ground rent

By a lease dated September 19 1966 the plaintiff landlord demised a site to S for a term of 42 years; S later assigned the term to the defendant tenant. The lease contained covenants by the tenant to erect a petrol-filling station and to pay certain specified stepped rents to the end of the 14th year of the term. From September 19 1980 and for the rest of the term, subject to the provisions of the rent review clause, the rent was to be £1,850 pa. Under the rent review clause the arbitrator was to determine whether, as at the date of the notice triggering the review, any rise had taken place since the commencement of the term, or since any last increase in the rent, in the annual rack-rental value of the property in the locality used for like purpose or purposes and, if so, by calculating the increase in the rent payable which would reflect the said rise. In the court below Judge Hague QC accepted the landlord's contentions: (1) that, because there has been no increase of rent under the rent review clause, the base date for calculation was the commencement of the term; and (2) the calculation of the rent increase should be made by assessing the rental value of the premises, including the buildings, the existence of a rise in values of other properties being only the arbitrary condition precedent for switching from a ground rent to current rack-rent. The tenant appealed.

Held: **The appeal was allowed.**

1. The base date from which the arbitrator must determine whether and to what extent the rents of other petrol stations have risen was the last increase to £1,850 pa.

2. The rise which the increase is to reflect is the proportion or percentage rise in rents of other petrol stations since September 19 1980. Although the £1,850 pa was a ground rent and the comparable to be used to determine the percentage increase is the rack-rental value of other petrol stations including their buildings, this comparable is merely to provide an index which can be applied to the ground rent payable under the lease.

No cases were referred to in this report.

This was an appeal by the defendant, Mobil Oil Co Ltd, from the decision of Judge Hague QC, sitting as a judge of the High Court, who had made a declaration in favour of the plaintiff landlord, British Railways Board, on the interpretation of a lease dated September 19 1966 held by the defendant.

David Neuberger QC (instructed by the solicitor to Mobil Oil Co Ltd) appeared for the tenant; Kirk Reynolds QC (instructed by the solicitor to British Railways Board) represented the landlord.

Giving the first judgment at the invitation of Sir Thomas Bingham MR, **Hoffmann LJ** said: This appeal concerns the construction of a rent review clause in a lease of a petrol-filling station constructed on railway land just north of King's Cross. By a lease dated September 19 1966 the British Railways Board demised the site to a Mr Harry Sibley for a term of 42 years. He subsequently assigned to the present tenant, Mobil Oil Co Ltd. The lease contains covenants by the tenant, first, to erect the filling station within one year and, second, to pay the following yearly rents: for the first year £350. This was a concessionary rent for the period during which the petrol station would be built and before the tenant commenced business. For the next six years, £1,325. From the end of the seventh to the end of the 14th years, that is, until September 19 1980, £1,470. For the rest of the term, subject to the rent review clause, £1,850. The rent review clause entitled the board to serve what is called a rent notice invoking the rent review mechanism at any time during the period of six months before the expiry of the 21st year, that is, before September 19 1987. It described a rent notice as:

> . . . a notice in writing . . . providing for the increase of the yearly rent of one thousand eight hundred and fifty pounds payable hereunder as from the expiration of the year of the term then current.

The board duly served such notice on Mobil seeking an increase

A from September 19 1987.

The rent review clause provided machinery for the parties to agree a new rent or for the appointment of an arbitrator to determine one if they could not. Para 4 states how the arbitrator should determine the new rent:

> The Arbitrator shall determine the question so referred to him by determining whether as at the date of the Rent Notice any rise has taken place since the commencement of the term hereby granted or if the rent payable hereunder shall previously have been increased since the date of the last increase in the annual rack rental value of the property in the locality used for the like purpose or purposes as the demised premises and the buildings thereon and if B so by calculating the increase in the rent payable hereunder which would reflect the said rise and the amount so calculated shall be the increase in the rent payable hereunder.

The arbitrator, therefore, has to determine whether, since a given date in the past, there has been a rise in the rents of other petrol stations in the locality and then to calculate an increase in the rent payable under the lease which would reflect that rise. This would seem a fairly simple and businesslike procedure, but the parties have been unable to agree on how it should be put into effect.

The first dispute concerns the base date from which the arbitrator must determine whether and to what extent the rents of other petrol C stations have risen. The board says that "since the date of the last increase" means since the date of the last increase agreed or determined under the rent review clause. As there has been no such increase, the base date is the commencement of the term. But, in my judgment, this makes no sense. The rent notice requested an increase in "the yearly rent of one thousand eight hundred and fifty pounds payable hereunder". The object of the arbitrator's determination is to calculate the increase in "the rent payable hereunder", which is the rent payable at the time of the rent notice. This was £1,850. The increase in that rent must reflect the rise in other rents since the base date. It would be very odd to calculate an increase in a rent of £1,850 D by reference to a rise in the rents of other petrol stations from a time when the rent under the lease was £350. One would not be comparing like with like.

In any case, the board's construction involves writing words into the lease. It says "since the date of the last increase" and the last increase was when the rent went up to £1,850. This makes good sense without having to imply words which disqualify that increase from consideration. The only argument against Mobil's construction is that it gives no effect to the conditional "if" which precedes the words "the rent payable hereunder shall previously have been increased". This is because, given the stepped rents provided for in the *reddendum*, it was E certain that the rent would have increased before any rent notice could be served. But this is easily accounted for. The rent review was a standard clause whereas the *reddendum* is custom-made for this particular lease. This is the kind of ruck in the texture of a document which frequently happens when the draftsman inserts a one-off clause without checking through the rest of the document to see what effect it is going to have on other standard clauses. It certainly is not enough to justify implying words into the clause which will produce a commercially absurd result.

The second dispute concerns the kind of calculation which the arbitrator must perform. What is the "rise in rents at other petrol F stations" which must be reflected in the increase under the lease? Is it a rise in absolute or percentage terms? In my view, it cannot possibly mean a rise in absolute terms. Petrol stations come in all shapes and sizes and it would be absurd to regard the absolute rise in the rent of, say, a station that is three times as big as being relevant to what should be the increase in the rent of this one. In any case, the clause contemplates considering more than one comparable and this also rules out the notion of any absolute figure. The rise which the increase reflects must therefore be the proportion or percentage rise in rents of other petrol stations since September 19 1980.

In my view, this produces a simple calculation. It is true that the £1,850 is a ground rent and the comparable being used to determine the percentage increase is the rack rental value of other petrol stations G including their buildings. But the rents of these other petrol stations are not being used to value the leasehold premises. They are being used only to provide an index which can be applied to the ground rent payable under the lease. It is in principle no different from, and in practice somewhat more appropriate than, using other indices such as retail prices or equity shares.

Before the judge, however, the board contended successfully for an altogether different approach, which Mr Kirk Reynolds QC has again urged upon us today. This is simply to calculate the current rental value of the demised premises, including the building, and to increase the rent to that figure. The rack-rents of neighbouring petrol stations H are on this basis no more than evidence of the market value of the demised premises and the extent of the rise in such rents since the base date is irrelevant. All that matters is that there should have been some rise, however small, since on this construction the existence of a rise is the arbitrary condition precedent for switching from a ground rent to a current rack-rent.

In my view, the clause cannot support this construction. It says that the increase in the rent under the lease must reflect the *rise* in the rents of neighbouring petrol stations since the base date and I have explained why this can mean only a percentage rise. The lease says nothing whatever about calculating the current rack-rental value of the J demised premises. This would not reflect the rise in other rents, but merely what those rents currently were.

I would therefore allow the appeal and for the declarations made by the judge substitute the following answers to the questions raised in the notice of motion. Under question 1, September 19 1980 and under question 2, that it should be answered in sense (a).

Steyn LJ agreed and did not add anything.

Agreeing, **Sir Thomas Bingham MR** said: On the second of the two points argued concerning the comparison on which any increase K in the rent is to be founded I have, with respect to the learned judge who thought otherwise, no doubt that the construction favoured by Hoffmann LJ is correct. It accords with the natural wording of the clause and not only gives better effect to the language of the lease than the judge's construction but also, as I think, gives a much more sensible commercial result.

I have, for my part, felt more doubt on the first point concerning the base date for the purposes of the comparison. The landlord, not surprisingly, has drawn attention to the words:

> . . . any rise has taken place since the commencement of the term hereby granted or if the rent payable hereunder shall previously have been increased L since the date of the last increase . . .

They submit that that language must apply to an increase taking place under rent review provisions because there will always, irrespective of any general increase in rents, have been a stepped increase in the rent payable under the lease before the date of the first rent review in the 21st year. Thus, they say, the clause must be read as applying only to rent review increases as opposed to stepped increases because otherwise the word "if" would be deprived of its conditionality since there could never be a comparison with a rent at the date of commencement of the term which the clause envisages as a possibility. That is an argument not without attraction. On the other M hand, even the landlord's interpretation involves, as I think, interpolation to some extent since the words "under the rent review provisions" need to be inserted after "increased".

Moreover, it seems to me that the landlord's construction involves a somewhat anomalous exercise of seeing whether there should be an increase in the rent of £1,850 by reference to an increase since the time when a rent of £350 was in force on September 19 1966. The overwhelming probability is, I think, that a mismatch arose in the course of drafting between the terms of subclause 4 and the provisions of the lease, which provide for payment of a stepped rent over the first 21 years of the term.

I accordingly agree with Hoffmann LJ on this point; also and furthermore agree that the question submitted for answer by the court should be answered as he proposes.

Appeal allowed. Declaration accordingly.

Chancery Division

September 9 1993

(Before Mr Justice RATTEE)

CHELSEA BUILDING SOCIETY v R&A MILLETT (SHOPS) LTD

Estates Gazette March 5 1994
[1994] 09 EG 182

Landlord and tenant — Rent review — Whether time of the essence of landlord's notice to implement the review

By a lease dated July 1 1976 the plaintiff landlord demised shop premises to the defendant tenant for a term of 25 years. The rent review clauses of the lease provided that "It shall be a condition precedent to the determination of the full market rental value of the premises that the lessor shall give the lessee during the first six months of the . . . fifteenth . . . year . . . of the said term notice in writing of its desire that such value shall be agreed or ascertained . . .". In respect of the rent review for the commencement of the 15th year, the landlord failed to give any notice seeking review within the first six months of that year expiring January 1 1992, but purported to trigger the rent review by two later notices. The landlord applied for declaration that the rent review had been implemented on the ground that time was not of the essence of the clause.

***Held*: The application was dismissed. The use of the words "It shall be a condition precedent" was a clear contrary intention to rebut the presumption that time is not of the essence of a rent review clause; the notice was essential to the landlord's right.**

The following cases are referred to in this report.

North Hertfordshire District Council v *Hitchin Industrial Estate Ltd* [1992] 2 EGLR 121; [1992] 37 EG 133

Phipps-Faire Ltd v *Malbern Construction Ltd* [1987] 1 EGLR 129; (1987) 282 EG 460

United Scientific Holdings Ltd v *Burnley Borough Council* [1978] AC 904; [1977] 2 WLR 806; [1977] 2 All ER 62; (1977) 33 P&CR 220; [1977] EGD 195; 243 EG 43 & 127, HL

This was an application by the landlord, Chelsea Building Society, seeking a declaration as to the validity of a notice implementing a rent review clause in a lease held by the defendant, R&A Millett (Shops) Ltd.

Kirk Reynolds QC (instructed by Bower Cotton & Bower) appeared for the plaintiff; Jonathan Karas (instructed by Titmus Sainer & Webb) represented the defendant.

Giving judgment, **RATTEE J** said: This is a dispute about the true effect of a rent review clause in a lease. The lease concerned was made on July 1 1976 between Chelsea Building Society (the plaintiff) as lessor, and R&A Millett (Shops) Ltd (the defendant), as lessee. It was a lease of certain shop premises in Cheltenham. It was expressed to be a lease for the term of 25 years from the date of the lease, that is to say from July 1 1976. It reserved certain specified rents during the initial period of the lease, and it then contained a fairly usual rent review clause entitling the landlord, the lessor, to have the rent periodically reviewed upwards only to what would be determined as a current market rent for the time being at the operative date of the review.

The relevant rent review provision commenced in these terms:

PROVIDED ALWAYS THAT IT IS HEREBY AGREED that the said yearly rent of £17,500 shall at the end of the fifth, tenth, fifteenth and twentieth years of the term respectively be increased to the then current market rental value of the premises as defined in the Schedule hereto and in this connection the following procedure shall be adopted:

(a) At least six months before the end of the fifth, tenth, fifteenth and twentieth years of the term respectively the Lessor and the Lessee or their duly authorised representatives will endeavour to reach agreement on the current market rental value of the premises and if they are able to agree such current market rental value they shall certify the amount of the then current market rental as agreed between them and a Memorandum of such current market rental will be annexed to these presents and signed by duly authorised officers of the Lessor and Lessee respectively

(b) In the event of the parties hereto failing to reach agreement as to the then current market rental value by a date three months prior to the fifth, tenth, fifteenth or twentieth year of the term as the case may be then the matter shall be referred to an independent Surveyor or Valuer to be appointed by the Lessor and Lessee in conjunction but failing agreement upon such appointment within a further period of one month to be appointed by the President for the time being of the Royal Institution of Chartered Surveyors for assessment and not by way of arbitration and the assessment of such independent Surveyor shall be communicated to the parties hereto in writing and shall be final and binding upon them

(c) Upon the signing of the Certificate by the duly authorised representatives of the Lessor and Lessee respectively or the receipt by the Lessor and the Lessee of the independent Surveyors' or Valuers' written communication as to the then current market rental value the rent hereby reserved shall as from the expiration of the fifth, tenth, fifteenth and twentieth years of the term be increased to the amount so certified

(d) Under no circumstances shall this clause operate to effect a reduction of the rent hereby reserved . . .

Then (e), which I need not read, refers to fees payable to the independent valuer.

Clause 4 of the lease, which is headed:

PROVIDED ALWAYS AND IT IS HEREBY AGREED AND DECLARED as follows: . . .

contains in subclause (6) the following:

It shall be a condition precedent to the determination of the full market rental value of the premises that the lessor shall give the lessee during the first six months of the fifth, tenth, fifteenth or twentieth year (as the case may be) of the said term notice in writing of its desire that such value shall be agreed or ascertained as at the expiration of such fifth, tenth, fifteenth or twentieth year in accordance with the provisions of the said Schedule hereto . . .

The Schedule contains the definition of the full market rental value for the purposes of the rent review provisions. Nothing turns for present purposes on that definition and I need not read it.

In order to comply with clause 4(6) of the lease in relation to any proposed rent review to take effect as from the end of the 15th year of the lease, that is to say the year ending on June 30 1991, it would have been necessary for the lessor to have given the notice specified in clause 4(6) not later than January 1 1992. In fact, as is accepted on behalf of the lessor, no notice was given within the meaning of clause 4(6) before that date, although in fact apparently two documents were subsequently served upon the lessee on each of which the lessor relies as having been a notice for the purposes of clause 4(6).

The question that arises in those circumstances on the originating summons before me is whether the six-month period specified in clause 4(6) of the lease is a period as to which time is to be treated as of the essence so as to disentitle the lessor from putting into operation the rent review provisions of the lease as from the end of the 15th year of the term, the lessor not having (as I have indicated) in fact given any notice within that six-month period. Needless to say the lessor's submissions are to the effect that time should not be regarded as of the essence of the provision in clause 4(6) and that, since notice was served before time had been made of the essence by any notice given by the lessee, the so-called condition precedent to the landlord's right to have the rent reviewed as from the end of the 15th year of the term is fulfilled. The lessee on the other hand equally inevitably submits, in the situation which I have described that the provision in clause 4(6) of the lease ought to be treated as one as to which time is of the essence, with the result that the lessor not having given notice within the six-month time-limit specified in that provision it has lost its right to have the rent reviewed as at the end of the 15th year of the term.

Counsel for the lessor, not surprisingly, based his submissions primarily on the decision of the House of Lords in *United Scientific Holdings Ltd* v *Burnley Borough Council* [1978] AC 904* in which the House of Lords held that in relation to timetables laid down by rent review provisions in leases there is a presumption that time is not of the essence, unless the parties have made it plain by the provisions of the lease concerned that time is indeed to be treated as being of the essence. There is no doubt that that principle is established by that decision of the House of Lords and the only question that arises in the present case is as to whether the terms of the lease entered into between the parties are such as to indicate sufficiently clearly an intention that time should indeed be of the essence in respect of the provision in clause 4(6) of the lease to rebut the presumption established by the *United Scientific Holdings* case.

Counsel for the lessor further relied upon a *dictum* of Warner J in *Phipps-Faire Ltd* v *Malbern Construction Ltd* [1987] 1 EGLR 129 in which at p131F-G Warner J said:

> The authorities seem to me to show that the presumption that time is not of the essence of a provision in a rent review clause is strong and that it will not be rebutted by any contra-indication in the express terms of the lease unless it is a compelling one.

The lessee's contention is that subclause (6) of clause 4 of the lease does indeed contain a compelling indication that the parties in this case intended time to be of the essence in relation to the six-month period specified for the service of the notice by the lessor required by that subclause. In particular, the lessee relies upon the opening words of subclause (6) of clause 4, namely:

> It shall be a condition precedent to the determination of the full market rental value of the premises that the lessor shall give the lessee during the first six months of [the relevant year] the [specified] notice in writing.

What clearer indication, submits the lessee, could there be of an intention that compliance with the time-limit set out in subclause (6) of clause 4 was intended to be regarded as essential to the lessor's right to have the rent review than to say that it should be a condition precedent of that right that notice should be given within the six-month period?

The lessor, on the other hand, contends that the use of the words "condition precedent" in subclause (6) of clause 4 are not sufficient to oust the presumption that time is not of the essence in respect of this particular provision of a rent review clause.

Not surprisingly, counsel on behalf of the lessor, relies strongly for that submission on a recent decision of Mr Evans-Lombe QC (as he then was), sitting as a deputy judge of this division, in *North Hertfordshire District Council* v *Hitchin Industrial Estate Ltd* [1992] 37 EG 133†. In that case the relevant provision being considered by the learned deputy judge was in these terms:

> It shall be a condition precedent to any such variation of rent as aforesaid that the Council or the Tenants shall be served twelve months' written notice on the Tenants or the Council as the case may require of their intention to vary the rent as hereinbefore provided and the parties shall be deemed to be in default of agreement if a rent shall not have been agreed in writing within four calendar months after the date of service of such notice.

I should say, as I think is probably obvious from that citation, that the reference there to the "council" is a reference to the lessor of the particular lease under consideration.

The learned deputy judge heard argument as to whether the effect of those words, and in particular the words "It shall be a condition precedent", was to make the giving of the 12 months' notice required by that provision an essential term of the lease, in the sense that time was of the essence in relation to the giving of that notice. The deputy judge held that the words used, and in particular the words "It shall be a condition precedent", did not indicate a sufficiently clear intention to make time of the essence in respect of the provision which those words introduced to oust the *United Scientific* presumption. The judge in that case referred, in the course of his judgment, to the fact that his attention had been drawn to certain passages in the speech of Lord Diplock in the *United Scientific* case in which the words "condition precedent" appeared and in particular to a passage appearing at p925F of the report of the *United Scientific* case in which Lord Diplock said:

> The first thing to be observed about each of these sub-sections [subsections of section 25 of the Supreme Court of Judicature Act 1873] is that they are concerned with matters in which before the Unifying Act came into force there had been a variance between the ways in which they were dealt with in courts of law and courts of equity respectively. Outside the field of mortgages and contracts for the sale of land, there were other kinds of contracts in which by 1875 some stipulations as to time were not treated in courts of law as being "conditions precedent" — which was then the common lawyer's way of saying that the particular stipulation as to time was not of the essence of the contract.

Counsel for the tenant in the *North Hertfordshire District Council* case relied upon that *dictum* of Lord Diplock as indicating that a reference to a time provision being a condition precedent was to be treated as synonymous with a reference to time being of the essence in respect of that provision. The learned deputy judge rejected that submission. He said in the course of his judgment with reference to it:

> I cannot accept this submission. Whatever may have been the meaning ascribed to the words "it shall be a condition precedent that", by a 19th century draftsman, I do not think that those words can be given a meaning equivalent to "time shall be of the essence" in a lease drafted in 1969. I do not think that the passages in the speech of Lord Diplock in the *United Scientific* case, in particular the passage I have cited

— I interject to say that that is the passage I myself have just cited —

> can be taken to establish that such would be the meaning of those words today.

Thus the learned judge in that case decided that the use of the words "It shall be a condition precedent" in respect of a giving of notice within a specified time was not sufficient to indicate a contrary intention to oust the *United Scientific* case presumption that time was not of the essence in the provisions of a rent review clause. However, the learned deputy judge went on to decide the case before him on an alternative basis, even if, he said, he was wrong in his first conclusion to the effect that the use of the words "It should be a condition precedent" did not make time of the essence; and the alternative basis on which he decided the case was that in any event the provision which I have cited from the lease that was being construed by him did not actually require service of the notice concerned within any particular time-limit. All that it required was that the notice should be 12 months' written notice, without any requirement as to the date by which that 12 months' notice should be given. However, it is plain that the learned deputy judge did adopt as one of the reasons for his decision in that case the conclusion that the use of the words "It shall be a condition precedent" in the provision being construed by him was not a sufficient indication of an intention to make time of the essence to oust the presumption.

That conclusion of the learned judge, of course, is not binding upon me. On the other hand, it is a conclusion which I treat with great respect and would differ from with reluctance. I cannot follow that conclusion in the present case because, in my judgment, the words of subclause (6) of clause 4 of the lease do indeed express what might almost be called the clearest possible intention on the part of the parties to the lease that the lessor should not have a right to have the rent reviewed at any given review date unless the lessor first gave notice, as required by subclause (6) of clause 4, within the first six months of the relevant year of the term. In my judgment, to say that

---

*Editor's note: Also reported at (1977) 243 EG 43 & 127.

†Editor's note: Also reported at [1992] 2 EGLR 121.

A the giving of such a notice within such a time is a condition precedent of the lessor's right to have a new market rent determined is equivalent to saying that that right of the lessor shall not arise unless he gives notice as required by subclause (6) of clause 4 within the period therein specified. In my judgment, to say that that right of the landlord is not to arise unless notice is given within the specified time does amount to saying that the parties intended that the giving of notice within the specified time was essential to the existence of the landlord's right. That, in my judgment, amounts to precisely the same thing as saying that time is of the essence in respect of that particular provision of the lease.

B There is no doubt, in accordance with the principles laid down by the House of Lords in the *United Scientific* case (and this is quite properly accepted on behalf of the lessor), that if in subclause (6) of clause 4 the parties had said "and time shall be of the essence as to the giving of this notice", the effect would indeed have been to oust the presumption that time was not of the essence in relation to that provision. In my judgment, to say that the giving of a notice within a specific time is a condition precedent to the lessor's right to have the rent reviewed is to say precisely the same thing as that time shall be of the essence in respect of the giving of the notice within the specific time. I reach that conclusion with all due respect to the decision of Mr Evans-Lombe QC (as he then was).

C There are, in my judgment, two other passages in the speech of Lord Diplock in the *United Scientific* case which suggest that Lord Diplock himself regarded the use of the term "condition precedent" in relation to a time stipulation as being equivalent to saying that time was of the essence in respect thereof. At p927E-F Lord Diplock said:

> Contemporaneously with this development of the rules of equity by the Court of Chancery, the courts of common law were in process of developing for themselves a not dissimilar rule in relation to stipulations as to time in other contracts, but were reaching their solution by a different route. They did so by a growing recognition of exceptions to the rule which had been fostered in the
> D early part of the 18th century by the necessity for the plaintiff under the then current rules of pleading to aver performance or willingness or ability to perform all stipulations on his part in the precise words in which they were expressed in the contract. This rule treated all promises by each party to a contract as "conditions precedent" to all promises of the other: with the result that any departure from the promised manner of performance, however slight that departure might have been, discharged the other party from the obligation to continue to perform any of his own promises. The history of the development by common law courts of exceptions to this rule is traced in the judgments of the Court of Appeal in *Hongkong Fir Shipping Co Ltd* v *Kawasaki Kisen Kaisha Ltd* [1962] 2 QB 26 from its origin in *Boone* v *Eyre* 1 Hy Bl 723n in 1779 to the judgment of Bramwell B in *Jackson* v *Union*
> E *Marine Insurance Co Ltd* (1874) LR 10 CP 125, 147 on the eve of the coming into force of the Supreme Court of Judicature Act 1873.
>
> My Lords, I will not take up time in repeating here what I myself said in the *Hongkong Fir* case, except to point out that by 1873;
>
> (1) stipulations as to the time at which a party was to perform a promise on his part were among the contractual stipulations which were not regarded as "conditions precedent" if his failure to perform that promise punctually did not deprive the other party of substantially the whole benefit which it was intended that he should obtain from the contract; . . .

I need not go on to refer to the remainder of the passage of Lord Diplock's speech in which he deals with other points arising in the
F *Hongkong Fir* case. But, a little later in his speech, at p928G, Lord Diplock said:

> Both in the Court of Chancery and in the courts of common law the rules that have been developed about particular stipulations not being of the essence of the contract or not being "conditions precedent" applied to synallagmatic contracts only. They did not apply to unilateral or "if contracts", of which the example most germane to the instant appeals is an option.

In my judgment, in particular in that last passage which I have cited from the speech of Lord Diplock, it is plain that Lord Diplock regarded a condition precedent as being something which is of the essence of the contract. In my judgment, with great respect, that must be right; and, by the same process of reasoning, by making the giving of notice within the specified six-month period a condition precedent G to the landlord's right to have the rent review provisions operated, subclause (6) of clause 4 of the lease in the present case made it quite plain that the giving of such notice within that time was of the essence of the contract. In other words, time was of the essence in respect thereof.

Counsel for the lessor put forward an alternative contention that if (as for the reasons I have endeavoured to explain in my judgment is the case) the use of the words in subclause (6) of clause 4 "It shall be a condition precedent" had the effect of making the giving of a notice by the lessor an essential provision of the contract, none the less time was still not of the essence in the sense that the words "It shall be a H condition precedent to the determination of the full market rental value of the premises" should not be construed as extending to the time, the six-month period, within which subclause (6) of clause 4 said that the notice should be given. All that was made, if anything was made, an essential term of the contract by clause 4(6), says counsel for the lessor, was the actual giving of the notice and time is still presumed not to be of the essence in respect of that provision. I do not accept that contention. It seems to me perfectly plain as a matter of natural construction of the words of subclause (6) of clause 4 that the parties were saying that it should be a condition precedent to the arising of the landlord's right to have the rent reviewed at the end of any relevant year of the term that the lessor should give the lessee the notice J required by subclause (6) of clause 4 within the time-limit there specified.

In the present case the lessor has failed to give a notice within the time there specified. In my judgment, the lessor's right to have the full market rental value of the premises determined as at the end of the 15th year of the term has not arisen and cannot now arise. Accordingly, the lessor's claim in the originating summons for declarations to the contrary must fail.

Application dismissed with costs.

K

Chancery Division

February 8 1994

(Before Mrs Justice ARDEN)

SCOTTISH AMICABLE LIFE ASSURANCE SOCIETY v MIDDLETON AND OTHERS

L

Estates Gazette March 12 1994
[1994] 10 EG 109

Landlord and tenant — Rent review — Whether direction to determine open market rent was a reference to rent after expiry of rent-free periods or concessionary rents referable to fitting-out or additionally to inducements — Whether disregard of rent-free periods and concessionary rents applied to inducements to be offered by hypothetical underlease to subtenants

By an underlease dated September 25 1987 the defendant tenants held the term of four floors at 3 Cloth Street, London EC1, from the M plaintiff landlord subject to provisions for the review of the rent. At review clause 3.2(b) of the underlease directed that the rent was to be determined ". . . the best yearly open market rent (at the rate payable following the expiry of any rent free period or periods at concessionary rents which might be granted on a new letting . . ." but disregarding, *inter alia*, "(d) any effect on rent of any initial rent free period or periods at concessionary rents or other inducements which might be offered in the open market . . .". The court was asked to determine the following questions: (1) whether the reference in clause 3.2(b) to rent-free periods or concessionary rents was to be interpreted as limited to fitting-out periods; (2) whether disregard (d) was

A concerned with the effect of the matters contemplated by clause 3.2(b) or the need for a landlord to offer rent-free periods, concessionary rents or other inducements to "prospective underlessees"; (3) whether disregard (d) applies to fitting-out periods or to all rent-free periods, concessionary rents and inducements; and (4) what was the true meaning of clause 3.2(b).

Held: **(1) The periods which might be granted following "a new letting", and referred to in clause 3.2(b), are restricted to fitting-out periods and do not include further inducement periods which may be given to increase the "headline" rent; (2) the word "underlessee" refers to subtentants from the underlessee and not**
B **the underlessee himself and therefore the cost to the hypothetical underlessee in giving inducements to his subtenants is disregarded; (3) having regard to the answer to question (1), this question was not answered; (4) under clause 3.2(b) the market rent is to be at the rate payable following the expiry of any rent-free or concessionary rents for fitting-out, and by disregard (d) any effect on rent attributable to the tenant's need to give rent-free or other inducements to a subtenant must be eliminated.**

The following cases are referred to in this report.

Basingstoke and Deane Borough Council v *Host Group Ltd* [1988] 1 WLR
C 348; [1988] 1 All ER 824; (1987) 56 P&CR 31; [1987] 2 EGLR 147; 284 EG 1587, CA

British Gas Corporation v *Universities Superannuation Scheme Ltd* [1986] 1 WLR 398; [1986] 1 All ER 978; (1986) 52 P&CR 111; [1986] 1 EGLR 120; 277 EG 980

Broadgate Square plc v *Lehman Brothers Ltd* [1994] 1 EGLR 143; [1994] 04 EG 135

City Offices plc v *Bryanston Insurance Co Ltd* [1993] 1 EGLR 126; [1993] 11 EG 129

Glofield Properties Ltd v *Morley (No 2)* [1989] 2 EGLR 118; [1989] 32 EG 49

London & Leeds Estates Ltd v *Paribas Ltd* [1993] 2 EGLR 149; [1993] 30 EG 89, CA

D *MFI Properties Ltd* v *BICC Group Pension Trust Ltd* [1986] 1 All ER 974; [1986] 1 EGLR 115; (1986) 277 EG 862

99 Bishopsgate Ltd v *Prudential Assurance Co Ltd* [1985] 1 EGLR 72; (1985) 273 EG 984, CA

This was an application by the landlord, Scottish Amicable Life Assurance Society, with the consent of the defendant tenants, William Middleton, Christopher Reginald Potts, David Nigel Godfrey and Derek Buckle West, and of the arbitrator, Geoffrey Barron Perkins, under section 2(1)(*b*) of the Arbitration Act 1979 to determine questions of construction arising in the course of an arbitration
E between the parties.

Jonathan Gaunt QC (instructed by Cameron Markby Hewitt) appeared for the plaintiff; Paul Morgan QC (instructed by Middleton Potts) represented the defendants.

Giving judgment, **ARDEN J** said: This is an application to determine points of construction on a rent review clause (clause 3.2) in an underlease dated September 25 1987 between the plaintiff ("the lessor"), the defendants ("the lessees") and others, of premises on four floors at 3 Cloth Street, London EC1. Dispute has arisen as to these points of construction in a pending arbitration. The arbitrator, Mr Geoffrey Barron Perkins [FRICS], and the parties consent to the
F determination of these questions by the court. The questions which I am asked to determine do not require me to find or assume any facts: indeed no evidence has been placed before me for this purpose. Moreover, I am told that, if the questions are decided now, there will be substantial costs savings. There will be no need for the arbitrator to sit with a legal assessor and (if the decision is in favour of the lessor) there would be no need for certain evidence to be called. In the circumstances I am satisfied that I should determine the questions on this application pursuant to section 2(l)(*b*) of the Arbitration Act 1979.

The rent review clause is in these terms:

> 3.2. The Existing Rent shall be reviewed on each Review Date and the New Rent (subject to the further review of rent on each succeeding Review Date and to the provisions of the sub-clauses hereinafter appearing) shall be G
> the greater of:
>
> (a) the Existing Rent to the intent that the Yearly Rent cannot decrease upon any rent review
>
> (b) such sum as shall be agreed by the Lessor and by the Lessee or determined as representing the best yearly open market rent (at the rate payable following the expiry of any rent free period or periods at concessionary rents which might be granted on a new letting of the Demised Premises or of comparable premises in the open market on the relevant Review Date) at which the Demised Premises might reasonably be expected to be let in the open market on the relevant Review Date without a fine or premium or value in the nature of a fine or premium:
>
> (i) as a whole by a willing lessor to a willing lessee or (if a sub-letting of H
> any Sublet Unit has been completed) in the Sublet Units by a willing lessor to a willing lessee of each Sublet Unit (whichever shall produce the higher rent or aggregated rents as the case may be)
>
> (ii) with vacant possession for a term equal to the original term of this Lease for the Permitted Use and otherwise subject to the provisions of this Lease (other than the amount of the Yearly Rent reserved but including provisions for rent review in accordance with this Clause 3 on every fifth anniversary of the commencement of such term) and
>
> (iii) on the assumption that at the relevant Review Date the Demised Premises are fit for immediate occupation and use for the purposes permitted by these Presents and that no work has been carried out thereon which has diminished the rental value of the Demised Premises and in case the Demised
> Premises or the Development have been destroyed or damaged on the J
> assumption that they have been fully restored before the relevant Review Date
>
> (iv) on the assumption that the Lessee's covenants herein contained have been fully performed and observed
>
> (v) on the assumption that the benefit of any planning or other consent or licence (whether or not granted hereunder) current at the relevant Review Date is also available for such willing lessee or lessees.
>
> But disregarding:
>
> (a) any effect on rent of the fact that the Lessee its sub-tenants or their respective predecessors in title have been in occupation of the Demised Premises
>
> (b) any goodwill attached to the Demised Premises by reason of the carrying on thereat of the business of the Lessee its sub-tenants or their K
> respective predecessors in title in their respective businesses and
>
> (c) any effect on the rental value of the Demised Premises attributable to the existence at the relevant Review Date of any improvement to the Demised Premises or any part thereof carried out not more than twenty-one years before the relevant Review Date with consent where required (or if the Court has determined that consent was unreasonably withheld) and otherwise than in pursuance of an obligation to the Lessor or its predecessors in title (except obligations requiring compliance with statutes or directions of local or public authorities) to the extent only that such improvement as aforesaid shall have been carried out without any liability on the part of the Lessor or its predecessors in title to reimburse or contribute towards the cost thereof and
> provided that such improvement was completed by the Lessee its sub-tenants L
> or their respective predecessors in title during the Term or during any period of occupation prior thereto arising out of an agreement to grant the Term.
>
> (d) any effect on rent of any initial rent free period or periods at concessionary rents or other inducements which might be offered in the open market to prospective underlessees of the Demised Premises or any part thereof or of comparable premises.

The questions for determination on this application may be stated as follows:

Question 1 — fitting-out question

In the direction in clause 3.2(b) that the rent to be determined is to M
be:

> the best yearly open market rent (at the rate payable following the expiry of any rent free period or periods at concessionary rents which might be granted on a new letting of the Demised Premises or of comparable premises in the open market on the relevant Review Date) at which the Demised Premises might reasonably be expected to be let in the open market . . .

is the reference to rent-free periods or concessionary rents to be interpreted as limited to such rent-free periods or concessionary rents as a willing lessor would give to reflect the fact that the willing lessee will not be able to make full beneficial use of the demised premises until after completion of his fitting-out works?

A I will call this kind of rent-free or concessionary rent period "a fitting-out period", and any other such period (see further below) an "inducement period". I will call the words in the passage which I have set out from clause 3.2(b) and which are in brackets "the bracketed words".

The lessees contend that the periods referred to in the bracketed words are fitting-out periods. The lessor contends that they are not so limited.

Question 2 — underlessee question

Is disregard (d) of clause 3.2(b) concerned with the effect on rent of:

B (a) rent-free periods, concessionary rents or other inducements which might be offered in the open market to the hypothetical willing lessee on the hypothetical letting contemplated by clause 3.2(b), or

(b) the need for a lessor of the premises, or of comparable premises, to offer such rent-free periods, concessionary rents or other inducements to prospective underlessees?

The lessees contend for (a) and the lessor contends for (b).

Question 3 — disregard (d) — fitting-out question

Does disregard (d) extend to fitting-out periods (and inducements for a like purpose) given to a willing lessee or does it apply to all
C rent-free periods, concessionary rents and inducements given to him?

This question is really an alternative to question 1 so far as the lessees are concerned. It arises if the lessees succeed on question 2. Consistently with its position on question 1 the lessor says that no limitation can be read into disregard (d) in this way and that it applies to fitting-out periods. The lessees contend the contrary. If the lessor is right, the result may be that disregard (d) cancels out the exclusion of rent-free periods in the bracketed words so that (contrary to the norm) the lessees will obtain the benefit of fitting-out periods even on mid-term rent reviews.

Question 4 — overall effect question

D What is the true meaning of clause 3.2(b) in relation to any rent-free periods or periods at concessionary rents or other inducements which might be offered or granted in the open market to the hypothetical lessee? This is really a recapitulation on the earlier questions.

In setting out the questions, I have used slightly different wording from that appearing in the originating motion with the aim of simplification.

There is little that I need from the underlease itself. On the front cover, it is described as an underlease, as indeed it is. Although, internally, it is called a "lease" and the parties are called "the lessor"
E and "the lessee", it is clear from, *inter alia*, the definition of "lease", and the references "to any superior lessor" and the headlease, that the terms "lessor" and "lessee" in this underlease mean the underlessor and underlessee respectively. The term of the underlease is 25 years from March 25 1987. There are five-yearly rent reviews. The lessees may not sublet less than an entire floor or, without the lessor's consent, the whole premises. The term "underlessee" is not defined. It is used only occasionally. It is used for instance in clause 1.6 and in clause 4.5 (which uses the term "prospective underlessee"). In these instances it means subtenants from the lessees (including perhaps subtenants from them).

F *Question 1 — fitting-out question*

Mr Jonathan Gaunt QC, for the lessor, submitted that there is nothing to limit in the words in brackets to fitting-out periods. Their object is to prevent a tenant obtaining additional rent-free periods during the term of his lease. As I understood it, it would be normal for a rent review clause of this nature to prevent a sitting tenant from getting further fitting-out periods. It was common ground that, once a lease was granted, it was inappropriate to claim further fitting-out periods that might have been given to a tenant on first occupation. Mr Gaunt pointed to the fact that the distinction between fitting out and other rent-free periods had been rejected in *City Offices plc* v *Bryanston Insurance Co Ltd* [1993] 1 EGLR 126, at p127L *per* Aldous J; and in *Broadgate Square plc* v *Lehman Brothers Ltd* G (November 29 1993, unreported*), *per* Harman J albeit on different rent review clauses. The result was to give the lessor a more favourable rent on review than would be obtained under some other rent review clauses.

Mr Paul Morgan QC, for the lessees, submitted that the commercial object of the bracketed words and/or of disregard (d) is to prevent the tenant from getting further fitting-out periods. He submitted that the bracketed words should not apply to other rent-free periods. Landlords from time to time give extended rent-free periods to maximise the "headline" rent (or rent actually specified in a lease following a rent-free period). This practice I am told improves the H valuation of the reversion. If the distortion in the headline rent is not corrected, the rent fixed on the rent review will be higher than it would otherwise have been. The commercial purpose of the bracketed words on the lessees' case is to prevent the open market rent being fixed by reference to the distorted figures. Moreover, in this case, the lessees are prevented from subletting the whole of the premises at a rent which is less than the rent they are liable to pay. The lessor's construction will thus make it difficult for them to sublet.

Mr Morgan further submits that the words "which might be granted on a new letting" in the brackets restrict the relevant rent-free periods to those attributable to the new letting. He also submits that the lessor's construction renders the words in brackets J after "at concessionary rents" otiose. Furthermore, the reference to concessionary rents suggests that the rent payable at the end of the relevant periods will be true market value, but this value may be distorted if periods other than the fitting-out period are taken into account as a result of the words in brackets. Alternatively, the words "value in the nature of a fine or premium" require the exclusion of inducement periods. The words "any initial rent free period or periods at concessionary rents or other inducements" in disregard (d) are wider than the equivalent in the bracketed words. This indicates that a narrower meaning is to be given to the bracketed words.

My duty is to find the intention of the parties as expressed in the K bracketed words bearing in mind the commercial purpose of the rent review clause. The leading authority in this field is *Basingstoke and Deane Borough Council* v *Host Group Ltd* [1988] 1 WLR 348†. This concerned the construction of a rent review clause in a lease of property comprising land and buildings restricted to use as a public house. The tenant had constructed the buildings at his own expense. The rent review clause required the valuer to assume that the premises were available for the greater of 20 years and the unexpired portion of the lease and that the property was a bare site. One of the questions which arose was whether in the absence of an express direction the valuer should assume that the site was available for development for L any lawful purpose. The Court of Appeal held that the valuer should assume that the notional letting was on the same terms, other than as to the amount of rent, as those subsisting under the existing actual lease. The reasoning is to be found in the following passages from the judgment of the court, which was given by Nicholls LJ [at p353D]:

> The question raised on this appeal is one of construction of a rent review clause in a lease. In answering that question it is axiomatic that what the court is seeking to identify and declare is the intention of the parties to the lease expressed in that clause. Thus, like all points of construction, the meaning of this rent review clause depends on the particular language used interpreted having regard to the context provided by the whole document and the matrix M of the material surrounding circumstances. We recognise, therefore, that the particular language used will always be of paramount importance. Nonetheless it is proper and only sensible, when construing a rent review clause, to have in mind what normally is the commercial purpose of such a clause.
>
> That purpose has been referred to in several recent cases, and is not in doubt. Sir Nicolas Browne-Wilkinson V-C expressed it in these terms in *British Gas Corporation* v *Universities Superannuation Scheme Ltd* [1986] 1 WLR 398, 401:
>
> "There is really no dispute that the general purpose of a provision for rent

*Editor's note: Also reported at [1994] 1 EGLR 143 (*ante*).

†Editor's note: Also reported at [1987] 2 EGLR 147.

A review is to enable the landlord to obtain from time to time the market rental which the premises would command if let on the same terms on the open market at the review dates. The purpose is to reflect the changes in the value of money and real increases in the value of the property during a long term."

To the same effect Dillon LJ said in *Equity & Law Life Assurance Society plc* v *Bodfield Ltd* [1987] 1 EGLR 124, 125:

"There is no doubt that the general object of a rent review clause, which provides that the rent cannot be reduced on a review, is to provide the landlord with some measure of relief where, by increases in property values or falls in the real value of money in an inflationary period, a fixed rent has become out of date and unduly favourable to the tenant. The exact measure of relief depends on the true construction of the particular rent review clause."

B The means by which rent review clauses afford landlords relief in respect of increases in property values or falls in the value of money is by providing, normally, for a valuer, in default of agreement, to assess the up-to-date rent for the demised premises at successive review dates. In making that assessment the valuer will be achieving the intended purpose of keeping the rent in line with current property values having regard to the current value of money if, but only if, he assesses the up-to-date rent on the same terms (other than as to quantum of rent) as the terms still subsisting between the parties under the actual, existing lease. If he departs from those terms, and assesses the up-to-date rent on the footing of terms materially less onerous to the tenant than those in the actual, existing lease, the rental at which he arrives will reflect in addition to the rental increases attributable to a rise in property values or a fall in the value of money, an additional element, viz, the increased rental C attributable to the fact that he is calculating the rent of a lease on terms more favourable to the tenant than the terms in the actual, existing lease. Conversely, if he assesses the up-to-date rent on the basis of terms materially more onerous to the tenant than those in the actual, existing lease, the rental figure at which the valuer arrives will not fully reflect the rise in property values or the fall in the value of money since the lease was granted or the rent was last fixed.

Of course rent review clauses may, and often do, require a valuer to make his valuation on a basis which departs in one or more respects from the subsisting terms of the actual existing lease. But if and in so far as a rent review clause does not so require, either expressly or by necessary implication, it seems to us that in general, and subject to a special context D indicating otherwise in a particular case, the parties are to be taken as having intended that the notional letting postulated by their rent review clause is to be a letting on the same terms (other than as to quantum of rent) as those still subsisting between the parties in the actual existing lease. The parties are to be taken as having so intended, because that would accord with, and give effect to, the general intention underlying the incorporation by them of a rent review clause into their lease.

We are fortified in this view by observations made in several cases . . .

Next there is *British Gas Corporation* v *Universities Superannuation Scheme Ltd* [1986] 1 WLR 398. There a rent review clause expressly provided for the notional letting at each review date to be a letting "containing the same provisions (other than as to the yearly rent) as are herein contained . . ." The E question arose whether the direction to leave out of account the provision in the actual lease "as to the yearly rent" was apt to exclude from the notional letting the fact that the actual lease contained provisions for future rent review. Sir Nicolas Browne-Wilkinson V-C held that the direction did not have that effect. Having referred to the general purpose of a rent review provision, as quoted above, he said, at pp401-402:

"Such being the purpose, in the absence of special circumstances it would in my judgment be wayward to impute to the parties an intention that the landlord should get a rent which was additionally inflated by a factor which has no reference either to changes in the value of money or in the value of the property but is referable to a factor which has no existence as between the actual landlord and the actual tenant, ie, the additional rent which could be obtained if there were no provisions for rent review. Of course, the lease may F be expressed in words so clear that there is no room for giving effect to such underlying purpose. Again, there may be special surrounding circumstances which indicate that the parties did intend to reach such an unusual bargain. But in the absence of such clear words or surrounding circumstances, in my judgment the lease should be construed so as to give effect to the basic purpose of the rent review clause and not so as to confer on the landlord a windfall benefit which he could never obtain on the market if he were actually letting the premises at the review date, viz, a letting on terms which contain provisions for rent review at a rent appropriate to a letting which did not contain such a provision."

. . . The fact that, unusually, the rent being reviewed is a ground rent, does not show or even suggest that in this case, where the landlord is to obtain an up-to-date rent for the site, unlike the ordinary case, where the landlord obtains an G up-to-date rent for the site plus the buildings, the parties intended that (to adapt the words of Sir Nicolas Browne-Wilkinson V-C in the *British Gas* case [1986] 1 WLR 398, 401) the landlord was to get a rent for the site additionally inflated by a factor which has no reference either to changes in the value of money or in the value of the site as demised but which is referable to a factor having no existence as between the actual landlord and the actual tenant: for instance, the additional rent which could be obtained for the site if the actual, existing lease did not contain user restrictions. To update the ground rent the valuer is to disregard the value of the buildings on the site. But we can see no more reason why, in doing so, he should disregard (or, more precisely, the parties can be supposed to have intended that he should disregard) the other subsisting terms of the actual, existing lease in this case than in any other case.

H We approach the construction of paragraph (vii), therefore, on the footing that, unless the paragraph otherwise requires, expressly or by necessary implication, or there is some context indicating otherwise, the parties are to be taken to have intended that the notional letting assumed for the purposes of the rent review assessment was to be on the same terms (other than as to quantum of rent) as those still subsisting under the actual, existing lease.

. . . There is no express direction that in thus assessing the reasonable current ground rental value the valuer is not to have regard to the other terms of the actual lease. Nor do we think that such a direction can fairly be spelled out as a matter of necessary implication. Quite the contrary: in the case of this lease, with its stringent user covenant, such a direction would be capable of working so unfairly on the tenant, in the manner we have explained, that we find it impossible to suppose that the original parties to the lease could have intended J this.

Nor are we persuaded by the contention that none of the covenants in the actual lease can be impliedly incorporated in the notional letting because many of the tenant's covenants are inappropriate to the notional site lease. The essential substance of the direction in paragraph (vii)(b) is that, for the purpose of computing the up-to-date ground rental value of the demised premises, the buildings actually there are to be disregarded. They are to be disregarded for that purpose because they were erected by the original tenant and at its own expense. No ground rental value, in other words, is to be attributable to the presence of the buildings. But although the tenant's covenants relating to buildings, such as covenants to repair and to insure, would be inappropriate to a simple lease of a bare site, those covenants will or may have some K materiality in assessing the up-to-date rental value of the premises demised by this lease (which includes buildings) even though no rental value is being attributed to the buildings themselves. They will or may have materiality in assessing that value, because those covenants impose obligations on the tenants. For this reason it is not inappropriate to carry those covenants into the notional lease, even though the buildings are being excluded therefrom. Paragraph (vii) is concerned with a valuation exercise, and the concept of a notional or hypothetical lease is a tool fashioned to that end. Where the subject of a lease is land and buildings, and the subject matter of the valuation is the up-to-date rental value of the leased property but excluding the buildings, it is not surprising if the notional lease thus predicated is a somewhat odd-looking creature.

L Mr Cullen pointed out that in paragraph (vii) the parties have plucked out of the lease two provisions only, those relating to the length of the term and to rent reviews. They have left unmentioned all the other provisions. He submitted that, in valuing the bare site, the only terms of fundamental importance to a valuer would be the terms relating to period, rent reviews and user. The parties have dealt expressly with two of these fundamental terms. If they had intended that the site should be valued on the basis of restricted user surely they would have said so. As arguments directed at the presumed intention of the parties we find these arguments of little weight when put in the scales against the contrary argument based on the manifestly unfair result to which the landlord's case leads.

M The Court of Appeal reached a similar result in *Glofield Properties Ltd* v *Morley (No2)* [1989] 2 EGLR 118. Hoffmann J (as he then was) had also reached a similar result in *MFI Properties Ltd* v *BICC Group Pension Trust Ltd* [1986] 1 All ER 974*. However, the *British Gas Corporation*† case was not cited in the *MFI Properties* case. Moreover, *MFI Properties* was decided before the *Basingstoke* case, but was not cited in that case. In view of the unanimous and recent decision of the Court of Appeal in the *Basingstoke* case, it is sufficient for me to limit my citation of authority to that case. Mr Gaunt urged the *MFI Properties* case on me, but, in so far as there is any difference

*Editor's note: Also reported at [1986] 1 EGLR 115.
†Editor's note: Also reported at [1986] 1 EGLR 120.

A between the approach in that case and the approach in the *Basingstoke* case, I am of course bound to follow the latter.

The exclusion of inducement periods results in a bonus to the lessor. If an inducement period has any effect on rent, it will be to increase the market rent artificially in the lessor's favour. The lessor could not obtain the increased rent on a new letting without suffering a rent-free period additional to a fitting-out period. Moreover, the ascertainment of the open market rent at the rate payable following expiry of an inducement period is not part of the process of reflecting changes in the value of money or real increases in the value of the property. In the case of this underlease it creates a difficulty for
B subletting. With all these factors in mind, I have come to the conclusion that, unless clause 3.2(b) otherwise requires, the bracketed words should be confined to fitting-out periods. This conclusion follows from the *Basingstoke* case which established that a rent review clause should be construed so as to give effect to the basic purpose of the rent review clause and not so as to confer a windfall on the landlord.

In my judgment, the words "on a new letting" are capable of more than one meaning. They can mean "in the event of a new letting", or "as a result of a new letting" and not as a result of (for example) the landlord's desire (to which the tenant has acceded) to maximise the
C headline rent. Applying the *Basingstoke* case I prefer the latter construction. Effectively I adopt Mr Morgan's submission that the words "on a new letting" mean "attributable to" (and only to) a new letting. I do not accept Mr Gaunt's criticism of this construction that an inducement period is also attributable to a new letting. It may be contemporaneous with a new letting, but causally it is distinct from it. A further reason for preferring Mr Morgan's construction is that it gives content to the words following the reference to "concessionary rents". These words otherwise for no apparent reason largely repeat directions given immediately after the brackets in clause 3.2. Finally, there is nothing in the clause to prevent this construction and it has not been suggested that it would be unworkable. In the circumstances I
D hold that the periods referred to in the bracketed words are fitting-out periods. It is unnecessary for me to deal with Mr Morgan's alternative or additional submissions.

Question 2 — underlessee question

Mr Gaunt submits that the words "prospective underlessees" are used in contradistinction to the words "lessee" and "the lessee's subtenants". Prospective lessees are persons to whom the hypothetical lessee may sublet individual floors (as is permitted by clause 4.5) and to whom the hypothetical lessee may envisage having to give rent-free periods or inducements, the cost of which he will reflect by reducing
E his own rental bid. The commercial purpose of disregard (d) is thus to negative the discount awarded on this basis in *99 Bishopgate Ltd* v *Prudential Assurance Co Ltd* [1985] 1 EGLR 72 and this purpose was acknowledged (*obiter*) as an aspiration of the draftsman of similar wording in *London & Leeds Estates Ltd* v *Paribas Ltd* [1993] 30 EG 89*. Mr Gaunt accepts, however, that such a discount would be unlikely in premises of this size.

Mr Morgan points out that the lease is in fact an underlease so that the hypothetical lessee contemplated by the opening paragraph of clause 3.2(b) will in fact be an underlessee. To disregard the hypothetical lessee makes no sense of the words "comparable
F premises". Disregard (d) should be given the same effect as disregard (iv) in *City Offices plc* v *Bryanston Insurance Co Ltd* [1993] 1 EGLR 126. This would mean that, where, for example, the market rent ascertained under clause 3.2(b) would in fact have been payable for the balance of a term following the expiration of a rent-free period, then, notwithstanding the bracketed words, it will be amortised over the whole of the term. This results in a lower open market rent.

I prefer Mr Gaunt's construction. In the context of this particular underlease, the word "underlessee" in my view more naturally refers to subtenants from the lessee and not to the lessee himself. Where the latter is referred to, the word "willing lessee" is used. The term "underlessee" is used in the underlease in the same sense as I have G given it here. The reference to "comparable premises" serves a purpose, *viz* the purpose of eliminating a discount for rent-free periods to subtenants found in comparables. My preferred construction prevents any question of any potential mismatch between the bracketed words and disregard (d). The words "to prospective underlessees" have no equivalent in the bracketed words, which provides some additional support for my construction. The result is not to confer a windfall on the landlord so the *Basingstoke* point does not arise.

Question 3 — disregard (d) — fitting-out question

Having regard to my answer to question 1, this question does not H arise.

Question 4 — overall effect question

For the above reasons, in my judgment:

(i) the effect of the bracketed words is that the market rent is to be at the rate payable following the expiry of any rent-free periods or periods at concessionary rents for fitting out;

(ii) disregard (d) requires the valuer to make any adjustment necessary to eliminate any effect on rent attributable to the tenant's need to give any rent free or concessionary period or other inducement to a subtenant.

I propose to ask counsel to agree a minute of order. J

*Editor's note: Also reported at [1993] 2 EGLR 149.

Chancery Division

November 24 1993

(Before Judge RICH, sitting as a judge of the High Court)

CO-OPERATIVE WHOLESALE SOCIETY LTD v NATIONAL WESTMINSTER BANK PLC K

Estates Gazette April 9 1994
[1994] 14 EG 130

Landlord and tenant — Rent review — Rent at review to be open market rental value after expiry of rent-free period of concessionary rent — Whether headline rent should be discounted

By a lease dated August 8 1986 the respondent tenant held a term of 25 years of premises from the applicant landlord. The lease contained provisions to review the rent to the rental value in the open market with vacant possession on the supposition, *inter alia*, that any rent-free period or concessionary rent or other inducement which may L be offered in the open market shall have expired or been given immediately before the relevant date of review. In his award determining the rent at review, the arbitrator found that the effect of the rent review clause was that the open market rental value should be the "headline rent". The applicant contended that if the object of the review clause is to arrive at the open market rental value let in the open market with vacant possession, and if it is the purpose to use comparable transactions in order to arrive at that conclusion, it is necessary to have regard to the practice of letting with rent-free periods; if what is truly a rental value is to be arrived at, then it is necessary to discount the headline rent to arrive at the true market M value of the premises let without a rent-free period.

***Held*: The application was refused. It must be assumed that at the review date any rent-free period has expired. That is the literal meaning of the clause and that is how the arbitrator construed it with the benefit of legal advice. There was no reason to depart from that literal meaning.**

The following cases are referred to in this report.

British Gas Corporation v *Universities Superannuation Scheme Ltd* [1986] 1 WLR 398; [1986] 1 All ER 978; (1986) 52 P&CR 111; [1986] 1 EGLR 120; (1986) 277 EG 980

A *City Offices plc* v *Bryanston Insurance Co Ltd* [1993] 1 EGLR 126; [1993] 11 EG 129

Daejan Investments Ltd v *Cornwall Coast Country Club* (1984) 50 P&CR 157; [1985] 1 EGLR 77; (1984) 273 EG 1122

This was an appeal by the tenants, National Westminster Bank plc, against an arbitrator's interim award in a rent review; the landlords were Co-operative Wholesale Society.

David Neuberger QC (instructed by Denton Hall, of Milton Keynes) appeared for the applicants; Kim Lewison QC (instructed by Linklaters & Paines) represented the respondent.

B Giving judgment, **JUDGE RICH** said: This is an appeal brought by agreement against the interim award of an arbitrator appointed to determine the rent review under a lease dated August 8 1986 in respect of premises in the City of London let by Co-operative Wholesale Society Ltd to National Westminster Bank at an initial rent of some very substantial sum.

The arbitrator's interim reward was concerned with the interpretation of a particular clause within that review found at clause 6(2)(d), which provided:

> The open market rental value of the demised premises shall be the rental value in the open market of the demised premises at the relevant date of review for a
> C term equal to the unexpired residue of the term commencing on the relevant date of review on the open market as between a willing lessor and a willing lessee with vacant possession without payment of fine or premium and having regard to the degree and terms of the actual availability of car parking for the tenant and on the suppositions if not facts: . . .
>
> (d) that any rent-free period or concessionary rent or any other inducement whether of a capital or revenue nature which may be offered in the case of a new letting in the open market at the relevant date of review shall have expired or been given immediately before the relevant date of review.

The arbitrator determined that the effect of the clause was, as he put it, that the open market rental value as defined in clause 6 shall be
D the "headline rental value" and that no discount shall be made for any rent-free periods or other inducements, whether they be of a capital or revenue nature. That determination involves an understanding of what the arbitrator meant by a headline rent, but it is in effect to be understood, in the case, for example, of the grant of a rent-free period, to be the rent at which the parties are agreed the premises shall be let and which is to be payable after the expiry of the rent-free period provided for in the lease. It is thus, in the treatment of comparable transactions for the purpose of determining the rental value at the relevant date, to be contrasted with what would otherwise be appropriate for the valuer to do, namely to consider the true rental effect over the period of the lease whether to the next review or for
E its whole term and to discount the so-called headline rent to reflect the value of the rent-free period, which is granted to the tenant in return for the payment of the headline rent over a period shorter than the term for which he will actually enjoy occupation of the premises.

Before turning to the words which the arbitrator was required to construe, it is, I accept, helpful to remind oneself of the purpose for which it is well understood that rent review clauses have now habitually been incorporated into leases granted for periods of years greater than three or five years, which are the common periods between reviews. This present lease was granted for a term of 25 years
F or thereabouts.

The familiar references for reminding oneself of that purpose are some observations of the then Vice-Chancellor, Sir Nicolas Browne-Wilkinson, in *British Gas Corporation* v *Universities Superannuation Scheme Ltd* [1986] 1 WLR 398*, at p401G where he said:

> There is really no dispute that the general purpose of a provision for rent review is to enable the landlord to obtain from time to time the market rental which the premises would command if let on the same terms on the open market at the review dates. The purpose is to reflect the changes in the value of money and real increases in the value of the property during a long term. G

Starting from that general purpose of a rent review, a question arises, if that is the objective of the parties, whether in the face of the common practice which has now arisen of granting rent-free periods at the outset of a lease, it is appropriate to assume at each review date the enjoyment of a further rent-free period at the beginning of the review period. The arguments as to how far such assumption would or would not succeed in properly reflecting changes in the value of money and real increases in the value of the property during a long term may often turn out to be circular because it is necessary, on the grant of the lease with its review periods and its review provisions, to have in mind whether the first rent-free period that is being granted is to be taken as H
the once for all rent-free period or whether successive rent-free periods are to be assumed for the purposes of each review and the appropriate initial rent will presumably reflect the purposes of the parties in so considering.

Mr David Neuberger QC, who, as I observed in indicating to counsel for the respondent to this appeal that I need not trouble him, has presented a case for saying that the arbitrator's construction was wrong in a way which turned what seemed to me, on the face of the words of the clause, almost unarguable into a case where at least it was clearly arguable. I hope I have understood his argument. The point that he makes immediately is that, if the object of clause 6(2) is J
to arrive at the open market rental value of the demised premises let in the open market with vacant possession as at the review date, and if it is the purpose to use comparable transactions in order to arrive at such conclusion, it is necessary to have regard to this practice of letting with rent-free periods; and if what is truly a rental value is to be arrived at, then it is necessary to discount the headline rent in the way that I have indicated in order to arrive at the true market value of the premises let without a rent-free period.

Accepting that that is true, does not, however, involve a conclusion as to what is the purpose of this particular review because, as I have indicated, even taking as the overall and general intention K
of review clauses to be that which was stated in the *British Gas Corporation* case, namely to reflect changes in the value of money and real increases in the value of the property during a long term, does not answer the question which I sought to pose as to what at the outset of the initial term the parties may regard as being the appropriate way of treating further rent-free periods that might in the reality of the market at each review date normally be expected to be granted.

It is for that reason that it appears to me that it is essential to look at the specific words of the clause. I am reinforced in my construction of the clause by the decision of Aldous J in *City Offices plc* v *Bryanston* L
Insurance Co Ltd [1993] 1 EGLR 126. In that case there was a clause which provided that the rent to be determined was:

> such as should reasonably be expected to be payable in respect of the demised premises after the expiry of any rent-free concession or fitting-out period which might be given to the tenant if a letting of the demised premises were negotiated in the open market between a willing landlord and a willing tenant upon a letting on the review date.

In spite of a written opinion of counsel, which was apparently before the arbitrator, and upon which the arbitrator had relied as to the meaning of those words, Aldous J felt constrained by those words to M
construe the clause as meaning that the rent has to be the best yearly rent payable after the expiry of any rent-free period whether provided as a fitting-out period or not, that is to say, he construed that clause as providing for the headline rent, as I have so far defined it.

The actual decision in that case was not, however, that the headline rent was payable, because in that case there was also a requirement for a disregard to be made which Aldous J construed as contradicting what would otherwise be the effect of the words which I have just read. The clause required that there should be disregarded any rent-free concession or fitting-out period for which allowance would or might be given to the tenant if the demised premises were let in the open market with vacant possession. Aldous J held that that

*Editor's note: Also reported at [1986] 1 EGLR 120.

A requirement to disregard such notional rent-free concession or fitting-out period contradicted what would otherwise have been the effect of the words which he construed in the sense which I have indicated as leading to the headline rent.

In the present case, there is only the requirement to determine the rental value on the particular supposition contained in clause 6(2)(d). Mr Neuberger points out that the supposition is, on its face, a supposition not as to matters of actual fact, but matters of a hypothetical nature.

That is partly true. It is only partly true in the sense that the conclusion as to the hypothetical transaction which the arbitrator is to
B assess is to be drawn almost inevitably from actual transactions in the market and it is, in my judgment, as to that with which the provision in clause 6(2)(d) deals. In referring to "... any rent-free period ... offered in the case of a new letting", it means offered not only in the case of the hypothetical new letting but also notionally likely to be offered, having regard to the practice of the market from which evidence is to be drawn as to rental value in the active market. It is the supposed rent-free period or concessionary rent which, in accordance with the practice of the market, would be offered in the case of the hypothetical new letting which is to be treated for the purpose of the assessment of rental value as if at the relevant date of review of such
C period it has already expired, and likewise that any inducements shall have been given immediately before the relevant date of review.

Mr Neuberger has very helpfully assisted me as to how that somewhat difficult supposition is to be made. Its difficulty lies in the requirement to consider rental value as at a particular date, namely the review date, on the basis of a transaction made between a willing lessor and a willing lessee at that same particular date for a tenancy to begin also at that same particular date, and yet to assume that the rent-free period has already been enjoyed. He drew my attention to the decision of Peter Gibson J (as he then was) in *Daejan Investments Ltd* v *Cornwall Coast Country Club* [1985] 1 EGLR 77 where at p80 he made this observation, with which I have the very greatest sympathy,
D at least in the context of that case, in respect of the arguments adduced in that case before him.

> In some respects the arguments between the parties, particularly in the pleadings, touched on the meta-physical, disputing as they were how to reconcile the hypothesis with dates in the real world. To my mind, much of such disputation misses the point. If the hypothesis requires one to assume that the casino user [which was there in issue, so far as valuation is concerned] is lawful, then it is necessary to assume that all preliminary steps, however complex, enabling such lawful user to be made will have been taken in a dateless continuum of time immediately before the letting.

E It is, in my judgment, into that dateless continuum of time that any rent-free period that would be offered in the case of a new letting in the open market is packed immediately before the review date, by reference to which the valuation was to be made in the present case; and the valuation is to be made on the basis that by that date of review such period shall have expired. That appears to me to be the literal meaning of this clause. It was so construed, for reasons with which I agree, by the arbitrator and by counsel advising the arbitrator. It does not appear to me that there is any reason to depart from that literal meaning, nor do I think there is any reason for me to assume that that literal meaning is inconsistent with the purpose of the parties when
F they subscribed to this particular form of words.

Chancery Division G

October 28 1993

(Before Judge RICH QC, sitting as a judge of the High Court)

GUILDFORD BOROUGH COUNCIL v COBB

Estates Gazette April 23 1994
[1994] 16 EG 147

Landlord and tenant — Rent review — Whether expression ground rent meant rent reflecting value of buildings erected at expense of H
tenant

By a building agreement dated September 11 1979 the defendant tenant erected at his expense industrial premises on land belonging to the plaintiff landlords. By a lease dated January 21 1980 the landlords granted the defendant a term of the land with the buildings at an initial ground rent of £6,000 pa subject to review thereafter. The rent review clause provided that the arbitrator was to determine the ground rental value of the demised premises assuming a value in the open market for use of the demised premises as a developed site for general industrial purposes. The landlords contended that the expression ground rental
value in the lease meant the rack-rental value. J

***Held*: The value of the buildings erected by the tenant are to be excluded. The rent review clause contained provisions which directed that assumptions which departed from reality should be taken into account. What is to be taken into account is the potentiality of the land for which the ground rent is to be assessed to carry out the original building works.**

No cases were referred to in this report.

This was an application by the landlords, Guildford Borough Council, for a declaration as to the meaning of a lease dated January
21 1980 held by the tenant, Mr Cobb. K

Joseph Harper QC (instructed by the solicitor to Guildford Borough Council) appeared for the plaintiffs; Gerald Moriarty QC (instructed by Mackrell Turner Garrett, of Woking) represented the defendant.

Giving judgment, **JUDGE RICH QC** said: In the proceedings, the plaintiffs seek a declaration as to the construction of a rent review clause in a lease granted by the plaintiffs to the defendant on January 21 1980. The premises that were thereby granted are described in the lease as:

> All that piece of land situated in Lissons Avenue Industrial Estate, Ashvale, L
> containing ... [and then it gives the areas] as there shown for identification on ... [a particular plan] together with the easements and rights specified in the first schedule hereto, which said land and any alteration thereto and easements and rights are hereinafter referred to as the demised premises.

I emphasise that the grant is all the land together with the rights and easements and "any alteration thereto" because, by the time that the lease was in fact granted, the land, as it first came into the possession of the defendant, had indeed been altered by the defendant in compliance with a building agreement that the defendant had entered into on September 11 1979. Under that agreement, the defendant
bound himself to construct on the land certain industrial premises and M
to do so within a limited period of time and the plaintiffs bound themselves to grant the lease that was in fact granted, which lease I now have to construe.

The rent reserved by the lease, which was for a term of 99 years from the date of the agreement to build, namely September 11 1979, was for the first five years £6,000 a year. The rent was payable as rent by virtue of the terms of the building agreement from the date of the completion of the buildings and the actual formal lease was entered into at that later date. Thus, what was taken by the defendant was the original land plus the alterations, namely the buildings erected on it by the defendant under the building agreement.

A The lease provided for the rent to be reviewed in accordance with the second schedule of the lease. That schedule is headed as containing provisions for the variation of rent and Mr Gerald Moriarty QC, on behalf of the defendant, has made the point that the original rent was described in the building agreement as a ground rent and for that reason I should treat the provisions for variation as being provisions to continue a ground rent properly so called. I do not regard that approach as helpful to the construction of the second schedule because, if the terms of the schedule did provide for a variation from what was initially a ground rent to what was to be a rack rent, then the varied rent would be a rack rent, and it is necessary to look at the
B provisions themselves to see whether that change was or was not to be made after the initial period of five years.

The schedule provides for the rent in successive periods of five years, after the initial period, are to be found by the surveyors for the parties reaching an agreement or, if no agreement is reached, it provides for a process of determination by an expert appointed for the purpose. The expert who is to be appointed for the purpose is required specifically by the terms of the schedule to be "one experienced in dealing with properties in similar use and of similar size in the County of Surrey". That is certainly a sensible provision, but I do not find it of great assistance in determining what the appointed surveyor was or was not to determine. Certainly, if he was to have experience of
C buildings, one might think that he was more certainly to be concerned with the valuation of buildings rather than merely of land. But since the value of land for the purpose of building is very much tied up with the valuation of the building which might be erected upon it, I do not find any great assistance from that requirement as to the kind of person who is to determine the rent in the absence of agreement between the parties.

The question of the basis upon which such surveyor is to determine the rent is, in my judgment, to be determined by reference to the particular words used by the parties in the provisions for variation of rent which I am called upon to construe.

D They are required by those provisions to agree a sum as being "the then current ground rental value of the demised premises hereinafter called the ground rental value". I have already drawn attention to the provision in the demise which defines the demised premises as being the land together with the alterations that have been effected upon the land during the currency of the building agreement. But what is to be found is the then current ground rental value of those premises. Ordinarily the word "ground rental value" imports, in my judgment, a valuation of the land as distinct from the land together with the buildings erected thereon. Mr Joseph Harper QC, who has appeared for the plaintiffs, has drawn my attention to the definitions in Stroud's
E dictionary of ground rent in these terms and, as I understand it, is willing to adopt it as the ordinary meaning of those words. The definition is:

> By the expression of ground rent, if unexplained, is to be understood a rent less than the rack rent of the premises. Its proper meaning is the rent at which land is let for the purpose of improvement by building.

That, Mr Harper concedes, is a definition which can be used to identify a ground rent even for land upon which a building has already been erected, because it is apt to identify a particular part of the rack rent which the ground rent is less than, in accordance with the
F definition, namely that part which is attributable to the land as opposed to the building.

There is, however, this difference between the ground rent of an empty and undeveloped site and the ground rent of a site upon which a building has already been erected, namely that in the latter case one has a ready answer to the question: what is one to assume is to be the pattern of development permitted for the land? If there were no provision in a rent review of a ground rent as to the assumed pattern of development, those charged with determining such ground rent would have to look exclusively to the guidance of the market for the particular piece of land. In the case of a review of ground rent, it would, however, commonly accord more closely with the reality of the commercial transaction that there should be considered not all G
potentiality of the particular site but only the value of the particular potentiality to erect the particular buildings which had in that particular transaction been erected by agreement of the parties.

As this matter developed — and it has taken a long time for it to develop and to come before the court, because the rent review that is being considered was due to take place in 1985 — the defendant tenant has understood the issue between the parties to be whether or not there is to be assumed in the valuation of the site a potentiality for the erection only of the particular buildings which are found upon the site, or a potentiality to erect whatever buildings might otherwise lawfully be placed upon it. But as the matter was presented to the H
court by Mr Harper in argument today, that was not in fact the issue which the plaintiffs wish to bring before the court.

The plaintiffs' contention was that, in the particular schedule which I have to construe, there is contained a peculiar — unusual, I mean by that — definition of ground rent which equates the ground rent so called, in this case, with what is actually a rack rent for the buildings demised by this lease.

It is essential to that argument, as Mr Harper conceded in the course of his submissions, to accept that the definition is concerned with the actual buildings actually found on the site and that the issue as it had been presented in an affidavit on behalf of the defendant, J
namely whether the potentiality of the site is to be assumed for the actual buildings or for some hypothetical buildings, if it arose at all, was to be resolved in the way in which the tenant contends, namely by reference to the actual buildings. It is necessary therefore for me, in order to construe the clause and resolve the differences between the parties, only to address the question whether or not the particular clause in the second schedule upon which Mr Harper relies does indeed contain a special definition of ground rent, to give to it in the context of the clause a meaning in effect equivalent to rack rent.

Mr Harper draws attention to the fact that the surveyors are required by the clause "to assume a value in the open market for use of the demised premises as a developed site for general industrial K
purposes" and he says that that is in fact a definition for the purposes of this clause of the ground rental value. In my judgment, it is nothing of the kind. The clause, after saying that what is to be assessed is the current ground rental value of the demised premises, goes on then to make a proviso as to how that ground rental value is to be assessed and it does so by identifying certain matters differing from the reality of the transaction between the parties which are, for the purposes of that assessment, to be assumed. Those additional provisions are, in my judgment, additional to the basic assumption imported by the use of the word "current ground rental value", namely that what is to be assessed by the valuers is the value of the land as distinct from the L
buildings placed thereon at the expense of the defendant.

The additional assumptions are set out as follows:

> Provided always that in assessing the ground rental value the surveyor shall assume a value in the open market for use of the demised premises as a developed site for general industrial purposes and otherwise upon the terms of this lease.

That is the first proviso and the first variation from reality, because it provides implicity that the terms of the lease are to be assumed save in respect of one matter, namely there is to be assumed the value in the open market for use of demised premises as a developed site for M
general industrial purposes. That differs from the covenant contained in clause 3.22 of the lease, which confines the use of the premises to use for the carrying on of the business of motor engineers and vehicle repairers and the Department of Transport vehicle-testing centre and the sale of vehicle spare parts, subject, however, to a right to change such use with the consent of the council in writing to another user, within the limits defined by class 3 of the Town and Country Planning Use Classes Order 1972, which is the use class not for general industrial purposes but for light industrial purposes. Thus that proviso requires, in my judgment, an assumption of a right of use of the demised premises for a use different from that contained in the user clause of the lease.

The other variations from reality which are required by the clause are the importation, not entirely appropriately, of disregards which are commonly imported into rent review clauses, first, of any goodwill attributable to the demised premises; second, any effects of any improvement to the demised premises carried out by the lessees or any underlessees otherwise and in pursuance of any obligation hereunder; and third, any effect on the ground rental value of the fact that the lessees or any underlessees may have been in occupation of the demised premises.

I had at first thought when I read the second of these disregards, namely of any effect of any improvement to the demised premises, which is said to be required to be disregarded without reference to the time at which the improvement was carried out, that those words in themselves required the disregarding of the effect upon the rent of the work carried out by the defendant tenant under the agreement to which I have already referred. But Mr Moriarty has not been prepared to rely upon that route to the result for which he contends and on reflection I think he was right not so to rely, for I accept that the reference to the words, "any improvement to the demised premises", are appropriately construed by considering what were indeed the demised premises at the date when the lease itself was granted and at that date the demised premises already included the buildings erected under the agreement made prior to the grant of the lease. Thus, that second disregard must have reference only to improvements to the demised buildings made after the date of the commencement of the lease.

Mr Harper would maintain that that is a clear indication that what is to be valued is in fact the buildings themselves. But I do not think that that consequence flows from that analysis of that clause, for this reason. As I indicated, in assessing a ground rent it is necessary, whether the assessment is made by the market or by a valuer assessing what the market would do, for those who make notional bids for the property to have some concept as to the development which they are proposing to make of the property. Thus, although I do not think that that second disregard was imported into this clause with as much consideration as I have since given to it — if I may be forgiven for so putting it — nevertheless I think that it is meaningful in the sense that it excludes from the assumed potentiality of development to be taken into account in assessing the development potential and thus the ground rent of the land, improvements in the demised premises made by the tenant subsequent to the grant of the lease unless they are made in pursuance of an obligation under the lease to the council.

That reasoning also has some value in the understanding of the effects of the clause, because I was puzzled why, when I looked for provisions in the lease as to the potential obligations of the tenant to the council to carry out works of improvement, the only obligation I could find was an obligation to rebuild in the case of destruction by fire. If the exception to the disregard does properly apply to that work of rebuilding which might arise in such circumstances — although there may be a question as to whether or not that would or would not be an improvement — it would certainly be appropriate to exclude it from being disregarded. What is to be taken into account, in my judgment, as the potentiality of the piece of land for which a ground rent is to be assessed, is the potentiality to carry out the original building works or, if those had had to be renewed following a fire, the renewed building works.

It is for these reasons that I do not construe the clause as giving a special definition to ground rental value. I regard the task of the valuers called upon to assess the ground rental value as being the task of assessing what is the value in rental terms for the period of the lease and, subject to the special provision as to assumed user upon the terms of the lease, of the potentiality of using the land, the subject of the lease, for the development that has actually been carried out under the agreement and which has in fact been erected on the land.

Thus, I would answer the questions put to me by making a declaration which will, as I follow it, conform almost exactly with the words used in the affidavit sworn on behalf of the defendant in para 8.1. But I will certainly hear argument as to whether or not that is the form of the declaration which follows from my judgment or whether there is some variation to the words there used that is appropriate in the light of my judgment.

NEGLIGENCE

A **Chancery Division**

December 3 1993

(Before Mr Justice JACOB)

CRANEHEATH SECURITIES LTD v YORK MONTAGUE LTD

Estates Gazette May 28 1994

[1994] 21 EG 124

B ***Negligence — Turnover valuation of restaurant premises for loan purposes — Whether valuer negligent in relying on figures and variables used in valuation***

In December 1989 and May 1990 the plaintiff bank advanced £500,000 and £100,000 respectively to T Ltd, the owner of a destination restaurant in a manor house at Kingsdown, Kent. The loans were secured as a second charge on the property. In making the loans the plaintiffs had copies of the audited and management accounts of T Ltd and a valuation made in September 1989 by the defendant firm of valuers in the sum of £5.25m. The property consisted of the main restaurant of 145 "covers", outbuildings, a gatehouse cottage, swimming pool and grounds. The valuation report recorded that he
C turnover for 1989 was likely to be in excess of £1m and net profitability of £330,000. The valuer, in making his going concern valuation, arrived at an estimation of turnover an efficient operator would achieve of £1.55m and a net profitability of 35%; taking 8 YP, he arrived at £4.34m to which he added the values of the other parts of the property. The restaurant later closed, T Ltd was wound up and the property was sold for £475,000, the plaintiffs losing the sum advanced and interest. The plaintiffs claimed the sum of £650,000 plus interest alleging that they relied on the defendants' valuation which was said to be negligent. The defendants counterclaimed, alleging contributory negligence.

D ***Held*: The claim was dismissed. It was accepted that the defendants' valuer should have used as a basis for his valuation the guidance notes on the valuation of assets prepared by the assets valuation standards committee of the RICS, as amended in September 1989. Where the property to be valued is to be used as security for a loan, the guidance notes are explicit that the basis of valuation is to be open market value. The open market value basis assumes that values will remain static. The valuer does not have to take into account a possible collapse of property prices after his valuation; that risk falls on anyone who lends or buys on the basis of the valuation. The "margin of error" approach to valuing**
E **depends upon evidence that there is such a margin and upon some notional "right" figure; there was no such evidence in this case. The valuation of the defendants was one which a competent valuer, using proper skill and care, could properly have reached. The plaintiffs knew that the valuation depended upon current turnover and profitability and they relied directly on the management accounts; there was no negligence in failing to take steps to substantiate the figures for turnover. The defendants had not been negligent. Had negligence been established, and the defendants' claim in contributory negligence been considered, it may be said that, where a valuation has been given to a person and that person has, and he knows he has, more information** F
affecting the valuation than the valuer had, he is very likely to find himself at least partly at fault if he seeks to act in reliance on the valuation without first giving the valuer that information for comment.

The following cases are referred to in this report.

Bolam v *Friern Hospital Management Committee* [1957] 1 WLR 582; [1957] 2 All ER 118

Corisand Investments Ltd v *Druce & Co* [1978] EGD 769; (1978) 248 EG 315, 407, 504

Mount Banking Corporation Ltd v *Brian Cooper & Co* [1992] 2 EGLR 142; G
[1992] 35 EG 123

Singer & Friedlander Ltd v *John D Wood & Co* [1977] EGD 569; (1977) 243 EG 212 & 295

This was a claim by the plaintiffs, Craneheath Securities Ltd, against the defendants, York Montague Ltd, for damages arising out of an alleged negligent valuation.

Jonathan Simpkiss (instructed by Allison & Humphreys) appeared for the plaintiffs; Richard Lynagh (instructed by Cameron Markby Hewitt) represented the defendants.

Giving judgment, **JACOB J** said: A manor house stands on the cliff H
overlooking the sea at the picture-postcard village of Kingsdown, Kent. It was built in 1820. By 1989 its ground floor was a large restaurant called Don Medi. Don Medi was the nickname of a Mr Medi Siadatan. He ran the business and lived on the first floor with four wives. Using the terminology of this case I shall call the manor house and its ancillary properties "Kingsdown". Kingsdown and the business was owned by a company called Traylodge Ltd ("Traylodge"). Traylodge was ultimately partly owned by a Mr Richard White, a chartered accountant based in Yorkshire and by Mr Siadatan. Mr White was the key financial figure in Traylodge.

J

The parties

The plaintiffs, Craneheath Securities Ltd ("Craneheath"), were, in 1989, a secondary bank and a subsidiary of Berisford International plc. Shortly before the events concerning this case, Craneheath had recruited from Lloyds Bank an experienced manager with a view to increasing their loan book substantially. It had until then been static for a number of years. The manager they recruited was Mr Christopher Stone and it was he who played a principal part in the two loans with which this case is concerned. He had a superior, Mr John Rule, who was an experienced accountant on secondment from Arthur Andersen, the large firm of accountants. Both of these men were witnesses before K
me, though neither is with Craneheath any more. Craneheath in 1991 called in all its loans and ceased further banking activity.

The defendants, York Montague Ltd ("YM"), are chartered surveyors. They were newly formed in mid-1989 but Mr Anthony Crabtree [FRICS IRRV], their managing director, was a man of very great experience, particularly in valuations in the hotel, leisure and restaurant industry. He had been with the well-known firm of Druce & Co for over 28 years. Also with YM was a Mr Derek Ames [FRICS FSVA]. He, also experienced in this field but less so than Mr Crabtree, had also come from Druce & Co. He was an old colleague of Mr

A Crabtree's. Mr Crabtree gave evidence (indeed was cross-examined for over a day), but Mr Ames did not.

Loans

On December 22 1989 Craneheath loaned £0.55m to Traylodge, secured by way of a second charge on Kingsdown. There was already a first charge in the sum of £2.4m in favour of Kreditforeningen Denmark Bank ("KD") who had loaned £3.15m altogether. On May 22 1990 Craneheath loaned a further £0.1m also subject to the second charge.

Claim

B On September 19 1990 joint administrators of Traylodge were appointed and on April 15 1991 a compulsory winding-up order was made. Kingsdown was eventually sold at auction on May 6 1992 for £0.475m. Craneheath lost all their money and KD most of theirs. Craneheath say that their two loans were made in reliance upon a evaluation of Kingsdown made by YM (and in particular by Mr Crabtree) in the sum of £5.25m. They say that valuation was negligent in a number of ways which I have to consider in due course. They add that, but for that valuation, they would not have made the loan at all. So they claim as their damage stemming from the negligence the sum of £0.65m plus interest — about £1m in all. YM say that the valuation
C was not negligent (or at least is not shown to be), and that even if it was, Craneheath are either wholly or partially the authors of their misfortune, if the latter that there should be a diminution of damages for contributory negligence.

Circumstances leading to the valuation and first loan

Mr Stone first met Mr White in 1988. He found him a very good prospective client — not only an FCA, but also a man with a variety of business interests outside Traylodge. Mr Stone learned that Mr White needed more finance for Traylodge. In the summer of 1989 Mr Stone himself went down four or five times to Kingsdown. He ate at the restaurant and saw over the premises. He was undoubtably very
D impressed with the operation, the state of decor of the restaurant and everything else about it. In a letter dated October 12 1989 to Mr White he eulogised:

> You have now shown me the site at Kingsdown, near Deal in Kent, which is most impressive and you and Medi are to be congratulated on developing such a magnificent concept, which really has to be seen to be believed.

The restaurant was extremely busy, with queues on occasions and people travelled a long way to it. It was a "destination" restaurant, having also the "advantage" (in terms of a large customer source) of a large permanent "mobile" home park nearby. Mr Stone set out his
E assessment of Mr White in a letter dated October 11 1989 to his superior, Mr Rule. After identifying a number of other substantial assets owned by Mr White, Mr Stone said:

> The balance sheets (of Traylodge) are enclosed which do not reflect the latest valuations of the Kingsdown and Canterbury sites, which exceed £10 million. Borrowing of £3.2 million to (KD) is the only long-term liability.
>
> We are dealing with a highly successful professional man who has substantial locked-in assets, who, despite funds due from the sale of his land in Scalby, finds himself in need of temporary bridging assistance. While this represents a very significant proportion of our lending book, I have no hesitation in recommending our approval, subject to the normal strict
> F conditions, which will provide us with significant fee income and solid interest contribution over a relatively short exposure period.
>
> There are numerous other propositions in line next year and I wish us to demonstrate our support to this blue chip client.

The reference to the Canterbury site is a reference to another Don Medi restaurant in Canterbury which had opened in about February that year. Mr White had told Mr Stone that the intention was to develop quickly a chain of Don Medi restaurants (either owned or franchised) in Kent and nearby. Kingsdown was to be the headquarters, providing training and facilities for these. The idea was, once these had been set up, to sell the lot as a going concern to a major brewing or hotel group. Mr White said he needed a further £0.55m for this purpose and for renovating a recent addition to the Kingsdown G
property called the Gatehouse Cottage. So the proposed loan was envisaged to be for a short term only.

There are a couple of other matters about this letter. First there is the reference to the latest valuations of the Kingsdown and Canterbury sites, "which exceed £10 million". This was inaccurate. While Kingsdown was valued at £5.25m there never was a valuation of Canterbury taking the total above £10m. I can think only that Mr Stone, in his enthusiasm to persuade his superior, inflated the value somewhat.

Second, the letter refers to balance sheets. On the evidence these were draft accounts of Traylodge to March 31 1989. They were H
prepared by Moore Stephens and had been sent to Mr Stone by fax on October 9. It is common ground that Mr Crabtree never saw these accounts until much later. The accounts show the following:

(i) that the turnover was just over £700,000 (this included only a small contribution from Canterbury, which had only just opened before the end of the accounting year);

(ii) that there were hire purchase payments for equipment;

(iii) that the property valuation (of both Kingsdown and Canterbury) was £5.425m;

(iv) that there was a loss for the year of £336,079.

Moore Stephens qualified these draft accounts in respect of one J
substantial item of about £1.25m in respect of building/restaurant renovations and furnishings. They could not independently confirm this expenditure, saying there was no system of control over it.

Third, at the time of the letter, Mr Stone did have at least one version of the YM valuation, addressed either to Craneheath or to KD or both. There are two versions, one addressed to KD (who were originally possibly going to make the loan) and one to Craneheath. Both are dated the same day, September 26 1989. Mr Crabtree said that both were done at the same time and that his firm would not backdate a valuation. Mr Stone (and the contemporaneous documents) suggest that the Craneheath version came into existence later. I do not K
think that this conflict matters, save potentially in relation to the reliability of Mr Crabtree's evidence. My own view is that the Craneheath version probably came into existence later, and that Mr Crabtree was mistaken. I do not think the mistake reflects generally on his evidence. The word processing of a fresh version for Craneheath (so they could rely upon it) would have been a matter of mere administration at the time. There was to be no further fee for it. There was no question of Mr Crabtree being asked to consider whether his valuation in fact made in September was still valid at a later date and particularly before the loan was made in December. The date of September was, of course, actually the appropriate date of the valuation, even if the document addressed to Craneheath was provided L
later.

Before the first loan was made, Craneheath adopted an internal policy under which Mr Stone was authorised to make loans of up to £0.25m on his own and that loans of above £0.5m required in addition to Mr Stone, Mr Rule or another Berisford board member. The projected loan fell within the latter class. On November 21 Traylodge, by a letter from Mr White, formally requested the loan. Mr Rule's copy of that letter contains his handwritten note "— current financial data". The evidence was clearly to the effect that Craneheath wanted such data before giving the loan. They already had the accounts to March 1989. I find that they did indeed get such data in the form of M
management accounts and that those accounts included up-to-date trading information (including in particular information as to turnover and profitability) and also, probably, future projections. Mr Stone said, and I accept, that the data was analysed in detail by his assistants. Only after this had been done was Craneheath prepared to make the loan. They also required that the YM valuation should be addressed to them (if indeed they had already not got such a copy). The requirement for management accounts appears to be explicit in a letter of November 30. It was suggested that there was a muddle in collating (so the requirement was in fact attached to a letter relating to a different matter). I do not think that this matters. Mr Stone said

A unequivocally in effect: if there had been no management accounts, there would have been no loan, just as Mr Rule had required in his handwritten note.

After the legal formalities had been completed, Craneheath arranged for the money, to be credited to Traylodge's account by a letter dated December 21. They did not check to see that the arrangement fee of £5,500 had been paid before draw down, as the terms of the loan required. It was never paid. That fee was originally claimed as damages in this action, but it is conceded that to that extent the claim must fail since it should have been paid before the advance was made.

B The upshot of the foregoing is that, at the time the loan was made, Craneheath had in their possession (and had read):

(i) Mr Crabtree's valuation made in September;

(ii) the Moore Stephens draft audited accounts to end of March 1989; and

(iii) the management accounts, giving current financial data.

Unaccountably item (iii) went missing before discovery was due. So also, according to Mr Stone, have several other discoverable documents. This may be because Craneheath moved premises, but it is difficult to imagine how just some documents went astray. I do not think this matters in relation to this item, for I have enough evidence

C of what was in it.

Second loan

All did not go well and by March 1990 Traylodge were asking for a further £0.25m. This arose because the National Westminster Bank ("NatWest") had a charge on the Canterbury property and were threatening to realise their security in that sum. Traylodge asked Mr Stone to buy NatWest out. Eventually a loan of £0.1m was made. Probably it did not have Mr Rule's authority as it should have done. Moreover at the time Traylodge had not paid the £5,500 arrangement fee nor an interest payment due to Craneheath and another to KD. Mr Robert Dibben, Craneheath's experienced and reliable banking expert,

D called the loan "imprudent" and I so find. I reject the suggestion that it was made in continued reliance upon the YM valuation which, by this time, was six months old. It was done out of panic in the hope of salvaging the situation.

What happened thereafter

Things went from bad to worse. There were some structural problems and Messrs White and Siadatan fell out, Mr Siadatan at one point even obtaining a *Mareva* injunction against Mr White, which was subsequently discharged. The restaurant was closed suddenly in midseason and stayed closed for some time. The details of this do not matter. It re-opened at one point and closed again. Naturally this must

E have harmed its reputation, generating "badwill". After Traylodge was wound up the liquidator's sale produced, as I have said, £0.475m. But this was in 1992 when the economy was very different. People were no longer willing to drive distances to destination restaurants or to pay the same amounts on anything like the scale they were in 1989.

Valuation: RICS guidance notes

I now turn to the valuation and the principles upon which it was made. It is common ground that Mr Crabtree ought to have been working in preparing his valuation on the basis of the guidance notes on the valuation of assets prepared by the assets valuation standards

F committee of the Royal Institution of Chartered Surveyors as it had been amended by September 1989. The valuation is explicit as to Mr Crabtree's terms of reference:

> Our instructions are to prepare a valuation of the Freehold interest in the property described, on an open market basis, as at today's date, on a going concern basis, including the benefit of goodwill, trade fixtures, fittings and contents used in connection with the business.
>
> We have provided our Valuation on an open market basis, and strictly in accordance with the Guidance Notes relating to the valuation of Property Assets published by the Royal Institution of Chartered Surveyors

He added "bearing in mind that the property will be offered for an advance secured by way of first mortgage". It was in fact, so far as Craneheath were concerned, a second mortgage. Mr Crabtree said he G thought at the time that there was to be a first mortgage, but nothing turns on this. For the method of valuation would be the same whether it was a first, second or nth mortgage, or indeed not a mortgage at all. The basis of valuation was, as Mr Crabtree said "an open market basis".

It is to what the RICS guidance notes say about such a basis that I must now therefore turn.

Guidance note GN 18 (revised January 1986) is headed:

Valuations of Assets for Investment Purposes and as Security for Loans

It contains the following paragraphs: H

> 4. When a property is valued as security for a loan by way of mortgage or debenture, the basis of valuation should be open market value. It would be usual for the valuer to be asked to express an opinion as to the suitability of the property as security for a loan. It is, however, a matter for the lender to assess the risk involved and express his assessment in fixing the terms of the loan, such as the percentage of value to be advanced, the provision for repayment of the capital and the interest rate. The valuer should refer in his Valuation Certificate or in his detailed report where one is to be provided, to all matters which are within his knowledge and which may assist the lender in his assessment of the risk.
>
> 5. It is not normally appropriate to value property to be used as security for a loan at forced sale value. Where, however, the Bank or other lending J institution requires the valuer to report forced sale value, either in addition to or in place of open market value, the valuer should ascertain from the client as a condition of providing such a valuation the period of time in which a sale is desired and whether or not such period is to include completion of the legal formalities.

The note is completely explicit therefore as to the basis of valuation when the property is to be used as a security. It is the "open market value". I find this somewhat surprising at least where, as here, the valuation is heavily dependent on the turnover or profitability of the business conducted from the property. If he has to resort to his security, the lender is unlikely in practice to be trying to realise it K when the business is running smoothly. It is much more likely that the business has collapsed wholly or partially. So the obvious implications of this method of valuation for a lender is that realisation of the security will not reach the valuation. However, that is what the guidance notes provide. There is no doubt that in this case at least the valuation was being given to an experienced and sophisticated lender who knew what the basis was.

I turn to the basis upon which the "open market value" is determined. This is set out in guidance note GN 22 (revised January 1989): L

> 1.1 "Open Market Value" means the best price at which an interest in the property might reasonably be expected to be sold at the date of the valuation assuming:
>
> (a) a willing seller;
>
> (b) a reasonable period in which to negotiate the sale taking into account the nature of the property and the state of the market;
>
> (c) that values will remain static during that period;
>
> (d) that the property will be freely exposed to the open market; and
>
> (e) that no account will be taken of any additional bid by a purchaser with a special interest.

This basis itself contains an important assumption from the point of view of this case — the assumption that values will remain static. The M valuer does not have to take into account a possible collapse of property prices after his valuation. The risk of that falls entirely on anyone who lends or buys on the basis of the valuation.

GN 22 also provides for a "forced sale value", which is the same as for the "open market value" save that there is an assumed time-limit for sale. Even that would not correspond well with what might be obtained when some kinds of business have collapsed. Take, for instance, a restaurant. Suppose the business collapses without explanation and the premises become idle. One would expect old customers to be wary of any re-opened restaurant on the same site — there would be a kind of "badwill" to be overcome.

A The RICS give further guidance as to open market valuations having regard to trading potential. This is contained in Background Paper no BP7 (page revised January 1986). The relevant paragraphs provide:

2. It has always been recognised that there are certain types of property designed or adapted for particular uses, which invariably change hands in the open market at prices based directly on trading potential for a strictly limited use. Examples of such properties which normally are sold as fully operational business units include hotels, public houses, cinemas, theatres, bingo clubs, gamin, clubs, petrol filling stations, betting shops, specialised leisure and sporting facilities.

B 3. Open market transactions which involve the sale of properties which come within this category, as opposed to those involving the sale of companies which own such properties, can provide evidence of value for use when valuing this type of property for balance sheet purposes. When analysing the prices paid for comparable properties and preparing a valuation of the subject property, the Valuer will normally have regard to the trading accounts for previous years, where these are available, and form an opinion as to the future trading potential and level of turnover likely to be achieved.

4. The example of hotels is typical of the type of property where these circumstances can apply, as they are usually sold as a fully operational business including fixtures, fittings, furniture and stock. The new owner will normally engage the existing staff and sometimes the management and would, of course, expect to take over the benefit of future bookings which are an C important feature of the continuing operation.

...

6. When problems are encountered in differentiating between the value of the trading potential which runs with the properly and the value of goodwill which has been created in the business by the present owner and which may be transferable to other properties in the event of the subject properly being sold, it is recommended that the Valuer when assessing future trading potential should exclude any turnover which would only be available to the present owner or management, but he should reflect any trading potential that might be realised in the hands of a more efficient operator.

...

11. (a) Advice on the suitability of the property as security for a loan

D This is generally covered by Guidance Note GN 18, but in the case of properties valued by reference to Trading Potential in addition to stating his opinion as to the suitability of the property as security for a loan at the date of valuation, the valuer should extend his advice to cover possible future fluctuations in the status of the property as security.

The valuer should state clearly the source and basis of the trading figures (actual and/or projected) and other trading information and assumptions upon which he has relied in arriving at his valuation. He should then add a rider to the effect that in the event of a future change in the trading potential or actual level of trade from that indicated by such information and assumptions, the open market value for existing use could also vary, in view of the importance the market for such properties attaches to trading figures actually achieved.

E As indicated in GN18 it is in any case a matter for the lender to assess the degree of risk involved and to reflect his assessment in fixing the terms of the loan including the percentage of open market value to be advised, the rate of interest to be charged and the provision for capital repayment.

Applicable legal principles

It was common ground that YM owed Craneheath a duty of care — that is why the valuation was addressed to them. The pleadings suggested a qualification to that duty, namely that it related only to a first mortgage. Wisely Mr Richard Lynagh [counsel for the defendants] abandoned that point at the outset.

F A number of authorities concerning negligence claims against valuers were cited to me. They were all helpful in their way, for they show how the courts have approached the problem in the past. The most difficult area is in assessing what amounts to negligence. There is a danger in unquestioningly taking an approach adopted on the basis of the facts in an earlier case as though that itself is a legal principle. For instance, in *Singer & Friedlander Ltd* v *John D Wood & Co* (1977) 243 EG 212, Watkins J had evidence that there was a permissible margin of error of +/-10% of a figure which could be said to be the right figure. This "margin of error" approach depends upon evidence that there is such a margin. It also seems to depend upon some notional "right" figure, though one could not ascertain this in many cases. The margin of error approach may well make sense when the property being valued has a number of good comparables (eg G houses on an estate, forestry or farmland). Then one can in effect take an "average" as the "right" figure and look at the deviation from it. That is one of the standard techniques in scientific measurement. In those circumstances one would expect different valuers to settle fairly closely upon the same sort of figure. But, as I say, the approach depends on evidence. I had no evidence before me which would begin to support a permissible "margin of error" for a property such as Kingsdown. The nearest I had was some evidence about possible margins of error in relation to certain of the factors (such as net profitability) adopted by Mr Crabtree, to which I shall return.

It is also of course vital, where it is said that a man has fallen short H of a particular standard, to identify the standard concerned. In this case the plaintiffs placed much reliance on what was said by Gibson J in *Corisand Investments Ltd* v *Druce & Co* (1978) 248 EG 315. The case was decided in 1978 when, it seems, there were no RICS guidelines as to valuation. Then, of course, the court was driven to standards of valuation as set forth in the evidence in that case. Thus, Gibson J had evidence that the going concern basis was inappropriate for a mortgage. He said:

In a valuation for mortgage purposes it is again common ground that the valuer cannot take the open market going concern valuation. The valuer cannot, in short, include in his valuation for this purpose any valuable part of J the going concern valuation which would not or might well not be there when the mortgagee attempts to realise his security.

By contrast in this case the standard of valuation was indeed to be "going concern", even though it was for a mortgage. From time to time Mr Crabtree's cross-examination included questions based on the express premise that he ought to have been "cautious" because the valuation was for mortgage purposes. While he ought to have exercised all proper skills in carrying out his valuation I reject the suggestion that he should have arrived at a lower figure simply because the valuation was for mortgage purposes. The practical effect of the RICS guidelines is to put any need to take extra caution where K the property is to be a security upon the lender, not the valuer. In some cases this might well lead to a misunderstanding, but not in this case. Mr Stone knew very well what a going concern valuation involved.

The key test, as it seems to me, is simply stated. It is this: was the valuation that which a competent valuer, using proper skill and care, could properly have reached: see *Mount Banking Corporation Ltd* v *Brian Cooper & Co* [1992] 2 EGLR 142. In answering this question the court should bear in mind what Watkins J said at p 213 in *Singer & Friedlander*, namely:

The valuation of land by trained competent and careful professional men is a L task which rarely, if ever, admits of precise conclusion. Often beyond certain well-founded facts so many imponderables confront the valuer that he is obliged to proceed on the basis of assumptions. Therefore he cannot be faulted for achieving a result which does not admit of some decree of error. Thus two able and experienced men, each confronted with same task, might come to different conclusions without any one being justified in saying that either of them has lacked competence and reasonable care, still less integrity, in doing his work.

As Gibson J noted there are parallels with medical negligence. For it is possible for different doctors properly and without negligence to adopt different treatments. One will not be negligent unless he has failed to act in accordance with a practice accepted as proper by a M responsible body of competent professional opinion, *Bolam* v *Friern Hospital Management Committee* [1957] 1 WLR 582.

Watkins J said in *Corisand* that a valuer should disregard any "speculative element". By this he meant he should ignore the possibility of a purchase by someone who was willing to pay more because the market might rise and he could sell at a quick profit. I think that is probably built into the RICS guideline requirement that the market (ie "values") will remain static. In any event there is no suggestion of a speculative element in this case.

One further qualification appears to me to be self-evident. Where a valuation is made by means of a written report, the lender can

A reasonably be expected to read the report and not simply the final figure. A valuation is not a guarantee of the figure reached — a "suable document" (to use Mr Rule's phrase) irrespective of what further or other information the lender may have. This is a matter of particular importance if the lender has significant further information which might well affect the valuation, as opposed to information which merely affects the creditworthiness of the lender. For primarily the lender is looking to the security of the property rather than the lender when he makes a secured loan. This is an important matter in this case, as will be seen.

Valuation: how Mr Crabtree arrived at £5.25m

B The property consisted of the main restaurant (with 145 "covers" as valuers call the seating capacity). In addition there were the upper floors and some outbuildings. The latter included the gatehouse cottage, recently purchased for £0.235m. The manor house had substantial first-floor accommodation which had a separate entrance. There was a swimming pool and grounds.

It was common ground that the property was unusual and accordingly difficult to value. Most of Mr Crabtree's detailed reasoning is to be found in his working notes. He not only adopted a primary method but sought to cross-check that by a secondary method.

The primary method of valuation of the restaurant was to estimate C its potential turnover and profitability. This, of course, started with its actual turnover and profitability. He said in his valuation:

> Although we have not been supplied with accounts for the restaurant it does, nevertheless, keep a careful record of all visitors in a Visitors' Book, and we attach in Appendix VII in *(sic)* an extract covering relevant dates during April through to early June. During our last inspection the attendance figures between June and September had shown a fairly significant increase on the earlier months.

Mr Crabtree said that he verified certain figures already noted by Mr Ames (who had prepared a valuation in June 1989 while he was with Druce & Co) and checked the later figures. It was suggested that D he had not in fact done this and had merely taken the Ames figures together with an average figure given by Mr White. I reject the suggestion. True he did not set out the figures for June, July and August (though he repeated the earlier Ames figures for March-May) but his working notes clearly show he must have made some check of the register. These notes were, he told me, themselves the result of transferring information from other pieces of paper which were not retained. This is highly likely, for Mr Crabtree's working notes are too neat and organised to be original notes for the whole valuation. Part of the criticism related to his reference to September when he could not have got those figures, his inspection being at the end of August. I E cannot think this matters. The valuation itself is dated September 26 and the inspection must obviously have been some time earlier.

Under the heading "Factors affecting Valuation" Mr Crabtree said:

> It is our understanding that the turnover during 1989 is likely to be slightly in excess of £1m, and that the net profitability is likely to be in the region of £.3m, due to efficient cost controls.

His working notes have a figure of £1.1m with the rider "figures provided by company estimated at £1.1—£1.2m". The first of the conditions of the valuation made it clear that:

> The valuation is largely based upon the information, data and plans supplied to F us. We have relied upon this information in arriving at our Valuation, which is made on the condition that there have been no matters undisclosed which could materially affect our opinion of value"

Mr Crabtree noted that the restaurant was closed on Sundays (he was told for religious reasons) and was not open for Saturday lunch. The restaurant sold pasta, hamburgers, pizzas, salads and the like. The menu range was short and there was a high turnover of customers. He considered that, given a high turnover, this is the sort of operation where there would be a net profitability of about 35% and this is the figure he took. Mr Crabtree considered that any purchaser would naturally open on Saturday for lunch in the summer and for Sunday all year — Kingsdown was, after all, a destination restaurant. He also considered that a price rise of 15% was achievable (the directors had G told him of a shortly planned 15% increase), though he used a figure of 10% in his calculations.

Mr Crabtree also thought that the upper parts and outbuildings could and would, by an efficient purchaser, be put to better use with an added effect on turnover for the restaurant.

Thus, he arrived at the turnover which in his estimation an efficient operator would achieve, namely £1.55m. Mr Crabtree considered what the net profitability would be (as I have said 35%). So he got to an annual net profit of £0.542m. He then, as is conventional, considered how many years' purchase would be appropriate. He took a figure of eight, thus arriving at a valuation of the restaurant valuation of H £4.34m. To this he added values for the upper floors at £0.5m, and the adjacent cottages at £0.25. He considered that the first and second floors had a potential for nine bedrooms/suites at £0.06m each giving a figure of £0.54m which he rounded down to £0.5m and the cottages had a potential of five letting bedrooms at £.05m each giving a figure of £0.25m. These were the same as Mr Ames' June valuation. He added £.235m for the gatehouse cottage, being the recent cost. The overall amount thus reached was £5.325 which he rounded down to £5.25m.

His cross-check method for the ancillary accommodation was to estimate the potential banqueting/wedding usage and conferencing J usage and prices, thereby arriving at an annual turnover of £0.22m. Again he took 35% net profitability. He applied a years' purchase of seven (not eight because this kind of use had not been established), deducted minor estimated conversion costs and arrived at £0.529m which he rounded down to £0.5m — the same sum as he had taken in his primary method and as had been reached by Mr Ames.

He also then considered the turnover as a whole, namely restaurant £1.55m and banqueting/conference (upper parts) £0.22m giving a total of £1.77m. Instead of the using the years' purchase method he then used another established method, namely considered the ratio of capital value/turnover. Taking the primary valuation of £5.325m this K gave a figure of £2.68m. He compared this with "comparables on restaurants" which he said "vary between 2.75/3.25".

It was an important part of Mr Crabtree's reasoning that he expected some big players to be anxious to acquire this kind of property — not only for its restaurant but for its hotel, conferencing and banqueting potential. He was very close to this kind of purchaser, he was experienced with the Kent market and I accept his evidence. No one as experienced as he gave evidence.

Was Mr Crabtree negligent?

It will be seen that Mr Crabtree's calculation was quite detailed and elaborate. Mr Jonathan Simpkiss [counsel for the plaintiffs], did not L attack the principles he had adopted, although originally there was an attack based on the valuation of the ancillary accommodation by Mr Clive Lickley [FRICS], Craneheath's expert witness. Mr Lickley was not prepared, under cross-examination, to say that no reasonable valuer would not use the method used by Mr Crabtree.

The attack began simply setting out other valuations of Kingsdown with the invitation to say that these made out a *prima facie* case. These valuations and dates were:

Date	*Valuation*	*Valuer*	*Basis*
29.1.88	£1.25m	Mr Spacey	Going concern
3.5.58	£1.25m	Edward Symmons	unstated
16.6.89	£4m	Mr Ames	Going concern
26.9.89	£5.25m	Mr Crabtree	Going concern
29.9.89*	£3.85m	Mr Crabtree	Bricks and mortar
2.6.93	£3.76m	Mr Spacey	Going concern
28.6.93	£2m	Mr Lickley	Going concern

M

*(Actually done by adjustment of September 1989 figure in March 1990 pursuant to a request from Moore)

The last two valuations were retrospective. They were prepared for this purposes of this case by Mr John Space [FRICS], YM's expert, and Mr Lickley.

A In the end Mr Spacey was not called. But since his report was referred to in part, the whole of it (and some supplementary letters) was before me. Although he reached a lower valuation (principally because he treated the ancillary accommodation differently, as had Mr Ames) he did not go so far as to say that Mr Crabtree's valuation was such that no competent valuer could have arrived at it. It was argued that Mr Crabtree's valuation was not only the highest but was way higher than the next lower figure, that of Mr Ames, even allowing for the fact that Mr Ames' valuation did not include Gatehouse Cottage. Moreover, it was pointed out that YM did not call a valuer prepared to reach anything like Mr Crabtree's figure. This latter point loses some
B force given that it seems that Cranheath themselves at an earlier stage in the litigation had engaged some other expert valuer whose report never saw the light of day.

I do not think that these figures alone make out a *prima facie* case of negligence. Mr Crabtree was, of course, conscious that his valuation was just over £1m higher than that of Mr Ames (allowing for the gatehouse). He had seen Mr Ames' valuation before he went down to make his own inspection. He discussed the matter with Mr Ames on his return. He had no ulterior motive for giving a higher figure. He told me that his principal reason for differing was in his assessment of what a major plc would be prepared to pay for the property given
C the possible uses of the upper parts. I accept that. If he had simply been lazy and unprepared to apply his own mind to the problem he could easily have just copied Mr Ames' views. He did not. Given that there is a clear and detailed explanation of how Mr Crabtree reached his conclusion, the question must be whether in that process Mr Crabtree was in any way negligent. The plaintiffs recognised this and sought to attack the valuation and the underlying assumptions.

I put aside what I regard as the trivial complaint about planning permission and begin with what lay at the heart of the attack. This was the figure for existing turnover and profits, which Mr Crabtree's notes
D indicate had been given to him. In opening Mr Simpkiss summarised the allegations as follows:

1. No serious attempt to substantiate turnover figures.

2. No or no adequate warning as to inadequate basis for figures those and reliance only on a director for them; especially since Mr Ames had been told the same thing. There might be something in that, if Mr Crabtree had not expressly made it clear that he had seen no accounts. But what makes the plaintiffs' position wholly untenable in this regard is the fact that they did have the management accounts. Moreover, without these (which Mr Stone told me were in accordance with the figures given to Mr Crabtree) Craneheath would not have
E gone ahead. Mr Stone said:

> If Mr Crabtree had obtained these management accounts I would have regarded it as a sufficient investigation by him. They were sufficiently consistent with Mr Crabtree's figures as to cause me no worry.

Lurking behind the scenes is a suggestion that the accounts were false, though noone ever said as much. I do not know whether they were. But Craneheath certainly had other material in their hands which put them on notice that there might be a problem. The Moore Stephens accounts spoke of a turnover of only £0.76m. They were of course historical. But, to my mind, if one had both the Moore Stephens
F accounts and Mr Crabtree's valuation one would naturally investigate more. I was not surprised to hear Mr Dibben to say so. I find it odd that Craneheath did not send these accounts to Mr Crabtree for comment. They knew the basis of the valuation was "profits" and here was something which might throw the calculation out. If they wanted to continue to rely upon the valuation it was only fair that they should have raised the matter with the valuer. If they had, not only would the disparity between the turnovers have emerged but also it would have been seen that what Mr Crabtree was told about fixtures and fittings (namely that they were fully owned by Traylodge) was inconsistent with payments of about £18,000 pa by way of hire purchase for equipment.

I think the fair conclusion is that Craneheath knew that the G valuation depended upon current turnover and profitability and relied directly on the management accounts and its own view of Mr White as a "blue chip" client. I do not see that Mr Crabtree can be faulted in this regard, or that, if he can, it would make him liable when Craneheath had more information than he had and knew they did.

It was faintly suggested by Mr Lickley in re-examination that even if Mr Crabtree had been given management accounts he ought to have gone further. I do not accept this. I cannot see why a valuer should, given proper management accounts by an established professional businessman, particularly an FCA, try to go behind these. A valuer is not an auditor. And what was good enough for Craneheath would have been good enough for the valuer. H

That disposes of the "crunch" attack. But there were other attacks upon various aspects of the valuation with which I must deal in turn. Collectively they were what was called in argument "bumping up".

Saturday/Sunday opening

The suggestion was that Mr Crabtree had allowed for an increased turnover without allowing for an increase in costs due to overtime and a possible need to close on Mondays. This was not made out on the evidence. Mr Crabtree pointed out how major companies indeed opened all day on Sunday (eg Harvester) and that there would not in the circumstances of Kingsdown (with plentiful part-time labour J available) no overtime would be necessary.

35% net profitability

Mr Lickley originally said that this was too high. And Mr Crabtree was criticised for drawing parallels with the "Travellers Fare" operation, the sale of which he had been concerned with in about 1987. Under cross-examination Mr Lickley was prepared to go to about 30%. However, this was not for an operation with the limited menu and large turnover such as Don Medi. He conceded that in those circumstances profitability might be higher than the kind of pub with large restaurant which he had been considering. Mr Crabtree pointed out that his choice of Travellers Fare as a comparable (which ranged K from station buffets, hamburger restaurants, to full restaurants) was based not on a parallel between the type of restaurant (which was obviously not the same), but on the profitability achievable where you have a large turnover and limited cheap-to-produce menu. I cannot see that the figure of 35% was negligent. It was in accordance with the anticipated price rise and increased usage.

Years' purchase of 8

Originally this was attacked by Mr Lickley as too high. He said that a purchaser would be cautious in the absence of accounts, a false basis since accounts existed. But when he was cross-examined some L "comparables" (necessarily not exact) were put showing higher figures for YP. In the end Mr Lickley accepted that the YP was a matter of judgment and was not prepared to say that a YP of 8 was wrong.

3 x turnover = value

The value/turnover factor was used by Mr Crabtree as a cross-check. Mr Lickley accepted that a factor of three was not unusual for a hotel. This property was not quite a hotel, but it was reasonable to look at hotel-cum-large restaurants as a comparable.

Average price per meal £12

It is here that I found myself unhappy with Mr Crabtree's M approach. What he did was to examine the menu and produce an average assuming three-course meals and half a carafe of wine between two. He also based his estimate on his own bill when he had dined with his family "incognito" a week before his official valuation visit. This does seem to me to be too rough and ready. However, I do not think it matters. He already had figures for turnover and profit and this figure affected only the extra profits from assumed increased turnover. Even if the right figure is £10 the valuation remains above £5m.

Alleged failure to take into account VAT

A very hostile attack was made here. I reject it. Mr Crabtree said

A that all valuers always ignore VAT and he always did. Mr Lickley accepted that valuations do not normally take VAT into account and that if it is not mentioned one would assume turnover figures to be less VAT.

I should say that, in my view, while Mr Lickley was evidently doing his best to help, his experience was comparatively limited and primarily concerned with public houses (including of course those with restaurants). Moreover, he had not seen the upstairs of the premises at all. He did concede that at the time the market was very buoyant and that there were a number of large operators anxious to buy substantial restaurant complexes. It is the existence of these (and
B Mr Crabtree was very familiar with them and the prices they were willing to pay) which lay at the heart of the valuation. This is not a "speculators price" of the kind contemplated in *Corisand*. These buyers were at the time interested in acquisition with the purpose of expanding operations and maximising turnover and profits.

Given the economic circumstances which emerged shortly after, these buyers simply disappeared at anything like the late-1989 prices. That is why the later valuations and the ultimate sale price is really of no great assistance.

Accordingly, I hold that it is not shown that Mr Crabtree was negligent. It is therefore not necessary for me to consider the arguments as to contributory negligence and do not propose to do so,
C save to say this: that where a valuation has been given to a man, and that man has, and knows he has, more information affecting the valuation than the valuer had, he is very likely to find himself at least partly at fault if he seeks to act in reliance on the valuation without first giving the valuer that information for comment.

Nor is it necessary for me to consider Mr Lynagh's alternative argument to the effect that if Mr Crabtree had given a value of £4.25m (approximately equal to Mr Ames' plus the gatehouse cottage) Craneheath were so anxious to loan that they would have loaned on that valuation. However, I have to say that I would have accepted Mr Stone's evidence, corroborated by a contemporaneous document, that
D the loan would not have been made below a £5m valuation.

The action therefore fails.

Queen's Bench Division

September 30 1993

(Before Mr C S C S Clarke QC, sitting as a deputy judge of the Queen's
E Bench Division)

ALLIED TRUST BANK LTD v EDWARD SYMMONS & PARTNERS

Estates Gazette June 4 1994

[1994] 22 EG 116

Negligence — Mortgage valuation — Whether valuer entitled to include element of hope value to reflect prospect of planning permission for conversion of property — Whether valuer arrived at a valuation within acceptable bracket

F In February 1988 C Ltd contracted to purchase Rackheath Hall, a Georgian country house, for £530,000. The company had proposals to develop the property by conversion and additional development to provide apartments and terrace houses and applied to the plaintiff bank for a loan of £370,000. The defendant firm provided a valuation report to the plaintiff in which it was stated that the open market value of the property was £595,000. In accordance with its policy, the plaintiff provided the loan based on 70% of the purchase price taking a legal charge as security. Completion took place in September 1988 and the loan was drawn down. C Ltd failed to pay interest or principal and the property was sold in 1991 for £220,000. The plaintiff claimed that: the defendants negligently valued the property; a proper valuation for mortgage purposes would have been £350,000; on such a valuation the G
plaintiff would have lent only £245,000 (being 70% of £350,000); and accordingly it had lost £125,000. The defendants' valuer, in using a residual valuation, assumed that it was likely that planning permission would be granted for the proposed conversion and development.

Held**: The claim was dismissed. The open market value of freehold property must necessarily include such additional amount (if any) as, in the open market, prospective purchasers would pay for the property on account of the prospect of obtaining planning permission for development, over and above that which they would pay if there was no hope that planning permission could be obtained. If the valuation is for mortgage** H
purposes, this adds little to open market value and accordingly the value of the prospect of planning permission should be included. The ISVA/RICS guidance notes for mortgage valuations were not appropriate for valuing property not intended for residential use by individuals. The *RICS Guidance Notes on the Valuation of Assets* (2nd ed) were a guide to the meaning of open market value for mortgage purposes. The observations on mortgage valuations in *Modern Methods of Valuation* were unlikely to be intended to refer to properties like Rackheath Hall. Although not published at the time, a valuer who followed the RICS *Statement of Asset Valuation Practice No 1* (July 1992) could not justifiably be J
criticised where he lacked specific instructions to the contrary. That note refers to "hope value" and provides confirmation that it was reasonable to include hope value in valuing the property. It was not wrong in principle in taking into account the prospect of planning permission. The use of the residual method of valuation was a valid and justified approach because the prospect of obtaining planning permission was sufficiently good for a developer to be likely to perform such an exercise in deciding to bid. The figures in the defendants' valuation justifying the valuation were not outside the range of figures which a competent valuer, using the residual method, might adopt. The defendants' K
valuer had not been negligent. However, if the property should have been valued as a residence only without the prospect of planning permission, its value would have been in the range £420,000-£470,000.

No cases were referred to in this report.

This was a claim for damages for negligence by the plaintiff, Allied Trust Bank Ltd, against the defendants, Edward Symmons & Partners.

Marion Smith (instructed by Freeman Pollard) appeared for the plaintiff; Stephen Worthington (instructed by Barlow Lyde & Gilbert) represented the defendants. L

Giving judgment, MR CSCS CLARKE QC said: Rackheath Hall is a Georgian country house situated at Rackheath, some 4 miles north-east of Norwich. The original building was completed in the late 18th century, but it was extensively remodelled between 1852 and 1854. It is listed in Grade II as a building of special architectural interest. In 1930 it was described as one of the prime estates in Britain. It has three floors and a basement. On the ground floor, in addition to the central hall, there is a suite of reception rooms, library, smoking room, domestic offices and a housekeeper's flat. On the first floor there are nine bedrooms, two-dressing rooms and three bathrooms. On M
the second floor there are nine bedrooms and two bathrooms. It could properly be described as a stately home. In 1988 it had not been completely occupied since the 1950s. The ground floor was used for storing and displaying antiques; and the owner lived in a flat over the billiard room.

In 1988 Mr Khalid Karrar was a property developer and a substantial customer of Allied Arab Bank Ltd ("the bank"), as the plaintiff was then named. In July 1988 companies owned or controlled by him had £1.6m by way of advances (for property development) outstanding from the bank. In February of that year Canterbrook Ltd, a Guernsey company under his control, contracted to purchase Rackheath Hall for £530,000 with completion due on August 1. Mr

A Karrar planned to develop the property by demolishing the Victorian additions, converting the hall into 10 self-contained apartments and erecting a new wing of eight terraced houses and three flats (one for a caretaker). He applied to the bank for finance and supplied them with a valuation report ("the report"), which he had obtained from Edwards Symmons & Partners, a firm of surveyors and valuers carrying on business at 56-62 Wilton Road, London SW1, who are the defendants. I shall refer to them as "the partnership". In 1988 they valued mainly commercial and industrial properties. The report was originally addressed to the Bank of Oman, but the bank required and later received a copy of the report addressed to itself dated July 28 1988 in **B** the same terms.

The author of the report was Mr William J Godfrey, then a salaried partner of the partnership. Para 15.00 of it reads as follows:

> 15.01. We are of the view that the open market value for mortgage purposes of the freehold property, free from encumbrances may be fairly expressed at the present time in the sum of: £595,000 (Five Hundred and Ninety Five Thousand Pounds)
>
> 15.02. We are further of the opinion that upon the grant of Planning Permission and Listed Building Consent for the conversion of the existing building into 10 self-contained apartments and the erection of a new wing of 8 terraced houses and 3 flats, the property will have a value in the order of: £950,000 (Nine Hundred and Fifty Thousand Pounds)
>
> **C** 15.03. We are further of the view that upon satisfactory completion of the building works, the apartments and houses should realise the sum of: £2,535,000 (Two Million Five Hundred and Thirty Five Thousand Pounds)

made up as shown on the attached schedule.

By a letter dated August 11 1988 the bank agreed to lend Canterbrook Ltd £370,000 to assist with 70% of the purchase price of Rackheath Hall. At that time it was the bank's policy to lend up to 70% of the purchase price or value of the property in question, whichever was the lower: £370,000 is 70% (strictly 69.8%) of the price of £530,000. The loan was expressed to be for a period of six **D** months to be "reviewed upon expiry, pending the outcome of planning application for conversion into flats and cottages". The loan was secured by a legal charge over the property and by the personal guarantee of Mr Karrar in the sum of £200,000 together with interest. Completion took place on September 20 1988 when the loan was drawn down. By early 1989 the market, which until the summer of 1988 had been buoyant, was turning and subsequently went into free-fall. Thereafter, Canterbrook failed to pay the interest due in full or on time. By a letter dated June 4 1990 the bank made formal demand for the £466,795.06 then due. On January 18 1991 Mr Karrar was adjudged bankrupt. In the summer of 1991 the property was sold **E** for £220,000.

The bank's principal (but not its sole) claim may be summarised as follows:

(1) The valuation of £595,000 was made negligently;

(2) A proper valuation made upon the same basis as that used by Mr Godfrey, namely "open market value for mortgage purposes", would have been £350,000;

(3) Upon such a valuation the bank would have lent only £245,000, being 70% of £350,000;

(4) Accordingly, it has lost £370,000 less £245,000, ie £125,000.

It is common ground that the bank's net receipts from the sale of **F** the property were less than £245,000 so that no deduction falls to be made for any excess. Since the bank's claim is based on the fact that it would lend only 70% of the lower of value or price, it follows that, if a competent valuation could have been as high as £530,000 the bank's claim would fail.

The issues

When Mr Godfrey valued the property at £595,000 he took the "open market value for mortgage purposes" to include such value as was attributable, in his view, to the prospect of obtaining planning permission for the development. The bank claim that this was a mistake and a negligent one. Since the valuation in clause 15.01 of the report was expressed to be for mortgage purposes he should, they say, **G** have ignored any value attributable to the prospect of planning permission and valued Rackheath Hall purely as a residence with no prospect of planning permission. They also claim that his report was either misleading or ambiguous and that he should have explained clearly that he was taking into account the prospect of planning permission. The partnership claim that the open market value for mortgage purposes rightly includes any value (if it exists) attributable to the prospect of planning permission. So, upon the assumption that the partnership owed to the bank a duty of care in compiling the report, the first main issue is whether the partnership was in breach of it, either because of the method of valuation used or because of a want of clarity in the report.

H The second main issue, which arises only if the partnership succeeds in respect of the first, is whether, even if the prospect of planning permission is to be, or may be, taken into account the figure of £595,000 was still wrong. The bank claims that, even if the prospect of planning permission is taken into account, the correct figure was £380,000. In that event their claim would be £104,000, namely £370,000 (70% of £380,000, ie £266,000).

The third main issue, which arises only if the partnership fails on the first, is whether, upon the assumption that the open market value for mortgage purposes is to be calculated on the basis of Rackheath Hall's use as a residence, any prospect of planning permission being **J** disregarded, the figure of £595,000 was wrong and, if so, what is the correct figure. As I have indicated, the bank say that it is £350,000. The partnership say that £595,000 is, even upon that assumption, not proved to be wrong.

Evidence

I heard evidence from: (1) the officers of the bank (other than Mr Geoffrey Emery) who sanctioned the loan — Mr Paul Stevens, Mr Steven Cook, Mr Allan James and Mr Anthony Wakelin; (2) Mr Edward Gilder, a former area planning officer of the Broadland District Council; and (3) Mr Godfrey. I also heard evidence from two **K** expert witnesses, Mr Nicholas J Bond [BSc ARICS] of Aldridge Landsell & Co, Norwich, for the bank, and Mr David J N Merrick [FRICS] of Savills, Norwich, for the partnership. Mr Bond had 10 years' experience in valuing residential property in the Norwich area for banks and other lending institutions. He does not specialise in large country houses or development sites (nor would he necessarily regard those as specialities), but he had prepared valuations for something like 10 such houses over this period and six or so such sites. By contrast Savills value three to five houses of the size of Rackheath Hall every year. In 1988 these valuations were principally for development purposes. I should say that I felt able to place confidence **L** in Mr Merrick's evidence as to what a reasonably competent valuer of Rackheath Hall would or could have done in July 1988; and where his opinion conflicted with Mr Bond's as to what such a valuer could reasonably do, I preferred his evidence. I should, also, record that it was common ground between the experts that Rackheath Hall was not easy to value, however you did it.

Sequence of events leading to the valuation

Mr Godfrey had acted for Mr Karrar before in respect of three developments: Broke Hall, Ipswich; 58 Fitzjohn Avenue, London NW3; and 13 Belsize Park, London NW3. In those three cases he had first been contacted by Mr Karrar and asked to provide a valuation for **M** the bank. Afterwards he had acted for the bank in monitoring the making of interim payments under the loan which the bank advanced. These developments had been satisfactory and had established for Mr Karrar a satisfactory track record in the eyes of the bank. In particular Broke Hall, which was an old country property near Ipswich, had been successfully converted into flats. In 1988 they were in the course of being sold and the bank was receiving repayments from the proceeds.

On July 5 1988 Mr Karrar contacted Mr Godfrey and told him that his company, Canterbrook Ltd, was purchasing Rackheath Hall, and that he wanted a valuation for the Allied Arab Bank to whom he was going for finance. Mr Godfrey understood (wrongly as it turns out) that the contract price was £595,000. Mr Karrar asked Mr Godfrey to

A value Rackheath Hall "as it stands" and to estimate the costs of development and what the finished units would be likely to fetch. On July 7 1988 Mr Godfrey spent a day at the property. He examined the condition of the property externally and internally. He also spoke to representatives of the local agents, Hockley & Co, who were the Ipswich agents in the course of selling the flats at Broke Hall, and William H Brown, agents in Norwich. These two firms were to be the joint selling agents for Rackheath Hall (with Hockleys as the main agent). The representatives of both firms believed that there would be good demand for both the flats and the houses in the development. The representative from Hockley & Co told Mr Godfrey that Broke B Hall was selling at much higher prices than Hockleys had originally anticipated. He estimated that, at July 1988 prices, the total of sales at Rackheath would be £2,765,000. A breakdown of these figures was supplied then or thereafter. The representative of William H Brown did not demur to these figures. Mr Godfrey also looked at the prices of properties advertised in estate agents' windows in Norwich, but the only flats which he could see were in the city centre and he did not regard these as comparable. Mr Godfrey thereafter produced, at Mr Karrar's request, his valuation report which was addressed to the Bank of Oman. No copy of the report in that form can now be found. At about the same time Mr Karrar telephoned Mr Paul Stevens, the lending manager at the bank, in the commercial lending division, C responsible for new commercial loans. Mr Karrar told Mr Stevens that Canterbrook had contracted to buy the property for £530,000 and that the partnership had valued it at £595,000. Mr Stevens told Mr Karrar that, if this was the case, the bank would consider providing a loan of £370,000 and that he should send the bank a formal request and a copy of the valuation. Mr Stevens mentioned the figure of £370,000 in accordance with the policy of the bank which I have described.

Mr Karrar supplied the bank with a copy of Mr Godfrey's report. He then rang Mr Stevens and asked to know whether the valuation was acceptable and whether the bank would lend £370,000. Mr Stevens told him that the bank would prepare an application for credit D to its sanctioning officials; but would require the valuation to be readdressed to the bank. Mr Karrar then got Mr Godfrey to readdress the report to the bank and he then supplied the readdressed report to the bank.

On July 29 1988 Mr Steven Cook, one of the assistant managers in the bank's commercial lending division, completed an application for credit at Mr Stevens' request. The relevant form recorded that the credit application was for £370,000; that the loan to value ratio was 62% (£370,000 as a proportion of £595,000); and that the proposed term was a six-month term followed by review pending the outcome of the planning application for permission for conversion into flats and E cottages. Under the heading "Financial Analysis and Recommendation Rationale" Mr Cook recorded the following:

(i) Past track record of K Karrar connection;
(ii) Well secured;
(iii) Good remuneration.

and he recommended the application. That recommendation was subject, *inter alia*, to a satisfactory structural survey of Rackheath Hall. The report accompanied the form as it passed up the sanctioning chain. By this time Mr Stevens was on holiday. At this stage Mr Karrar was preparing to provide the moneys due on completion, then F scheduled for August 1, either from his own funds or from some other source of finance.

On August 8 1988 Mr Geoffrey Emery, the manager of the credit department, noted on the application the following:

Recommended subject to KM's notes. The maximum exposure to the Group should not exceed £2m.

The reference to the "group" was to the group of companies owned or controlled by Mr Karrar. "KM" stands for Keith Mason. Mr Mason was a manager in the credit department. His job was to provide Mr Emery with notes on the pros and cons of granting any facility that was requested.

Mr Mason's notes on this proposal included the following: G

NAME OF ACCOUNT: K Karrar connection.
DATE: 2/8/88

Observation:
Obviously an important connection but the amount proposed takes our exposure over £2m, hence my comments tend to reflect some of the political nuances...

Canterbrook Limited
This new proposal to lend £370K against FD (ie Freehold Deeds) £595K L/V 62% takes us over the £2m figure. The deal appears speculative in that customer (Mr K) wishes to buy a Grade II listed Georgian Mansion, to H demolish part of the house and to do it up. Planning permission has not been obtained and completion of the purchase was due yesterday 1/8. Timing is of the essence herewith the reduction monies on Brookside coming in during August. The main problem is will Planning Consent be given for the demolition/refurbishing. Edward Symmons appeared to think so in their Report in about six weeks. With Mr K's G(uarantee) we would not lose money and:

Conclusion and Recommendations
...I recommend we indicate interest in assisting with 70% of the purchase if he can find the completion monies subject to:
(i) Planning permission being granted for the whole scheme and satisfactory full structural survey. J
(ii) Brookside completion monies being received as projected thereby containing the connection exposures under £2m.

Brookside Ltd was the company developing Broke Hall. As appears from this note, Mr Mason recommended that the loan should be granted only subject to the grant of planning permission.

On August 9 1988 Mr James, the bank's general manager, noted on the application the following:

Recommended as above ie (1) Group exposure Ltd to £2m and (2) Subject to planning permission being held. Refer also to my comments on Brookside Limited. K

On the same day Mr Wakelin, the bank's chief executive, noted on the application:

Sanction. I just have a feeling £2m may be tight. He must agree this limit.

On August 11 1988 the bank wrote to Canterbrook Ltd offering to make the loan. On August 17 1988 Canterbrook Ltd accepted this offer. By an error the offer was not made subject to the condition that planning permission be granted.

On August 18 1988 Mr Cook wrote a memorandum to the "sanctioning chain", which reads as follows:

L *Canterbrook Limited — Purchase of "Rackheath Hall"*
The Facility of £370,000 was sanctioned for this Company on the basis that we would only lend when Planning Consent for the proposed conversion and extension were in place.

The Mortgage valuation by Edward Symmons and Partners valued the property at £595,000 in its existing state and at £950,000 with the benefit of the Planning Consent. Therefore, we would be lending at 39% L/V percentage.

The condition for planning consent being in place is somewhat harsh being (sc. seeing) that we would only be lending at 62% without planning consent and when you take into account the track record of the Karrar Connection combined with the background wealth...

Whilst the Edward Symmons Report on "Rackheath Hall" gives the M opinion that the planning consent would be readily forthcoming for the conversion and extension even if this was delayed for a considerable length of time we are confident there are the resources available to meet interest payments on a loan of £370,000 to simply "hold" the property until such time as planning consent is in place or a sale of the property is achieved if the planning consent is not forthcoming.

We therefore seek a revised Sanction for Canterbrook Limited, deleting the stipulation that Planning Consent must be in place prior to our advance for £370,000...

PS: Despite the foregoing, it would appear that our Offer Letter of 11th August did not stipulate that Planning Consent must be in place prior to drawdown and even spells out that the period and repayment of the loan

A relates to the outcome of the Planning Application. This oversight has meant that the Bank is committed anyway to the proposed revision.

Mr Stevens added by way of manuscript comment on this note:

I totally agree with these comments regarding the condition of planning being held prior to our loan being made. Imposing this condition would have not helped him at all.

On August 23 Mr Mason noted the following:

MR EMERY

Mr Karrar is committed to accept the imposed ceiling of £2m for the "Group".
B In view of the expected receipt of reductions on Brookside (see note 22/8 herewith) and the delayed completion for Rackheath Hall now on 1/9 this should be easily achieved. In view of the wording on the Offer Letter we have no option other than to withdraw the Planning Consent requirement and suggest we can be quite relaxed about it.

Mr Emery initialled this memorandum "OK". Mr James noted "Recommended". Mr Wakelin wrote:

OK but this is the second facility letter which has gone out against sanction. How did this happen? Let me have a report please.

The bank had no direct contact with the partnership. It was Mr
C Karrar who commissioned Mr Godfrey to produce the valuation; he who paid them and he who supplied the report to the bank. The bank did not give the partnership any instructions as to the method of valuation which they should adopt. Mr Godfrey knew, however, that his report would be relied on by the bank in deciding whether to lend and, if so, how much. Indeed, the report was readdressed to the bank in order that they might rely on it. In those circumstances the partnership owed to the bank a duty of care in respect of the valuation contained in the report. The bank did in fact rely upon that report to tell them what the value of the property was: because, if 70% of it was less than £370,000 they would not have lent the excess.

D The method of valuation which Mr Godfrey used can be described by reference to an appendix to his report. This appendix is described as a "Preliminary financial appraisal". He took the estimated sales of the completed development as £2,535,000 (the figure which appears at para 15.03 of his report). This figure he reached by reducing the figure of £2,765,000 given to him by Hockley & Co to take account of his own assessment of the value of individual units. He believed that the price which Mr Karrar was going to pay for the property was £595,000. He estimated the building costs, finance charges and fees on acquisition at, in total, £1,303,000. £2,535,000 less £595,000 and £1,303,000 is £637,000. £637,000 is 25% of £2,535,000. Mr Godfrey
E reckoned that 16% of gross sales was the minimum return which was appropriate for a viable development.

By this means Mr Godfrey was satisfied that the property was worth the £595,000 which Mr Karrar (as he understood) intended to pay for it. This method of valuation represents in substance, but not in form, or (in so far as it is set out in the appendix) description, what is known as the residual value method. By that method the value of a property intended to be developed is derived by raking the estimated sales proceeds upon the completion of the development and deducting from them: (i) the cost of building; (ii) financing charges; (iii) fees; and (iv) a figure for overheads and profit leaving, as a residue, a figure which represents the value. Mr Godfrey would have produced a text
F book valuation by this method if he had produced, as the bottom line of his calculations, the figure of £595,000, having deducted £637,000 instead of £595,000 at the earlier stage. But the fact that he deducted £595,000 and then reached a figure of £637,000 as a surplus does not, in my judgment, invalidate his calculations *if* he was entitled to use the residual method.

What the method of presentation used in the appendix does do, however, is to obscure (I do not mean intentionally) the fact that, in reaching his valuation, Mr Godfrey had started with a figure of £595,000 and then seen what surplus it produced, having regard to the other figures used in the calculation. Thus, it is not wholly clear from the appendix, where the £595,000 figure is described as "Freehold property at valuation", that the valuation of the £595,000 is reached by G
taking that figure as the value to be tested, and then seeing whether, using the residual value method, it stands up. A valuer, or a layman familiar with the residual value method, would probably realise that that was what Mr Godfrey had done. But someone else might not.

Mr Godfrey told me that he reached the £950,000 figure in para 15.02 of his report by adopting the higher price for sales of £2,765,000 and taking a lower percentage for surplus of 16%. He cannot have done this without adjustment because the difference between £2,765,000 and £2,535,000 is £230,000. The difference between £637,000 (25% profit) and £405,600 (16% profit) is £231,400 and, if these two figures are added to £595,000, the product is £1,056,400. Mr H
Godfrey did perform that calculation and then scaled the result down to £950,000. He told me that he did so because it was his view "that a much higher figure would have been achieved in the market-place if planning consent and listed building consent had been in place and a clear development opportunity without any outstanding matters and associated risks was available". In those circumstances he thought that developers would have taken a more bullish view over the likely sales prices which could have been achieved and would have, perhaps, adopted the figures put forward by Hockley & Co. He also said that the availability of a "clean proposition" would have appealed to a larger market which could well have included local builders with lower overheads and in those circumstances lower building cost rates J
and profits would have been accepted. I can well see that, if all consents are in place, a higher price may be obtainable than if they are outstanding and that it may, therefore, be appropriate to use something like a figure of 16% (which would produce a valuation of £825,000). But I do not follow why a developer would estimate the eventual sales prices as higher or the building costs as lower simply because planning permission was in place. I do not, however, pursue the question as to whether the £950,000 valuation was too high because it is the £595,000 figure which was in issue.

It is plain that Mr Godfrey's method of valuation leading to the £595,000 figure takes into account the prospect of planning K
permission being granted: indeed it assumes that, after an interval, it is likely to be granted.

Open market value for mortgage purposes

Open market value

In my judgment, the open market value of freehold property must necessarily include such additional amount (if any) as, in an open market, prospective purchasers would pay for the property on account of the prospect of obtaining planning permission for development, over and above that which they would pay if there was no hope that planning permission could be obtained. If there are purchasers, who L
would pay the extra, a valuation which excludes them from consideration is not an *open* market valuation. Rackheath Hall was a property which would appeal either to a wealthy private purchaser, who intended to use it as a grand residence; a developer; or to a purchaser who wished to use it as an office, hotel, club, nursing home or institution. In those circumstances, if there was a prospect of planning permission being granted for use other than as a residence, this would necessarily affect the property's open market value. Any wealthy individual who sought to purchase Rackheath Hall for use as a residence would be in competition with anyone who wanted it because he thought that it could be developed; and he would have to M
match whatever enhanced price such a developer would be prepared to pay on account of the prospect that planning permission for development would be granted.

For mortgage purposes

If, as I hold, the open market value includes any value properly attributable to the prospect of planning permission, the addition of the words "for mortgage purposes" does not, in my judgment, make any significant difference. The purpose of the valuation does not *prima facie* alter the open market value. I accept that upon this analysis the words "for mortgage purposes" add little to "open market value" other than to invite a degree of conservatism in valuation having regard to

A the fact that the valuation is to be used for security purposes. But, unless these words have some special meaning to valuers, I see no basis upon which they can be regarded as excluding from an open market valuation such part of the value as participants in such a market would attribute to the prospect of obtaining planning permission. I turn, therefore, to consider whether the words "for mortgage purposes" would necessarily signify to a surveyor or valuer of reasonable competence that he should value Rackheath Hall as a residence and ignore any prospect of planning permission, however likely.

In December 1985 the Incorporated Society of Valuers and B Auctioneers ("ISVA") published jointly with the Royal Institution of Chartered Surveyors ("RICS") a pamphlet entitled *Mortgage Valuations: Guidance Notes For Valuers*. It contains the following:

> *Guidance Notes for Valuers on the inspection of residential property undertaken for mortgage purposes on behalf of building societies, banks and other lenders.*
>
> The unabated discussion on the extent of a valuer's duty to be undertaken in the preparation of a valuation report for mortgage purposes and the differences of opinion expressed have given rise to the need to compile a set of guidelines which can be accepted as an agreed basis of approach. The guidelines which follow have been drawn up by the Incorporated Society of Valuers and Auctioneers and the Royal Institution of Chartered Surveyors and C published after consultation with the Building Societies Association. . .
>
> THE NOTES WHICH FOLLOW ARE NOT INTENDED AS A DEFINITE STATEMENT OF THE NATURE AND EXTENT OF THE INSPECTION AND SHOULD BE ACCEPTED AS COMPLEMENTARY TO, AND NOT IN SUBSTITUTION FOR, THE INITIATIVE AND JUDGMENT OF THE VALUER WHO MUST BE THE FINAL ARBITER.
>
> 1. (a) The valuation of residential property for mortgage purposes is, unless the lender instructs otherwise, open market value as at the date of inspection . . .
>
> (b) Open market value for mortgage purposes is the amount at which a qualified Valuer believes the property would sell in the open market as at the D date of its inspection, allowing for reasonable time to find a buyer in the prevailing market conditions and disregarding any exceptional reasons the buyer may have for purchasing, including, if appropriate, the fact that it is a new property. . .
>
> (d) Planning potential will be disregarded, unless otherwise instructed.

I do not regard these notes as applicable to the valuation of Rackheath Hall. As their heading makes clear they are notes on the inspection of residential property and they are, in my judgment, intended to apply to the valuation of properties which are being bought for use as a residence by private individuals. Rackheath Hall was not being purchased for that purpose, as the bank well knew; indeed much E of Mr Godfrey's report deals with the prospects of obtaining planning permission and the viability of development, neither of which are relevant if the property is being purchased simply as a residence. In any event, the notes are notes for guidance and, as the passage in capitals which I have set out above makes clear, they are not intended as a substitute for the valuer's judgment.

Mr Godfrey's judgment was that planning prospects could not be ignored, and that it would be unrealistic to value the property, on an open market basis, as a dilapidated country mansion without hope of development. I agree.

In 1988 there were, also, in existence *Guidance Notes on the* F *Valuation of Assets* (2nd ed) published by the RICS. Those notes were described as applicable to the valuation of all fixed assets including valuations for security purposes: see para 1.1 of guidance note 1. Para 2 of guidance note 1 provided as follows:

> *Bases of Valuation*
>
> 2.1. ***"Open Market Value" (OMV)***
>
> 2.1.1. Valuations of Land and Buildings will normally be on the basis of open market value and will have regard to evidence of open market transactions in similar property.
>
> 2.1.2. Such valuations may reflect either:
>
> (a) The use of the property for the same or similar purposes as hitherto (Open Market Value for Existing Use). This includes properties normally sold as fully operational business units and valued with regard to trading potential, G such as hotels, public houses, cinemas and many other properties;
>
> (b) The prospective use of the property for other purposes (Open Market Value for Alternative Use);
>
> 2.1.3. The purpose of the valuation will determine the choice between (a) and (b) above, as explained in the Background Papers. *"Open Market Value" without the limitation of the words "for existing use" or "for alternative use" will have regard to both existing and alternative uses to the extent to which these are reflected in the price obtainable on the open market.*
>
> **3. *Extent of Instructions***
>
> 3.1. Members are advised to discuss and agree with the client and his advisers the properties to be valued, the basis of the valuation, and the Plant and Machinery to be included with the buildings. H
>
> . . .
>
> 6. *Special Assumptions*
>
> In arriving at Open Market Value, the Valuer will have to take into account *the assumptions which the market would generally make in the particular circumstances.* He should, however, exercise extreme caution in making assumptions which might be unrealistic eg that land in which no interest is owned could be acquired or that consents for development would be forthcoming. The Valuer may be required to justify the use of such assumptions.

(Emphasis supplied.)

Guidance note no 18 provided as follows: J

> VALUATION OF ASSETS FOR INVESTMENT PURPOSES AND AS SECURITY FOR LOANS
>
> 1. Where the property being valued comprises an investment, or a property held for, or in course of development, on completion of which it will be held as an investment, or is to be charged as security for a loan by way of mortgage or debenture, the under-mentioned considerations apply. . .
>
> *Security for Loans*
>
> 4. *When a property is valued as security for a loan by way of mortgage or debenture, the basis of valuation should be Open Market Value.* It would be usual for the Valuer to be asked to express an opinion as to the suitability of the property as security for a loan. It is, however, a matter for the lender to assess the risk involved and express his assessment in fixing the terms of a K loan, such as the percentage of value to be advanced, the provision for repayment of the capital and the interest rate. The Valuer should refer in his Valuation Certificate or in his detailed report where one is to be provided, to all matters which are within his knowledge and which may assist the lender in his assessment to the risk.
>
> 5. It is not normally appropriate to value property to be used as security for a loan at forced sale value. . .

(Emphasis in para 4 supplied.)

I regard these guidance notes as a surer guide to the meaning of "Open Market Value for Mortgage Purposes" in the present case. In L accordance with that guidance Mr Godfrey was entitled, in my judgment, to have regard to the prospect of planning permission being granted for Rackheath Hall and to include in his valuation such amount as he properly thought attributable, by way of value, to such prospect. I accept and prefer Mr Merrick's evidence to that effect.

I was also referred to a passage in a well-known text book, *Modern Methods of Valuation*. In dealing with valuations for mortgage purposes the authors write (at p442) as follows:

> . . . Valuations should be on the basis of existing use value unless planning permission has been granted for development. Even then (the Valuer) should disregard any development value: if an owner wishes to raise capital to carry M out development then he will enter into a funding agreement which is different from a straightforward mortgage.

I am not convinced that the authors of the book intended these observations to be directed at a property such as Rackheath Hall which was being purchased for development, or that they intended to suggest that, in such circumstances, the valuer must ignore any prospect of planning permission if would-be purchasers in the market would take it into account. I note that shortly afterwards, on the same page, there appears the following passage:

> Any future element of value which is reasonably certain in its nature, such as reversion to a full rental value on the expiration of an existing lease at a low

A rent, may properly be taken into account, but anything which is purely speculative in its nature should be disregarded . . .

I further note that, on p445, there is to be found the following paragraph:

> On the approach to mortgage valuations of residential properly, the Valuer is referred to Guidance Notes for Valuers published jointly by the Incorporated Society of Valuers and Auctioneers and the Royal Institution of Chartered Surveyors and on definitions of value to the latter's *Guidance Notes on the Valuation of Assets* (2nd edition).

I have already referred to both these notes. If there is any conflict B between what is said at p442 in *Modern Methods of Valuation* and the RICS' guidance notes, it is, in my opinion, to be resolved by referring to the latter. At the very least it seems to me that Mr Godfrey was entitled to base himself upon them. I am confirmed in that conclusion by the fact that the residual value method is referred to and explained in chapter 12 of the book, which includes the observation:

> . . . In commercial situations, such as where a valuer is preparing a valuation for the purchase of the development site, a realistic view must be adopted and the residual method is accepted as a proper approach.

Planning permission

C In 1990 the RICS published the third edition of the guidance notes under the heading *Statements of Asset Valuation Practice and Guidance Notes*. *Statement of Asset Valuation Practice No 1* ("SAVP 1"), as revised as at June 1992, reads as follows:

> GENERAL PRINCIPLES TO BE OBSERVED IN PREPARING ASSET VALUATIONS
>
> 1. *General*
>
> 1.1. *Application of the Statements of Asset Valuation Practice.*
>
> The Statements of Asset Valuation Practice (SAVP) and Information Papers (IP) in this Handbook apply to the valuation of all fixed assets, that is, land and buildings of all kinds . . .
>
> 1.2. The SAVPs and IPs apply to all valuations which will be, or may be, D included or referred to in any public or published document. They apply whether or nor the Valuer has endeavoured to contract out of responsibility to Third Parties (that is, parties other than those to whom the valuation was addressed or for whom it was intended) and whether or not there is any likelihood that Third Parties will place reliance on the valuation. . .
>
> 1.3. The SAVPs and the IPs do not apply to valuations undertaken only for the private information of the client. Accordingly they do not apply in any of the following cases:
>
> . . .
>
> (c) valuations of domestic dwellings where the owner or prospective owner of a dwelling is borrowing from a lender such as a bank, or similar organisation;
>
> (d) valuations for mortgages or loans secured on commercial property for E the private purposes of the lender or the borrower;. . .
>
> In such cases, Valuers are strongly advised to make clear in their report that it is for the private information of the client and not suitable for publication nor to be disclosed to any Third Party without consent. . .
>
> 3. *Open Market Bases of Valuation*
>
> 3.1. Open Market Value (OMV) (see SAVP No 2) is the normal basis of valuation used in preparing valuations of fixed assets. Open Market Valuations are generally based on the evidence of open market transactions in similar property in the locality.
>
> 3.2. The use of the expression "Open Market Value" (which is defined in SAVP No 2), not qualified by any reference to Existing Use or Alternative Use, implies the value for any use to the extent to which that value is reflected F in the price obtainable in the open market. It is the only basis of valuation appropriate for properties held as investments, properties held for future development and properties declared surplus to the requirements of an undertaking.

The philosophy behind SAVP 1 was, I infer, to prescribe uniform principles to be observed in all valuations likely to be made public and thus relied on by persons other than those who commissioned the valuation and gave instructions as to how it was to be prepared. In this manner anyone who read such a report would (or could) know the basis upon which it had been prepared. The guidelines do not apply — see para 1.3. — when the transaction to which the valuation relates is wholly private. In such a case the client may be expected to give whatever instruction he chooses (of which, by definition, he will be G aware) as to the basis upon which the valuation is to take place. I doubt that the bank (which did not commission the report and had no contract or contact with the partnership) is to be regarded as the client for the purposes of these notes; but, even if it should be, the bank in fact gave no instructions as to the basis upon which the report was to be made. Mr Karrar gave instructions to Mr Godfrey to value the hall "as it stands". There is no evidence before me that Mr Godfrey's report did not match Mr Karrar's expectations.

In those circumstances, even if this valuation had been given in July 1992, I would not regard SAVP 1, although in terms inapplicable, as irrelevant. I reach this superficially paradoxical conclusion H because, in my judgment, a valuer who, lacking specific instructions to the contrary, adopted in respect of a wholly private transaction a method of valuation prescribed for valuations which were to be made public could not justifiably be criticised for having acted in accordance with the guidance provided by the SAVP. The effect of the amended SAVP, which reflected developing practice, was to permit a wider latitude in respect of private valuations than that applicable to valuations which might be made public and to confine the mandatory application of the guidance notes (in so far as any guidance notes are mandatory) to the latter. That does not, however, in my opinion, mean that the standard for "public" valuations lost all relevance. J

Additionally, in June 1992 SAVP 2 was published. It provides, *inter alia*:

> *Statement*:
>
> 1. Open Market Value (OMV)
>
> 1.1. Open Market Value means the best price at which the sale of an interest in property might reasonably be expected to have been completed unconditionally for cash consideration on the date of valuation, assuming:
>
> (a) a willing seller;
>
> (b) that, prior to the date of valuation, there had been a reasonable period (having regard to the nature of the property and the state of the market) for the proper marketing of the interest, for the agreement of price and terms and for K the completion of the sale;
>
> (c) that the state of the market, level of values and other circumstances were, on any earlier assumed date of exchange of contracts, the same as on the date of valuation; and
>
> (d) that no account is taken of any additional bid by a purchaser with a special interest.
>
> 1.2. The use of the expression "Open Market Value" not qualified by any reference to Existing Use or Alternative Use, implies the value for any use to the extent to which that value is reflected in the price obtainable in the open market.
>
> *Guidance Notes*:
>
> . . . L
>
> 1.4. . . . the best price . . . Open Market Value should equal the best price which could reasonably be expected to have been obtained in the market on the date of valuation (subject to the exclusion of any additional bid by a purchaser with a special interest) not a "fair" price, nor an average price, nor the price which the Vendor thinks ought to be achieved.
>
> 1.5. Hope Value. Open Market Value includes such element of "hope value" (if any) as the property may have for uses other than the existing use,. . . but limited to the extent that the expectation of realising such "hope value" would, in practice, be reflected in offers that would be made in the open market by prospective purchasers (other than the additional bid of "a purchaser with a special interest"). Open market valuations should be made only on such assumptions as the Valuer reasonably and realistically could M expect to be made by prospective purchasers in the open market (other than "a purchaser with a special interest").

This SAVP, like SAVP 1, post-dates the July 1988 valuation by two years; and is not, therefore, directly in point. But it does, in my judgment, provide some confirmation of the conclusion which I have already reached as to the reasonableness of including "hope" value in any valuation of Rackheath Hall.

Mr Bond told me that, unless specifically instructed otherwise, he always valued on existing use basis for mortgage purposes and that he understood the phrase "open market value for mortgage purposes" as excluding "hope value", although he described the phrase as not

A often used in reports. It is to be noted, however, that his experience is overwhelmingly in the residential field and not with development properties. Mr Merrick, on the other hand, saw no difference between open market value and "open market value for mortgage purposes" material to the present case. I prefer the opinion of Mr Merrick. Accordingly, I conclude that Mr Godfrey was not wrong in principle in taking into account the prospect of planning permission in valuing Rackheath Hall for mortgage purposes. To do so was consistent with a practice adopted by competent valuers and he was not negligent in following it.

I next consider whether, given that it was open to Mr Godfrey to B take into account the prospect of planning permission, it was legitimate to use the residual value method as the means of valuation. Mr Bond's opinion was that the residual value method had too many uncertainties to be used for a valuation for security purposes. It depended upon too many assumptions, eg as to the likely sale value of the completed development, the time that the development would take, financing charges and the like; and changes (even if relatively modest) in the assumptions could substantially alter the end result. He accepted that the method was useful or, at any rate, was used, as a means of enabling a developer to estimate what surplus would be left if a development were to be carried out, so that he might estimate how C much money was available to pay for an undeveloped site and leave a reasonable profit. But he regarded it as too risky for mortgage purposes.

If it is right to take into account "hope value" there is no easy way to do so. There may be in existence a comparable property with the same prospect of obtaining planning permission for the same sort of development as the property to be valued. But this would be rare. Alternatively, it might be possible (although I doubt whether developers would be likely to do so) to take the value of the property in question with planning permission and its value without hope of planning permission and derive an intermediate figure which reflects D the percentage likelihood of planning permission being granted. Last, a valuer could perform the residual value calculation which a prospective developer would or might employ. I would regard this as a valid approach provided that the prospect of obtaining planning permission was sufficiently good for a developer to be likely to perform such an exercise in deciding what to bid.

The drawback to using the residual value method is that it assumes that the development will go ahead whereas, *ex hypothesi*, that is not certain. But to treat this logical imperfection as a bar to any use of the residual value method would be to ignore the practical reality that developers, if they think that planning permission is likely to be E granted, are prepared (as was Mr Karrar) to purchase without knowing that they will be bound to get planning permission, and to calculate the price that they are prepared to pay by the residual value method or something like it. In such circumstances they would, however, if prudent, be likely to pitch their price at such a level as would produce a greater surplus than 16%, to allow for the lack of finality in respect of planning.

Planning position

In January 1980 Broadland District Council had granted planning permission for the hall to be used either for office accommodation, country club, restaurant and hotel, or as a nursing home. These F permissions had lapsed five years later. An application for Mr Karrar's development was lodged with the council on July 13 1988. In his report Mr Godfrey expressed the view that the application (which, when he drafted the report, he knew was about to be made) would receive favourable consideration. He said that he expected that all statutory consents would be obtained by October 1988. In the event, on November 1 1988 the council authorised the head of development and planning services to approve the conversion of Rackheath Hall after the completion of a section 52 agreement (under the then town and country planning Act) controlling the occupation of the new development until the hall had been restored. On April 10 1989 listed building consent was given.

In my judgment, the likelihood of planning permission for the G development being granted was, at the end of July 1988, high. Indeed, I think that Mr Godfrey could properly regard it as reasonably certain to be granted and in the short term, ie something like three months. The case for planning permission was considerably more than marginal. The ground had been prepared. Between March and July 1988 a number of meetings had taken place between Mr Colin Huntley of Adams Huntley Associates, who were responsible for the design of the new building, and Mr Christopher Gorniak, responsible for the renovation of the hall and officers of the council. Mr Gorniak had been in contact with the council prior to March 1988. On July 7 1988 a planning officer of the council had told Mr Godfrey, when he H telephoned, that he thought that the proposals would be favourably considered by the relevant committee and confirmed, as was the case, that English Heritage had been consulted and approved the proposals in principle. Although the hall was located outside the village of Rackheath and there was, thus, in accordance with policy H of the Norfolk County structure plan, a presumption against new residential development taking place, policy E8 of the same plan recorded that the council would:

> seek to ensure the restoration of historic buildings and their settings by:. . .
>
> (2) permitting and promoting new uses for old and valuable buildings to assist in their survival, providing the adoption can be achieved without J detrimentally changing the historic or visual character of the building.

The council could be expected to be, and in the event were, anxious to ensure the restoration and secure the future of the hall, which was deteriorating. I do not, however, consider that there was any real likelihood that the council would grant planning permission without some form of section 52 agreement. But I do not think that the fact that a section 52 agreement would be required should have been regarded as having a significant effect on a proper valuation.

In those circumstances I conclude that Mr Godfrey was justified in using the residual value method (which he told me he had used once or twice a month over the previous seven years in valuing K development properties). I accept the evidence of Mr Merrick to that effect. He told me that he, himself, would have valued it in exactly the same way: and I do not think that either he would have been, or that Mr Godfrey was, negligent in doing so.

I next have to decide whether Mr Godfrey's figure of £595,000 was a figure which, in July 1988, a reasonably competent valuer could have reached using the residual value method. In this connection Mr Bond put forward a residual valuation on what he described as a "more reasonable basis than that adopted by the defendants". His valuation was as follows:

L

	£	£
Estimated Sales		2,155,000
Deduct		
Building Costs		
(£555,000 + £575,000)	1,130,000	
Finance on building costs:		
12% x 1 year x 33.3%	45,000	
Agents and Legal fees on sale of units (@ 2%)	43,000	
Allowance for overheads and profit:		
20% of estimated sales	431,000	
Contingency allowance —		
6% of building cost	61,000	1,719,000
		445,000
Fees & Stamp Duty on Acquisition (@ 2%)	8,000	
Finance on Residual Value:		
12% x 15 months	58,000	66,000
		Residual Value = 379,000

M

The first major difference between the two sets of figures lies in the value of the estimated sales. Mr Godfrey takes £2,535,000. Mr Bond's figure is 15% less at £2,155,000. Mr Godfrey's figure was itself an 8% reduction from Hockley's figures of £2,765,000. Mr Bond deducted a further 15%, basing himself upon a comparison with what he regarded as similar units in the centre of Norwich. I do not think that Mr Godfrey was at fault in using the figure of £2,535,000. It was reasonable for him to accept advice from Hockleys and William H Brown. He did not follow their advice slavishly, but in fact reduced their figures. I accept the evidence of Mr Merrick that Mr Godfrey's figures were a reasonable estimate. Mr Merrick told me that he would have come up with the same sort of figures himself; indeed that he would have considered Hockley's higher figures themselves reasonable.

The fact that some flats in the centre of Norwich were selling at appreciably less per square foot than the proposed flats at the hall does not, in my judgment, establish that the selling prices proposed by Mr Godfrey in respect of the hall were unreasonably high. The two types of development are quite different in type and location and the prospective purchasers are different. I have considered Mr Bond's comparison of Mr Godfrey's estimate of prices at Rackheath Hall with the prices in fact obtained at Gunton Hall. That is a property some 20 miles north of Norwich and 10 miles from the coast. It was converted into 19 houses or cottages, but with no new buildings. By comparing the prices obtained for 3, 5 and 6 Brewery Yard and the Old Kitchens at Gunton Hall (adjusted to July 1988) to the prices estimated for flats at Rackheath Hall with a similar square footage, it appears that some of the flats at Rackheath Hall were estimated by Mr Godfrey to fetch a higher price than was in fact obtained at Gunton Hall. I accept the evidence of Mr Merrick that this comparison does not mean that Mr Godfrey's estimate was unreasonable. In particular, square footage is not everything. For example, while the adjusted price for 5 Brewery Yard was £204,957 and the flat nearest in size to 5 Brewery Yard at Rackheath Hall was flat 1, which Mr Godfrey valued at £225,000, 5 Brewery Yard is in a comparatively gloomy position. Flat 1 at Rackheath Hall would be the best and of considerable grandeur.

The second difference is that Mr Bond has assumed that one-third of the building costs of £1,130,000 would have to be financed at any one time; Mr Godfrey has assumed a figure of a quarter which Mr Merrick told me he would, himself, have chosen. (Mr Bond's figure is £45,000. The figure in Mr Godfrey's report is £31,000, but the building costs are estimated at £1,130,000 (not £1,030,000 as mistyped) and the financing charge, should therefore, be £33,900 and not £31,000.) I accept Mr Merrick's evidence that both one-third and one-quarter are tenable figures. Both Mr Bond and Mr Godfrey took agency and legal fees as 2% of gross sales. This produces, in Mr Bond's case, £43,000 and in Mr Godfrey's case, £59,000; 2% of £2,535,000 is, in fact, £50,700 and not £59,000 which appears in Mr Godfrey's valuation.

The next difference is that Mr Godfrey's calculations produce a profit margin of 25% of estimated sales. Mr Bond's allowance is 20% of £2,155,000, ie £491,000. In this respect Mr Godfrey's deduction is higher in amount, being a higher percentage (25%) of a higher figure (£2,535,000). Mr Godfrey did not, of course, choose the figure of 25%: that was the surplus which resulted from his calculation. But the effect of his calculation is to allow for a considerably higher surplus than that which a developer would demand: 16-17% would be the textbook figure. In July 1988 many developers were working on considerably less, in the belief, unfounded as it turned out, that the market would continue to rise. A figure of 25% thus incorporates a reasonable allowance to allow for the want of finality in respect of planning permission. If the amount to be deducted is reduced (as in Mr Bond's calculation) the residual value necessarily increases. Mr Godfrey's more conservative figure of 25% deserves no criticism. Mr Bond then deducts an extra 6% of the building costs of £1,130,000, less the £110,000 of contingency which Mr Godfrey had included in those costs. This is a figure of £61,000. Such an additional deduction produces an overall contingency figure of 16.7%. Mr Bond was of the opinion that Mr Godfrey's contingency figures of £30,000 for roof repairs, £50,000 for damp and rot and £30,000 for stonework constitute special contingency figures to which a further general contingency figure should be added. I do not, however, regard Mr Godfrey's contingency figure, which works out at 10%, as unreasonable. I accept Mr Merrick's evidence that it was reasonable.

Mr Bond's figure for fees and stamp duty is 2% of his residual value of £379,000 compared with Mr Godfrey's 2% of his residual value of £595,000, ie £12,000. Mr Godfrey's figure for finance on the residual value of £379,000 is taken over 15 rather than 12 months. Again, I would regard both periods as tenable.

In short, I am not persuaded that Mr Godfrey's calculations justifying the figure of £595,000 are outside the range of figures which a competent valuer, using the residual value method, might adopt. *A fortiori* figure of £530,000 would not have been outside that range. I draw attention to the fact that the residual value is critically affected by the figure taken for estimated sales and the allowance for overhead and profit. If, for instance, the figure for estimated sales is £2,765,000 (Mr Godfrey's figure), then the equivalent figure to the £445,000 in Mr Bond's calculation is, by my mathematics, £920,700 and the residual value is £787,000. There are of course other permutations.

I am confirmed in the view which I have reached by the point made on behalf of the partnership that if, as Mr Bond's calculations suggest, the value of the whole site was £380,000, then if, as Mr Merrick thought and Mr Bond agreed, the valuation of the site for new buildings was £200,000, the value of Rackheath Hall itself was £180,000. This seems remarkably modest.

I also take into account the fact that, in February 1988, Mr Karrar had bought Rackheath Hall for £530,000. Mr Karrar was a property developer of some experience. I have no reason to suppose that he was acting imprudently or recklessly in purchasing at £530,000; or that he paid an enhanced price for any special reason. Nor did the market dip between February and July 1988. Indeed it rose. In those circumstances the fact that Mr Karrar purchased at £530,000 itself provides some support to a valuation of at least that much.

Other comparables

Mr Merrick also relied upon a number of comparable values, which he accepted were not direct comparables, as supporting Mr Godfrey's valuation of £595,000. They are set out at appendix 1 to his first report. In order to convert the value of his comparables to a July 1988 date he used the Savills' index. This is an index compiled by the Savills' research department. It charts the movement in prices of, *inter alia*, small Norfolk houses and large Norfolk houses. Savills take 20 small and 20 large houses (small being four bedrooms and one bathroom or below, large being four bedrooms and two bathrooms and above). They revalue them every six months either by taking actual sales figures (where sales have taken place) or by making a valuation. Unlike the Nationwide Anglia East Anglia index, the Savills' index includes sales where the property has not been bought upon mortgage but for cash. I regard this index as a useful guide, but no more, to the general trend in residential prices.

Mr Bond did not agree that Mr Merrick's comparables supported a valuation of £595,000. The respective contentions of both sides in relation to these comparables were set out in a helpful schedule. In the light of the conclusion which I have already reached as to the validity of the £595,000 figure, I do not propose to embark upon a detailed analysis of these comparables. I have considered all of them and I have taken into account Mr Bond's criticism of them. They do not, for the most part, provide any sure guide to the proper valuation of Rackheath Hall (including therein the value of the prospect of planning permission) on July 28 1988. I have, however, found two of the comparables of some confirmatory assistance. Thickthorn Hall, Hethersett, is situated approximately 4 miles south-west of Norwich. By October 1985 the hall had been converted into eight flats on three floors. The style and size of the converted units was similar to that which was proposed at Rackheath. A sale was agreed (though not completed) in October 1985 at £310,000, with no planning permission for further development. The prospective purchaser was a licensed victualler who wished to purchase the property as an alcoholic reform centre. The sale did not proceed because of restrictive covenants against commercial use. If the £310,000 is indexed to July 1988 using the Savills' house price index, it equates to a value of £775,794 as at July 1988. Mr Bond pointed out that the proposed use was not residential and that to use a house price index was inappropriate. In addition, the relevant planning permission was in place. Nevertheless the figure of £775,794, which relates to a conversion of only eight flats (as opposed to the 10 at Rackheath Hall, itself), provides some support to the figure of £595,000. Morningthorpe Hall, Norfolk, is situated 11 miles south of Norwich. It was sold for £450,000 in November 1988 with planning permission for conversion to seven units. If this figure is indexed to July 1988 using the Savills' house price index it equates to a figure of £500,000 or £71,000 per unit. Since there were to be 10 units at Rackheath Hall, Mr Merrick reckons that this equates to a figure of £710,000 at Rackheath Hall. Mr Bond points out that, immediately after the purchase, the new owner entered into negotiations with the South Norfolk District Council and was granted, in September 1989, planning permission to increase the number of units from seven to nine. If one adopts Mr Merrick's analysis, but using nine rather than seven units, a figure of £55,555 per unit is obtained, giving a figure for the hall at Rackheath of £555,555. Again this seems to me some confirmation that £595,000, *a fortiori* £530,000, was within the right range.

Was the report misleading?

Miss Marion Smith on behalf of the bank forcefully submitted that it was Mr Godfrey's duty to produce a clear and unambiguous report; that this report was unclear because it left undefined what was meant by "for mortgage purposes" so that the report was liable to be and was in fact misinterpreted. This submission derives support from the fact that clause 15.02 of the report values the property at £950,000 "upon the grant of Planning Permission and Listed Building Consent" whereas clause 15.01 values it "at the present time". The contrast between clause 15.01 and clause 15.02 is, in terms, a contract between a valuation with and without planning permission. The wording is silent as to whether, in the latter case, any account is to be taken of the prospect of planning permission. But the juxtaposition of the two paragraphs, together with the substantial difference between £950,000 and £595,000, is capable of being understood as meaning that, in the case of clause 15.01, no account is being taken of planning permission, actual or potential.

I appreciate the force of these submissions. But I have come to the conclusion that it would be wrong to find Mr Godfrey guilty of negligence because of the wording of clause 15.01.

Mr Godfrey there expresses his opinion of the open market value of the property for mortgage purposes. In so doing he was entitled, as I find, to follow the RICS guidance notes and to value the property on the basis of its value with the prospect of planning permission. Indeed the guidance notes indicated that he should. He was also, in my judgment, entitled to use the residual value method and, by that method, to reach a figure of £595,000. In those circumstances I do not think that he can fairly be regarded as having acted in a manner in which no reasonably competent surveyor or valuer would act by expressing himself as he did. He in fact produced what he said he was producing, namely an open market value for mortgage purposes, on a basis acceptable to the profession. If I had concluded that the words "for mortgage purposes" had a defined meaning which excluded, in the present context, the value of any prospect of planning permission, or if there had been specific instructions from the bank (which, in my judgment, it was for the bank to give and not him to seek), my conclusion would have been different.

What did the bank understand from the report?

The bank's officers gave evidence of how they understood the report. I make the following findings as to their understanding.

Mr Stevens: took it for granted that the valuation in clause 15.01 was a valuation of the hall as a residence, ie without considering the improvement in value attributable to any possible grant of planning permission. But, while he did not address his mind to the point at the time, he agreed, in my view inevitably, that, if a purchaser is looking to develop the property in question, as Mr Karrar was, the open market value must include a valuation of the prospect of planning permission.

Mr Cook: assumed that the valuation would follow the relevant RICS' guideline (although he was hazy as to what that guideline was). Otherwise I am left in some doubt as to what view, if any, he took, at the time, of the meaning of clause 15.01. He said in cross-examination that he did not expect that Mr Godfrey would value Rackheath Hall as a residence, but in re-examination that he understood that the clause 15.01 valuation was for residential purposes. He did not, I find, realise that clause 15.01 included "hope value". He, too, agreed that a true valuation of the property could not ignore the prospect of planning permission.

Mr Mason: did not give evidence. His notes show that he thought that the loan of £370,000 against a valuation of £590,000, should be subject to planning permission because the deal was speculative, given that planning permission had not been obtained. Logically this suggests that he may have thought that the £595,000 valuation took some account of the prospect of planning permission since, if the hall was worth £595,000 even if planning permission was refused, the £370,000 was well secured. I do not, however, conclude that this is what Mr Mason thought since I doubt that the apparent logic of his notes was as well thought through at the time as Mr Worthington, for the partnership, submitted that it was. If he did think this, he did not communicate that conclusion to others in the sanctioning chain, of which he was not a member. I believe that this note simple reflects a concern not to begin lending money for the purposes of a speculative project until the essential planning permission was secure.

Mr James: thought that the £595,000 valuation disregarded any value attributable to planning potential because of the use of the words "for mortgage purposes"; but he agreed with the proposition that the open market value on a willing buyer/seller basis necessarily included the value attributable to the prospect of planning permission.

Mr Wakelin: thought that "for mortgage purposes" meant the value of the property in its existing use, disregarding the prospect of planning.

Causation

If, in July 1988, it had become apparent to the bank that the £595,000 figure took into account the prospect of planning permission, the likelihood is that they would have lent only such sum as represented 70% of the value of the property ignoring the possibility of planning permission, ie its value solely as a residence. I accept the evidence of the bank's officers in that respect. I do not conclude that, if 70% of the value was less than £370,000, the bank would, nevertheless, have lent at least £370,000 after November 1988 or May 1989, when the section 52 agreement was close to finality. If they had been prepared to lend further at all they would, in my judgment, have waited until planning permission was in fact granted and the section 52 agreement in place. They would not have repeated the error which they made in July 1988 of lending without planning permission being in place.

Value of Rackheath Hall as a residence

Mr Bond's valuation of Rackheath Hall as a residence, ignoring any prospect of planning permission, was £350,000; Mr Merrick's was £555,000. In his written statement of August 1991 Mr Godfrey expressed the view, which he was to describe in evidence as "off the cuff" that its value was of the order of £200,000-£300,000. He plainly did not consider the value of Rackheath Hall purely as a residence in July 1988.

Rackheath Hall was both rare and unusual, even for Norfolk. Norfolk has many country houses, including a few that are large and palatial. But houses such as Rackheath Hall do not change hands with great frequency. When they do they do so by private treaty: and the sale price is not always known. Further, Rackheath Hall was not in good condition and the site which it occupied was relatively small for its size. This makes the task of valuing Rackheath Hall difficult. No exact or nearly exact comparable was put before me and the difference between such comparables as were available and Rackheath Hall is such that it is difficult to assess the appropriate adjustment to be made.

Mr Bond relied principally upon two comparables: (i) Home Farmhouse, Burlingham; (ii) Fritton Grange.

Home Farmhouse

Home Farmhouse lies approximately 9 miles east of Norwich. Like Rackheath Hall it is a Grade II listed building. It is about (in fact slightly less than) a quarter of the size of Rackheath Hall, with about half the rooms: but they are much smaller than the rooms at Rackheath Hall. It was, when sold in November 1987 at public auction, in poor condition. It fetched £110,000. The land with it was some 2.42 acres, approximately 60% of the land of Rackheath Hall. At the time of sale it had been converted into three separate cottages, but it was sold as one unit.

Fritton Grange

Fritton Grange is approximately 9 miles south of Norwich. It is listed Grade II. Most of it is Georgian or Victorian. It was modernised and improved in 1984 and reroofed in 1987. It, too, is about a quarter the size of Rackheath Hall. In October 1987 it needed a damp course. There was possibly some rot on the ground floor, isolated wet rot to external joinery and some rewiring work was needed. Its site was some 3.4 acres including well-established gardens. It was sold in October 1987 for £225,000.

Both of these houses were, thus, Grade II, within travelling distance of Norwich and sold with land of less than 5 acres. But neither of them is a stately home and it is obvious from the photographs of them that they are quite different in character to Rackheath Hall.

Mr Bond reached his figure of £350,000 by the following route:

(i) He multiplied the sale price of Fritton Grange (£225,000) by four, thus producing £900,000. He took four as the multiplier upon the ground that, although Rackheath Hall is more than four times as big as Fritton Grange, the upper floor of Rackheath Hall was, in his opinion, less valuable than the lower floors.

(ii) He then took the difference between £225,000, the cost of the improved Fritton Grange, and £110,000, the cost of the unimproved Home Farm, namely £115,000, and used a multiplier of five to determine the repair and renovation costs for Rackheath Hall at £575,000. He took a multiplier of five on the ground that, in his opinion, the cost of repair to the upper parts of Rackheath Hall might be proportionately higher owing to the need for scaffolding and having to bring items up from the ground floor.

(iii) £900,000 less £575,000 is £325,000. This figure he adjusted to allow: (a) for the relatively small site at Rackheath Hall but, also, (b) for the improvement in the market between October/November 1987 and July 1988. Thus, he reached a figure of £350,000, which looked right to him.

While I accept that, for a valuation exercise as difficult as this one, it may be necessary, in the absence of satisfactory direct comparables or nearly comparables, to take such comparables as are available and apply adjustments which are a matter of judgment, I am highly sceptical of the validity of Mr Bond's exercise, which is one of his own devising, at any rate if it is intended to fix an upper limit on value beyond which a reasonably competent valuer could not go. Not only does it assume that the difference between the price of Home Farm and Fritton Grange was solely attributable to want of repair but, more important, it also assumes that the value of Rackheath Hall in good condition may satisfactorily be reached by multiplying the value of Fritton Grange by four: a multiple based on floor area. Given the difference in character between the two properties I do not think it satisfactory simply to use a multiplier drawn from a straight size comparison. (I note that, if a multiplier of five was used for Fritton Grange, the resultant value for Rackheath Hall would be £550,000 (£225,000 x 5 - £575,000).) I also doubt the validity of using the £115,000 as a "cost of repair" multiplicand; and am not satisfied that the multiplier of five is necessarily correct. It is not self evident that repair costs necessarily increase in direct proportion to size. Indeed, Mr Bond's selection of a multiplier of five, as opposed to four, assumes that they do not, but also discounts any possibility of economies of scale. I decline to adopt this method of valuation as a reliable means of reaching a figure of £350,000.

Mr Merrick put forward as a more satisfactory, but far from direct, comparable: Shotesham Park Estate. The Shotesham Park Estate is described, in my opinion rightly, as a particularly fine residential and sporting estate at Shotesham St Mary, some 6 miles from Norwich. It is a Grade I listed hall designed by Sir John Soane in 1784. In all it is about 763 acres in extent, including a home farm of 575 acres with a listed farmhouse and modern grain storage building, 188 acres of woodland, rough and river, 12 cottages, a shoot and gravel deposits. The estate was sold as a whole in early 1989 for approximately £2,500,000. It was not in good condition, having dry and wet rot.

The method by which Mr Merrick reached his figure of £555,000 is set out at pp212A and B of trial bundle 5. Essentially Mr Merrick amended a valuation carried out by Mr Hugh Coghill [FRICS FAAV] of Savills' in 1989 to reflect the more accurate data available to him. That valuation was made up as follows:

(i) Land less 10% acres with house:	£745,570
(ii) House:	£750,000
(iii) Cottages:	£240,000
(iv) Farm buildings:	£30,000
(v) Shoot	£40,000
	£1,805,000
Plus 15% premium to buy:	£270,750
	£2,075,750

The 15% premium to buy reflects the fact that the value of an estate such as this is greater than the sum of its parts.

From that Mr Merrick derived the following percentage, before application of the premium:

Hall:	£750,000	41.5%
Remainder:	£1,055,570	58.5%

Mr Bond reached a similar figure as follows:

(i) Land:	£827,000
(ii) Houses:	£750,000
(iii) Cottages:	£240,000
(iv) Farm Buildings:	£30,000
(v) Shoot & sporting rights to adjoining 150 acres:	£42,755
	£1,889,753
Plus 15% premium to buy:	£283,463
	£2,173,218

From this Mr Bond derived the following percentages:

Hall	£750,000	39.7%
Remainder	£1,139,755	60.3%
	£1,889,755	

The differences between these two sets of figures are apparent from a comparison of trial bundle 5/212A and B and the schedule produced by Mr Bond. I find that in this respect Mr Bond's figures are the more reliable. On the basis of those figures the proportion of the actual sale price of £2,500,000 to be attributed to the hall is £992,500 (£2,500,000 x 39.7%).

Thereafter, the difference between the two valuations is as follows. Mr Merrick would deduct 10% from the value of the hall to allow for its being split from the main part of the estate and a further 10% because Shotesham is a better selling area than Rackheath Hall. He

A would then deduct £275,000 on the grounds that, although the budget for Shotesham Hall showed that £500,000 needed to be spent on the hall, the purchaser was able to use direct labour from the estate, spending approximately £300,000. So the extra cost of repair at Rackheath Hall was £275,000. If a base figure of £992,500 is used with those deductions, the resultant value is £992,500 x 80% — £275,000, which equals £519,000.

Mr Bond agreed that 10% should be deducted because Shotesham is a better selling area. He also agreed the deduction of £275,000, but not that it was generous. But Mr Bond would:

(i) make a 15%, and not a 10%, deduction to reflect the reduction
B in value on account of the split of the hall from the estate; Mr Bond described 10% as a minimum. Mr Merrick described 15% as "over the top" on the ground that it was quite possible that there might be more buyers for the hall without the estate;

(ii) make a further deduction of 10% to reflect the superior quality of Shotesham Hall; and

(iii) make a further reduction of £51,285 to allow for the fact that Shotesham Hall had 13.15 acres more land retained with the hall. He valued these acres at £3,900 per acre: three times the value of agricultural land. Mr Merrick would make no adjustment because, in his view, 4 acres was the amount of land that Rackheath Hall needed and no discount was necessary on account of the fact that it did not
C have 17.

By this calculation he produced a figure of £318,840 as opposed to £519,000.

I am thus left in the position where I do not regard Mr Bond's figure of £350,000 as reliable because of the method which he used to reach it; and where the figures produced by the respective experts, as derived from Shotesham Park, vary by no less than £200,000. This, perhaps, illustrates the great difficulty that would have existed in July 1988 of valuing Rackheath Hall purely as a residence. Mr Worthington, for the partnership, submits that, on the evidence as a whole, it was simply not established that £595,000 was wrong if
D Rackheath Hall fell to be valued as a residence only. I do not accept that submission. Doing the best I can on the evidence before me, I have come to the conclusion that, if Rackheath Hall has to be valued in July 1988 without regard to the prospect of planning permission, a proper valuation would be in the range of £420,000-£470,000. I have reached that conclusion by the following route. I start with the base figure of £992,500 referred to above and deduct the agreed £275,000. That makes £717,500. Thereafter, Mr Bond would deduct a further 40% of £992,500 to allow for the split of the hall from the estate (15%), the better selling area (10%), the superior quality of Shotesham Hall (10%) and the difference in acreage (which amounts to 5%). In
E my judgment, the aggregate deduction of 40% is too high. I do not think that the figure should be more than 30%. Mr Merrick, by contrast, would deduct only a further 20%. I think that that figure is marginally too low and that the minimum deduction should be no less than 25%. I reach this conclusion particularly because of what I regard as the higher architectural quality of Shotesham. This, as well as being a personal view, is reflected in the different grading for the property. The range of £420,000-£470,000 assumes either a 25% or a 30% deduction.

Accordingly, if I had held that Rackheath Hall should have been valued as a residence only, I would have awarded the plaintiffs the
F sum of £41,000 being £370,000, the amount lent, less 70% of £470,000, ie £329,000. To that there would have to be added a sum for interest which, had it been necessary, the parties contemplated that they would agree or which, in default, would have to have been determined by a master.

As it is, for the reasons which I have given, this claim, in my judgment, fails and there will be judgment for the defendants. I should like to record my appreciation of the clarity of presentation by both counsel of their respective cases and of the cogency of their submissions: particularly their closing written submissions.

Judgment for the defendants.

Queen's Bench Division G

February 18 1994

(Before Judge MARR JOHNSON, sitting as a judge of the High Court)

AXA EQUITY & LAW HOME LOANS LTD v GOLDSACK & FREEMAN

Estates Gazette June 11 1994

[1994] 23 EG 130

Negligence — Whether valuation of £155,000 for mortgage purposes in 1988 negligent overvaluation H

On September 16 1988 W, of the defendant firm, provided the plaintiffs with a valuation of £155,000 for mortgage purposes of a leasehold flat in Brighton. The plaintiff then advanced sums to Mr and Mrs B amounting to £108,500 to enable them to redeem an existing mortgage and carry out improvements to the flat. On the default of Mr and Mrs B in repaying the loan, the plaintiffs obtained possession of the flat and sold it in October 1991 for £75,000. The plaintiffs claimed damages from the defendants alleging a negligent over-valuation and that a proper value of the flat in September 1988 would have been £120,000. W had qualified as a building surveyor rather than as a J
valuation surveyor and became an associate of the RICS in 1986. In making his valuation he consulted two local estate agents. One of the agents consulted informed him that the flat would sell for £160,000.

***Held*: The plaintiff's claim was dismissed. The figure of £160,000 could be justified and there was no difference between this and open market value assessed under joint RICS/ISVA Guidance Notes GN22. W pitched his valuation at the top end of the bracket of values of £140,000 to £155,000. There was no negligence. Had the defendants been liable, the plaintiffs were not contributorily liable in lending to Mr and Mrs B. Any loss attributable to a fall in property values would not likely to have K
been caused by any negligence of the defendants had negligence been established.**

The following cases are referred to in this report.

Mount Banking Corporation Ltd v *Brian Cooper & Co* [1992] 2 EGLR 142; [1992] 35 EG 123

Singer & Friedlander Ltd v *John D Wood & Co* [1977] EGD 569; (1977) 243 EG 212 & 295

This was a claim for damages in negligence by the plaintiffs, Axa Equity & Law Home Loans Ltd, against the defendants, Goldsack & L
Freeman.

Nigel Thomas (instructed by Vizards) appeared for the plaintiffs; Christopher Gibson (instructed by Davies Arnold Cooper) represented the defendants.

Giving judgment, JUDGE MARR JOHNSON said: The plaintiffs are a limited company, which began trading in June 1988, at a time when the property market in this country was extremely buoyant. As their name indicates, their business was to lend money to private borrowers who wished to purchase their own home.

In September 1988 a married couple, Clive and Ludmilla Barnes, M
applied to the plaintiffs for a loan in order to remortgage their leasehold flat at 9 Lewes Crescent, Brighton. Initially, the application was for a loan of £93,750. In accordance with their usual practice, the plaintiffs instructed the defendants, a local firm of estate agents, to value the property. On September 16 1988, Mr Kevin Woudman [ARICS], of the defendants' firm, provided a written valuation of the flat for mortgage purposes at £155,000. That was on an open market basis. Relying, it is said, on this valuation, the plaintiffs advanced sums to the borrowers which eventually totalled £108,500. These sums were secured by way of legal mortgage over the property. In due course the borrowers defaulted on their obligations under the mortgage and the plaintiffs obtained possession of the property on February 11

1991. When the plaintiffs obtained further valuations of the property, it was said to be worth only about £75,000 to £80,000 on the open market or £65,000 to £70,000 on a forced sale. It was finally sold in October 1991 for £75,000.

The plaintiffs contend in this action that the valuation of £155,000 provided by the defendants in September 1988 was a negligent over-valuation and that a proper value for the flat at that time would have been £120,000. On any showing, therefore, there had been a substantial drop in value since September 1988. It is common ground that this was due in part to a severe fall in the property market between 1988 and 1991. It is also said that by 1991 the property was less attractive to a prospective purchaser than it had been in September 1988. The defendants deny negligence, but in the event that they are held liable, they allege contributory negligence on the part of the plaintiffs in a number of respects. In the event of a finding of liability, there are also interesting questions which arise on the issue of damages, including the currently fashionable question whether the defendants should be held liable to compensate the plaintiffs for loss caused by a fall in the property market.

Much of the evidence which I heard went to the issue of damages. However, the first question to decide is, of course, whether the valuation of £155,000 was a negligent over-valuation.

The flat in question is a ground-floor flat in a Regency terraced house forming part of Lewes Crescent, which is on the sea front at Brighton, to the north side of Kemp Town Enclosures. Lewes Crescent lies just south of Sussex Square. All the houses in the area constituted by the square, the crescent and two terraces running off the crescent, Chichester Terrace and Arundel Terrace, are Grade I listed buildings. They form part of the Kemp Town conservation area, which is said by those who know to be particularly fashionable and a most desirable area in which to live.

Mr and Mrs Barnes had purchased the lease of the flat for £69,000 in 1986. In the summer of 1988 the property was subject to a mortgage with the Skipton Building Society and a second charge in favour of Maxvale Securities Ltd. In early September 1988, Mr and Mrs Barnes submitted a written form of application to the plaintiffs in which they asked to borrow the sum of £93,750 in order to redeem the existing charges and effect home improvements. They gave their estimate of the present value of the property as £135,000. By letter dated September 8 1988, the defendants received instructions to inspect and report on the property. The letter was sent on behalf of the plaintiffs by Mortgage Systems Ltd, a company through whom the plaintiffs carried out matters of daily administration. It was in standard form and included a requirement that the report should be completed and signed by a member of at least three years standing of either the Royal Institution of Chartered Surveyors or the Incorporated Society of Valuers and Auctioneers.

Mr Woudman went to inspect the flat on September 16 1988 and made his report on the same day. A rough note of his inquiries and findings survives and was referred to by Mr Woudman in the course of his evidence. In his written report he made comments on the construction of the property, its marketability, the state of repair (which was good) and various other matters. The leasehold interest was for a term of 99 years from November 2 1976. Mr Woudman concluded his report by stating that his valuation for mortgage purposes was £155,000.

The plaintiffs sent a copy of Mr Woudman's report to the Barnes. When the Barnes learned that the property was worth more than they had supposed, they promptly applied to the plaintiffs for a larger mortgage. In the event, the plaintiffs lent the Barnes £99,000 initially and then further sums totalling £9,500 under a facility which was described as an equity account. Thus the total amount advanced as capital amounted to £108,500. This was the maximum amount which could be advanced under the plaintiffs' rules governing the particular type of loan involved, being 70% of the surveyor's valuation of £155,000.

Mr Woudman gave evidence before me. Although he was obviously, and understandably, on the defensive, I saw no reason to disbelieve his evidence and I accept what he told me. He qualified as a building surveyor rather than as a valuation surveyor, but no point is taken on that. He was an associate partner of his firm at the relevant time, having become an associate of the Royal Institution of Chartered Surveyors in October 1986. It follows that at the time he made the valuation he had been an associate of that institution for less than the minimum period of three years stipulated by the plaintiffs in their letter of instruction. Mr Woudman told me, however, that in the course of his work for the defendants he carried out a large number of valuations.

It is a point made by counsel for the plaintiffs that throughout his evidence Mr Woudman repeatedly used the conditional tense, telling me what he would have done rather than what he actually did. This is a justified comment, but I am satisfied that Mr Woudman was not deliberately prevaricating. I believe he was simply attempting to underline the fact that he had in some respects an imperfect recollection of his actions at the time, but where his memory was deficient he was satisfied that he would have taken all proper steps and made all proper inquiries.

He inspected the flat for probably about half an hour and noted its condition. He was given a plan of the layout of the rooms in the flat by Mr Barnes. When he got back to the office, he consulted his firm's database where he was unable to find any suitable comparables. He then telephoned at least two local firms of estate agents to inquire as to their views on the probable value of the flat. First and foremost he telephoned Mr Paul Lewis who worked for a firm then called Bernard Thorpe & Partners. That firm has since changed its name to Jonathan Rolls. Mr Lewis was, according to Mr Woudman, by far the most knowledgeable person whom he could have consulted. His firm's office was very close to Lewes Crescent and they were instructed in most of the sales of property in and around that area. Mr Woudman thought they handled as much as 95% of the business around the Kemp Town enclosures and Mr Lewis was the expert to consult within that firm. Mr Lewis told Mr Woudman that he knew the flat and he reckoned he could sell it for £160,000. This was, of course, a great deal more than the Barnes had paid for it in 1986, but it is common ground that there had been a significant upward shift in the market between 1986 and 1988. The peak of the property market was between June and September 1988 when the Government abolished double MIRAS tax relief. Mr Woudman thought, but was not sure, that Mr Lewis mentioned two other properties in the neighbourhood which were said to be worth about £135,000 and £150,000. He was reasonably sure that in addition to speaking to Mr Lewis he had telephoned the firm of Fox & Sons and he thought he might also have spoken to Hayler & Sawyer, who were another firm of estate agents in the locality. In addition, Mr Woudman thought he spoke to two of the partners in his firm. Having considered all the available evidence, he concluded that a proper value for mortgage purposes would be £155,000. This valuation was made on the basis of open market value, which is not necessarily the same as the price which an estate agent might hope to obtain on a sale of the property. I will return to the difference between these two concepts later.

The plaintiffs' attack on this valuation is based on the evidence of Mr Stewart Gray of the firm of Austin Gray, who are a firm of local estate agents and valuers. Their office is only about one and a half miles from Lewes Crescent.

Mr Gray is an extremely experienced valuer and an associate member of the Incorporated Society of Valuers and Auctioneers. He first visited the flat in February 1991 and it was his opinion then that a proper valuation for the flat in September 1988 would have been £115,000. He subsequently adjusted this figure upwards to £120,000 when he realised that the internal condition of the flat had changed since Mr Woudman made his inspection in September 1988. When Mr Woudman saw the flat it was in good repair and good decorative condition. The little room behind the main bedroom was a dressing room and the kitchen was at the rear of the premises. Subsequently, it seems that the Barnes executed certain building works and also, or perhaps as a result, allowed the condition of the flat to deteriorate.

A They changed the main bedroom into a rear reception room by knocking down an internal wall. They also ripped out the original kitchen at the back in order to make that room the second bedroom and installed a kitchenette in the former dressing room. These alterations were probably undesirable in principle, but in any event it seems they were incompletely carried out. Mr Gray justified his figure of £120,000 by reference to a number of comparables or, as I think the defendants would prefer to say, nearer or more remote comparables. It is accepted that for historical reasons the houses in Lewes Crescent are internally distinctive one from another and thus true comparables are difficult to obtain.

B Lewes Crescent and Sussex Square were originally laid out and constructed by Bushy & Wilds during the period 1820 to 1830. The houses were designed for single-owner occupation. In the years that followed, extensions of varying kinds were built at the back of most of the houses and in the present century most of the houses were converted into flats. Thus the main ground-floor reception room would form the nucleus of one flat, the first-floor main bedroom would form the nucleus of another flat, and so on. The practical result is that no two flats are identical even if on the same floor in adjacent houses.

Mr Gray put before the court a schedule of nine comparable properties, consisting in the main of ground- or second-floor flats in or C near Lewes Crescent or Sussex Square. He had not himself personally viewed any of the comparable properties. Three of them had been sold by the firm of Jonathan Rolls during 1988 for prices ranging between £69,000 and £110,000. The highest price agreed, although the sale subsequently proved abortive, was £135,000 in August 1988 for a first-floor flat at 6 Lewes Crescent. Mr Gray claimed that all these comparables supported his figure of £120,000 for the flat in question.

For the defence, two expert witnesses were called: Mr Paul Lewis, to whom reference had already been made, and Mr Jonathan Rolls [ARICS] who is the sole principal of the firm which now bears his name. Mr Lewis has practised as an estate agent for about 25 years. For the past 10 years he has managed the agency department at his D firm which specialises in the sale of flats in the area with which I am concerned. Mr Lewis is not a qualified valuer. Nevertheless he used to receive one or more telephone calls a day from local valuers wishing to make use of his special expertise in relation to the valuation of flats in this area. He told me quite frankly that he had no recollection of the conversation spoken to by Mr Woodman. It would have been a routine casual conversation over five years ago, which he would have had no reason to remember at the time. However, he equally had no reason to doubt Mr Woodman's word that the conversation had taken place and that he had said in September 1988 that he could sell the flat for £160,000. His view was that £155,000 was a proper open market E valuation of the flat at that time.

Mr Lewis had personally visited every one of Mr Gray's comparable properties at one time or another. In his written statement he said he regarded some of them as being so remotely comparable as to be of no real assistance. In cross-examination he modified his stance to some extent by accepting that they were of limited value, but he was at pains to stress the unique quality of the flats in Lewes Crescent and the fact that no two properties were the same. Some of the comparables, for instance the first-floor flat at 6 Lewes Crescent where a sale was agreed at £135,000, he regarded as actively supporting Mr Woodman's valuation of £155,000 for the ground-floor F flat at 9 Lewes Crescent. He had personally inspected both properties and, contrary to Mr Gray's view, he regarded the ground-floor flat at no 9 as being more desirable and more valuable than the first-floor flat at no 6.

Mr Jonathan Rolls gave evidence to similar effect. He is another extremely experienced valuer, having been an ARICS since 1971. Since that time he has specialised in the valuation, management and sale of flats in the Kemp Town area. He made the point that the ground-floor flat at 9 Lewes Crescent was a particularly fine example of its kind. It was converted to a high standard by the former freeholder of the house for his own use and included in the lease was the use of a very pleasant patio garden at the rear. Mr Rolls agreed with Mr Woodman's valuation of £155,000 in September 1988 and said that he found Mr G Gray's valuation of £120,000 surprisingly low. He supported Mr Woodman's valuation by reference to two comparables of his own — a ground-floor flat at 5 Lewes Crescent, which was sold by his firm for £146,000 in August 1989, and a ground-floor flat at 19 Lewes Crescent, which was sold for £140,000 in May 1990. By the time these later sales were made, the market was somewhat weaker than it had been in the summer of 1988. The plaintiffs disputed that either of these flats were truly comparable and I heard a considerable body of evidence about that.

In summary, each side challenged the relevance of the other's comparables or, alternatively, claimed that they supported the case H being advanced by their own side. My own conclusion on one aspect of this matter may be shortly stated. I accept that in September 1988 Mr Lewis did tell Mr Woodman that he could sell the flat for £160,000 and that was his honest opinion as the most experienced estate agent dealing in that particular type of property in the locality.

In his final submissions, it seemed to me that Mr Nigel Thomas, for the plaintiffs, came close to conceding that £160,000 was, if not the definitive estate agent's selling price, at all events a price which a competent estate agent might well hope to achieve on a sale of the property. His submission was to the effect that an estate agent's selling price was of only limited assistance here and Mr Woodman should J still have gone through the process of reviewing other comparables which was undertaken by Mr Gray. Such a process, it was argued, would have led Mr Woodman to Mr Gray's figure of £120,000 or thereabouts. This in itself prompts the question: if it is accepted that a proper selling price for the flat was around £155,000 to £160,000, how could the proper open market valuation come to be only £120,000, which is some £40,000 less? The answer, according to the plaintiffs, lies in the true meaning of the expression "open market valuation".

I was referred to the current guidance notes for valuers published jointly by the ISVA and the RICS, which give guidance, among other things, on the matters which should be taken into account when K valuing a property for mortgage purposes. As it happens, Mr Woodman said that he had in mind another document altogether, which he referred to as the "Red Book" and which I believe to be a reference to the RICS Guidance Note GN22. However, the defendants were content to fight the case on the wording of the notes* referred to by the plaintiffs. The guidance notes are for valuers on the inspection of residential property and they include comments which are designed to assist a valuer at arriving at a proper open market value for the property. Note 1 is as follows (I read only the relevant parts):

> (a) The valuation of residential property for mortgage purposes is, unless the lender instructs otherwise, open market value as at the date of inspection, . . .
>
> L (b) Open market value for mortgage purposes is the amount at which a qualified valuer believes the property would sell in the open market as at the date of inspection, allowing for reasonable time to find a buyer in the prevailing market conditions and disregarding any exceptional reason the buyer may have for purchasing, including, . . .
>
> (c) Among the relevant factors to be taken into account will be:
>
> (i) age, type, accommodation, fixtures and features of the property;
>
> (ii) construction and general state of repair: subsidence, liability to flooding and other risks;
>
> (iii) siting and amenities of the locality;
>
> (iv) easements and burdensome or restrictive covenants if known or apparent.

M Then there are further comments set out in the notes which I need not trouble to read.

Mr Thomas submits that, in order to arrive at a proper open market value, a valuer may have to take into account various considerations which would not necessarily affect the mind of an estate agent charged with selling the property for the best price obtainable. For instance, the estate agent might be contemplating a vendor client who was prepared to wait indefinitely in order to achieve the price sought, whereas the valuer must contemplate the existence of a buyer prepared to purchase

*Editor's note: Extracted from a booklet published by the RICS – *Guidance notes for valuers*.

A within a reasonable time. There may be structural defects which would affect the value of a property more in the mind of a prudent valuer than an enthusiastic purchaser, or the purchaser may have some special reason for paying a higher price as, for instance, where he already owns the property next door. These are all considerations which may lead to the open market value of a property being substantially lower than a particular estate agent's selling price. On behalf of the defendants, Mr Christopher Gibson does not challenge the principle behind the submission. He simply denies its applicability in the circumstances of the present case. He submits that there is no reason in this case for the open market value to be any lower than the
B most experienced local estate agent's selling price and that, on all the evidence before the court, it is impossible for the open market value of the property to have been as low as £120,000 in September 1988. In reply the plaintiffs say that the evidence of Mr Lewis and Mr Rolls ought to be discounted because they are, in effect, giving evidence in support of Mr Lewis' own valuation, albeit unremembered, in September 1988.

At this point I pause briefly to consider the law. In *Singer & Friedlander Ltd* v *John D Wood & Co* (1977) 243 EG 212, Watkins J said [at p213]:

> The valuation of land by trained, competent and careful professional men is a
> C task which rarely, if ever, admits of precise conclusion. Often beyond certain well-founded facts so many imponderables confront the valuer that he is obliged to proceed on the basis of assumptions. Therefore, he cannot be faulted for achieving a result which does not admit of some degree of error. Thus, two able and experienced men, each confronted with the same task, might come to different conclusions without any one being justified in saying that either of them has lacked competence and reasonable care, still less integrity, in doing his work. The permissible margin of error is said by Mr Dean, and agreed by Mr Ross, to be generally 10% either side of a figure which can be said to be the right figure, ie so I am informed, not a figure which later, with hindsight, proves to be right but which at the time of valuation is the figure which a competent, careful and experienced valuer
> D arrives at after making all the necessary inquiries and paying proper regard to the then state of the market. In exceptional circumstances the permissible margin, they say, could be extended to about 15%, or a little more, either way. Any valuation falling outside what I shall call the "bracket" brings into question the competence of the valuer and the sort of care he gave to the task of valuation.

Then, bearing in mind that Mr Woodman made his valuation at what proved in retrospect to be the very peak of the market, I think it is worth quoting a further passage from the judgment.

> If a valuation is sought at times when the property market is plainly showing signs of deep depression or of unusual buoyancy or volatility, the valuer's task
> E is made more difficult than usual. But it is not in such unusual circumstances an impossible one. As Mr Ross said, valuation is an art, not a science. Pinpoint accuracy in the result is not, therefore, to be expected by he who requests the valuation. There is, as I have said, a permissible margin of error, the "bracket", as I have called it. What can properly be expected from a competent valuer using reasonable skill and care is that his valuation falls within this bracket. The unusual circumstances of his task impose upon him a greater test of his skill and bid him to exercise stricter disciplines in the making of assumptions without which he is unable to perform his task; and I think he must beware of lapsing into carelessness or over-confidence when the market is riding high. The more unusual be the nature of the problem, for no matter what reason, the greater the need for circumspection.

F

The principles to be deduced from this and other cases were helpfully examined in *Mount Banking Corporation Ltd* v *Brian Cooper & Co* [1992] 2 EGLR 142. In that case the learned judge took the view that the proper approach is not to try to ascertain a single ideally correct valuation for the property in question and then ask whether the impugned valuation differs from it by more than 10% or 15%, or whatever is said to be the permissible limit of deviation. Rather the question should be where the proper bracket of valuation lies and then whether the impugned valuation lies within a permissible limit of that bracket. I find that an attractive approach and it is one which I adopt in the present case. I think it is particularly appropriate to have regard to a bracket of valuation in this case for two reasons: first, as has already been explained, this was not a case where a valuer G could have had access to any true comparables, so that a greater than usual element of informed estimation was required; and, second, it now transpires that Mr Woodman was required to make his valuation at the very peak of the property market. At such a time it is, of course, particularly necessary to exercise caution. But two equally competent valuers might well have different ideas about the degree to which such caution should affect the open market value of the property.

I have already stated that I accept that in September 1988 Mr Lewis, the most experienced estate agent in the district, genuinely believed that he could sell the flat for £160,000. I conclude that he was justified in that belief and in the circumstances of this case I am unable H to see that there was any great difference between Mr Lewis' selling price and the open market value. Bearing in mind the plaintiffs' warning that Mr Rolls is not truly an independent expert but is, in effect, seeking to support a valuation made by his own firm, I none the less conclude that Mr Rolls' evidence is to be preferred to that of Mr Gray and that Mr Woodman's valuation of £155,000 lies within a proper bracket of valuation. Strictly speaking, I do not think I need state what the bracket was. However, doing the best I can, I would say that it probably included valuations lying between £140,000 and £155,000. It follows that I consider that Mr Woodman pitched his valuation at the very top end of the bracket. It also follows, in my J opinion, that Mr Rolls is right when he says that Mr Gray's considered valuation of £120,000 is distinctly low. One possible explanation for this may lie in the consideration that the property was bought for £69,000 and subsequently sold for £75,000. The leap from there to £155,000 in just two years is a very substantial leap, and it is difficult to remain unaffected by the knowledge that after 1989 the market subsided almost as rapidly as it had risen. Another possible explanation may involve the distinctive nature of the flats around the Kemp Town enclosures, which makes them unusually difficult to value.

I conclude, therefore, that the plaintiffs' case of negligence has not been made out. In the circumstances, I propose to deal with the K remaining issues very briefly.

The defendants contended that the plaintiffs were not shown to have relied on Mr Woodman's valuation when they advanced the mortgage money. I reject this ground of defence. Both Mr Hadley and Mr Armstrong, two witnesses called by the plaintiffs, testified as to the importance of the valuation. Furthermore, a number of documents in the third court bundle, amply demonstrate, to my mind, that the plaintiffs placed great reliance on the valuation supplied by the defendants.

The defendants also made allegations of contributory negligence arising out of the fact that, on the documents disclosed to the L plaintiffs' solicitors by previous lenders, the Barnes were potentially unsatisfactory borrowers. It is said that the plaintiffs were fixed with constructive notice of what the solicitors ought to have realised was evidence tending to show that the Barnes were in arrears with both their existing mortgages. If the solicitors had reported that information to the plaintiffs, so the argument goes, the plaintiffs would have made further inquiries, the loan would not have been made and thus the loss would have been avoided. I reject this argument for the reasons set out in Mr Thomas' helpful skeleton argument.

Briefly, the solicitors would not have known what information the plaintiffs already had or did not have, and it was not their function to M act as detectives on the plaintiffs' behalf. Their duty was simply to ensure that title was in order to carry out the necessary conveyancing and see that the charge was properly taken. If, therefore, I had found negligence proved, I would not have made any finding of contributory negligence on the part of the plaintiffs.

The final issue is that of damages, upon which a number of interesting arguments were raised. However, in view of my finding on liability, I do not propose to say anything more on the subject, save to mention one answer given by Mr Hadley in cross-examination. When asked whether the plaintiffs' home loan scheme was designed to protect against a fall in the value of the market, he replied: "I do not

A think any mortgage product is designed to cope with a fall in the market." In the light of that answer it may be that the plaintiffs would in any event have experienced difficulty in showing that any part of their loss attributed to a fall in property values was also caused by B negligence on the part of the defendants.

In the result there will be judgment for the defendants.

REAL PROPERTY AND CONVEYANCING

A
Chancery Division

February 18 1993

(Before Mr J SHER QC, sitting as a deputy judge of the division)

NATIONAL SCHIZOPHRENIA FELLOWSHIP v RIBBLE ESTATES SA AND OTHERS

Estates Gazette January 22 1994

[1994] 03 EG 132

B ***Restrictive covenant — Covenant against any trade or pursuit which could be a nuisance, annoyance or danger to neighbouring property — Whether use of property for housing schizophrenics in quiescent stage breach of covenant***

On March 25 1991 the plaintiff, a registered charity working for the welfare of people with schizophrenia and allied disorders, purchased a property which had previously been administered by the local health authority on behalf of the Secretary of State for Health. The Secretary of State had acquired the property by a conveyance dated March 11 1983 from the first defendant. By clause 2 of, and the third schedule to, the conveyance the purchaser covenanted that no
C "noisy offensive or dangerous trade or pursuit shall be carried on on the property hereby conveyed nor any trade or pursuit which may be or become in any way a nuisance annoyance or danger to the vendor or its successors in title or to the owners or occupiers of any neighbouring property which may tend to depreciate the value of the property" . The plaintiff's proposals for the property was to convert it to provide living accommodation for 12 persons who had in the past suffered from severe mental illness, but whose condition had stabilised on medication. The plaintiff sought a declaration that these proposals would not involve a breach of the restrictive covenant. The defendants, who were principally adjoining owners, did not, in the
D end, oppose the relief sought.

***Held*: The proposed activity would not cause nuisance, annoyance or danger to the defendants within the meaning of the covenant: there being no evidence of a reasonable apprehension of risk of injury. The words of the covenant were not wide enough to restrict or control the owners of the property as to the identity or personal characteristics of the individuals they chose to permit to occupy it.**

The following cases are referred to in this report.

C&G Homes Ltd v *Secretary of State for Health* [1991] Ch 365; [1991] 2
E WLR 715; [1991] 2 All ER 841; (1990) 62 P&CR 69; [1991] 1 EGLR 188; [1991] 06 EG 70, CA

Tod-Heatley v *Benham* (1888) 40 ChD 80

This was the hearing of an originating summons issued by the plaintiff, National Schizophrenic Fellowship, against nine defendants seeking a declaration as to the meaning of a covenant in a conveyance dated March 11 1983 by which the first defendant, Ribble Estates SA, sold 73 Scarisbrick New Road, Southport, Merseyside, to the Secretary of State for Health.

David Van Hee (instructed by Lawrence Graham) appeared for the plaintiff; the defendants did not appear and were not represented. F

Giving judgment, MR J SHER QC said: This is an originating summons taken out by National Schizophrenia Fellowship, a registered charity working for the welfare of people with schizophrenia and allied disorders. As part of its work the fellowship plans to provide housing for persons who have been receiving in-patient treatment in national health service mental hospitals for schizophrenia and who are ready to be discharged back into the community, but who need a period of accommodation in stable surroundings because they do not have homes and families of their own to which to return or are not yet ready to live independently on their own. The need for such accommodation in the Southport and G
Hornby area will be greatly increased in the near future by the imminent closure of Greaves Hall Hospital, a national health mental hospital in Southport.

On March 25 1991 the fellowship completed the purchase of a property known as 73 Scarisbrick New Road, Southport, Merseyside, a property that had been administered by the local health authority on behalf of the Secretary of State for Health. I shall refer to it throughout this judgment as "the property" .

The freehold in the property had been acquired by the Secretary of State for Social Services by a conveyance dated March 11 1983 from the first defendant company, Ribble Estates SA, a Panamanian H
company. By clause 2 of that conveyance the Secretary of State covenanted as follows:

> The purchaser for the benefit and protection of the remainder of the land belonging to the vendor of which the property hereby conveyed forms part or any part thereof and so as to bind the said property into whosesoever hands the same may come hereby covenants with the vendor that the purchaser and the persons deriving title under him will at all times hereafter observe and perform the restrictions and stipulations set out in the Third Schedule hereto.

The third schedule reads as follows: J

> No noisy offensive or dangerous trade or pursuit shall be carried on on the property hereby conveyed nor any trade or pursuit which may be or become in any way a nuisance annoyance or danger to the vendor or its successors in title or to the owners or occupiers of any neighbouring property which may tend to depreciate the value of the property.

The originating summons asks for a declaration that the fellowship's proposed user of the property will not infringe this restrictive covenant. As it was thought by the fellowship that the first defendant company might still own land in the vicinity of the property capable of benefiting from the restrictions on the property, that company was joined as a defendant to the summons. Also joined were K
defendants two to nine, all but two of whom (those being seven and eight) are owners of properties which were on March 11 1983 undoubtedly part of the retained land vested in the first defendant for the benefit of which the covenant was taken.

Those defendants all own houses near the property. The nearest is Kathleen Marland Leach, defendant three, at 1a Balfour Road immediately next door to the property (which occupies a corner plot between Scarisbrick New Road and Balfour Road). Defendants eight, seven, five and four are further up Balfour Road. The second defendant and defendants six and nine are much further away from the

A property. Defendants three, four and five have not acknowledged service and have played no part in the proceedings. Defendants one, seven and eight did acknowledge service, but have taken no further part, although some of the evidence put in on behalf defendants two, six and nine was said to be put in on behalf of defendants seven and eight as well.

The most active defendants were the remaining ones, namely defendants two, six and nine, who put up a vigorous fight in affidavit evidence and exhibits running in all to some 194 pages, but who have not appeared before me either by counsel or in person. Their solicitors have written to the court explaining that the day before the hearing B was due to commence the legal aid board in Liverpool discharged their clients' legal aid certificates and they were not proposing to attend the hearing.

All the defendants have written saying that they now do not resist the relief asked for in the summons provided that no order for costs is made against them in these proceedings. Mr David Van Hee, counsel for the fellowship, has confirmed that no such order will be asked for by the fellowship. In the event I have had no one from the defendants' side to put the case against Mr Van Hee's arguments for the fellowship, but as the court's ruling on the construction of the covenant is still required I have had to look at the matter critically and C do the best I can under the circumstances.

I return to the fellowship's proposals for the property. The property is in the course of being converted so as to provide living accommodation for 12 people, who have in the past suffered from severe mental illness, but whose condition has been stabilised on medication and who are ready to be discharged back into the community.

These 12 residents will be chosen from a particular group of patients now in a rehabilitation ward, known as Patterson Unit, Hescott Centre, which provides 25 in-patient beds for patients, many of whom are ready to be discharged back into the community. The fellowship's proposals are supported by the district health authority, D which will take a major part in selecting the residents. There will be a selection panel, which will include members of the staff of the Patterson Unit and members of the management staff of the health authority.

In order to provide individual care and support for the residents and to comply with the wishes of the social services department of the local authority, the staff will consist of one project manager, one senior care manager, two senior care officers, four care assistants, an administration officer, two part-time cooks and domestic staff plus two waking night staff. This will mean that at all times there will be a senior member of staff available on the premises.

E It is intended that the staff who will be recruited will include present members of the staff of Greaves Hall Hospital and Patterson Unit, from which the residents will have come. Medical attendance will be available from a general practitioner, sympathetic to the fellowship's work, whose surgery is opposite the property.

When the originating summons was issued the fellowship's plans were to include facilities in the property for training the residents and non-resident daytime trainees as receptionists, waiters, waitresses, chambermaids, chefs etc along with re-establishing the ordinary skills of daily living.

Following anxiety expressed on behalf of some of the defendants F these proposals have been abandoned. The current proposals are simply to provide a home for 12 patients with the necessary support staff and the declaration sought in the summons is to be amended accordingly.

Those defendants who have put in affidavit evidence have expressed their concern and anxiety about the fellowship's proposals and have forcefully objected to those proposals. I can summarise their concerns under a few general heads.

First, they are concerned about the impact on the value of their houses. It is interesting that this concern is a common and consistent feature of complaint and I have little doubt that it was a major, perhaps the major, motivation for defending these proceedings. Whether the implementation of the fellowship's plans for the property will have G this effect of reducing property values is, of course, another matter. There is much surmise in this regard by non-expert witnesses on both sides, but not one scrap of admissible evidence on which I can reasonably make a finding.

If the point is of importance it will have, in my judgment, to be resolved against the party who had the onus of proof, ie in my judgment, the fellowship, which has come to this court asserting that their proposals will not infringe the covenant.

I should add here that, although there is considerable difficulty in identifying the property meant by the last two words of the covenant, the common sense of the covenant is that it must refer to the retained H land of the vendor for the benefit of which the covenant was taken or, at least, the neighbouring property referred to in those specific terms in the covenant. In my view, it does not, despite the valiant efforts of Mr Van Hee, refer to the property itself.

In the event, as will be seen, the issue as to whether the implementation of the fellowship's proposals will tend to depreciate surrounding property values does not have to be decided because the requirement in the covenant to this effect is additional to and separate from the requirement that the activity should cause annoyance. As I have decided that that activity will not cause annoyance within the meaning of the covenant the fellowship will succeed despite the failure to show that property values will not be adversely affected. J

The second broad head of complaint is the fact that there appears to be a discrepancy between the fellowship's proposals as contained in early documentation prepared long before these proceedings had started and the proposals as formulated in the evidence of Mr Alan Bolton, national development officer of the plaintiff, filed in support of the originating summons. There is, however, nothing sinister in that. The fellowship's proposals have undergone several changes over time and discrepancies are simply attributable to such changes from time to time. Indeed, as I have indicated, there have been further significant changes in those proposals during the course of these proceedings themselves. I need say nothing more about this head of K complaint.

The defendants exhibit a concern about the training facilities, in particular, in respect of non-residents who would be attending at the property during the day time. Those aspects of the proposals have been abandoned and I need say no more about this concern.

That leaves the last head of complaint and that is the defendants' concern for the general safety of their children: see, for example, Mrs Maxine Diane Sullivan's affidavit of June 9 1992 filed in support of the cases of the defendants two, six, seven and eight in which she says at para 3 and 4:

> 3. I have genuine fears for the safety of my children . . . in particular in view of the day care facility and the outpatients who will be coming and going at all times. L
>
> 4. My fears are based upon personal knowledge obtained whilst working at Greaves Hall Hospital, Southport from 1980 until 1985 as a nursing assistant. My first year at the hospital was working with mentally handicapped patients and schizophrenics and I also worked for two years with other psychiatric patients. I confirm that I also had contact with outpatient schizophrenics at the day unit and on the psychiatric unit when I was employed on night shifts. I am concerned with regard to the security arrangements for the residents and also for the day centre attendants. I am aware from my own experience that such patients are susceptible for example to taunts from children and other people who will be aware of them attending the unit and they therefore may act aggressively. I am also aware of the fact that such a facility will affect the M value of the property in the immediate area of 73 Scarisbrick New Road.
>
> 5. The property 3 Balfour Road [which is her property] is a family residence and any prospective purchaser with young children will, no doubt, be deterred by the presence of such an establishment as is proposed at 73 Scarisbrick New Road.

I pause to notice that Mrs Sullivan's concern is largely associated with the day care facility which, of course, has now been abandoned. She does, however, include in her expression of anxiety the residents themselves and says that they are susceptible to taunts from children and that they therefore may react aggressively.

Mr Blewitt is another witness who filed an affidavit in support of the defendants' case. He said in para 4 of his affidavit:

> I am particularly concerned for my wife and my 12-year-old son. I have had first hand experience of people suffering with mental instability in my employment as a police officer. For that reason I would submit that my fears are genuine and well-founded. I cannot preclude the possibility that something may happen to my wife or my son particularly in view of the day centre facility and the attendants who will be using such a facility.

Two points need to be made as to this. First, Mr Blewitt's experience as a police officer has been with mental instability on occasion no doubt in an acute phase of mental illness. I am, however, concerned in this case with a much more specific type of mental illness, namely schizophrenia and, more important, with such illness in a quiescent stage when the patient has been fully treated and stabilised and is ready to go back into the community.

The other point I should make about Mr Blewitt's evidence is that the day centre facility has been abandoned, so that particular anxiety of his has been allayed.

Josephine Barnes, the second defendant, expressed concern in a letter written to Mr Bolton on November 10 1990. She said:

> I have first hand knowledge of schizophrenic behaviour. From past experience in my own family I know how unstable and dangerously violent schizophrenic people can be. For this reason and because my family includes two young children it is my intention to take legal action to oppose the proposed use of no 73 Scarisbrick New Road.
>
> I am not unmindful of the problems and current needs of this group of people, but to place them in circumstances which expose them to their own personality inadequacies and may lead to (and have in other places) terrible incidents which criminilise them is, I feel, both lacking in wisdom and true long term concern.

Again, her concern relates to patients in the acute face of the illness, which is not what this case is about.

Another witness, Maureen Llewellyn, had this to say:

> 9. During the course of acting on behalf of the Scarisbrick New Road and Balfour Road Residents' Association, I have amassed a file of newspaper reports which, sadly, recount tragic incidents, sometimes fatal, involving schizophrenic sufferers and their families. While I have to accept that the reporting of such cases by the national media tend to be melodramatic, I have to say that these incidents have enforced my own personal fears of problems arising not only with residential patients but, in particular, with those outpatients who would be attending 73 Scarisbrick New Road and the day centre facilities which would be available.

Again a witness who refers to the disease in its acute phase and who refers, in particular, to concern about the visit to the property daily of outpatients, a feature of the proposals which has now been abandoned.

All the defendants who put forward evidence draw attention to a pamphlet called *Schizophrenia Notes for Relatives and Friends* written by, among others, Julian Leff, assistant director, Medical Research Council, Social Psychiatry Unit in London. These notes were an attempt to give relatives of those suffering from this disease some information about the disease, its effects, causes and treatment. The defendants rely on selected excerpts from this document in order to show how serious the condition can be. Dr Leff is now Professor of Social and Cultural Psychiatry at University of London and Honorary Consultant in Psychiatry at Bethlem Royal and Maudsley Hospital and director of the Medical Research Council, Social and Community Psychiatric Unit at the Institute for Psychiatry, De Crespigny Park, London SE5. He is plainly a highly qualified witness in this area of medicine.

The fellowship has put forward an affidavit by him in which he points out that, while the pamphlet was intended to inform relatives about the symptoms to be expected in acute episodes of schizophrenia, the persons who are to be moved into the property are not in an acute phase of the disease. They have had treatment for many years and are in a quiescent stage of the illness.

Professor Leff goes on in his affidavit to say:

> 4. I am a director of the team for the Assessment of Psychiatric Services at Friern Hospital, London, and in connection with this work I have studied the transfer into community homes of 500 long stay patients from Friern and Claybury Hospitals in north London. These patients were very similar in mental state and behaviour to the patients currently in Greaves Hall Hospital from whom the residents of 73 Scarisbrick New Road will be chosen.
>
> 5. Where patients from Friern and Claybury Hospitals were placed in staffed homes in the community, then in their first year of living in staffed homes, none of the patients was involved in aggressive incidents involving neighbours or members of the public. There was no deterioration in their mental state or behaviour.
>
> 6. My team conducted a survey of the neighbours of one home for 20 patients in north London six months after they moved there from the hospital. The great majority of these neighbours were unaware that the residents were suffering from mental illness. A number of neighbours wrongly identified the home as being an old people's home. This is understandable since the average age of discharged long stay patients is 60.
>
> 7. In general, the neighbours found the residents of these staffed homes to be quieter than the average person. A significant proportion of the 500 patients we studied made friends with their neighbours in the course of their first year in the community.
>
> 8. I am sure that the neighbours of 73 Scarisbrick New Road have no cause to fear for their own safety or that of their children.

I now turn to my task which is to construe the covenant and decide whether the present proposals of the fellowship would infringe it.

In *Tod-Heatley* v *Benham* (1888) 40 ChD 80 a house known as 6 Gloucester Terrace, Brompton, had been let for 99 years from June 25 1820. The lessee covenanted that he would not (at p81):

> . . . do or wittingly or willingly cause or suffer to be done any act, matter, or thing in or upon or about the said premises, which shall or may be or grow to the annoyance, nuisance, grievance or damage of the lessor, her heirs and assigns, or the inhabitants of the neighbouring or adjoining houses.

The adjoining houses, 7 and 8 Gloucester Terrace, had formed part of the original lessor's estate and had been demised in plots for building purposes subject to covenants similar to those in the lease. Dr Benham became the occupier of no 6 with notice of the covenants in the original lease. In February 1887 the premises were opened by him as a hospital for treatment of various diseases.

By the time of the hearing at first instance in May 1888 some 12,000 persons had attended at the hospital (which indicates the extent of the activity on the property in that case). The action for injunctive relief was brought by Mr Tod-Heatley, the owner of the reversion expectant on determination of the lease in no 6 and by the lessees and occupiers of 7 and 8 Gloucester Terrace.

The case is important because it gives guidance as to the meaning of the word "annoyance" — far and away the most relevant and important word in the covenant I have to construe in this case.

At p98 Bowen LJ said:

> "Annoyance" is a wider term than nuisance, and if you find a thing which reasonably troubles the mind and pleasure, not of a fanciful person or of a skilled person who knows the truth, but of the ordinary sensible English inhabitant of a house — if you find there is anything which disturbs his reasonable peace of mind, that seems to me to be an annoyance, although it may not appear to amount to physical detriment to comfort. You must take sensible people, you must not take fanciful people on the one side or skilled people on the other; and that is the key as it seems to me of this case. Doctors may be able to say, and, for anything I know, to say with certainty, that there is no sort of danger from this hospital to the surrounding neighbourhood. But the fact that some doctors think there is makes it evident at all events that it is not a very unreasonable thing for persons of ordinary apprehension to be troubled in their minds about it. And if it is not an unreasonable thing for an ordinary person who lives in the neighbourhood to be troubled in his mind by the apprehension of such risk, it seems to me there is danger of annoyance, though there may not be a nuisance. As the Lord Justice Cotton has said, the very opinions of the doctors called on behalf of the Plaintiffs to the effect that there may be a risk, even if they are wrong in thinking there is risk, shew that it is not at all events unreasonable for less skilled and competent persons to think so. The measure is what is reasonable enjoyment of a man's premises — not according to the estimate given by skilled doctors as to the effect of an

A hospital, but according to what ordinary people, who are not skilled doctors, may think it.

In short, the question is whether there is a reasonable apprehension of risk. In *Tod-Heatley* the risk identified was that of infection. In the present case the risk is that of injury, particularly to the defendants' children. It is not a complete answer for doctors to be called by the person sought to be restrained to the effect that there is no risk because those complaining might still reasonably apprehend risk based on their own medical evidence. In the *Tod-Heatley* case it appears that the line-up of evidence was doctor against doctor. Cotton LJ said at p94:

> B I do not go into or criticise the evidence of the doctors. Some have been called on behalf of the Plaintiffs, who state their views and point out what they think may be a case of nuisance. Even if they do not make out that there is really risk, yet the expression of their opinions shews that persons residing in the neighbourhood . . . and the inhabitants of adjoining houses in *Gloucester Terrace*, might reasonably apprehend risk in consequence of those patients being brought there. It is said that in walking along the *Strand* there is just as much risk as is run by the people here. Possibly that may be so. I do not know. In *London* (speaking for myself) one does not think so much of the risk as people living in the country do. But this is bringing to a centre in the immediate neighbourhood of these houses people, some of whom have contagious diseases, some of whom, there is reason to believe, having regard to the evidence of the doctors on both sides, it cannot be said that the
> C apprehension of risk from this hospital being carried on is unreasonable. There has been, therefore, a violation of the covenant, and the appeal must be dismissed.

I have not had the benefit of submissions on behalf of those who assert an apprehension of risk of injury in this case. That would, of course, have been helpful to me as would an opportunity of seeing them cross-examined. Attendance for cross-examination has never, so far as I know, been requested by the fellowship in this case.

However, I have, in the course of outlining the evidence of those who have complained, already pointed out that those who have
D expressed anxiety have been largely concerned about the day centre facility, which has now been abandoned, and have been influenced by their non-expert perception (apart from Mrs Sullivan who has some expertise in this area) of schizophrenia in its acute, rather than quiescent, phase. Moreover, it is plain that a major reason for their concern is their apprehension of a diminution in the market value of their respective homes.

Against this I have the clear evidence of Professor Leff, who concludes that the defendants have no cause to fear for their own safety or that of their children.

In my judgment, on the evidence before me, I find that the
E defendants have not exhibited a reasonable apprehension of risk of injury to justify treating the fellowship's present proposals as something which, in the words of the covenant, will be or become in any way a nuisance, annoyance or danger to the defendants.

I bear particularly in mind that in *Tod-Heatley* there was medical evidence on the part of those complaining, suggesting that there was a risk; and if those experts thought there was a risk, it was not unreasonable that the plaintiffs should think so too. Here there is no expert evidence put forward on the part of the defendants.

I bear in mind also, it should be said, that the defendants have now indicated that they do not resist the relief being sought, albeit against
F the understanding and expectation that no order for costs will be made against them.

I should before leaving this point and in fairness to the defendants note that at a meeting of the residents in or about 1990 Mr Bolton is said to have stated that the worst that the defendants could expect from the residents of the property was a brick through the window. It has troubled me that that statement by Mr Bolton might of itself have induced in the defendants a reasonable apprehension of risk. Mr Bolton has left the employment of the fellowship and has not himself responded to this evidence, although Ms Jane John, director of the fellowship's northern region who has taken over responsibility for the property, has said:

> 16. I accept that Mr Bolton may well have said that the worst you could expect was a brick through a window. I am sure that he intended thereby to G
> emphasise that nothing worse should be expected, not that this was a likely occurrence. The residents will all be allowed into the property on terminable interests and if they cause a nuisance they will be removed. The residents will understand the terms of their agreements and that the consequences of any nuisance will be that they will be removed.

I am not entirely satisfied that I have had all the evidence relevant to this statement made by Mr Bolton. The context in which it was said is plainly of considerable importance. I would have wished to see some further explanation from Mr Bolton himself. I am not, however, satisfied that it is enough to make me change my view that, on the evidence before me and as matters now stand, the fellowship's present H
proposals will not, on the true construction of the covenant, amount to an infringement of it.

That would be sufficient to dispose of the case, but it has been submitted by Mr Van Hee that in any event the activity targeted by the covenant is not the use by the fellowship of the property, but the activity of the occupants in the property. It would be as well to look back at the covenant itself.

First, in my judgment, there is no question here of an infringement of the first few words of the covenant: there is no question of any noisy, offensive or dangerous trade or pursuit being carried on on the property. The issue lies in the next few words "nor any trade or J
pursuit which may be or become in any way a nuisance annoyance or danger".

In short, therefore, the covenant, so far as relevant to the facts of this case, is that no trade or pursuit shall be carried on on the property which may be or become in any way a nuisance, annoyance or danger to the surrounding properties. In my view, the word "trade" is inapposite to encompass the fellowship's proposed user of the property.

I turn to the word "pursuit". This word is an unusual word in this legal context. It is not a term of art and means, so the *Concise Oxford Dictionary* tells us: K

> a profession, employment, recreation, that one follows.

Certainly, it conjures up the idea of a specific and identifiable activity that may be carried on in the property such as, for example, playing the violin at unsociable hours or, as in Mr Van Hee's example, late-night carpentry, which can cause legitimate annoyance to the neighbours. The word "pursuit" is more appropriate to refer to an activity carried on in the premises than to the actions of the fellowship in setting up and running the supervised home for expatients.

When coupled with the reference to the words "shall be carried on on the property" the covenant seems to me to be directed to the L
activities of the occupants (whoever they may be) in or upon the property rather than to the identity of the occupants themselves or to the activity of the fellowship in selecting the occupants and providing a home for them.

No complaint could be made about the activities of the occupants themselves on or in the property. They will merely be living there. The burden of complaint relates to the personal characteristics of persons whom the fellowship has chosen or will choose to live in the property.

The matter may be tested by an extreme example. Suppose from time to time the fellowship simply granted licences to occupy the M
property as a residence for expatients who had at some time suffered from severe schizophrenia but were ready to attempt to face life in the community; can it really be said that because of a reasonable apprehension of risk of injury to children the neighbours could stop the granting of these licences on the grounds that it was a pursuit carried on on the property which may become an annoyance to them? I think not. The words of this covenant are not, in my judgment, wide enough to restrict or control the owners of the property as to the identity or personal characteristics of the individuals they choose to permit to occupy it.

I do not say that the matter is clear. It may be different if the scale

A of the fellowship activity was particularly great; but I do gain some comfort from the case of *C&G Homes Ltd* v *Secretary of State for Health* [1991] Ch 365* where, in a rather similar situation of the provision on a residential housing estate of supervised housing for a group of former mental in-patients (albeit a case where the only complaint was that neighbouring property would be reduced in value), Lord Donaldson of Lymington MR held that the restrictive covenant in that case was directed to conduct in or upon the premises which may cause annoyance rather than the overall use being made of the premises by the Secretary of State, namely that of supervised housing.

This was not the *ratio* of the case, although the observations of
B Nourse LJ at p386C were supportive of the view expressed the Master of the Rolls. The covenant concerned in that case read (at p385):

> 20. Not to cause or permit or suffer to be done in or upon the property any act or thing which may be or become a nuisance, annoyance, danger or detriment to the transferor or owners or occupiers for the time being of other parts of the estate.

There was in that case, in addition, a separate covenant against user for trade or business purposes or any purpose other than those incidental to the enjoyment of a private dwelling-house. The existence of that separate covenant assisted the construction which the Master of the Rolls put upon the covenant in clause 20.

C That is a feature absent from the present case but one which, in my assessment, is not critical to the fellowship's point. Overall, what impresses me most in this case is that the fellowship's proposals merely involve a residential (albeit supervised) user of the property. There can be no complaint as to the activity proposed to be carried on in the property itself; that activity is merely residential. I do not think that the words of the covenant are apt to control the identity or personal characteristics of the persons whom the fellowship chooses to house in the property. I will make the declaration requested in the originating summons.

D *Declaration accordingly.*

---

*Editor's note: Also reported at [1991] 1 EGLR 188.

Chancery Division

May 18 1993

(Before Mrs Justice ARDEN)

E MID KENT WATER PLC v BATCHELOR AND OTHERS

Estates Gazette February 19 1994

[1994] 07 EG 197

Valuation — Transfer of land — Provision for additional consideration upon grant of planning permission — Whether value without benefit of grant of planning permission should include hope value

By a transfer dated February 4 1986 the plaintiff acquired a 20-acre field. Clause 2 of the transfer provided that in the event of any
F development permissions being obtained in respect of the land during the following 20-year period, the plaintiff would pay to the first and second defendants a sum of 50% of the amount by which the value of the land in respect of which development permission has been granted has thereby been increased. On June 20 1989 the local planning authority resolved to grant planning permission to the plaintiff to construct a concrete reservoir on the land, and on June 28 notice of the grant of planning permission was given. The plaintiff contended that, in arriving at the first valuation, with which the value with the benefit of planning permission which was granted had to be compared, the valuer should take into account the prospects of obtaining planning permission, including that ultimately obtained, ie hope value.

G ***Held*: In order that clause 2 may fulfil what objectively was its aim — that of sharing development gain — its meaning must be that the value of the land in question, without the benefit of planning permission, must be ascertained on the same basis as the contract price appears to have been ascertained. The inference to be drawn from the inclusion of clause 2 is that the contract price did not include any expectation of development gain, but reflected its then existing use. The valuation with which the increased valuation is to be compared is a valuation performed on the same date as the increased valuation is performed of the land, but with the use at the time of the transfer.**

H

The following case is referred to in this report.

Prenn v *Simmonds* [1971] 1 WLR 1381; [1971] 3 All ER 237, HL

This was an action concerning the meaning of a transfer dated February 4 1986 by which the plaintiff, Mid Kent Water plc, covenanted to pay additional consideration to the first and second defendants.

Gordon Nurse (instructed by Gouldens) appeared for the plaintiff; Mark Cunningham (instructed by Hallett & Co) represented the first and second defendants; the third defendant did not appear and was not represented. J

Giving judgment, **Arden J** said: This action concerns a dispute which has arisen over the construction of clause 2 of a transfer dated February 4 1986 made between Mr and Mrs May, of the first part, the first and second defendants referred to therein as the "nominee" in the second transfer respectively of the second part and the plaintiff referred therein as the "transferee" of the third part.

By the transfer the plaintiff acquired a field comprising some 20 acres adjoining Hermitage Lane, Maidstone, Kent, for £76,000. Clause 2 of the transfer provided in material part as follows:

> The transferee hereby covenants with the second transferor and the nominee K that in the event of any development permission or permissions being obtained in respect of the land hereby transferred or in part thereof during the period of 20 years from the date hereof, the transferee will on a subsequent sale of the said land or upon implementing that development permission, but in any event within two years of the grant of such permission, pay to the second transferor and the nominee a sum equivalent to 50 per cent of the amount by which the value of the land in respect of which development permission has been granted has thereby increased, subject only to an allowance of any tax liability which the transferee or his successor in title may prove to the satisfaction of the second transferor and the nominee to be created either as a result of the grant of the development permission on such subsequent sale or the implementation of the permission. The amount of such increase in value shall unless agreed between the relevant parties within one month of the liability L arising be decided by Messrs Cobbs of Sevenoaks, or such other agent as the second transferor and the nominee may nominate.

On June 28 1989 the plaintiff obtained planning permission to construct a concrete reservoir on the land in question and a dispute has arisen as to exactly how the valuer should proceed under clause 2. The dispute does not relate to the valuation of the land in respect of which planning permission has been granted with the benefit of planning permission, but rather with the approach which the valuer should take to the valuation with which that valuation should be compared. Mr Gordon Nurse, for the plaintiff, has argued that the valuer should, when doing this valuation, take into account the prospects of obtaining M planning permission including that ultimately granted. Mr Mark Cunningham, for the defendants, contends that this would defeat the commercial purpose of clause 2. He invites me to dismiss the plaintiff's claim for a declaration on this basis.

At the hearing before me, Mr Nurse has contended in the alternative that the valuer should disregard any value attributable to the hope of obtaining the development permission that was in fact obtained. Mr Cunningham has not sought to oppose this on its merits, but has reserved his position on costs. I should add that the third defendant is the agent whom the defendants would wish to use for the purpose of any proposed valuation to clause 2 arising out of the grant

A of planning permission on June 28 1989. The third defendant has not, however, taken any separate part in the hearing before me. None of the parties has sought to say that the court should leave the question in dispute to the chosen valuer. I should add that the question has been argued before me purely on the basis that it is a question of construction, and I have not been concerned with any question of quantification.

As a matter of the literal construction of clause 2, the amount which the valuer has to quantify is the amount by which the market value of the land is increased simply as a result of the grant of planning permission. But on this basis the valuer would be entitled to
B take into account the fact that an application had been made or, if it had occurred, the fact that the planning authority had resolved to grant it. In the usual course of events the planning authority will not change their mind after the passing of the resolution, but some days will pass before their decision is notified to the applicant. Mr Cunningham has argued that planning permission is granted when formal notification is given to the applicant and not before: see vol 46 of *Halsbury's Laws*, 4th ed (reissue 1992), pp 338 to 339, para 448, note 6.

Mr Nurse has not sought to disagree with this proposition of law, but accepts that the valuer has to disregard the fact that the planning authority has resolved to grant permission. This shows that the parties
C cannot have intended clause 2 to bear a purely literal construction.

It is indeed now well established by the highest authority that a contract must be construed against the factual background known to the parties at or before the date of the contract: see *Prenn* v *Simmonds* [1971] 1 WLR 1381 at p1385H *per* Lord Wilberforce, with whom the other members of the House agreed. Accordingly, I next turn to consider that background.

The plaintiff called two witnesses, Mr David William Walker, the present company secretary of the plaintiffs, and Mr John Hoyle [FRICS], a chartered surveyor. The defendants did not call any witnesses. Mr Walker gave evidence as to the circumstances surrounding the transfer and some later events, although he himself
D joined the plaintiffs only in February 1992 and was relying upon his researches into this matter. However, no exception was taken to this.

Mr Hoyle gave evidence as to the valuation practice and as to the plaintiffs' applications in 1988 and 1989 for planning permission: see below. I now set out the facts relevant to the question of construction in the events that happened as I find such facts to be.

1. The plaintiff is a statutory water undertaker. It was previously a statutory company, but has since registered under the Companies Act 1985 as a public limited company. The plaintiff acquired the land in question in this case for the purposes of its business as a water
E company.

2. On February 4 1986 the transfer with which these proceedings are concerned was signed. The first defendant was a nominee for the second defendant, who is the second transferor for the purposes of clause 2. The land in question was previously used for a "pick your own fruit" farming enterprise and had no planning permission. The auction particulars refer to the existence of further medium- or long-term development potential in the area. The price paid by the plaintiff amounts to approximately £2,708 per acre, but there is no evidence as to the precise relationship between this figure and agricultural land values for this type of land in the area at the time. However, the inclusion of clause 2 suggests to me that the price paid
F must have been on the basis of the existing use of the land and it is difficult to see why if the position were otherwise, the plaintiff would have agreed to the inclusion of clause 2.

3. The broad purpose of clause 2 of the transfer was to enable the plaintiff and the first and second defendants to share development gains after tax if planning permission were obtained. It appears that it was assumed by the parties to the transfer that there would be such an increase in value.

4. In June 1988 there was a proposal, which did not in the event proceed, that the plaintiff would sell the land to Maidstone Football Club for a sum in excess of £0.5 m on terms that they — that is the club — met the liability to the second defendant or an offer in excess of £1m with the plaintiff making any payment to the second G
defendant. The defendants say that this proposal, while not strictly relevant to the construction of clause 2, supports the commercial purpose of clause 2, which I have already outlined and found as fact on the basis of Mr Walker's evidence.

5. Later in 1988 the plaintiff made an unsuccessful application for planning permission for the construction of divisional offices, workshop, district depots and a service reservoir with a booster pump station. This application was unsuccessful.

6. By an application dated April 3 1989 the plaintiff applied to Maidstone Borough Council for planning permission to build a concrete reservoir. The planning authority approved the application on H
June 20 1989. Notice of the grant of planning permission was given on June 28 1989. The permission has now been implemented.

Mr Hoyle, who, as I have explained, is a chartered surveyor, gave evidence as an expert. In the course of his evidence he gave his opinion on a number of matters relevant to quantification of any increase in value as a result of the grant of development permission in this case. However, questions of quantification are for determination by the agent appointed for the purposes of clause 2 and Mr Hoyle was not cross-examined on these matters. For the record, I would add that Mr Hoyle was of the opinion that while existing local planning policies remained in force there was no likelihood of further development being permitted on the land during the period of the Kent J
deposit draft plan. He was further of the opinion that the development which had taken place on the land was of a very specific nature which could be used only by a public water authority because any increase in the value of the land would not have to be taken into account if the plaintiff was to use its compulsory powers, and for the above reasons in particular there had been no increase in the value of the land as a result of the grant of planning permission even if hope value were excluded.

Mr Hoyle also gave evidence as to the current guidelines of the Royal Institution of Chartered Surveyors. These guidelines were adopted since the date of the transfer, but required a chartered K
surveyor for ascertaining the open market value of the property without restriction as to its use to take into account any hope value. In the words of Mr Hoyle:

> Such value includes such element of hope value, if any, as the property may have for uses other than existing use or for the realisation of such marriage value, if any, which the property may have for merger with another property or which an interest in the property may have for merger with another interest in the said property but limited to the extent that the expectation of realising such hope value would in practice be reflected in offers that would be made in the open market by prospective purchasers other than the additional bid of a purchaser with a special interest. L

Mr Hoyle concludes that, "Hope value is thus part of the valuation exercise in this case", and it is on the basis of his evidence that the plaintiff asks the court to declare that the value of the land with which the benefit of development permission is to be compared includes hope value of both the development permission in question and any other development permission. Mr Hoyle, however, accepted that he was not required to follow the RICS guidelines, to which I have referred, in ascertaining the market value of property if the party's contract required him to do otherwise. As a professional matter, however, he would in that event be bound to point out the differences between his valuation on the basis agreed by the parties and a M
valuation in accordance with the RICS guidelines in his valuation. This point aside, Mr Hoyle confirmed, as one would expect, that there was nothing in the RICS guidelines which would prevent effect being given to the party's contract if it required some other form of valuation.

The difficulty in the present case has arisen because clause 2 does not expressly state the basis on which the value of land, prior to the grant of development permission, is to be ascertained. I do not myself think that it is a question of the date at which the land is to be valued. Indeed, I would accept Mr Nurse's submission that in order to ascertain the amount of the increase in value owing to the grant of

A planning permission there should be the shortest practical interval in time between the valuation with the benefit of planning permission and the valuation without such benefit. But, as I have said in my judgment, the real question is not the date of such valuation; as I see it, the real question is the basis of valuation. What the plaintiffs say is that when the valuer ascertains the market value of the land without the benefit of development permission which has triggered clause 2 he should, applying the principles of valuation, of which Mr Hoyle has given evidence, take into account the hope of obtaining development permission or, alternatively, the hope of obtaining development permission other than the permission which was in fact granted. If this
B happens then, according to the defendants, clause 2 will not achieve the purpose that any development gain is shared. On the contrary, on the plaintiff's construction, the greater the expectation that some permission will be obtained the less the amount of increasing value to be shared.

At this point I return to *Prenn* v *Simmonds*. In that case Mr Prenn, a purchaser of a majority shareholding in the capital of a company called RTT, a holding company, proposed to sell an equity interest in RTT to a valued employee, called Dr Simmonds, of a trading subsidiary of RTT called Airmec Ltd. Under the party's contract, Mr Prenn was to transfer those shares to Dr Simmonds when he himself had paid deferred instalments of the purchase price for his majority
C shareholding out of the proceeds of redemption of certain redeemable preference shares issued by RTT or, alternatively, when RTT had sufficient profits to effect this redemption. The alternative was included because Mr Prenn, as the new controller of RTT, could prevent the redemption of the preference shares. The difficulty arose because Mr Prenn said that the alternative which used the words "the profits of RTT", meant the profits of RTT as an entity and not the consolidated profits of it and its subsidiaries. The House of Lords held that the profits must be the consolidated profits. As Lord Wilberforce put it at p1388H:

> D . . . no purpose can be discerned why the reference should be to the separate profits of RTT, which in fact means such part of the group profits as the board of that company, effectively Mr Prenn, decided to pass up to the parent company.

Thus, the contract in that case was construed by reference to the objective aim of the transaction, so in this case I must construe clause 2 by reference to the objective of the parties in agreeing to it.

Having considered the contentions on either side, in the light of *Prenn* v *Simmonds*, I am satisfied that the defendants are right. In order that clause 2 may fulfil what objectively was its aim — that is, the sharing of development gain — its meaning must, in my judgment,
E be that the value of the land in question, without the benefit of planning permission, must be ascertained on the same basis as the contract price appears to have been ascertained, but such valuation is to be performed at a later date. The inference which I have drawn from the inclusion of clause 2 is that the price did not include any expectation of development gain, but reflected its then existing use. Thus, in my judgment, the true construction of clause 2 is that the valuation with which the increased valuation is to be compared is a valuation performed on the same date as the increased valuation is performed of the land, but with the use at the time of the transfer. In argument, the example was given of successive developments of the
F same part of the land. I do not, on reflection, think that this poses a particular difficulty. To give effect to the purpose of clause 2 any profits already shared under clause 2 would have to be deducted from the amount of any subsequent increase in profit which was also liable to be shared between the parties.

Thus, subject to any further submissions by counsel as to the detail of the declaration, the declaration which I propose to grant is a declaration on the terms of clause 1(a) as regards the first paragraph of that proposed declaration, but continuing in lieu of the alternatives (i), (ii) and (iii) in the draft provided by Mr Nurse of some such words as "being the value which such land would then have if valued on the basis of its use for agricultural purposes". I therefore propose to hear submissions by counsel as to the declaration which should be made. G

Declaration accordingly.

Chancery Division

September 17 1993

(Before Mr David NEUBERGER QC, sitting as a deputy judge of the division) H

MARCHANT AND OTHERS v ONSLOW

Estates Gazette April 2 1994
[1994] 13 EG 114

School Sites Act 1841 — Whether site reverted to adjoining land of grantor at date of grant — Whether proceeds of sale held in trust for grantor's successor

By a conveyance dated April 18 1848 and made pursuant to the School Sites Act 1841, Elizabeth Foley and Richard Onslow conveyed J
a parcel of land to the vicar and church wardens of Newent Parish to be held by the grantees on trust for the purposes of the Act. The premises were used as a school until its closure in July 1984. In September 1987 the plaintiff trustees, the successors in title to the original grantees, sold the premises for £60,100. The defendant claimed that by virtue of section 2 of the 1841 Act and section 1 of the Reverter of Sites Act 1987 the proceeds of sale was hers. The premises should have reverted to her because: (1) they were freestanding and not part of any manor or state; (2) even if they were part of such manor, section 2 should be construed to give effect to reverter; or (3) section 2 should be so construed having regard to the K
1987 Act.

Held: **The plaintiffs hold the sale proceeds on trust for the defendant. As a matter of impression section 2 of the 1841 Act requires a school site to revert and be rejoined to the grantor's neighbouring land. There was insufficient evidence to show that on April 18 1848 the premises were anything other than freestanding. It is not right that section 1(2) of the 1987 Act alters the provisions of reverter in section 2 of the 1841 Act so that a site which reverts pursuant to the proviso to the section always reverts to the grantor of his successors in title irrespective of whether the site formed part of a manor or other lands. Section 1(2) merely transfers the rights of reverter from the site to the proceeds of** L
sale.

The following cases are referred to in this report.

Cawston's Conveyance, Re [1940] Ch 27
Dennis v *Malcolm* [1934] Ch 244

This was an application by the plaintiffs, Rev Canon Iain William Marchant, Fred Harry Passant and Elaine Joy Frampton, by an originating summons to determine questions arising out of the School Sites Act 1841 and the Reverter of Sites Act 1987 in relation to a claim by the defendant, Susan Margaret Onslow. M

Nigel Gerald (instructed by Treasures & Rivers Wyatt, of Gloucester) appeared for the plaintiffs; Timothy Jennings (instructed by Stoneham Langton & Passmore, of Croydon) represented the defendant.

Giving judgment, **Mr David Neuberger QC** said: This case raises a point of some difficulty as to the proper construction and effect of the School Sites Act 1841 ("the 1841 Act").

On April 18 1848, Elizabeth Foley and Richard Onslow conveyed to the vicar and church wardens of the Parish of Newent a "close piece or parcel of arable land or garden ground containing by admeasurement 1 rood and 7 perches (being the same more or less)

A situate lying and being at Picklenash in the Parish of Newent". The conveyance was expressly stated to be pursuant to the 1841 Act. The land thereby conveyed ("the premises") was to be held by the grantees "for the purposes of the [1841] Act and upon trust to permit the said premises and all buildings thereon affected to be . . . used as and for a school . . . And for the residence of the Schoolmaster and Schoolmistress of the said School and for no other purpose . . .". The premises devolved from the original grantees presumably through various subsequent owners to the plaintiffs. The premises were continuously used as a school until the end of July 1984, when the school was closed down, the council having built a new school nearby.

B On September 3 1987, the plaintiffs sold the premises for conversion to residential use for £60,100. That sum has been deposited in an interest-bearing account in a bank, and I am informed that the sum in that account is now in the region of £80,000. The issue between the parties is whether this money is, by virtue of the provisions of section 2 of the 1841 Act ("section 2") and section 1 of the Reverter of Sites Act 1987 ("the 1987 Act"), the property of the defendant. The issue arises in this way.

Section 2 provides:

> Any person, being seised in fee simple, fee tail, or for life, of and in any manor or lands of freehold, copyhold, or customary tenure, and having the beneficial C interest therein, . . . may grant, convey, or enfranchise by way of gift, sale, or exchange, in fee simple or for a term of years, any quantity not exceeding one acre of such land, as a site for a school for the education of poor persons, or for the residence of the schoolmaster or schoolmistress, or otherwise for the purposes of the education of such poor persons in religious and useful knowledge; provided that no such grant made by any person seised only for life of and in any such manor or lands shall be valid, unless the person next entitled to the same in remainder, in fee simple or fee tail, (if legally competent) shall be a party to and join in such grant: Provided also, that where any portion of waste or commonable land shall be gratuitously conveyed by any lord or lady of a manor for any such purposes as aforesaid, the rights and interests of all persons in the said land shall be barred and divested by such D conveyance; Provided also, that upon the said land so granted as aforesaid, or any part thereof, ceasing to be used for the purposes in this Act mentioned, the same shall thereupon immediately revert to and become a portion of the said estate held in fee simple or otherwise, or of any manor, or land as aforesaid, as fully to all intents and purposes as if this Act had not been passed, anything hereincontained to the contrary notwithstanding.

The long title of the 1987 Act is "An Act to amend the law with respect to the Reverter of Sites that have ceased to be used for particular purposes; and for connected purposes". Section 1 is headed "Right of Reverter replaced by Trust for Sale". The provisions of subsections 1(1) and (2) thereof are as follows:

> E 1.—(1) Where any relevant enactment provides for land to revert to the ownership of any person at any time, being a time when the land ceases . . . to be used for particular purposes, that enactment shall have effect . . . as if it provided (instead of for the reverter) for the land to be vested after that time, on the trust arising under this section, in the persons in whom it was vested immediately before that time.
>
> (2) Subject to the following provisions of this Act, the trust arising under this section in relation to any land is a trust to sell the land and to stand possessed of the net proceeds of sale . . . upon trust for the persons who but for this Act would from time to time be entitled to the ownership of the land by virtue of its reverter.

F The issue between the parties may conveniently be taken from the originating summons as being:

> Whether upon the true construction of section 2 and in the events which have happened, the Plaintiffs hold the net proceeds of sale of the [premises] . . . on trust for
>
> (a) the successors in title to the grantors of the said conveyance [as the defendant contends] or
>
> (b) the successors in title to the grantors' land of which the said site once formed part [as the plaintiff contends].

It is common ground that if the answer to this question is in sense (a), then the moneys are held on trust for the defendant, and it is right to record that evidence has been put before the court which seems to show that the defendant is indeed the successor in title of the grantors, G Elizabeth Foley and Richard Onslow. On the other hand, if the answer is in sense (b), then, pursuant to certain provisions of the 1987 Act to which I have not referred, the plaintiffs, having comprehensively advertised for any claimants to the proceeds of sale of the premises, and no one having come forward other than the defendant, the plaintiffs are entitled to approach the charity commissioners for the approval of a charitable scheme in relation to the moneys.

It is contended on behalf of the defendant that the premises should revert to her essentially on three different grounds. The first is that, as a matter of fact, the premises were in 1848 freestanding and not part of any manor or estate or any other land. The second ground is that, even H if the premises were in 1848 part of any such manor, estate or other land, the proper construction of section 2 none the less leads to the conclusion that the premises should revert to the grantors' successors in title. Third, it is said that, even if that result does not accord with the construction of section 2 as originally enacted, the effect of that section has been changed by the provisions of the 1987 Act.

I propose, first, to set out the relevant facts relied on by the parties and the inferences which I have been invited to draw from them. Then I propose to consider the proper construction and effect of section 2. I will then turn to the proper conclusion as to the effect of section 2 on the facts as I find them. Finally, I will turn to the effect of the 1987 Act. J

In the conveyance, the premises are described as being bounded in part by "a garden belonging to the said Elizabeth Foley". The Newent Parish Tithe Map 1841 and surveyor's report to the tithe commissioners apparently show that Elizabeth Foley was in 1841 the largest owner in the parish possessing at least 735 acres including three fairly large farms and dozens of smaller holdings as well as a number of very small plots, as well as a further 700 acres of woodlands described as "in hand". In the parish records of 1848 she was described as "the Lady of the Manor". Richard Onslow is shown in the 1841 records as owning 16 acres in the parish. The tithe map has a plot (numbered 1009), which seems to approximate to the premises, K and is recorded as containing 1 rood and 12 perches, and is also recorded as being in the occupation of a Hannah Lewis and a Philip Owen. Immediately adjoining this site was another site, numbered 1008, containing 9 perches, and described as a cottage in the occupation of John Guest, and the other side of parcel 1008 was a further site of 15 perches, described as 1007, a cottage in the occupation of Lydia Need. All three sites are recorded as owned by Elizabeth Foley. In the conveyance of April 18 1848, the land is described as "late in the several occupations of John Child and James Child or their undertenants and now in the possession of the . . . Vicar and Church Wardens of Newent". L

On the basis of this evidence, it is suggested on behalf of the plaintiffs that it would be right to conclude either that the premises formed part of the landholding or estate of Elizabeth Foley, possibly in her capacity as lady of the manor, or that it formed part of a landholding, being site 1009 of 1 rood and 12 perches, or that it formed part of a landholding including site 1008 (and possibly 1007) and that accordingly the reverter provisions of section 2 lead to the conclusion that the reverter is not to the successor in title of the grantors, the defendant, but to the person or persons to whom the remainder of the landholding (be that all or some of the property owned by Elizabeth Foley in Newent in 1848) has devolved. It is M suggested that the fact that Richard Onslow was a party to the conveyance of April 18 1848 may well be explicable on the basis that the property held by Elizabeth Foley was enjoyed by her as tenant for life, and that Richard Onslow who, it is common ground, was her nephew, the remainderman, and that therefore he had to join in the conveyance as required by section 2.

On behalf of the defendant, it is submitted that the evidence relied on by the plaintiffs is insufficient, and may well be inadmissible, in that, when considering whether the premises were part of an estate manor or another piece of land one must confine oneself to the four corners of the conveyance. Even if that is not right, it is said that the

A points relied on by the plaintiffs are insufficient to establish that the premises were not, as it were, freestanding in 1848. There is no explanation why Richard Onslow was a party to the conveyance, and the plaintiffs' explanation for this is said to be conjecture. In addition, it is pointed out that there is no acknowledgement by the grantors in the conveyance of the grantees' right to production and delivery of copies of documents (although the conveyance does contain covenants for type and for further assurances), which one would have expected if the premises were part of other land.

Section 2 is not a satisfactory piece of drafting, and its infelicity is particularly apparent in the final proviso, with which I am principally
B concerned in this case. The stipulation that on cesser of school use the site shall "immediately revert to and become a portion of the said estate held in fee simple or otherwise or of any manor or land as aforesaid" is particularly obscure, not least because despite the "said" there is no previous reference to an "estate". It seems to me that the strictures in the law commission report *Property Law Rights of Reverter* (Law Commission's Report No 111) are well founded. At para 29, one finds the following:

> What Parliament actually had in mind is a matter of pure speculation but the phraseology . . . suggests that it was expected that sites provided under the Act . . . would always constitute small parts of landowners' existing estates; and,
> C moreover, that it was not anticipated that those estates would be broken up. If those expectations had been fulfilled it would be a matter of substantial indifference whether the site reverted to the ownership of the *grantor* (or his successors) or was rejoined to the *grantor's neighbouring land*; and the fact that the Act . . . expressed [itself] in the latter manner would not give rise to problems. Unfortunately the conditions necessary for avoiding problems have not been satisfied.

Considering the matter free of authority, it seems to me that section 2 does indeed envisage that the site to be provided would be a small part of a substantial estate, and that if and when the site conveyed pursuant to the 1841 Act ceased to be used for school purposes, the site should revert to being part of the estate. In other words, subject to
D what I have to say about the precise formulation, I would have thought that the site, rather than reverting "to the ownership of the grantor (or his successors)" would be "rejoined to the grantor's neighbouring land".

In the first place, that is how section 2 strikes me as a matter of impression. With such an ill-drafted provision, one is a little hesitant of entering into a more detailed analysis, but I find it particularly difficult to give any meaning to the words "or of any manor or land as aforesaid" in the second proviso to section 2 unless the reverter provisions have this meaning. Indeed, it is hard to see what the point of the reference at the very beginning of the section to "any manor or
E lands of freehold, copyhold, or customary tenure" may be, unless it is to emphasise that when the site reverts pursuant to the second proviso, it reverts to being part of that manor or other lands.

Section 2 may well not have envisaged the possibility of a grantor conveying what I have called a free-standing site pursuant to the provisions of that Act. In those circumstances, still considering the matter free of authority, it seems to me that one would have to conclude that Parliament must have intended that the effect of the second proviso to the section was not that the site would simply fail to revert but that, there being no other land which it could, as it were, rejoin, it should revert to the original grantor or his successors in title.
F Indeed, I consider that one can read the second proviso to section 2 as having that express effect if one construes the reference therein to "the said estate" rather loosely as meaning the estate of the grantor. It does not appear to me, however, that such a result in relation to the reverter of the free-standing site vitiates the conclusion I have reached in relation to a site which is part of other land at the date of grant.

However, I think that it is also necessary to consider in what circumstances a site would be treated for the purposes of section 2 as part of "any manor or lands of freehold, copyhold, or customary tenure". In this connection, the mere fact that the grantor of the site happens to own other land in the vicinity, or even other land which adjoins the site does not appear to me to be in any way conclusive that the site, was, on the date of its conveyance pursuant to section 2, part G
of other land. Equally, the fact that the grantor has no other land which actually adjoins the site would not be conclusive that the site was not part of other land. In each case, it would be a question of fact as to whether or not the site was part of other land. I am conscious that in describing the site as being "part of other land" I am departing from the express language of section 2 itself, but it seems to me to be a permissible way of describing in vernacular terms the sort of concept which the section has in mind particularly where it refers to a "quantity . . . of such land". An obvious case where the site would be treated as part of other land might be where the site was included in a large self-contained parcel of agricultural land in one ownership, H
farmed as a single farm by the owner. If, subsequent to granting such a site, the owner sold to a third party, and the site then ceased to be used as a school, it would seem at least sensible that it should revert to the third party as to the original grantor. On the other hand, where a site is one of two houses in their own grounds with a common boundary it would not, in the absence of other evidence, be right to describe it as a "quantity . . . of such land" on the basis of "such land" being the two houses in their grounds.

It is suggested that one of the problems with such a construction of section 2 is that, in the event of the estate of which the unit formed part being sold off in lots, one would arrive at the site being owned J
beneficially by a large number of people in very small shares. While it would be wrong to reject such an argument wholly, I think that there are two answers to it. The first is that contained in the paragraph which I have cited from the law commission report, namely that it was not a situation envisaged in 1841. The second is that, if the sort of practical approach I have suggested is applied in full, such a problem will not occur very frequently. If 1 acre of a 200-acre estate were the subject of a grant under section 2, and thereafter the estate owner conveyed, say, three 5-acre plots for the purpose of building and selling off a number of houses, it seems to me that, if the site were to revert, a practical application of section 2 would result in the site reverting in full to the owner of the balance of the 184 acres on the basis that these are in K
reality the lands that the estate referred to in the proviso.

It is suggested that the decision and reasoning of the Court of Appeal in *Re Cawston's Conveyance* [1940] Ch 27 is inconsistent with the conclusion that I have reached. I do not accept that. First, it was a case clearly concerned solely with what I have called a "self standing" site: there was no question of the site being part of any manor or other lands at the date of grant. The only question, therefore, arising in relation to section 2 was that to which I have already referred, namely whether in those circumstances the reverter provisions contained in the second proviso to section 2 applied at all. L
Accordingly, any observations in the judgment of Sir Wilfred Greene MR (with whom the other members of the Court of Appeal agreed) on the instant point are strictly *obiter*. Second, I have come to the conclusion that, in any event, this is not a case where there is anything said, even *obiter*, in the Court of Appeal which is inconsistent with the conclusion which I reach in the absence of authority. It is true that there are one or two sentences in the judgment of the Master of the Rolls which, if taken on their own, might be said to indicate a view different from that which I have reached, but, when one reads his judgment as a whole, I find nothing in it which calls into question my conclusion. In particular, I consider that the last paragraph beginning on p38 is, if anything, supportive of my conclusion. Further, as is M
pointed out in para 34 of the law commission's report, Clauson LJ, who was party to the decision in *Cawston*, made observations when sitting at first instance in *Dennis* v *Malcolm* [1934] Ch 244 at p251, which seem to indicate that, at any rate when he was hearing that case, he took the same view of the effect of the reverter provisions in section 2 as I do.

Having reached my conclusion as to the proper construction and effect of the second proviso to section 2, I turn to apply them to the facts of this case. There has been some argument about whether the onus of proof is on the defendant to establish that the premises were free-standing or whether it is on the plaintiffs to establish that it was part of some other lands. I am not at all sure that this is a case which should be decided on the basis of the onus of proof, but, if it is, it seems to me that the defendant is entitled to say that there is nothing in the conveyance which indicates that the premises were part of other lands owned by the grantors, and that that is enough to shift the onus of proof, if it was initially on the defendant, to the plaintiffs.

Whatever the right analysis, I have reached the conclusion that, even bearing in mind the various indications on which the plaintiffs rely, there is insufficient evidence to satisfy me that the premises were, as at April 18 1848, anything other than freestanding. On the basis of my construction of the effect of the proviso to section 2, it is not sufficient that the grantor owned other land in the parish, and even owned land which adjoined the site which is conveyed. The force of the plaintiffs' argument on the point is further weakened when one bears in mind that there were two grantors, and there is no evidence of any land in the parish being owned by both of them. Of course, it is quite possible that the premises, and some or all of the other land in the parish recorded as being owned by Elizabeth Foley, were held by her as tenant for life with Richard Onslow as remainderman. However, while that is one possible explanation, it is purely a matter of speculation. The fact, if it is one, that Elizabeth Foley was lady of the manor does not appear to be of great significance: there is no evidence either way as to whether the premises were part of the manor.

The fact that site 1009 is shown in the 1841 records as being somewhat bigger than it is recorded as being in the conveyance may be significant, but as counsel for the plaintiffs said, tithe records are not always particularly accurate, and I note that in the conveyance the area is described as being "the same more or less".

The fact that the conveyance contains no covenant for production is a mild factor supporting the defendant's contention that the land was freestanding; it may well be that the grantors took the view that, as the conveyance was a voluntary and charitable one, and they did not wish to have to produce their documents of title, they were not prepared to give such a covenant. That, too, is a matter of speculation.

Accordingly, there being nothing in the conveyance to show that the land was part of other lands or of a manor as contemplated by section 2, and there being insufficient evidence in the circumstances relied on by the plaintiffs to satisfy me that the premises were part of other lands or of the manor, I find in favour of the defendant and answer the question raised in the originating summons in sense (a).

I ought to refer briefly to the argument raised on behalf of the defendant to the effect that the provisions of the 1987 Act, and in particular the closing words of section 1(2) thereof, altered the effect of section 2 so that a site which reverts pursuant to the proviso to that section always reverts to the grantor or his successors in title irrespective of whether the site formed part of a manor or of other lands. It is suggested that, in this connection, the provisions of section 1(2) of the 1987 Act are ambiguous, and it is therefore open to me to have regard to what was said by the Lord Chancellor, Lord Hailsham of St Marylebone, in the House of Lords when the 1987 Act was a Bill before Parliament.

I reject that argument. First, I cannot see how section 1, in particular the closing words of subsection (2) thereof, of the 1987 Act can be said to be ambiguous. It seems to me clear that all that it is purporting to do is to transfer the rights of reverter from the site to the proceeds of sale of the site. It virtually states in terms that it is not intending to alter the identity of the person to whom the site reverts, because it refers to the trust created by the 1987 Act as being "for the persons who *but for this Act* would from time to time be entitled to the ownership of the land by virtue of its reverter" (emphasis supplied). It seems to me that those words refer one straight back to the 1841 Act. If confirmation of this view were needed, I think that it is to be found in the long title of the 1987 Act and indeed the title of section 1 thereof. Second, I do not think that what was said by the Lord Chancellor when introducing the Bill is anything like clear enough to assist the defendant's argument on this point, even were I satisfied that there was some sort of ambiguity in section 1 of the 1987 Act which entitled me to look at what the Lord Chancellor had to say.

In these circumstances, I answer the first question in the originating summons in sense (a). The second question, which seeks a declaration that, if the answer to the first question is indeed in sense (a), whether the moneys are held on trust for the defendant, I answer in the affirmative.

County Court

September 24 1993

(Before Judge Roger COOKE sitting at Mayors and City of London County Court)

DEAKINS v HOOKINGS

Estates Gazette April 9 1994
[1994] 14 EG 133

Right of light — Erection of building extension — Whether interference with right of light — Whether appropriate to grant mandatory injunction — Basis of damages if mandatory injunction wrongly ordered

The plaintiff had lived at 2 Ashbourne Terrace, London SW19, part of a terrace of cottages, for 40 years. At its rear the property had a small courtyard. Windows of the property facing the courtyard had acquired easements of light. The defendant owned the adjoining property, 3 Gladstone Road, and in 1986, after the grant of planning permission, had constructed a rear extension which affected light to the windows to the plaintiff's property. In August 1986 the plaintiff commenced proceedings alleging interference with her right to light to the living room and kitchen.

Held: **There had been an actionable interference to right light to the living room, but not to the kitchen of the plaintiff's property. Notwithstanding the failure to obtain an interlocutory injunction, there had been protestations on behalf of the plaintiff at an early stage and a mandatory injunction would be granted, the plaintiff's loss being sufficiently serious to justify the inconvenience and expense of the necessary work to remove the interference to light. Should the case go further, the judge assessed the damages he would have awarded in the alternative to an injunction in the sum of £4,500 being 15% of the benefit of the development to the defendant.**

The following cases are referred to in this report.

Allen v *Greenwood* [1980] Ch 119; [1979] 2 WLR 187; [1979] 1 All ER 819, CA

Bracewell v *Appleby* [1975] Ch 408; [1975] 2 WLR 282; [1975] 1 All ER 993; [1976] EGD 190; (1974) 237 EG 731

Carr-Saunders v *Dick McNeil Associates Ltd* [1986] 1 WLR 922; [1986] 2 All ER 888; (1986) 53 P&CR 14; [1986] 2 EGLR 181; 279 EG 1359

Colls v *Home & Colonial Stores Ltd* [1904] AC 179, HL

Ough v *King* [1967] 1 WLR 1547; [1967] 3 All ER 859; (1967) 19 P&CR 40; [1967] EGD 692; 204 EG 251, CA

Price v *Hilditch* [1930] 1 Ch 500

Pugh v *Howells* (1984) 48 P&CR 298, CA

Shelfer v *City of London Electric Lighting Co Ltd* [1895] 1 ChD 287; [1891-4] All ER Rep 838, CA

Whitwham v *Westminster Brymbo Coal & Coke Co* [1896] 2 Ch 538

Wrotham Park Estate Co Ltd v *Parkside Homes Ltd* [1974] 1 WLR 798; [1974] 2 All ER 321; (1973) 27 P&CR 296

This was the hearing of a claim by the plaintiff, Miss Lillian Ellen Winifred Deakins, for an injunction or alternatively damages for wrongful interference with a right of light by the defendant, Diane Hookings.

Stephen Bickford-Smith (instructed by Evill & Coleman) appeared for the plaintiff; Phillip Reed (instructed by Partridge & Wilson, of Norwich) represented the defendant.

Giving judgment, **Judge Roger Cooke** said: This is a right to light case. The plaintiff, Miss Deakins, has lived at 2 Ashbourne Terrace, London SW19, for nearly 40 years. In 1955 she and her mother and sister bought the property as joint trustees. Mother and sister are now deceased and she is the sole proprietor.

Ashbourne Terrace is, as its name suggests, a residential row of terraced cottages. Its appearance suggests construction well before 1900. Ashbourne Terrace runs into Gladstone Road. Immediately adjacent to no 2 is the corner property, which has frontages on to both Ashbourne Terrace and Gladstone Road and consists of 3 Gladstone Road. At all material times the facades on both sides of the corner consisted of projecting shop fronts. Like many 19th-century properties in London the scheme of construction was that each pair of terraced houses had projecting rearwards from their point of junction a wing (referred to as the back addition), half of which belonged to each of the pair. So the ground plan of each cottage is an L shape having a common wall with the reverse L shape next door. There are windows on the side face of the back addition and, in the case of the present property, there would once have been a window at ground-floor level, on the rear face, but it has now been replaced by a glass door, which leads into a single-storey addition, which has a window to the rear. The part of the rear wall of the property not comprised in the back addition has a window to the rear.

So the general plan is that there is a narrow courtyard to the rear of each property with windows facing into it both from the rear of the property and the back addition. The remaining side of the courtyard in all properties but this are formed by a fence fencing off next doors' courtyard.

This property is different to its neighbour's because the courtyard abutted directly not on to any courtyard at the back of 3 Gladstone Road (for there was none), but with the blank wall of the narrow (and different) back addition of 3 Gladstone Road lying almost immediately behind the fence with only a foot or two's space in between. So there was only about a 6 ft width between the two properties. At the rear of this back addition was a single-storey wooden garage, which projects forward of the building line towards Gladstone Road. The Gladstone Road facade of the back addition had ordinary windows giving on to Gladstone Road and simply matched the remainder of the Gladstone Road facade of no 3. Apart from the garage projection, which was clearly later, the back addition of no 3 seems from surviving photos to have been built as one with no 3.

The back addition of no 3 had two features of particular relevance to this case: first, the garage, as I have said was single-storey only. Second, the roof was a single slope of 30° with the high side at the road facade and the low side facing no 2.

Subject to whatever limitations on light were caused by the original back addition of no 3 as described, the windows of no 2 had clearly stood and enjoyed the use of light for many years. Despite indications to the contrary on the pleadings, it was treated as common ground before me that the windows facing on to the courtyard of no 2 and more especially the ground-floor windows had acquired an easement of light by prescription and were ancient lights.

Of these windows the window facing the courtyard accommodated the kitchen and the window on the rear wall of the main building accommodated the living room.

So the matter stood in 1985. No 3 had had a chequered history. It had latterly been a health food shop, "Nature's Corner". But this had closed and the premises had been bought for use as a betting shop, a licence had been refused and the betting shop people decided to dispose of what by now was a derelict property. Ms Hookings, now Mrs Van den Berghe, bought it. Her intentions were to use the downstairs as a hairdressers' salon and to construct a flat upstairs. Her plans involved the partial or total demolition of the back addition and the garage.

In summary what she wished to do was:

(a) To extend at ground level the entire ground-floor facade to line up with the shop-window projection.

(b) At first-floor level the ground-floor facade in the rear section would be in line with the ground floor, but the part nearer the shop was set back so as to form a small open balcony.

(c) The new building would extend rearwards to the end of where the garage had stood.

(d) On the no 2 side the building line would be to where the rear of the garage had been, ie virtually to the boundary fence a matter of 10 ins to 1 ft for the whole of its length thus coming a foot or two closer to no 2 than had the original back addition.

(e) That wall would be one single vertical wall up to eaves height and the roof would be flat.

Thus, on the proposals the relevant changes as regards no 2 would be:

(i) the back addition was closer to no 2 than its predecessor had been;

(ii) it now went up to eaves height on the near side instead of there being a slope as formerly;

(iii) instead of the facade in the garage area being only just over 6 ins in height as formerly it was now of the same height as the rest of the back addition.

So there was something substantially different and it was quite reasonable for anyone living in no 2 to suppose that if built as designed it would adversely affect the already rather precarious supply of light to the rear ground-floor windows.

On May 21 1985 the defendant applied for planning permission. In July the local authority refused it precisely because of the possible adverse effect on light.

The defendant was concerned; she had the bank loan, the builders and was keen to get on. She discussed it with the council. They said she could appeal to the Secretary of State, but they suggested that she come up with a modified plan, cutting out part of the building over the old garage site so as to allow rather more light from there. So she could start building and if the appeal succeeded easily build in the missing section.

So this is what she did. On July 22 she submitted an amended planning application. The plaintiff got wind of it and on July 30 she put in a letter of objection on grounds of interference with light. The letter was drafted for her by her friend Mr Underwood and was clear in its terms.

The council considered the matter and on October 3 gave conditional planning permission. It seems they must have been satisfied that the modification reasonably met the objection on light. Nobody appears, however, to have considered it in close technical detail. The defendant fairly clearly thought she now had no problem. Nobody had explained to her (and from her evidence I conclude that until much later nobody did) that the meeting of planning objections had no effect as regards any infringement of the plaintiff's private rights.

Meanwhile, the appeal proceeded. Representations were put in in January. A number of objections were received both from the plaintiff and neighbours and on March 13 the appeal was dismissed. So the defendant never attempted to build the full scheme, but she did build the modified one.

There is considerable dispute, if only as to detail, about precisely what contacts were made by the defendant to the plaintiff in the Autumn of 1985 and who said what to whom. After nearly eight years it is hardly surprising that recollections differ and that it is even unclear whether there was one meeting or two. It would I think be unprofitable to reach a firm conclusion even if I could. For present purposes I need only find as follows:

(1) In early October the defendant let the plaintiff have a copy of the plans.

(2) At least once in October there was a meeting at which there were present the plaintiff, her friend Mr Underwood, the defendant and an unnamed male friend of defendant's.

(3) It is not at all clear whether the meeting was really directed to getting the plaintiff to consent to the development or merely to

A allowing access to her property. In hindsight, the plaintiff and Mr Underwood think the former; I am not wholly sure they are right.

(4) What is clear is that the plaintiff declined to co-operate in whatever she was being asked to do. Although the defendant behaved pleasantly and courteously throughout (she is a lady accustomed to dealing with the elderly), her friend showed an unfortunate lack of patience, which may have soured the atmosphere.

(5) Probably then the defendant said things to the plaintiff which on her side may have been intended to show what a good neighbour the hairdressing business would be and which the plaintiff and Mr Underwood construed as inducements.

B At this stage the plaintiff had no solicitor. But she soon consulted one. He did not act all that effectively. There was doubt in his mind whether planning permission had been granted or not, and although he made inquires that doubt remained well into 1986 despite a letter in answer to his inquiries from Merton London Borough Council of November 6 1985 which was perfectly clear. He did however write to the defendant on October 31 1985, asking if planning permission had yet been granted and objecting to the obstruction of the plaintiff's right to light. This produced a letter from the defendant's solicitors asking for the precise objections and appearing to base their confidence in their position (mistakenly) on the grant of planning C permission.

At the end of November scaffolding went up, but the purpose appears to have been to facilitate boarding up and nothing else happened beyond that.

The plaintiff's solicitors instructed a local surveyor at this time, and he recommended the well known expert Mr John Anstey [FRICS], who gave evidence before me. Unfortunately because of the solicitors' confusion regarding planning permission, having fairly quickly instructed Mr Anstey they then told him to hold his hand, which he thus did throughout the most critical period.

Nothing further happened *inter partes* until on February 19 the D defendant put in some footings and her workmen took down the boundary fence, telling the plaintiff as they did so that the wall would be as high as the guttering.

This provoked the plaintiff's solicitors to action. They complained (but without benefit of expert assistance) in some detail (and in response to the November letter) of the obstruction to light. This produced a denial from the defendant's solicitors and a proposal to paint the wall with white paint. On March 10 the plaintiff's solicitors threatened proceedings, but then did nothing about it. Further correspondence and a stalling reply followed in April.

Further correspondence produced a reply from the defendant's E solicitors indicating that the work was going well ahead and offering a personal discussion between the lay parties which was not taken up.

In May Mr Anstey was at last asked to inspect. He also seems to have told the plaintiff something she did not know, that she was eligible for legal aid (the solicitors have appeared to have been reluctant to do a legal aid case). This prompted a change of solicitors and in July her present solicitors were instructed, applied for emergency legal aid, instructed counsel who viewed the premises and advised. By now the building was up and counsel advised as anyone would have done, that the time for an interlocutory injunction was past.

F So no injunction was obtained restraining the building, which was duly completed and occupied. Miss Hookings started her business on the ground floor and moved into the flat. As built the building was the smaller version for which planning permission had been obtained. The flank facing no 2 was faced in white brick and the upper storey of the set back part was faced in, *inter alia*, white plastic so it was hoped as to mitigate the effect on the plaintiff's light.

The plaintiff started her action by writ issued on August 18 1986 (I infer as soon as legal aid had been obtained) and the action has taken incredibly almost seven years to come to trial. On one occasion the defendant faced a debarring order in respect of incomplete discovery and even at trial it was apparent that discovery had not been properly done. The defendant is now heavily in dispute with her former solicitors, who have been discharged. I cannot of course adjudicate G between them but I do not approach this case on the basis that any of this was the defendant's personal fault.

Just to complete the story, the defendant's personal circumstances have changed. She met and married her husband, Mr Van den Berghe in, I think, 1988. He is a Belgian. Their intentions were that she should carry on with her home and business and he would move to the UK and get a job here. But a close family bereavement caused a change of plan. In the result the defendant disposed of her business to Mr O'Sullivan and let the ground floor to him on January 6 1989. She now lives mostly in Germany, but retains the flat as her home in the UK and occasionally stays there.

With that history I come to the issues. The first and critical issue is H whether there has been actionable interference with plaintiff's light.

Interference with an easement sounds in nuisance and not in trespass. For some purposes the two are scarcely distinguishable, but the critical difference is that it is not any interference with an easement of light that gives a cause of action. The interference must be such as to cause a legal nuisance.

The starting point is *Colls* v *Home & Colonial Stores Ltd* [1904] AC 179 where the House of Lords considered the whole basis of liability. The best-known passage, paraphrased in the headnote and really encapsulating the decision of the House of Lords, is in Lord Davey's speech at p204 where he says: J

> . . . the owner or occupier of the dominant tenement is entitled to the uninterrupted access through his ancient windows of a quantity of light, the measure of which is what is required for the ordinary purposes of inhabitancy or business of the tenement according to the ordinary notions of mankind.

It is not limited to any particular use to which a room has been put in the past — *Price* v *Hilditch* [1930] 1 Ch 500, though what is ordinary must depend on the nature of the building and to what it is ordinarily adapted: *Allen* v *Greenwood* [1980] Ch 119 at pp130-131. The building should be left adequately lighted for all ordinary purposes to which they were or might reasonably be expected to be K used in the future: *Carr-Saunders* v *Dick McNeil Associates Ltd* [1986] 1 WLR 922*. Furthermore, what is in question is the interruption of the direct light, special arrangement made by the servient owner to ensure a degree of reflection by, as has been done here, colouring the obstruction white will not do because there is no certainty that this can be maintained in the future: *Price* v *Hilditch* [1930] 1 Ch 500.

Pausing here it is obvious that what the court is looking at is what is the light that is left and whether that light so left is adequate in all the circumstances according to the objective standards of a reasonable man: see *Gale on Easements* 15 ed — now sadly I suppose a work of L authority — at pp268 and 387.

What are the standards of a reasonable man? The well-known expert work of Mr Waldram in the 1920s led for many years to the belief that there was almost a rigid rule that if half the room was adequately lit, ie had not less than one lumen at table level, the light left was sufficient. This was held not to be a rigid rule by the Court of Appeal in *Ough* v *King* [1967] 1 WLR 1547† at pp1552-3. The Master of the Rolls saying that in some cases a higher standard may be reasonably required.

It is important to read some of the detail in *Ough* v *King* because it is quite apparent that what the Court of Appeal had in mind (following something the judge below had said) was that the notions of mankind M as to what was adequate had changed and continued to change, ie people in modern conditions were no longer prepared to put up with the dingy lighting standards of their ancestors. So *Ough* v *King*, I think, really means not so much that one disregards the 50/50 rule, but that it is a bare minimum.

In the present case there is the particular problem that: (i) the living room was badly designed from the start with the window in one end of the long side so that one side of the room was never particularly well

*Editor's note: Also reported at [1986] 2 EGLR 181.
†Editor's note: Also reported at (1967) 204 EG 251.

A lit — the effect of the obstruction is to cause the frontier between the well- and ill-lit areas to advance by a small, but arguably critical distance; and (ii) that in the kitchen the ill-lit area is a circulation area rather than a working area. I will develop this more fully later.

It seems to me that having regard to the authorities I ought to approach the problem on these bases:

(i) In a room that is already ill-lit every bit of light is precious.

(ii) Save in an extreme case it would be difficult to say that once a living room (contrast a store) fell below 50/50 that the light left was adequate.

(iii) In considering whether a room where more than 50% remained **B** well-lit regard should be had to the use to be made of the remainder and how bad, *vis à vis* that use, the remaining light was.

(iv) The test is not merely a statistical one; test (ii) provides a pretty irreductible minimum.

I have had the benefit of expert evidence from two distinguished experts, who disagree about very little and I have also myself inspected the premises. As I have said the experts are largely on common ground and from their evidence and the helpful profiles that they have drawn I derive the following material:

The living room

(a) Before the building the well-lit area was 65 sq ft, equivalent to **C** 51% of the floor area. It is now 52 sq ft and therefore equivalent to 41%.

(b) What has happened is that the ill-lit area which is the end of the room furthest from the window has increased on a bow-shaped profiled strip extending the whole width of the room, broadest at the window wall and narrowest at the rear. The broader part of the strip is somewhat over 1 ft wide.

(c) The part of the room nearest the window and the fireplace where at present the principal sitting arrangements are remains unaffected.

D *The kitchen*

(a) Prior to the building the kitchen was well lit as to 88% (47.5 sq ft).

(b) Now it is to be taken as at best 31 sq ft and at worst 29.75 sq ft representing respectively totals of 57.4% and 55.1% so that the loss is on the one hypothesis (33.3 sq ft) equivalent to 18% and on the other 31 sq ft equivalent to 16.75%. I shall have in a moment to decide between the two versions.

(c) The profile of loss is complex. There is a small peak-shaped area of loss to the right of the window and near to the window wall. But the main area of loss is to the rear of the kitchen near the door that gives on to the living room. The kitchen is L-shaped so if one views it **E** per the plan in the defendant's report it is the right-hand projection of the L that has been almost totally lost. The working part of the kitchen, that is the area nearer the window where at present the sink, the working surface and the cooker are placed is largely unaffected.

It is at this point I should deal with the issue between the experts. It is in practical terms small, but raises a point of principle. There is in addition to the window a small supply of borrowed light throughout the glazed door to the lean-to structure, which itself has a window. Mr Anstey said that if this had simply been a window he would have given full effect to it, but because it was twice removed he did not. Mr Paul Wiseman [RIBA Hon ARAM], for the defendant, gave it full effect. **F** The difference as said is 2.3%, not decisive. But so far as necessary I would prefer Mr Wiseman, simply because on principle I do not think you can ignore any part of the effect of any other source of natural light, given the test that I have to apply, ie "what light is there now".

My own inspection took place at about 3 pm on an early September day when the weather conditions could be described as "cloudy bright", ie there was plenty of light, but it was diffuse and there was no direct sunlight and therefore no hard shadows. As seems to have been the experience of the county court judge in *Ough* v *King*, I too did not find the rooms as dark as I had expected, though I should remember that on a really overcast day things could be appreciately worse and there might well be a difference later in the year when the sun is lower, but I had no evidence of what that difference would be. **G**

There is an obvious contrast in the living room between the window end and the door end. The window end is moderately well lit though not particularly bright, the door end is sombre, but not dark. It was just possible to perform Mr Waldram's celebrated test with a copy of a newspaper at table level. It could be read, but not without some strain.

In the kitchen the sink and working area appeared fairly bright and though the circulation area was sombre it did not appear to be so dark as, in the conditions in which I saw it, to cause real difficulty in finding one's way and avoiding obstacles (the most notable obstacle being the step down from the living room). It is also, however, **H** important to see what the plaintiff herself says about the conditions as, not least, because I saw the property only in one comparatively favourable state, while plaintiff of course lives there all the time. As to the living room she said:

> The amount of light coming through the downstairs windows has been severely diminished by the extension

She then complains about her loss of the open view and says:

> I never had any view from the downstairs window but there was a feeling of openness and there was a lot more light than there is now. There has been a hugh loss of sunlight. The whole room used to be filled with sunlight in **J** summer. Now a very thin strip of light travels along the wall during the day. Since the extension was built I have suffered severe depression.

— which she attributes to loss of light though there is no medical evidence. She adds that she switches her lights on earlier than she did and her electricity bill has gone up in consequence. She confirmed and elaborated on this in the witness box.

I find it difficult to say that even though the expert has stated that the actual area of loss is comparatively small that this loss is merely a small injury. I take care to discount some of what she says (i) part of her complaint is as to the loss of openness, which is not a matter for **K** liability; (ii) I think it more than likely that the depression may relate more to (i) than to loss of light. But having done that discounting I am left with the firm impression that though the loss of light is limited in scope it is none the less of real significance to somebody who is to live in that room.

As far as the kitchen is concerned the principal complaint is as to loss of sunlight. On the evidence and my observations I have little doubt as to the conclusions I should reach.

The living room

The cottage has only two living rooms, the other being the front room facing the street and it is eminently reasonable to treat the back **L** room as the room that most people living there would treat as their principal living room as indeed the plaintiff does. So viewed it was poorly lit to begin with and a drop below 50%, even though it is a small drop (and I shall have to consider the effect of this on relief) really does push it below what is acceptable. But it is not merely a question of statistics. The plaintiff's evidence makes it obvious to me that there is an appreciable drop in the comfortable occupation of the room going beyond the comparatively slight drop that the expert evidence alone would suggest. While the expert evidence alone would be enough to say there was an actionable interference but a slight one, the plaintiff's evidence goes to show that it is real and deleterious in **M** terms of beneficial occupation and goes beyond the merely slight. I conclude that there is an actionable interference with the right to light. The fact that the rear wall of the back addition to the defendant's premises has been advisedly constructed in a white material is as I have explained irrelevant as a matter of principle.

The kitchen

The drop is substantial, but what is left is over 50%. Here I think the use to which the room could be put is critical. If the affected area could be used for anything other than circulation then one would be very much inclined to say that the fact that room is over 50% is in modern conditions not good enough. But realistically, constructed as

A the room is, that area can never be used for anything else and the area of the room that can be used is adequately lit. As I have said the circulation area is not itself so dark as to render the room uncomfortable. Had the extension been constructed as originally envisaged and not in the modified form required by the planners the situation could I think be critically different. I do not disregard the plaintiff's evidence as to the drop in sunlight, but I think it is less critical in this room than in the living room and that all in all there is adequate light. I conclude there is no actionable interference in respect of the kitchen.

Relief

B It is now necessary to consider the question of relief, should it be by mandatory injunction or will damages in lieu of an injunction suffice? I propose, despite the conclusion to which I have come on liability, to make all necessary findings of fact in case the matter should go further.

Mr Anstey, the expert for the plaintiff, says that in order to abate the nuisance in regard to the living room what needs to happen is for a corner of the first floor of the defendant's structure to be shaved off — I will refer to as the "chamfer". In order to abate any nuisance to the kitchen there would need to be a substantial lowering of the roof over the staircase at the rear of the property substituting a sloping roof for C the present flat one and probably involving some lowering of the street frontage at that point. I do not understand Mr Wiseman seriously to disagree with this.

The defendant's valuer flippantly said in his report that all the plaintiff needed to do was to clean her windows. He accepts having sat through the expert evidence that this was unfortunate and it is the sort of comment that ought not to have been given in evidence.

No evidence was given as to what would need to be done to the defendant's premises to achieve these results, but in the course of the view I was invited to inspect and did inspect the interior to the defendant's premises. It is apparent that if the chamfer is carried out a D corner will need to be cut off the defendant's bathroom. This is a shower room with toilet, and cutting off the corner will remove the toilet, which is the only one in the premises. Obviously the bathroom area will need substantial redesign and as it is next to the kitchen I would have little doubt that it would be a tricky and expensive operation in order to obtain the necessary consent. I suspect, however, that it could be done somehow.

The alteration of the roof over the staircase is even to the non-expert an obviously very difficult operation because it would leave insufficient headroom at the staircase head and would make it impossible to make a sufficiently high doorway into the kitchen into E which the staircase leads directly. The redesign would be major and expensive if, as to which I have reservations, it could effectively be done at all. But in view of my conclusion, having once found this fact I can leave it there.

Although a mandatory injunction is always up to a point discretionary the usual rule is that where a plaintiff establishes an actionable interference he is entitled to a mandatory injunction at trial, the law in general setting its face against the compulsory purchase of an aggrieved party's rights, which is what the refusal of an injunction would mean.

Therefore, the plaintiff would be entitled to ask for an injunction to F effect the chamfer unless there was some particular circumstance rendering the grant of an injunction inappropriate. I turn to consider what those circumstances might be:

Failure to obtain an interlocutory injunction

I have set out the history in some detail. Seen from the defendant's point of view there was a lot of huffing and puffing and no action. The plaintiff's solicitors kept on saying there would be an application for injunction and then did nothing effective. Behind the scenes what happened, as the plaintiff fully satisfies me, was that (a) the solicitors initially and not very reasonably allowed themselves to be misled into thinking there was not yet any planning permission, so they delayed fatally instructing Mr Anstey and despite the fact that the plaintiff was an elderly pensioner with no income but her old age pension and no G property but her cottage they showed a marked disinclination (possibly compounded by ignorance) to advise her to obtain legal aid. When she changed her solicitors they did the obvious thing and immediately applied but by then it was too late to obtain an interlocutory injunction. One must be moderate in criticising solicitors who have had no opportunity to explain their conduct but on the evidence I have heard I am not impressed — indeed if one looks at the way both firms handled the matter their clients were not well served. Sensible advice and action at an early stage — possibly involving immediate instructions to surveyors — could have saved all the expense and agony of this action. H

Mr Phillip Reed, for the defendant, submits that the failure to obtain an interlocutory injunction and the accompanying delay in issue of the writ is a critical factor in deciding whether to refuse relief. While the defendant went on in the face of a warning she did not go on in the face of a writ, there was no writ until far too late.

This type of difficulty received attention in *Wrotham Park Estate Co Ltd* v *Parkside Homes Ltd* [1974] 1 WLR 798 at p810 where Brightman J says:

> . . . it is no answer to a claim for a mandatory injunction that the Plaintiffs, having issued proceedings, deliberately held their hand and did not seek the assistance of the court for the purpose of preserving the status quo. On the J other hand, it is, in my view, equally true that a Plaintiff is not entitled "as of course" to have everything pulled down that was built after the issue of the Writ.

What I get from that passage is really this: that a failure to move the court will not of itself disentitle you to final relief, but that if you do not move the court and the building is erected and there is some other ground for refusing a mandatory injunction you will not get the building pulled down, whereas if you had moved *quia timet* the building would not have been erected at all (the discretionary ground not being relevant at the interlocutory stage). Thus here if one were to come to the conclusion that the chamfer came within what I will call K (and elaborate on) as the *Shelfer* exception a mandatory injunction would be refused, but the *Shelfer* exception would have been irrelevant to a negative injunction, which at the interlocutory stage would almost certainly have been made.

One must remember that Brightman J was dealing with the case actually before him and that the relevant date was in fact there the issue of the writ, all the houses having been built after it. The passage is therefore necessarily silent about what the position is if the right is asserted before any building takes place, but no writ is issued until the building is substantially up. For my part, I would have thought on general principles that if it is known that the right is asserted and L nothing done by the plaintiff to make it appear that right has ceased to be asserted then the builder takes the risk of going on the failure to apply interlocutorily will not of itself stop final relief being granted, although as I have said circumstances may be at the trial stage which prevent a mandatory injunction which would not have prevented an interlocutory injunction had one been applied for. I am fortified in this view by *Snell's Equity*, 29th ed, at p658 and cases there cited.

On the particular facts of this case rights were first asserted by letter of October 31 1985. They were asserted again on February 20 1986, March 10 and after a fashion, May 13. True it is that they were M under-particularised and the threats of proceedings were not carried out, but any recipient, especially reviewing the correspondence, would have known of a consistent assertion of interference with a right to light. None of this in any way indicates that the plaintiff has forgone her rights or acquiesced in the tort.

I therefore would not refuse a mandatory injunction on the ground of failure to move. I would add for completeness that though I sympathise with the position in which the plaintiff found herself through no fault of her own I do not think her difficulty with her solicitors are relevant to the question. I think as in most other places in litigation the actions of the solicitors and the actions of the client

A cannot be separated, ie it does not matter that a party's delay is the fault of their solicitor; it is still, as regards the merits of the action, delay.

I come now to what to my mind is the much more serious ground of opposition to a mandatory injunction. The starting point is the well known case of *Shelfer* v *City of London Electric Lighting Co Ltd* [1895] 1 ChD 287. As summarised usefully by *Snell* at p657 what *Shelfer* decided that was a good working rule that an injunction (as I read it whether prohibitory or mandatory) would be refused and damages awarded only where there was: (1) a small injury to the plaintiff; (2) that injury was capable of being estimated in money; (3) B that injury would be adequately compensated by a small sum; and (4) the grant of an injunction would be oppressive.

Shelfer is I think really an authority on the general principles of awarding damages in lieu of an injunction and while it comprehends mandatory injunctions is not the sole ground for refusing one. It is well established that although in general a mandatory injunction follows the event its grant is always discretionary taking into account such factors (by no means exclusive) as to the extent of damage which the plaintiff would suffer, and conduct of the defendant in rushing up a building in face of objections: see generally *Snell* at p659. This last factor is a factor which if one read the judgments in *Shelfer* is a C criterion that provides an exception to what would otherwise be a *Shelfer* result.

I was referred to the decision of the Court of Appeal in *Pugh* v *Howells* (1984) 48 P&CR 298. In that case the county court judge had come to the conclusion that the *Shelfer* criteria applied in a right to light case and awarded damages — refusing a mandatory injunction. The Court of Appeal reversed him. They did not in terms criticise the application of *Shelfer* to a right to light case (see Fox LJ), but it was considered that (i) on the facts it was dubious whether injury could be said to be small and (ii) there was no valuation evidence as to the effect in money so the judge just took an educated guess. (The latter is as will appear not the case here.) But what really caused the Court of D Appeal to allow the appeal was the conduct of the defendants in rushing up the building in the face of the plaintiff's assertions, described as taking a risk that they would get away with it. One is left with distinct flavour that the Court of Appeal thought the defendants' conduct more than a little disgraceful.

I would first consider whether *Shelfer* criteria would apply at all. There is at first sight a degree of intellectual tight-rope walking between saying that an interference with light is a substantial interference with comfort and saying that it is small and within the *Shelfer* criteria. But I observe that although in *Pugh* the Court of Appeal were against the judge on the facts they did not rule out the E principle. In this case the expert evidence viewed alone allows the interference to be precisely measured. Its real significance is that it reduces the room below the acceptable minimum. But what it does is very precisely to take out a comparatively narrow band leaving what is on the design of the room the more useful area still substantially useful on its own. That would suggest the change is small, but ought to consider the plaintiff's own evidence, which I have reviewed at some length earlier in this judgment.

I am left with the firm impression that though the loss of light is limited in scope it is none the less of real significance to somebody who is to live in that room and therefore: (1) it cannot be dismissed as F small; and (2) damages are not really an adequate remedy for that sort of loss of amenity. Had I considered otherwise, I would have found that the loss could be valued, because there was evidence before me (to which I will refer later *de bene esse*) on which damages could properly be assessed and had the injury merely been small then the effect of the injunction on the defendant would have been disproportionate and therefore oppressive. But as I now find, hard on the defendant though the result is, it is justified by the fact that the injury to plaintiff is not merely small.

It follows that not only do I reject the application of the *Shelfer* criteria, but I also conclude that if one balances the plaintiff's loss against the defendant's loss, although the plaintiff's loss is less devastating none the less it is sufficiently significant to justify the G expense and inconvenience of the work which it follows the defendant must be called on to do.

I add for the sake of completeness in case this goes elsewhere that had I found *Shelfer* satisfied I would not have found that this was a case of the *Pugh* v *Howells* type. Although the plaintiff went on complaining and threatening, the defendant did nothing hurried, her solicitors asked for details which they did not get, she was perfectly open about what she did and her personal conduct was impeccable. Sadly she did not receive the advice she might have had which was to have a surveyor look at the effect of what she was going to do. She honestly believed the planners that by modifying her plans had met the H plaintiff's objections. She really and honestly thought it was a matter of planning and not of private rights. Her conduct was quite different from that attributable to the defendants in *Pugh* and should not be stigmatised.

The result thus far is that I would grant a mandatory injunction in the case of the living room so as to compel the defendant to execute the works which I have referred to as the chamfer. She cannot do it overnight; plans will have to be prepared, building regulation consent sought, finance raised and workmen instructed. It is essential that she is given a reasonable time in which to comply. I will hear submissions as to what such time should be. J

Strictly speaking that disposes of the action. But if it goes elsewhere and the Court of Appeal were to consider that there ought to be an award of damages in lieu of an injunction I ought to make findings as to what those damages might be. I have heard both expert evidence, evidence of the parties and submissions. There are two ways of approaching the damages: (i) on a simple basis of loss of market value; or (ii) on what is called the "user principle". It was on (i) that the experts expressed their views. The defendant's expert gave evidence first. It was he who had expressed the flippant view about cleaning windows to which I referred earlier in this judgment. It was not only for that reason that I had doubts about him, but it hardly K helped. He seemed to think there was no real difference in market value at all which as a proposition I do find a little surprising.

Mr Martin Higgins [FRICS], the plaintiff's expert thought there was. As I analyse his evidence I would not have thought the difference was high, but it was there and based on that evidence had I thought it right to approach the case this way I would have said the figure was £1,500.

I turn now to (ii) the so called user principle. This was originally derived from cases of trespass in the 19th century of which a prominent example was *Whitwham* v *Westminster Brymbo Coal & Coke Co* [1896] 2 Ch 538. In *Wrotham Park* it was applied to cases of damages in lieu for breach of restrictive covenant where no actual L damage suffered by the owner at all, but he had lost the right of veto given by his restrictive covenant by the refusal of the mandatory injunction, so Brightman J's approach was to apply *Whitwham* by parity of reasoning and fix damages at what he would have had to pay for a wayleave — the plaintiffs set their sights high at substantial proportion of the developers' profit but the court only gave 5%, compare the *Stokes* v *Cambridge* type of case where the percentages are high.

The principle has also been applied to easements: see *Bracewell* v *Appleby* [1975] Ch 408*. The basis of valuation is I would have thought to take what a willing vendor would have paid a willing M purchaser.

In my judgment, user principle damages would be appropriate to a case like this where what one would be looking at is damages to replace an injunction that might otherwise have been granted. An injunction can be granted in circumstances where no or no conventional common law damage has been suffered at all, see *Wrotham Park* where there was no damage to the dominant tenement. What was important to the plaintiffs and what they had a right to uphold was the covenant simpliciter. It may be here that the damage to the value of the house on the market is slight but the person entitled to

*Editor's note: Also reported at (1974) 237 EG 731.

A the right of light has a right, not itself always easily measured in money terms, to uphold that right specifically. If they are to lose that right they should be compensated for losing it on some basis that represents a fair price for being bought out.

In *Wrotham Park* the method adopted was to look at what profits the developer would have made. This is a familiar concept in compulsory purchase cases, classically also in the development world in "ransom strip" cases. The question is how much of the development value is the person bought out reasonably entitled to demand. Such percentage is really a matter of valuation for the court, but the sort of factor that would bear on it seems to me to be:

(i) is it a ransom situation, ie is the development stuck without it or merely inconvenienced; B

(ii) does the loss of the right inconvenience the seller?

In *Wrotham Park* there was a ransom situation, but no loss. This is not a ransom case, but 5% is too low in this case. I think the right figure is 15% of the benefit from the development. Mr Reed says the only benefit which should be taken into account is that attributable to the flat. It is not right to leave out the salon. This was a single development, based on an enhancement of £30,000, I would assess the damages at £4,500.

TOWN AND COUNTRY PLANNING

A
House of Lords

October 28 1993

(Before Lord Keith of KINKEL, Lord TEMPLEMAN, Lord Jauncey of TULLICHETTLE, Lord BROWNE-WILKINSON and Lord MUSTILL)

BRITISH RAILWAYS BOARD v SECRETARY OF STATE FOR THE ENVIRONMENT AND OTHERS

Estates Gazette January 15 1994

B [1994] 02 EG 107

Town and country planning — Whether planning permission should be refused on ground that a desirable condition appears to have no reasonable prospects of fulfilment

In May 1988 British Rail applied to Hounslow London Borough Council for outline planning permission to develop for housing and ancillary purposes part of a former marshalling yard. The plan annexed to the application covered land belonging to British Rail and an area of land owned by the council over which it was proposed to provide a vehicular access to the housing development. The council
C failed to determine the application within the statutory period and British Rail appealed to the Secretary of State for the Environment. Following the holding of a public local inquiry, the inspector reported on September 13 1989 recommending that the application be granted subject to the completion of an agreement with the council under section 52 of the Town and Country Planning Act 1971 and subject to a condition that the main access road should be completed prior to the commencement of the construction of the residential development. Because the council objected on environmental grounds to the use of their land for the access, no agreement was entered into. By a decision letter dated June 5 1990 the Secretary of State refused planning permission on the basis that he was precluded in law from
D granting the permission subject to conditions relating to the provision of access over the council land which had no reasonable prospect of fulfilment within the five-year life of the permission. British Rail's application under section 288 of the Town and Country Planning Act 1990, to quash the Secretary of State's decision, was dismissed by Mr Gerald Moriarty QC (sitting as a deputy judge of the Queen's Bench Division) and that decision was affirmed by the Court of Appeal.

Held: **The appeal was allowed and the matter remitted to the Secretary of State. A planning application may be made by an applicant in respect of land he does not own. If the owner of that**
E **land objects to the grant of planning permission for unsound reasons, the mere fact that the owner objects and is unwilling that the development should go ahead cannot in itself necessarily lead to a refusal. The function of the planning authority is to decide whether the proposed development is desirable in the public interest. The answer to that question is not to be affected by the consideration that the owner of the land is determined not to allow the development so that permission for it, if granted, would not have reasonable prospects of being implemented. That does not mean that the planning authority, if they decide that the proposed development is in the public interest, is absolutely disentitled from taking into account the improbability of** F
permission for it, if granted, being implemented. There is no absolute rule that the existence of difficulties, even if apparently insuperable, must necessarily lead to refusal of planning permission for a desirable development. A would-be developer may be faced with difficulties of many different kinds, in the way of site assembly or securing the discharge of restrictive covenants. If he considers that it is in his interest to secure planning permission notwithstanding the existence of such difficulties, it is not for the planning authority to refuse it simply on their view of how serious the difficulties are. In the present case the proposed condition relating to the provision of the road over G
land owned by the council, if imposed, would not derogate from the planning permission if granted. If the condition is of a negative character and appropriate in the light of sound planning principles, the fact that it appears to have no reasonable prospects of being implemented does not mean that the grant of planning permission subject to it would be irrational in the *Wednesbury* sense so that it would be unlawful to grant it. The mere fact that a desirable condition appears to have no reasonable prospects of fulfilment does not mean that planning permission must necessarily be refused.

H

The following cases are referred to in this report.

Bradford City Metropolitan Council v *Secretary of State for the Environment* (1986) 53 P&CR 55; [1986] 1 EGLR 199; 278 EG 1473; [1986] JPL 598, CA

Grampian Regional Council v *City of Aberdeen District Council* (1983) 47 P&CR 633; 1984 SLT 197; [1984] JPL 590, HL

Jones v Secretary of State for Wales (1990) 61 P&CR 238; [1990] 3 PLR 102, CA

Newbury District Council v *Secretary of State for the Environment* [1981] AC 578; [1980] 2 WLR 379; [1980] 1 All ER 731; (1980) 78 LGR 306; 40 P&CR 148; [1980] JPL 325, HL

J

This was an appeal by British Railways Board against the decision of the Court of Appeal who on October 6 1992 affirmed the decision of Mr Gerald Moriarty QC (sitting as a deputy judge of the Queen's Bench Division), who on December 20 1991 dismissed British Rail's application to quash the Secretary of State's decision refusing planning permission upon an application made by British Railways Board to the second respondents, Hounslow London Borough Council.

Jeremy Sullivan QC and Nicholas Huskinson (instructed by the solicitor to British Railways Board) appeared for the appellant; John Howell QC and Nathalie Lieven (instructed by the Treasury Solicitor) K
appeared for the first respondent; the second respondents, Hounslow London Borough Council, did not appear and were not represented.

In his speech, LORD KEITH OF KINKEL said: In May 1988 appellants, the British Railways Board ("British Rail"), applied to Hounslow London Borough Council ("Hounslow") for outline planning permission to develop for housing and ancillary purposes part of the former Feltham marshalling yard. The site shown on the plan annexed to the application covered land belonging to British Rail, but also an area of land owned by Hounslow, lying between British

Rail's property and the A314 road. It was proposed that a vehicular access to the housing development should be formed over the land so owned by Hounslow. Hounslow failed to determine the application within the statutory period and so British Rail appealed to the Secretary of State for the Environment ("Secretary of State") against the deemed refusal of it.

The Secretary of State appointed an inspector, Mr David Fenton, to hold a public local inquiry into the appeal and to report to him. Mr Fenton duly held the inquiry in July 1989 and reported on September 13 1989. He recommended that the application providing for access from the A314 be granted, subject to the completion of an agreement with Hounslow under section 52 of the Town and Country Planning Act 1971 and to certain conditions set out in the report. These conditions included:

> 10. The residential development hereby approved shall not exceed 440 dwelling units
>
> . . .
>
> 13. The works to provide the main access road shall be completed to base course level prior to the commencement of the construction of the residential development hereby approved, and shall be fully completed prior to the occupation of buildings.

The Department of the Environment, on November 7 1989, sent copies of Mr Fenton's report to British Rail and to Hounslow with a letter stating that the Secretary or State agreed generally with the inspector's view but that before making a decision he desired confirmation of the completion of a legal agreement under section 52 within 28 days. Thereafter British Rail made strenuous efforts to persuade Hounslow to enter into such an agreement, but without success. Hounslow's objection was based on environmental grounds. The former marshalling yards, disused for 30 years, had become the habitat for various interesting species, and the land belonging to Hounslow over which the proposed access road was to pass, consisted in the Pevensey Road Open Space and former sewage pits, which the inspector found to be areas with their own nature conservation interest and public amenity value. Various correspondence followed until, on March 21 1990, the Department sent to British Rail a letter stating, *inter alia*:

> 2. There are, of course, similar difficulties about your alternative suggestion of imposing a *Grampian* condition or conditions. A House of Lords decision in 1984 was to the effect that a condition prohibiting the carrying out of development or the occupation of buildings until such time as works (or other action) have been carried out by a third party is a valid condition provided its imposition is reasonable in the circumstances of the case. The "reasonableness" of the condition is to be judged on the basis of whether there is at least a reasonable prospect of the works in question (or other action to be taken) being carried out within a reasonable time. The relevant timescale to be considered is that within which the permission must be implemented.
>
> 3. Given the present circumstances there appears to be no reason to believe that the council would be willing to co-operate. That being so, it is felt that there is no reasonable prospect of the works in question being carried out within a reasonable time. A *Grampian* condition or conditions would therefore not seem appropriate.

A *Grampian*-type condition, which takes the name from the decision of this House in *Grampian Regional Council* v *Aberdeen District Council* (1983) 47 P&CR 633, is one which provides that approved development shall not be commenced until some event, in that case the closure of a section of public road, has taken place. The proposed condition 13 in the inspector's report was a *Grampian*-type condition. It will be seen that the Secretary of State took the view that the adjection to the planning permission of this condition would not be valid unless there was a reasonable prospect of its being fulfilled. Hounslow, on March 28 1990, wrote to the Secretary of State stating that the council at a meeting on March 27 had decided not to enter into a section 52 agreement with British Rail. By decision letter dated June 5 1990 the Secretary of State refused planning permission. He did so on the basis that he was precluded in law from granting the permission subject to conditions which appeared to have no reasonable prospect of fulfilment within the five-year life of the permission. The question in the case is whether he was correct in that view of the law.

British Rail applied to the High Court, under section 288 of the Town and Country Planning Act 1990, to quash the Secretary of State's decision. The application was heard by Mr Gerald Moriarty QC, sitting as a deputy judge of the Queen's Bench Division, who on December 20 1991 dismissed it, and his decision was affirmed by the Court of Appeal (Dillon, Kennedy and Hirst LJJ) on October 6 1992. British Rail now appeals to your Lordships' House.

The power to adject conditions to the grant of planning permission is contained in section 29(1) of the Act of 1971, which provides in its latter part that a local planning authority: "(*a*) subject to sections 41 and 42 of this Act, may grant planning permission, either unconditionally or subject to such conditions as they think fit: or (*b*) may refuse planning permission."

The power, though widely expressed, is not unrestricted. In *Newbury District Council* v *Secretary of State for the Environment* [1981] AC 578, Lord Lane said at p627:

> Despite the breadth of the words "subject to such conditions as they think fit," subsequent decisions have shown that to come within the ambit of the Act and therefore to be intra vires and valid a condition must fulfil the following three conditions: (1) it must be imposed for a planning purpose; (2) it must fairly and reasonably relate to the development for which permission is being given; (3) it must be reasonable; that is to say, it must be a condition which a reasonable local authority properly advised might impose . . . The third test is probably derived from *Associated Provincial Picture Houses Ltd* v *Wednesbury Corporation* [1948] 1 KB 223, and ensures that the Minister, if he is asked to review the actions of a local authority, may, even if tests (1) and (2) are quite satisfied, nevertheless allow an appeal on much broader grounds, if the effect of the condition would be to impose an obviously unreasonable burden upon the appellant. Decisions of the local planning authority should not, however, lightly be set aside on this ground.

It is clear from the latter part of this passage that Lord Lane had in mind that where a local planning authority adjected to a planning consent a condition which was unreasonable in the *Wednesbury* sense an appeal by the would-be developer would be allowed to the effect of rendering the consent free of the condition, as happened in *City of Bradford Metropolitan Council* v *Secretary of State for the Environment* (1986) 53 P&CR 55*. However, it appears to have come about that the law is viewed, particularly in the Court of Appeal, as being to the effect that if a condition which on its merits is reasonable and necessary on planning grounds has no reasonable prospects of fulfilment then that condition cannot validly be imposed and planning permission must be refused. That view is apparently thought to be supported by a passage in the speech which I delivered, with the concurrence of my colleagues, in *Grampian Regional Council* v *Aberdeen District Council* (1983) 47 P&CR 633. That was a case where the regional council had applied to the district council and another local planning authority for planning consent to industrial development on a site which spanned the boundary between the authorities. The reporter, to whom the Secretary of State for Scotland had delegated the decision on the application, considered that the development, while desirable on planning grounds, would generate a volume of traffic likely to cause a road safety hazard. This hazard could, however, be eliminated if a section of public road leading to a certain junction were to be closed. The reporter took the view that it would not be competent to grant permission for the development subject to the condition that the road in question should be closed, since it was not in the power of the applicants to close the road at their own hand. An appeal by the applicants to the Court of Session was allowed, and the decision was affirmed by this House, the permission being granted subject to the condition that the development should not commence until the relevant section of road had been closed. It was argued for the district council before this House that there was no

*Editor's note: Also reported at [1986] 1 EGLR 199.

A practical distinction between a condition requiring a result which it was not within the power of the applicant to bring about and one providing that no development should begin until that result had been achieved. In either case the practical effect was to require the applicant to bring about something which was not within his power. In rejecting this argument I said, at pp636-637:

> My Lords, in my opinion there is no substance in the appellants' contentions. In the first place, there is a crucial difference between the positive and the negative type of condition in this context, namely that the latter is enforceable while the former is not. In the second place, the reasonableness of any condition has to be considered in the light of the circumstances of the case. In this case the proposals for development put
> B forward by the first respondents were found by the reporter to be generally desirable in the public interest. The only aspect of them which he regarded as disadvantageous was the traffic problem they would be likely to cause at the Findon House junction. That problem was capable of being solved by the closing of the southern part of Wellington Road, something which had at least reasonable prospects of being achieved under statutory powers to that effect. In the circumstances, it would have been not only not unreasonable but highly appropriate to grant planning permission subject to the condition that the development was not to proceed unless and until the closure had been brought about. In any event, it is impossible to view a condition of that nature as unreasonable and not within the scope of section 26(1) of the Act if regard is had to the provisions of section 198. Subsection (1) provides: "The Secretary
> C of State may by order authorise the stopping up or diversion of any highway if he is satisfied that it is necessary to do so in order to enable development to be carried out in accordance with planning permission granted under Part III of this Act, or to be carried out by a government department."
>
> A situation where planning permission has been granted subject to a condition that the development is not to proceed until a particular highway has been closed is plainly one situation within the contemplation of this enactment, though no doubt there are others. The stopping up of the highway would very obviously be necessary in order to enable the development to be carried out. So it is reasonable to infer that precisely the type of condition which is in issue in this appeal was envisaged by the legislature when enacting section 26(1). As it happens, the first respondents have themselves power,
> D under section 12 of the Roads (Scotland) Act 1970, to promote an order for the closure of Wellington Road. But that is an accident, though it may perhaps make the case an *a fortiori* one. Section 198 is entirely general and is apt to favour strongly the reasonableness of negative conditions relating to the closure of highways in all appropriate cases.

In *Jones* v *Secretary of State for Wales and Ogwr Borough Council* (1990) 61 P&CR 238*, a decision of the Court of Appeal which that court followed in the present case, Purchas LJ regarded this passage in my speech as laying down that a negative condition was necessarily unreasonable in the *Wednesbury* sense unless there was a reasonable prospect of it being fulfilled. The case was similar to the present in
E respect that the applicant for planning permission to build a house required vehicular access to it over land which formed part of a common. The owners of the common refused to consent to this so long as the association of commoners and the board of conservators were in opposition to it. The inspector refused planning permission on the ground that a condition regarding the vehicular access could not be imposed unless there was a reasonable prospect of the condition being satisfied. His decision was quashed by Judge Marder QC, but restored by the Court of Appeal. Purchas LJ said, at p241, after quoting from the *Grampian* case:

> The issue with which we are concerned is whether that part of the speech in
> F particular and the authority in general is to be read as putting a restriction upon the use of *Grampian* conditions to those circumstances in which there is some evidence that there was at least a reasonable prospect of the circumstance or the removal of the obstacle to which the condition is applied coming to pass within a reasonable time, which it seems to be common round would be the time envisaged in the planning statutes as the life of a planning consent.

And later at p245:

> In my reading of the speech of Lord Keith, he clearly had in mind that a condition of this kind would be reasonable and, in certain circumstances, more than reasonable, but only if and in so far as it was established that there was a reasonable prospect of the removal of the obstacle being achieved.

*Editor's note: Also reported at [1990] 3 PLR 102.

Nourse LJ did not base his judgment so directly on the *Grampian* G case. He said, at pp246-247:

> If planning permission had been granted subject to a condition that the development should not commence until such a right had been obtained, it would necessarily have followed that the condition, being one with no reasonable prospect of fulfilment, could not reasonably be imposed. The condition would thus have failed to meet the third of the requirements for a valid condition which were authoritatively stated by the House of Lords in *Newbury District Council* v *Secretary of State for the Environment*. It must be unreasonable to impose a condition which at once shuts out any reasonable prospect of the permission's being implemented.
>
> In order to arrive at this conclusion it is not really necessary to rely on the later decision of the House of Lords in *Grampian Regional Council* v *City of* H *Aberdeen District Council*, where the precise question with which we are confronted was not in point. But there is certainly nothing in the speech of Lord Keith of Kinkel which is inconsistent with the conclusion and, on a reading of what he said, I think it highly likely that he and the others of their Lordships would have agreed with it.

The third member of the court, Glidewell LJ said, at p248:

> For the reasons which have already been explained in my Lords' judgments, the condition, if it is to be imposed, has itself to be a reasonable one and it cannot be reasonable to impose a condition where there is no reasonable prospect of the problem being overcome since this would inevitably stultify J the permission.

In the present case (1992) 65 P&CR 401 Dillon LJ, who delivered the leading judgment in the Court of Appeal said, at p408, in dealing with the argument that it made a difference that the proposed access was within the site covered by the application:

> But to my mind there is no logic in that position at all. The fundamental flaw in the *Grampian* case is imposing a condition in granting a planning permission which, on the information available, the applicant has no prospect of satisfying within the life of the permission. That is so however the plan on the planning application happens to have been drawn, and I cannot see that there is any justifiable logical distinction between the one case and the other. K It is on either view a nonsense to grant a planning permission subject to a condition which the planning authority, or the Secretary of State on an appeal, knows perfectly well there is no reasonable prospect of the applicant being able to satisfy.
>
> It was submitted to us by Mr Sullivan that the decision in *Jones*, in so far as the approach road in question was part of the appeal site in that case, must be treated as having been given *per incuriam*; but I do not take that view. It is a considered decision of this court with which I respectfully agree. It would be a nonsense to grant planning permission, subject to the suggested condition in the present case, in view of the attitude firmly adopted by the Hounslow Council in the correspondence that followed the inspector's report.

Kennedy LJ said, at p408: L

> It is clear from the speech of Lord Keith in *Grampian*, and from the other authorities to which my Lord has referred, that a negative condition can be imposed on land which an applicant seeks to develop, which condition can only be satisfied by, for example, getting control of adjacent land required for access; and *Grampian* is also authority for the proposition that planning permission may be refused if there are no reasonable prospects of the applicant being able to fulfil the condition because, for example, the owner of the adjacent land simply will not sell.

Hirst LJ agreed with both judgments.

As I observed in the *Grampian* case, the question whether or not a certain condition is unreasonable depends on the circumstances of the M case. The circumstances of the *Grampian* case itself were such that the proposed condition was not only reasonable but highly appropriate. One of these circumstances was that the closure of the southern part of Wellington Road had reasonable prospects of being brought about. It is a mistake, however, to turn the decision on its head to the effect of treating it as carrying the necessary implication that a condition is unreasonable and invalid if it does not have such reasonable prospects. The 1971 Act contemplates that an application for planning permission may be made by a person who does not own the land to which it relates. This is made plain by section 27(1)(*b*), which provides that an application for planning permission is not to be

A entertained unless, in the appropriate case, the applicant accompanies it with:

a certificate stating that the applicant has given the requisite notice of the application to all persons (other than the applicant) who, at the beginning of the period of twenty-one days ending with the date of the application relates, and setting out the names of those persons, the addresses at which notice of the application was given to them respectively, and the date of service of each such notice.

Section 29(3) provides that where the application is accompanied by such a certificate the local planning authority:

B (*a*) in determining the application, shall take into account any representations relating, thereto which are made to them, before the end of the period mentioned in subsection (4) of that section, by any person who satisfies them that he is an owner of any land to which the application relates or that he is the tenant of an agricultural holding any part of which is comprised in that land; and

(*b*) shall give notice of their decision to every person who has made representations which they were required to take into account in accordance with the preceding paragraph.

The owner of the land to which the application relates may object to the grant of planning permission for reasons which may or may not be sound on planning grounds. If his reasons are sound on planning
C grounds no doubt the application will be refused. But if they are unsound, the mere fact that the owner objects and is unwilling that the development should go ahead cannot in itself necessarily lead to a refusal. The function of the planning authority is to decide whether the proposed development is desirable in the public interest. The answer to that question is not to be affected by the consideration that the owner of the land is determined not to allow the development so that permission for it, if granted, would not have reasonable prospects of being implemented. That does not mean that the planning authority, if they decide that the proposed development is in the public interest, are absolutely disentitled from taking into account the improbability of
D permission for it, if granted, being implemented. For example, if there were a competition between two alternative sites for a desirable development, difficulties of bringing about implementation on one site which were not present in relation to the other might very properly lead to the refusal of planning permission for the site affected by the difficulties and the grant of it for the other. But there is no absolute rule that the existence of difficulties, even if apparently insuperable, must necessarily lead to refusal of planning permission for a desirable development. A would-be developer may be faced with difficulties of many different kinds, in the way of site assembly or securing the discharge of restrictive covenants. If he considers that it is in his
E interests to secure planning permission notwithstanding the existence of such difficulties, it is not for the planning authority to refuse it simply on their view of how serious the difficulties are.

In the present case British Rail have applied for a planning permission which would cover their own land and also land belonging to Hounslow. Hounslow's land is to be the site of the access road which they seek. The proposed condition relates simply to the stage which construction of the access road must have reached before the construction of the houses starts and before the houses are occupied. The condition, if imposed, would not derogate from the planning permission if granted. So the position is essentially that British Rail have applied for planning permission affecting land not in their
F ownership, a common state of affairs specifically contemplated by the Act. The proposed condition does not relate to land outside the ambit of the permission applied for. Even if it did, the relevant considerations would be the same as those to be applied where an application for planning permission relates to land not in ownership of the applicant. If the condition is of a negative character and appropriate in the light of sound planning principles, the fact that it appears to have no reasonable prospects of being implemented does not mean that the grant of planning permission subject to it would be irrational in the *Wednesbury* sense so that it would be unlawful to grant it. If it is irrational to grant planning permission subject to a condition which has no reasonable prospects of being implemented, then it must be no less irrational to refuse planning permission on the G
ground that a desirable condition has no reasonable prospects of implementation and therefore cannot be imposed. In truth, neither course is irrational. What is appropriate depends on the circumstances and is to be determined in the exercise of the discretion of the planning authority. But the mere fact that a desirable condition appears to have no reasonable prospects of fulfilment does not mean that planning permission must necessarily be refused. Something more is required before that can be the correct result.

My lords, for these reasons I am of the opinion that the Secretary of State misdirected himself in law in considering that imposition of the proposed condition regarding the access road to the development H
would be invalid. I would therefore allow the appeal and remit the matter to the Secretary of State for reconsideration in the light of your lordships' judgment. *Jones* v *Secretary of State for Wales and Ogwr Borough Council* (1990) 61 P&CR 238 was wrongly decided and must be overruled.

LORDS TEMPLEMAN, JAUNCEY OF TULLICHETTLE, BROWNE-WILKINSON and MUSTILL agreed that the appeal should be allowed, and did not add anything.

J

Queen's Bench Division

DIVISIONAL COURT

July 12 1993

(Before Lord Justice MANN and Mr Justice SEDLEY)

WYCOMBE DISTRICT COUNCIL v MICHAEL SHANLY GROUP LTD

K

Estates Gazette January 15 1994
[1994] 02 EG 112

Town and country planning — Advertisement regulations — Display of advertisements in contravention of regulations — Whether owner and occupier of land upon which advertisements displayed without express consent not guilty of offence by proving advertisement displayed without his consent

The respondent owner and occupier of land had given no express consent to the display of advertisements on hoardings erected on its land of MD Ltd. Although the owner had had negotiations with MD L
Ltd, these had not come to fruition and upon the respondent becoming aware of the advertisement hoardings, it removed them as soon as it could. On those findings the justices concluded that the advertisement hoarding was erected without the knowledge or consent of the respondent who, by virtue of section 224(5) of the Town and Country Planning Act 1990, was not guilty of an offence of displaying an advertisement. The appellant local planning authority appealed by way of a case stated contending that it could not be a proper defence to a charge under section 224(3) of the 1990 Act for a defendant owner or occupier of land upon which advertisements have been displayed with its knowledge that they were displayed without its knowledge or M
consent.

***Held*: The appeal was dismissed. The Divisional Court in *Merton London Borough Council* v *Edmonds* The Times July 6 1993 decided that in the context of the defence in section 224(5) of the 1990 Act the expression "knowledge or consent" must be construed disjunctively so that even though the person in question has, or acquires, knowledge of a display, it is still open to him to seek to establish, the burden being on him, that he did not and does not consent to it. That decision would be applied to this appeal. On the facts found by the justices it could not be said that the respondent had not established absence of implicit consent.**

A The following case is referred to in this report.

Merton London Borough Council v *Richard Edmonds* The Times, July 6 1993

This was an appeal by way of a case stated by the applicants, Wycombe District Council, from a decision of the justices for the county of Buckinghamshire acting in and for the petty sessional division of Wycombe, who on October 20 1992 dismissed informations preferred by the appellants against the respondent, Michael Shanly Group Ltd, relating to the display of advertisements upon hoarding at Queen Alexander Road, High Wycombe.

Charles Mynors (instructed by the solicitor to Wycombe District
B Council) appeared on behalf of the appellants; Philip Astor (instructed by Harold Benjamin & Collins, of Harrow) represented the respondent.

Giving judgment, **MANN LJ** said: There is before the court an appeal by way of case stated. The appellants are Wycombe District Council in their capacity as district planning authority. The respondent is Michael Shanly Group Ltd in its capacity as owner and occupier of land at Queen Alexander Road, High Wycombe. The case is stated by the justices for the county of Buckinghamshire acting in and for the petty sessional division of Wycombe in respect of their
C adjudication as a magistrates' court sitting at High Wycombe on October 20 1992. On that day, the justices had before them three informations which had been preferred by the appellant authority against the respondent company. Each information related to the display of advertisements upon a hoarding located on the land owned and occupied by the respondent company.

The informations related to three different dates. That is to say, March 26, April 8 and June 2 1992. The display of advertisements, such as were these, was subject to planning control. It is an offence to display an advertisement in contravention of regulations made under section 220 of the Town and Country Planning Act 1990. The offence
D is constituted by section 224(3) of that Act.

The regulations currently in force are the Town and Country Planning (Control of Advertisements) Regulations 1992 (SI 1992 No 666). Those came into force on April 6 1992, replacing regulations of the same name made in 1989, but which were in very similar terms. Strictly, the offence alleged on March 26 was a contravention of the earlier regulations, but nothing turns upon that.

By virtue of regulation 3, paras 2 and 5, either deemed or expressed consent is required for the display of an advertisement unless it comes within one or more of the classes set forth in the Schedule to the regulations. None of those classes is relevant
E in the present case. Nor are the provisions in relation to deemed consent.

The advertisements displayed on March 26, April 8 and June 2 had no express consent. Their display was, accordingly, a display in contravention of the regulation.

Section 224(4) of the 1990 Act provides that:

> Without prejudice to the generality of subsection (3), a person shall be deemed to display an advertisement for the purposes of that subsection if —
>
> (a) he is the owner or occupier of the land on which the advertisement is displayed; or
>
> F (b) the advertisement gives publicity to his goods, trade, business or other concerns.

Section 224(5) provides that:

> A person shall not be guilty of an offence under subsection (3) by reason only —
>
> (a) of his being the owner or occupier of the land on which an advertisement is displayed; or
>
> (b) of his goods, trade, business or other concerns being given publicity by the advertisement,
>
> if he proves that it was displayed without his knowledge or consent.

The phrase "knowledge or consent" is the phrase which has agitated argument in this appeal.

Upon the facts as found by the justices, the respondent knew of the G display of advertisements on March 27 when the display was bought to its attention by the appellant's enforcement officer.

The justices found that the respondent had given no express consent to the display of advertisments in that negotiations with an advertising company named Motivation Display Ltd had not come to fruition. Despite the absence of fruition, Motivation Display Ltd erected a hoarding and posted the advertisments. They have been separately prosecuted in respect of those activities.

The justices found that the respondent company, "did everything in their power to have Motivation Display Ltd, the owners of the sign to remove it" and also, that "The respondent took the hoarding down as H soon as they could". In the light of those findings, the justices concluded that:

> The advertising hoarding was erected without the knowledge or consent of the respondent and that the respondent took all proper steps, in the circumstances, to remove the advertisements once their attention had been drawn to them, and accordingly dismissed the informations.

The case was stated at the request of the planning authority and the question posed by the justices was whether:

> In view of the production of apparently uncontested evidence:
>
> (a) as to status of the defendant or owner or occupier of the relevant land; J
>
> (b) as to the display on the land of advertisements in contravention of the Town and Country (Control of Advertisements) Regulations 1992, and
>
> (c) as to the knowledge of the defendant as to the existence of that display, it was a proper defence to charge under section 224(3) of the Town and Country Planning Act for the defendant to argue that those advertisements were displayed without its knowledge or consent, and
>
> whether there was any or any sufficient evidence adduced to support any other defence justifying the acquittal of the defendant.

Mr Charles Mynors, on behalf of the local planning authority, had two submissions. First, that the defence was not established unless the defendant could establish that he did not know and did not consent. The K second argument was that if, despite knowledge, lack of consent was a defence then in this case there was consent.

The first argument can be attractively deployed and is plainly a tenable one. However, since the case was stated, this court in *Merton London Borough Council* v *Richard Edmonds*, reported in The Times, July 6 1993, has decided the matter. The judgment was given on June 23 1993. We have the advantage of an unapproved transcript of the leading judgment which was given by Buckley J. In it he refers to the argument on which Mr Mynors sought to rely. He said:

> As a pure matter of language, I confess my first thought was, that the L proper construction was conjunctive, but one must look at the overall purpose of the Act, one must consider whether that would be manifest and see how it would work. Various examples were canvassed in argument, but it seems to me that someone to become in these circumstances, guilty of a criminal offence simply on acquiring knowledge of a particular state of affairs for which they were not responsible would be quite contrary to the fundamental principles of our criminal law.

A little later he says:

> My conclusion is that in the context of this particular Act, "knowledge or consent" as it appears in section 224(5) must be construed disjunctively so that even though the person in question has, or acquires, knowledge of a display, it M is still open to him to seek to establish, the burden being on him, that he did not and does not consent to it.

Kennedy LJ is expressed as agreeing. This court, while not strictly bound by its own previous decisions, will not depart from them unless they are plainly wrong.

I find it impossible to say that that decision is plainly wrong. It effectively disposes of Mr Mynors' first argument. I turn to his second argument which is that in this case the respondent had not established absence of consent. In regard to consent, it seems to me that consent may be either expressed or implicit from circumstances.

There was, here, no question of express consent because the

A negotiations with Motivation Display Ltd did not come to fruition. Mr Mynors suggested that there was implicit consent which the respondent had failed to disprove, the onus being upon it by virtue of section 101 of the Magistrates' Court Act 1980. However, it seems to me that an impossible obstacle to this submission is presented by the justices' findings of fact. In particular the finding that the hoarding was taken down as soon as the respondent could.

It follows that, in my judgment, the justices were entirely right to conclude that the defence was available on the basis that the respondent had successfully established the fact, albeit, it had knowledge, that it did not consent either expressly or impliedly to the
B display of the advertisement. I would accordingly answer the question posed by the justices "Yes" and dismiss this appeal.

Agreeing, **SEDLEY J** said: I would add this. Since, as Mann LJ has said, the issue is certainly arguable either way, the recent decision of this court in *Edmonds* v *Merton London Borough Council* concludes the present appeal.

That decision, as appears from the draft judgment which we have seen, is founded on the holding that an interpretation of section 224(5) which treats knowledge alone as sufficient to establish guilt runs contrary to the fundamental principles of criminal law. From this flows the argument that the subsection requires that there be both
C knowledge and consent before an offence is committed. Therefore an absence of either offence affords an offence. D

The contrary argument, which, in my view, is a respectable one, is that once a person is deemed by section 224(3) to have displayed an advertisement without planning consent by virtue of his being the owner or occupier of the land for the special defence under section 224(5) the defendant must then show that the display occurred on the date charged without his knowledge and without his consent.

On this view, neither absence of knowledge alone or absence or consent alone would be a defence. Both must be proved by the defendant.

The objection that this would mean that a person loses the defence the moment he learns of the display, and that it is wrong that a person
should be found to be a criminal without time to act, would on that E
view be met by the ample power which the justices have in terms of disposal as to costs to reflect the justices' view of an overhasty prosecution. However, the decision of this court to the contrary means that this appeal must fail.

In any event, even if the reverse were the case, the finding of the justices that the respondent took the hoarding down as soon as it could even if it was generous on the evidence would be practically bound to lead to an absolute discharge on the particular facts. For those reasons, I respectfully agree with Mann LJ's conclusions.

Appeal dismissed. F

LANDS TRIBUNAL

COMPENSATION

A DONOVAN v DWR CYMRU — WELSH WATER PLC AND ANOTHER

(REF/42/1992)

March 4 1993

(Judge MARDER QC, the President)

Estates Gazette February 5 1994

[1994] 05 EG 163

B ***Compensation — Water Act 1989 — Sewer laid by contractors in land outside the notified route by private agreement — Whether Lands Tribunal had jurisdiction to hear claim — Whether contractors acted as agents for water company***

On October 3 1989 the first respondents served a notice under section 153 of the Water Act 1989 of their intention to lay a sewer through the claimant's land at Lime Kiln Farm, Rogerstone, Newport, Gwent. The second respondents were employed as contractors to lay the sewer and, because of an existing pipe, in laying the new sewer they uncovered or disturbed 5,117 m of land adjoining the strip depicted on the plan which accompanied the original notice of intention to enter. On the first respondents denying liability for paying
C compensation in respect of the area of additional land disturbed by the second respondents, the president ordered two questions to be heard as preliminary points of law: (1) whether the Lands Tribunal has jurisdiction to hear a claim relating to the uncovered land outside the original easement strip as specified in the notice of entry; and (2) if yes, whether the second respondents acted as agents for the first respondents in forming a private agreement with the claimant to strip soil from the uncovered land and/or to deposit materials on it. Entry under the private agreement was said to be unauthorised and not attributable to the exercise of statutory powers by the first respondents.

***Decision*: The tribunal does have jurisdiction to determine the**
D **claimant's claim. The second respondents entered the claimant's land as agents for the first respondents and in no other capacity, and that *vis-à-vis* the claimant whatever work was done on his land was done by the first respondents by their agents and entitled him to compensation. Any dispute as to the authority of the second respondents was a matter between the first and second respondents and did not affect the first respondents' liability.**

No cases were referred to in this report.

This was the hearing of preliminary points of law in a claim by the
E claimant, Edward John Donovan, against the respondents, Dwr Cymru —Welsh Water plc and Alfred McAlpine Homes Midlands Ltd, brought under the Water Act 1989.

Malcolm Bishop (instructed by Hornby Baker Jones & Wood, of Newport) appeared for the claimant; Jonathan Walters (instructed by the Chief Solicitor to Welsh Water plc) represented the respondents.

Giving his decision on a preliminary point of law, **JUDGE MARDER QC**, the President, said: This reference stems from a claim by Mr Edward John Donovan for compensation arising from the laying of a sewer across his land at Lime Kiln Farm, Rogerstone, Newport, Gwent. The claimant lodged a reference to the tribunal in January F
1992.

The facts of the matter do not emerge clearly from a document lodged with the tribunal in September 1992, apparently with the agreement of the parties, and purporting to set out those facts agreed and those in issue between the parties. However, it appears not to be in dispute that on October 3 1989 the first respondents, Welsh Water plc, served notice pursuant to section 153 of the Water Act 1989 of their intention to lay a sewer or drain across the claimant's land along a line depicted on a plan attached to the notice; Welsh Water employed the second respondents, McAlpine, as contractors to lay the pipe on their behalf; McAlpine entered on the land in about G
mid-August 1990; and in the course of carrying out the work McAlpine "uncovered" or "disturbed" land belonging to the claimant adjoining the strip depicted on the plan, but not included within that strip; the reason for that additional "disturbance" appears to have been an operational difficulty caused by the discovery in the land of another water pipe in the course of laying the new pipe. The area of land thus "uncovered" or "disturbed" was 5,117m.

The parties apparently agreed most items of the claimant's claim, but could not agree as to the liability of the respondents in respect of this extra area of "uncovered" land. The first respondents repudiated any liability, apparently arguing that this extra work was unauthorised. H

Subsequently, at the request of the first respondents, the former president set down for hearing as preliminary points of law the following questions:

(1) whether the Lands Tribunal has jurisdiction to hear a claim relating to the uncovered land outside the original easement strip as specified in the Notice of Entry;

(2) if yes, whether the second respondents acted as agents for the first respondents in forming a private agreement with Mr Donovan to strip soil from the uncovered land and/or to deposit materials on it.

I held a hearing at Cardiff on February 9 1993 for the purpose of J
determining these preliminary points of law. At the hearing the claimant was represented by Mr Malcolm Bishop. Both respondents were represented by Mr Jonathan Walters. Mr Bishop indicated at the outset that counsel had agreed that the second question did not call for determination at this stage. He then outlined the relevant statutory provisions.

For the purposes of the Water Act 1989 it is clear that the first respondents are statutory undertakers, exercising powers under section 153(1) and Schedule 19. The relevant provisions of Schedule 19 are paras 4 and 6. Para 4(1) confers power:

to lay a relevant pipe (whether above or below the surface) in any land which K
is not in . . . a street . . .

Para 6 contains the provisions for compensating the landowner on whose land the power to lay pipes has been exercised. So far as material, that paragraph provides:

6(2) Where the person entitled to an interest in any relevant land sustains loss or damage which —

(a) is attributable to the exercise by . . . any water undertaker . . . of any power conferred by virtue of paragraph 4 or 5 above;

(b) does not consist in depreciation of the value of that interest; and

A (c) is loss or damage for which he would have been entitled to compensation by way of compensation for disturbance, if his interest in that land had been compulsorily acquired . . .

he shall be entitled to compensation from . . . the undertaker in respect of that loss or damage . . .

(3) where any damage to, or injurious affection of, any land which is not relevant land is attributable to the exercise by . . . any water undertaker . . . of any power conferred by virtue of paragraph 4 or 5 above, the . . . undertaker shall pay compensation in respect of that damage for injurious affection to every person entitled to an interest in that land.

(4) Any question of disputed compensation under this paragraph, shall be referred to and determined by the Lands Tribunal . . .

B (9) In this paragraph "relevant land", in relation to any exercise of a power conferred by virtue of paragraph 4 or 5 above, means the land where the power is exercised or land held with that land.

Mr Bishop suggested that the first respondents' objection to the jurisdiction rested on the proposition that the "uncovered land" which had not been indicated on the plan accompanying the notice of intention was not "relevant land" within the meaning of Schedule 19.

At that point, Mr Walters rose to say that that was not the position of the respondents. It was accepted that the land on which work had been done was "relevant land", but the submission was that the second respondents, McAlpine, had made an agreement with the claimant C which concerned and affected land that had not been included or referred to on the plan attached to the notice of entry and therefore the work that had been carried out on that land was not authorised and was not attributable to the exercise of statutory powers by Welsh Water.

In my view, this submission does not raise a preliminary point of law going to the jurisdiction of the Lands Tribunal. If it discloses any point of law at all, then it is a point that can be determined only on the basis of evidence to be given at a substantive hearing.

I made plain at the hearing my view that, on the facts disclosed and apparently agreed, the second respondents had entered the claimant's land only as agents for the first respondents and in no other capacity D and that *vis-à-vis* the claimant whatever work was done on his land was done by the first respondents by their agents and entitled him to be compensated. If there was a dispute as to the authority of McAlpine, that appeared to me to be a dispute between Welsh Water and McAlpine, and could not affect Welsh Water's liability to compensate the claimant in respect of work carried out by Welsh Water's agents upon his land. The only possible answer to question (1) raised in the preliminary point of law had to be "Yes", the Lands Tribunal does have jurisdiction to determine the claimant's claim.

The respondents will pay the claimant's costs of and incidental to the hearing of the preliminary point of law, such costs if not agreed to E be taxed by the registrar of the Lands Tribunal on the High Court scale.

TOYE v KENSINGTON AND CHELSEA ROYAL LONDON BOROUGH COUNCIL

(REF/84/1991)

June 9 1993

F (J C Hill TD FRICS)

Estates Gazette March 5 1994

[1994] 09 EG 186

Compensation — Refusal of planning permission for enlargement of dwelling-house — Determination of value after proposed works — Costs where award above sealed offers of both parties

The claimant was the owner of 90 Palace Gardens Terrace, Kensington, London W8, a mid-terrace house on lower-ground, ground and two upper floors under a slate roof, standing in an area which was subject of a direction under article 4 of the General Development Order 1977. The claimant's appeal to the Secretary of State for the Environment, against the decision of the compensating authority to refuse planning permission on February 29 1988 for the G extension of the subject property within 10% of its volume by the addition of a new mansard roof, was dismissed on June 29 1988. In support of the claimant's claim for compensation under section 169 of the Town and Country Planning Act 1977 (*Note:*— consolidated into section 114 of the Town and Country Planning Act 1990, but now repealed in respect of applications made on or after November 16 1990 by section 31(2) of the Planning and Compensation Act 1991), it was contended that the value of the property on the agreed valuation date of June 29 1988 was £1.1m on the assumption that the proposed development had been carried out. Having regard to the agreed costs of the proposed works of £92,000 and the agreed value of the property H as it existed of £645,000, the compensation claim was £363,000. The compensating authority's valuer, who advanced a different scheme for improvement works, supported a compensation sum of £91,998 after analysing sales on a value per square foot and having regard to building society indices of house values.

Decision: **The compensation was £263,000. The use of building society indices by their general nature makes them unsuitable to value individual properties. In relation to floor areas, houses were valued directly experience rather than by a mathematical approach. The practical approach of the claimant's valuers was to be preferred to the theoretical approach of the compensating J authority's valuer. The value of the subject property had the proposed works been carried out would have been £1m from which had to be deducted the agreed costs of £92,000 and its agreed value as existing of £645,000. Because the award was above the sealed offers of both parties, section 4(3) of the Land Compensation Act 1961 applied and the compensating authority was ordered to bear their own costs and pay the costs of the claimant so far as they were incurred after his sealed offer was made.**

The following case is referred to in this report. K

Segama NV v *Penny Le Roy Ltd* [1984] EGD 74; (1984) 269 EG 322

This was the determination of a claim for compensation under section 169 of the Town and Country Planning Act 1971 by Mr B E Toye following the refusal of planning permission against the compensating authority Kensington and Chelsea Royal London Borough Council.

Nicholas Dowding (instructed by Ashurst Morris Crisp) appeared for the claimant; Timothy Corner (instructed by the solicitor to Kensington and Chelsea Royal London Borough Council) represented the compensating authority.

L Giving his decision, MR J C HILL said: This is a reference to determine the amount of compensation payable to the claimant as a consequence of the refusal of planning permission for the extension of the subject property by the addition of a new mansard roof containing dormer windows. Although the proposed extension was within 10% of the existing cubic content of the property, the latter was situated in an area which was the subject of a direction under article 4 of the General Development Order 1977 (SI 1977, No 289). Planning permission was refused on February 29 1988 and the claimant's appeal against the refusal was dismissed on June 29 1988 by the inspector appointed by the Secretary of State for the Environment. M Compensation therefore falls to be assessed in accordance with section 169 of the Town and Country Act 1971*. The valuation date is June 29 1988.

Mr Nicholas Dowding, of counsel, appeared for the claimant and called Mr Tony Johnson BSc FRICS and Mr Jonathan Hunt. Mr Timothy Corner, of counsel, appeared for the compensating authority and called Mr Richard Carroll MSc BSc FRICS.

---

*Editor's note: Consolidated into section 114 of the Town and Country Planning Act 1990, but now repealed in respect of applications made after November 16 1990, by section 31(2) of the Planning and Compensation Act 1991.

A The following matters were agreed either prior to or during the hearing:

1. The subject property was built in the second half of the last century and is a mid-terraced house of brick construction faced with stucco on the front elevation and having a pitched-slate roof. The property comprises lower ground, ground and two upper floors.

2. The compensation payable is to be the difference between the value of the subject property if consent for the proposed extension had been granted and work carried out (valuation A), less the estimated cost of carrying out the development and the value
B following the refusal. The post-refusal value is agreed to be the greater of:

(a) the value of the property as existing (valuation B1), or

(b) the value of the property if alterations were made which would not require planning permission (valuation B2) and after making allowance for the cost involved in such alterations.

3. Initially the cost of both the planning application scheme and the non-application scheme improvements was agreed to be £65,000 exclusive of fees, VAT, finance charges, accommodation costs and contingencies. Subsequently, however, the cost was agreed to be £92,000.

C 4. Floor areas of the subject property:

(a) existing: gross internal area 2,794 sq ft
effective floor area 2,133 sq ft

(b) additional floor areas:

(i) planning application scheme:
effective floor area 468 sq ft

(ii) non-application scheme:
effective floor area 216 sq ft

5. The value of the property as existing: £645,000.

6. Details of comparable properties:

(a) sales of four-storey houses:

D (i) 55 Palace Gardens Terrace: sold for £665,000 on April 29 1988

(ii) 86 Palace Gardens Terrace: sold for £640,000 on July 14 1988

(iii) 86 Palace Gardens Terrace: sold for £950,000 in December 1989. (Following improvements to provide usable accommodation in the attic area.)

(iv) 63 Palace Gardens Terrace: sold for £515,000 on August 22 1988

(v) 84 Palace Gardens Terrace: sold for £665,000 on August 11 1988

(b) Sales of five-storey houses:

E (i) 98 Palace Gardens Terrace: sold for £1,185,000 on January 15 1988

(ii) 98 Palace Gardens Terrace: sold for £1,475,000 on July 11 1988

(iii) 6 Palace Gardens Terrace: sold for £975,000 on August 18 1989

Case for the claimant

Mr Johnson said that, having considered transactions involving other properties in the area and the report by Mr Hunt, he was of the opinion that there was a marked differential between the values of five-storey and four-storey houses. The accommodation provided was
F a key factor and although the subject property after conversion would not be a full five-storey house he had taken this into account. In his view the value of the subject property in June 1988 and on the assumption that the proposed development had been carried out would have been £1,100,000. In arriving at the latter figure he had had particular regard to the sale of 98 Palace Gardens Terrace in January 1988 for £1,185,000. This property had been sold again in July 1988 for £1,475,000, which was in excess of the asking price and was an indication of competition in a buoyant market. Deducting the agreed costs he arrived at a figure of £1,008,000 as the value of the property if planning permission had not been refused (valuation A).

The value of the property as it existed in June 1988 was agreed at £645,000 (valuation B1). Taking account of potential for creating bedrooms in the roof space and rearranging the accommodation on the lower floors, he put the value of the property, if improved, at £750,000; deducting the agreed costs produced a value of £658,000 (valuation B2). He regarded the latter figure as being largely hypothetical and since it was higher than the value as existing he thought that in reality the value of the property at the valuation date was £645,000. He regarded the sale of 86 Palace Gardens Terrace in July 1988 for £640,000 as providing support for the agreed £645,000 "as existing" value of the subject property. G
No 86 Palace Gardens Terrace was sold again in December 1989 following alterations to provide bedroom accommodation in the attic and he agreed in cross-examination that no 86, after conversion, would be very similar to no 90 if the attic of the latter had also been converted to provide bedroom accommodation (valuation B2). H

In his view, the compensation payable was the difference between £1,008,000 and £645,000, ie £363,000.

He disagreed with Mr Carroll's floor area comparison approach. Throughout his (Mr Johnson's) career, he had never known houses to be valued for the purpose of sales in the open market on an area basis
except in the limited case of very high-value penthouses. He was also critical of the use by Mr Carroll of indices. Indices normally related to the sales of average houses, not £1,000,000 houses in Notting Hill Gate which were not affected by the same mortgage considerations, the value of the higher-value properties having been maintained after the market for smaller houses had crashed. J

Mr Hunt said that until August 1988 the demand for family houses in Kensington and Chelsea was very active and although the abolition of dual mortgage tax relief in August 1988 caused a collapse in the residential market generally, the effect on large family houses was not fully felt until 18 months later. If the subject property had been
marketed in the summer of 1988 in fair condition it was his opinion that it would have achieved a price of about £650,000, which would have included any potential for alterations. In his view, the addition of a mansard-roof floor to the subject property would have created an infinitely more saleable family house than the existing four-storey house. A mansard floor would enable the top-two floors to be used as bedrooms, with a through lounge on the first floor and dining room and kitchen on the ground floor. It would change the character of the house and a decent-sized mansard floor would effectively make it into a five-storey house; this would not be the case if rooms were created within the existing roofspace. With the addition of the proposed K
mansard floor the market value of the subject property in summer 1988, on the assumption that it was finished to interior design standard ie in outstanding decorative order internally and externally, was £1,200,000. If in reasonable condition only he put the value at £1,100,000. He was of the opinion that at the relevant date five-storey houses in the immediate area had a value of between £1,400,000 and £1,600,000. He had made what he considered to be an appropriate discount to arrive at the value of the subject property as a four-storey house with a mansard floor. He had never used floor areas as a basis for the valuation of houses although the area was sometime quoted in the case of modern houses. Never during his career had he used indices for valuation purposes; the use of indices was not a recognised L
method of valuation. M

Case for the compensating authority

Mr Carroll said that he had analysed the sale of 6 Palace Gardens Terrace and the sale in January 1988 of no 98 (both five-storey houses) on an effective floor area (EFA) basis. The average price per square foot was £393; if this figure was applied to the EFA of the subject property the resulting value was £1,022,193. The five-storey houses in Palace Gardens Terrace were however of superior construction and on a grander scale than the four-storey houses like the subject property and he thought that a more correct approach would be to use the analysis of the sale in December 1989 of no 86 following the improvements made subsequent to the sale of the same property in July 1988. The Nationwide Anglia Building Society House Price Indices suggested that property values had increased by 13.2% between the third quarter of 1980 and the fourth quarter of 1989. The rate per square foot derived from the December 1989 sale of no 86 was £413.22; adjusting this figure back to June 1988 by using the index produces a rate per square foot of £365.03. Applying the latter figure to the area of the subject property on the assumption that the proposed mansard scheme had been carried out resulted in a figure of £949,443. We adopted the latter figure as the value of the planning application scheme (valuation A).

He said that the existing layout of the subject property was less than satisfactory with the four bedrooms split between the lower ground and second floors and with the kitchen in the front of the ground floor providing an unimposing and unattractive entrance vista. The property was capable of considerable improvement without requiring planning permission by converting the attic into bedrooms and rearranging the accommodation on the lower floors. Again by applying the index-adjusted figure of £365.03 per sq ft derived from the later sale of no 86 he arrived at a figure of £857,445, which he adopted as the value of the subject property if a limited scheme of improvement had been carried out (valuation B2). From both his valuation A and valuation B2 he deducted £96,000 being the agreed cost of works plus the addition of £4,000, which he regarded as the cost likely to be incurred by the owner on other accommodation while the works were being carried out. The difference between the resulting figures (£853,443 and £761,445) ie £91,998 was, in his view, the compensation payable. He had prepared calculations which showed that if the figure of 13.2% from the Nationwide index was different from the actual increase in values between June 1988 and December 1989 any increase in the percentage would reduce the compensation and vice versa.

The respective valuers' figures can be summarised as follows:

Claimant

		£
Planning application scheme (valuation A)		1,100,000
Less costs		92,000
		1,008,000
Value as existing (valuation B1)		645,000
Compensation		363,000

Compensating authority

	£	£
Planning application scheme (valuation A)		949,443
Less costs		96,000
		853,443
Non-planning application scheme (valuation B2)	857,445	
Less costs	96,000	
		761,445
Compensation		91,998

Reference was made during the hearing to the following case: *Segama NV* v *Penny Le Roy Ltd* (1984) 269 EG 322.

Subsequent to the hearing I inspected the subject property and visited the other properties referred to during the hearing.

Decision

Mr Johnson put the value of the subject property, on the assumption that the planning application scheme had been implemented, at £1,100,000. He considered that the sale of 98 Palace Gardens Terrace in January 1988 for £1,185,000 supported his figure and Mr Hunt's valuation was also £1,100,000. Both Mr Johnson and Mr Hunt regarded the subject property with the benefit of the proposed new mansard floor as being more akin to the adjoining five-storey houses than the other four-storey houses.

Mr Carroll arrived at his valuation A by reference to the sale in December 1989 of no 86, a four-storey house, first by analysing the sale price in terms of floorspace and then by adjusting for the rise in values between June 1988 and December 1989. Mr Carroll admitted in cross-examination that he had no personal practical experience of the residential property market and also conceded that valuers of residential properties would rarely use an approach based on a price per square foot although his valuations were in fact wholly dependent on that approach. He used the Nationwide index to adjust for different sale dates, but was unable to produce the specific edition of the index on which he had relied and was unable to say whether the index covered the whole of London or central London only.

I would comment in passing that although indices may be a useful guide to overall trends in value their general nature must make them highly unsuitable for use in the valuation of individual properties, which may differ widely in size, type and quality from the properties on which the indices are based and where local movements in value may vary dramatically in the same period from the movements over the wider area covered by the indices. Although from the index Mr Carroll identified an increase in value of 13.2% between June 1988 and December 1989 the actual increase in value of 98 Palace Gardens Terrace between January 1988 and the second sale in July 1988 was 24.5%. He relied on a hearsay opinion that the second sale was in excess of market value, but I prefer the views expressed by Mr Johnson and Mr Hunt that the price obtained was a production of competition and rising values. In any event the result underlines the unreliability of indices in specific cases. Although Mr Carroll accepted that floor areas were not normally used when houses were being valued for open market purposes, he justified his floor area approach in the present case on the grounds that for compensation purposes he considered that he had to adopt a calculated approach to arrive at as precise a figure as possible.

Mr Timothy Corner argued that the claimant's valuers "hunch" approach was not sufficient and that it was necessary for them to be able to show how they had arrived at their figures. It seems to me, however, that although floor areas may have played a part in the valuation of houses for purposes such as rating (at least prior to 1990) the valuation of houses for private occupation is a highly-subjective process, values being arrived at directly from experience or from comparison with the sale prices of other houses with *ad hoc* adjustments being made for any apparent differences such as location, age, quality, construction, appearance, size, state of repair, potential for improvement, etc on an overall rather than a mathematical approach. Both Mr Johnson and Mr Hunt said that floor areas normally played no part in the valuation of houses and this view is supported by the absence of floor areas in any of the estate agents' particulars referred to during the hearing, size being dealt with entirely in terms of the dimensions of the principal rooms. On balance I prefer the practical approach of the claimant's valuers to the largely theoretical basis adopted by Mr Carroll.

Although I accept that the addition of a mansard floor would have materially improved the subject property and would have set it apart from the four-storey houses, the fact remains that not only do the five-storey houses have an extra floor, they also have greater room heights as is apparent both from the photographs submitted by Mr Carroll and from an external inspection; they are in Mr Carroll's words "on a grander scale". The evidence is far from conclusive, but on balance I would put the value of the subject property on the assumption that the planning application scheme had been carried out at £1,000,000. This is substantially more than the sale prices of any of the four-storey houses (with the exception of the second sale of no 86 in December 1989 for £950,000 following the attic conversion; it is common ground however that prices rose between June 1988 and December 1989), but is below the lowest of the sale prices of the five-storey houses, which from the evidence are clearly more valuable than the subject property would be after the mansard conversion, regardless of how well the latter conversion was carried out.

A It was agreed that £92,000 should be deducted for the cost of the proposed works and Mr Carroll deducted a further £4,000 for the cost to the owner of alternative accommodation during the works. We are concerned here, however, with the open market value in June 1988 and what an existing owner might have done is irrelevant. If sold in the open market work could not start until the sale was completed and provision would probably need to be made for any interest charges during the period of works. I have not been provided with the analysis of the agreed cost figure of £92,000 and do not therefore know if it included any provision for interest. I note that some of the earlier cost figures did in fact include an interest charge and in the absence of B evidence to the contrary I assume that £92,000 is an inclusive figure and that no further deduction need be made.

I accept that the layout of the subject property as it existed at the valuation date was not ideal and it is agreed that improvements could be made without planning permission being required. Different figures were put forward during the hearing as the values of the property as existing and as if a limited conversion scheme had been carried out. In reality, however, the property could only have one value at the valuation date as any potential for improvement would have been reflected in the purchase price. Mr Carroll's "non-application scheme" value (£857,455) was wholly dependent on the assumption that no 86, in its post-improvement state in December C 1989, was directly comparable with the subject property if the attic conversion had been carried out and that the Nationwide index could be applied to produce an accurate equivalent value 18 months earlier. I find both these assumptions unacceptable.

Mr Carroll had not seen the interior of no 86, but Mr Johnson had inspected after the conversion and was of the opinion that the latter was of a higher standard than that proposed for the subject property; this is borne out by the estate agent's particulars, which referred to the property have been "totally renovated to an extremely high standard". Even allowing for the rise in values between the first sale of no 86 in July 1988 at £640,000 and the second sale in December 1989 at D £950,000 the difference of £310,000 suggest that the property had been substantially improved in the interim. I have already said that I regard the use of the Nationwide index to be inappropriate, but in any event it seems to me unnecessary to look at the post-conversion value of no 86. This property sold in its unimproved state in July 1988, one month after the valuation date for the subject property, for £640,000 with the benefit of what in retrospect was obviously at least as much development potential as existed in the subject property at the same time. Nos 55 and 84 Palace Gardens Terrace were also sold in 1988, the sale price in both cases being £665,000. The comparatively narrow range of sale prices wholly supports the figure of £645,000, E which the parties agreed as the value of the property in its existing state at the valuation date and I see no reason to substitute a wholly theoretical figure for the latter figure.

I therefore determine the compensation in the sum of £263,000 arrived at as follows:

Valuation with planning application scheme completed	£1,000,000
Less costs	£ 92,000
	£ 908,000
Value as existing	£ 645,000
Compensation	£ 263,000

F I received written submissions upon costs on behalf of the parties. The award is in excess of the sealed offers made by both parties. In the circumstances I consider that the provisions of section 4(3) of the Land Compensation Act 1971 apply and I direct that the compensating authority shall bear their own costs and shall pay the costs of the claimant so far as they were incurred after his offer was made, such costs, if not agreed, to be taxed by the registrar of the Lands Tribunal on the High Court Scale.

GLASSHOUSE PROPERTIES LTD v DEPARTMENT OF TRANSPORT G

(Ref/009/1991)

April 23 1993

(Before T HOYES Esq FRICS)

Estates Gazette April 23 1994 and April 30 1994

[1994] 16 EG 150 and 17 EG 156

Compulsory purchase — Section 8 of the Compulsory Purchase Act 1965 — Whether claimant's counternotice must be served before entry by acquiring authority — Whether such notice served H

The claimant is the owner of Station House, a 20-storey office building let in floors or suites to numerous occupiers on a site of 3.4 acres fronting the North Circular Road. The curtilage of the site has 409 car-parking spaces for use by the occupiers and by the occupier of a filling station on the site. The acquiring authority served a notice to treat and a notice of entry on March 22 1989, with entry to be taken on or after March 30 1990, and took possession of an area of land resulting in the permanent loss of 154 car-parking spaces. A notice of claim was lodged on May 4 1989 seeking compensation for the land taken and for injurious affection, although the claim was not J quantified. The parties agreed that the following questions should be determined as preliminary issues: (1) what is the land the acquiring authority is to be taken to have entered on March 30 1990; (2) whether the claimant may apply to the Lands Tribunal for a determination under section 8 of the Compulsory Purchase Act 1965 if no notice was given to the authority by the claimant that such an application would be made, or that the whole of its property would be required to be purchased, before the acquiring authority entered the land; (3) whether the claimant gave any such notice to the acquiring authority prior to entry; and (4) whether the acquiring authority is estopped from denying that the claimant may apply to K the Lands Tribunal for a determination under section 8 of the 1965 Act.

***Decision*: (1) The date of entry for the purposes of question 1 was agreed as March 30 1990 having regard to *Chilton* v *Telford Development Corporation* [1987] 1 EGLR 12. (2) Any notice or application by a claimant seeking to operate the provisions of section 8 of the 1965 Act with the intention of requiring the acquiring authority to take the whole of Station House had to be served prior to the agreed date of entry of March 30 1990. In the absence of such notice or application the claimant is therefore not entitled to seek a determination from the tribunal under section 8. (3) On the evidence the claimant did not require the acquiring L authority to purchase and stated its willingness to sell the whole of Station House prior to March 30 1990 by a notice in the manner required as appropriate in *Gardner* v *Charing Cross Railway Co* (1862) 31 LJ Ch 181. No submissions were addressed as to the fourth question.**

The following cases are referred to in this report.

Ashton Vale Iron Co Ltd v *Bristol Corpn* [1901] 1 Ch 591; 17 TLR 183, CA
Binney v *Hammersmith & City Railway Co* (1863) 8 LT 161
Chilton v *Telford Development Corporation* [1987] 1 WLR 872; [1987] 3 All ER 992; [1987] 1 EGLR 12; (1987) 281 EG 1443, CA M
Gardner v *Charing Cross Railway Co* (1862) 31 LJ Ch 181
Gonty & Manchester, Sheffield & Lincolnshire Railway Co, Re [1896] 2 QB 439; [1895-9] All ER Rep 651; 45 WR 83; 12 TLR 617, CA
Grierson v *Cheshire Lines' Committee* (1874) LR 19 Eq 83
Harvie v *South Devon Railway Co* (1874) [XXXI NS] LT 424; (1875) 32 LT 1
King v *Wycombe Railway Co* (1860) 29 LJ Ch 462
Marson v *London, Chatham & Dover Railway Co* (1869) LR 7 Eq 546
R v *London & South Western Railway Co* (1848) 12 QB 775
Royal Life Insurance v *Phillips* (1990) 61 P&CR 182; [1990] 2 EGLR 135; [1990] 43 EG 70
Spackman v *Great Western Railway Co* (1855) 1 Jur NS 790

A *Sparrow* v *Oxford, Worcester & Wolverhampton Railway Co* (1852) 2 Dc GM & G 94

This is an application under rule 49 of the Lands Tribunal Rules 1975 for the hearing of preliminary issues in a claim for compensation by the claimant, Glasshouse Properties Ltd, against the acquiring authority, Department of Transport, arising out of the compulsory acquisition of land.

Christopher Lockart-Mummery QC and David Elvin (instructed by Binks Stern) appeared for the claimant; David Holgate and Nicholas Taggart (instructed by the Treasury Solicitor) represented the acquiring authority.

B

Giving his decision, **Mr T Hoyes** said: These are preliminary issues of fact and law arising under a reference by the claimant dated January 21 1991 seeking the determination of the Lands Tribunal as to whether the acquiring authority can be required to purchase the freehold interest in the whole of Station House, Wembley, rather than only in those parts identified in the notice to treat dated March 22 1989.

Following a pre-trial review, the preliminary questions for determination, as agreed between the parties, were contained in an order of the registrar dated April 6 1992. These questions are:

C 1. What land of the claimant the Secretary of State is to be taken as having entered on March 30 1990 (pursuant to a notice given under section 11 of the Compulsory Purchase Act 1965 dated March 22 1989) (such land is hereafter referred to as "the land").

2. Whether the claimant may apply to the Lands Tribunal for determination under section 8 of the Compulsory Purchase Act 1965 if no notice was given to the Secretary of State by the claimant that such an application would be made, or that the whole of its property would be required to be purchased, before the Secretary of State entered the land.

3. Whether the claimant gave any such notice to the Secretary of D State prior to such entry.

4. Whether the Secretary of State is estopped from denying that the claimant may apply to the Lands Tribunal for a determination under section 8 of the Compulsory Purchase Act 1965.

The date of entry in this reference has been a matter of contention; the claimant in January 1991 identified March 30 1990, whereas in June 1991 the acquiring authority were expecting to enter in November 1991. However, at the outset of the hearing it was agreed between counsel that, as a matter of law following *Chilton* v *Telford Development Corporation* [1987] 1 EGLR 12, the acquiring authority had entered upon all the parcels of land identified in the notices to treat and to enter (both dated March 22 1989) on March 30 1990. E Second, it was agreed that this date of entry was to be applied as appropriate to matters within questions 2, 3 and 4, albeit that physical entry upon the majority of the identified land did not take place until a later date: on the facts disclosed in evidence during late 1991. Question 1 is therefore disposed of by agreement of the parties.

In order to put the questions posed into a factual context it is necessary to outline briefly the scheme of the acquiring authority, describe the land of the claimant, indicate how the land is likely to be affected by the scheme and list the procedural steps in the acquisition taken thus far. These matters will be dealt with by reference to a statement of facts dated April 28 1992 lodged by the claimant with the F tribunal, an agreed bundle of documents (mainly correspondence) and non-contentious matters in the oral evidence.

1. The scheme of the acquiring authority is the improvement of the A406, North Circular Road, between Hanger Lane and a point east of Harrow Road (A404).

2. Station House is a 20-storey office building on a site of about 3.4 acres fronting the North Circular Road at its junction with Harrow Road from which it enjoys vehicular access.

3. The curtilage of Station House provided 409 car-parking spaces at surface level for use by the occupiers of the office building; a small number were used by the occupiers of a petrol-filling station, car showroom and repair workshop also sited within the curtilage.

4. The office building was constructed in the mid-1960s and is let G in floors or suites to numerous occupiers.

5. It had a long history of underoccupation until the late 1980s, but it is recognised as affording generous car parking and has now become attractive to occupiers generating or attracting car traffic.

6. The freehold interest in Station House was owned by Colman Consolidated Properties Ltd (Colman) until April 3 1989, when it was purchased by Glasshouse Properties Ltd (Glasshouse), an investment company which was put into administrative receivership in October 1991 by a bank mortgagee.

7. The principal effect of the acquiring authority's scheme upon Station House is the permanent loss of 154 car-parking spaces, about H 37.5% of the total, to the proposed road works together with disruption to occupation during the construction period. The petrol-filling station will also become incapable of operation owing to the loss of the forecourt area.

8. At all material times the owner has maintained that the loss of the car-parking spaces and the proposed works will adversely affect its ability to continue letting offices in Station House and also the rents obtainable from occupiers; both matters are considered fundamental to, *inter alia*, the capital value of Station House.

9. Colman from about mid-1987 and Glasshouse for most of its ownership sought to overcome and mitigate this prospective damage J by the construction within the curtilage of a multi-level car-park structure for which planning permission was obtained on March 6 1990.

10. This structure has not been erected because of the alleged inability of Glasshouse to fund the cost, estimated at £5,000,000, owing to the absence of an offer by and receipt from the acquiring authority of compensation of that order.

11. The making of the A406 London North Circular Trunk Road (Hanger Lane to Harrow Road Improvement) Compulsory Purchase Order was first published on September 15 1983.

12. A public local inquiry into objections, including that of K Colman, was opened on September 26 1984 and on June 11 1986 the Secretaries of State for the Environment and Transport indicated that they were minded in principle to confirm the order subject to some modifications, involving matters of detail rather than of substance for Station House.

13. On December 22 1988 the compulsory purchase order was confirmed, notices to treat and of entry were served on Colman on March 22 1989, entry to be taken on or after March 30 1990.

14. A completed claim form dated May 4 1989 in answer to notice to treat was lodged with the acquiring authority by surveyors instructed by Glasshouse. This claim was forwarded by the Department of Transport to the district valuer for the Ealing area on L May 11 1989 with instructions that negotiations upon compensation be commenced on its behalf.

15. The claim form identified the land in respect of which compensation was claimed by reference to the parcels numbered on the plan attached to the compulsory purchase order, ie part of the curtilage of Station House, and, in terms, sought compensation for the land to be taken and for injurious affection (temporary and permanent) to the value of land to be retained. A possible loss of rental income was claimed under "disturbance" and reference was made to four other items under "any other matter".

16. The claim was not quantified in total nor as to any particular M item of claim, *quantum* was to be advised later.

Mr Christopher Lockhart-Mummery QC and Mr David Elvin of counsel appeared for the claimant and called Mr A Hardiman FNAE MBIM, a former director of Colman, who spoke to the position of that company, Mr M R Turner FRICS, a director of Glasshouse, who spoke to the position and the material activities of that company, and Mr A B Crabtree BSc FRICS IRRV, who gave evidence concerning his discussions with the office of the district valuer.

Mr David Holgate and Mr Nicholas Taggart of counsel appeared for the acquiring authority and called Mr J Echlin ARICS and Mr D A Russell-Smith FRICS, both of the district valuer's office Ealing, since

July 1991 merged with the Harrow office.

Reference was made to the following authorities;

Chilton v *Telford Development Corporation* [1987] 1 EGLR 12, CA
Binney v *Hammersmith & City Railway Co* (1863) 8 LT 161
Spackman v *Great Western Railway Co* (1855) 1 Jur NS 790
Gardner v *Charing Cross Railway Co* (1862) 31 LJ Ch 181
R v *London & South Western Railway Co* (1848) 12 QB 775
Sparrow v *Oxford, Worcester & Wolverhampton Railway Co* (1852) 2 De GM & G 94
King v *Wycombe Railway Co* (1860) 29 LJ Ch 462
Marson v *London, Chatham & Dover Railway Co* (1869) LR 7 Eq 546
Grierson v *Cheshire Lines' Committee* (1874) LR 19 Eq 83
Harvie v *South Devon Railway Co* (1874) [XXXI NS] LT 424; on appeal (1875) 32 LT 1
Gonty & Manchester, Sheffield & Lincolnshire Railway Co, Re [1896] 2 QB 439, CA
Ashton Vale Iron Co Ltd v *Bristol Corpn* [1901] 1 Ch 591, CA
Royal Life Insurance v *Phillips* [1990] 2 EGLR 135

Question 2

Section 8 of the Compulsory Purchase Act 1965 is in the following form:

> (1) No person shall be required to sell a part only —
> (*a*) of any house, building or manufactory or
> (*b*) of a park or garden belonging to a house, if he is willing and able to sell the whole of the house, building, manufactory, park or garden, unless the Lands Tribunal determines that —
> (i) In the case of a house, building or manufactory the part proposed to be acquired can be taken without material detriment to the house, building or malufactory, or
> (ii) In the case of a park or garden, the part proposed to be acquired can be taken without seriously affecting the amenity or convenience of the house, and if the Lands Tribunal so determine, the Lands Tribunal shall award compensation in respect of any loss due to the severance of the part proposed to be acquired, in addition to its value; and thereupon the party interested shall be required to sell to the acquiring authority that part of the house, building, manufactory park or garden.

Subsections (2) and (3) of section 8 have no relevance to the matter in issue. In the instant case this section entitles the owner, part of whose land is being acquired, to require the acquiring authority to take the whole unless the Lands Tribunal determines that the part "can be taken without material detriment" to the whole. The questions here are, first, whether Glasshouse or its receivers in order to have the benefit of the section needed to serve a notice upon the acquiring authority before March 30 1990, the agreed date of entry, to the effect that it was required by the claimant to take the whole of Station House and its curtilage. Second, if the claimant intended to refer the matter of whether a part can be taken without material detriment to the remainder to the Lands Tribunal, its first jurisdiction under section 8, as the claimant in fact did on January 21 1991, it needed to give notice of that intention before March 30 1990. These questions hinge on whether the agreed date of entry is, in effect, a barrier beyond which the claimant ceases to have any right to invoke section 8 to secure any relief or benefit that it may afford.

Mr Christopher Lockhart-Mummery QC submitted that there is no requirement in section 8 of the Compulsory Purchase Act 1965 that a notice invoking its provisions should be given prior to entry by the acquiring authority or indeed at any particular time. A claim under section 8 can therefore be made by notice before or after entry upon the part of the land authorised to be acquired and subject to notices to treat and entry. His reason for this position was that the legality of entry now depends upon notice of entry being given in accordance with section 11(1) Compulsory Purchase Act 1965 and not, as formerly under the Lands Clauses Consolidation Act 1845, upon the prior assessment of compensation and the deposit of a bond for the assessed sum: see *Garner*. Furthermore, he said compensation is now generally assessed after entry and if an owner does not lodge a quantified claim, a common situation, an acquiring authority will have no warning of the extent of a compensation claim until after they have entered upon the land and thereby become obligated to pay compensation.

A section 8 claim now concerns only the *quantum* of compensation whereas formerly under the 1845 Act it went also to the adequacy of the bond and the legality of entry. He urged that an acquiring authority is well able to assess the issue of material detriment under section 8, sufficient to inform themselves of the effect upon compensation in the same way that they have to inform them of the amount of compensation generally. The current notice of entry procedure in section 11 of the 1965 Act renders inapplicable the established authorities relating to "counternotices" under the 1845 Act. Mr Lockhart-Mummery referred to and distinguished the specific requirements for the service of a notice at a particular time under the Compulsory Purchase (Vesting Declarations) Act 1981, because that Act operates on the basis of deemed entry rather than upon the more common and longer-established notions of legal and physical entry following notice by the acquiring authority under section 11 of the 1965 Act.

In closing, Mr Lockhart-Mummery submitted that as section 8 of the 1965 Act was wholly silent as to the time for serving notice by the landowner upon an acquiring authority, in contrast to the 1981 Act, the statute should be construed strictly against the acquiring authority in favour of the landowner following *King*.

For these reasons the fact that the acquiring authority entered the land of the claimant on March 30 1990 is irrelevant to the question of the time when notice is to be served and the determination sought from this tribunal on January 21 1991 can and should be granted.

In response, Mr David Holgate said that while notice or counternotice provisions were absent from section 8 of the 1965 Act they were likewise absent from section 92 of the 1845 Act. The 1965 Act was a consolidation, which must have taken account of the case law on the service and the timing of counternotices. Therefore, the established authorities remain relevant and binding. The only difference in the wording of section 92 and section 8 is the deletion of the words "at any time" and the insertion of the "material detriment" provisions, which provide a possible defence to an acquiring authority against a claim to purchase the whole. Neither assist the claimant in the instant case nor do the material detriment provisions alter the need for a time-limit by which a counternotice should be served; because the acquiring authority have to decide whether to raise a lack of material detriment as a defence, and if that is decided against them, then the acquiring authority can elect to withdraw the notice to treat. He cited *Gonty* as illustrating that there was no inconsistency between material detriment provisions and a time-limit.

The introduction of section 11 of the 1965 Act as a replacement for sections 84 and 85 of the 1845 Act, payment of compensation prior to entry or the giving of security for full compensation, does not affect the question of when a counternotice should be served because the counternotice was not only needed by the acquiring authority for operating section 85 (deciding what money to deposit) but also for making an election as to whether to abandon the intended purchase of the land: see *Ashton Vale Iron Co*. That requirement remains as much today as in the 19th century. If section 8 can be relied upon by a claimant at any time, before or after entry, it would be penal in effect because after entry it would enhance the anticipated cost of schemes without affording a "retreat" for the acquiring authority. He urged that this was the main purpose of a counternotice and that, in effect, it set in motion a procedure for the determination of how much land the acquiring authority will take: see *Gardner* and *Spackman*.

Mr Holgate submitted that for these reasons at least a notice must be given before the acquiring authority enters on to the land because after this time they are bound to take the whole and they cannot withdraw their notice to treat in respect of the part: see *King* and *Marson*. An invalid counternotice entitles an acquiring authority to proceed with acquisition of the part only: see *Harvie*.

Mr Holgate also referred to the Compulsory Purchase (Vesting Declarations) Act 1981, section 12 and Schedule 1, which he regarded as enshrining the principles of *King* by providing specific time-scales for both the landowner and the acquiring authority. In effect, para 4 of Schedule 1 includes the statutory enactment of the election procedure

A available to the acquiring authority. He submitted that it was inconceivable in 1968, when general vesting declarations were first introduced, that Parliament should have imported the principles of the case law into the then new vesting procedure if it thought that entry under section 11 of the 1965 Act would be and was being operated differently. Parliament has recognised that owners cannot force an acquiring authority to take the whole of their land after entry and there should be no procedural difference between normal compulsory purchase and that where the vesting declaration route is followed.

In essence, the case of the claimant relies on the proposition that B the provisions and procedures in section 8 of the 1965 Act are free standing and self-contained so as to render irrelevant the established authorities which require that any notice or counternotice to take an entire premises, instead of a part, shall be served before entry by the acquiring authority. While superficially this preposition has attractions I do not find it convincing. The 1965 Act was a consolidation of earlier legislation relating to procedural matters for compulsory purchase and, had Parliament intended to depart from established procedures and practice, I consider that it would have specifically so provided.

I recognise the importance of the submission by Mr Holgate C concerning the opportunity and the obligation residing in the acquiring authority to make an election, following receipt of a notice, whether to proceed with their scheme, but involving a larger and probably a more expensive area of land or retreat from the whole proposal. The authorities clearly establish that the time of entry is central to the issue of election and I am not persuaded that the procedure for entry in section 11 of the 1965 Act, although different from that under the 1845 Act, detracts from the established position in any way. An acquiring authority are entitled to know the extent of the compensation claim they will be required to pay and to this end they invite the claimant to notify a claim in response to the notice to treat. This D entitlement is unaffected by any suggestion that an acquiring authority are well able, and frequently do, make their own estimate of the likely sum of compensation payable.

To follow the route sought by the claimant involving notice at any time up to the time of the agreement or determination of compensation would be contrary to the objectives of the introduction into the "modern" legislation of the concept of "material detriment", which affords a defence to an acquiring authority, subject to safeguards for the landowner, against being obliged to take more land than they require for their scheme. The historic remedy was to retreat immediately upon receiving a counternotice whereas section E 8 of the 1965 Act entitles an acquiring authority to retreat only after this tribunal has found against them on the issue of material detriment. The absolute right of the landowner to force the taking of the whole has thus been reduced in certain circumstances to the advantage of the acquiring authority in the control and conduct of their affairs. If the position of the claimant in this reference is accepted it would be contrary to this trend of affording a second line of defence and a degree of certainty to the acquiring authority prior to entry.

For these reasons, and in answer to question 2, I find that any notice or application by the claimant seeking to operate the provisions F of section 8 of the Compulsory Purchase Act 1965 with the intent of requiring the acquiring authority to take the whole of Station House had to be served prior to the agreed date of entry, namely March 30 1990. In the absence of such a notice or application the claimant is therefore not entitled to seek a determination from this tribunal under section 8 of the 1965 Act.

Question 3

"Whether the claimant gave any such notice to the Secretary of State prior to such entry"

It was common ground between counsel that no particular form of notice is prescribed by section 8 of the 1965 Act, nor indeed by section 92 of the 1845 Act. The position concerning practice under the 1845 Act is described in *Cripps on Compulsory Acquisition* 11th G ed, at para 2.046 on p359 as follows:

> The counter-notice, . . . is not required to be in any particular form. It may be verbal . . . It is sufficient for the owner to give notice by any method of description which accurately specifies the premises, the whole of which he requires to be taken.

There was contention between counsel as to the ingredients and nature of the required notice, although both accepted that it could be either written or verbal. Mr Lockhart-Mummery said, if it was found that notice was required to be served before entry, that it had to be "an adequate warning of a potential claim prior to entry", see *Spackman*, H or put another way, "a warning sufficiently clear on an objective test to put the acquiring authority on notice of risk", it could be verbal, see *Binney*, or qualified or conditional, a "backstop" position, see *Gardner*.

Mr Holgate submitted that notice of a potential claim was inadequate as the owner has to be willing and able to convey the whole premises and the notice must require the acquiring authority to take the whole. The notice must be unequivocal and not contingent since a contingent notice does not confirm the willingness of the owner to sell. Furthermore, a contingent notice does not enable the acquiring authority to serve an objection as a step in a process of having the issue of material detriment determined by the Lands J Tribunal. Any without prejudice notice which the claimant may have given is equivocal and therefore offends the concept of willing and able to sell found in section 8 of the 1985 Act.

Evidence was called by both parties to establish their respective positions in discussions, negotiations and correspondence which had been conducted since about 1985, more particularly details of these matters were to be found in an agreed bundle of correspondence and in the submitted statements of the witnesses upon which they were cross-examined. The documents and evidence can usefully be divided chronologically into the following periods:

1. Before notice to treat on March 22 1989, essentially the period K of ownership by Colman.

2. From March 22 1989 to the agreed date of entry on March 30 1990.

3. After March 30 1990.

Mr Holgate said, in respect of the first period, that it was too early and irrelevant as section 8 of the 1965 Act can be of relevance only in response to notice to treat. In respect of the third period, having regard to his contentions (and now my findings above) that date of entry is a watershed, this material was too late to be of any significance in answer to the question in issue.

In broad terms I accept this analysis of the assistance to be derived L from the early and later material and it will be referred to only briefly for the sake of order and completeness.

Evidence

In evidence Mr Hardman said that he first became acquainted with Station House following his appointment as group property manager to the Colman companies in early 1985. In about 1987 it was perceived that the problem which would arise from the loss of 154 car-parking spaces to the road works could largely be overcome by the construction of a multi-storey car park of about 300 spaces within the curtilage of Station House.

This would make good the physical loss of spaces to the improved M highway and the further spaces to be lost due to building a structure on part of the existing car park. Moreover, if the new car park could be available for use before the loss of the 154 spaces, it would avoid a period of reduced car parking and a consequent loss in revenue from rents etc for the offices. The long-term detriment to Station House would be overcome and the short-term losses likely to be inflicted would be mitigated or eliminated. Looked at broadly from the claimant's point of view, the compensation payable could be measured by the cost of providing the new multi-storey car park — at that time the estimated cost was of the order of £2,500,000, a sum later increased to about £5,000,000.

The proposal was discussed, albeit informally prior to notice to treat, with the Department of Transport and the district valuer, then a Mr Daxendaie, whom it is recorded in minutes of a meeting on October 15 1987 expressed reservations that the taking of 154 car-parking spaces would cause detriment to Station House. The matter was also put in writing on May 11 1987 by the chairman of Colman to a then minister in the Department of Transport, the letter contained, *inter alia*, the following statement, "we do not wish to invoke the powers of Section 8 of the Compulsory Purchase Act 1965 although, of course, we must reserve the right to do so". Mr Hardiman also wrote to the minister on May 14 1987 giving various figures relating to anticipated losses and concluded by stating, "I do not want Section 8 as I simply cannot pay the capital gains tax (consequent upon the sale of the whole premises)."

During the hearing the Colman approach became known as "the car park and backstop position" and was last referred to by Mr Hardiman in a letter of February 10 1989 to Mr Baxendale of the district valuer's office urging action in discussions on compensation and concluded in the following terms:

> Please can we take this matter on board now. If not we will have to pursue section 8 as we will have been given no alternative. The cost to the Department of Transport will then be the full value plus an incalculable sum for disturbance (presumably to occupying lessees) and leave them with a half empty building.

In cross-examination Mr Hardiman said that it was his practice to serve a notice under section 8 in response to notice to treat, but it was premature for Colman to serve such a notice as it was to complete the sale of its freehold interest to Glasshouse as an investment on April 3 1989. Nevertheless he considered that the position of the building owner (Colman) had been made clear in that a warning had been given to the Department of Transport and the district valuer. He ceased to be involved with Station House very soon after the service of notice to treat on March 22 1989.

Mr Turner, a director of Glasshouse, said that the company purchased Station House as an investment in the knowledge of both notice to treat in respect of part of the car park and the proposal by Colman to build a multi-storey car park to make good the loss of parking spaces. Glasshouse expected to pay for the cost of the new car park out of compensation moneys from the acquiring authority, and accordingly instructed consultants to design the structure and seek planning permission for it; the permission was granted on March 6 1990 and detailed design work etc continued thereafter.

Some time in April 1989 Mr Turner instructed Mr Crabtree to undertake negotiations relating to compensation on the basis of the Colman "car park and backstop position". On May 4 1989 Mr Crabtree lodged with the Department of Transport a claim in answer to notice to treat on the usual proforma, which was signed by Mr Turner. The claim was unquantified in amount and plainly sought compensation for land taken and injurious affection to the residue of Station House; neither the claim form nor the covering letter made any reference to section 8 of the 1965 Act as a "longstop" or otherwise. Negotiations between Mr Crabtree and the district valuer were opened and proceeded until September 1990 (see later).

Perturbed at the lack of progress and with conditions worsening at Station House for the occupiers, Mr Turner sought clarification and action from the Department of Transport. On March 28 1990, he wrote and included in his letter the following:

> Due to the disastrous effect on the building of your department's works, I cannot fund the building of the alternative car park. I need to know that the funds will be provided by your department for its construction before I can let the (building) contract.
>
> The alternative cause of action is for us to consider serving a purchase notice on the whole property, due to its blighting by the works being carried out. This is an alternative on which I am seeking Counsel's advice.

In oral evidence Mr Turner accepted that he was not very familiar with technical terms in the field of compensation and that the last paragraph cited above was intended to refer to the use of section 8 of the 1965 Act and not to provisions within the Town and Country Planning Acts.

In cross-examination Mr Turner accepted that when writing the letter of March 28 1990 (two days prior to the now agreed date of legal entry) the company had not decided that it wanted the acquiring authority to take the whole of Station House. He was hopeful at that time, based on general advice received, that the compensation payable and the cost of the proposed car park would approximately equate, a conclusion which the company had reached in both 1989 and early 1990. He agreed that Mr Crabtree had acted within his instructions and that he had no authority from the company (as its agent) to "trigger" section 8 by verbal notice to the Department of Transport or the district valuer. Finally, Mr Turner fully accepted that during 1989 the dispute as to compensation could have been referred to the Lands Tribunal, on the same basis as the current reference, but he said the company was reluctant to do this as it was then hopeful that agreement would be reached.

In his evidence Mr Crabtree said that one of the advantages of Station House to occupiers was the generous provision of car parking and he saw this as an aid to letting space and underpinning the capital of the freehold interest. On purchase Glasshouse wished to hold the property as an investment and, when the formal claim for compensation was prepared, "the possibility of serving a Section 8 notice was discussed, but the directors of Glasshouse made the point that they would only wish to pursue the Section 8 claim in the event that they could not, for whatever reason, construct the car park and have the cost refunded out of any compensation from the Department of Transport": see report, para 8.

His first meeting with the district valuer was on June 13 1989, by then a Mr Echlin. The object of the meeting was to seek to establish the "likely *quantum* of the injurious affection claim in order to establish that there would be sufficient compensation to cover the cost of the new car park". A sketch plan of a multi-level car park was produced and it was indicated that design work was moving forward with a view to construction starting in early 1990, it then being anticipated that the Department of Transport would physically enter in late 1990. This was the practical Colman proposal of replacing before losing car-parking spaces and hopefully eliminating consequential loss. The district valuer expressed a view that the proposal would yield a measure of betterment to the claimant. Mr Crabtree said that he put a figure of compensation to the district valuer in the range of £3,500,000 to £4,000,000 and that the estimated cost of the car park was then about £2,000,000. The district valuer "advised that he could not guarantee that the compensation for injurious affection would cover the cost of the car park".

In his report (para 13) Mr Crabtree says: "I stated to him that it was still open to Glasshouse to amend the basis of the original claim if sufficient compensation was not forthcoming and to take the total loss route at this point, ie to make a Section 8 claim."

Events then took an unexpected turn because Mr Echlin made a verbal offer on behalf of the Department of Transport to purchase the whole of Station House subject to contract and ministerial approval. Mr Echlin repeated the offer of £13,000,000 in writing on June 16 1989 and added the term that "each side should be responsible for its own fees and legal costs". Mr Crabtree discussed the offer with the claimant and upon instructions declined it in writing on June 20 1989. Mr Echlin replied to that letter on July 26 1989 noting the refusal and stating: "I remain of the opinion it may well be in the interests of both parties for such a sale to take place. I should therefore be obliged if you would advise me of the figure your clients would consider for the sale of Station House." For reasons explained, that letter was delayed in reaching Mr Crabtree, but he replied on October 30 1989 in terms that the claimants regarded its ownership of Station House as a long-term investment and that it had in mind refurbishment on expiry of leases and the completion of the road works. He expressed the view that, having regard to these matters, it was "unlikely that they would consider an offer for the freehold of less than £15,000,000,

A subject to existing tenancies". Mr Crabtree said that he invited Mr Echlin to make a further and improved offer because (report, para 19):

> Glasshouse Properties were now reaching the conclusion that the car park could not be constructed prior to the land being taken, as a result of which the building would be incapable of beneficial use.

The claimant attaches considerable importance to events at and arising from the meeting on June 13 1987 between Mr Crabtree and the district valuer. First, it is asserted that a verbal notice under section 8 was given to the acquiring authority's agents in the manner of *Gardner*; and, second, that the offers made by Mr Echlin on behalf of
B the acquiring authority were in furtherance of such notice, albeit they were declined. Mr Echlin in his evidence has a different explanation wholly unrelated to section 8 of the 1965 Act (see later).

Negotiations with the district valuer lost momentum and Mr Crabtree wrote again on April 9 1990, following a meeting with the Department of Transport on April 4 jointly with Mr Turner (of Glasshouse), indicating that the contractor carrying out the road works would enter upon the reference land on or about November-December 1990 and stating:

> the fact that the building is affected by a compulsory purchase order is going to make funding of the car park a virtual impossibility for my clients, unless
> C there is a formal agreement that the cost of construction will be met by the Department of Transport as part of the compensation claim.

An inconclusive meeting took place on May 3 1990 between Mr Crabtree and Mr Turner with the district valuer (then a Mr Russell-Smith), which was followed by a joint site visit on May 11 1990. On July 10 1990 Mr Crabtree wrote to the district valuer reiterating the proposal to construct the car park, the cost of which was then estimated at "in excess of £3,000,000". With that letter Mr Crabtree enclosed, for discussion only, "a suggested approach to the valuation exercise". This document was clearly based upon a land taken and injurious affection approach in the sum of £4,900,000.

D Meetings took place with the district valuer on July 11 and September 11 1990 on the basis that Glasshouse was still able to build the car park, but no progress was made in agreeing compensation as the district valuer remained unconvinced that there would be significant injurious affection or detriment to Station House.

In cross-examination Mr Crabtree agreed that section 8 was seen as a "backstop" up to the end of March 1990 and that in September 1990 he was not negotiating with the district valuer on the basis of a claim under section 8; the basis of claim was not amended from May 1989 to late September 1990, when he ceased to be concerned with the claim. He accepted that the cost of building a new car park was not "a
E recognised head of claim" in compensation for compulsory purchase and he agreed that office values in Wembley reached a peak during the period June to September 1990. In re-examination he said he thought that the 1989 offer by the district valuer to purchase the whole of Station House meant acceptance on behalf of the acquiring authority that the operation of section 8 was appropriate to the claim.

The evidence of Mr Echlin related to the period May 1989 to April 1990. He received the claim form from the Department of Transport on May 12 1989 together with the minutes of a meeting on October 15 1987 between the Department of Transport and Colman concerning the multi-storey car park proposal. The claim was clearly for
F compensation for land taken and injurious affection and he said neither the claimant nor its advisers ever made any written or oral claim to him involving section 8 of the 1965 Act.

The offer to purchase the whole of Station House made by him was not in the context of section 8 but as "an opportunity purchase" upon private-treaty terms, as he believed a profit could accrue to the Department of Transport from such a transaction at £13,000,000. He considered that the car-park proposal would accord a measure of betterment to the claimant and for this and other reasons he did not agree that its cost should, or could, be set by the Department of Transport. In his opinion the injurious affection to Station House was not as much as the estimated cost of the car park, put to him at £7,000-£9,000 per space, giving a total, including professional fees, of G
about £2,300,000.

In cross-examination it was put to Mr Echlin that documents in the agreed bundle indicated that if the car park was not built section 8 would be pursued. He said that was not his reading of the documents and that he regarded section 8 as the "longstop" position of the claimant.

In his evidence Mr Russell-Smith said that he first became aware of matters relating to Station House in February 1989 from correspondence directed to Mr Baxendale, one of his colleagues, with whom (as superior officer) he discussed the case. Up to May 11 1989 when formal instructions to negotiate compensation were received H
from the Department of Transport all discussions were of necessity informal.

He said he was very surprised that Mr Echlin had made the offer to purchase the whole of Station House on June 16 1989, which he regarded as outside the authority and practice of the valuation office. He took over direct responsibility for the case in May 1990 and declined to give any undertaking that the Department of Transport would "fund", directly or indirectly, the cost of the proposed car park.

At his meeting with Messrs Crabtree and Turner on May 3 1990 there was no mention of section 8 of the 1965 Act. In his view the diminution in value of Station House was nowhere near the J
£4,400,000 put forward under that head by Mr Crabtree at the meeting on July 11 1990 and he regarded Mr Crabtree's figure as driven by the cost of the car park; his "before and after" approach, not section 8, was appropriate in principle, but his figures were unacceptable. In September 1990 there had been no physical entry by the Department of Transport on to the reference land and negotiations upon the compensation claim rested until January 11 1991 when a Mr Langley replaced Mr Crabtree as the valuer for the claimant and sought a meeting to discuss compensation.

By letter dated January 11 1991 Mr Langley indicated that a reference to the Lands Tribunal was in preparation "based on material K
detriment to the property caused by the proposals". This reference was dated January 21 1991 and a copy was served on the Department of Transport on the same day by solicitors for the claimant. Mr Russell-Smith said he met Mr Langley, who confirmed that the approach required by the claimant was now under section 8 with an entry and valuation date of March 30 1990. This was the first occasion on which it was indicated that the section 8 route was to be followed; previously he had seen nothing in writing which he regarded as a notice and nothing had been said to him which could be a verbal notice, nor had he been advised of any such approach direct to the Department of Transport. The claimant served a section 8 notice by L
virtue of its reference to the Lands Tribunal.

In cross-examination, the letter from Mr Hardiman to Mr Baxendale of the district valuer's office dated February 10 1989 was put to Mr Russell-Smith and he agreed that Colman was saying, that if they were not put in funds to build the car park prior to entry a section 8 notice would follow. He saw this early reference as a warning that a section 8 notice might well be served in the future if the condition were not met.

Findings on the evidence

1. If it is correct in law that correspondence and representations prior to the date of notice to treat (March 22 1989) are irrelevant, the M
position adopted by Colman can have no bearing on the question in issue.

2. Glasshouse from April 4 1989 up to about January 20 1991 continued the approach of Colman, namely it looked to the compensation for injurious affection to provide funds from which to construct the replacement car park. In the cited statements to this effect (and in others not cited but of similar import, see annexure 1 [not reproduced here] the invoking of section 8 was a "longstop" position.

3. These representations following the Colman approach were, at their simplest, a mere statement of the rights residing in the claimant

and when considered against the detailed factual background of the instant case, threats or warnings of possible action if the precondition of adequate finance being available, sufficiently in advance of physical entry, were not met.

4. The representations in the correspondence and made at meetings with the Department of Transport and the district valuer did not comprise a notice under section 8 upon which the acquiring authority could act until the reference to this tribunal on January 21 1991.

5. There is no evidence that Glasshouse was "willing and able to sell the whole". The evidence was all the other way until January 21 1991, namely that it wished to retain Station House as an investment. Indeed, it declined an offer to purchase made by Mr Echlin in mid-1989 on behalf of the Department of Transport.

6. The only notice served upon the acquiring authority seeking to invoke the provisions of section 8 was the reference to the Lands Tribunal on January 21 1991. This finding is subject to consideration later as to whether, at any earlier time, a notice in the manner of *Gardner* was served.

In response, Mr Holgate said that the position of the acquiring authority could be only clarified by unequivocal written or verbal notice. No correspondence or recorded oral statements in the agreed bundle constitute a sufficient notice under section 8. There is no point in a notice served before the power of acquisition is put in motion by the service of notice to treat; *per* the authorities the notice is to be served in response to notice to treat.

He stated that Glasshouse did not lodge its claim for compensation in May 1989 on the basis of section 8 nor did it seek to amend its claim until January 1991. Up to September 1990 the claimant's surveyor was still discussing a claim on the basis of injurious affection and that claim was neither agreed nor rejected; this position contrasted with that in *Gardner,* where the claim in respect of the part was rejected.

He urged that the claimant could have referred the matter to the Lands Tribunal before legal entry on March 30 1990. It missed this date due to its own fault and it should not now be permitted to oblige the acquiring authority to take the whole premises, subject to proof of material detriment under section 8.

In closing, Mr Lockhart-Mummery said that the approach adopted by the acquiring authority was too narrow and highly technical; the merits of the case were with the claimant and it had been working with the acquiring authority for years to find a practical solution to the problem created by the road improvements. He accepted that the acquiring authority were entitled to some warning of the risk inherent in entering upon part of the land, namely that they may be required to take the whole, but Mr Holgate's approach was too rigid. Mr Russell-Smith had agreed that a warning was sounded early in 1989 that a claim under section 8 might arise later. The Colman stance had been continued by Glasshouse and there was an agreement in November 1989, made verbally with Mr Turner, to delay physical entry from April 1991 until November 1991 to facilitate the construction of the car park. The acquiring authority had been aware of the position of the claimant throughout and a conditional or "backstop" notice in the form given is sufficient in law: see *Gardner*.

All my above findings on the evidence are against the claimant save for the *Gardner* reservation at 6. Subject to this reservation, I therefore accept the primary submission of Mr Holgate that there was no documentary or oral evidence before me sufficient to constitute a notice under section 8 of the 1965 Act and served on the acquiring authority between the dates of notice to treat, March 22 1989, and the agreed date of legal entry, March 30 1990. If I am right in law in answer to question 2 it is an unavoidable conclusion that, *Gardner* apart, the provisions of section 8 of the 1965 Act are not available to the claimant, but if I am wrong the claimant can rely upon the reference to this tribunal made on January 21 1991.

In respect of question 3 there remains the issue of whether *Gardner* can assist the claimant. The essence of the judgment in *Gardner* is aptly summarised at para 1 of the headnote to the report as follows:

> a landowner who, upon being served by a company with notice to treat for part of his property, replies by claiming a given sum for the part, is not thereby precluded, if that sum be refused, from asserting his rights, under the 92nd section of this Act, to require the company to take the whole.

The facts in *Gardner* were briefly as follows; the owners delivered a written claim for the part of their land sought by the acquiring authority on the basis of land taken plus injurious of section to the residue in the sum of £800. The surveyor for the acquiring authority rejected this sum as excessive and the solicitor for the claimants replied on September 14:

> Messrs Gardner consider that their premises will be damaged to the extent of £800 — the sum claimed for a portion required by the railway — and are not disposed to agree to accept less. If you cannot advise this amount to be given, they will require the company to take all the premises, and for which they will want £3,000.

The acquiring authority took the formal steps prior to entry required by the 1845 Act and deposited the sum of £150 for part of the land. Upon receiving notice of this action the solicitor served a formal counternotice requiring the acquiring authority to take the whole, "and not to enter upon the part specified in the notice to treat until they had obtained a valuation of the whole of the premises . . .; or until they agreed to accept the terms proposed on behalf of the owners for the purchase of the portion of the premises specified in their notice to treat". Despite this counternotice the acquiring authority entered upon the part and the owner "filed a bill" requiring the whole to be taken as they were willing, and offering, to sell and convey the whole.

At p1053 Sir W Page Wood V-C said:

> The question, therefore, is reduced to this — whether, the company having served the Plaintiffs with notice to treat for part of the property in question, the Plaintiffs, by claiming £800 for that part, have precluded themselves, thenceforth and forever, from insisting that, if £800 be not given for the part, they would require the company to take the whole.
>
> In my judgment it would be unreasonable to hold anything of the kind; £800 may be a very good price for part of a property, and the owner may well be disposed—he may even prefer—to retain the rest provided he can obtain £800 for the part required; but the moment an attempt is made to compel him to sell the part for a less sum than £800, he may say with reason that, unless he can obtain £800 for the part, he would prefer to sell the whole.

He further said that, "I take that letter [of 14 September] to be a sufficient counternotice under the 92nd section of the Act [of 1845]".

It seems to me that if the substance of that letter is applied to the facts in the instant case the following issues are raised:

(a) Is it essential that the amounts respectively required for the part and the whole be quantified in pounds sterling?

(b) Can the stated objective of the claimant to secure compensation at least sufficient to construct a new car park be realistically equated with the £800 in *Gardner*?

(c) Did the Department of Transport by their refusal to directly take on or underwrite the cost of the car park effectively reject that cost as being excessive compensation?

(d) Did the district valuers in negotiations with the claimant's surveyor, by making reference to betterment (Mr Echlin) and that injurious affection was not so high as Mr Crabtree claimed (Mr Russell-Smith), thereby reject the consideration sought by the claimant for the part?

(e) Had that rejection become apparent prior to entry by the acquiring authority on March 30 1990?

(f) Must the terms of any notice specifically express a requirement to take the whole to satisfy the words "willing and able to sell" in section 8 of the 1965 Act?

(g) Assuming that the claimant was able to sell the whole, was it willing before March 30 1990 or did that willingness arise only in or about January 1991?

Mr Lockhart-Mummery's submission was that a *Gardner*-style notice had been served on the acquiring authority by the claimant, time and time again, by virtue of the position adopted that, if it cannot have the money for the car park, it would revert to section 8.

Mr Holgate said in response that any notice which may be regarded as having been served by the claimant was only contingent and that at the time of entry the claimant did not want to rely on section 8. It was pursuing a claim for land and injurious affection and when the Department of Transport made clear that they would not underwrite the cost of the car park the claimant did not amend the claim to section 8, at least not until January 1991 when it then required the acquiring authority, subject to the issue of material detriment, to take the whole.

Reverting to the issues (a)-(g) above, I find that the respective amounts were sufficiently quantified in mid-1989 as a result of the discussions between Mr Crabtree and Mr Echlin. After this date the stated estimated cost of the car park was in effect equivalent to the £800 in *Gardner*. For their own administrative and legal reasons the Department of Transport did refuse, as accepted by Mr Holgate, to underwrite the cost of the car park. The district valuers rejected the estimated cost of the car park as a proper measure of the payment which should, and could, be made for land taken and injurious affection under the compensation code and that rejection occurred before March 30 1990.

The notice held to be adequate in *Gardner* was clear that it required the acquiring authority to purchase the whole if the precondition as to the stated price for the part was not met. In the instant case that clarity was not present in any of the claimant's "notices", nor was the position made express by variation of any notice and or the claim when negotiations between Mr Crabtree and Mr Echlin pointed to the precondition of financing the car park not being met. Mr Turner in cross-examination stated that Mr Crabtree had no authority to invoke section 8 on behalf of the claimant, that is, in the *Gardner* situation, positively to require the acquiring authority to take the whole of Station House. Likewise, Mr Turner himself as a director of the claimant company took no steps in this direction before March 30 1990: see document dated March 28 1990. On May 3 1990 Mr Turner and Mr Crabtree attended a meeting with Mr Russell-Smith and the record of that meeting discloses no clear requirement to purchase the whole of Station House: see also document by Mr Crabtree dated July 10 1990, and documents by Mr Turner dated July 17 and September 6 1990.

Having regard to these documents and taking a balanced view of all the evidence I do not find that Glasshouse required the acquiring authority to purchase and stated its willingness to sell, in the manner of *Gardner*, the whole of Station House before entry on March 30 1990. I therefore answer question 3 in the negative.

I heard no submissions in relation to question 4 and it is accordingly taken as abandoned by the claimant.

I have received written submissions upon costs from the parties and it is ordered that the claimant pay to the acquiring authority its costs in the reference thus far. In the absence of agreement the costs to be taxed by the registrar of the Lands Tribunal upon the High Court standard basis.

SHIMIZU (UK) LTD v WESTMINSTER CITY COUNCIL

(Ref/14/1992)

May 21 1993

(T HOYES FRICS)

Estates Gazette May 21 and 28 1994

[1994] 20 EG 154 and 21 EG 131

Compensation — Listed building — Whether listed building consent refused for works of demolition or for alteration

The claimant is the owner of a building at the corner of Piccadilly and Old Bond Street, London W1, occupied at the ground floor by shops and as to the upper floors by offices. The building is grade II listed, save for 45, 46 and 47 Old Bond Street. The listed part was reconstructed in conjunction with the redevelopment, by the erection of an entirely new building, of 45, 46 and 47 Old Bond Street in 1990-91. The steel structural frame of the two buildings is integral and internally the parts are virtually indistinguishable. Planning permission was originally granted for the proposals in July 1988, but in June 1990 an application was made for listed building consent for the removal of chimney breasts and the creation of additional floorspace. The parties sought the determination of the following preliminary issues: (1) whether the works, which were the subject-matter of the application for listed building consent, which was refused by the Secretary of State for the Environment on June 4 1991, constitute alterations to or extension of a listed building under section 27(1)(a) of the Planning (Listed Buildings and Conservation Areas) Act 1990, with the result that compensation is accordingly payable by the compensation authority to the claimant; and (2) if compensation is payable, the date at which the amount of compensation is to be assessed.

***Decision*: (1) Although many alterations to buildings will inevitably involve an element of demolition, to categorise the total activity and the works as demolition would so reduce the scope of section 27 of the 1990 Act that refusal of consent for alterations would rarely attract compensation. The issue of whether proposed works are alterations to or demolition of part of a listed building falls to be considered on the facts of each case having regard to the nature of the works and all the surrounding circumstances. The presence of an element of demolition in the proposals of an applicant for listed building consent, and how that application is regarded by the planning authority, is a factor to be put in the balance when considering the nature or character of the overall proposals. On the evidence the works constituted an alteration to a listed building and the decision of the Secretary of State on June 4 1991 gave rise to a potential liability for compensation. (2) The date for the purposes of the second preliminary issue was agreed by the parties to be June 4 1991.**

The following cases are referred to in this report.

Antoniades v *Villiers* [1990] 1 AC 417; [1988] 1 WLR 912; [1988] 3 WLR 1205; [1988] 3 All ER 1058; [1988] 2 EGLR 78; [1988] 47 EG 193, HL

Ashland Oil & Refining Co v *Commissioner of Internal Revenue* 1938 99 Federal Reports, 2nd series 588, US Circuit Court of Appeals

Britt v *Buckinghamshire County Council* (1962) 60 LGR 430; 14 P&CR 332; [1962] EGD 557; 184 EG 167; on appeal [1964] 1 QB 77; [1963] 2 WLR 722; [1963] 2 All ER 175; (1963) 61 LGR 223; 14 P&CR 318; [1963] EGD 566; 185 EG 683, CA

Cambridge City Council v *Secretary of State for the Environment* (1992) 90 LGR 275, CA

Craven v *White (Stephen)* [1989] AC 398; [1988] 3 WLR 423; [1988] 3 ALL ER 495, HL

Customs and Excise Commissioners v *Viva Gas Appliances Ltd* [1983] 1 WLR 1445; [1984] 1 All ER 112; [1983] STC 819, HL

Furniss v *Dawson* [1984] AC 474; [1984] 2 WLR 226; [1984] 1 All ER 530, HL

Gisborne v *Burton* [1989] QB 390; [1988] 3 WLR 921; [1988] 3 All ER 775; [1988] 2 EGLR 9; [1988] 38 EG 129, CA

Glenwood Lumber Co v *Phillips* [1904] AC 405, PC

Hagee (London) Ltd v *Co-operative Insurance Society Ltd* (1991) 63 P&CR 362; [1992] 1 EGLR 57; [1992] 07 EG 122

Long v *Secretary of State for the Environment* [1981] JPL 886 QB

Manchester, Sheffield & Lincolnshire Railway Co v *North Central Wagon Co* [1888] 13 AC 554, HL

R v *North Hertfordshire District Council, ex parte Sullivan* unreported 1981

Sainty v *Minister of Housing and Local Government* (1964) 62 LGR 179

Schuler (L) AG v *Wickman Machine Tool Sales Ltd* [1974] AC 235; [1973] 2 WLR 683; [1973] 2 All ER 39; [1973] 2 Lloyd's Rep 53, HL

Shepherd Homes Ltd v *Sandham (No 2)* [1971] 1 WLR 1062; [1971 2 All ER 1267

Stephens v *Cuckfield Rural District Council* [1960] 2 QB 373; [1960] 3 WLR 248; [1960] 2 All ER 716, CA

Street v *Mountford* [1985] AC 809; [1985] 2 WLR 877; [1985] 2 All ER 289; [1985] 1 EGLR 128; (1985) 274 EG 821, HL

Wells v *Minister of Housing and Local Government* [1967] 1 WLR 1000; [1967] 2 All ER 1041; (1967) 65 LGR 408; 18 P&CR 401, CA

A *Windsor and Maidenhead Royal Borough Council* v *Secretary of State for the Environment* (1987) 86 LGR 402; [1988] 2 PLR 17

This was an application under rule 49 of the Lands Tribunal Rules 1975 for the determination of preliminary issues in a claim by the claimants, Shimizu (UK) Ltd, for compensation under section 27 of the Planning (Listed Buildings and Conservation Areas) Act 1990 against the compensating authority, Westminster City Council.

David Holgate (instructed by Freshfields) appeared for the claimant; Michael Barnes QC (instructed by the solicitor to Westminster City Council) represented the compensating authority.

B Giving his decision, MR T HOYES said: These are preliminary issues of fact and law under a reference claiming compensation under section 27 of the Planning (Listed Buildings and Conservation Areas) Act 1990 (the 1990 Act) in respect of a building situated at the corner of Piccadilly and Old Bond Street in London W1.

The building comprises shop units at street level, currently vacant, and offices on four upper levels, plus a fifth level as to part. The offices are now occupied by the claimant, the owner of the freehold interest. The building is grade II listed save for 45, 46 and 47 Old Bond Street. The listed part (formerly known as Qantas House) was reconstructed in conjunction with the redevelopment, by the erection of an entirely new building, of 45, 46 and 47 Old Bond Street in C 1990-91. The steel structural frame of the two buildings is integral and internally the parts are virtually indistinguishable.

Preliminary issues

The preliminary issues for determination are:

(a) whether the works which were the subject-matter of the application for listed building consent, which was refused by the Secretary of State for the Environment in his decision letter dated June 4 1991, constitute alterations to or extension of a listed building under section 27(1)(a) of the Planning (Listed Building and Conservation Areas) Act 1990, with the result that compensation is accordingly D payable by the compensating authority to the claimant;

(b) if compensation is payable, the date at which the amount of compensation is to be assessed.

In respect of issue (b) it is now agreed that the relevant date is June 4 1991.

Statutory provisions

The marginal note to section 27 of the 1990 Act reads, "Compensation for refusal of consent to alterations, etc of listed building" and the section is in the following terms:

> (1) This section shall have effect where —
>
> E (a) an application is made for listed building consent for the alteration or extension of a listed building;
>
> (b) the works do not constitute development or they do so but the development is such that planning permission for it is granted by a development order; and
>
> (c) the Secretary of State, either on appeal or on the reference of the application to him, refuses such consent or grants it subject to conditions.

Section 7 of the 1990 Act, in the absence of authorisation, prohibits works

> for the demolition of a listed building or for its alteration or extension in any manner which would affect its character as a building of special architectural or historic interest.

F

It was common ground that in section 7 a distinction is made between works of demolition and works of alteration and extension and, as the former is not mentioned in section 27, no entitlement to compensation arises in an owner if consent is refused for demolition. It was further agreed that section 27(1)(b) is not material in this reference and that a refusal of consent, on appeal, as required by section 27(1)(c), had been issued by the Secretary of State on June 4 1991.

The dispute centres upon the nature of the works which the claimant wished to carry out and for which its application for consent was refused. For the claimant it is said that those works were for the alteration of a listed building and for the compensating authority G it is said that they were for the demolition of part of a listed building. The one falls within section 27 and the other is outside it and for this reason the council denies they have any liability to pay compensation.

Mr David Holgate, of counsel, appeared for the claimant and called Mr J Mitham BA (Hons) Dip Arch RIBA, an associate of Frederick Gibberd Coombes & Partners. By arrangement between counsel a factual report prepared by Mr Erik Pagano BSc DMS, a development manager for the claimant, was put in evidence as an agreed document.

Mr Michael Barnes QC appeared for the compensating authority and called Mr M Lowndes BA (Hons) Dip TP Dip Cons (AA) MSC, a principal H area conservation and design officer in the employ of the council.

An agreed statement of facts was lodged and a brief summary is attached to this decision as annexure 1.

ANNEXURE 1

Extract from agreed statement of facts

1. The site comprises Qantas House (56-60 Piccadilly and 48-50 Old Bond Street) and 45, 46 and 47 Old Bend Street, originally they were four individual buildings.

2. Qantas House is Grade II listed, was designed by Read & McDonald J (architects) and was built in 1905 in Portland stone and roofed in Westmorland slate. The facade is a fine example of Edwardian baroque and the building is one of the architect's best-known works.

3. The entire site lies within the Mayfair conservation area.

4. The interior of the buildings was of poorer quality comprising inter-connecting rooms with low ceiling heights.

5. The chimney stacks at roof level form an integral part of the external appearance. The internal chimney-breasts were wholly devoid of mantlepieces, fireplaces or surrounds and the openings had been bricked up.

6. For the town planning history see pp3 and 4 of decision.

7. Demolition commenced on the site in March 1990, and the facade, chimney stacks and chimney-breasts comprising parts of Qantas House were K given structural support during the demolition works.

8. Rebuilding works began in September 1990 and a steel frame was erected to, *inter alia*, give support to the retained facade and chimney stacks of Qantas House. The chimney-breasts ceased to have a structural function; they have been "boxed in" and finished in the same manner as the adjacent walls.

9. If the chimney-breasts were removed at a future date the works scheduled in annexure 2 would be required to integrate the space they occupy into adjacent office areas.

10. The new building constructed on the site has a floor area of accroximately 30,000 sq ft.

The chimney-breasts (with boxing) occupy 327 sq ft and (excluding boxing) 276-299 sq ft; about 1% of the total. L

It was agreed that the reports of Mr Pagano and Mr Lowndes, with their supporting documents, were in effect amplifications of the agreed statement, although Mr Lowndes was briefly cross-examined.

Town planning history

The following applications have been made and permissions or consents have been granted.

1. On November 7 1986 application for planning permission for the "redevelopment of existing premises (with retention of the facade of M Qantas House) for office and retail uses". This application was deferred in February 1987 as it was likely to be refused due to the extent of demolition proposed.

2. On July 6 1988 a revised proposal for "the demolition of all except the facades, chimney-breasts and stacks of Qantas House and redevelopment to provide a retail arcade on the ground floor, ancillary retail, and office accommodation and a winebar/restaurant in the basement and offices on first to fifth floors" was given conditional planning permission and listed building consent.

3. On September 6 1989 the above proposal was given conservation area consent. (The applicant and the freeholder up to November 2 1989 was the Prudential, thereafter Shimizu.)

A 4. On December 4 1989 an application to amend the 1988 planning permission was made and this along with a revised listed building consent was granted on May 25 1990.

5. Further amendments were sought on July 20 1990 and September 18 1990. These were approved on January 29 1991 and the redevelopment has been carried out in accordance with these permissions and consents.

6. On June 22 1990 application was made for listed building consent for "the removal of internal chimney-breasts at ground, first, second, third and fourth floor levels behind the Piccadilly and Old Bond Street facades".

B 7. On July 2 1990 the above application was amended to include the chimney-breasts at fifth-floor level.

8. On August 2 1990 a further amendment was made to include temporary and permanent support to the chimney stacks over.

9. No decision was made by the council within the statutory eight-week period and the claimant appealed to the Secretary of State on October 22 1990.

10. On June 4 1991 the appeal was dismissed on behalf of the Secretary of State.

Application for listed building consent

The relevant application for listed building consent was made on C the standard form provided by the council and the description of the proposed works at question 3 was stated to involve "the demolition of internal chimney-breasts *and* alteration by the removal of the chimney-breasts". In respect of the materials to be used in relation to the alterations they were stated to be as approved by the planning permission (Ref PT/TP/18090) dated July 6 1988. The covering letter dated June 19 1990 from the claimant's town planning consultants, Fibbens Fox, referred to carrying out the development in accordance with the planning permission and listed building consents dated May 25 1990 (registered numbers 896618B and 896619, which in turn referred to the consents of July 6 1988). The letter stated,

D we hereby make application for listed building consent for the demolition and removal of those parts of the chimney-breasts

— and later —

In terms of the layout of the proposed new building retention of the chimney-breasts will be extremely undesirable providing unnecessary and damaging interruptions to the floor space available at each level.

Upon receipt of the application it was necessary for the council to decide whether it was purely for demolition and nothing else in which case they needed to have regard to the direction made under section E 15(5) of the 1990 Act, concerning notification to six specified national bodies: see Circular 8/87, para 81. The alternative course, chosen by the council, was to notify English Heritage (formerly the Historic Buildings and Monuments Commission for England), an action only necessary because of the extensive demolition of the listed building already approved and then taking place: see Circular 8/87 para 86 II (2)(b)(iii).

The Secretary of State on appeal, but effectively at first instance in this case, also had to decide whether the proposal of the claimant was demolition or alteration to a listed building in order to apply the respective national policies relevant to the decision in Circular 8/87: para 89/91, demolition, and para 92, alterations. The reference in the F decision letter (at para 6) to para 92 of the circular, indicates that the Secretary of State did not regard the application as wholly demolition although clearly aware that an element of demolition or the removal of a feature within the listed building was involved.

Subsequent to the application in June 1990 and for the proceedings before the Lands Tribunal the claimant and the compensating authority have agreed a schedule of structural and other works which would be necessary were the chimney-breasts to be removed: see annexure 2.

These are works, other than demolition, essential to the objective of the claimant in enlarging the office floorspace and removing obstructions to render the existing floorspace more useful.

Evidence

The evidence of Mr Mitham was that demolition of the former G Qantas House commenced in April 1990 and that the approved scheme of redevelopment, involving reconstruction and new building, began in September 1990 and continued thereafter so that by June 1991 the new fabric was water tight and ready for finishing internally. The structural steel frame was designed and erected as an entity to serve both the site of the listed buildings and 45, 46 and 47 Old Bond Street, which were wholly demolished. Incorporated into that frame was support for the chimney stacks over in the event that the June 1990 application for the removal of the chimney-breast was successful or indeed any similar application in the future. Removal could thus be H undertaken as a minor work without unacceptable disruption to occupation.

ANNEXURE 2

Structural works to be carried out if the chimney-breasts were removed today

1. The space created by the removal of the chimney-breasts would require infilling so that a continuous structural floor area would be created.

These alterations would serve structural functions:

(a) The creation of additional floor space such that it would be capable of supporting imposed loads upon it (ie office equipment) to the same J specifications as that required for the remainder of the floor area; and

(b) allowing this additional floor area to form part of the overall structural floor slab, adding to the lateral bracing effect imparted to the structural steel frame of the building.

The above alterations would require horizontal holes to be drilled into the existing surrounding structural concrete slab into which new and additional reinforcing bars would be linked. Once completed, hollow-rib decking (a corrugated steel sheet used to support the wet concrete) would be inserted into the empty floor areas. Concrete would then be placed on to the hollow-rib.

2. *Other works*

(i) installation of a new area of raised flooring (not structural in the sense of providing lateral bracing to the steel frame, but structural in the sense of K carrying the imposed floor loadings);

(ii) reposition of the air conditioning units and associated ductwork including the heated and chilled water supply pipework;

(iii) repositioning of and addition to the electrical wiring serving both the air conditioning units and ceiling lighting system;

(iv) installation of a new area of suspended ceiling complete with appropriate light fittings.

(v) extension of the ceiling perimeter bulkhead (ie the junction of the ceiling with the outside walls of the building).

(vi) extension of the perimeter air conditioning linear diffuser grill in line with (v) above;

(vii) the installation of new finishes to the external wall of the building L (inside face) including; a thermal barrier, a plasterboard lining and decorative applications, eg wallpaper or paint.

Both parties hereby agree that the above works would be necessary if the chimney breasts were removed today following the granting of the appropriate listed building consent.

Signed
Freshfields
Westminster City Council
21st of December 1992

Ref: RL/CPD/079710-0180
(140014C) M

In cross-examination Mr Mitham said that in June 1990 the retained facade, the new steel frame, and the retained chimney stacks were clearly visible on the site.

The report prepared by Mr Pagano (accepted as agreed and not subject to cross-examination) disclosed that Shimizu purchased the freehold interest from the Prudential on November 2 1989. A new architect, Mr Mitham, was appointed following which revisions to the approved 1988 scheme of development were put in hand, including the proposal to remove the chimney-breasts. In order not to delay the

A start of demolition works on the site, it was decided to deal with the variations in stages, for example the abandonment of an approved ground-floor shopping arcade in favour of larger units, the relocation of the lift and service core in the new building and the removal of the chimney-breasts. As a result there was an overlap between specific town planning and listed building consent applications and discussions with the council upon prospective applications. In effect the council appear not to have required a consolidation of the applications so as to include the proposals relating to the chimney-breasts with other modifications.

The evidence of Mr Lowndes related primarily to the history of the
B planning applications, recorded above. In cross-examination he agreed that it was common for applications for planning permission, etc to be acceptable in principle but not in detail and that it was also common for applicants to amend their proposals so as to attract a favourable recommendation from a planning officer. He stated that most developer applicants accepted an agreed compromise and it was somewhat rare for revisions to be sought after formal approval of an agreed scheme, as in the instant case. He regarded subsequent applications and possible appeals as a "waste of the time of the planning authority" although of possible advantage to a developer. He accepted that since July 6 1988 the owner had a planning permission which could be implemented at the subject site.

C

Authorities

The following authorities were referred to:

Wells v *Minister of Housing and Local Government* [1967] 1 WLR 1000, CA

Cambridge City Council v *Secretary of State for the Environment* (1992) 90 LGR 275, CA

Hagee (London) Ltd v *Co-operative Insurance Society Ltd* (1991) 63 P&CR 362, Ch*

Windsor and Maidenhead Royal Borough Council v *Secretary of State for the Environment* (1987) 86 LGR 402, QB†

Sainty v *Minister of Housing and Local Government* (1964) 62 LGR 179, QB

D *R* v *North Hertfordshire District Council, ex parte Sullivan* unreported 1981, QB

Ashland Oil & Refining Co v *Commissioner of Internal Revenue* 1938 99 Federal Reports 2nd Series 588, US Circuit Court of Appeals

Stephens v *Cuckfield Rural District Council* [1960] 2 QB 373, CA

Britt v *Buckinghamshire County Council* [1964] 1 QB 77, CA

Furniss v *Dawson* [1984] AC 474, HL

Craven v *White (Stephen)* [1989] AC 398, HL

Gisborne v *Burton* [1988] 3 WLR 921, CA‡

Street v *Mountford* [1985] AC 809, HL§

Glenwood Lumber Co v *Phillips* [1904] 1 AC 405, PC

E *Antoniades* v *Villiers* [1990] 1 AC 417, HL¶

Manchester, Sheffield & Lincolnshire Railway Co v *North Central Wagon Co* [1888] 13 AC 554, HL

Shepherd Homes Ltd v *Sandham (No 2)* [1971] 1 WLR 1062, Ch

Long v *Secretary of State for the Environment* [1981] JPL 886, QB

Schuler (L) AG v *Wickman Machine Tool Sales Ltd* [1974] AC 235, HL

Submissions

The following matters were common ground between counsel:

1. Alteration to a listed building can include an element of demolition provided it is *de minimis* in extent.

F 2. Listed building consent was required for the proposals of the claimant, despite the opinion expressed by Mr Mitham to the contrary in his report.

3. In terms of processing the application of June 22 1990, the council treated it as being for alterations, likewise the Secretary of State in his decision on appeal dated June 4 1991.

4. Demolition is purely negative in function.

5. There was no deliberate scheme by the claimant, involving the G
fragmenting of its applications, so as to secure and maximise compensation.

For the claimant

Mr Holgate said that there was very little case law to differentiate the words demolition and alteration. He submitted that the 1990 Act does not indicate that demolition and alteration are exclusive categories; an alteration need not involve making an addition, it could relate just to the removal of part of the building. This position is contemplated by para 92 and Appendix IV of Circular 8/87 and also see *Hagee* p731 (albeit the alteration consequent upon removal in that case was *de minimis*) and *Windsor and Maidenhead* at pp407 and 408 H
where the following words of Lord Diplock in *Customs and Excise Commissioners* v *Viva Gas Appliances Ltd* [1983] 1 WLR 1445 at p1451 (a VAT appeal case) are cited and approved.

> I can see no ground on which the meaning of the ordinary English word "alteration" qualified by the adjectival phrase "of any building" should be construed as excluding any work upon the fabric of the building except that which is so slight or trivial as to attract the application of the *de minimis* rule.

Mr Holgate submitted that demolition cannot involve anything more than removal of the building or a part thereof; it is entirely negative in concept. If the works involve an addition to the fabric as J
well as removal, they constitute an alteration: see *Sainty* where the demolition and replacement of a wall, in contrast to a complete dwelling, was considered to be an alteration for the purposes of the general development order by Parker CJ at p180. In the instant case the works involved demolition of the chimney-breasts and the addition of floorspace, therefore they constitute an alteration. Mr Holgate went further and contended that situations may arise where proposals can be both demolition and alteration due to the overlapping of the two non-exclusive categories. Section 27(1)(a) does not provide that the works must not constitute demolition as well as alteration, it does not refer to "only" alteration, nor does it provide that if the works proposed K
include demolition of part then compensation is not payable. In essence the Act requires that the application, to attract compensation on refusal, must involve alteration (or extension) of a listed building and that it does not relate solely to the demolition of the whole or a part; only pure demolition is excluded from compensation.

A further submission by Mr Holgate related to the issue of whether there may be some works involving solely removal which can only fall within demolition. He said if that question does arise the only possible distinction drawn in the Act is in section 7, which provides that not all alterations or extensions require consent but only those which materially affect the character of the building as a building of L
"special architectural or historic interest" ("the special character test"). The Act does not require the special character test to be met when deciding whether consent is needed for "demolition". Clearly there is no need for such a test where the demolition relates to the whole; the special character is completely lost. Similarly, the word "demolition" only covers removal of such a substantial part of the building that there is no need for the special character test to be considered when determining whether an application for listed building consent is necessary. Unless demolition of part is construed in that way, ie substantial and affecting character, the word demolition "would cover any removal of any part of the building no matter how small, *de* M
minimis apart, and irrespective of whether that removal had any significance for the preservation of the listed building".

However, he contended that this alternative interpretation cannot be correct because it would result in listed building consent being required for the removal (demolition) of a modern office partition, which bore no relationship to the original form of the building and was of no significance to listed building control. He said it would be curious if on the special character test, the planning authority would have concluded no consent was required but that position had to be reversed because of the works being designated demolition and thereby attracting more stringent procedural requirements: see *North Hertfordshire*.

---

*Editor's note: Also reported at [1992] 1 EGLR 57.

†Editor's note: Also reported at [1992] 2 PLR 17.

‡Editor's note: Also reported at [1988] 2 EGLR 9.

§Editor's note: Also reported at [1985] 1 EGLR 128.

¶Editor's note: Also reported at [1988] 2 EGLR 78.

A The definition of "building" in section 336(1) of the 1990 Act applies throughout town planning legislation "except insofar as the context otherwise requires". The context of section 7 of the 1990 Act indicates that demolition does not apply to a part of a building however small. In the present case, even if the works of removal alone are considered, those works were too insubstantial to be considered as falling exclusively within demolition and wholly outside alteration. In considering whether any works of demolition are substantial the judgment needs to be applied, where section 27 is involved, at the date of the Secretary of State's decision refusing consent because by section 27(c) that decision is a precondition to the entitlement to B compensation and also because the Secretary of State decides the appeal as a *de novo* application, see section 22(l)(a) of the 1990 Act, having regard to all circumstances at the date of his decision. In the instant case, at June 4 1991, the listed building comprised a redeveloped structure ready for internal finishing *per* Mr Mitham's evidence.

Mr Holgate submitted that it was an incorrect approach for the council to suggest that the June 1990 application was in reality only an application for demolition because the issue concerning the chimney-breasts in earlier applications had been how much of the original building should and could be allowed to be demolished: the evidence C see of Mr Lowndes and the position of English Heritage. First, the previous listed building consents granted covered both demolition and everything erected behind the retained facade, thus these consents included alterations to the listed building. Second, the 1990 application related not just to the removal of the chimney-breasts, but also to the creation of additional floorspace, namely how the listed building should be further altered. Third, if the 1990 application related to an alteration that position should not be affected by the fact that at the date of application the listed building then only comprised the retained facades, the chimney stacks and breasts and their temporary restraints. The nature of the application giving rise to D compensation should not be treated as varying according to the time it is made.

For all these reasons, Mr Holgate considered that the totality of the proposals of the claimant, forming the 1990 application, should be regarded as constituting an alteration to a listed building within section 27 and not solely the demolition of part of the building.

For the compensating authority

Mr Barnes in reply said that the basic question is did the works proposed amount to alteration or the demolition of part of Qantas House, a listed building? He submitted that only two possibilities were E open with respect to the application of section 336 Town and Country Planning Act 1990 and the reservation in section 336(1), "except insofar as the context otherwise requires", it either applies or it does not. Building includes any part of a building subject to *de minimis* or building means the whole and not any part. He contended that there was no possibility of approaching the matter in some truncated form, ie disregarding limited demolition and no authorities had been cited by Mr Holgate for that approach. The council say that the statutory definition of building is not to be disregarded in this case; the guidance in Circular 8/87 indicates an awareness of the problem and it was so treated in *North Hertfordshire*.

F Mr Barnes said that the correct approach was to look at the application and ask, "What is it an application for?" In his submission it was an application for the demolition of part of a listed building, there is a demolition of part of a building if any part of the structure is demolished. Whether the removal of the chimney-breasts denotes demolition must be judged by reference to the listed building alone, by what it comprised at the date of the application (then a skeleton), by reference to all physical facts including the aesthetic value of the chimney-breasts and the financial consequences of their removal and to the fact that the chimney-breasts were to be removed and not replaced. He referred to three rules or "common-sense principles" relevant to the identification of demolition; works can be *de minimis*, alteration always involves some change to the existing and where removal and replacement are involved works can be categorised as G either demolition or alteration, he cited *Windsor and Maidenhead*, *Hagee*, *Sainty* and *North Hertfordshire*, in the latter case Comyn J regarded demolition as the dominant factor in proposals to demolish and replace.

Mr Barnes contended that since the listed building was a skeleton at the application date, the chimney-breasts were an important part of that which then existed and therefore the proposal of the claimant must have been a demolition, because "that is how an ordinary man would see it". He attached little importance to the proposed increase in floorspace as indicating more than mere demolition.

The covering letter with the application (from Fibbens Fox on June H 19 1990), in terms, referred to "partial demolition"; this was the application and it is improper in law to use subsequent documents and actions such as the council's planning committee report and the Secretary of State's decision letter, to determine the meaning of the application, he referred to *Schuler* as the authority. He said it is fundamentally wrong to contend, as Mr Holgate did, that some works can be both demolition or alteration they must be one or the other, there may be borderline cases but that is nothing to the point; it merely makes the decision more difficult. As the legislation intended demolition to be treated differently, in the ways already identified by Mr Holgate, there cannot be a hybrid or overlapping situation. This J differentiation is common in town planning law, for example section 74 of the 1990 Act concerning demolition in conservation areas and section 55(1)(A)(a), demolition, and section 55(2)(a), other alteration, in the Town and Country Planning Act 1990. This is the significance of the "dominant" point in *North Hertfordshire*; where there is doubt it is the demolition category which is paramount.

He submitted that the 1988 listed building consent for Qantas House was for demolition and that replacement by a new structure did not make it alteration, likewise in the instant case making good the floors and ceiling, etc (the agreed works at annexure 2) cannot change what is a demolition, there will frequently be making good after K demolition of part of a building. The construction of a statute, as a matter of principle, cannot be influenced by guidance and direction from the Secretary of State in Circular 8/87: see *Stephens* and *Britt*.

A further principal submission by Mr Barnes was that the question should be considered not solely by reference to the application of June 22 1990 but also with the first application of November 7 1986, which resulted in the consent of July 6 1988 to demolish and reconstruct Qantas House. He contended that this was, "the application of the general principle of law that in construing a statute or in ascertaining the nature for legal purposes of any transaction the courts would have regard to the substance and reality of the transaction and not to its L outward form". He said an aspect of this principle is that when a transaction can naturally and appropriately be effected by one step, but two or more steps are employed with the result that some advantage under statute is gained, the court will ignore the additional steps and treat the transaction as effected by a single step. He added that the approach of different areas of the law would vary within the principle and he emphasised that in this submission he was not contending that there was an elaborate scheme to maximise compensation. He cited *Furniss* and *Craven* and *Ashland Oil* in taxation, *Gisborne* in agricultural holdings, *Street*, *Glenwood* and *Antoniades* in leases or licences, *Manchester Railway Co* in mortgagees and *Shepherd Homes* M in restrictive covenants. He said the claimant in the instant case, while genuine in its application, had made a "series" of applications to get to the best end result, presumably the most useful and valuable building after reconstruction, the declared objective of the claimant.

He said the principles in the authorities cited were relevant to the instant case because, in his submission, there was nothing to be gained by making a separate application for the removal of the chimney-breasts, there was no purpose which had a "commercial standing on its own". It may well be that the staged approach arose from a change in the ownership of Qantas House in 1989, but to fragment applications is to enable owners to fragment schemes; the instant case should be decided on the reality and substance, the form of the application should not yield large compensation to the claimant.

In conclusion he said that if the only scheme had been the removal of the chimney-breasts it may well have been an alteration, but the true scheme had always been something bigger with "the sticking point" being the chimney-breasts. The claimant has not shown in its evidence any commercial purpose in the later and separate application.

In closing Mr Holgate accepted that the issue fell to be decided by reference to the listed building only but that its physical state was to be taken as that existing on June 4 1991 when the application for consent was refused on appeal, in support he cited *Wells*. He said structural alterations can include the demolition of part of a building, see *Cambridge City*, and it was inappropriate to seek to "shoehorn" such proposals into the demolition category. The approach of Mr Barnes was, if in doubt to call it demolition, but this was not a fair reading of the principle and approach in *North Hertfordshire* as the judge was not in any doubt on the facts of that case; in the instant case the removal was not so substantial as to take the proposed works outside alteration. It is, and always was, accepted by the council as planning authority that new floorspace was to be created by the claimant's proposals which had a clear commercial objective going beyond pure demolition.

As there is no evidence of any preordained sequence of events, accepted by Mr Barnes, his submission that the claimant should lose its right to compensation having regard to the principles in *Furniss* and other authorities is ill founded. These cases were concerned with the evasion of taxation and the denial of security of tenure by the use of sham transactions, etc; the avoidance of mandatory statutory requirements. In the instant case there was ample commercial purpose in the June 1990 application for listed building consent and equally there were good commercial reasons for not including these works in the modified application of the approved major scheme; namely to avoid delay and its associated cost *per* the evidence of Mr Pagano. If the tribunal is satisfied that the proposed works are properly classified as alterations the claimant should not be deprived of compensation by the excursion of Mr Barnes into *Furniss* and other cases.

At the joint request of the parties I inspected the building and identified the chimney-breasts the subject of the refusal of consent and considered on site the agreed works detailed in annexure 2. I also observed the effect of the chimney-breasts, as now boxed in and finished in similar manner to the adjacent walls, upon the layout and use of the office space.

Decision

Many alterations to buildings will inevitably involve an element of demolition. In such circumstances to categorise the total activity and the works as demolition, as Mr Barnes sought to do, thereby placing them outside section 27 would so reduce the scope of the section that refusal of consent for such alterations would rarely, if ever, *de minimis* apart, attract compensation. Likewise, with extensions where the negative function of demolition is involved at the outset of the total scheme of works.

In my judgment, the issue of whether proposed works are alterations to or the demolition of part of a listed building, falls to be considered on the facts in each case having regard to the nature of the works and all the surroundings circumstances.

The starting point must be the information contained in and given at the time of the application, whether on a particular application form, in supporting plans and sketches, etc and in covering correspondence: see *Long*. The applicant describes and defines the proposal in this manner and often indicates the purpose in pursuing the proposals or scheme. From this information the planning authority are able and are required to consider the overall character of the proposal and process the application, in terms of the extent of notification and consultation, accordingly. That judgment should relate to the overall proposals comprising the application and no doubt frequently must be one of impression. It will concern such matters as the scale of the proposals relative to the total building, whether there is demolition and replacement or remodelling of a part of the building and the nature and scale of these activities.

The presence of an element of demolition in the proposals is a factor to be put in the balance when deciding the nature or the character of the overall proposals, but unless it is predominant it cannot dominate and dictate the final impression of the nature of the proposed works: see *North Hertfordshire*. To conclude otherwise would effectively involve disregarding the overall "flavour" of the proposals by seizing upon one particular ingredient and considering the proposals in a fragmented manner. Demolition of the whole or a part of a building is an identifiable function or activity compromising purely destructive works: see *Cambridge* at p294. The alteration of a building is wider, involving other functions or operations among which may be included an element of demolition. Only when demolition is the dominant function does it sensibly inform the impression of the nature or character of the proposals as a whole.

In this case a specific free-standing application for further works to a listed building which had already undergone extensive demolition, and was undergoing reconstruction, was made. An earlier permission and or consent was plainly being implemented and therefore the resulting form of the reconstructed building determines the background against which the later application falls to be considered; the scheme being implemented for this purpose is deemed to be completed. This was the context of the decision by the Secretary of State refusing the consent for the works proposed by the claimant.

In the accepted absence of any scheme to secure or increase compensation by the timing and form of the application for consent, I see no relevance in the principles derived from *Furniss* and the other cases cited. Even if on the face of things they had relevance by virtue of a particular objective I find that there was ample good commercial purpose in the particular application, in short to improve the building. There was also a separate commercial purpose in proceeding earlier upon the larger scheme as the consent for it gave the claimant authorisation to create 99% of the possible physical space within the envelope of the building, albeit with some disability to interior layout: see annexure 1, item 10. For these reasons I reject the submissions of Mr Barnes that the proposals of the claimant should be viewed in aggregate and that when so considered the works for which consent was refused should be categorised as demolition.

Taking a balanced view of the facts and the evidence relating to the background of the proposals, also being mindful of how the application was categorised upon receipt by the council and subsequently treated by the Secretary of State, his refusal being fundamental to any compensation being payable, I reach the conclusion that the works constituted an alteration to a listed building rather than the pure negative function of the demolition of a part of the building. This is the position I arrive at if I follow the routes urged by Mr Barnes, "the view of the ordinary man" and by Mr Holgate, to treat demolition and alteration as plain ordinary words *per Windsor and Maidenhead*.

There is undoubtedly an element of making good in the agreed works scheduled at annexure 2, but, in my judgment, they go beyond matters purely incidental to or consequent upon making good after removal of the chimney-breasts. Furthermore, the identified objective of the claimant, specific throughout, was to improve the building being created by increasing its useable floorspace and enhancing the usefulness of the space existing around the chimney-breasts, namely to improve the internal layout. This is an opportunity to further alter the building which the Secretary of State has determined should be denied to the claimant in the cause of retaining listed buildings in their existing form for the public good.

In answer to the first question I therefore find that the proposed works for which consent was refused were an alteration to a listed building and that a potential liability to pay compensation arises in the council as compensating authority.

The answer to the second question is agreed to be June 4 1991.

Written submissions were invited and have been received from the parties upon costs. Following consideration of those submissions it is

A ordered that the compensating authority do pay to the claimant its costs in the preliminary issue, including in the application leading to the order of November 9 1992, such costs if not agreed to be taxed by the registrar of the Lands Tribunal upon the High Court standard basis. Any other costs incurred thus far in the reference are reserved. B

Note: The next case appears on p 229 of the green section

SUPPLEMENT

CASES NOT REPORTED IN "ESTATES GAZETTE"

COMPULSORY PURCHASE AND COMPENSATION

A County Court

June 10 1993

(Before Judge HAGUE QC, at Mayors and City of London Court)

MERCURY COMMUNICATIONS LTD v LONDON & INDIA DOCK INVESTMENTS LTD

Compensation — Wayleave agreement — Telecommunications code B *— Whether compulsory purchase compensation principles or ransom value proper approach to determination of consideration for deed of grant*

The plaintiff, Mercury Communications Ltd, proposed to lay 230m of additional cable ducts under Eastwood Wharf Road, a private road owned by the defendant, to link the eastern end of the Canary Wharf development, London Docklands, with one of Mercury's earth stations; these receive and transmit international telecommunications by means of satellite. The parties agreed the terms of a deed of grant of rights to lay the cable duct save for the determination of the consideration. By para 7 of the C Telecommunications Code (Schedule 2 to the Telecommunications Act 1984) the county court may determine "such terms with respect to the payment of consideration . . . as it appears to the court to be fair and reasonable if the agreement had been given willingly". It was contended, on behalf of Mercury, that compulsory purchase compensation principles were applicable and that the consideration was nil or nominal; the consideration should be limited to a small capital payment similar to the capital payments agreed with the National Farmers' Union and Country Landowners' Association. For the defendant landowner it was argued that the rights to be conferred by the deed of grant were equivalent to a "ransom strip" and that an annual payment of £24,175 was appropriate.

D **Held: The direction in the code that the terms are to be "fair and reasonable" involves an element of subjective judicial opinion; the court's determination is not necessarily the same as the result in the market would have been if the grant had been given willingly. Compulsory purchase principles, including the *Pointe Gourde* principle, are not applicable. A ransom value approach, having regard to the profit which Mercury was forecasting from the use of the cables, was inappropriate. The evidence of the terms of aerial agreements with telecommunications operators was of general assistance, in** E **showing the relative bargaining powers of grantors and grantees, but the agreements with the NFU and CLA were of no value because they related to agricultural land and did not include anything for the bargaining power of the grantor. Two agreements in 1987 and 1988 relating to the laying of cables in the immediate area for annual payments of £3,000, and arrived at with the code in the background, contained the best evidence. Adjusting that figure for inflation and the additional number of cable ducts, the appropriate annual payment was £9,000. *Re Naylor Benzon Mining Co Ltd* [1950] Ch 567 and *BP Petroleum Developments Ltd* v *Ryder* [1987] 2 EGLR 233 not followed.**

The following cases are referred to in this report. F

Batchelor v *Kent County Council* (1990) 59 P&CR 357; [1990] 1 EGLR 32; [1990] 14 EG 129; (1989) RVR 181; [1990] JPL 571, CA
BP Petroleum Developments Ltd v *Ryder* [1987] 2 EGLR 233; [1987] RVR 211
Colchester Estates (Cardiff) v *Carlton Industries plc* [1986] Ch 80; [1984] 3 WLR 693; [1984] 2 All ER 601; [1984] EGD 461; (1984) 271 EG 778
Consett Iron Co Ltd's Application, Re [1938] 1 All ER 439
Delaforce v *Evans* (1970) 22 P&CR 770; (1970) 215 EG 315, LT
Evans (FR) (Leeds) Ltd v *English Electric Co Ltd* (1977) 36 P&CR 185; (1978) EGD 67; 245 EG 657
Fraser v *City of Fraserville* [1917] AC 187; 116 LT 258; 33 TLR 179, PC
Hertfordshire County Council v *Ozanne* [1989] 2 EGLR 18; [1989] 43 EG 182; [1989] RVR 179, CA G
Hertfordshire County Council v *Ozanne* [1991] 1 WLR 105; [1991] 1 All ER 769; (1991) 89 LGR 345; [1991] 1 EGLR 34; [1991] 13 EG 157, HL
Inland Revenue Commissioners v *Clay* [1914] 3 KB 466; (1914) 83 LJKB 1425; 111 LT 484; 30 TLR 573
Lambe v *Secretary of State for War* [1955] 2 QB 612; [1955] 2 WLR 1127; [1955] 2 All ER 386, CA
Lloyd-Jones v *Church Commissioners for England* (1981) 261 EG 471, LT
Lucas and Chesterfield Gas & Water Board, Re [1909] 1 KB 16
Lynall v *Inland Revenue Commissioners* [1972] AC 680; [1971] 3 WLR 759; [1971] 3 All ER 914, HL
Naylor Benzon Mining Co Ltd, Re [1950] Ch 567; [1950] 1 All ER 518; (1950) 1 P&CR 69 H
Pointe Gourde Quarrying & Transport Co Ltd v *Sub-Intendent of Crown Lands* [1947] AC 565, PC
Raja Vyricherla Narayana Gajapatiraju v *Revenue Divisional Officer, Vizagapatam* [1939] AC 302
Sidney v *North Eastern Railway Co* [1914] 3 KB 629, DC
South Eastern Railway Co v *London County Council* [1915] 2 Ch 252
Stokes v *Cambridge Corporation* (1961) 13 P&CR 77; [1961] EGD 207; 180 EG 839, LT
Whitwham v *Westminster Brymbo Coal & Coke Co* [1896] 2 Ch 538

Richard Aikens QC and John Nicholls (instructed by Bird & Bird) appeared for the plaintiff; John Hobson (instructed by Allen & Overy) J represented the defendant.

Giving judgment, JUDGE HAGUE QC, said: This is an application made by Mercury Communications Ltd ("Mercury") for the grant of certain rights to lay and use ducts under a private roadway in the Isle of Dogs owned by London and India Dock Investments Ltd ("LIDI"), a subsidiary of the Port of London Authority ("the PLA"). The application is made under the "Telecommunications Code" contained in Schedule 2 of the Telecommunications Act 1984 ("the 1984 Act"). It has been agreed that these rights should be granted by an appropriate deed of grant, but the parties are in dispute as to the K financial terms on which the grant should be made. That is now the sole matter for my decision. At the time of issue of the application there were other matters in dispute, but these were resolved by agreement before the hearing. All the terms and conditions, apart from the financial terms, are incorporated in an agreed deed of grant.

I have been told that this is the first application under the telecommunications code which has reached the stage of a judicial decision, and that the result of this application may have important consequences in other cases between Mercury and landowners generally. That is no doubt so, but of course I can only deal with the facts and circumstances of this case, which in some respects are rather special. However, as will appear, I deal with several points of interpretation of the telecommunications code which are likely to be of general application.

Background

It is convenient first to describe briefly the statutory and factual background to the application.

Prior to 1981, the Post Office had a monopoly in the provision of telecommunication services, as a result of various Acts some of which I shall refer to later. Under the British Telecommunications Act 1981 ("the 1981 Act") a new statutory corporation, called British Telecommunications and defined in the 1981 Act as "the corporation", was established to take over the telecommunications part of the business of the Post Office. By section 10 of the 1981 Act, all the relevant property, rights and liabilities of that business were transferred to the corporation. By section 12, the corporation was given the exclusive privilege of running a telecommunications business in the British Isles; but by section 15, the Secretary of State was empowered to grant a licence to any other person to operate a telecommunications service, as specified in the licence. Mercury was granted such a licence.

The 1984 Act made provision for the corporation's property, rights and liabilities to vest in a new company of the same name ("BT"), in order to facilitate its privatisation. The 1984 Act also abolished BT's statutory monopoly, but made it necessary for any other operator to obtain a licence from the Secretary of State. Under section 7, such a licence could authorise the operator to connect to and use the telecommunication system and apparatus of any other operator, including BT. Under section 10, the Secretary of State was given power to apply the telecommunications code to an operator who was granted a licence.

Mercury was duly granted such a licence on November 5 1984 for the period of 25 years from November 8 1984. By para 2, the licence applied the telecommunications code to Mercury (subject to certain exceptions and conditions, none of which are relevant). Mercury was given power to connect to and use the systems of BT and any other licenced operator and provision was made for agreements whereby the charge for any call is divided between the operators whose lines are used. (There are similar agreements in force as to international calls.) Mercury is the only operator licenced to operate nationally, so that nationally there is effectively a duopoly with BT. There are, however, some operators who are licenced to operate locally, including East London Telecommunications Ltd, which as its name indicates operates in the east London area, including the Isle of Dogs.

Mercury's telecommunications network is based on an optical fibre trunk cable network, which at present connects over 90 of the cities and major towns in the UK. Over the last 10 years, Mercury has invested £1.7 bn in building a national and international trunk network with other associated activities. In the main, the trunk cables run alongside British Rail tracks, but in some instances also alongside canals or in London Underground tunnels. Many short stretches are laid under public highways under the Public Utilities Street Works Act 1950 (which enables public utilities to lay installations under public highways without payment of consideration to the highway owner). Within each city or town, access to and from the network is gained via a "trunk access node". Customers are connected via local distribution cables and a "Cable Distribution Node" (CDN) to the trunk access nodes and thus to the trunk network. A CDN performs a similar function to an exchange. Optical fibre cables connect the various trunk access nodes and CDNs to each other.

One such CDN, known as the Thameside CDN, comprises part of the Docklands Telecommunications Centre, at Eastwood Wharf Road, on the Isle of Dogs. This is also one of Mercury's three "earth stations" which receive and transmit international telecommunications by means of satellite. It is also an international cable station. In addition, it handles virtually all of Mercury's telecommunications traffic between Docklands and the rest of London. The Thameside earth station, as I shall call the whole centre, is thus a vital part of Mercury's national network. It is built on land owned freehold by LIDI under a lease granted in 1984 pursuant to an earlier building agreement entered into granted shortly after Mercury received its licence under the 1981 Act.

Canary Wharf

As is well known, a major part of the plans for the redevelopment of London's Docklands on the Isle of Dogs comprises the mainly commercial development at Canary Wharf by Olympia & York Canary Wharf Ltd ("O&Y"). This redevelopment was originally intended to be in three phases. As I understand it, phase 1 is now more or less complete; but the future of phases 2 and 3 is, to say the least, uncertain. Mercury is very anxious to provide telecommunication services to the Canary Wharf properties, for obvious business reasons. Most of the anticipated commercial occupiers will be heavily dependant on full telecommunication services for the purposes of their businesses. Also, some occupiers will be existing customers of Mercury who are relocating, and Mercury is anxious to retain their custom. In order to obtain internal authorisation for the capital expenditure involved, a so-called "business case" was prepared by the city region of Mercury as the sponsoring region. The purposes and contents of the business case were described by Mr John Redgate (currently the controller, capital expenditure, for Mercury), in his statement and oral evidence and I do not propose to go into further detail about it. The business case contained a financial summary, which has now been updated and really overtaken by a revised forecast, prepared under Mr Redgate's supervision and completed in July 1992.

The business case was duly accepted by those responsible in Mercury. It showed an attractive commercial anticipated return on investment, but in addition it is, I think, clear that Mercury regarded a full telecommunications service for Canary Wharf as a prestige project which it was essential for Mercury to be involved in. Mercury have accordingly negotiated suitable arrangements with O&Y. As part of those arrangements, O&Y have constructed an underwater tunnel to Blackwall Place which will carry Mercury's cable ducts. From the end of the tunnel, it is proposed to lay the ducts under a private roadway owned by LIDI to connect with Mercury's Thameside earth station.

Proposed rights

The private roadway in question is Eastwood Wharf Road, which is a tarmac estate road of secondary quality. It runs off a public highway to the east called Prestons Road. It is a no-through road which comes to an end at West India Dock. It serves a number of adjoining commercial properties on the Wood Wharf Business Park, the freeholds of which are owned by LIDI or associated companies. Immediately to the north is the elevated Cartier Circle, a road junction constructed at the eastern end of Canary Wharf.

One of the adjoining commercial properties is the Thameside earth station. The right to lay and use ducts is required to expand Mercury's telecommunications link from the earth station to the eastern end of the Canary Wharf development. There are existing cable ducts laid under Eastwood Wharf Road by Mercury running from the Thameside earth station to Canary Wharf under two agreements, the first being with the PLA dated January 29 1987 ("the 1987 agreement") and the second with LIDI dated June 22 1988 ("the 1988 agreement"). I shall have to refer to these two agreements in due course. But the existing ducts are quite inadequate for Mercury's purposes.

Mercury therefore proposes to lay further cable ducts under Eastwood Wharf Road. The total length of the new duct system will be 230m. Of this length, 177m will comprise 24, 90mm pvc ducts, set three wide by eight deep in a reinforced concrete block, laid at a minimum depth of 600mm. The remaining 53m will comprise six 90mm pvc ducts laid directly in the ground, also at a minimum depth of 600mm. The proposed works will also include four manholes. These will be underground concrete chambers with external dimensions of 2,950mm x 1,650mm x 2,530mm, accessed by vertical shafts, 900mm^2, of varying lengths and each topped by a single metal cover 610mm^2 level with the road surface.

A It is common ground between the parties that Eastwood Wharf Road is much the most convenient route for the proposed new ducts, and that the only possible alternative routes would all lie across LIDI's land let to others. LIDI's agreement to the grant of the proposed right is therefore essential.

As well as Mercury's existing ducts which I have mentioned, there are also a number of service and public utility ducts, drains and other conducting media currently laid under the verges and carriageway of Eastwood Wharf Road. An illustrative diagram of these was produced in evidence. In addition, O&Y have a right to lay further drains there. There was some disagreement as to whether the installation of
B Mercury's proposed new ducts would restrict the ability to lay further services in the road, and there was no real evidence on the point. I do not think it really matters one way or the other, but in the absence of clear evidence to the contrary I think I should assume in favour of Mercury that other services could be laid without undue difficulty.

Agreed statement of facts

Much of the foregoing is taken from an agreed statement of facts prepared for the purposes of the hearing by the surveyor expert witnesses on each side, Mr David J Room FRICS of Grimley J R Eve, on behalf of Mercury, and Mr Edward M Sheard FRICS of Matthews & Son, on behalf of LIDI. After preparing and exchanging their initial
C reports and supplementary reports pursuant to an interlocutory order dated May 15 1992, Mr Room and Mr Sheard very sensible met together to endeavour to agree as many matters of fact and valuation as they could. They were happily able to agree virtually all the material facts, including the facts concerning the comparables, and also to come to a compromise (strictly for the purposes of this litigation only) as to all but a few of the relevant figures and percentages, in particular those arising in connection with Mr Sheard's valuation. All these matters are set out in the agreed statement, which is dated December 23 1992. This must have involved each of them in a great deal of work and negotiation. I have found that document of
D the greatest assistance; and without the sensible compromises which they managed to reach this already difficult case would have been even more complicated and the length of the hearing would have been much extended. I should like to express my gratitude to both of them for their achievements in this regard and also for the clarity of their various reports.

I am also grateful to the solicitors on each side for the convenient and careful way in which the numerous documents in the case have been prepared and presented and to counsel, Mr Richard Aikens QC and Mr John Nicholls, on behalf of Mercury, and Mr John Hobson on behalf of LIDI, for the lucidity and helpfulness of their arguments.
E They summarised their arguments in written "outline submissions" which I have found of great assistance.

Deed of grant

As I have mentioned, there is no dispute that the proposed rights should be granted and the parties finally agreed a form of deed of grant, apart from the financial terms, shortly before the start of the hearing. The deed is a detailed document granting Mercury the rights it requires, subject to various terms and conditions. It is not necessary for me to refer to it in any detail, but I should mention the following. (1) The grant is for a term ending in the year 2009. (2) All the
F installation and maintenance costs are to be paid by Mercury. (3) Mercury gives LIDI a full indemnity against any liability arising as a result of the installation of the ducts and the operation of the cables, whether or not there is any default on the part of Mercury. (4) On the footing that the consideration for the grant should be an annual sum (and not a capital payment as Mercury contends), there is provision for an annual review in line with inflation: this includes, if necessary, a reference to arbitration, which is permitted by para 7(4) of the telecommunications code set out below.

Telecommunications code

I turn now to the telecommunications code set out in Schedule 2 of the 1984 Act, which I will refer to as "the code". It is necessary for me to set out a few of its provisions in full. Under the definitions in para 1 G and in relation to the facts of this case, "the operator" means Mercury, "the operator's system" means Mercury's telecommunication system, and "the statutory purposes" means the purposes of establishing and running Mercury's system. Para 2 reads as follows, so far as material:

> 2.—(1) The agreement in writing of the occupier for the time being of any land shall be required for conferring on the operator a right for the statutory purposes—
>
> (a) to execute any works on that land for or in connection with the installation, maintenance, adjustment, repair or alteration of telecommunication apparatus; or
>
> (b) to keep telecommunication apparatus installed on, under or over that land; or H
>
> (c) to enter that land to inspect any apparatus installed (whether on, under or over that land or elsewhere) for the purposes of the operator's system.
>
> (2) A person who is the owner of the freehold estate in any land or is a lessee of any land shall not be bound by a right conferred in accordance with subparagraph (1) above by the occupier of that land unless—
>
> (a) he conferred the right himself as occupier of that land; . . .
>
> (5) A right falling within subparagraph (1) above shall not be exercisable except in accordance with the terms (whether as to payment or otherwise) subject to which it is conferred; and, accordingly, every person for the time being bound by such a right shall have the benefit of those terms. J

In this case, as LIDI is the freehold owner as well as the occupier of the private road, no difficulty arises from subpara (2) of this paragraph.

Para 5 of the code gives the court power to dispense with the agreement of any person in quite wide circumstances, and by a court order to confer on the operator the rights which he requires. It is this power which Mercury is asking me to exercise in this case. There is no dispute that the conditions for the grant of such an order are satisfied; as I have indicated LIDI has always been ready to grant to Mercury the rights it requires on terms, and the only term now outstanding concerns the consideration. Subpara 4 reads: K

> (4) An order under this paragraph made in respect of a proposed right may, in conferring that right or providing for it to bind any person or interest in land and in dispensing with the need for any person's agreement, direct that the right shall have effect with such modifications, be exercisable on such terms and be subject to such conditions as may be specified in the order.

The crucial paragraph on which this case really turns is para 7. This enables the court to fix financial terms where agreement is dispensed with, and reads, so far as material, as follows:

> 7.—(1) The terms and conditions specified by virtue of subparagraph (4) of paragraph 5 above in an order under that paragraph dispensing with the need L for a person's agreement, shall include—
>
> (*a*) such terms with respect to the payment of consideration in respect of the giving of the agreement, or the exercise of the rights to which the order relates, as it appears to the court would have been fair and reasonable if the agreement had been given willingly and subject to the other provisions of the order; and
>
> (*b*) such terms as appear to the court appropriate for ensuring that that person and persons from time to time bound by virtue of paragraph 2(4) above by the rights to which the order relates are adequately compensated (whether by the payment of such consideration or otherwise) for any loss or damage sustained by them in consequence of the exercise of those rights.
>
> (2) In determining what terms should be specified in an order under M paragraph 5 above for requiring an amount to be paid to any person in respect of—
>
> (*a*) the provisions of that order conferring any right or providing for any right to bind any person or any interest in land, or
>
> (*b*) the exercise of any right to which the order relates, the court shall take into account the prejudicial effect (if any) of the order or, as the case may be, of the exercise of the right on that person's enjoyment of, or on any interest of his in, land other than the land in relation to which the right is conferred.
>
> . . .
>
> (4) The terms specified by virtue of sub-paragraph (1) above in an order under paragraph above may provide—

(*a*) for the making of payments from time to time to such persons as may be determined under those terms; and

(*b*) for questions arising in consequence of those terms (whether as to the amount of any loss or damage caused by the exercise of a right or otherwise) to be referred to arbitration or to be determined in such other manner as may be specified in the order.

The court

"The court" is defined by para 1 of the code as meaning the county court. Presumably, Parliament thought that cases under the code would be relatively straightforward and could be accommodated in the normal county court listings without difficulty. The hearing before me extended over seven full days. The papers are contained in eight lever arch files, some of them quite bulky. As well as considering the several reports from each expert and hearing their oral evidence, I have read statements from seven other persons and four of them also gave oral evidence. Counsel made their submissions to me with economy, but their written outline submissions together covered 60 pages. Further, the valuation issues which I have considered are of the kind which are familiar to the Lands Tribunal, but not to most county court judges.

Bearing these matters in mind, I would suggest that in any future application under the code which is likely to be of substance, consideration ought to be given at an early stage to: (a) transferring the application to this court (Mayors and City of London Court), as happened in this case (the application having been commenced in Bow County Court); and (b) making arrangements so that the application is heard by a judge who has some experience in valuation matters.

Matters agreed on the code

Certain matters arising on para 7 of the code are clear and were common ground between the parties.

(1) The valuation date is the date of the court order.

(2) The consideration must be assessed on the basis of the terms and conditions of the deed of grant.

(3) As a result of those terms and conditions, there will be no loss or damage to LIDI as a result of the deed and the exercise of the rights thereby granted. No compensation for "loss or damage" is therefore payable under para 7(1)(*b*) of the code.

(4) Similarly, there will be no loss or damage to LIDI as a result of severance or injurious affection. No compensation is therefore payable under para 7(2) of the code.

It follows that the only financial term which can be imposed is the payment of "consideration" under para 7(1)(*a*). That word, to my mind, suggests that the rights to be granted have a price, irrespective of any detriment to the grantor. The word is used in contrast to "compensation" in paras 7(1)(*b*) and 7(2), which suggests loss or damage to the grantor for which he should be compensated.

Outline of rival arguments

The sole point at issue is therefore what, if any, consideration is payable under para 7(1)(*a*) of the code, ie what "would have been fair and reasonable if the agreement had been given willingly". The whole argument centres round the meaning and effect of those words in the circumstances of the case.

The contentions of Mercury, put forward by Mr Room as expert and supported by Mr Aikens and Mr Nicholls as counsel, are (very briefly) as follows: (1) the general principles of compulsory acquisition of land apply in the assessment of the consideration; (2) under those principles, (a) only the value to LIDI can be considered, and (b) any increase in value to LIDI due to the scheme underlying Mercury's acquisition must be ignored ("the *Pointe Gourde* principle"); (3) the consideration is therefore nil or nominal; and (4) if not nil or nominal, the consideration should be limited to a small capital payment calculated by reference to similar capital payments which have been negotiated by Mercury with the National Farmers' Union and the Country Landowners' Association.

The contentions of LIDI, put forward by Mr Sheard as expert and supported by Mr Hobson as counsel on behalf of LIDI, are (again very briefly) as follows: (1) the general principles of compulsory acquisition do not apply; (2) the rights conferred by the deed of grant are equivalent to a "ransom strip" of the kind identified in *Stokes* v *Cambridge Corporation* (1961) 13 P&CR 77*; and (3) LIDI can thus negotiate for and so should be entitled to a percentage share of Mercury's anticipated profits from its Canary Wharf operation. In his original report, after studying Mercury's business case and the financial statement which I have mentioned, Mr Sheard put forward a claim for the consideration to be £140,000 pa. However, after receiving Mr Redgate's revised forecast (the figures in which are agreed) and discussing the matter and for the purposes of the case agreeing certain figures and percentages with Mr Room, Mr Sheard has reduced this figure to £24,175 pa.

Preliminary points on para 7(1)(a) of the code

Before considering the rival arguments of the parties there are a number of general points arising on para 7(1)(*a*) of the code which it is convenient to deal with.

The words "willingly given" clearly refer to the grantor, who must therefore be a "willing" grantor. However, in my judgment, a "willing" grantee must also be assumed, that is to say a grantee who is willing to take on "fair and reasonable" terms. After all, it is the grantee who is applying to the court for a grant on those terms, and it necessarily follows that he is a "willing grantee" in that sense. Statute frequently refers only to the "willing" vendor, lessor or grantor, without mention of the purchaser, lessee or grantee, but it is always assumed that the latter too will be "willing": see eg r2 of section 5 of the Land Compensation Act 1961; section 25(1) of the Finance (1909-10) Act 1910, mentioned below; section 34 of the Landlord and Tenant Act 1954; and section 9(1) of the Leasehold Reform Act 1967.

In my judgment, I have to consider what would be "fair and reasonable" terms as between these parties, ie LIDI and Mercury, as opposed to a hypothetical grantor and a hypothetical grantee. That appears to me to be the consequence of the wording of para 7(1)(*a*), particularly in the light of the earlier parts of the code which assume that there have been prior negotiations between the parties. Mr Aikens argued to the contrary, saying that para 7(1)(*a*) requires the court to consider a hypothetical situation. That is to a certain extent true, but I see no reason to depart from reality more than is necessary or is required by the relevant wording of the legislation. The wording of para 7(1)(*a*) is significantly different in this respect from other legislation, eg section 8(2) of the Mines (Working Facilities and Support) Act 1966, which refers to "a willing grantor" and "a willing grantee".

What I have to decide is what would have been "fair and reasonable" if the grant had been willingly given. That, in my view, necessarily involves an element of subjective judicial opinion, for there can be no proof or objective determination of what is fair and reasonable. To a certain extent, the answer must depend on the judge's own perception of what is "fair and reasonable". There is thus a distinction between this case and the more common category of case where the legislation or the agreement in question requires a determination of "market value" or "market rent" or the like; such a determination involves an objective assessment of a factual matter not involving any discretion or subjective opinion. This distinction may perhaps account for the fact that the code directs the consideration to be determined by a judge rather than the Lands Tribunal.

Following on from that, it is, in my judgment, clear that what I have to determine is not the same as what the result in the market would have been if the grant had been given willingly. That is, however, far from saying that the market result is irrelevant or can afford no guidance. Indeed, in my view, the market result is the obvious starting point; and in most cases it will come to the same thing as what is "fair and reasonable", because *prima facie* it would be neither fair nor reasonable for the grantor to receive less than he would

*Editor's note: Also reported at (1961) 180 EG 839.

in the market or for the grantee to have to pay more than he would in the market. But there may be circumstances, of which the absence of any real market may be one, in which a judge could properly conclude that what the evidence may point to as being the likely market result is not a result which is "fair and reasonable".

In my view, the word "willingly" in para 7(1)(*a*) cannot be taken in isolation. For it is a meaningless concept if considered apart from the financial terms of the grant. Whether or not a vendor or grantor is "willing" must necessarily depend on the price or consideration he is to receive. Thus, a vendor cannot sensibly be stigmatised as being unwilling because he refuses to accept an offer at an inadequate price. "Willing" must mean "willing to sell at a fair price". Likewise "given willingly" in para 7(1)(*a*) must, in my judgment, mean "given willingly for a consideration which is fair and reasonable". In his submissions, Mr Aikens argued that, *ex hypothesi*, LIDI was not willing because it would not give consent to the proposed grant except on what he described as its own prohibitive terms. In my judgment, that does not advance the argument and is logically fallacious, because it begs the real question and indeed really assumes what it is setting out to demonstrate.

Authorities on "willing"

In my judgment, the meaning and effect of the word "willingly" in its context is critical; and I consider it helpful, and indeed essential, to refer at some length to three important authorities in which the meaning of the word "willing" in a similar context has been considered.

(1) *Inland Revenue Commissioners* v *Clay* [1914] 3 KB 466 (CA), arose from the first attempt to tax increases in land values made by the famous Lloyd George budget which resulted in the Finance (1909-10) Act 1910. For the purpose of "increment value duty", it became necessary to determine the "gross value" as at April 30 1909 of a house. That house had a value at that date for a person wishing to use it as a residence of £750. But it adjoined a nurses' home which the trustees of the home desired to extend. It was advantageous to them to pay at least £1,000 for the house for that purpose, and they did in fact purchase it for that sum on September 29 1910. The "gross value" was defined by section 25(1) of the Act to mean:

> . . . the amount which the fee simple of the land, if sold at the time in the open market by a willing seller in its then condition, free from incumbrances, and from any burden, charge, or restriction (other than rates or taxes) might be expected to realise.

At first instance before Scrutton J at [1914] 1 KB 349, it was argued by the Solicitor-General on behalf of the Revenue that the "gross value" was £750, and that (to quote Scrutton J at p348) it "must exclude the price which one particular buyer will give because of his particular need; that this is not the price in the 'open market'; and that the willing seller must be willing to sell at a market price, not a fancy price". In support of his argument, the Solicitor-General relied on some *dicta* of Fletcher Moulton LJ in *Re Lucas and Chesterfield Gas & Water Board* [1909] 1 KB 16 that in assessing compensation on a compulsory purchase you cannot consider the special needs of the compulsory purchaser. Scrutton J rejected that argument, and held the "gross value" to be £1,000, saying in the course of his judgment at p348:

> The seller is not to be assumed to be making a forced sale at any price he can get, however low. He must be willing to sell, not demanding compensation for a forced sale, but he is not required to exclude the principal bidder from his market, because that principal bidder wants the house more than any one else and will therefore give more for it.

In the Court of Appeal the Solicitor-General repeated his argument before Scrutton J, but in addition put forward what I will refer to as "the one more bid argument". This was rejected by all three judges in the Court of Appeal. Cozens-Hardy MR said at p472:

> An "open market" sale of property "in its then condition" presupposes a knowledge of its situation with all surrounding circumstances. To say that a small farm in the middle of a wealthy landowner's estate is to be valued without reference to the fact that he will probably be willing to pay a large price, but solely with reference to its ordinary agricultural value, seems to me absurd. If the landowner does not at the moment buy, land brokers or speculators will give more than its purely agricultural value with a view to reselling it at a profit to the landowner.

Swinfen Eady LJ said at p476:

> The fact that No 83 [the house] had a potential value, by reason of its proximity to the nurses' home, and the very rapid increase of the work of that institution, pointing to a necessary extension at an early date, ought properly to be taken into consideration in arriving at the value it might be expected to realize.

Later he said:

> It was then urged by the Solicitor-General that if the probability of this special buyer purchasing, above the price, which but for his needs would have been the market price, could be taken into consideration at all, then only one further point or bid could be allowed, and it must be assumed that this special buyer would have become the purchaser upon making this one extra bid. Such an assumption would ordinarily be quite erroneous. The knowledge of the special need would affect the market price, and others would join in competing for the property with a view of obtaining it at a price less than that at which the opinion would be formed that it would be worth the while of the special buyer to purchase.
>
> A sale by a willing seller is distinguished from a sale which is made by reason of compulsory powers, where the vendor frequently obtains an addition to the price by reason of being under compulsion to sell. It does not mean a sale by a person willing to sell his property without reserve for any price he can obtain.

Pickford LJ said at p478:

> . . . I think a willing seller means one who is prepared to sell, provided a fair price is obtained under all the circumstances of the case. I do not think it means only a seller who is prepared to sell at any price and on any terms, and who is actually at the time wishing to sell. In other words, I do not think it means an anxious seller.

I should add that in relation to compulsory purchase, the effect of *Inland Revenue Commissioners* v *Clay* was reversed by r(3), section 2 of the Acquisition of Land (Assessment of Compensation) Act 1919 (now section 5 of the Land Compensation Act 1961), which provided that there should be left out of account any value "for which there is no market apart from the special needs of a particular purchaser". Those words have now been repealed by para 1, Schedule 15 to the Planning and Compensation Act 1991, thus restoring the effect of *Inland Revenue Commissioners* v *Clay*.

(2) In the well known "Indian case", *Raja Vyricherla Narayana Gajapatiraju* v *Revenue Divisional Officer, Vizagapatam* [1939] AC 302, the facts were briefly as follows. A harbour was being constructed at Vizagapatam, and land had been acquired compulsorily for that purpose. Part of that land had been allocated for use by oil companies and other industrial concerns, but it was malarious. The appellant's land immediately to the south of this land contained a spring which yielded a constant and abundant supply of good drinking water which could easily be made available for the oil companies and those engaged in the harbour works, but not others. The appellant's land was compulsorily acquired for the purpose of executing antimalarial works. The main question which came before the Privy Council was the extent to which the water supply should be taken into account in assessing the appellant's compensation. Under the relevant Indian legislation the same principles applied as those under the Lands Clauses (Consolidation) Act 1845 before the coming into operation of the 1919 Act.

Lord Romer delivering the judgment of the Privy Council said at p312:

> The compensation must be determined, therefore, by reference to the price which a willing vendor might reasonably expect to obtain from a willing purchaser. The disinclination of the vendor to part with his land and the urgent necessity of the purchaser to buy must alike be disregarded. Neither must be considered as acting under compulsion. This is implied in the common saying that the value of the land is not to be estimated at its value to the purchaser.

But this does not mean that the fact that some particular purchaser might desire the land more than others is to be disregarded. The wish of a particular purchaser, though not his compulsion, may always be taken into consideration for what it is worth.

A little later on he said in relation to "market value":

But sometimes it happens that the land to be valued possesses some unusual, and it may be, unique features, as regards its position or its potentialities. In such a case the arbitrator in determining its value will have no market value to guide him, and he will have to ascertain as best he may from the materials before him, what a willing vendor might reasonably expect to obtain from a willing purchaser, for the land in that particular position and with those particular potentialities.

At p314, Lord Romer dealt with the argument that the arbitrator must hold any imaginary auction at which the persons present would include those only interested in the land itself and not interested in the potentialities; those persons would pay only what was referred to as the "poramboke" value (which I take to be the rough equivalent of existing use value). Lord Romer accepted that such purchasers would drop out when the bidding reached the "poramboke" value, which would leave the auction to the "potentiality" buyers. He continued at p315:

But at what figure will this bidding stop? As already pointed out it cannot be imagined as going on until the ultimate purchaser has been driven by the competition up to a fantastic price. For he is *ex hypothesi* a willing purchaser and not one who is by circumstances forced to buy. Nor can the bidding be imagined to stop at the first advance on the "poramboke" value. For the vendor is a willing vendor and not one compelled by circumstances to sell his potentiality for anything that he can get.

Lord Romer then expressed the view that an imaginary auction was of no help, because the arbitrator still had to estimate the true value to the vendor of the potentiality. He continued at p316:

Upon the question of the value of the potentiality where there is only one possible purchaser, there are some authorities to which their Lordships will have to refer. But dealing with the matter apart from authority it would seem that the value should be the sum which the arbitrator estimates a willing purchaser will pay and not what a purchaser will pay under compulsion. It was contended on behalf of the respondent that, at an auction where there is only one possible purchaser of the potentiality, the bidding will only rise above the "poramboke" value sufficiently to enable the land to be knocked down to that purchaser. But if the potentiality is of value to the vendor if there happen to be two or more possible purchasers of it, it is difficult to see why he should be willing to part with it for nothing merely because there is only one purchaser. To compel him to do so is to treat him as a vendor parting with his land under compulsion and not as a willing vendor. The fact is that the only possible purchaser of a potentiality is usually quite willing to pay for it.

Lord Romer then referred with approval to *Inland Revenue Commissioners* v *Clay*, and expressed the opinion that if the house, in that case, had been compulsorily purchased under the Land Clauses (Consolidation) Act 1845 before its purchase by the trustees, it should have been valued at £1,000 and not £750.

Lord Romer a little later referred to *Re Lucas and Chesterfield Gas & Water Board supra*, and expressly disapproved the *dicta* of Fletcher Moulton LJ (relied on by the Solicitor-General in *Inland Revenue Commissioners* v *Clay*) mentioned above. Lord Romer then referred to *Sidney* v *North Eastern Railway Co* [1914] 3 KB 629 and some *dicta* of Rowlatt J in that case and said at p322:

If and so far as this means that the value to be ascertained is the price that would be paid by a willing purchaser to a willing vendor, and not the price that would be paid by a "driven" purchaser to an unwilling vendor, their Lordships agree. But so far as it means that the possibility of the promoter as a willing purchaser being willing to pay more than other competitors, or in the cases where he is the only purchaser of the potentiality, more than the value of the land without the potentiality is to be disregarded, their Lordships venture respectfully to differ from the learned judge.

For these reasons, their Lordships have come to the conclusion that, even where the only possible purchaser of the land's potentiality is the authority that has obtained the compulsory powers, the arbitrator in awarding compensation must ascertain to the best of his ability the price that would be paid by a willing purchaser to a willing vendor of the land with its potentiality in the same way that he would ascertain it in a case where there are several possible purchasers and that he is no more confined to awarding the land's "poramboke" value in the former case than he is in the latter.

(3) *FR Evans (Leeds) Ltd* v *English Electric Co Ltd* (1977) 36 P&CR 185* concerned a rent review clause under which the rent was to be "the rent at which the demised premises are worth to be let with vacant possession on the open market as a whole between a willing lessor and a willing lessee" for the remainder of the term of the lease and upon the same covenants and conditions. The demised premises in question were "the Walton Works", which were exceptionally large factory premises. Their size made them, for practical purposes, unlettable to anyone except English Electric. (There was thus effectively a monopoly position on either side, akin to that in the present case.)

Donaldson J, on an appeal from an arbitrator, stressed that both the lessor and the lessee were hypothetical persons. At p190, under heading (c), he said:

The fact that it is very likely that the English Electric Co Ltd would have been the only potential lessee is relevant, but its relevance is indirect. It does not matter whether the only potential lessee was this company or the XYZ Co Ltd. What matters is that in the state of the market there was not likely to be more than one willing lessee. The effect of this fact is not, however, decisive because this single potential lessee is to be assumed to be a willing lessee — neither reluctant nor importunate, but willing. Just as the hypothetical lessor cannot rely too much on the fact that no property similar to the Walton Works is available on the market, so the hypothetical lessee cannot rely too much upon the fact that he has no competitors — he is, and is known to be, a willing lessee. Furthermore, it is known that he will remain a willing lessee so long as the willing lessor does not press his demand for rent beyond the point at which he is ceasing to act as a willing lessor and at which a willing lessee would cease to be such.

Donaldson J then mentioned *Inland Revenue Commissioners* v *Clay*. Later he said at p192:

The rent for which each is negotiating is that which is high enough to be acceptable to a willing lessor and low enough to be acceptable to a willing lessee.

Finally, he said at p193:

I also agree that the possibility of the parties' failing to reach agreement is to be disregarded; to borrow and adapt an immortal phrase, "We are not interested in the possibilities of failure to reach agreement. They do not exist." As the negotiations proceed, however, each will be considering whether it would not be better at a given level of rent to break off the negotiations. True it is that they will resist these temptations, but the extent to which they operate on their respective minds will be reflected in the rent which will, notionally, be agreed in the end.

There are further passages in the judgment of Donaldson J which are in my view of assistance. His comments are, it seems to me, very much in line with those stated in the "Indian case". The report does not say whether he was referred to that case; but (particularly having regard to the experienced counsel involved) I would be surprised if he was not.

The *FR Evans* case went to the Court of Appeal, but there is no report of the judgments there. It appears, however, that Donaldson J's reasoning and decision were approved: see the note at (1978) 245 EG 657 at p662.

Mercury's submissions as to compulsory purchase principles

As I have indicated, Mr Aikens submitted that, in interpreting and giving effect to para 7(1)(*a*) of the code, an analogy should be drawn with the principles applicable on the compulsory purchase of land and the same principles applied. Those principles are now embodied in the Land Compensation Act 1961 as amended, which replaced and broadly re-enacted the relevant parts of the 1919 Act. It was not suggested that either of those statutes could be applied directly, but

*Editor's note: Also reported at (1978) 245 EG 657.

that the main principles which were developed by decisions of the courts under section 63 of the Land Clauses (Consolidation) Act 1845 should be. The relevant part of that section provides that in estimating the compensation payable, regard should be had:

> . . . not only to the value of the land to be purchased or taken by the promoters of the undertaking, but also to the damage, if any, to be sustained by the owner of the lands by reason of the severing of the lands taken from the other lands of such owner, or otherwise injuriously affecting such other lands . . .

The section thus provided for compensation for severance and injurious affection, but did not go into any detail about the "value" of the land.

The principles evolved by the courts were summarised by Eve J in *South Eastern Railway Co* v *London County Council* [1915] 2 Ch 252 in a much-cited passage at p258 where he states them under six heads, the material ones of which read as follows:

> (1) the value to be ascertained is the value to the vendor, not its value to the purchaser; (2) in fixing the value to the vendor all restrictions imposed on the user and enjoyment of the land in his hands are to be taken into account, but the possibility of such restrictions being removed for his benefit is not to be overlooked; . . . (4) increase in value consequent on the execution of the undertaking for or in connection with which the purchase is made must be disregarded; (5) the value to be ascertained is the price to be paid for the land with all its potentialities and with all the use made of it by the vendor.

Head (4) is now commonly called "the *Pointe Gourde* principle" from the case in which it was affirmed by the Privy Council, *Pointe Gourde Quarrying & Transport Co Ltd* v *Sub-Intendent of Crown Lands* [1947] AC 565. I will return to this principle in due course.

I should deal with one point which arises out of the *South Eastern Railway* case. The facts were that the railway company owned two adjoining pieces of land near Charing Cross Station. The LCC compulsorily acquired a frontage strip of one piece for the purposes of widening The Strand. The arbitrator found that the value of the strip taken, on its own, was £18,330. But he added that the aggregate value of the two pieces of land together had been reduced by only £3,465. The LCC argued that the compensation payable was thus only £3,465, that being the loss to the owner. Eve J and the Court of Appeal rejected that argument, holding that the compensation was £18,330. Eve J at p259 of his judgment commented that the LCC's argument:

> . . . would bring about some startling results; the most obvious one being that it would upset all uniformity of value, inasmuch as the value of the identical piece of land in the hands of one vendor might be assessed at many times it value in the hands of another, and this, not from any intrinsic distinction, but by reason solely of extraneous considerations. Moreover, it would be calculated to work injustice in that a vendor compelled to sell, and who the Legislature intended should be compensated for being compelled to sell, might have to accept from the undertakers a price far less than he would have obtained from any other purchaser, and out of all proportion to the true value of the land had it been ascertained without reference to the fortuitous circumstances of his being also interested in the contiguous land.

Eve J's reference to "uniformity of value" in this passage has been seized on in later cases as affording support for the proposition that there should be uniformity of value whoever is *the purchaser*. But it is clear from the facts and passage cited above that Eve J was dealing simply with uniformity of value in the hands of different *vendors*, and that his remarks have been taken out of their context in the two later cases to which I now refer.

In support of his argument, Mr Aikens referred to and relied on two decisions on section 9(2) of the Mines (Working Facilities and Support) Act 1923 and its successor section 8(2) of the Mines (Working Facilities and Support) Act 1966. Those Acts provide for the compulsory acquisition of rights to work minerals. The sections each provide that "compensation or consideration" is to be assessed:

> on the basis of what would be fair and reasonable between a willing grantor and a willing grantee, having regard to the conditions subject to which the right is or is to be granted.

The resemblance between these words and those of para 7(1)(*a*) of the code is obvious. The only difference of any possible significance is that the willing grantor and the willing grantee are hypothetical persons, not the actual grantor and grantee (as I have held above to be the case under para 7(1)(*a*) of the code). In both of the decisions, the court held that the compulsory purchase principles outlined above applied by analogy. I must consider each of them in some detail.

In *Re Naylor Benzon Mining Co Ltd* [1950] Ch 567, the applicants owned land owned by one Dunkerley. They had mined their own land for ironstone, but it was almost worked out and they negotiated with Dunkerley for the right to mine the ironstone in his land. It was clearly convenient for the applicants to work Dunkerley's land. Wynn-Parry J said (in a passage omitted from the Law Reports, but appearing at [1951] 1 All ER 521A) that it was "common ground between the applicants and Mr Dunkerley that the applicants are in a particularly favourable position to develop the land . . ., both as regards continuity of work and as regards a saving of expense". The applicants were thus clearly in the position of special purchasers.

The applicants therefore put forward a proposed draft mining lease setting out certain royalty payments. It was clear from the evidence, said Wynn-Parry J at p574 of the law report:

> first that the applicants were anxious to come to an amicable agreement with the respondent Dunkerley if possible; secondly that they considered what from their point of view would represent a good commercial transaction, taking into account that, for the reasons to which I have already referred they were in a particularly favourable position to conduct mining operations on Dunkerley's land.

But Dunkerley did not accept the draft lease and its royalties, and so the applicants applied to the court for compulsory rights under the 1923 Act. Matters then took a curious turn, because at an early stage of the hearing Dunkerley indicated that he was now willing to accept the draft lease with its royalties as it stood. The applicants, however, had by then been advised that the royalties proposed in the draft lease were higher than those prevailing in the area, and said that, although they did not desire to resile from the terms which they had put forward, in view of certain sections of the Iron and Steel Act 1949 which had by then come into force, they could no longer properly consent to them.

Counsel for the applicants argued that the compulsory purchase principles were applicable, and that under those principles it would be wrong to take into account the special value which the property might have to a special purchaser who had obtained powers of compulsory purchase (such as the applicants); on the latter point he relied on the *dicta* of Fletcher Moulton LJ in *Re Lucas and Chesterfield Gas & Water Board*, mentioned above. Unfortunately, counsel for Dunkerley failed to draw Wynn-Parry J's attention to the fact that those *dicta* had been expressly disapproved in the "Indian case", although he did refer to *Inland Revenue Commissioners* v *Clay*. Wynn-Parry J accepted both of those arguments, citing at some length and placing considerable reliance on the disapproved *dicta* of Fletcher Moulton LJ. Although this is not entirely clear, Wynn-Parry J seems, following and applying those *dicta*, to have ignored the special value of the minerals to the applicants and to have set the scale of royalty payments by reference only to the general level of such payments in the area.

I have to say that I find the *Naylor Benzon* case a difficult one. Wynn-Parry J did not discuss the limitations inherent in the grant being on the basis of a willing grantor and a willing grantee (discussed under "Preliminary points on para 7(1)(*a*)" and in the authorities dealt with in "Authorities on 'willing'" above), and seems to have thought that the alternative to applying the compulsory purchase principles would be an entirely open negotiation. In applying the authorities he referred to the "uniformity of value" *dicta* of Eve J in the *South Eastern Railway Co* case set out above, but with all respect applied them out of context and wrongly. His other reason for applying the compulsory purchase principles stemmed from the disapproved *dicta* of Fletcher Moulton LJ. Finally, even if those principles are applicable, in my view, the exclusion of the applicants' own offer from being taken into account was contrary to the "Indian case" and to *Inland Revenue Commissioners* v *Clay*.

BP Petroleum Developments Ltd v *Ryder* [1987] RVR 211*, on which Mr Aikens placed considerable reliance, concerned the Wytch Farm oilfield in Dorset. By the Petroleum (Production) Act 1934, petroleum was nationalised (without compensation), but under section 2 power was conferred on the responsible minister to grant licences to search, bore for and get petroleum. By section 3, the 1923 Act, and any Act replacing it was made applicable to petroleum with certain modifications, and available to any person granted a licence. One such modification is that, by section 3(2)(*a*), an additional allowance of not less than 10% has to be made to the compensation on account of the acquisition of the working rights being compulsory. Another is that by section 6, model clauses prescribed by regulations must be incorporated in any licence. The model clauses prescribed provide for the Crown alone to receive royalty payments, and preclude the licensee from entering into arrangements to share the proceeds of sale of the petroleum with any other person.

BP had been granted a licence to exploit the Wytch Farm oilfield, and applied under the 1966 Act for the grant of working rights. A good deal of the decision of Peter Gibson J relates to the question of whether any, and if so what, rights should be granted and to other matters (including compensation for disturbance and severance) with which I am not concerned. But one of the main issues was that of the "compensation or consideration" payable under section 8(2) of the 1966 Act. Counsel for BP argued that the compulsory purchase principles were applicable, relying on the *Naylor Benzon* case. Counsel for the respondent landowners argued the contrary, pointing out: (i) that a *dictum* of Finlay J in *Re Consett Iron Co Ltd's Application* [1938] 1 All ER 439 at p447 that no assistance was to be derived from the Land Clauses (Consolidation) Act 1845 in considering compensation under the 1923 Act had not been drawn to Wynn-Parry J's attention; and (ii) that the *dicta* of Fletcher Moulton LJ had been disapproved in the "Indian case". Counsel also pointed to analogous situations, including that identified in *Stokes* v *Cambridge supra*, a case to which I shall return, to justify a different approach. But Peter Gibson J was unpersuaded and, following the *Naylor Benzon* case held that the compulsory purchase principles were applicable. I will return later to the manner in which he applied those principles, and in particular in relation to the *Pointe Gourde* principle and the assessment of the potentiality to the landowners. I would also mention that the landowners lodged an appeal, which was settled by agreement before it reached the Court of Appeal.

Had the matter been free from authority, I would not myself have considered that the compulsory purchase principles were applicable to the 1923 and 1966 Acts. I would have held that the relevant principles were those derived from the phrases "willing grantor", "willing grantee" and "fair and reasonable" discussed above. These are of course in many respects very similar to, and to large degree overlap, the compulsory purchase principles. In most cases, the practical result is likely to be the same. But I cannot myself see any justification for going beyond the former principles and introducing further concepts which Parliament refrained, presumably intentionally, from incorporating into the legislation. Nevertheless, had the present case been one under the 1966 Act, as the *Naylor Benzon* case had been strongly challenged but followed in the *BP* case, I would have considered myself bound to follow them in that respect (even assuming that as a county court judge I am not technically bound by the decision of a High Court judge): cf *Colchester Estates (Cardiff)* v *Carlton Industries plc* [1986] Ch 80† at p85.

Mr Aikens submitted that it necessarily follows, particularly from the *BP* case, that I should likewise apply the compulsory purchase principles to para 7(1)(*a*) of the code. He said that there was no significant difference between the wording of that paragraph and the wording of section 8(2) of the 1966 Act, and that Parliament must have had that section and its interpretation in the *Naylor Benzon* case in mind when enacting para 7(1)(*a*) of the code.

*Editor's note: Also reported at [1987] 2 EGLR 233.
†Editor's note: Also reported at (1984) 271 EG 778.

Mr Hobson pointed out a number of differences which he said distinguished the *BP* case from the present one. These were briefly: (1) the legislative provisions use different wordings; (2) the case concerned the exploitation of petroleum which had been nationalised without compensation by the 1934 Act; (3) a licence under the 1934 Act confers a monopoly; (4) there are no model clauses in the present case; (5) licensees under the 1934 Act can acquire rights compulsorily (but so in effect can an operator under the code); (6) the criteria for the grant of rights are rather different under the 1966 Act and para 5 of the code; (7) in the present case there is no addition for the acquisition being compulsory; and (8) the particular phrases "compensation or consideration" and "a willing grantor" and "a willing grantee" in section 8(2) of the 1966 Act are materially different. However, I agree with Mr Aikens that these differences would not, by themselves, be of sufficient significance to justify me in drawing a distinction between the *BP* case and the present case. But the matter does not end there.

LIDI's submissions as to compulsory purchase principles

Mr Hobson submitted that there are three reasons why the compulsory purchase principles cannot properly be applied to para 7(1)(*a*) of the code.

First, he drew attention to certain provisions of the code where the statutory provisions as to compulsory acquisition are expressly incorporated into the code. These are: (i) para 4, and in particular subpara (8) which expressly incorporates some of the rules in section 5 Land Compensation Act 1961; and (ii) para 16, which deals with injurious affection and which (in my view significantly) incorporates the relevant provisions of the 1961 Act into the assessment of that part of the compensation for a grant specified in para 7(2). Mr Hobson also drew attention to section 34 of the 1984 Act, which also directly incorporates the 1961 Act. These express references, he submitted, preclude any unexpressed incorporation of compulsory purchase principles by analogy. It is, in my view, also significant that where the 1961 Act is incorporated, the Lands Tribunal (which of course is familiar with compulsory purchase law) is given jurisdiction, as one would expect. The fact that the county court is otherwise given jurisdiction suggests that Parliament did not consider that compulsory purchase law would otherwise be applicable.

Second, Mr Hobson drew attention to the code concerning telecommunications under previous Acts, which was replaced by the code under the 1984 Act. The relevant parts of the previous Acts provided as follows. Section 6 of the Telegraph Act 1863 gave any telegraph company power to execute certain works under, over along or across any street (defined as a public way) or public road, and also power under subpara (4) thereof to:

> place and maintain a telegraph and posts under, in, upon, over, along or across any land or building, or any railway or canal, or any estuary or branch of the sea, or the shore or bed of any tidal water . . .

By section 7 it was provided, so far as material, as follows:

> In the exercise of the powers given by the last foregoing section the company shall do as little damage as maybe, and shall make full compensation to all bodies and persons interested for all damage sustained by them by reason or in consequence of the exercise of such powers; the amount and application of such compensation to be determined in manner provided by the Land Clauses (Consolidation) Act 1845 . . . and any Act amending those Acts, for the determination of the amount and application of compensation for lands taken or injuriously affected.

By the Telegraph Act 1868, which enabled the Postmaster-General to acquire the undertakings of telegraph companies, the 1863 Act was extended to the Postmaster-General. By the combined effect of sections 3 and 4 of the Telegraph Act 1878 and the Telegraph (Construction) Act 1916, it was provided that differences between the Postmaster-General and other persons should be determined by a county court.

The important point stressed by Mr Hobson was that until the 1984 Act the telecommunications code thus expressly incorporated the compulsory purchase legislation (and in particular the 1845 Act), as well as conferring jurisdiction on the county court. He submitted that Parliament, in replacing the previous code with a new code which did not (apart from the specific instances I have mentioned) incorporate the compulsory purchase legislation, was deliberately not applying the compulsory purchase principles. To construe the new code as nevertheless continuing to apply those principles would, said Mr Hobson, be flouting the plain intention manifested by Parliament. (I should mention that both counsel agreed that there was no assistance to be gained from *Hansard* on the matter.)

Third, Mr Hobson referred to other statutes by which public utility companies have been given powers to acquire rights compulsorily. In all such cases, the compulsory purchase legislation, and in particular the Land Compensation Act 1961, had been expressly incorporated. Mr Hobson instanced section 9 and Part II, Schedule 3, Gas Act 1986; section 10, Part II, Schedule 3 and para 7, Schedule 4, Electricity Act 1989; and section 155, para 1, Schedule 9 and para 8, Schedule 11, Water Resources Act 1991. It is not necessary to go through these in any detail. They essentially repeated similar provisions in previous Acts. Mr Hobson said that the contrast with the code was obvious, and reinforced his argument that Parliament had not intended to apply compulsory purchase principles to the code.

Conclusion on application of compulsory purchase principles

I prefer the arguments of Mr Hobson. It seems to me that the idea that Parliament must have had the wording of section 8(2) of the 1966 Act and the *Naylor Benzon* decision in mind when enacting the code is rather far-fetched anyway, but made quite untenable by the other parts of the code, the provisions of the code it replaced, and the comparable legislation. I therefore hold that the *Naylor Benzon* and *BP* cases are distinguishable in this respect, and that para 7(1)(*a*) of the code must be applied without regard to compulsory purchase principles. The principles applicable are those derived from the words "willingly" and "fair and reasonable" discussed above.

Pointe Gourde principle

It follows from the foregoing that the *Pointe Gourde* principle in compulsory purchase, on which Mr Room and Mr Aikens placed reliance, has no application. However, I heard a good deal of evidence and argument relating to it, and for the sake of completeness I propose to make some comments and findings about it in relation to the facts of this case.

The *Pointe Gourde* principle and its relationship to a *Stokes* v *Cambridge* situation were considered by the Court of Appeal in *Batchelor* v *Kent County Council* (1990) 59 P&CR 357*. That case concerned the compulsory acquisition for highway purposes of a "ransom" strip of land giving access to a substantial area of housing land. Mann LJ, with whom the other members of the court agreed, referred to the short statement of the *Pointe Gourde* principle in the *Pointe Gourde* case itself *supra* at p572:

> It is well settled that compensation for the compulsory acquisition of land cannot include an increase in value which is entirely due to the scheme underlying the acquisition.

Mann LJ also cited another short passage from *Fraser* v *City of Fraserville* [1917] AC 187 at p194:

> . . . the value to be ascertained is the value . . . excluding any advantage due to the carrying out of the scheme for which the property is compulsorily acquired, the question of what is the scheme being a question of fact . . . in each case.

Mann LJ then referred to *Stokes* v *Cambridge* saying [at p360]:

> In that case there was an assessment of compensation for land which had an agreed potential for industrial use subject to the provision of satisfactory access. Satisfactory access should be secured across an adjoining strip of land which was in separate ownership. The Lands Tribunal held that the value of the subject land should be diminished in the open market to take account of what the owner of the ransom strip would require before granting access across his strip . . . *Stokes* v *Cambridge* was not doubted before this court. Nor was it suggested that the open market value of acquired land cannot embody a ransom component if it holds a key to development elsewhere . . .

Counsel informed the court that the case was the first to come before it in which the relationship between the *Pointe Gourde* principle and *Stokes* v *Cambridge* had to be considered. On that, Mann LJ said at p361:

> I find no difficulty with the relationship. If a premium value is "entirely due to the scheme underlying the acquisition" then it must be disregarded. If it was pre-existent to the acquisition it must in my judgement be regarded. To ignore the pre-existent value would be to expropriate it without compensation and would be to contravene the fundamental principle of equivalence . . .

What is the "scheme underlying the acquisition" in the present case? Various suggestions were put forward on behalf of Mercury, including Mercury's licence and national network or the provision of telecommunication services to the whole docklands area. I agree with Mr Hobson that these suggestions are too wide. In my judgment, the "scheme" must be confined to the provision of telecommunication services to the Canary Wharf development and I so find as a fact.

However, in my judgment, the premium value of the right to lay cable ducts was pre-existent to the acquisition of the right. The Thameside earth station and the proposal to develop Canary Wharf both existed before the acquisition. Following and applying the *dicta* of Mann LJ in the *Batchelor* case at p361 set out above, even if the compulsory acquisition principles had been applicable, I would have found that the *Pointe Gourde* principle was not relevant and so account had to be taken of the special value to Mercury of the proposed grant.

I should mention that although Peter Gibson J in the *BP* case mentioned the *Pointe Gourde* principle, as I read his judgment he did not consider it applied on the facts of that case: see at p231 (left hand column) where he said:

> . . . I do not see why the fact that the presence of oil and gas in the Wytch Farm oilfield gives to the licensee (but only the licensee) a special interest in acquiring the rights should be ignored.

That is clearly correct and in accordance with the *dicta* of Mann LJ set out above. The existence of the oilfield and the value of the surface rights to exploit it plainly were pre-existing to the acquisition of those rights. In my view, the situation is the same in the present case.

Factors affecting quantification of the special interest

In the *BP* case, evidence was given of annual sums paid for similar rights in comparable cases, which were greatly in excess of, and bore no relationship to, agricultural values (which were the existing use values). Although as mentioned above he considered that the consideration for the grant should properly reflect the special interest of the licensee, Peter Gibson J felt able to disregard that evidence. Instead, he returned (at p232) with some enthusiasm to the concept of an imaginary auction, notwithstanding its rejection by Lord Romer in the "Indian case". He said:

> In the real world the willing special purchaser would bid a price sufficient to make sure he would beat all who were acquiring the lands for their existing use. I cannot see the logic of a purchaser, however willing, being assumed to pay a price far in excess of other bidders. The willing grantor is to be assumed to be willing to sell at the best price he can obtain and he cannot be assumed to insist on a reserve price measured by reference to some value other than what is fairly and reasonably obtainable in the market.

On that basis, Peter Gibson J decided that the consideration should be just "something more" than the agricultural value of the land, and he fixed it at £45 per acre, an increase of 8% over the agricultural value of £40 per acre. (This compared with the landowners' valuer's figure, based on the comparables of £2,500 per acre.) Mr Aikens relied on this part of Peter Gibson J's judgment and argued that I should adopt a similar approach.

I am unable to accept that argument and I must respectfully decline to follow this part of Peter Gibson J's judgment, for the following reasons:

*Editor's note: Also reported at [1990] 1 EGLR 32.

A (1) In my judgment, it is inconsistent with the reasoning of the Court of Appeal in *Inland Revenue Commissioners* v *Clay* and that of Lord Romer in the "Indian case". I prefer those cases, by which I am bound.

(2) It appears simply to resurrect the "one more bid" argument which was decisively rejected in those cases. It has also been rejected in other contexts (see, eg *Lloyd-Jones* v *Church Commissioners for England* (1981) 261 EG 471), and is inconsistent with the actual decision in other cases (see, eg *Lambe* v *Secretary of State for War* [1955] 2 QB 612).

(3) The fallacy of that argument is exposed if one postulates a B property which is of interest to only one purchaser or grantee, ie a property with a negligible existing use or "poramboke" value, such as a factory constructed or adapted for the special use of a particular tenant or of such a size as to be of interest to only one tenant (as in the *FR Evans* case). On the "one more bid" argument, the vendor or grantor would sell or let at only a nominal or very small figure, without regard to the value to the purchaser or grantee. In my view, on that basis he would plainly be a driven or forced vendor or grantor, and not a willing one.

(4) Similarly, the fallacy of the argument is also exposed by considering the case of a grant of a right of way or a wayleave C agreement, which will cause no detriment to the landowner. If there is only one possible user of the way in question, runs the argument, the person wishing to acquire the right of way will only have to offer a small consideration which the landowner will therefore be obliged to accept. The logical consequence of this, it seems to me, is that the consideration will be the same whether the right of way is for the purposes of a small cottage or a large mansion or a factory. That is plainly absurd and contrary to all experience. In practice the landowner is able to negotiate an appropriate consideration, which can take one of several forms, eg a capital payment, an annual ascertained rent or a wayleave rent, depending on the circumstances. The landowner is able to do so because he is able to take into account in D his negotiating stance the importance and value of the proposed right to the grantee. If he is forced to disregard these factors, he will, in my view, again cease to be a "willing" grantor and will become a driven or forced one.

(5) On the facts of the *BP* case, it was held that a grantor would be "willing" to grant on fair and reasonable terms at £45 per acre even though other grantors were receiving £2,500 per acre for similar grants. It may well be that the latter figure was a somewhat false one and reflected a strong element of anxiety on the part of the grantee. But making every possible allowance for that, I have to say that in my view the disparity borders on the absurd, and I find it impossible to see E how the grantor would be considered "willing" at the former figure.

(6) In his comments on *Stokes* v *Cambridge* at p231, Peter Gibson J expressed the view that that decision could only be justified on the footing that there was more than one possible purchaser of the "ransom strip"; this view again appears also to depend on the "one more bid" argument. With respect, that is a novel suggestion, and it finds no support in the decision itself nor in the recent cases in the Court of Appeal in which the *Stokes* v *Cambridge* principle has been considered by the Court of Appeal, ie the *Batchelor* case *supra* and *Hertfordshire County Council* v *Ozanne* [1989] 2 EGLR 18 (reported in the House of Lords on a different point at [1991] 1 All ER 769*).

F

Conclusion on Mercury's main submissions

For these reasons I reject Mercury's main submissions that the compulsory purchase principles are applicable, that the *Pointe Gourde* principle is applicable and that the consideration should be only a nominal or small one.

LIDI's main submissions

I turn to the main arguments advanced on behalf of LIDI. Although the installation of the ducts in Eastwood Wharf Road would involve only a very small addition to Mercury's network, it was said that the benefit to Mercury would be considerable. The use of the ducts (and G thus of LIDI's land) would unlock the benefits which Mercury expects to gain from the Canary Wharf extension to its network, as originally set out and quantified in the business plan, but now adjusted in the revised forecast. This is a highly relevant factor which LIDI would be able and entitled to take into account in any negotiations with Mercury. LIDI would not be acting "willingly" unless proper (but not undue) account is taken of the "potentiality" of the land and LIDI's bargaining position in this respect. For the reasons which I have given above, and following *Inland Revenue Commissioners* v *Clay*, the "Indian case" and the *F R Evans* case, thus far and in principle I agree with this submission. But it of course leaves open the crucial question which is what would be "fair and reasonable" if LIDI's agreement to H the deed of grant had been given willingly.

Mr Sheard in his evidence, supported by Mr Hobson in his submissions, argued (as I have indicated) that the situation is analogous to the acquisition of an essential right of access to a development site. The well known principle in such a situation, which is based on market transactions and which has been accepted by the Lands Tribunal and by the courts, is that the owner of the access land which of itself is of little value (or "ransom strip" as it is often pejoratively, and perhaps misleadingly, referred to) can negotiate and is entitled to a percentage share in the resulting increase in the development value of the development site. The quantification of the J percentage share depends on the facts and circumstances of the particular case. In *Stokes* v *Cambridge supra*, there was a compulsory acquisition of a development site. The Lands Tribunal held that one-third of the development value should be deducted from the compensation payable because that amount would have to be paid to the owner of the access land. The *Batchelor* case *supra* concerned the compulsory acquisition of the access land. The Lands Tribunal awarded compensation at 15% of the development value of the development site. In the *Ozanne* case *supra*, another case of compulsory acquisition of the access land, the award of 50% of the development value, which was on the footing that it would be split K between different owners of the access land (a point stressed by Mr Hobson, because ducts would have to be laid elsewhere than in Eastwood Wharf Road in this case).

This principle is not necessarily confined to the valuation of access land. It may be equally applicable in any other situation where a development value is unlocked. As an example, Mr Sheard in his evidence mentioned a case where a right to light had been released enabling a development to take place.

On the footing of that principle, Mr Sheard suggested that the proper consideration to be paid by Mercury should be an annual sum representing a percentage of the anticipated net profit to Mercury from L its Canary Wharf network "loop". On the basis of the business case (and assuming average annual depreciation of capital at 10% and interest on capital expended at 15%), in his original report he arrived at an annual sum of £140,000; this was based on a percentage share attributable to the rights to be granted of 15%.

As I have mentioned, as a result of the figures in the revised forecast and the agreement reached at their meeting, Mr Sheard and Mr Room have agreed all the figures and percentages except the appropriate deduction for operating costs. The average annual revenue forecast was agreed at £3,150,000; the revised capital expenditure estimate at £6,586,000; the appropriate rate of depreciation on capital M expenditure at 10%; the appropriate rate of return on capital expenditure at 15%; and the appropriate percentage attributable to the deed of grant (making allowances to reflect certain agreed factors) at 4.33%. This agreement and the annual consideration derived by each party therefrom can be summarised as follows:

Description	*Mercury*	*LIDI*
Estimated revenue	£3,150,000	£3,150,000
Deduct annual costs		
Operating costs on revenue	1,417,500 (45%)	945,000 (30%)
Depreciation on £6,586,000	658,600 (10%)	658,600 (10%)

*Editor's note: Also reported at [1991] 1 EGLR 34.

A Return on capital do	978,900 (15%)	978,900 (15%)
Net benefit	86,000	558,500
Share attributable to deed of grant, agreed at 4.33%	3,724	24,183
but say	£3,725	£24,175

The only difference is the percentage of operating costs on revenue. Mr Room was of the opinion that the deduction for this factor should reflect all Mercury's operating costs, and on that basis the appropriate percentage was agreed at 45%. Mr Sheard was of the opinion that it should reflect only the incremental costs attributable to the Canary
B Wharf loop, and on that basis the appropriate percentage was agreed at 30%. In their evidence, both gave detailed reasons for their opinions on this aspect of the matter.

Conclusion on LIDI's main submissions

I do not propose to choose between these two opinions, because I altogether reject this method of arriving at a "fair and reasonable" consideration. In my judgment, it is not an appropriate method to use for any situation except where: (1) there is a single capital payment to be made; and (2) the benefit to the developer/payer can be relatively easily quantified, as in the typical *Stokes* v *Cambridge* situation. I have heard no evidence of this kind of "profit" calculation being employed
C in practice in any other situation, and in particular in any situation involving an annual payment. In none of the comparable transactions referred to in evidence (including the "aerial agreements" mentioned below) was there any suggestion of it being used. I cannot think that, except in very special and unusual instances, it would be so used where its basis is an estimated future profit on an annual basis.

Mr Sheard in his evidence and Mr Hobson in his argument, sought to draw support for LIDI's argument and the application of the *Stokes* v *Cambridge* principles by reference to wayleave rents. The sharing of the benefit, said Mr Sheard in para 7.5 of his first report

D is a principle adopted not only where a development is made accessible by the grant of access rights by agreement, but also in relation to the grant of easements or wayleaves to transport minerals across lands; in such cases the owner of the land being crossed effectively charges a toll for the right to transport material by reference to the value or amount of material transported. In doing so, it is frequently the case that an estimate is made in advance of the amount likely to be transported, and an annual sum fixed by reference to that estimate without subsequent adjustment to reflect actual movements.

That, in my view, is an accurate description of a wayleave rent, and I would add that royalties are similarly ascertained by reference to the quantities of minerals extracted from the ground. But, in my judgment, this does not support LIDI's claim to share in the profits at all. The
E whole point of wayleave and royalty payments is that they are related solely to the relevant quantity of materials, which reflects the use made by the grantee of the rights granted, not the profit made. The payments have to be made whether the grantee operates at a profit or a loss.

In the course of his argument on this aspect of the matter, Mr Aikens referred me to *Whitwam* v *Westminster Brymbo Coal & Coke Co* [1896] 2 Ch 538, in which the Court of Appeal considered the question of the proper measure of damages where "one man runs trucks on rails on another man's land [when] it does not do any harm whatever, and there is no pecuniary damage": see at p542. Mr Aikens
F stressed that in such a situation the damages are not assessed by reference to the profits derived by the defendants from such a user. As Rigby LJ pointed out at p543, "in fact it is a matter of indifference whether the defendants made a profit out of the transaction". In such a case, the damages are assessed on the basis of an appropriate wayleave rent. In my judgment, the *Whitwam* case clearly supports Mercury's argument against LIDI's main submission that the consideration should be based on Mercury's anticipated profit, although as mentioned below I think it may be something of a two-edged sword from Mercury's point of view.

There are, in my view, further cogent reasons for not assessing the consideration by reference to Mercury's profit.

It seems to me that whether Mercury makes a profit or not on its G
Canary Wharf loop is not really material. Mercury might decide for business and prestige reasons to operate the loop at a nil profit or even at a loss. It is difficult to see why that should affect the consideration. The most material matters, in my view, are the amount of use of the ducts and cables, and their overall importance to Mercury and its operations.

Although Mr Room and Mr Sheard have come to an agreement as to the manner in which the profit benefit to Mercury of the Canary Wharf "loop" should be calculated for the purposes of these proceedings, there is obviously great scope for argument on many of the component factors. In the normal case, this would make LIDI's H
method of arriving at the appropriate consideration very difficult to apply in practice, and I think it is most unlikely that the market would attempt to do so.

It will be apparent from the calculations set out above in relation to the figures agreed by Mr Room and Mr Sheard that some quite minor variations in some of the estimated figures and/or the appropriate percentages to be applied can and do lead to enormous differences in the resulting consideration figure. This is vividly illustrated, first by the reduction in Mr Sheard's original figure of £140,000 pa to £24,175 pa, and second by the fact the relatively small difference between them as to the proper percentage deduction for operating costs results J
in Mr Sheard's figure being over six times Mr Room's figure. Other similar examples could be given by taking quite small adjustments to the figures or percentages; Mr Room's schedule DJR 13 shows how sensitive the results are to small changes in the figures used. This, to my mind, demonstrates that this method certainly does not accord with what the market would do and, in my judgment, does not produce a "fair and reasonable" consideration.

All the above figures and percentages have been arrived at after disclosure of Mercury's business case and the revised forecast, and as the result of negotiations between Mr Room and Mr Sheard. In the normal way in an ordinary transaction, the relevant information would K
not be available to the proposed grantor. The market would therefore not take the figures and percentages into account, and I do not consider that it is right for me to take them into account: see the estate duty case of *Lynall* v *Inland Revenue Commissioners* [1972] AC 680, mentioned by Peter Gibson J in the *BP* case at p230.

Basis of determination of consideration

Having now rejected the main submissions of both sides, I must consider how otherwise to determine what would be the "fair and reasonable" consideration for the deed of grant in accordance with the principles discussed above. One suggestion was that it should be L
related to the development value of Eastwood Wharf Road. In relation to that, there was evidence of the development value of sites in the area. I do not propose to discuss that evidence, for two reasons. In the first place, the road itself is not part of a development site and I am unable to see how that evidence is of any assistance in determining the value of the road. I agree with Mr Room that the road's value is only as an access and for carrying services. Second, and more importantly, the value of the road for present purposes is the bargaining power it gives to LIDI, which is quite unrelated to either any development value of neighbouring sites or the intrinsic value of the road simply as a piece of land.

The nearest analogous situation, in my view, is that of a wayleave, M
as discussed under "Conclusion of LIDI's main submissions" above. The *Whitwam* case *supra* gives a good illustration of the appropriate method of fixing an appropriate consideration for a right of importance and value to the grantee but which causes no detriment to the grantor, which is why I have described that case as something of a two-edged sword for Mercury to employ; and, as I have already said, wayleave payments automatically reflect the use made of the right granted and its importance to the grantee. In principle, it seems to me that the equivalent of a wayleave rent would be the most "fair and reasonable" method of determining the consideration. However, that is plainly not practical in the present case, just as it is not practical in the

A majority of right of way cases. There must necessarily be instead a capital payment or annual rent, but that should, in my view, reflect the anticipated use of the right and thus its importance and the value to the grantee.

The only material on which this can be based is, I consider, the evidence of comparable transactions and I must now turn to these.

Aerial agreements

At an early stage of the hearing before me, Mr Hobson made an application for leave to adduce evidence of transactions relating to the grant of rights to fix and use aerial antennae on radio masts and
B elsewhere. Mr Aikens objected to the introduction of this evidence on the grounds: (1) that it was irrelevant; and (2) that it was too late. As to (1), I considered that the evidence could well be of assistance, in that it might show how the market fixed the compensation for the grant of a not dissimilar right of value to the grantee which caused little or no detriment to the grantor, and that it was not therefore irrelevant. As to (2) however, I took the view that in general terms the evidence was much too late and so there would be unfair prejudice to Mercury if such evidence were to be allowed in. The evidence would effectively be of comparable transactions which Mercury's witnesses would be unable to investigate and thus put themselves in a position to be able to comment upon. But it appeared that some of the transactions
C involved organisations associated with Mercury. I took the view that Mercury's witnesses would not have undue difficulty in ascertaining sufficiently the facts and circumstances of those transactions. I therefore allowed evidence of those transactions, but not others, to be adduced. Mercury was in fact able to produce evidence as to those transactions, and the circumstances in which they were negotiated.

I summarise that evidence as follows. The transactions in question concerned the grant of licences to Mercury Personnel Communications Network Ltd ("MPCN") or Mercury Personal Communications ("MPC"), a partnership with which Mercury is connected, albeit indirectly, and of which MPCN became a part.
D MPCN was, and MPC is, involved in the mobile telephone business, and required the use of suitable radio masts. Negotiations took place for this purpose with the PLA, which owned Cory's Radio Mast at Gallion's Reach near the River Thames. There was an initial agreement in early 1991 for MPCN to be entitled to fix two antennae (with associated equipment) to the mast at an annual rent of £3,000, subject to an annual upwards-only RPI adjustment. Shortly afterwards, MPCN's requirements were increased first to four installations and subsequently to varying larger numbers of installations. It became clear that the PLA required £1,500 per installation and was not prepared to concede any discount for the larger numbers. The licence
E which was eventually granted to MPC permitted six aerials and eight microwave dishes at an annual rent of £21,000 (ie £1,500 per installation) subject to yearly upwards-only RPI reviews.

The affidavit evidence of Mr Rodney J F Gillington [FRICS], a chartered surveyor experienced in this field, is of interest. Towards the end of his affidavit he says:

> It is not unusual to express rental on a "per antenna" basis especially on existing radio masts or on sites operated by telecommunications companies. Owners of existing radio masts and telecommunications companies are aware of the value to an operator of a radio mast or an established telecommunications site. There is a market in radio masts because the benefit
> F of such a site to an operator like MPC is known. As regards this mast, firstly, there is no real alternative in the area, especially to an operator faced with the time constraints which faced MPC in this case. Secondly, the mast is available for immediate use by the operator. Thirdly, such a mast involves the operator in no capital cost and limited planning constraints. All these must have been weighed up by MPC before they agreed to accept the rental.

This evidence is not, in my view, of any assistance as regards the *quantum* of the consideration in the present case, for the rights granted and the circumstances generally are wholly different. But nevertheless the evidence is, in my judgment, of value because it provides illustrations and cogent support for some of the matters I have discussed above, as follows:

(1) In the real world of the market, a grantor who is in a position to G
grant a right of value to a grantee can and does obtain a not insubstantial consideration for the grant. There is no question of his accepting a nil or nominal amount because the grant of the right involves him in no loss or detriment. The fact is that he is able to use his bargaining strength to negotiate a realistic, but not excessive, figure. How much he is able to negotiate depends on the importance of the right to the grantee, the special suitability of the site to the grantee and the availability of the alternatives.

(2) Mr Room, in his second supplementary report dealing with the aerial agreements, accepted that Eastwood Wharf Road, like Cory's
Radio Mast, has a special suitability to Mercury by reason of its H
location. But he argued that the circumstances were different in that the mast was an expensive capital investment of value to the PLA and it was not surprising that the PLA should expect a payment by way of contribution to the costs from any one who wished to share its facilities. In my judgment, this argument is fallacious, for the expenditure of the grantor is not really relevant to the critical factor, which is his bargaining power. The expenditure would count for nothing if there was no demand for the use of the mast, and an absence of expenditure does not affect the value to the grantee of a facility and therefore the bargaining power of the grantor.

(3) However, there was never any question of the consideration J
being ascertained by reference to the profits of MPCN or MPC. The consideration is an index-linked annual sum assessed, indirectly at any rate, by its value to the grantee and the market generally.

(4) For these reasons, in my judgment, the evidence of the aerial agreements affords no support for, and indeed is evidence at variance with, the main contentions of both Mercury and LIDI.

(5) It was argued on behalf of LIDI that the evidence shows that there should be no discount for bulk allowed in the assessment of consideration, but, in my view, the situations are not comparable in that respect, for these reasons: (a) It appears that there are plenty of
would-be grantees for rights to use Cory's Radio Mast, so that a K
grantee has little bargaining power in that respect; (b) Cory's Radio Mast can accommodate only a finite number of antennae, and the increase in the number of antennae correspondingly reduces the amount of space available. The position is different in the case of the installation of cables encased together, occupying only the surface area covered by three cables although going relatively deep; it was accepted by Mr Sheard in evidence that there would be little prospect of other services being laid under Mercury's cables; (c) I think there may well be substance in Mr Room's view that the course of the negotiations may have affected the availability of a discount, and that if facilities for the full 14 antennae had been requested at the outset,
some reduction from the per antenna figure might have been obtained. L

Agreements with NFU and CLA

Mr Room gave evidence, supported by the documents in his appendix DJR 5, of standard wayleave agreements negotiated by Mercury with the National Farmers' Union and the Country Landowners' Association. The figures applied during the three years ended December 31 1991, but are currently under renegotiation. The agreements provide for capital payments for the grant of 25-year terms, calculated at £3 per metre for single cables, £3.50 per metre for double cables, £285 for surface manholes and £10 for deep buried
manholes. (These are the figures for owner-occupiers, and are M
apportioned in other cases between the owner and the tenant.) On a similar basis, Mr Room calculated that the consideration in the present case should be a capital sum of £2,000, which he put forward as a possible alternative to a nominal figure.

I am unable to accept this evidence as being of any value in the present case. To start with the obvious point, the figures were in respect of non-urban land and the documents show that they were to some extent arrived at by reference to the value of agricultural land. Further, the fact that the standard figures are on a per metre and per manhole basis show that they are really only, or at any rate mainly, concerned with compensation for actual depreciation and

A inconvenience. It is clear that the figures do not include anything to reflect any bargaining power individual owners may have due to special circumstances. They appear to cover, for instance, cases where it would be relatively easy (if less convenient) for Mercury to lay ducts under adjoining public highways at no consideration, and do not provide for any addition where the opposite is the case. They are thus dealing with an entirely different situation, and can afford no help in the present case where (as I have held) the considerable bargaining power of LIDI is the dominant factor, and indeed the only factor of any consequence, to be taken into account.

Schedule and analysis of comparables

B Mr Room and Mr Sheard agreed a schedule and analysis of 16 transactions in the City of London. Owing to the confidentiality of BT's transactions, all the transactions in the schedule were bargains struck by Mercury with landowners in private negotiations without any application to the court under the telecommunications code. All were grants (mostly for a 25-year term) permitting the installation of between one and four 90, 95 or 100mm cables, ie cables of similar width to those in the present case, but many fewer of them. Mr Room and Mr Sheard omitted from the schedule a number of transactions in which the grantor had received direct benefit from the grant (by the landowner's building being connected to the Mercury system); in my C judgment, Mr Room and Mr Sheard were clearly right to omit these, because they could not be comparables of any evidential value in relation to a transaction in which the grantor receives no benefit apart from the monetary consideration itself.

The amounts paid by way of consideration for the grants were a mixture of capital and annual payments. The capital payments have been converted into annual payments by taking a seven years' purchase, and the figures adjusted in accordance with the Retail Price Index to October 1992 (at the time, that being the most recent date for which figures were available), in order that the transactions can be compared on the same basis.

D Unfortunately, there are special factors affecting most of the transactions which, in my judgment, renders them of no real evidential value, as follows:

(a) In each of the transactions numbered 1, 4, 5, 10, 11, 13, 14 and 15 in the schedule, it is apparent that the grantor's tenants would obtain benefit from the installations. Mercury's services being connected to the relevant buildings. This might well increase the rental and/or lettability of the buildings and thus benefit the grantor. In my view, these transactions really fall into the same category as those omitted by Mr Room and Mr Sheard mentioned above.

(b) In three other transactions, numbered 1, 9 and 12, the cables E had been laid across private land in preference to breaking up the adjoining public highway (when, as I have indicated, no consideration is payable). This factor plainly affected adversely the bargaining power of the grantor (and indeed may well have removed it altogether), and probably accounts for the consideration in nos 1 and 12 being nil. As regards no 9 (647 Commercial Road), the evidence was that it had not been appreciated that Mercury's cables, which had been laid under a pavement, had in fact been laid in private land until a sale of that land had been agreed. The grantor was anxious to regularise the position as quickly as possible in order not to jeopardise his sale.

F (c) In the transactions numbered 2 and 3 (Throgmorton Avenue), it appears that the services had to be connected to adjoining buildings, which puts them in the same category as (a) above. But in addition the grantor has the right to determine the agreement on six-months notice.

(d) In the transaction numbered 16 (Marks and Spencer car park), the cable serves only one customer.

The transaction numbered 6 (Billingsgate Market) was an early transaction for 10 years only, the two 100mm cables being intended as only temporary until replaced, as they now have been. The consideration was £1,500 pa, subject to RPI increases, and in addition Mercury agreed to make certain other payments during the laying of the cable and to pay the grantor's costs of £1,250. These factors, in my judgment, render the transaction of some evidential value (in contrast G to those mentioned in paras (a) to (d) above, albeit limited value.

The best comparables, purely as transactions, are those numbered 7 and 8 in the schedule, which are "the 1987 agreement" with the PLA and the "1988 agreement" with LIDI which I mentioned under "Proposed rights" above. Under the 1987 agreement, Mercury was granted the right to lay two 100mm cables over a length of 700m, and under the 1988 agreement the right to lay four 95mm cables over a length of 340m. However, the evidence as to the negotiations leading up to these two agreements shows that there was no direct consideration of the telecommunications code or of the principles on which the consideration should be determined. What happened as to H the 1987 agreement, which was in fact negotiated in October 1985, was that Mr Graham D Hall [FRICS] (who gave evidence) had a meeting with two representatives of Mercury. There had not been much preparation for the meeting. Mercury's representatives suggested an annual rent of £1,000, Mr Hall suggested £5,000, and the parties split the difference by agreeing the figure of £3,000 pa. It was, said Mr Hall, really a "horse deal". As to the 1988 agreement, Mr Desmond Collier on behalf of Mercury and Mr Hall merely took the 1987 agreement figure. The number of cables was to be double but the length of them was to be about half. Both Mr Collier and Mr Hall said that there was very little discussion as to the consideration, which was J considered as being almost ancillary to other matters.

Mr Aikens submitted that, having regard to the evidence about the negotiations, little reliance could be placed on these agreements as comparables. He also stressed that there were important practical circumstances which must have had the effect of inducing Mercury to agree excessive amounts. Although it would have been theoretically possible to invoke the code and have the consideration determined by the court, this was not an option in practice open to Mercury by reason of (a) the inevitable delay and the time constraints on Mercury to get its network into operation; (b) the inherent uncertainty involved; and (c) the expense of any litigation, which would be out of all proportion K to the amounts involved (although this factor would also affect LIDI). Mercury thus had a considerable "anxiety to settle". In my view, there is plainly great force in these points, and they have been recognised by the Lands Tribunal in other contexts where there is a grantee's "anxiety to settle" as affecting the reliability and evidential value of comparables: see eg *Delaforce* v *Evans* (1970) 22 P&CR 770*. If there had been more reliable comparables available, I would not have placed put much weight on them.

However, that is far from saying that the 1987 and 1988 agreements (and transaction no 6) are of no evidential value. Although the telecommunications code was not directly considered, it was in the L background. The cables were of equivalent dimensions. Importantly, to my mind, both parties were undoubtedly "willing" to enter into the agreements. Furthermore, the figures agreed, albeit after an unsophisticated "horse deal", seem to me to be, if a little high, at any rate in the right sort of bracket, and of a kind that appears to be fair to both parties and reasonable. The agreements are of course at total variance with the main arguments of either party before me. But I have to say, that in my view, the unsophisticated first thoughts and reactions of Mr Hall on behalf of LIDI and Mr Collier and the other representatives of Mercury in the initial negotiations produced results which are much more in accord with what was envisaged by para 7(1) of the code than the alternative results produced by the main M arguments of the parties before me.

Conclusion

Having, for the reasons set out above, rejected the main arguments of both parties, and also Mercury's reliance on the NFU/CLA agreements, I must do the best I can with the sparse material available. In the circumstances, and in default of anything better, the starting point must, I think, be the 1987 and 1988 agreements. At current money values, these can be taken as providing in round figures for annual payments of £4,000. I adjust this figure downwards by about

---

*Editor's note: Also reported at (1970) 215 EG 315.

A one-third to take account of Mercury's "anxiety to settle". I then use it to determine whatI think would be a corresponding figure for three cables laid side by side, as the uppermost cables. This is not easy, because the two £3,000 annual payments in the 1987 and 1988 agreements took into account different lengths of cable, which I consider in this case (depending as it does on the parties' bargaining strengths) as a factor of little importance. This determination is necessarily rough and ready and a matter of judgment, but I arrive at a figure of £2,000 pa. I must then take account of the seven layers of three cables underneath the uppermost three cables. I reject the suggestion that these should be ignored, which seems to me to ignore B LIDI's bargaining strength and to be quite unrealistic. But, in my judgment, it is appropriate to make a fairly substantial discount for these layers, and I determine a discount of 50%. That adds up to a total annual figure of £2,000 + 50% of 7 x £2,000 = £9,000. I do not C consider it appropriate to add anything for the manholes etc, as they are simply component parts of the installation of the 24 cables and the figures I have determined already take them into account.

I now consider that figure objectively and ignoring the method of arriving at it. In my judgment, it is the right sort of figure given the importance of the ducts to Mercury and the bargaining strengths of both sides, and I consider that it is fair to both parties and reasonable on the basis that the deed of grant was to be willingly given.

Accordingly, I hold that the figure to be inserted in the deed of grant should be an annual figure (and not a capital sum), and should be £9,000 pa. I will hear counsel as to the precise form of order and as to D costs and any other matters.

For further cases on this subject see p 33

LANDLORD AND TENANT
GENERAL

A **Court of Appeal**

June 10 1988

(Before Lord Justice WOOLF and Sir John MEGAW)

TWYMAN v CHARRINGTON

Landlord and tenant — Repairs — Service charges — Whether external walls and roof within expression "mutual or party walls" or "mutual or party structures"

B The plaintiff landlord occupied the first and second floors of a dwelling-house; the defendant occupied the ground floor and basement by way of an underlease. By clause 3(4) of the underlease the defendant covenanted to repair the demised premises. By clause 3(7) of the underlease the defendant covenanted to contribute and pay a rateable or due proportion of the costs and expenses expended by the landlord in repairing, *inter alia*, all "mutual or party walls and fences mutual or party structures and other items which may belong to or be used for the demised premises in common with other premises near or adjoining thereto". The plaintiff landlord incurred expenditure in repairing the roof and upon its non-payment by the
C defendant obtained an order for possession in the court below following forfeiture proceedings. The defendant appealed contending that the roof did not fall within the terms of clause 3(7).

***Held*: The appeal was dismissed. Looking at the terms of the underlease as a whole, the external walls are not included in the demise. Considering clause 3(7) on the general basis that the clear intent of the underlease as a whole was that the tenant should be responsible for "internal repairs" of his demised premises, that he should have no obligation with regard to the internal parts of the premises occupied by his landlord, but that the expenses of repairing the external parts of the premises are**
D to be shared in common with a landlord. The roof can be regarded as being a "mutual structure". The same approach can be adopted in the case of external walls, which are mutual structures. If there was any doubt about the position and the roof was not a mutual structure it was certainly an "other item". It was largely a question of impression for the court. The word "mutual" has the meaning of "in common". The fact that the roof was not physically adjacent to the demised premises did not prevent it from being a "mutual structure".

The following case is referred to in this report.

F *Douglas-Scott* v *Scorgie* [1984] 1 WLR 716; [1984] 1 All ER 1086; (1984) 48 P&CR 109; [1984] EGD 325; 269 EG 1164, CA

This was an appeal from Judge Edwards (sitting at Brentford County Court), who on March 11 1988 made an order for possession against the defendant, Robert David Charrington, in respect of 37 Church Street, Twickenham, in proceedings for forfeiture and arrears of service charges brought by the plaintiff, Brian Charles Twyman.

Edward Cousins (instructed by Camerons, of Harrow) appeared for the appellant; David Elvin (instructed by Coleman & Betts, of Kingston upon Thames) represented the respondent.

Giving the first judgment, **WOOLF LJ** said: This is an appeal from G
a judgment of Judge Edwards, sitting at Brentford County Court, which he gave on March 11 1988. On that occasion he held that the plaintiff was entitled to recover possession of premises which were part of 37 Church Street, Twickenham, and ordered that a sum of money totalling £129.40 together with certain costs were to be paid. Although the proceedings took the form of forfeiture proceedings and for possession, in fact the purpose of the proceedings was to resolve a dispute between the plaintiff and the defendant as to the proper interpretation of the lease pursuant to which the defendant was in possession of the ground floor and basement of 37 Church Street, Twickenham. H

The plaintiff occupies the first and second floors of those premises. He was an assignee of the lease of the whole premises and the underlease was one entered into by the plaintiff's predecessor in title with the defendant's predecessor in title. By the underlease the defendant was entitled to occupy the ground floor and basement, as already indicated, and the underlease which gave rise to the dispute contained a number of covenants. Fortunately, however, it is only necessary for the purposes of this appeal to refer to a few of those provisions because in fact the dispute is confined to determining whether the defendant tenant is responsible for making a contribution to the costs of repairs to the roof. J

By the underlease it was provided that there was demised "unto the Tenant ALL THOSE premises more particularly described in the Schedule hereto (hereinafter referred to as 'the demised premises')".

The schedule is in these terms:

> ALL THOSE premises comprising the ground floor and basement of the building known as 37 Church Street Twickenham . . . as the same are for the purpose of identification only shown coloured pink on the attached plan.

The court has looked at the attached plan, but the attached plan provides no assistance in resolving the question which is of some relevance as to whether or not the external walls are included in the demise. K

The underlease contains a covenant — covenant 3(4) — pursuant to which the tenant is under an obligation to repair. That is not the covenant which is primarily involved in the issue on this appeal but it is, however, relevant and I should read it. It is in these terms:

> THE Tenant hereby COVENANTS with the Landlord . . .
>
> (4) to repair maintain cleanse and keep in good and substantial repair and to replace or rebuild where necessary the demised premises and all additions and improvements thereto and all Landlord's fixtures and fittings thereunto belonging and all appurtenances forming part of the demises premises (damage by any insured risk only excepted) AND ALSO when and so often as any Landlord's fixtures belonging to the demised premises shall so require to L
> substitute other fixtures of a similar description quality and value to the satisfaction of the Landlord.

The covenant which is directly relevant to the issue on this appeal is at clause 3(7), and that is the only other covenant to which I need refer. It provides:

> The Tenant hereby COVENANTS with the Landlord . . .
>
> (7) to contribute and pay a rateable or due proportion of the costs and expenses (to be determined by the Landlord's Surveyor) expended by the Landlord or the Superior Landlord in making laying repairing maintaining

renewing rebuilding and cleansing all ways roads pavements sewers drains pipes watercourses mutual or party walls and fences mutual or party structures and other items which may belong to or be used for the demised premises in common with other premises near or adjoining thereto and to keep the Landlord and the Superior Landlord indemnified against such proportion of such costs and expenses.

The judge came to the conclusion that the effect of clause 3(7) was that the tenant was under an obligation to pay a contribution towards the costs of repairing roof to the building. He said, in the note of his judgment:

> It seems to me that repairs to the roof are within the covenant (clause 3(7)) as being repairs of a mutual structure used for the demised premises . . . with other premises. Or it is an "other item" which maybe so used within the wording of the covenant. The arguments adduced by the Plaintiff are stronger than those by the Defendant.

It is common ground that sadly this lease is not drawn with the precision that is desirable. There is a dispute, first of all, between the landlord and the tenant as to whether the external walls of the ground floor and basement are included in the demise. As to that, looking at the terms of the lease as a whole, I am quite satisfied that the external walls are not included. Having come to that conclusion, I approach the construction of clause 3(7) on the general basis that the clear intent of the lease as a whole is that the tenant should be responsible for what I will describe in general as "internal repairs" of his demised premises, that is, the ground floor and basement; that in general he should have no obligation with regard to the internal parts of the premises which are occupied by his landlord; but with regard to the external parts of the premises the expenses of repairing these is to be shared in common with his landlord. Such an approach fits in with the headlease, pursuant to which the plaintiff is under an obligation to keep in repair the whole of the premises. These premises being now in the form of two flats, the convenient way for the cost of keeping them in repair to be met when they are in dual occupation is for the costs in relation to the external parts and what I will call in general the "common parts" to be shared by the two occupiers. That, in my view, is the general intent of clause 3(7).

When one condescends to look at the terms in detail of that clause, one finds that there is such an obligation clearly established with regard, for example, to sewers and drainpipes; and indeed in relation to mutual or party walls. That, in my view, helps when considering the vital words which were considered by the judge in coming to his conclusion in favour of the landlord. First of all, the judge came to the conclusion that the roof could be regarded as a mutual structure. In the context of this covenant I would agree with the learned judge's conclusion in that regard. It is important that the word "mutual" is followed by "structures or party structures". Clearly the intent here is that structures which are immediately adjoining the demised premises shall be regarded as party structures, and those parts which are used for their common benefit which are not immediately contiguous should be regarded as mutual, and I would so regard the roof. It provides benefit to both demised premises and it is within the general intent of the clause that the costs of repair should be contributed to by the tenant.

So far as the external walls are concerned, that is clearly dealt with by the preceding words "mutual or party walls", and I would adopt the same approach with regard to structures.

If there were any doubt about the position, then, in my view, the doubt is removed by the reference which follows "structures" to "other items which may" (and here I quote the relevant words) "be used for the demised premises in common with other premises near or adjoining thereto". To talk about a roof as an "other item" might not be the normal language which one would apply, but as "other items" is clearly intended to embrace all that goes before and to be therefore of general import, the fact that it goes on to identify those items as being ones which are used in common with the other premises means that if the roof was not a mutual structure it was certainly an "other item".

In the very helpful arguments which were advanced on behalf of both sides by Mr Edward Cousins and Mr David Elvin, we were referred to references in the *Oxford Dictionary* — the shorter and concise dictionaries — as to the meaning of the word "mutual". We were also referred to the decision of this court in the case of *Douglas-Scott* v *Scorgie* [1984] 1 All ER 1086*. We were also referred to an authority dealing with the general approach to the *contra proferentum* rule. However, I do not propose to make any further references to those authorities because, in my view, they do not provide any real assistance on the issue which arises on this appeal. As Mr Cousins readily accepted in the course of argument, this is largely a question of impression for the court. Clearly in this case it is open to the court to form a view as to whether the roof is covered by the obligations placed upon the tenant by clause 3(7) and I find the language of that clause sufficiently clear to enable this court to uphold the decision of the learned judge without reference to other assistance.

I accordingly would dismiss this appeal.

Agreeing that the appeal should be dismissed, **Sir John Megaw** said: I would add my agreement with the tribute which Woolf LJ has paid to the very concise and well-expressed arguments put by counsel on both sides. They also have my sympathy in being faced with the formidable task of construing a document, in particular subclause (7) of clause 3 of the lease, which is, at the best, very unfortunately phrased from the point of view of clarity. The principal difficulty, I think, is the use of the word "mutual". That is a word which in its original and proper meaning involves reciprocity, but it has long since escaped from that narrow meaning in its ordinary everyday use in the English language. Very often now it is used to mean no more than "in common", such as "our mutual friend", which has got nothing to do with reciprocity, but simply refers to somebody who is a friend of each of the two people concerned.

I think that the word "mutual" in clause 3(7) really has the meaning of "in common", and a "mutual structure" includes in this case the roof because that structure provides a service, a benefit, an advantage, to the demised premises as well as to the rest of the building, even though it is not physically adjacent to the demised premises, being separated from it by the other floors of the building. But I do not think that in the circumstances here the fact that it is not physically adjacent prevents it from being a "mutual structure" for the purposes of the clause.

I would agree also that, if there is doubt about the meaning of "mutual structures" in this context, that doubt ceases to be of importance because of the following words "other items which may belong to or be used for the demised premises". It is agreed that the words "may belong to" are not of relevance here. It is submitted with some force by Mr Elvin that it is difficult to describe the roof as being "used for" the demised premises in common with adjoining premises, but to my mind it is not very difficult to see that the real intention is that the roof should be regarded as being "used" for the demised premises just in the same way as a sewer or a drain or a downfall water-pipe would be properly described as being "used for the demised premises". Here the roof is used for the demised premises because, if the roof were not there, the demised premises would ultimately suffer from the effects of rainfall coming through an unroofed building.

I agree that the appeal should be dismissed.

Appeal dismissed with costs.

*Editor's note: Also reported at (1984) 269 EG 1164.

A
Court of Session

INNER HOUSE

October 8 1993

(Before Lord McCLUSKEY, Lord WEIR and Lord CLYDE)

SCOTTISH WHOLEFOODS COLLECTIVE WAREHOUSE LTD v RAYE INVESTMENTS LTD AND ANOTHER

B *Option — Option in lease to purchase leased subjects — Price to be current open market value and at a price to be mutually agreed — Whether option unenforceable by being agreement to agree*

In 1984 the pursuers and respondents became the tenants of leased subjects at 23 Fleming Street, Glasgow, under a contract of lease. Clause ninth, of the contract of lease, contained an option for the tenants to purchase the leased subjects at the current open market price between a willing buyer and a willing seller. The clause provided: ". . . and as said price shall be mutually agreed between the Tenants and the Landlords". The tenants sent a timeous notice exercising the option and included a certificate prepared by
C surveyors stating the current open market value of the subjects to be £90,000. The landlords asserted the open market value to be £250,000. The tenants raised the present action seeking declarator that they were entitled to purchase the subjects at a price of £90,000 "or at such other price as the Court may determine". The dispute between the parties was whether the option clause created an agreed means for determining the price, being a means which was not dependent upon some future agreement between the parties. The defenders and appellants appealed from the sheriff principal who had refused an appeal from the sheriff who repelled the first defenders', plea to the relevancy holding that there was an enforceable
D obligation.

***Held*: The clause must be looked at as a whole to discover the objective intention of the parties from the whole words used. The words in the clause ". . . and as said price shall be mutually agreed between the Tenants and the Landlords" did not detract from the clear intention that the tenant could exercise the option at the current open market value. Failing agreement by the parties to the price, there was no difficulty in having the matter determined objectively by the court. The option was valid and enforceable.**

E The following cases are referred to in this report.

Didymi Corporation v *Atlantic Lines & Navigation Co Inc* [1987] 2 Lloyd's Rep 166; [1988] 2 Lloyd's Rep 108

Foley v *Classique Coaches Ltd* [1934] 2 KB 1; [1934] All ER Rep 88, CA

Malpass, decd, In Re [1985] Ch 42; [1984] 3 WLR 372; [1984] 2 All ER 313

May & Butcher v *King* [1934] 2 KB 17

Michael Richards Properties Ltd v *St Saviour's Parish, Southwark* [1975] 3 All ER 416

Nicolene Ltd v *Simmonds* [1953] 1 QB 543; [1953] 2 WLR 717; [1953] 1 All ER 822, CA

Sudbrook Trading Estate Ltd v *Eggleton* [1983] 1 AC 444; [1982] 3 WLR
F 315; [1982] 3 All ER 1; (1982) 44 P&CR 153; [1982] EGD 392; 265 EG 215, HL

Talbot v *Talbot* [1968] Ch 1; [1967] 2 All ER 920; [1967] 3 WLR 438, CA

Trade Development Bank v *David W Haig (Bellshill) Ltd* 1983 SLT 510

This was an appeal from the decision of the sheriff principal who had refused an appeal from the sheriff who had repelled the appellants and defenders, Raye Investments Ltd and Barclays Bank plc, plea to the relevancy in the action raised by the respondents and pursuers, Scottish Wholefoods Collective Warehouse Ltd.

Donald Macfadyen QC and Stephen Woolman (instructed by Bird Semple Fyfe Ireland, of Glasgow appeared for the second defenders and appellants; Matthew Clarke QC (instructed by J&RA Robertson of Edinburgh, agents for Stone & Co, of Glasgow) represented the G
pursuers and respondents.

Giving his opinion, **LORD McCLUSKEY** said: In 1984 the pursuers and respondents became the tenants under a contract of lease with Pickford Removals Ltd, the then owners of the leased subjects, at 23 Fleming Street, Glasgow. The first defenders later succeeded to the landlords' interest, by which time the term of the lease had been extended. The contract of lease was in an unremarkable form. It was professionally drawn up. It contained a clause from which the dispute between the parties to this appeal derives. As the issue in the appeal turns upon the construction of this clause it is essential to notice its H
whole terms. It reads:

NINTH The Tenants will have the option to purchase the leased subjects at any time during the currency of this Lease which option they shall be bound to declare by six months prior written notice to the Landlords. The price shall be the current open market price pertaining at the time when the notice is served as between a willing buyer and a willing seller and on the assumption that vacant possession can be granted to the purchaser and no account being taken of any goodwill attaching to the premises or any fixtures and fittings therein (other than those belonging to the Landlords) and as said price shall be mutually agreed between the Tenants and the Landlords. In the event of the Tenants exercising their rights hereunder, the Tenants will be responsible for all expenses incurred in connection therewith including but without prejudice J
to the generality of the foregoing, all legal expenses of the Landlords incurred in connection with the conveyance to them.

On the understanding that this clause gave them an option to purchase the leased subjects, the tenants sent the landlords a timeous written notice declaring their intention to exercise that option. With the written notice they sent a certificate of value, prepared by a firm of surveyors, stating the current open market value of the subjects to be £90,000.

Correspondence between the parties followed, but the landlords disputed the valuation and asserted the open market value to be K
approximately £250,000. As the parties came to no agreement on the figure for open market value the landlords adopted the position that there was "no enforceable obligation" upon them to sell the subjects to the tenants. The tenants then raised the present action seeking declarator that they are entitled to purchase the subjects at a price of £90,000 "or at such other price as the Court may determine". They also sought a court order to make the landlords, the first-named defenders, sell the subjects to the pursuers. No issue has been taken in these proceedings about the form of the declarator or the order sought or about any other matter of detail.

After hearing debate on the first-named defenders' plea to the L
relevancy, the sheriff held that there was an enforceable obligation, repelled that plea and allowed a proof on the single issue of the open market value of the subjects at the material time. The sheriff principal refused an appeal against the sheriff's interlocutor. In each case, the sheriff and the sheriff principal rested his decision upon his construction of clause ninth. They gave somewhat different reasons for holding that it contained an enforceable obligation; but each held that the pursuers were entitled to purchase the subjects at the open market price as determined at a proof. The first-named defenders are no longer parties to the cause, but the second-named defenders and appellants have been sisted as parties to the appeal. They have adopted M
the grounds of appeal originally lodged for the first-named defenders. The title and interest of the second-named defenders to pursue the appeal upon these grounds is not challenged.

The issue between the parties is an issue as to the meaning and effect of clause ninth quoted above. Both parties to this appeal accept that the option conferred on the tenants by clause ninth was an option which, if properly exercised in terms of the clause, could result in the sale and disposition of the subjects by the landlords to the tenants; they were agreed that the price of the subjects would be an essential element in such a sale and that, in the circumstances here, clause ninth did not confer on the tenant any enforceable right to purchase the subjects unless the terms of the clause itself enshrined agreement as to price.

Neither party to this appeal suggested that it was necessary for the clause to contain a clear agreement as to a certain price as specified in the contract of lease; an agreement which specified a workable and agreed method of ascertaining a price would be sufficient; and both parties accepted the principle stated by Viscount Dunedin in *May & Butcher Ltd* v *King* (reported as a note in *Foley* v *Classique Coaches Ltd* [1934] 2 KB 1 at p17):

> To be a good contract there must be a concluded bargain, and a concluded contract is one which settles everything that is necessary to be settled and leaves nothing to be settled by agreement of the parties. Of course it may leave something which still has to be determined, but then that determination must be a determination which does not depend upon the agreement between the parties. In the system of law in which I was brought up, that was expressed by one of those brocards of which perhaps we have been too fond, but which very often express very neatly what is wanted: *certum est quod certum reddi potest*. Therefore, you may very well agree that a certain part of the contract of sale, such as price, may be settled by some one else.

It was not in dispute that an agreement to make a future agreement on an essential element of the bargain was not enforceable. Accordingly, the dispute between the parties here was whether or not the clause created an agreed means for determining the price, being a means which was not dependent upon some future agreement between the parties. That being so, it is not necessary to attempt to analyse in any detail the many cases which were referred to in the debates in the Sheriff Court. It is enough in this context simply to mention the *May & Butcher* case and *Trade Development Bank* v *David W Haig (Bellshill) Ltd* 1983 SLT 510.

Mr Donald Macfadyen QC, for the appellants, argued that, although the rest of the clause could be read as envisaging that the tenants would have an option to purchase the leased subjects on certain terms, the clause fell to be read as a whole; the 14 words "and as said price shall be mutually agreed between the Tenants and the Landlords" could not be read out of the clause. They formed an integral part of it and, properly understood, meant that there was no agreement within the clause either as to the price to be paid or as to any mechanism independent of the parties for determining that price. When the clause was read as a whole the only mechanism provided for was that the tenants and the landlords would seek to reach agreement when the "option" was sought to be exercised as to what the then current open market price was. Unless such an agreement could then be reached the price would remain indefinite. In short, clause ninth encapsulated only an agreement to agree, not an enforceable obligation.

While it was acknowledged that the second sentence of the clause pointed towards an intention to transact at the current open market price and also defined the factors which would determine precisely the character of what was to be marketed — subjects which were vacant and in respect of which there was no good-will element and no addition in respect of fixtures and fittings — all upon the basis that the transaction was to be regarded as one between two parties at arm's length ("a willing buyer and a willing seller"), none the less there was no express agreement as to what that price was to be. Nor was there any agreement upon a mechanism for ascertaining the price, other than the agreement that it was to be "mutually agreed" in the future. Accordingly, both sheriffs were in error and the appeal should be allowed; the action should be dismissed.

In reply, Mr Matthew Clarke QC, for the pursuers and respondents, submitted that there was no failure by the parties to agree; there was simply a dispute about what "the current open market price" was. The distinction between a dispute and a failure to agree could be found in the speech of Viscount Dunedin in *May & Butcher*: at p22 he had said:

> In no proper meaning of the word can this be described as a dispute arising between the parties; it is a failure to agree, which is a very different thing from a dispute.

In the present instance the parties had an agreement to allow the tenants to purchase the subjects at the current open market price. They were simply in dispute as to what that price was and it was appropriate that the court should allow a proof in order to ascertain the price. In these circumstances, the sheriffs had reached the correct conclusion. Looked at as a whole, clause ninth plainly envisaged that the parties to the lease were intending that the tenants should have a right to purchase upon certain terms. The 14 words founded upon by the appellants did not have the effect of rendering the agreement incomplete, inchoate and unenforceable. That part of the clause referred back to "said price" and therefore the notion of current open market price was clearly brought in again at the stage in the clause where the 14 words appeared. Reading the clause as a whole, it was clear from its terms that the parties were agreed upon the price — admittedly an essential element in such a contract — in that they were agreed upon means of determining it. The clause no doubt envisaged that they would attempt to reach agreement as to the sum which represented the open market price; but they were agreed that the price was to be the open market price; that figure could be objectively ascertained without further agreement between the parties. The approach of the sheriff in this case was to be preferred to that of the sheriff principal and that approach, to the effect that the quoted words contained something akin to a mandatory injunction upon the parties to agree, was adopted.

Reference was also made to *Didymi Corporation* v *Atlantic Lines & Navigation Co Inc* [1987] 2 Lloyd's Rep 166 and [1988] 2 Lloyd's Rep 108. In that case, the parties had agreed that the amount of the hire of a vessel, would in certain circumstances be decreased or increased "by an amount to be mutually agreed between owners and charterers". Hobhouse J and the Court of Appeal considered that the clause containing these words contained what were clearly intended to be words of obligation and that the function of the court was to extract from the words used the objective intention of the parties. The ability of the court to give effect to such a clause depended upon whether or not the clause provided "a sufficient criterion to enable the appropriate reduction of increase in hire to be determined". In the present case the position was similar. The words relating to mutual agreement did not introduce such uncertainty into the contract as to deprive the rest of the clause of meaning and effect. Given the clear agreement on current open market price then it was clear that if the parties failed to reach agreement as to what that current open market price was then there would arise a simple justiciable issue of the kind which courts were well accustomed to resolving. If, however, the court came to the view that the drafting of the clause was so clumsy and awkward that it was plain that it contained some accidental or unintended contradiction then the words creating that contradiction could be read out. This was the approach of the court in *Nicolene Ltd* v *Simmonds* [1953] 1 QB 543 and also in *Michael Richards Properties Ltd* v *St Saviour's Parish, Southwark* [1975] 3 All ER 416. Reference was also made to *Sudbrook Trading Estate Ltd* v *Eggleton* [1983] 1 AC 444*.

In my opinion, the submissions of the pursuers and respondents are to be preferred. The clause must be looked at as a whole to discover the objective intention of the parties from the whole words used in the clause. I think it is abundantly plain that the parties intended that rights and obligations should be created by the clause. This is particularly clear from the use of the words in the final sentence:

> In the event of the Tenants exercising *their rights hereunder*, the Tenants will be responsible for all expenses incurred in connection therewith including . . . all legal expenses of the Landlords incurred in connection with the conveyance to them

(Emphasis supplied.)

It is the words "their rights hereunder" which are telling; but even the provisions as to expenses in connection with the conveyance are indicative. The opening words of the clause are also very clear. The expression they contain is, "the option to purchase". If one were to

*Editor's note: Also reported at (1982) 265 EG 215.

read the clause in the way which the appellants contend the tenants would have no rights and no contractual option to purchase at all. They would simply have a would need clause ninth in order to have the right to make an offer; they would have *that* right without any such provision. Yet this clause not only confers "the option to purchase"; it actually obliges them, if they wish to exercise it, to do so within a limited, specified time scale and also by a specified method, "written notice".

Furthermore, it appears to me to envisage that the written notice will contain reference to "the current open market price, at the time when the notice is served . . .". The clause goes into some detail to ensure that the nature of the market is specified and the precise character of what is to be marketed is defined. Accordingly, everything which is necessary to define or ascertain all the elements of a contract to give the tenant an enforceable right to exercise an option to buy is to be found in this clause. Indeed counsel for the appellants did not suggest that if the last 14 words of the second sentence had been omitted the clause would have failed to create an enforceable obligation.

I agree with counsel for the appellants that these words must, if possible, be given meaning in the context, but I do not read them as detracting from the clear intention of the clause, which was to confer upon the tenants a right, namely a right to exercise an option to purchase the leased subjects at the current open market price pertaining at the time when the written notice was served. I do not find it a strained reading of the 14-word phrase to construe it as indicating that the parties envisaged or hoped that they would be able to agree what the open market price was without having to submit their dispute on that matter to the arbitrament of some other tribunal; but I do not doubt that they had in mind that the right to purchase at the open market price could be exercised even in the absence of such agreement. I also regard the reference to "*said* price" as being very telling.

On my reading of the clause, the first step in the exercise of the tenants' right would be to serve the written notice which would contain reference to the current open market price. That would allow a sum representing the open market price to be agreed after the service of the notice, but would not exclude agreement before that. Failing such agreement, I can see no difficulty whatsoever in having the matter determined objectively by the court. That is the kind of task which courts frequently have to undertake, for example in relation to certain types of indemnity insurance claim. In the whole circumstances, I consider that the clause as a whole falls to be read as stating the intention of the parties, namely to confer upon the tenants a right, an option to purchase the subjects at a price, the current open market price, an objective concept, the money sum of which, if not agreed, would be determined by the court.

In arriving at this view, I cannot avoid the conclusion that the effect of inserting the 14 words at the end of the second sentence resulted in the intention of the parties being expressed clumsily and awkwardly. Had it been necessary, I should have been prepared, given the strength and clarity of the expression of that intention elsewhere in the clause, to treat these 14 words as being sufficiently meaningless, ambiguous or mistaken to warrant their being ignored: in respect of that I should have been prepared to follow the reasoning to similar effect contained in the opinion of Denning LJ in *Nicolene Ltd*.

For these reasons, I would refuse the appeal.

Lord Weir agreed and did not add anything.

Agreeing, Lord Clyde said: This case concerns the construction of a contract of lease entered into between the pursuers, as the landlords, and the first defenders, as the tenants. The lease was of warehouse premises in Glasgow. The lease was entered into in 1984 and extended and varied in 1986. After the action left the Sheriff Court, Barclays Bank plc were sisted into the case as minuters and second defenders. The appeal to this court was presented by counsel on behalf of the second defenders adopting the grounds of appeal which had been lodged by the first defenders. It is evident that the second defenders have a legitimate interest to pursue the appeal and no objection was taken to their appearance at this stage. The only clause of the contract of lease which is material for the purposes of the appeal is clause ninth. Its full terms have been set out in Lord McCluskey's opinion.

The pursuers have averred that on September 14 1988 they sent written notice to the landlords of their intention to exercise their option to purchase the property, together with a certified valuation of the property. The first defenders have averred that they dispute the valuation, that there is no mutual agreement on the price, and that there is no enforceable obligation on them to sell the subjects to the pursuers. The pursuers have raised the present action for a declarator that they are entitled to purchase the property from the first defenders at the price stated in their own valuation or at such other price as the court may determine. The sheriff, and on appeal, the sheriff principal, have each held that there should be a proof to determine the price. The first defenders have appealed to this court and the second defenders have developed the submissions which were presented in the lower court to the effect that the defenders' first plea-in-law which is a general plea to the relevancy should be sustained and the action dismissed.

The dispute in the case is one concerning the proper construction of the clause and in particular the phrase "and as said price shall be mutually agreed between the Tenants and the Landlords". Counsel for the second defenders and appellants submitted that the effect of that phrase was that there was no effective obligation to convey the property to the tenants unless there was an agreement identifying a mutually agreed open market price. In the absence of an averment of such agreement the pursuers' case was irrelevant. He referred to *Trade Development Bank* v *David W Haig (Bellshill) Ltd* 1983 SLT 510 to support the proposition that there was no effective exercise of an option to lease where the duration of the lease was not settled. It was not in dispute that the agreement on the price is one of the essential elements of a sale and accordingly there was no effective exercise of the option in the present case where that essential remained unresolved.

He accepted that a contract of sale could be effectively made where a mechanism had been agreed for fixing the price, and if machinery broke down or one party failed to operate it the court could intervene to resolve that matter. In that connection he made reference to *Talbot* v *Talbot* [1968] Ch 1, *Sudbrook Trading Estate Ltd* v *Eggleton* [1983] 1 AC 444 and *Lloyds Bank plc* v *Malpass* [1985] Ch 42*. But as was clearly stated in *May & Butcher Ltd* v *King* [1934] 2 KB 17 at p21 if the price is left to be agreed between the parties there is no contract. The sheriff had taken the view that the word "shall" was a mandatory injunction. Counsel for the second defenders criticised that, pointing out that such a mandatory injunction was antithetical to a mutual agreement and in any event was not enforceable. The sheriff principal adopted a construction whereby the critical phrase was to be understood as the first of two alternative ways of defining the price, either a price agreed by the parties or the contract market price. Counsel for the second defenders submitted that that construction does too much violence to the structure and language of the clause and indeed neither party before us adopted it. What the parties had done in the submission of counsel for the second defenders was to make an agreement to agree, each reserving their position so as not to be committed to the sale.

Counsel for the pursuers and respondents submitted that the court should try to construe the clause in a way which would make the arrangements made by the parties workable. The clause was designed to provide for a future sale and sought to provide a formula for the exercise of the option. The agreement of the parties was that the price should be the current market price at the time of the exercise of the option. Under reference to the case of *Didymi Corporation* v *Atlantic Lines & Navigation Co Inc* [1987] 2 Lloyd's Rep 166 and [1988] 2

*See *Malpass, decd, In Re* [1985] Ch 42; [1984] 3 WLR 372; [1984] 2 All ER 313.

A Lloyd's Rep 108, he submitted that the critical phrase in the present case was of a "directory or mechanical" character as opposed to the substantive provision that the price was to be the current market price. Under reference to the case of *Sudbrook* he submitted that the making of the agreement was subsidiary and non-essential. If parties failed to agree then the court could intervene to establish the current market price. As an alternative approach counsel for the pursuers submitted that the phrase in dispute was meaningless and could be severed and disregarded. He referred in this connection to the cases of *Nicolene Ltd* v *Simmonds* [1953] 1 QB 543 and *Michael Richards Properties Ltd* v *St Saviour's Parish, Southwark* [1975] 3 All ER 416. The phrase
B in his submission was unnecessary verbiage stating the obvious in that the parties had already agreed that the "said price" should be the current market price.

I am not persuaded that the phrase in question is altogether without meaning. It indicates one method by which the criterion of current market value is to be applied so as to produce a specific sum of money. But it does seem to me that counsel for the pursuers was well-founded in the main submission which he presented that the phrase should be understood as subsidiary and non-essential. Here, as in the *Didymi* case, the parties had agreed upon an objective standard which was the prescribed measure of the price. Here, as there, it seems to me,
C that that is the substantive provision of the clause which enables the parties to identify the price. The case is not simply one where the parties have left the price to be agreed so as to create simply a mere agreement to agree. On the contrary they have identified the criterion to be applied and the later reference to their mutual agreement is not an essential or substantial element so as to peril the operation of the clause on such agreement being reached.

Counsel for the second defenders sought to distinguish the case of *Didymi* from the present on the ground that it concerned a charter party and not a sale or purchase of heritage. He suggested that Scottish law recognised a greater flexibility with regard to the former category of case. But I do not find that feature weighing prominently in the
D decision. The case seems to have been treated in the same way as any commercial contract and the reliance placed on the case of *Sudbrook* E illustrates the absence of any particular considerations peculiar to mercantile matters. In *Didymi* there was an arbitration clause which could save the position of a failure to agree but in the present case while there is no arbitration clause it is accepted that if agreement was not an essential for the operation of the clause then failing agreement the matter could properly be resolved by the court.

The main objection to the second defenders' approach, to my mind, is that it tends to deprive the clause of any real effect. The achievement by the tenants of a purchase is made to depend upon the consent of the landlords to the price as being the current market value. Not only does that leave the way open for the landlords to avoid F parting with the property if they so desired, but it leaves the tenants not significantly better placed than if they were to make an offer to buy the property without reference to this clause at all. But the clear intention of the clause was to create rights to be exercised by the tenants. That is evident from the final sentence of it.

The first sentence plainly provides for the tenants to have an option to purchase the property. That should mean that the landlords are under an obligation to sell it if the notice is duly given. But the second defenders' construction of the central sentence fits uneasily with the first and the last when it imposes a qualification virtually destructive both of the right of the tenants and the obligation of the landlords. Counsel for the second defenders submitted that the parties could have G wished to hold themselves back from any absolute commitment to a sale. But that is not what is indicated by the clear terms of the first and the last sentences of the clause. I am not persuaded that properly construed this clause was nothing more than an agreement to agree. On the approach put forward by counsel for the pursuers, it seems to me, that the intention indicated in the clause to give rights to the tenant and to impose an obligation on the landlord can be realised and made effective.

On the whole matter, in my view, the appeal should be refused and the interlocutor of the sheriff principal affirmed.

H

For further cases on this subject see p 67

LANDLORD AND TENANT
BUSINESS TENANCIES

A **County Court**

March 21 1983

(Before Judge Aron OWEN, at Clerkenwell County Court)

M CHAPLIN LTD v REGENT CAPITAL HOLDINGS LTD

Landlord and Tenant Act 1954 — Renewal opposed by landlord on ground (d) of section 30(1) — Whether landlord's offer of suitable alternative accommodation must be made before service of notice
B ***under section 25 of the Act***

The applicant tenant held a tenancy dated July 17 1978 from the respondent landlord, the contractual term of which expired on March 25 1982. The landlord served a notice dated July 17 1981 on the tenant under section 25 of the Landlord and Tenant Act 1954 terminating the tenancy on March 25 1982 and stating that they would oppose the grant of a new tenancy on, *inter alia*, ground (*d*) of section 30(1) of the Act, namely that the landlord has offered alternative accommodation. The landlord's offer of alternative accommodation, on another floor in the same building as the holding, was made in a letter which accompanied the section 25
C notice. At trial the tenant contended that, having regard to the use of the past tense in para (*d*) of section 30(1), the offer of alternative accommodation must be made before the service of the section 25 notice, and further, the accommodation which has been offered did not satisfy the requirements of para (*d*).

Held: **The tenant's application for a new tenancy was dismissed. The requirement in para (*d*) that the landlord *has* offered alternative accommodation is satisfied by an offer of accommodation before the issue is joined in the pleadings, namely before the landlord files his answer in the county court or his affidavit in reply in the High Court. The offer of**
D **alternative accommodation in the letter of July 17 1981 was therefore a valid offer. On the evidence the accommodation offered by the landlord was suitable for the tenants.**

The following case is referred to in this report.

Betty's Cafés Ltd v *Phillips Furnishing Stores Ltd* [1959] AC 20; [1958] 2 WLR 513; [1958] 1 All ER 607; [1958] EGD 92; (1958) 171 EG 319, HL; affirming [1957] Ch 67; [1956] 3 WLR 1134; [1957] 1 All ER 1, CA

This was an application by the applicant tenant, M Chaplin Ltd, for a new tenancy under the Landlord and Tenant Act 1954, the renewal of which was opposed by the respondent tenant, Regent Capital
E Holdings Ltd.

John Dyson QC (instructed by Fooks & Co) appeared for the applicant; Christopher Priday (instructed by Brecher & Co) represented the respondents.

Giving judgment, **JUDGE ARON OWEN** said: In this application, M Chaplin Ltd, the applicant (to whom I shall hereinafter refer as "the tenants") apply to the court for an order for the grant to them of a new tenancy, pursuant to the provisions of Part II of the Landlord and Tenant Act 1954 ("the Act"), by Regent Capital Holdings Ltd, the respondent (to whom I shall hereinafter refer as "the landlords"). The tenants' premises, in respect of which they seek a new tenancy, consist F
of rooms on the ground floor at 21/22 St Cross Street, Hatton Garden, London EC1, which are in the building called St George's House on the corner of 44 Hatton Garden. The current lease is that dated July 17 1978 made between the landlords and the tenants for a term which expired on March 25 1982.

The proceedings in this application were initiated by the landlord serving a notice under section 25 of the Act, such notice being dated July 17 1981 and terminating the tenancy on March 25 1982, which was also the contractual date of expiry. The tenants have since stayed on in the premises under the provisions of the Act.

In their section 25 notice the landlords stated that they would G
oppose the grant of a new tenancy on two grounds. The two grounds set out in the notice are those provided by section 30(1)(*d*) and (*f*) of the Act, namely the provision relating to alternative accommodation (which is para (*d*)) and the provision relating to demolition or reconstruction (para (*f*)). The strict timetable laid down by the Act was then operated. The tenants' counternotice was by their solicitor's letter dated September 4 1981, which stated formally that the tenants were not willing to give up possession at the date of termination.

The tenants' application to the court was made by originating summons dated October 29 1981. That summons set out the various particulars as required by the County Court Rules (the relevant rule is H
now r6 Ord 43 in the County Court Rules 1981), including the tenants' proposals as to the period, rent and other terms of the new tenancy for which they were applying.

The landlords' answer is dated November 13 1981 and the two grounds of opposition to the grant of a new tenancy, which had been set out in their section 25 notice, are pleaded. They also pleaded that if a new tenancy was granted they objected to the terms proposed by the tenants.

So battle was joined between the parties. The line of battle was, however, altered when by letter dated February 1 1982 the solicitors for the landlords wrote to the solicitor for the tenants stating that the J
landlords had decided to abandon their opposition under ground (*f*), the demolition/reconstruction objection. This left para (*d*), the alternative accommodation provision, as the only ground upon which the landlords oppose the grant of a new tenancy to the tenants.

The trial before me was concerned with the true construction of section 30(1)(*d*) of the Act and evidence relating to the requirements thereof and with the determination of the figure for rent in respect of any new tenancy granted by the court. I was not concerned with the duration or other terms of any new tenancy which might be granted, the parties having agreed those other matters themselves. I was informed by Mr John Dyson QC and Mr Christopher Priday, counsel K
for the tenants and landlords respectively, and to both of whom I am greatly indebted for their helpful submissions, that, so far as their researches revealed, there is no direct authority on how para (*d*), section 30(1) of the Act is to be construed. The Act has now been in operation for nearly three decades, but it appears that the implications of this paragraph have never had to be fully and directly considered.

In the County Court Practice 1982 ed at p1487 there is this note:

Alternative accommodation — this expression has the same meaning as in the Rent Act and therefore part of the existing accommodation may afford

A alternative accommodation for the purposes of sub-s (1)(*d*): *Lawrence* v *Carter* (1956) LJo 259 (County Court).

That however does not deal with the problem with which we are concerned in the present application.

There are *obiter dicta* in the well-known House of Lords' decision *Betty's Cafés Ltd* v *Phillips Furnishing Stores Ltd* [1959] AC 20* which both counsel prayed in aid and to which I shall presently refer in my judgment. That case concerned the construction of para (*f*), but reference was made to the other paragraphs in section 30(1) in the course of their lordships' speeches.

B Counsel told me that the facts in the present application raise for the first time, it would appear, a direct consideration by the court of the true construction of para (*d*). Let me, therefore, first read in full section 30(1)(*d*) of the Act. It provides:

> 30. Opposition by landlord to application for new tenancy
>
> (1) The grounds on which a landlord may oppose an application under subsection (1) of section 24 of this Act are such of the following grounds as may be stated in the landlord's notice under section 25 of this Act or, as the case may be, under subsection (6) of section 26 thereof, that is to say:
>
> . . .
>
> (*d*) that the landlord has offered and is willing to provide or secure the provision of alternative accommodation for the tenant, that the terms on
> C which the alternative accommodation is available are reasonable having regard to the terms of the current tenancy and to all other relevant circumstances, and that the accommodation and the time at which it will be available are suitable for the tenant's requirements (including the requirement to preserve goodwill) having regard to the nature and class of his business and to the situation and extent of, and facilities afforded by, the holding;
>
> . . .

The construction of that paragraph submitted by Mr Dyson was as follows:

D (1) The landlord must have made an offer of alternative accommodation before the time of the section 25 notice.

(2) That offer must have been of alternative accommodation which satisfies reasonableness and suitability as set out in para (*d*).

(3) The offer must not be withdrawn and must remain open for acceptance until the hearing at which time reasonableness and suitability must still be satisfied.

(4) Terms which are reasonable include whether the rent upon which the alternative accommodation is available is a reasonable rent. But it must be a reasonable rent on which to agree and not a reasonable rent to ask as an "opening shot"; it must be not a
E reasonable asking rent, but a reasonable concluding rent.

In support of his first submission, Mr Dyson quoted Lord Denning in the *Betty's Cafés* case *op cit* at p50:

> If a landlord opposes on the ground that: "I have offered you alternative accommodation and am willing to provide it" he clearly means that *in the past*, at some time before the notice, he has offered alternative accommodation, and that *in the present*, at the time of giving the notice, he is willing to provide it.

The importance of that construction for Mr Dyson's submission on the facts in our case is this: the landlords' offer of alternative
F accommodation is contained in the letter dated July 17 1981, which was the letter sent with the section 25 notice. Mr Dyson said that offer, sent at the same time as the notice, would not satisfy the words "has offered", which refer to an offer in the past, at some time before the notice (as *per* Lord Denning *obiter*). And, said Mr Dyson, if he was wrong on that construction as to time, and the offer of alternative accommodation could accompany the section 25 notice, then he relies on his second and fourth submissions: the offer when put forward must there and then satisfy the requirements of para (*d*) in respect of reasonableness and suitability and the landlord cannot subsequently amend the terms of that offer in an attempt to make it conform.

*Editor's note: Also reported at (1958) 171 EG 319.

The effect of such a construction of para (*d*), when applied to the G facts in our case would, Mr Dyson submits, be fatal for the landlords. Because one of the terms included in the landlords' offer of July 17 1981 was an offer to pay for the tenants' costs of removal, but only up to a sum not exceeding £500. That sum has since been conceded by the landlords to be unreasonable; heavy safes would have to be moved and resited and the reasonable figure for moving costs is accepted as being in the region of £5,000.

Further, it was submitted, the landlords having put forward in their offer a definite figure for rent for the alternative accommodation are stuck with it and it does not avail them to pass a resolution, as in fact they did at their company meeting on January 18 1983, that their offer H as to rent is the figure proposed "or such other rental as the court may determine".

Mr Dyson submitted that such departures from the original offer are not permitted by the requirements of para (*d*). A landlord offering alternative accommodation, he said, would have to devise some formula in which to couch his offer to avoid such pitfalls and the landlords in our case had failed to do so.

Mr Priday, needless to say, does not accept that construction of para (*d*). Though in any event he says, on the facts of our case, as to timing, since no time whatsoever is stipulated in the Act as to when in the past the landlord must have made his offer, the fact that the offer (in the instant case) was contained in the letter which would be read J first by the recipient, that is before reading the section 25 notice enclosed with the letter, was good enough.

I might add that on the facts of the present case there was an offer of alternative accommodation in the same building which was the subject of correspondence between the parties in 1979. And thereafter there were some informal meetings between the parties (between Mr EN Goodman for the landlords and Mr AD Chaplin for the tenants) when the landlords offered orally various "packaged deals" of alternative accommodation on various floors, including the second floor, in the same building.

Mr Priday, however, submitted that, on a true construction of para K (*d*), the time for the operation of the words "has offered" is not restricted to an offer which must precede the giving of the section 25 notice. And he also disagreed with Mr Dyson's submissions as to the immutability of the terms proposed in the landlords' original offer.

On the question of time. Mr Priday quoted the *obiter* views of Viscount Simonds in the *Betty's Cafés* case *op cit* at p36. The passage he prays in aid reads as follows:

> Perhaps a brighter light is thrown by (*d*) which opens with the words "that the landlord has offered and is willing", etc. Here the perfect and the present tense are used. Leave out the perfect and look only at the present tense. "The L landlord is willing". It would be a hardship and worse on the tenant, if the relevant date were any other than that of the hearing: it is to his advantage that the opportunity of accepting an offer of alternative accommodation should be open to the last moment, and it is inconceivable that the landlord should at the hearing be permitted to say that, though no longer willing, he had been willing at an earlier date, and therefore could validly oppose the application. Nor would it be reasonable to reduce the time within which the landlord should have the opportunity of finding and offering alternative accommodation. If the tenant complains that he has had too little time to consider its suitability, his grievance can be met by an appropriate adjournment. In ground (*d*), therefore, I find support, if it be needed, for the view that the word "intends" in ground (*f*) means "intends at the date of the hearing".

M Mr Priday (who, it so happens, was one of the team of counsel who appeared in the *Betty's Cafés* case) submitted, and in my view correctly, that the key to the understanding of Viscount Simonds' remarks in the passage I have quoted (and which Mr Dyson confessed he found difficult) was to be found in the fact that their lordships were considering a section 26 case.

In that case it was the tenant who initiated proceedings by his request for a new tenancy under section 26 of the Act. And section 26(6) reads:

> Within two months of the making of a tenant's request for a new tenancy the landlord may give notice to the tenant that he will oppose an application to the

A court for the grant of a new tenancy, and any such notice shall state on which of the grounds mentioned in section 30 of this Act the landlord will oppose the application.

Thus a landlord is here given a statutory time of up to two months in which to put forward his ground or grounds of opposition and to serve the notice setting out such ground or grounds and those grounds include para (*d*). But if the opening words of that paragraph, "the landlord has offered", mean that the offer must have been made before the service of the notice, then in order to be able to put forward this ground of opposition, the landlord will have been required to have offered alternative accommodation in a shorter time than the two
B months allowed by the Act as the statutory time in which the Act, in terms, permits him to put forward this ground of opposition.

Viscount Simonds thought it would not "be reasonable to reduce the time within which the landlord should have the opportunity of finding and offering alternative accommodation". With respect, I agree. Such a construction infers that a different time scale is to be ascribed to the words "has offered", one that is *not* restricted and confined to a time before service of the notice.

As to the submissions that the terms of the offer of alternative accommodation must be a kind of "once and for all", "take it or leave it" type of offer as suggested by the tenants' counsel, Mr Priday drew
C my attention to a number of matters, which he submitted militated against any such construction.

The statutory timetable is such that a considerable period of time, usually many months, inevitably elapses between the giving of a notice by the landlord under section 25 or section 26(6) of the Act and the hearing of the application in court. During that time, circumstances can change enormously. Yet (on the tenants' construction of the paragraph) the terms of the landlord's offer would not be permitted to be altered to fit in with changed circumstances. It would be an impossible, or almost impossible feat for a landlord to make an offer which is "spot on" from inception and will remain so throughout the
D lengthy period that is going to elapse.

A further possible extension of that period is postulated by section 31 of the Act. That section, which is the only other section in the Act where para (*d*) is specifically mentioned (paras (*e*) and (*f*) are also referred to), contains a provision in subsection (2) which, in the circumstances and time-limit set out, has the effect of prolonging the date, at which the court is to be satisfied that the ground has been established, to a future date when that ground will be satisfied.

Apart from those considerations, some of the matters which have to be established under para (*d*), particularly the requirement to preserve goodwill, are of a subjective nature and all the evidence relating thereto could hardly be in the landlord's possession right from the
E start. The only statutory provision as to what preliminary information the landlord can obtain from the tenant is that set out in section 40 of the Act. That information would not be of much assistance to the landlord in formulating his alternative accommodation offer.

Another requirement of para (*d*) is that regard must be had, *inter alia*, to the terms of the current tenancy when considering the terms on which the alternative accommodation is available. Both counsel accept that "terms of the current tenancy" in this context refer not to the old terms under which the tenant holds his existing tenancy, but to such terms as the parties agree or the court would determine in a grant of the tenant's application for a new tenancy of his holding.

F Thus the requirement in para (*d*) to have "regard, etc" incidentally encompasses consideration of that trio of sections in the Act, namely section 33 (duration of new tenancy), section 34 (rent under new tenancy) and section 35 (other terms of new tenancy). Each of those sections is worded so as to give the parties the opportunity to reach an agreement themselves and only "in default of such agreement" does the court step in.

That, if I may say so, has been one of the great strengths of Part II of the Act and why, on the whole, it has worked so well. Within the framework of a tight procedural schedule, the Act nevertheless provides a flexibility, giving full rein to the parties themselves to negotiate right up to the last moment, to the very door of the court. The court is a long stop if the parties themselves cannot reach G agreement.

So here you have incorporated, as it were, into para (*d*) a requirement which relates to "terms" which, under other sections of the Act, are permitted to be negotiated and agreed between the parties or, in default, are determined by the court. Why then should a similar flexibility not apply to the terms of an offer of alternative accommodation?

How then should para (*d*) be construed? Mindful of the fact that I am treading new ground and should, therefore, tread warily, I put forward the following construction:

(1) The opening words of subsection (1) of section 30 indicate, to H my mind, that there should be no difference, when considering any of the grounds in para (*a*) to (*g*), whether the landlord's opposition arises as a result of the service of a notice under section 25 or from the service of a notice under section 26(6). And neither of those sections and the appropriate notices thereunder require the landlord to state the grounds upon which he *opposes*; the requirement is for him to state the grounds on which he *would* (section 25(6)) or *will* (section 26(6)) oppose.

(2) Dealing specifically with para (*d*), that paragraph has three separate limbs each of which is preceded by the word "that", namely:

(*a*) "that the landlord has offered" etc; J

(*b*) "that the terms" etc;

(*c*) "and that the accommodation" etc.

(3) It will be seen that the past tense occurs only once in the whole paragraph and that this is in the first part of the first limb, "has offered", and even there it is coupled with the present tense, "is willing", which forms the second part of the first limb. The tense used in the second and third limbs is the present and there is also a future tense, "the time at which it (the alternative accommodation) will be available".

(4) The first limb is expressed in language of a general nature, a statement about the provision of alternative accommodation which is K not particularised in any way. There is no definite article preceding "alternative accommodation" mentioned in the first limb. The definite article does appear in relation to the alternative accommodation referred to in the second and third limbs.

(5) In my judgment, the first limb is in the nature of an "opening shot" by the landlord, telling the tenant the ground of opposition to be relied upon. And the requirement of the landlord having offered in the past (what I have described in military terms as the "opening shot"), is satisfied if that has taken place before "battle is joined", namely if the offer has been made before the issues are joined in the pleadings. This would be before the landlord files his answer (if the tenant's L application is to the county court); and before the landlord files his affidavit in reply to the tenant's affidavit (if the tenant's application is to the High Court). The times are set out in the rules which are to be found respectively under Ord 43 of the County Court Rules 1981 and under Ord 97 in the Rules of the Supreme Court.

(6) Take, for example, what happens after the landlord has served a section 25 notice. Until the tenant serves a counternotice, the landlord does not know if there is going to be any issue. If no counternotice is served, the landlord does not need to rely on any ground of opposition (provided by section 30) to oppose the grant of a new tenancy. Again, after the tenant has made his application to the court, until the landlord has filed his answer, the tenant does not know if the landlord is going M to pursue his opposition; there is nothing to prevent the landlord withdrawing his opposition. But once the answer is filed, the issues between the parties are set out and joined. So if in that period between the service of the counternotice and the service of the answer the landlord makes an offer of alternative accommodation that, in my judgment, will satisfy the words "the landlord has offered".

I interpose here to say that I am glad to be able to put forward such a construction for those words in the first limb of para (*d*). Because it seems to me, the words, "the landlord has offered" are of little worth to the tenant if, as is the fact, the words do not specify at all the time in the past when such offer must have been made, whether that offer had to be in writing or could have been made orally and precisely what that past offer (made prior to the notice) should have contained.

How will it assist the tenant if, for example, the landlord makes an offer by writing it to the tenant the day before he serves his section 25 notice specifying ground (*d*) as the ground for opposition; or by sending the offer by first class post and the said section 25 notice in a separate envelope sent later on the same day by second class post; or by telephoning an offer to the tenant the day before the tenant receives the said notice?

In this context I adopt the words used by Romer LJ, in the course of his judgment in the *Betty's Cafés* case in the Court of Appeal, and say: "it is permissible to inquire whether the requirement" — I refer to a requirement of a prenotice offer — "would, on the one hand, be of advantage to the tenant and, on the other hand, would or might result in hardship or inconvenience to the landlord; for if it would be a disadvantage to the landlord without securing any corresponding benefit to the tenant, it is surely a legitimate inference that the legislature did not intend to impose it".

Another comment I make is this. I appreciate that the provisions of para (*d*) differ materially from the suitable alternative accommodation provisions in section 98 of the Rent Act 1977 and Schedule 15, Part IV thereto. Yet, it seems to me, that whether the person affected is a residential tenant who has to have a home in which to live, or is a business tenant who has to have premises in which to conduct his business, the real concern of that person, in an alternative accommodation case, is what is available to him now, when the matter is before the court or when the order of the court takes effect, not was or might have been available to him in the past.

The relevant Rent Act provisions make no reference to any past offer. The wording in section 98(1)(*a*) of the Rent Act 1977 reads:

> the Court is satisfied that suitable alternative accommodation is available for the tenant or will be available for him when the order in question takes effect.

However, the words, "the landlord has offered" are in para (*d*) and effect must be given to them and that is achieved by the construction I have put forward.

I am fortified in that construction by the view expressed by Viscount Simonds in the *Betty's Cafés* case *op cit* at p36. And it is also interesting to note the way in which Birkett LJ considered para (*d*) in the course of his judgment in that case when it was in the Court of Appeal. His views on this paragraph are, of course, *obiter* but this is what he said [1957] Ch 67 at pp83-84. Having quoted para (*d*) in full he went on:

> At the hearing of the application, the landlord must establish to the satisfaction of the Court that he has filled the various conditions laid down in the subsection. He must therefore show that the alternative accommodation is available at the hearing, and must then be in the position to express his willingness to provide it, or to secure its provision for the tenant. The time at which the alternative accommodation will be available must be suitable to the tenant's requirements, and the court must be satisfied on all these points. *At the date of the notice the landlord might very well be unable to offer the requisite alternative accommodation*, though he was negotiating in order to be able to do so at the hearing. If at the hearing the landlord was able to fulfil all the conditions of the subsection, and established this to the satisfaction of the court, there would be no discretion left in the court to order a new tenancy; and, provided that the landlord had stated the ground in his notice, the mere fact that it was not possible at the date of the notice to fulfil all the requirements of the section would not enable the court in my judgment to say that the landlord was debarred from establishing the ground at the hearing.

(Emphasis supplied)

I now return to the construction of the remainder of para (*d*); see paras (1) to (6) above.

(7) Having offered alternative accommodation, the offer must be kept open from the time made and throughout and must not be withdrawn. The present tense, "is willing", in the second part of the first limb connotes a continuing willingness.

(8) The second limb of para (*d*) deals with the requirements of reasonableness in respect of the matters therein set out. In my judgment, the time when the landlord has to establish that requirement to the satisfaction of the court is at the hearing. There is a passage in *Woodfall's, Law of Landlord and Tenant* (28th ed vol 2 p2417 para 2-0715) which reads:

> Thirdly, that the alternative accommodation is available on terms which are reasonable having regard to the terms of the current tenancy and to all other relevant circumstances. The accommodation must have been offered on these reasonable terms, and it must still be available on those terms.

With respect to the learned editors of *Woodfall*, I do not consider that passage to be correct in stating "the accommodation must have been offered on these reasonable terms . . .", a past occasion, linked with a present occasion "and it must still be available on those terms".

The statutory wording of this limb of para (*d*) is: "the terms on which the alternative accommodation is available are reasonable", not "was and is available and were and are reasonable". The construction suggested by *Woodfall* would entail putting in those extra words of a past tense which do not appear in the Act.

(9) The third limb of para (*d*) deals with suitability. In respect of that requirement also, the time when it has to be established is at the hearing.

Again, I draw attention to the relevant wording of the Act which is here couched only in the present, with a future included, "the accommodation and the time at which it will be available are suitable"; not "the time at which it was and will be available are and were suitable".

So much for the construction of para (*d*). I turn now to the evidence and whether the landlords have, on the evidence, established all the requirements of that paragraph. I find first of all, that the offer was made timeously within the meaning of the words "has offered". Even if the construction I have put forward, as to the time to which those words refer, is held to be wrong, in this case the letter (dated July 17 1981) was intended to be read before the notice and that would suffice and as I have said, there was some evidence of an earlier oral offer or offers.

I have next to consider whether the terms on which the alternative accommodation is available are reasonable having regard to the statutory requirements. The terms offered (apart from rent which I shall, hereinafter, consider separately) are precisely the same as the terms of the current existing lease under which the tenants are in occupation. As to duration, I understand that a period of five years is acceptable to both parties. The draft of the proposed lease is exhibited in bundle B.

Since it is my determination of the rent for the alternative accommodation which the landlords will accept, I find, having considered the terms on which the alternative accommodation is available and all other relevant circumstances, that the landlords have proved the requirements of reasonableness.

Next the requirement of suitability as set out in the third limb of para (*d*). This involved the consideration of quite a number of matters. The nature and class of the tenants' business, as described to me by Mr Alfred D Chaplin, are these:

(i) The tenants (the company, M Chaplin Ltd) deal in new jewellery, which they buy from manufacturers and sell to retail jewellers and occasionally to other wholesalers.

(ii) There are two associated companies also trading from the same premises: M Marston & Co (Gold Coins) — which appears to be a firm name and not that of a separate limited company — which deals in gold coins. This company or firm buys from bullion houses, retail shops, dealers and members of the public. It sells to bullion houses, retail jewellers and other dealers. The other company is Turner & Leveridge Ltd, which trades in secondhand and antique jewellery and silverware. It buys from auction rooms, retail jewellers, brokers, other dealers and members of the public. It sells to customers within the trade and to retail jewellers.

The tenants' trading activities must, of course, come within the permitted user under the lease. The relevant clause states:

> (13) Not to use the demised premises or any part thereof or permit the

> A same to be used in any manner except as offices for the purposes of the tenant's business of watch and clock importers and wholesale jewellers . . . and not to use or permit or suffer the use thereof for the sale by retail of any goods or merchandise . . .

There is a further clause in the lease which provides as follows:

> (ii) Subject to the provisos hereinafter contained not to underlet or part with provided
>
> (1) the tenant may share the possession and occupation with any other company or firm for the time being beneficially owned and/or controlled by Mr Alfred Douglas Chaplin and/or Mr Maurice Chaplin.

B As to the situation and extent of, and facilities afforded by, the tenants' holding, they are as follows. The situation is on the ground floor at 21/22 St Cross Street, Hatton Garden, London EC1, which is part of the building St George's House, 44 Hatton Garden.

The extent and layout are shown on the plan ASB appendix B and I inspected the tenants' premises and saw all the facilities of their holding. The alternative accommodation is situated in the very same building, but on the second floor, the entrance to the building being the same entrance as now used by the tenants. A new lift has been installed and new toilet facilities.

I can do no better than quote the description given of the tenants' present accommodation and of the alternative accommodation as set
C out in the report of Mr Eric Frank Shapiro [FRICS], the tenants' surveyor. After referring to the extent of usable floor area as being a total of 758 sq ft and to his plan, Mr Shapiro describes the tenants' present accommodation as follows:

> . . . the accommodation is most irregular in shape and as such it constitutes uneconomical space. It is totally unmodernised, but it is fully burglar alarmed and has a security lobby. The standard of finish is substantially below that provided on the second floor having no suspended ceilings, no modern decorations or fitted carpets. The bulgar alarm and security lobby are tenants' improvements. The accommodation has no internal WC facilities or water supply as these are located on the top, second and third floor of the building so
> D that a visit to the lavatory is most inconvenient. Although there is central heating in the building as a whole there is no heating in the private office or entrance lobby.

Mr Shapiro's description of the alternative accommodation reads as follows:

> The alternative accommodation offered is part of the second floor of 44 Hatton Garden which comprises approximately 1,400 sq ft. The actual size of the accommodation is to be agreed, but it is not less than 800 sq ft and it has been modernised by the provision of a suspended ceiling with fluorescent lights and is fully carpeted.

E The tenants' objections to the alternative accommodation were a general allegation that goodwill would not be preserved and a number of specific matters were relied on. These included:

(i) the alternative accommodation was on the second floor and not on the ground floor;

(ii) the special facility which Mr Chaplin now enjoyed of being able, when standing in his office on the ground floor, to see out of the window and to be seen by passers-by in the street;

(iii) the notice boards on the front of the ground floor premises advertising the tenants' two associated companies;

(iv) a special burglar alarm system on the ground floor.

F Any objections which may have existed in relation to layout and cost of removal I find are no longer tenable in view of the landlords' offer to provide whichever of the various layouts the tenants wish to choose from among the alternatives offered, and their offer to pay all reasonable removal costs, the figure of £5,000 being put forward.

On the question of goodwill, it has to be borne in mind that the tenants' business, as permitted by their lease, is a wholesale business carried on in premises which can be used only as offices for such business. The business cannot and must not be a retail business and the premises are not to be used as a shop or showroom.

When I inspected the Hatton Garden area, I could not help but notice that very many wholesale jewellery businesses were carried on in upper floors and not on the ground floor. In the newest building G
erected in the area, "The Hatton Garden Centre", a building specially designed, I understand, for the diamond and precious stone trade and including the wholesale jewellery trade and containing the London Diamond Bourse, all the accommodation offered is on the first and second floors. It is true that Mr Chaplin can see out of the window into the street and be seen from the street when he stands in the room he uses as his office. His head in the window is clearly shown in one of the photographs produced in evidence and I saw the position for myself when I inspected the site. But again I stress the tenants' business is not retail and the tenants would be precluded from dealing with passers-by in the street in any retail capacity. Mr Chaplin told me H
that the only persons whose attention he tries to attract, or who attract his attention through the window, are established customers in the trade, not strangers. There is no question whatsoever of touting for business from the public through this window.

The facility amounts to no more than this: someone in the trade who knows Mr Chaplin and who wanted to contact him and who happened to be passing by would not need to telephone to find out if Mr Chaplin was in; he would see his head in the window. Similarly, someone in the trade whom Mr Chaplin knows and wanted to contact and who happened to be passing by could be contacted by Mr Chaplin attracting that person's attention by tapping on the window. In the J
alternative accommodation there would be nothing to prevent Mr Chaplin having his desk by the corner balcony window facing St Cross Street and being seen if he stands there. True he would not be able to attract the attention of a person in the street below, but if that person wished to know if Mr Chaplin was in, and he was passing by, he would look up to the window and would be able to see Mr Chaplin, if Mr Chaplin was standing by the window. In any event, in my judgment, this matter will not in any way detract from the preservation of goodwill. It is a matter which was not referred to at all, and neither was any alleged loss of goodwill mentioned at all, during the negotiations which took place in 1979 between the parties when K
alternative accommodation was offered on the third floor in the same building.

In the present proceedings it is a fact that, after July 17 1981, the landlords' solicitors wrote on several occasions asking for the tenants' comments on the alternative accommodation offered. It was not until May 14 1982 that the tenants' solicitor wrote, objecting for the first time in the terms that "an upper part is not the same as a ground floor" and that "in addition", any move by the tenants "must entail serious loss of goodwill".

I accept that an upper floor is not the same as a ground floor — and L
that would be a factor of special significance when dealing with a retail shop or showroom business — but I do not find that a move by the tenants (in the present case) from the ground floor to the second floor will entail any loss of goodwill.

As to the notices on the premises advertising the two other companies, the landlords have said that the name of these companies will be shown on the board at the main entrance doors of the building where the names of all the occupiers of the building are listed. This is the very same entrance which leads to the tenants' present premises.

Mr Priday submitted, rightly in my view, that alternative accommodation does not have to mirror exactly the existing M
accommodation. And what would appear to be the main source of advertising for these two companies will continue precisely as before. The tenants are connected with a company called AB Davis Ltd, who trade as jewellers, silversmiths and pawnbrokers and have a retail shop at 89-91 Queensway, London W2, and another retail shop at 18 Brook Street, London W1.

In those shops, I understand, a printed card advertising the tenants' business and those of Turner & Leveridge Ltd and M Marston & Co, with the address St George's House, 44 Hatton Garden, London, and telephone numbers, is available for distribution. The card, which is printed in blue and red colours was produced in evidence. In addition,

A customers are also found and contacted by travellers employed by the tenants. Those matters will remain exactly as before. And, most importantly, the address of the tenants and its two associated companies will not be changed.

On the question of protection against burglary, I have no hesitation in finding that the alternative accommodation on the second floor would be at least as safe, if not safer, than the tenants' present accommodation on the ground floor.

The special precautions now taken must be viewed in the light of the circumstances pertaining, namely that the accommodation is on the ground floor with a door opening directly on to the pavement in St B Cross Street. That door as I saw for myself, could easily be broken open and the premises entered directly from the street. Accordingly, special precautions have been taken to deter any would-be criminal from making such breaking and entry. Those special precautions, other than the usual, sensible precautions for jewellers' premises in Hatton Garden, will not be required on the second floor.

There is no issue between the parties as to the time at which the alternative accommodation will be available. I find that the landlords have proved all the requirements in para (*d*) relating to suitability.

I now return to the question of rent, which is a matter very much in issue. I had the advantage of visiting, on Friday January 28 1983, the Hatton Garden area, traditionally the home of wholesale jewellers, of C inspecting the tenants' present accommodation, the alternative accommodation and most of the comparables. I also had the reports of, and heard evidence given by, the experts: Mr HAE Cox for the landlords (supported by Mr AS Browne who produced the plans to which I have referred, and who gave evidence in place of Mr Hartford, the architect, who has retired), and Mr EF Shapiro for the tenants.

The figures for rent, put forward by the respective parties, for a new tenancy of the tenants' existing premises on the ground floor and for the alternative accommodation on the second floor, were as follows:

Landlords' figures: D
Ground floor: £15 per sq ft
Second floor: £12.50 per sq ft
Tenants' figures:
Ground floor: £10.50 per sq ft
Second floor: £9.50 per sq ft

Many considerations were canvassed before me on behalf of the parties. They included inflation, recession, the state of the market generally, availability of accommodation, user clauses, differences between ground floors and upper floors, detailed analysis and breakdown of the "comparables", renovations carried out to the St George's House building and so on. A great deal of "fine tuning" has E to be done to arrive at the proper figure for rent under a new tenancy bearing in mind what are the statutory requirements prescribed by section 34 of the Act.

I must say that, as between the two expert surveyors, Mr Cox and Mr Shapiro, I greatly preferred the evidence given by Mr Shapiro who is a highly qualified surveyor and with great experience in valuing and whose expert evidence I found impressive. In my judgment, his valuations and "fine tunings" are much nearer the proper mark than those given by Mr Cox.

Having considered all the evidence and the circumstances, I determine the new rents as follows:

For a new tenancy of the ground floor: £11.75 per sq ft F
For a new tenancy of the second floor: £10.50 per sq ft

To sum up, I find that the landlords' opposition to the tenants' request for a new tenancy succeeds on the ground set out in section 30(1)(*d*) of the Act which they have established to my satisfaction. I am thus precluded by section 31(1) of the Act from making an order for the grant of a new tenancy to the tenants. But, of course, the alternative accommodation is available to the tenants on the basis of all the various matters (including the rent I have now determined) as set out in my judgment.

For further cases on this subject see p 129

NEGLIGENCE

A

Official Referee's Business

December 13 1993

(Before Judge BOWSHER QC)

EZEKIEL AND ANOTHER v McDADE AND OTHERS

Negligence — Valuation for building society — Liability in tort — Whether valuer negligent — Damages for reduction in value — Whether additional damages for worry, distress and inconvenience

B In October 1986 the plaintiffs purchased a dwelling-house for £56,000 partly financed by a loan from a building society. The property was part of a terrace built by the "Bison system". At the request of the building society, the first defendant provided a valuation report. An attempt to sell the property in 1988 failed because defects were discovered by an intending purchaser. The first plaintiff lost his job and failed to pay mortgage instalments resulting in the building society obtaining possession in August 1990. It was eventually sold at auction in March 1992 for £29,000, the building society claiming £67,397.59 from the plaintiffs for arrears and costs. The plaintiffs were put into temporary accommodation before
C obtaining a council house tenancy. It was agreed that the first defendant owed a duty of care in tort to the plaintiffs. The plaintiffs claimed damages for negligence alleging that the first defendant failed to observe that a concrete purlin observable in the roofspace was unevenly positioned so that at one end was a gap of 40mm. In the event of liability the plaintiffs claimed in addition to a reduction in value of the property, damages for worry, distress and inconvenience.

***Held*: Judgment was given for the plaintiffs. The first defendant was negligent; a reasonably competent surveyor making a "head and shoulders" inspection of the roofspace
D would have observed the unacceptable 40mm gap and would not have reported, as the first defendant did, that there was no evidence of any serious defect in the examined parts of the property. Damages totalling £43,364.65 were awarded which included: damages based on the difference in value of the property in 1986 if it had been in the condition represented by the first defendant's report and its actual value were awarded in the sum of £36,000; damages for worry, distress and physical inconvenience of £6,000; and agreed damages of £1,364.65 for legal expenses and other costs.**

E The following cases are referred to in this report.

Smith v *Eric S Bush (a firm)* [1990] 1 AC 831; [1989] 2 WLR 790; [1989] 2 All ER 514; (1989) 87 LGR 685; [1989] 1 EGLR 169; [1989] 17 EG 68 & 18 EG 99, HL

Watts v *Morrow* [1991] 1 WLR 1421; [1991] 4 All ER 939; [1991] 2 EGLR 152; [1991] 43 EG 121; 26 Con LR 98

This was a claim by the plaintiffs, David Albert Ezekiel and Margie Chrysanthemus Ezekiel, against the first defendant, Ian McDade, the second to fifth defendants, Roy Burgess FRICS, Joseph Leckie FRICS FSVA, David Giles ARICS and John Crosbie FRICS (formerly trading as Brendons), the sixth defendant, Britannia F
Brendons (a firm), and the seventh defendant Britannia Estate Agents Ltd, for damages arising out of a valuation report prepared by the first defendant.

Brian Leech (instructed by Burch Phillips & Co) appeared for the plaintiffs; Anna Guggenheim (instructed by Cameron Markby Hewitt) represented the defendants.

Giving judgment, **Judge Bowsher QC** said: The plaintiffs claim damages for loss arising from an allegedly negligent survey made by the first defendant in August 1986 of a house at 40 Armytage Road, Heston, Middlesex.

All the defendants are jointly represented, and it has been agreed G
that no distinction is to be made between them, although as a matter of law some point might have been taken that at least the sixth and seventh defendants were not liable for any default on the part of the first defendant. By agreement, if there is to be judgment against the first defendant, there should be judgment against all defendants.

This is a sad story of our time. While denying liability on behalf of her clients, Miss Anna Guggenheim expressed sincere sympathy with the plaintiffs.

Mr Ezekiel, who is now aged 52, was working in 1986 as an agent behind the ticket counter of El Al Airlines at Heathrow Airport. He lived with his wife and three children (they have since had a fourth H
child) in a two-bedroomed house in Wood Lane, Isleworth. Mr Ezekiel gave evidence in this trial. He is a sensible and prudent man. The misfortunes which he has suffered in the last three or four years are not uncommon today. In Mr Ezekiel's case, I am satisfied that the misfortunes were not of his own making. Whether the defendants are responsible is a matter for me to decide.

To get more room for their growing family, in 1986 the plaintiffs moved from Wood Lane to a house at 40 Armytage Road. The house at Armytage Road was bought for £56,000 financed by £16,000 net proceeds from their previous house and a loan of £40,000 from the Halifax Building Society. The purchase was completed on October 27 J
1986.

As a condition of making their loan, the Halifax Building Society obtained a valuation report on the house from the first defendant in August 1986. The plaintiffs paid for that report and were sent a copy of it. In reliance on that report they bought the property without obtaining any more detailed survey of their own. It is submitted on behalf of the plaintiffs and not disputed on behalf of the defendants that in the circumstances the defendants owe a duty to the plaintiffs in tort pursuant to the principles set out in *Smith* v *Eric S Bush (a firm)* [1990] 1 AC 831*.

In February 1988, the first plaintiff was made redundant. After K
three months he was able to find part-time work with a travel agency. He also worked as an office cleaner in the evening to make enough money to keep up his mortgage payments. In mid-1988 the plaintiffs decided to sell their house. In straitened circumstances, they yet had the courage to plan to buy a larger house with family help, intending to let out a room to someone working at the airport.

At the peak of the property market, the house at 40 Armytage Road was put on the market with an asking price of £82,950. There was

*Editor's note: Also reported at [1989] 1 EGLR 169.

reasonable interest, and a Mr Farooq agreed, subject to contract, to buy at £79,000. Mr Farooq had the house surveyed and as a result of the survey withdrew from the purchase. He said that the house had defects and suggested to the plaintiffs that they should take legal advice, which they did. This development made the plaintiffs anxious and depressed. The first plaintiff's work was affected and he was dismissed. As a result, the plaintiffs fell into arrears with their mortgage payments and when the arrears amounted to over £3,500, the Halifax Building Society issued proceedings. A possession order was made in March 1990, but as a matter of grace not executed until August 1990.

After their eviction from their home, the plaintiffs and their three children were housed by the council in one room in bed and breakfast accommodation at a cost of £57.63 per week. To support themselves, they sold everything they had, including their wedding rings. After about eight or 10 weeks they were moved to an unsatisfactory two-bedroomed house and then to their present council house at 237 Great West Road, Hounslow, a three-bedroomed house. By borrowing from friends, they managed to get the house centrally heated for the comfort of Mrs Ezekiel and one of her daughters who suffer from asthma and for the care of their fourth child, who had then arrived. After their eviction, the council provided the plaintiffs with two garages in which to store their belongings, but the garages were damp and many of their belongings were spoilt.

From August 1990, the Halifax Building Society tried to sell the property without success and in July 1991 suspended advertising the property while awaiting a report from a structural engineer. The property was put back on the market, but not sold by private treaty and was eventually sold by auction on March 13 1992 for £29,000. On October 21 1991, the Halifax Building Society claimed that the plaintiffs owed them £67,397.59 plus some further legal fees. That will have been reduced by the £29,000 (less expenses) received at auction to about £38,000, but there will no doubt be additional interest accrued since October 1991. I do not know what are the intentions of the Halifax Building Society, and they could not be criticised for exercising their full rights in favour of their investors. The amended claim in this action is for £37,364.65 plus general damages for inconvenience and distress so that it seems likely that even if I find the defendants liable to the plaintiffs and award to the plaintiffs everything that they are entitled to claim under common law as recently enunciated by the Court of Appeal (*Watts* v *Morrow* (1991) 26 Con LR 98*), the plaintiffs are likely to be still dependent on council accommodation instead of living in their own home because the maximum fruits of this action will provide little for the plaintiffs after settlement of the obligations to the building society and the Legal Aid Board.

With the assistance of legal aid, the plaintiffs issued proceedings in the Queen's Bench Division on May 29 1992. The action was transferred to an official referee on January 2 1993. On March 18 1993, a date was set for trial to begin on July 12 1993. The plaintiffs continued to be dogged by misfortune. Shortly before trial, the plaintiffs' expert witness sadly died and a fresh date was set for trial for December 6 1993, when the trial began, the plaintiffs having instructed a partner of their deceased expert to give evidence on their behalf.

The plaintiffs' former home is one of a terrace of four houses on the edge of a council estate near Heathrow airport and it is very convenient for people working at the airport. When walking around the estate with counsel, I observed that the estate is well tended and the buildings well decorated. There are no graffiti on the buildings. One can see aircraft landing regularly at Heathrow, but the flight path is a little way away and there is no substantial nuisance by noise. Although one could smell aircraft fuel outside Hounslow West underground station when travelling to a view of the property, there was no similar smell on the estate.

40 Armytage Road is one of the two middle houses of a terrace of four. The terrace was built in a large panel precast concrete Bison Unit system. This particular house had been bought from the council pursuant to statutory rights. Other adjacent properties were not so bought.

The report by the first defendant of which the plaintiffs complain is dated August 29 1986. The report accurately described the house as being a former council house on the edge of a council estate, the house being of "non-traditional construction with the exterior panels being of reinforced pre-cast concrete panels with a lightweight inner skin". The most important sentence of the report was as follows:

> There is no evidence of any serious defect in the examined areas of the premises and it has been modernised and improved to a good standard throughout.

After pointing out some remediable defects, the report concluded:

> However, we believe the current price is realistic and fairly reflects the present condition.

I bear in mind that the report was a valuation report and not a structural survey. It is important to distinguish between the duties of a surveyor making a valuation report and the duties of a surveyor making a structural survey. There is a most helpful practice note of the Royal Institution of Chartered Surveyors describing what is required on a structural survey of residential property. By contrast there is an RICS document entitled *Guidance notes for valuers* first issued in December 1985, which includes the following:

> Guidance notes for valuers on the inspection of residential property: . . .
> 1(a) The valuation of residential property for mortgage purposes is, unless the lender instructs otherwise, open market value as at the date of inspection . . .
> (c) Among the relevant factors to be taken into account will be: . . . .
> (ii) construction and general state of repair . . .
> 2. The following general principles should be adopted . . .
> (d) If the valuer suspects that hidden defects exist, he may, if he thinks it necessary, recommend that further investigation be carried out.

It was common ground between the parties that in 1986, a surveyor making a valuation survey of this property would be expected to make what is called a "head and shoulders" inspection of the roofspace, that is, he should find any trap door into the roof and at least put his head and shoulders through the trap door and make a visual inspection.

The technical evidence in this case has concentrated on a heavy concrete purlin in the roofspace. The agreed evidence is that that purlin was constructed with lugs at each end and would have been lowered into place by a crane so that the lugs fell into slots in other parts of the structure. It is also agreed that this particular purlin was unevenly positioned so that at one end there was a gap of 40 mm (just over 1.5 ins) between the bulk of the purlin and the main member of the wall, and at the other end the gap was 10 mm or less. The expert evidence led on behalf of the plaintiffs was that this imbalance was dangerous, it ought to have been noticed and reported on a valuation survey and it was negligent not to report it. That evidence was given by Mr Stephen Whitley [FRICS], who broadly supported the contents of the expert's report written by his partner, the late Mr Victor Oster [FRICS].

The first point taken on behalf of the defendants was that a reasonably competent surveyor looking into the roofspace when making a valuation survey could not have been expected to see the gap of 40 mm. That point was put forward in evidence by Professor Malcolm Hollis on behalf of the defendants. Professor Hollis supported his evidence with coloured photocopies of photographs which he took of the interior of the roof from the trap door into the roof. Mr Whitley produced coloured photocopies of photographs which he took which he contended showed the contrary position. After some pressure, I managed to obtain from the parties glossy prints of the photographs. The photographs produced did indeed appear to support the contrary positions taken by each expert. A photograph taken by the expert for the plaintiffs showed the gap clearly, another photograph taken by Professor Hollis for the defendants showed only darkness over the position of the gap. I strongly urged the parties to

*Editor's note: Also reported at [1991] 2 EGLR 152.

A arrange for me to have a view of the premises and on December 7 the second day of the trial, by courtesy of the present occupant tenants of the house, I climbed a collapsible surveyor's ladder and looked through the trapdoor into the roofspace of the house. First using a torch, the gap of 40 mm at the end of the concrete purlin was immediately obvious. Within easy reach when standing on the ladder was a light switch. After switching on the light and without using the torch that gap at the end of the purlin was also obvious. During the trial there was some discussion (which I found extremely odd) of the extent to which a surveyor should project his head into the roof space to satisfy his duty of care. I have to say that although eventually I
B stood on the ladder so that my waist was level with the ceiling (which I did not find a great imposition) my view of the important gap of 40mm was plainly visible even when only my head was above the level of the trapdoor. In her closing speech, Miss Guggenheim suggested that from a distance of 5 or 6 ft a gap of 20mm which might have been acceptable was not appreciably different from 40mm. I simply do not accept that that submission is correct in point of fact. The submission is plausible when demonstrated by a line on a piece of paper in court, but the appearance of a gap at the end of a purlin is a different matter entirely.

In his written statement, Mr Ian McDade [ARICS] did not state that
C he did inspect the roof void of this property. He wrote:

> During the course of my mortgage valuation inspection I *would* have carried out a head and shoulders inspection of the roof void.

As a result of a question from me, Mr McDade said in oral evidence that he did carry out such inspection, but I very much doubt whether he did do so. Certainly he cannot recall any details of any such inspection and he did not record them. Mr McDade was quite clear that it was his firm's practice that a head and shoulders inspection of the roofspace should have been carried out and he was required to carry a collapsible ladder and a very high-powered halogen torch for that purpose. In fairness to the firm of Brendons, I must make
D it plain that the system of work which they required of their former employee, Mr McDade, could not be criticised on the basis of any evidence given in this trial.

Professor Hollis' evidence went far beyond his photographs. In his first report, verified on oath, he stated:

> There are no readily visible defects within the roof space. Any unusual gap to a purlin end cannot be seen from the roof hatch. . . . There may be two gaps at the joist end . . . Even if these gaps are regarded as being a defect, which I doubt, they were not readily visible looking into the roof void from the hatch.

In a report written on December 1 1993 in response to the written
E report of the plaintiffs, expert Mr Whitley, and also verified on oath, Professor Hollis stated:

> Mr Whitley states that Mr McDade's head and shoulder inspection of the roof void should, without question, have revealed the unacceptable joint of the concrete purlin. This is a matter which is central to the case. In my view, the test is whether the defective joint, if that joint is defective, was readily visible from a head and shoulders inspection. It is not a question as to whether it could have been revealed but whether it was readily visible. My photographs show what one would see generally when looking into a roof space at the time of an inspection and those show that the purlin joints in question are not necessarily visible.

F In his oral evidence, Professor Hollis sought to discount the photographs taken by Mr Whitley. He said that those photographs concentrate on one element rather than on the roof generally and that they were taken with a high-speed film which, according to Professor Hollis, tends to record information one cannot see with the naked eye. As I have no greater claim than Professor Hollis to express an expert view on photography, I only say that I do not accept his qualification to express a view on photography generally. What I do say most emphatically is that visual inspection from the trap door either by torch light or by the installed electric light reveals the existence of the gap at the end of the purlins far more clearly than is shown even by Mr Whitley's photographs.

If I had had to try this case by reference to photographs and G
conflicting oral evidence, I would have been in some doubt. Having had a view, I am in no doubt. I entirely accept the evidence of Mr Whitley that Mr McDade's "head and shoulders" inspection of the roofspace (if he made one) should have revealed the unacceptable joint of the concrete purlin with the party wall to no 42 on the front-roof slope. That is in accordance with the view very strongly expressed by the late Mr Oster in his report. The only excuse for failing to see and report on the joint would have been if there had been some obstruction in the way. Mr McDade did not suggest that there was any obstruction in the way at the time. He did not suggest that there was anything stored in the roofspace at the time. He stated in H
evidence:

> I have difficulty recalling any details of my inspection. I cannot therefore say whether a gap between the roof purlin and party wall was present at the time of my inspection or visible from the vantage of the roof hatch, or even possibly obscured by the presence of stored goods and other furniture.

I find that any surveyor of reasonable competence ought to have seen, noted, and reported the gap at the end of the purlin. If there had been goods stored in the roof which made a reasonable "head and shoulders" inspection ineffective, he should have reported that and qualified his report. Instead, Mr McDade reported without J
qualification, "There is no evidence of any serious defect in the examined areas of the premises". The readers of the report were entitled to understand that the "examined areas" included the roofspace from a head and shoulders inspection.

Whether Mr McDade ought to have seen and noted the gap at the end of the purlin is only the first point in the case, but my finding in relation to it affects the whole of the credibility of Professor Hollis' evidence, including his evidence in relation to other issues. He was firm in his evidence that the photographs he produced gave a fair picture of what could be seen on a head and shoulders inspection of the roof space. Having rejected that evidence, I approach the K
remainder of his evidence with great caution and I only accept it where it is independently supported, despite his impressive qualifications.

The next issue raised by Miss Guggenheim for the defendants was whether a reasonably competent surveyor looking into the roofspace should have concluded from what he saw that there was a possibility of a serious defect in the property so that the house should not be accepted as security for a loan or that a structural engineer should inspect. Professor Hollis said that the gaps at the end of purlins do not represent defects in construction. I accept from him that there is no sign in the property that the gaps have resulted in any distress or other L
damage to the property. It may also be the case, as he says, that repairs carried out by the council to the adjacent properties 38 and 42 (still owned by the council) have strengthened the support to no 40. But all of that is to a considerable extent beside the point. The questions are, what does the potential buyer think about this property and what effect does that have on its value? The proof of the pudding is in the eating. The plaintiffs had a willing buyer at an acceptable price, Mr Farooq, who withdrew after a survey. The plaintiffs failed to find another purchaser although they tried to do so. The Halifax Building Society tried unsuccessfully to sell the property for 11 months from August 1990 to July 1991 and only succeeded in selling it after a further eight M
months (after getting a structural engineers report) at auction for far less than the plaintiffs paid for it.

The history of the attempted sales of the property does, of course, have to be viewed in the light of the developing knowledge of the Bison type of construction. In considering that history one has to distinguish between high-rise buildings (including the notorious Ronan Point) and low-rise buildings.

In this regard, Professor Hollis stated:

> In 1986, building societies were lending on this form of concrete frame. Since that time, the character of lending has changed and it is now the case that a mortgage is difficult to obtain on this form of construction.

I accept that statement as correct so far as it goes, but it does not tell the whole truth. Professor Hollis' statement should be read in the light of the comment made upon it by Mr Whitley:

> My view is that it is now impossible to obtain a mortgage where this form of construction is involved and, in 1986, it is quite likely that not all building societies would have accepted this form of construction as representing a suitable security.
>
> In any event a prudent surveyor should in 1986 have applied additional care when inspecting a building involving any type of concrete construction due to suspicions held generally in the surveying profession towards any form of concrete dwelling.

I accept the evidence of Mr Whitley that surveyors practising in the area knew of problems with the Bison system on this estate before 1986. Mr Oster annexed to his report a standard form letter dated August 9 1984 addressed to council tenants on the estate from the chairman of Hounslow London Borough Council housing committee referring to defects inherent in the Bison system. While Mr Whitley did not see the letter at the time, he knew of the problem because it was a matter of local politics reported in the local press. The council set out on a programme of what they described as "essential repairs". It is not clear when that programme began, but the third and last phase began in July 1987. Mr McDade and Professor Hollis accepted that the works on phase 1 probably began before the survey made by Mr McDade. A local surveyor ought to have been put on his guard about the Bison system in mid-1986 even in respect of two-storey buildings both from general knowledge of the building system and from local knowledge of particular problems in the locality.

I am forced to find that Mr McDade was negligent in the conduct of the survey of this property and that therefore there should be judgment for the plaintiffs against all the defendants.

I turn to the question of damages.

It is agreed between the parties that on the authorities the main item of damage should be the difference between the value the house in 1986 if it had been in the condition represented by the surveyor's report and its actual value. It is also agreed that the value of the house in 1986 if it had been in the condition represented would have been what was paid for it, namely £56,000.

The actual value of the house in 1986 was stated by Mr Oster in his report to be £4,960. Mr Oster could find no comparable sale value and he made a calculation designed to show what would be the value of the property to an investor buying the property at auction seeking to gain a return on his money by renting out the property. Mr Whitley did not seek to support Mr Oster's calculations. Mr Oster's calculations in any event do not take into account that at the end of the life of this building there will be a residual site value. There was no Civil Evidence Act 1968 notice served in respect of Mr Oster's report and it is not of itself evidence and only becomes evidence if adopted by Mr Whitley.

The best evidence of the value of a property is usually what someone is prepared to pay for it. What was paid for this property was £29,000 at auction on March 13 1992, six years after the date which I have to consider. Mr Whitley gave his opinion that the value of the property in its true condition in 1986 was £20,000. His reasoning was not based on any difference between 1986 and 1992 values, nor did he support his evidence with sales of comparable property, nor did he put forward any calculation of what could be made out of an investment in the property. Mr Whitley relied on his experience as a local estate agent as well as a surveyor and he suggested that the price of £29,000 paid at auction was attributable to "overzealous bidding".

How the auction price in 1992 relates to values in 1986 has not been considered in evidence. I have evidence that the market price was at its peak in 1988 and that at that time the plaintiffs received an offer for £79,000 for the property in good condition. The property market has fallen since then, but I have no evidence whether the fall is such that a price fetched at auction in 1992 should be regarded as equivalent to a price which the property might have been expected to fetch at auction in 1986. It would not be right to assume that the value in March 1992 was the same as the value to be expected in mid-1986. It is possible that that might be the case, but I have no evidence to that effect.

The defendants, case was stated by Professor Hollis as follows:

> I believe that the maximum reduction in value that may have occurred would have been £12,000. I believe that this takes into account the risk of having to face the possibility of spending about £8,500.

In 1987, and probably earlier, Hounslow London Borough Council were offering to undertake remedial work to the houses for £9,000 each. A letter to that effect was put in evidence. In making their offer, the council stated, "Matters such as the subsequent mortgage ability of your home will have to be ascertained for yourself. The council cannot give an opinion." When considering a terrace of four, putting one house right would not eliminate fears of defects unless the other houses in the short terrace also were put right: but that was done and there was every reason to believe in August 1986 that Hounslow London Borough Council as a caring landlord would put their own houses in order. However, the work would involve the provision of angle irons, which would draw to the attention of any purchaser the fact that remedial work had been done. No remedial work has in fact been done to this property and it seems likely that the property will be supported by the strengthened properties on each side, but there will always remain a question mark about its condition.

I do not think it right to assess the reduction in value by reference to remedial work which has not been carried out and which, even if carried out, would not remove the question mark which stands over this property. The only other evidence which I have as to value is the evidence of Mr Whitley. That evidence was not strongly tested in cross-examination, perhaps because it was considered to be reliable.

I therefore assess the damages attributable to reduction in value at £36,000, that is the difference between £56,000 and Mr Whitley's valuation of £20,000.

The amount claimed for legal and other costs was not disputed and I award those in the sum claimed, £1,364.65.

The plaintiffs also claim damages for worry, distress and inconvenience. I bear in mind what was said by the Court of Appeal in *Watts* v *Morrow*. As in that case, any damages awarded are totally unrelated to the strain and stress of litigation. The question is, what damage in terms of inconvenience has been caused by the negligence of the defendants? I have already described what has happened to the plaintiffs by way of changes to their circumstances. It is true that if the first plaintiff had not become less efficient at work as a result of worry, he would have been able to keep up his payments to the building society and he would have been able to stay in his home. In that event, the plaintiffs would not have suffered so much distress and inconvenience even though their intention had been to move. But they would still have had the knowledge that they were trapped in a house which could not be sold for a sum sufficient to pay off the purchase price. However, worry leading to unemployment and repossession of property is such a common result of this type of circumstance that one has to say that what has happened to the plaintiffs should have been reasonably foreseeable to Mr McDade and is a natural consequence of his negligent survey. Taking a very modest view of the damage suffered by the plaintiffs, which is still continuing, I assess their damages for worry, distress and physical inconvenience at £6,000.

I therefore award to the plaintiffs by way of damages the sums of £36,000, £1,364.65 and £6,000 totalling £43,364.65. I invite argument on the questions of interest and costs.

For further cases on this subject see p 159